1 MONTH OF
FREE
READING

at

www.ForgottenBooks.com

By purchasing this book you are eligible for one month membership to ForgottenBooks.com, giving you unlimited access to our entire collection of over 1,000,000 titles via our web site and mobile apps.

To claim your free month visit:
www.forgottenbooks.com/free172268

ISBN 978-0-266-17779-1
PIBN 10172268

Department of the Interior:

U. S. NATIONAL MUSEUM.

—— 34 ——

PROCEEDINGS

OF THE

UNITED STATES NATIONAL MUSEUM.

Vol. V.
1882.

PUBLISHED UNDER THE DIRECTION OF THE SMITHSONIAN INSTITUTION.

WASHINGTON:
GOVERNMENT PRINTING OFFICE.
1883.

ADVERTISEMENT.

The extension of the scope of the National Museum during the past few years, and the activity of the collectors sent out in its interests, have caused a great increase in the amount of material in its possession. Many of the objects gathered are of a novel and important character, and serve to throw a new light upon the study of nature and of man.

The importance to science of prompt publication of descriptions of this material led to the establishment, in 1878, of the present series of publications, entitled "Proceedings of the United States National Museum," the distinguishing peculiarity of which is that the articles are published in signatures as soon as matter sufficient to fill sixteen pages has been obtained and printed. The date of publication being plainly expressed in each signature, the ready settlement of questions of priority is assured.

The articles in this series consist: First, of papers prepared by the scientific corps of the National Museum; secondly, of papers by others, founded upon the collections in the National Museum; and, finally, of interesting facts and memoranda from the correspondence of the Smithsonian Institution.

The Bulletins of the National Museum, the publication of which was commenced in 1875, consist of elaborate papers (monographs of families of animals, &c.), while the present series contemplates the prompt publication of freshly acquired facts relating to biology, anthropology, and geology; descriptions of restricted groups of animals and plants; the settlement of particular questions relative to the synonymy of species, and the diaries of minor expeditions.

This series of publications was commenced in 1878, with volume I, under the title "Proceedings of the United States National Museum," by the authority and at the expense of the Interior Department, and under the direction of the Smithsonian Institution.

The present volume, constituting the fifth of the series, has been prepared under the editorial supervision of Dr. Tarleton H. Bean, curator of the department of fishes.

SPENCER F. BAIRD,
Director of the U. S. National Museum.

UNITED STATES NATIONAL MUSEUM,
Washington, June 20, 1883.

(II)

TABLE OF CONTENTS.

LIST OF ILLUSTRATIONS.

1.—PLATES.

UNIO CLINOPISTHUS (sp. nov.).

FIG. 1. Left side view; natural size.
FIG. 2. Dorsal view of the same example.

ANOMIA MICRONEMA Meek.

FIG. 3. View of the under valve, showing the byssal plug.
FIG. 4. Exterior view of an upper valve.
FIG. 5. Similar view of another example, showing coarser radiating lines.
FIG. 6. Interior view of a very large upper valve, showing muscular scars
    and process beneath the umbo. All of natural size.

CAMPELOMA PRODUCTA (sp. nov.).

FIG. 7. Lateral view of type specimen; natural size.
FIG. 8. Opposite view of the same.
FIG. 9. Lateral view of a more robust example.

PYRGULIFERA HUMEROSA Meek.

FIG. 10. Lateral view of type specimen; natural size.
FIG. 11. Opposite view of the same.
FIG. 12. Similar view of a smaller example.

LIST OF ILLUSTRATIONS.

2.—CUTS.

    a, larva; *b*, pupa; *c*, imago—enlarged, hair-lines showing nat. size; *d*, front wing of a pale var.; *e*, seed, nat. size, with empty pupa skin; *f*, do., showing hole of exit. (After Riley.)

LIST OF CORRECTIONS.

Page 10, line 27, is should read in.

Page 16, line 11 from bottom, *sanctœ-luciœ* should read *sanctœ-luciœ*.

Page 89, line 21, *Oncorhyncus* should read *Oncorhynchus*.

Page 95, line 21, *berthondi* should read *berthoudi*.

Page 9⁴, line 9, *humrosa* should read *humerosa*.

Page 99, line 15, view should read view.

Page 114, line 19 from bottom, *vov* should read *nov*.

Page 115, line 6 from bottom, Siniahmoo should read Semiahmoo.

Page 115, line 7 from bottom, Hiniahmoo should read Semiahmoo.

Page 122, line 4, Street should read Streets.

Page 131, line 8 from bottom, should read: 12.—UPENEUS VANICOLENSIS C. & V.

Page 131, line 16 from bottom, *trifa sciatus* should read *trifasciatus*.

Page 136, line 16 from bottom, Swian should read Swain.

Page 141, line 17 from bottom, Ba toë should read Batoë.

Page 222, line 18, URANIDBA should read URANIDEA.

Page 239, line 12 from bottom, Culpea should read Clupea.

Page 255, line 19, Fundnlus should read Fundulus.

Page 263, line 19, *Exocœtus* should read *Exocœtus*.

Page 266, line 23, sal and rays should be separated by more space.

Page 267, line 23, *Atherinia* should read *Atherina*.

Page 281, line 12, Scæna should read Sciæna.

Page 285, line 12, *Spctted* should read *Spotted*.

Page 290, line 20, OPISTOGNATHIDÆ should read OPISTHOGNATHIDÆ.

Page 293, line 2 from bottom, retained is not properly spaced.

Page 297, line 1, Blocb should read Bloch.

Page 308, line 26, *Corythoichthys* should read *Corythroichthys*.

Page 315, line 6 from bottom, MOLUSCA should read MOLLUSCA.

Page 376, last line, olclection should read collection.

Page 386, line 7 from bottom, *migratorius* should read *migratoria*.

Page 412, line 14, LETHARCUS should read LETHARCHUS.

Page 413, line 3, Baiostoma brachialis should read Bæostoma brachiale.

Page 437, line 3, Letharcus should read Letharchus.

Page 456, line 17, *Turdinœ* should read *Turdinœ*.

Page 485, line 2, (Plate VI) should read (Plate XI).

Page 486, last line, *Anarhichas* should read *Anarrhichas*.

Page 520, line 15, Chondropteygii should read Chondropterygii.

Page 523, line 25, larnal should read larval.

Page 524, line 6 from bottom, *Burmeister* should read Burmeister.

Page 524, line 5 from bottom, Soc should read Soc..

Page 529, line 9 from bottom, Poliptila should read Polioptila.

Page 548, line 15, Nyctherodias should read Nyctherodius.

Page 549, line 21, Virosylvia should read Vireosylvia.

Page 552, line 1, *Himantololophus* should read *Himantolophus*.

Page 563, line 26, it should read its.

Page 564, line 6 from bottom, Plate VII should read Plate XII.

Page 570, line 10, should be in the usual title caps.

PROCEEDINGS

OF THE

UNITED STATES NATIONAL MUSEUM.

1882.

INFORMATION CONCERNING SOME FOSSIL TREES IN THE UNITED STATES NATIONAL MUSEUM.

By Lieut. Col. P. T. SWAINE, U. S. A., and Lieut. J. T. C. HEGE-WALD, U. S. A.

[Letter to General William T. Sherman.]

SIR: I have the honor to furnish you the following information with regard to the two fossil trees procured from "Lithodendron" at the request of Lieut. Col. P. T. Swaine, Fifteenth Infantry:

On or about the middle of May, 1879, the honor was conferred upon me to carry out written instructions received from General Sherman, with regard to procuring several specimens of fossil trees from "Litho-dendron" for the National Museum. A sergeant, ten men, and two drivers, rationed for twelve days, with teams and two heavy stone wagons, were ordered to accompany me on the expedition, taking with them such tools as would be necessary to procure and handle the specimens. We made the usual drives, stopping at a forage agency each night until we arrived at Navajoe Springs, Arizona.

The country passed over was very dry and dusty, though the road was in good condition, being the regular mail route to Prescott.

At Navajoe Springs we left the road, cutting diagonally across the country about 20 miles, arriving at Bear Spring near the head of Litho-dendron in the evening. We had to cross several aroyas, but being in the dry season, we had nothing to fear from water or marshy soil. The country traversed was desolate and barren, sage-brush and pinon trees abounding, good grazing and water being very scarce. Here and there mountain peaks stood out in bold relief like great sign-posts to guide the traveler on his way. The water, when found, was in small quantities and alkaline.

Near the head of Lithodendron, and about Bear Spring, grazing was good, the Navajoes having thousands of heads of sheep there which they drove to the spring every morning and evening; being always on the *qui vive* for news, they thought it strange the "Great Father in

Washington" should want some of the bones of the "Great Giant" their forefathers had killed years ago when taking possession of the country, the *lava beds* being the remains of the blood that ran from his wounds.

Camping at Bear Spring, I turned the mules out to graze and left the men to prepare an early dinner whilst I rode down the valley to examine the thousands of specimens that lay scattered on each side of the valley along the slopes, which were perhaps 50 feet high; the valley of the Lithodendron, at its widest part, being scarcely a half mile. Along the slopes no vegetation whatever was to be seen, wood being very scarce; the soil was composed of clay and sand mostly, and these petrifactions, broken into millions of pieces, lay scattered all adown these slopes. Some of the large fossil trees were well preserved, though the action of the heat and cold had broken most of them in sections from 2 to 10 feet long, and some of these must have been immense trees; measuring the exposed parts of several they varied from 150 to 200 feet in length, and from 2 to $4\frac{1}{2}$ feet in diameter, the centers often containing most beautiful quartz crystals.

I encountered considerable difficulty in trying to procure two specimens answering to the General's description, and which I thought would please. After finding the larger of the two fossils sent, I could find no mate, the remainder being of a different species, and the exposed part broken in segments too short to answer. Finally, I concluded to unearth part of the same specimen, which entered the ground at an angle of about 20°.

Bringing back men and teams, I dug along some 30 feet, finding the second dark specimen, which made a good match, and which saw the light, perhaps, for the first time for ages, though both were parts of the same tree. This was on the right bank or slope of Lithodendron, one mile and a quarter from Bear Spring. I got both fossils loaded on the wagons, and camped at the Spring that night.

Next morning we left quite early, encountering some difficulty in getting over the rough country, frequently stopping to make a road to get on a mesa or over some aroya; late the same evening we arrived at Navajoe Springs.

From here we encountered no further difficulties. Arriving at the post I reported my return and the result of the expedition. (The post was Fort Wingate, N. Mex.)

These specimens remained at the post until Colonel Bull, in September, 1879, had them boxed up and sent to Santa Fé, New Mexico. From there they were shipped east to Washington, I believe.

Very respectfully, your most obedient servant,
J. T. C. HEGEWALD,
(*Late*) *Second Lieutenant, Fifteenth Infantry.*

NEW ALBANY, IND.,
September 21, 1881.

HISTORY OF THE TWO SPECIMENS OF FOSSIL TREES IN THE SMITH-
SONIAN INSTITUTION, WASHINGTON, D. C.

The General of the Army, General W. T. Sherman, while on a tour across the continent in the fall of 1878, suggested to Lieut. Col. P. T. Swaine, Fifteenth United States Infantry, then in command of the post of Fort Wingate, N. Mex., the expediency of procuring two of the petrifactions of the country in that vicinity of reasonable dimensions for transportation, yet sufficiently large to be worthy of a place in the Smithsonian Institution. Acting upon this suggestion, an expedition was organized early in the spring of 1879 to proceed to the Lithodendron (stone trees) in Arizona. Thomas V. Kearns, a gentleman of long residence in that part of the country, and familiar with the locality to be explored, kindly volunteered his services, and success was, in a great measure, due to his efforts in carrying out the wishes of the General. The military detail consisted of Second Lieut. J. T. C. Hegewald, one sergeant, and twelve soldiers, all of the Fifteenth United States Infantry, and the party was well supplied with army wagon running gears specially arranged for hauling stone, and with tools and appliances complete. Lieutenant Hegewald has furnished a detailed and comprehensive statement of the events connected with this expedition, which is interesting as an appendix to this paper.

Only one of the two specimens obtained from the Lithodendron by Mr. Kearns and Lieutenant Hegewald was forwarded to Washington. This is the large dark-colored one. In the place of the second one brought in from the locality of the Lithodendron a better specimen was found on the Mesa to the north of and adjacent to Fort Wingate, about two miles from the flag-staff. This is the smaller and lighter colored one.

First. Lieut. S. R. Stafford, regimental quartermaster, Fifteenth United States Infantry, had a strong platform made of plank spiked together, and rolled each fossil on separately, fastening them in place with strap iron, and hauled them to Santa Fé, N. Mex., where they were detained in the government corral awaiting the collection of enough other curiosities to make up a car load, when they were shipped to Washington under the direction and care of agents of the Smithsonian.

P. T. SWAINE,
Lieutenant-Colonel Fifteenth Infantry, Brevet Colonel, U. S. A.

A STUDY OF THE PHRONIMIDÆ OF THE NORTH PACIFIC SURVEYING EXPEDITION.

By THOS. H. STREETS, M. D., U. S. N.

The identification of the *Phronimæ* has been attended with difficulty on account of the absence of properly-defined characters. Claus, who gives the most detailed account of them, combines in his description of *P. sedentaria* more than one species. I have had no opportunity to examine *P. sedentaria*. The following article is the result of close study, and comparisons of a number of specimens of each species; and the specific characters here presented and figured were found to be constant, and apply to all sizes.

The family characteristics are as follows:

Head broad and rounded above, tapering below to the oral apparatus. Eyes on the dorsal and lateral surfaces of the head. Both pairs of antennæ present in the male, and long; in the female the inferior pair obsolete, and the superior pair short. Thorax broad anteriorly, and tapering posteriorly. The first and second pairs of thoracic feet short; the extremity of the fourth joint being more or less produced, and the fifth joint with a pair of wing-like appendages on either side of its apex. The fifth pair of thoracic feet developed into a stout, prehensile organ. The remaining pairs of feet simple. Abdomen narrow. The caudal appendages slender, cylindrical, and two-branched.

There is a very marked resemblance among the *Phronimidæ*. The family characters are many; the generic and specific characters are few, but constant.

The eggs of the female are carried in an incubatory pouch between the posterior thoracic feet. Females with the young in every stage of development within the eggs may be found swimming free; yet when the young leave the eggs, they are always found, I believe, inside the body of a *Pyrosoma*, a *Beroe*, or a *Medusa*, which the female amphipod appropriates as a home for her immature species. The parent and young are usually found inclosed in the same case. The former by this action manifests, apparently, a maternal solicitude for the welfare of her offspring. This is interesting as appearing in animals so low in the scale of being as the amphipods.

There was observed a great disparity between the number of males and females collected in any locality. In the preparation of this article there were examined forty-five specimens belonging to the different genera of the family, and the proportion of males to females was found to be as 1 to 8. Until quite recently the male form—being so different—was not recognized as belonging to the same species. The discovery was made by Claus.

PHRONIMA, Latreille.

Head, thorax, and abdomen as described under *Phronimidæ*. The first and second pairs of thoracic feet short and slender, with the fourth, or carpal joint *broadly produced;* the third and fourth pairs long, simple, and subequal. The fifth pair stoutly developed, and provided with a *strong prehensile organ, resembling the claw of some of the Cancridæ.* The last two pairs of legs shorter than the preceding, and subequal. The three pairs of caudal appendages long and slender, each furnished with two lanceolate branches. Telson short.

Sexual differences.—Males smaller than the females. In the female the inferior antennæ are absent. In the position of these organs—beneath the lateral eye—is a broad, rounded prominence, slightly projecting beyond the anterior margin of the head. The apex of this prominence usually bears a single short hair. The superior antennæ are short and three-jointed, the last joint being beset with a few auditory hairs. In the male both pairs of antennæ are present, and are provided with long, flexible flagella; the last joint of the peduncle of the superior pair long, as in the female, but much more robust, and densely furnished with hairs; the peduncle of the inferior pair three-jointed. The abdomen of the male is stouter, and the bases of the swimming feet more nearly rounded; in the female the basal portion of these feet are oblong-ovate, and the last segment of the thorax is longer and narrower than the corresponding part in the male.

PHRONIMA ATLANTICA, Guérin.

(Plate I, Fig. 1, 1a, 2.)

Phronima atlantica, Guérin-Méneville, Iconogr., pl. 25, fig. 4; Mag. Zool., 1836, cl, vii, pl. 18, fig. 1.—Milne-Edwards, Hist. des Crust., 1840, iii, p. 93.—C. Spence Bate, Catalogue Amphi. Crust., 1862, p. 319, pl. 51, fig. 4.—Dana, U. S. Explor. Exped., 1852, p. 1001.

Female.—The first and second joints of the peduncle of the superior antennæ short; the last more than twice the length of the first two. The first and second pairs of thoracic feet with the carpal joint produced antero-inferiorly, and the produced portion evenly set with sharp spines along its anterior edge; the following joint, which antagonizes with the produced portion of the preceding, slightly arched and spinous along its inferior edge; the last joint notched below the end, and furnished with a ribbed, pectinated appendage on either side of its base; the third joint prolonged anteriorly below, truncated, and set around with short, sharp bristles or spines. The second pair of legs longer than the first. The third and fourth pairs with the basal joint armed behind, at its extremity, with a sharp spine; the basal joint of the fifth pair armed in the same manner as the two preceding, but the spine is much larger in the former; there is likewise a spine on the middle of the following joint, in front. The third joint of the fifth pair enlarged, arched above, and lengthened; the fourth joint, or palm, long, attenuated at its articulation with the third, and gradually broadening to its junction with the fifth joint, arched above, the inferior angle produced anteriorly into a long and stout point, corresponding to the immovable finger of the *Cancridæ*, the anterior border with two stout, prominent teeth, the upper the larger, tuberculated on the edge towards the movable finger, and beset with a few bristles or hairs; the fifth joint, or movable finger, longer than the anterior border of the palm, arched above, and with a broad prominence on the middle of the inferior margin; the last joint very small, and in old subjects fused with the preceding joint. The basal joint of the sixth and seventh pairs of legs armed at the extremity, in front, with a short spine; and the second and third joints of the last pair with a prominent, rounded projection on the anterior surface, that on the second joint more pointed. The first pair of caudal appendages extending almost as far backward as the extremity of the third pair; the second pair falling short of the articulation of the rami of the third pair, and terminating about opposite the articulation of the rami of the first pair. Telson minute, unguiform.

Male.—The fifth pair of thoracic legs relatively shorter in the male; all the joints of the leg individually shorter and stouter than the corresponding parts in the female. The produced portion of the fourth joint, corresponding with the immovable finger of a crab, more produced downward, and less anteriorly, and arises from about the middle of the inferior surface. The fifth joint is more curved at its proximal extremity, so as to antagonize with the produced portion of the fourth joint. These sexual characters of the fifth pair of legs are only developed in the mature male; in the young of this sex, the fifth pair partakes of the characters, more or less, of the young female.

There were examined twenty-eight specimens of this species, coming from many different localities in the Pacific Ocean, varying in length from 4 to 21^{mm}, and there was found no material variation in the structural character of the prehensile organ, dependent upon age (presuming the size of the specimen to be dependent upon its age); that of 4^{mm}, as well as that of 21^{mm}, presenting all the essential characteristics of the species as described and figured by Guérin. The shape of the hand varies somewhat with size, but not sufficient to lead to a mistaken identity of the species. In the young of from 4 to 6^{mm}, the hand is almost as deep posteriorly as anteriorly, and all the joints are relatively shorter and stouter. As the animal increases in size the parts become lengthened, and the hand is much narrower posteriorly than anteriorly. In one specimen only, did the teeth on the anterior surface of the hand show any variation; in that, the detached tooth, nearest the produced portion, was wanting. *P. custos*, probably, represents this occasional variation. In another example, the prominence on the concave surface of the movable finger was very prominent, almost tooth-like. With these exceptions, I found no tendency to variation in these parts, which is contrary to the researches of Claus. According to this authority, *P. atlantica* is nothing more than the immature female form of *P. sedentaria*. I think, however, that the validity of the species will no longer be questioned, now that the male form of *P. atlantica* is presented.

Locality (of those examined): Pacific Ocean, north and south of the equator, from latitude 30° 42′ south to 37° north; and from longitude 81° 40′ west to 160° 25′ west. The temperature of the water varied from 60° to 79° Fahr.

PHRONIMA PACIFICA, Streets.

(Plate I, Fig. 3, 3*a*.)

Phronima sedentaria, Claus, Zeitschrift wissen. Zoologie, Leipzig, 1872, XXII, pls. xxvi, xxvii, fig. 1–12.

Phronima pacifica, Streets, Bulletin of the National Museum, No. 7, Washington, 1877, p. 128.

Female.—The first and second joints of the superior antennæ short (the first narrow, the second broad); the last joint about twice the length of the first and second combined. The structure of the first and second pairs of thoracic feet similar to those of *P. atlantica*. The spine on the posterior extremity of the basal joint of the third and fourth pairs is wanting in the present species, and in its place is a bristle-like hair. The fifth pair of legs are relatively shorter, when compared with those of *atlantica;* a prominent spine on the posterior extremity of the basal joint, but none on the following joint, in front; the third joint short, broad, and considerably arched above; the fourth joint (palm) broadly quadrate, almost as broad as long, the superior border rounded posteriorly to the articulation of the third joint, the lower border slightly curved, the character of the dentition on the anterior border similar to that of *atlantica* in the general arrangement of the teeth, but the teeth are not nearly so prominent, or pointed, the lower, single tooth but slightly separated from the larger crenulated tubercle; the prolonged inferior angle more curved upward, and shorter than in the former species. The fifth joint curved, about as long as the anterior margin of the palm, a low convexity on the inferior margin. The first pair of caudal appendages do not reach as far backward as the third pair, extending to, or slightly beyond, the middle of the rami of the last pair; the second pair extends to, or slightly beyond, the point of articulation of the rami of the third pair, and more than half way the length of the branches of the first pair.

The characters of the fifth, or prehensile pair of legs, and the relative length of the second pair of caudal appendages are sufficient to readily distinguish this species from *P. atlantica*.

In the young of 3^{mm} the shape of the hand is the same as in the adult. On the anterior margin there are, in the place of the dentated tubercle, two or three pointed teeth, springing from a slightly elevated base. The hand of the male is similar to that of the female, except that the immovable finger rises from a more receding angle, which, however, is less receding than that observed in *P. atlantica*.

Claus confuses this species with *P. sedentaria*. (*Vide* Zeitschrift wissen. Zoologie, Leipzig, 1872, xxii, pls. xxvi, xxvii, fig. 1–12.)

The number of specimens examined was ten—nine females and one male. Their lengths varied from 3 to 12^{mm}.

Locality.—Pacific Ocean, north and south of the equator—from latitude 40° north to 30° 42′ south; and from longitude 97° 14′ west to 157° 37′ west. The temperature of the water of the localities whence the specimens were obtained varied from 66° to 73° Fahr.

The following facts may be deduced by comparison with *P. atlantica*. The present species is smaller in size, less numerous in the localities given, and a relatively larger proportion of those in the collection came from localities south of the equator.

PHRONIMELLA, Claus.

The shape of the head and antennæ, and the general form of the thorax and abdomen very similar to *Phronima*. The third pair of thoracic feet *long*—much longer than the succeeding pair. The fifth pair enlarged, and used for prehension; *the extremity, or claw, resembling that of the Squilla*—the movable finger (fifth joint) flexing against the anterior aspect of the palm, which is furnished with teeth. Three pairs of styliform caudal appendages;* the second, or middle pair short, or rudimentary.

Sexual differences.—Males smaller than the females, and more robust. In the females the second pair of caudal appendages are rudimentary, almost obsolete; in the males well developed.

In respect to the antennæ and other parts of the body the sexual differences are similar to those observed in *Phronima*.

PHRONIMELLA ELONGATA, Claus.

(Plate I, Fig. 4, 4a, 5, 5a.)

Phronima elongata, Claus, Würzburger naturwissen. Zeitschrift, Würzburg, 1862, III, p. 247, pl. vi, fig. 6–11 (male and female).—Zeitschrift f. wissen. Zoologie, Leipzig, 1863, XII, p. 193, pl. xix, figs. 2, 3, 7 (female).

Phronimella elongata, Claus, Zeitschrift f. wissen. Zoologie, Leipzig, 1872, XXII, pp. 333, 336, 3 17.

Anchylonyx hamatus, Streets, Bulletin of the National Museum, No 7, Washington, 1877, p. 131 (female).

Female.—The first joint of the superior antennæ short; the second long and with a few auditory hairs at its apex. The first and second pairs of thoracic feet shorter than the succeeding pairs; the first shorter than the second, with the fourth joint hardly produced at its posterior distal extremity, the produced portion spine-like; the second pair with the fourth joint elongate and slender, and with the spine on the posterior distal extremity often wanting; where it is present it is much smaller than that on the corresponding joint of the first pair. The third pair of thoracic feet extremely elongate, nearly as long as the animal, the excessive lengthening being in the last two joints; the bases of the third and fourth pairs of feet spinous along the posterior edge. The base of the fifth, or prehensile, pair longer than that of the preceding pairs, and spinous on the anterior edge, two or three spines on the posterior edge near the distal extremity; the anterior edge of the second, third, and fourth joints spinous; the fourth joint enlarged at its extremity, and armed with four or five large teeth, against which the following joint, or finger, impinges; the lowest of the teeth the largest, and touches the finger about its middle; the fifth joint about one-third the length of the fourth, arched; the claws of all the pairs of feet anchylosed with the fifth joint, and fixed at a right angle to it, forming a hook, and the apex of the fifth joint slightly produced as a straight, acute spine. The bases of the last two pairs somewhat club-shaped,

---

* Claus states that there are "only two pairs of styliform caudal appendages." This is true of the female, but not of the male. In one of his plates, where the caudal extremity of a male is given, the three pairs of styliform appendages are very clearly represented.

PLATE I.

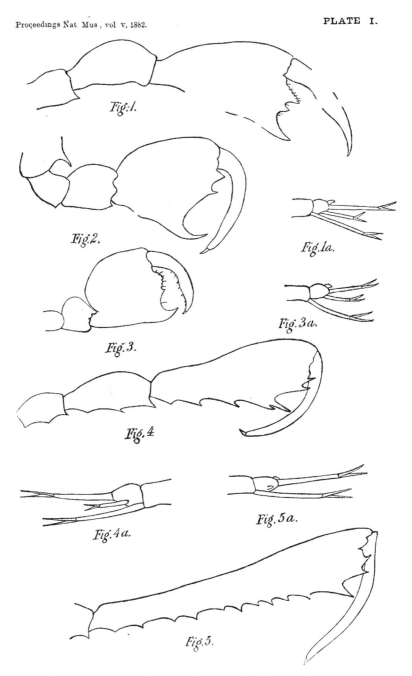

Fig:1.

Fig.2.

Fig.1a.

Fig.3.

Fig.3a.

Fig.4.

Fig.4a.

Fig.5a.

Fig.5.

PHRONIMIDÆ OF NORTH PACIFIC SURVEYING EXPEDITION.

and apex armed with a spine in front; a spine on the anterior edge ot the following joint. The first pair of caudal appendages terminate half way the rami of the third pair; the second pair rudimentary, represented only by a projecting tubercle.

Male.—The base of the superior antennæ stouter than in the female, the first joint broad, the second long and straight, with its inferior apex produced, and its lower edge densely hairy; the first and second joints of the flagellum subequal, and together about as long as the third; the third and fourth subequal, the remainder of the flagellum lost. The inferior antennæ more slender than the superior; peduncle three-jointed, and bent upward at the third joint; the first joint broad, the others successively diminishing in breadth; flagellum very long, one-half, or more, than the length of the body, filamentous, joints elongate, the first the longest, the remainder subequal. The under surface of the flagella of both pairs furnished with long, equidistant hairs. The body of the animal smaller and stouter than the female; the last two joints of the third pair of feet relatively shorter, and all the feet shorter and more robust; the fifth joint of the fifth pair about one-half the length of the fourth joint, and impinges on the large tooth anterior to its middle. The second pair of caudal appendages well developed, and extend to the commencement of the rami of the first pair.

The number of specimens examined was seven—six females and one male—varying in lengths from 9 to 15^{mm}, and coming from localities in the Pacific Ocean north and south of the equator, from latitude 34° 00′ north to latitude 30° 40′ south, and from longitude 102° 43′ west to longitude 150° 00′ west. Claus first describes the species as coming from the Mediterranean Sea. The length of the male specimen, 10^{mm}.

EXPLANATION OF PLATE I.

FIG. 1. *Phronima atlantica* (female). Fifth thoracic foot; 1 *a.* Caudal appendages.
FIG. 2. *Phronima atlantica* (male). Fifth thoracic foot.
FIG. 3. *Phronima pacifica* (female). Fifth thoracic foot; 3 *a.* Caudal appendages.
FIG. 4. *Phronimella elongata* (male). Fifth thoracic foot; 4 *a.* Caudal appendages.
FIG. 5. *Phronimella elongata* (female). Fifth thoracic foot; 5 *a.* Caudal appendages.

WASHINGTON, D. C., *March* 1, 1882.

DESCRIPTION OF SEVERAL NEW RACES OF AMERICAN BIRDS.

By ROBERT RIDGWAY.

1. METHRIOPTERUS CURVIROSTRIS OCCIDENTALIS.*

CH.—Similar to *M. curvirostris*, Swains., but tail much longer, colors darker and browner, spots of lower parts better defined and regularly

---

* METHRIOPTERUS CURVIROSTRIS OCCIDENTALIS Ridgw., MS.
"*Harporhynchus curvirostris*" LAWR. Mem. Boston Soc. N. H. II. pt. iii, No. 2, 1874, 267 (Tepic and Mazatlan).

cuneate or deltoid on the breast, the posterior lower parts suffused with much deeper fulvous, and the tail spots pale isabella-color or brownish white, instead of pure white.

Adult: Above grayish brown, the remiges and tail more brownish; middle and greater wing-coverts sometimes narrowly tipped with dull white, but these markings occasionally quite obsolete; three or four outer tail-feathers tipped with dull brownish white or pale isabella-color, the spots about .35–.40 of an inch wide on inner web of lateral feathers, successively much more restricted on the others.   Lower parts pale isabella-color, paler on chin and throat, which are nearly white, as is sometimes also the breast and middle of the abdomen, the color gradually deepening into brownish ochraceous or fulvous on the flanks, anal region, and crissum.   Jugulum marked with distinct, regularly cuneate or deltoid, spots of grayish brown, like the color of the upper parts; breast and sides marked with roundish, elliptical, or tear-shaped spots of the same, the spots largest on the breast, where sometimes more or less blended.   Bill black, the basal portion of the mandible more brownish; legs and feet dark brownish.   Wing, 4.45–4.70 (4.56); tail, 5.00–5.20 (5.10); culmen, 1.12–1.30 (1.20); bill from nostril, .90–1.15 (1.02); gonys, .70–.85 (.77); tarsus, 1.40; middle toe, 1.00–1.10 (1.05).*

Hab.—Coast region of western Mexico, in the vicinity of Tepic and Mazatlan ("common resident").

2. MIMUS GILVUS LAWRENCEI.

Ch.—Differing from true *M. gilvus* in much longer wing and tail, decidedly smaller and slenderer bill, decidedly lighter and browner gray of upper parts, much less distinct light superciliary stripe, and other details of coloration.   From var. *gracilis* is much less distinctly black wings, with less sharply contrasted light markings, upper parts browner, the bill smaller and more slender, etc.

Adult: Above uniform brownish gray (much as in *M. polyglottus*, but rather browner); wings and tail dusky (not black), the greater coverts and remiges broadly edged with brownish gray (like the back), the middle and greater wing-coverts distinctly tipped with white (forming two narrow bands), and the extreme base of the primaries white, usually, however, concealed by the primary coverts; three to five outer tail-feathers abruptly tipped with white, this 1.40–1.65 inches in extent on the outer feather, which has the outer web mostly or entirely white; the middle rectrices narrowly and indistinctly whitish or pale grayish at extreme tips.   A very indistinct paler superciliary stripe, strongly contrasted only with the dusky lores; an indistinct dusky post-ocular streak; eyelids pure white.   Lower parts dull white, purer on the throat and belly, the jugulum shaded with pale grayish, the flanks and anal-region, sometimes the crissum also, more or less strongly tinged with buff.   Bill,

---

* Extreme and average measurements of 4 adults.

legs, and feet, black;. iris "gray," "light olive," or "yellow" (SUMI-CHRAST, MS.). Wing 4.30–4.50 (4.40), tail 4.90–5.20 (5.02), culmen .65–.67 (.66), tarsus 1.20–1.35 (1.27), middle toe .80–.85 (.82).

Hab.—Isthmus of Tehuantepec (Tehuantepec City: F. Sumichrast).

Types in U. S. Nat. Mus. (Nos. 59678, ♂, and 59677, ♀, Tehuantepec City, October 8 and 29, 1869: F. Sumichrast).

The bird described above is a well-marked race, apparently referable to *M. gilvus*, though possibly (with *M. gracilis*, Cabanis, of Yucatan, Guatemala, and Honduras) distinct specifically. A considerable number of specimens of the various forms referred by authors to *M. gilvus*, representing many localities, have been examined in this connection, and the result appears to justify the subdivision of that species into several races, as follows:

A. Whitish superciliary stripe very distinct.

    a. gilvus. Above dark brownish gray, general outer surface of the wings not distinctly darker. Wing 3.85–4.40 (4.08), tail 3.90–5.00 (4.39), culmen .70–.80 (.73), tarsus 1.18–1.32 (1.27), middle toe .80–.90 (.83). *Hab.*—Guiana, Tobago, Grenada, Sta. Lucia, St. Vincent, and Martinique.* (10 specimens examined.)

    β. melanopterus. Above much lighter gray, the wings distinctly darker, by reason of narrower paler edgings. Wing 4 60–4.75, (4.67), tail 4.80–5.30 (5.12), culmen .72–.80 (.77), tarsus 1.35–1.38 (1.36), middle toe .88–1.00 (.92). *Hab.*—Venezuela and Colombia. (4 specimens.)

B. Superciliary stripe very indistinct.

    γ. gracilis. Above deep gray (about intermediate in shade between *gilvus* and *melanopterus*), the wings pure black, in abrupt and very conspicuous contrast, and with the clear white markings very sharply defined. Wing 4.15–4.80 (4.49), tail 5.00–5.80 (5.38), culmen .70–.75 (.72), tarsus 1.20–1.37 (1.31), middle toe .80–.90 (.87). *Hab.*—Guatemala, Honduras, and Yucatan. (5 specimens.

    δ. lawrencei. Above decidedly brownish gray, the wings about as in *M. melanopterus.* Wing 4.30–4.50 (4.40), tail 4.90–5.20 (5.02), culmen .65–.67 (.66), tarsus 1.20–1.35 (1.27), middle toe .80–.85 (.82). *Hab.*—Southern Mexico (Isthmus of Tehuantepec). (3 specimens.)

The synonymy of the several forms is as follows:

α. GILVUS.

Turdus gilvus VIEILL. Ois. Am. Sept. ii, 1807, 15, pl. 68 *bis* (Guiana); Nouv. Dict. xx, 1818, 296; Enc. Méth. 1823, 678.

---

*A very young bird, unquestionably of this species collected by Ober (Nat. Mus., No. 75125; orig. No. 793; "August").

Mimus gilvus JARDINE Ann. N. H. ser. 2, xx, 1847, 329 (Tobago).—SCL. P. Z. S., 1859,
    342.—TAYLOR Ibis 1864, 80 (Trinidad).—SEMPER, P. Z. S. 1871, 268 (Sta. Lucia,
    W. I.); ib. 1872, 648 (do.).—SCL. & SALV. Nom. Neotr. 1873, 3 (part).—BOU-
    CARD, Cat. Av. 1876, 146 (Guiana).—LAWR. Pr. U. S. Nat. Mus. i, 1878, 187
    (St. Vincent, W. I.); ib. 1879, 268 (Grenada, W. I.).—SALV. & GODM. Biol.
    Centr. Am. Aves, i, 1879, 36 (part).
"*Mimus melanopterus*" (part) SCL. P. Z. S. 1859, 342 (spec's from Trinidad and To-
    bago); Cat. Am. B. 1861, 9 (Trinidad).

β. MELANOPTERUS.

Mimus melanopterus LAWR. Ann. Lyc. N. Y. 1849, 35, pl. 2 (Venezuela).—SCL. P. Z.
    1859, 342 (part: spec's from New Granada and Venezuela); Catal. Am. B.
    1861, 9 (Rio Negro and Bogota).—SCL. & SALV. P. Z. S. 1868, 1866 (Vene-
    zuela).—WYATT, Ibis, 1871, 320 (Sta. Marta, New Granada).
Mimus columbianus CABAN. Mus. Hein. i, Jan. 1851, 82 (Colombia; Venezuela).

γ. GRACILIS.

Mimus gracilis CABAN. Mus. Hein. i, Jan., 1851, 83 (Honduras?).—SCL. & SALV. Ibis,
    1859, 5 (Belize; Vera Paz).—SCL. P. Z. S. 1859, 343 (Guatemala; Honduras),
    Cat. Am. B. 1861, 9, No. 58 (Salamá, Guatamala; Honduras).—MOORE, P. Z.
    S. 1859, 55 (Belize).—TAYLOR, Ibis, 1860, 110 (Comayagua, Honduras).—
    OWEN Ibis, 1861, 60 (San Geronimo, Guat.; fig of egg, pl. ii, fig. 2).—BAIRD,
    Review, 1864, 54 (San Geronimo).—LAWR. Ann. Lyc., N. Y., ix, 1869, 199
    (Merida, Yucatan).—FRANTZIUS, Jour. für Orn. 1869, 290.
"*Mimus gilvus*" (part) SCL. & SALV. Nom. Neotr. 1873, 3 ("Central America to Guat-
    emala").—SALV. & GODM. Biol. Centr. Am. Aves, i, 1879, 36 (Merida, Yuca-
    tan; Belize, Comayagua, Light-house, and Glover's reefs, Honduras; Salamá,
    San Geronimo, plain of Zacapa, upper Montagua valley, Dueñas, and Jutiapa,
    Guatemala).

δ. LAWRENCEI.

"*Mimus gracilis*" LAWR. Bull. U. S. Nat. Mus. No. 4, 1876, 12 (Tehuantepec City).

3. MERULA FLAVIROSTRIS GRAYSONI.*

CH.—Above grayish brown, slightly grayer on the nape, decidedly
ashy on primaries, upper tail-coverts, and outer webs of tail-feathers,
the wing-coverts and scapulars yellowish brown or raw-umber-brown;
lores dusky. Malar region, chin, and throat, white, streaked (except
on chin) with brownish dusky; jugulum light grayish brown, or brownish gray, indistinctly streaked with darker; breast, sides, and flanks,
plain light brown or grayish ochre; axillars and lining of wings deeper,
more reddish, ochraceous; abdomen, anal region, and crissum, white;
tibiæ light dingy grayish. Bill yellowish, dusky at tip and on basal
portion of culmen; "iris reddish brown" (GRAYSON); legs and feet
light brown (dull yellowish in life?). Wing 4.80–4.85, tail 3.90–4.00,
culmen .80–.85, bill from nostril .60, tarsus 1.35, middle toe .80–.90.
Hab.—Tres Marias Islands, off coast of Western Mexico.

Types, Nos. 37322, ♂, and 37323, ♀, U. S. Nat. Mus.; Tres Marias,
Jan. 1865; Col. A. J. Grayson.

---

* MERULA FLAVIROSTRIS GRAYSONI, Ridgway, MS.
    "*Turdus flavirostris*" LAWRENCE, Proc. Boston Soc. N. H. 1871, 276; Nat. Hist.
Tres Marias and Socorro, 1871, 17; Mem. Bost. Soc. N. H. ii. pt. 3, No. 2, 1874, p. 266.

4. SIALIA SIALIS GUATEMALÆ.*

Ch.—Similar to *S. sialis* of the eastern United States, but with decidedly longer wing and tail, the cinnamon of breast, etc., paler; ♀ with the back decidedly bluish.

♂ *adult*: Above uniform rich cobalt blue (exactly as in *S. sialis*), the shafts of the rectrices and remiges deep black, and the ends of the primaries dusky black. Chin, throat, breast, sides, and flanks, pale cinnamon; abdomen white; anal region and lower tail-coverts white, the latter tinged with blue, and with dusky shafts. Bill and feet deep black; iris brown. Wing 4.15–4.40, tail 2.80–3.00, culmen .50, tarsus .80–.85, middle toe .62–.65.

♀ *adult*: Above dull grayish blue, more brownish across the nape; feathers of pileum and back with blackish shaft-streaks (obsolete in winter plumage); rump, upper tail-coverts, and tail, bright blue, lighter and more greenish than in the ♂; wings dull blue; throat, jugulum, breast, sides, and flanks, pale dull cinnamon; abdomen, anal region, and crissum, white. Bill, tarsi, and toes, black; iris, brown. Wing 4.00–4.10, tail 2.70–2.80.

Hab.—Highlands of Guatemala and Honduras.

A considerable series of specimens of Guatemalan Bluebirds are quite uniform in their characters, as described above. It is somewhat strange that this extreme southern form should resemble much more closely in colors the true *S. sialis* of the eastern United States than the *S. azurea* of eastern Mexico, but such is nevertheless the case. Specimens in the National Museum collection are from central Guatemala ("Coban to Clusec"), and Vera Paz (Tactic and Coban). I have not seen a specimen from Honduras, but the birds of that country are probably identical with those from Guatemala.

5. CHAMÆA FASCIATA HENSHAWI.

Ch.—Differing from *C. fasciata* of the coast district of California in very much paler and grayer colors. Above brownish gray, becoming decidedly ashy on sides of head and neck, the tail showing very indistinct narrow transverse bars of a darker shade (quite obsolete in some specimens). Beneath pale vinaceous-buff, more or less tinged with pale ashy, especially on the sides. Wing 2.20–2.50, tail 3.20–3.70, culmen .40–.45, tarsus .95–1.05.

---

* SIALIA SIALIS GUATEMALÆ, Ridgw., MS.

"*Sialia wilsoni*" SCL. & SALV. Ibis, 1859, 8, (highlands of Guatemala; "El Azulejo"); Nom. Neotr. 1873, 4 (part).—SALVIN, Ibis, 1860, 29 (Coban and Dueñas; resident).—TAYLOR, Ibis, 1860, 15, 110 (highlands of Honduras, pine region, alt. 5,000 ft.).—OWEN, Ibis, 1861, 60 (Guatemala; descr. nest and eggs).—SCL. Cat. Am. B. 1862, 10 (part).

"*Sialia sialis*" SALV. & GODM. Biol. Centr. Am. Aves, i, 1879, 45 (part).

"*Sialia azurea*" BAIRD, Review, 1864, 62 (part).—SEEBOHM, Cat. B. Brit. Mus. v, 1881, 331 (Guatemala).

Hab.—Interior districts of California, including west slope of Sierra Nevada; north to Sacramento, south to Walker's Basin, Tejon Mts., and San Diego.

The differences in coloration between this interior form and the coast race (true *fasciata*) are very striking on comparison of specimens, and may be briefly tabulated as follows:

Var. FASCIATA. Above deep umber-brown, more grayish on side of head and neck; beneath deep cinnamon-buff, or light cinnamon, the throat and jugulum more or less distinctly streaked with dusky or grayish. Wing 2.20–2.60, tail 3.20–3.70, culmen .40–.45, tarsus 1.00–1.10. *Hab.*—Coast of California, south to Sta. Clara, north to or beyond Nicasio.

Var. HENSHAWI. Above brownish gray or grayish brown, the sides of head and neck decidedly ashy; beneath pale cinnamon-buff, or pale vinaceous-buff, usually more or less suffused with pale ashy, the darker streaks on jugulum, etc., nearly or quite obsolete. Wing 2.20–2.50, tail 3.20–3.70, culmen .40–.45, tarsus .95–1.05. *Hab.*—Interior of California, including western slope of Sierra Nevada.

As may be seen from the above measurements (taken from seven adult specimens of *fasciata* and eight of *henshawi*), the dimensions of the two forms are essentially identical. The extreme development of the characters distinguishing *C. henshawi* is seen in specimens from Walker's Basin and the Tejon Mts., collected by Mr. H. W. Henshaw, to whom this new form is dedicated. Specimens from Sacramento are darker, but still not enough so to make them referable to the coast form, to which all specimens from Stockton seem to belong. The darkest examples of *C. fasciata*, as restricted, come from the coast district north of San Francisco Bay (Nicasio, Marin Co., C. A. Allen).

1. CHAMÆA FASCIATA (typica).
    Parus fasciatus GAMB. Proc. Phil. Acad. Aug. 1845, 265 ("California").
        Chamæa fasciata GAMB. Proc. Phil. Acad. Feb. 1847, 154; Jour. Philad.
        Acad. 1, 1847, 34, pl. viii, fig. 3 (adult).—CABAN. Weigm. Archiv. 1848,
        i, 102.—CASS. Illustr. 1853, 39, pl. 7 (adult).—BAIRD, B. N. Am. 1858,
        370 (part); Review, 1864, 76 (part).—COOPER, B. Cal. i, 1870, 39 (part;
        "Coast of California, north to lat. 38°").—B. B. & R. Hist. N. Am. B. i,
        1874, 84, pl. vi, fig. 8.—BELDING, Proc. U. S. Nat. Mus. i, 1879, 402 (Stock-
        ton; constant resid.).

2. CHAMÆA FASCIATA HENSHAWI.
    Chamæa fasciata BAIRD, B. N. Am. 1858, 370 (part; specs. from Sacramento and Ft.
        Tejon); Review, 1864, 76 (specs. Sacramento Valley, Ft. Tejon, and San Diego).
        XANTUS, Proc. Phil. Acad. 1859, 191 (Ft. Tejon).—B. B. & R. Hist. N. Am.
        B. i, 1874, 84 (part).—COOPER, Orn. Cal. i, 1870, 39 (part; specs. from San
        Diego and foot-hills of Sierra Nevada).—NELSON, Proc. Boston Soc. N. H.
        xviii, 1875, 356 (Nevada, Cal.).—HENSHAW, Rep. Wheeler's Exp. 1876, App.
        J. J. p. 228 ("Chamoea"; Tejon Mts. and Walker's Basin, Aug.-Nov.).—
        BELDING, Proc. U. S. Nat. Mus. i, 1879, 402 (part; Marysville, Yuba Co., and
        Murphy's,* Calaveras Co.; constant resid.).

---

*Altitude, 2,400 feet.

6. PERISOREUS CANADENSIS NIGRICAPILLUS.

CH.—Similar to *P. canadensis fumifrons* in darkness of coloration, but forehead, lores, chin, throat, and sides of neck distinctly white, in marked and abrupt contrast with the dark color of adjacent parts; crown, occiput, and upper part of auricular region decidedly black, with little or no admixture of slaty anteriorly. Differing from true *canadensis* in much darker coloration throughout, much blacker crown, black auriculars, less extensive white area on forehead, and more marked contrast of the white portions of head and neck, with adjacent darker colors.

♂ *adult:* (No. 85950, U. S. Nat. Mus. Labrador, Apr. 2, 1880; "Schneider"; presented by Dr. L. Stejneger.) Whole forehead (back to about .75 of an inch from the anterior points of the nasal tufts), lores, malar region, chin, throat, and sides of neck soiled white, many of the feathers of the chin and throat having black shafts; crown and occiput, with upper and posterior portions of auricular region, deep black, somewhat mixed with slaty anteriorly and posteriorly. Upper parts dark dull slate, lighter and more grayish on the nape, and changing to plumbeous on the secondaries and tail-feathers, all of which are narrowly bordered at ends with white, which is about .25 of an inch wide on lateral rectrices; primaries edged with grayish white beyond their sinuations. Lower parts dark brownish gray, quite abruptly defined against the soiled white of the jugulum. Bill and feet deep black. Wing, 5.40; the primaries 1.10 longer than secondaries; tail, 5.30, its gradation only .75; culmen, .85; tarsus, 1.40; middle toe, .65.

It is only after very careful comparison with numerous specimens of the true *P. canadensis* from Maine, Nova Scotia, New Brunswick, Minnesota, and various localities in the interior of British America, and of an even larger series of *P. canadensis fumifrons* from Alaska, that I have concluded to base a new geographical race of this species upon the single specimen described above. That I am fully justified in doing so is evident from the fact that not one specimen among nearly 100 adult birds of this genus resembles very closely the specimen in question. In all probability the form to which the present specimen belongs inhabits the coast-district of Labrador, and would thus represent on the Atlantic side the littoral race of Alaska, called *P. canadensis fumifrons*.

SYNOPSIS OF THE WEST INDIAN MYADESTES.

By LEONHARD STEJNEGER.

Having had occasion to examine the various species of Myadestes* in connection with a study of the genera of Turdidæ, certain differences in the wing-structure among species of the West Indian group, typified by

---

* So the name is originally spelt by SWAINSON, and as μύα is found besides μυῖα, I have preferred the older form to AGASSIZ's restoration.

M. genibarbis SWAINS., led me into a further examination of the species of this section of the genus, with the aid of additional material. The inspection of the fine series of specimens, which, through the kindness of the authorities of the U. S. National Museum, I have been enabled to bring together, has resulted in a discovery of such interesting relationships between the forms in question, that I have concluded to put my notes into the shape of a monograph of all the West Indian species.

The National Museum collection, while probably more complete than any other, is still lacking in specimens from a large number of the West Indian Islands.† Mr. GEORGE N. LAWRENCE, of New York City, has kindly placed at my disposal his entire collection of species of this genus. Mr. J. A. ALLEN, of the Museum of Comparative Zoology, Cambridge, Mass., has loaned me seven specimens from the island of Sta. Lucia, while Mr. C. B. CORY, of Boston, has generously put in my hands the unique type of his *M. montanus*, from Haiti. These, together with the collection of the National Museum, make a series of 35 specimens, which represent very satisfactorily all the forms herein described, with the exception of *M. montanus*.

I desire to express my obligations to my friend ROBERT RIDGWAY for the kindness with which he has rendered me assistance in the preparation of these pages.

WASHINGTON, D. C., *February* 10, 1882.

SYNOPSIS OF THE SPECIES.

a^1. Throat and crissum orange-brown, abruptly defined; breast slaty blue, a patch of white on under eyelid.

   b^1. Upper parts sooty black, back and breast different in color.

      1. *M. sibilans* LAWR.

   b^2. Upper parts slaty blue; back and breast of the same color.

      c^1. Legs light yellow; no yellow armilla on tibia.

         d^1. Ears streaked with white; a white, or white and brown stripe along lower part of cheeks, bordered beneath by a blackish line.

            e^1. Chin of same color as throat, not white; whole abdomen like the crissum.

               2. *M. genibarbis* SWAINS.

            e^2. Chin white, abruptly defined; upper abdomen like the breast.

               f^1. Only the fore half of the malar stripe white, the hind part brown; tail-feathers not shorter than wing.

                  3. *M. sanctæ-luciæ* STEJNEGER.

               f^2. Almost the whole malar stripe white, only a few feathers at the lower end tinged with brown; tail feathers not longer than wing.

                  4. *M. dominicanus* STEJNEGER.

----

† Of the West Indian Islands inhabited by a species of *Myadestes*, but from which the National Museum possesses no specimens, are St. Domingo and Sta. Lucia. No species are known to occur upon the islands of Porto Rico, Guadeloupe, or Grenada, but as these islands are mountainous and resemble in other physical features those upon which species of *Myadestes* are known to occur, it is altogether probable that each one of these also possesses its peculiar species of the genus.

d^2. Ears blackish, not streaked; extreme point of base of lower mandible with an almost inappreciable white or brown spot.

e^1. Whole chin, and the spot on the malar apex brown.

5. *M. montanus* CORY.

e^2. Extreme point of chin, and malar apex, each with a white spot.

6. *M. solitarius* BAIRD.

? c^2. Legs brown; a yellow armilla round lower end of tibia.

? 7. *M. armillatus* (VIEILL.).

a^2. Whole under surface uniform whitish; a white ring round the eye.

8. *M. elisabeth* (LEMB.).

1. MYADESTES SIBILANS LAWR.

[Plate II, Fig. 6.]

1847.—*Ptilogonys armillatus* GOSSE, Birds of Jamaica, p. 198 (*nec* Vieill.) (*part*).

1878.—*Myiadestes sibilans* LAWR. Ann. N. Y. Ac. Sc. I, p. 148; Pr. U. S. Nat. Mus. 1878, p. 188.—OBER, Camps in the Caribbees (p. 199).—LISTER, Ibis, 1880, p. 39.

U. S. NAT. MUS. No. 74062 (δ *ad. St. Vincent, Nov.* 3, 1877.—F. A. OBER).

First primary about half the 2d, not falcate; 2d shorter than 7th, normal in shape; 3d, 4th, 5th, and 6th longest; tail much shorter than wing, and double rounded; 3d and 4th pairs the longest.

Above smoky black, forehead, crown, nape and sides of head more intense; lower back, rump, and upper tail-coverts more slaty, with a distinct tinge of olivaceous. Chin and the upper third of the malar stripe white, as also the lower eyelid, and a narrow stripe along the shaft of each ear-covert; throat and the lower two-thirds of the malar stripe bright orange-rufous, a well defined black line separating the malar stripe from the throat; breast, upper part of abdomen, and flanks clear ash-gray, many feathers, especially on the flanks, edged with rufous, remaining underparts of the same color as the throat, only a little paler; tibia gray, each feather tipped with rufous. Wings black with the edge, and a large patch at the base on the inner web of each of the six inner primaries, pure white, on the three innermost primaries also extending on to the outre web, and thus forming a very distinct white speculum; base of outer web of the inner secondaries dark ash forming an obscure band; the innermost secondaries with a narrow edge of faint olivaceous; under wing coverts and axillars whitish gray, several feathers being edged or tipped with rufous. The innermost pair of tail-feathers grayish-black at the base, becoming pure and deep black towards the tip; the following three pairs uniform black; the fifth pair has a large wedge-shaped white spot on the inner web along the outer two-thirds of the shaft, outer web also tipped with white; on the outermost pair the white spot extends further towards the base, only leaving a small portion at the base of both webs black, the terminal third of the outer web being dusky ash. Bill black; legs clear pale yellow, claws horny brown. "Iris bright hazel" (LAWR. l. c.).

As to the dimensions see the table below.

Another male (*No.* 74065, *U. S. Nat. Mus.*) has an irregular white spot at the tip of the third of the tail-feather from the outside, which is not to be found in any of the other specimens examined.

The ♀ does not differ from the ♂ either in color or size.

Table of dimensions.

Collection.	Museum number.	Collector's number.	Collector.	Locality.	Sex and age.	When collected.	Total length.*	Wing.	Tail-feathers.	Tarsus.	Middle toe with claw.	Exposed culmen.
							mm.	mm.	mm	mm	mm.	mm.
U.S. Nat.M.	74061	423	F. Ober	St. Vincent	♀ ad.	Nov. 1, 1877	190	88	71	25	22	11
Do......	74062	433	...do ...	....do	♂ ad.	Nov. 3, 1877	180	89	73	25	21	11
Do.....	74065	435	...do ...	....do	♂ ad.	Nov. 9, 1877	177	83	74	23	21	11
Lawrence...	......	425	.. do ...	....do	♂ ad.	Nov. 1, 1877	187	89	72	25	21	11
Do......	......	424	...do ...	....do	♀ ad.	Nov. 1, 1877	190	87	78	24	22	11
Average measurements of the above five specimens							185	87	74	24	21	11

* Fresh.

HAB.—St. Vincent. Mr. F. A. OBER states (Pr. U. S. Nat. Mus. 1878, p. 188) that this bird "is an inhabitant of all the high ridges containing deep woods and ravines." He procured several specimens from the top of the volcan Souffrière (about 3,000′ from the sea) and one from "High Woods, Sandy Bay, Carib Country." LISTER met with it in every part of the high woods that he explored.

REMARKS.—Strangely enough, the "Souffrière Bird" is the most distinct and remote species of the whole rufous-throated group, although the distance between St. Vincent and Sta. Lucia is not greater than between Martinique and Dominica, not to mention the forms of St. Domingo and Jamaica, which, in spite of their remote habitat, are more nearly related to the Sta. Lucia bird than the St. Vincent species is. *M. sibilans* is easily distinguished by its proportionately shorter tail and longer tarsus, the normal second primary, the black color of the upper surface, and the white speculum on the wing. Besides, the rufous color on the under surface is mixed with orange, and totally different from the brownish tint of the other species.

2. MYADESTES GENIBARBIS SWAINS.

[Plate II, Fig. 3.]

†1818.—*Muscicapa armillata* VIEILL. N. Dict. d'Hist. Nat. xxi, p. 448 (*juv., nec* 1807).
1837.—*Myadestes genibarbis* SWAINS. Nat. Libr., XIII Ornith. Flycatch., p. 134, pl. 13.—BAIRD, Rev. N. A. Birds, I, 1866, p. 423.—LAWR. Pr. U. S. Nat. Mus. 1878, p. 352.

U. S. NAT. MUS. NO. 75136 (♂ ad. *Martinique, July*, 1877, F. A. OBER).

Second primary about two and a half times the 1st, which is attenuated, but not falcate; 2d also attenuated towards the tip, but not sinuated; 3d normal; 2d equal to the 8th; 3d shorter than 6th; 4th, 5th, and 6th largest. Tail considerably graduated and less emarginated, the middle pair being equal to the 2d pair from the outside; tail-feathers a little shorter than wing.

Upper surface pure slaty-plumbeous, forehead slightly washed with olivaceous; lores black; also a stripe below the white patch on the under eyelid, assuming the color of the back on the ear-coverts, each feather of which and the above-mentioned stripe having a narrow, well-defined white central streak behind, very faintly washed with brownish. From the base of lower mandible a well-defined malar stripe runs backwards, the anterior third of which is white, while the lower two-thirds have the color of the throat, from which the malar stripe is separated by a narrow, but distinct, black stripe, reaching close to the lower edge of the mandible. Throat and chin chestnut-rufous, the white bases of the feathers on the latter showing somewhat through. Breast and upper sides of abdomen lighter than the back, almost clear ash-gray, becoming gradually lighter towards the abdomen; remaining underparts of the same color as the throat, only somewhat paler, and assuming a faint olivaceous shade on the upper abdomen; tibia like the back, a few feathers being tipped with rufous. Wings blackish, with pale edges on the primaries and two ash-gray bars across the secondaries, leaving between them a deep black patch; wing-coverts, except the primary coverts, broadly edged with gray like the back; innermost secondaries almost entirely so; inner web of the quills white at the base, forming a broad bar on the under surface of the wing; edge of wing grayish white. Middle tail-feathers uniform slate-gray; the following pairs black, the three outmost with a wedge-shaped white spot on the inner web at the end, making on the innermost only one-fifth of the length of the quill, on the middle one about one-half, and on the outermost about two-thirds, the outer webs being light slate-gray for the same extent from the tip. Bill black; legs pale brownish yellow.

The female seem to differ from the male in having the gray color of the breast less pure, this part being somewhat suffused with rufous-olive.

A young bird in the collection of Mr. GEO. N. LAWRENCE (*Martinique*, *July*, 1877, F. A. OBER), which has begun to assume the adult plumage, has the underparts dull orange-rufous, each feather with blackish edges, except on the throat and under tail-coverts, which are almost unicolor; upper parts and small wing-coverts much darker, with small rufous spots before the black terminal edge; greater and middle wing-coverts edged at the tip with rufous. Wing-feathers elsewhere and tail almost identical with the same parts in the adults.

Table of dimensions.

Collection.	Museum number.	Collector's number.	Collector.	Locality.	Sex and age.	When collected.	Total length.*	Wing.	Tail-feathers.	Tarsus.	Middle toe with claw.	Exposed culmen.
							mm.	*mm*	*mm.*	*mm*	*mm.*	*mm.*
U. S. Nat. M.	75136	716	Ober ...	Martinique .	♂ ad..	July, 1877	190	87	85	21	19	11
Do......	75137	714	..do	....do	♀ ad..	...do	190	86	78	22	20	11
Do	75138	734	..do	....do	♀ ad..	...do	196	85	83	21	19	11
Lawr	....	711	..do	....do	Juv...	...do	184	84	81	21	20	11
Average measurements of the above three *adult* specimens							192	86	82	21	19	11

*Fresh

HAB. Martinique. The label on No. 75136 states that the species is "abundant in high valleys."

REMARKS.—I have applied SWAINSON'S name to this species with some hesitation, because Mr. P. L. SCLATER (P. Z. S. 1871, p. 269) states, that he has "compared the Santa Lucia skins of this bird with two examples of *M. genibarbis* in the Swainsonian collection at Cambridge (which, although not so marked, are in all probability typical specimens), and find them agree." On the other hand, the Martinique bird agrees much better with the figure and description of SWAINSON, which give the chin as having the same color as the throat. And as it is not quite clear from the statement of Mr. SCLATER—who expressly mentions, that the St. Lucia skins do not agree with the said figure and description—to perceive, whether the birds in the Swainsonian collection differ in the same manner, I have preferred to give the name in question to the form which best agrees with the plate and the description, and to which Prof. S. F. BAIRD, in his admirable review (l. c.) already has applied the name. From Professor Baird's description it is evident that he has had before him specimens of this species, and that the determination of the locality, "Martinique," in the Lafresnaye collection was right. Besides, it is more probable that SWAINSON has had specimens from Martinique than from Sta. Lucia, since birds from the former island were common in collections, while it is very doubtful whether any skins at all had been brought to Europe from the latter at the time when SWAINSON described his species.

3. MYADESTES SANCTÆ-LUCIÆ STEJNEGER.

[Plate II, Fig. 4.]

1871.—*Myiadestes genibarbis* SCLAT. Proc. Zool. Soc. Lond. 1871, p. 269.—SEMPER, Proc. Zool. Soc. Lond. 1872, p. 649.—SCL. and SALV. Av. Neotrop. (1873) p. 4.

MUS. COMP. ZOOL. CAMBR. NO. 29582. (*Ad. Sta. Lucia.* JOHN SEMPER.)

Second primary two and a half times the 1st, which is attenuated and very slightly falcate; 2d attenuated and slightly sinuated at end; 3d normal; 2d intermediate in length between 7th and 8th, 3d equal to 6th,

3d, 4th, 5th and 6th longest.   Tail as in *M. genibarbis*; tail-feathers equal to or a little longer than the wing.

Whole upper parts slaty plumbeous with a conspicuous olivaceous wash, becoming more intense on the lower back, but lacking on the rump and upper tail-coverts.   The pattern of the head that of *M. genibarbis*, except that the black stripe below the eye extends further back on the auriculars, and that the white part of the malar stripe occupies the forward half.   Chin pure white, this color abruptly defined against the throat, which is rufous-chestnut.   The remaining underparts like those of the Martinique bird, except that the gray of the breast extends more backward on the abdomen.   Wings and tail also have the same general appearance as in the above-mentioned species; on the wings, however, the black speculum of the secondaries is more reduced, the adjacent grey cross-bands being broader, and on the tail the white is more extended, especially on the outer pair, in which the middle third of the outer web is white; besides, the outer webs of the three outermost rectrices are broadly tipped with white, and the following two pairs have also very distinct white tips.   Bill black, feet pale yellow.

In none of the seven specimens before me is the sex indicated; but as they show no differences from the specimen described above, I presume there is no difference between the male and female.

Table of dimensions.

Collection.	Museum number.	Collector's number.	Collector.	Locality.	Sex and age.	When collected.	Total length.	Wing.	Tail feathers.	Tarsus.	Middle toe with claw.	Exposed culmen.
							mm.	*mm.*	*mm.*	*mm*	*mm*
Mus C. Z C.	26714	.....	Semper .	Sta. Lucia ..	— ad.	....'.....	87	90	22	21	11
Do.....	27388	......	...do ...	....do	— ad.	....	92	92	22	21	11
Do....	27389	......	... do ...	....do	— ad.	....	88	93	22	20	11
Do....	27390	......	.. do ...	....do	— ad.	....	86	90	21	21	11
Do.....	27391	......	...do ...	....do	— ad.	....	89	94	22	22	11
Do.....	27392	......	.. do ...	....do	— ad.	....	87	92	22	20	12
Do.....	29582	......	...do ...	....do	— ad.	....	89	89	21	21	11
Average measurements of the above seven specimens							88	91	22	21	11

HAB.—Santa Lucia.   Mr. SEMPER (l. c.) states these birds are "generally found in the virgin forest or near it," and that "they are fond of cool shady places on the hills and high lands."

REMARKS.—Although very nearly allied to the foregoing species, the *M. sanctæ-luciæ* is easily distinguishable by the well defined white chin, the greater amount of white in the malar stripe, the darker, more chestnut shade of the rufous of the throat, the greater extent of the gray on the lower parts, and by the olivaceous tinge of the back in front of the rump.

The differences from the next form, *M. dominicanus*, will be pointed out under the head of the latter.

4. MYADESTES DOMINICANUS STEJNEGER.

[Plate II, Fig. 5.]

1878.—*Myadestes genibarbis* LAWR. Pr. U. S. Nat. Mus. 1878, p. 53.

U. S. NAT. MUS. NO. 77801. (♂ ad. *Dominica*. F. A. OBER.)

Second primary two and one-third times the 1st, which is acute and somewhat falcate; 2d attenuated but scarcely sinuated at end, intermediate in length between 7th and 8th; 3d equal to 6th, normal; 3d, 4th, 5th, and 6th longest. Tail somewhat graduated,* the longest feathers equal to or a little shorter than the wing.

Above slaty plumbeous, with a very faint tinge of olivaceous on head and back; lores and a narrow stripe above the eyes conspicuously suffused with olivaceous; almost the whole malar stripe whitish, the feathers the lower end tipped with chestnut; chin white, throat pure chestnut; breast, flanks, and abdomen, except the lower middle part of the latter, ash-gray, duller on the breast, more whitish on the abdomen, and very faintly washed with olivaceous, especially on the flanks, where more tinged with rufous; lower middle of abdomen, crissum, and under tail-coverts chestnut-rufous; wings and tail as in *M. sanctæ-luciæ*, the light basal spot on the outer web of the innermost primaries being very conspicuous and well defined; the black speculum on the secondaries larger, and the amount of white on the outer tail feathers rather less than in that bird; bill black, feet pale yellow.

The ♀ differs only in having a stronger wash of olive on the back, as Mr. Lawrence has already remarked (*l. c.*).

A young ♀ in the first plumage, shot the 18th of September by Mr. Ober (U. S. Nat. Mus., No. 77803) resembles very much the young bird of *M. genibarbis* from Martinique, but may be easily distinguished by the deeper tinge of the rufous, by less well-defined edgings on the under surface, and by the rufous tips of the wing-coverts being larger and better defined, forming two very distinct bands across the wing. Besides, the tail shows the same differences as in the adults, the 4th and 5th pair being tipped with white in the Dominica bird, while those feathers are uniform black in the typical *M. genibarbis*.

Table of dimensions.

Collection.	Museum number.	Collector's number.	Collector.	Locality.	Sex and age.	When collected.	Total length.†	Wing.	Tail-feathers.	Tarsus.	Middle toe, with claw.	Exposed culmen.
							mm.	*mm.*	*mm.*	*mm.*	*mm.*	*mm.*
U. S. Nat. M.	77801	153	Ober....	Dominica.	♂ ad..	....	......	89	86	21	19	11
Do.......	77802	60	....do ..	....do	♀ ad..	....	......	92	88	23	20	11
Do.......	81780	32	... do ...	....do	♀ ad..	....	......	91	90	23	21	12
Lawr.......	....	104	.. do ...	....do	♂ ad..	....	......	92	87	23	20	11
Do.......	....	105	....do ..	....do	♀ ad..	....	......	88	87	23	19	12
U S. Nat M	77803	366	....do ...	....do	♀ juv.	Sept. 18	190	87	82	23	21	.....
Average measurements of the above five adult specimens						......	90	88	23	20	11

* In the other specimens the middle tail feathers are shorter, the tail thus being emarginated, as in the foregoing species.

† Fresh.

HAB.—Dominica. "Frequents the most gloomy and solitary mountain gorges. . . . Never found below 1,000 feet altitude." (OBER, P. U. S. N. M., 1878, p. 53.)

REMARKS.—Compared with the two foregoing forms, the Dominica bird differs in having the throat of a much richer and deeper tint, being beautiful chestnut without any mixture of rufous; the rufous of the abdomen and crissum is still more restricted than in *M. sanctæ-luciæ*, and is also of a deeper shade, agreeing with the color of the throat in the latter. It also differs from both in having almost the whole of the malar stripe whitish as described above. With the Santa Lucia bird it agrees in having the chin white, and the fourth and fifth pair (counting from outside) of the tail-feathers tipped with white, differing in both these respects from the typical *M. genibarbis*.

In the tint of the throat the Martinique form is exactly intermediate between the other two, as might be expected on account of the intermediate position of this island between Sta. Lucia and Dominica; but it is a strange fact that the birds from these latter islands agree in other respects much better than either of them do with the bird from the island between them.

The three forms here discussed are very closely allied, but as the differences mentioned above hold good through the extensive series of skins which I have been able to examine, I have not hesitated to describe them as separate forms. The singular relation between their mutual resemblances and the situation of the islands in which they occur, have convinced me that they, although originally grown out from the same parent stock, have how become distinct.

5. MYADESTES MONTANUS CORY.
[Plate II, Fig. 1.]

1881.—*Myiadestes montanus* CORY, Bull. Nutt. Orn. Club, 1881, p. 130.—Id. ibid. p. 151.

MUS. C. B. CORY, Boston, No. 1253 (♀ *ad., neighborhood of Fort Jacques, Haiti. March* 3, 1881).

Second primary two and two-thirds times the 1st, which is acute and somewhat falcate, equal to the 7th, strongly sinuated and somewhat attenuated at the tip; 3d longer than the 6th, normal; 3d, 4th, and 5th longest. Tail gratuated and emarginated; middle pair equal to the 2d pair (from outside); tail-feathers equal to the wing.

Above slaty plumbeous, with a very faint tinge of olivaceous on the middle of the back; lores, cheeks, and auriculars black, unstreaked; lower eyelid brownish (?) white; chin, throat, and a small patch on the malar apex, rufous-chestnut, or the same color as the throat in *M. sanctæ-luciæ*; chin without any white spot; breast, flanks, and abdomen (except the middle portion of the latter) ash-grey, as light as in *M. sibilans*, many of the feathers tinged with rufous; middle and lower abdomen, crissum, and under tail-coverts rufous, exactly like the same parts in *sanctæ-luciæ*; tibia slaty plumbeous without rufous. Wings and tail marked as in the allied species, with the exception that the gray on the outer web of the outer tail-feathers is more restricted and lighter in shade; fourth and fifth pair without white tips. Bill black; legs yellow; claws a little more dusky.

Total length (fresh) 177^{mm}; wing 85^{mm}; tail feathers 85^{mm}; tarsus 23^{mm}; middle toe with claw 20^{mm}; exposed culmen 10^{mm}.

HAB.—*Haiti.*—The only specimen which is yet known was procured by Mr. CHARLES B. CORY in the neighborhood of Fort Jacques, Haiti, He states (*l. c.*) that it is " an apparently rare species, frequenting the summits of the highest mountains."

REMARKS.—This species may be distinguished from the nearly related *M. solitarius* from Jamaica by the absence of the white spot on the extreme chin angle, and by having the malar spot rufous instead of white. The color of the throat is less chestnut, being considerably mixed with rufous; the gray color of the breast and upper abdomen is much clearer and more mixed with rufous; the rufous on the abdomen and crissum extends farther forward on the former, and is much lighter than in the Jamaican bird. Besides, the species under consideration seems to be of smaller size.

Although only the one specimen has been examined, I have very little doubt that the species will prove to be well founded. The individual variation among these birds seems to be very limited, and the differences, pointed out above, are trenchant enough to make the two forms readily distinguishable.

6. MYADESTES SOLITARIUS BAIRD.

[Plate II, Fig. 2.]

1847.—*Ptilogonys armilattus* GOSSE, Birds of Jamaica, p. 198, cfr. VIII (*nec* Vieill.).—SCLATER, Proc. Zool. Soc. Lond. 1861 (p. 73).—(*Myiadestes*) MARCH, Proc. Ac. Nat. Sc. Phila. 1863, p. 294.

1866.—*Myiadestes solitarius* BAIRD, Rev. Amer. Birds, I, p. 421.—(*Myiadectes*) A. and E. Newton, Handb. of Jamaica for 1881, p. 107.

U. S. NAT. MUS. No. 30285 (♂ ad., *Port Royal Mountains, Jamaica, March,* 1863. W. T. MARCH).

Second primary about two and two-thirds times the 1st, which is acute, and very falcate,* considerably shorter than 7th, sinuated and somewhat attenuated at end; 3d shorter than 6th, conspicuously attenuated toward the tip; 4th, 5th and 6th longest. Tail graduated, and slightly emarginated, middle pair being equal to the 3d pair from outside; longest tail-feathers about equal to the wing.

Upper surface pure slaty plumbeous, without any olivaceous wash, except on the forehead; lower cheeks and auriculars black, graduating into the plumbeous of the neck on the latter, the auriculars unstreaked; a large patch on lower eyelid, and a smaller one on malar apex, and on the extreme point of the chin-angle, white; chin and throat pure chestnut, exactly as in *M. dominicanus;* breast, flanks, and abdomen pure slaty

---

* More so than in the other 6 specimens, which I have had opportunity to examine.

plumbeous on the upper breast, almost of the same shade as the back, but becoming much lighter on the lower parts towards the belly; anal region, crissum and under tail-coverts rufous-chestnut; tibiæ like the back. Wings and tail as in the foregoing species, the edge of the wing being purer white.† Bill black, legs yellow, claws blackish brown.

The *females* seem not to differ materially from the males.

Mr. GOSSE states that the irides are hazel, or dull orange.

Table of dimensions.

Collection.	Museum number.	Collector's number.	Collector.	Locality.	Sex and age.	When collected.	Total length.	Wing.	Tail-feathers.	Tarsus.	Middle toe, with claw.	Exposed culmen.
								mm	*mm*	*mm*	*mm*	*mm*
U.S Nat M.	30285	.....	March	Pt Royal Mts , Jamaica	♂ ad.	Mar.,1863	....	91	96	22	20	11
Do.....	74602	2307	Bryant	Moneague, Jamaica....	♂ ad.	Feb., 1865	....	91	90	21	20	10
Do.....	74603	2314	Bryant	......do	♀ ad.	Feb., 1865	....	95	94	22	20	10
Do. ..	78216	.....	....	Jamaica	— ad.	Oct , 1878	....	89	87	21	20	10
Do....	78217	.....	....	......do	♀ ad	Mar.,1879	....'	91	91	23	20	10
Lawrence..	....	2313	Bryant	Moneague, Jamaica....	♀ ad.	Feb., 1865	....	92	89	23	21	10
Do.....	....	.....	March	......do	♂ ad	Apr ,1866	....	91	94	21	20	10
Average measurement of the above seven specimens........................							....	91	91	22	20	10

HAB.—*Jamaica.* "It is entirely restricted to the dense highland woods; it is at times very common about the woods, above New Castle, in Port Royal Mountains, and along the ridges between that parish and Saint George's, as well as about Abbey Green, one of the approaches to the Blue Mountains." (MARCH, *l. c.*)

? 7. MYADESTES ARMILLATUS (VIEILL.).

1807.—*Muscicapa armillata* VIEILL. Ois. Amer. Sept. I, p. 69, pl. 42.
1866.—*Myiadestes armillatus* BAIRD, Rev. Amer. Birds, I, p. 422.—SCLATER, Proc. Zool. Soc. Lond., 1871, p. 270.—LAWRENCE, Ann. N. Y. Acad. 1878, p. 149.

The description of VIEILLOT (l. c.*) does not agree with any of the West-Indian *Myadestes* yet known. That it is not the *genibarbis* from Martinique is evident from the description, although VIEILLOT in 1818 gives that island as the especial habitat of his bird. It may, however, be, that the description of the young bird, which he gives for the first time in N. Dict. d'Hist. Nat. xxi, p. 448 (1818), belongs to the Martinique species, and hence the statement of the habitat. Mr. SCLATER (l. c.) thinks "It is possible that *M. armillatus verus* may be the species from St. Domingo," but the bird detected in that island by Mr. CORY agrees less with VIEILLOT's description than any of the other

† In this specimen, Prof. BAIRD's type, two or three feathers on each edge are tipped with rufous, which is not to be seen in the other specimens.

* And N. Dict. d'Hist. Nat. xxi, p. 448 (1818), where a few phrases are changed, and the breast given as "more blackish" (*plus noir*) than the back, instead of "paler" (*plus clair*) of the original description.

known forms, and in view of the peculiarity in their geographical distribution, that each of the mountainous islands has its own distinct species, it seems very improbable that another form is still to be found in St. Domingo. The most perplexing features of VIEILLOT'S bird are the brown feet and the beautiful yellow bracelet on the lower part of the tibia, and I am inclined to indorse the view of Mr. ROBERT RIDGWAY, that it is one of the known species,* poorly described, from a specimen supplied with legs and feet belonging to a quite different bird. The strong scutellation of the tarsus, as shown in the plate, seems to indicate that this suspicion is well founded. On the other hand, it should not be overlooked that *M. sibilans* has the tibiæ colored somewhat like the bird in question, and that the West-Indian islands are not yet so satisfactorily explored that anything can be said with certainty.

I therefore here reprint Professor BAIRD'S translation (l. c.) of VIEILLOT'S description of the adult:

"Bill blackish; a white spot on the sides of the throat, and at its origin (the chin) immediately below the lower mandible (the two continuous); the eye surrounded by the same color. Head, back, rump, two intermediate tail-feathers, and the breast of a grayish-slate, paler below. Wing and tail feathers blackish, bordered externally by gray, the three lateral on each side of the tail more or less white. Belly and hinder parts brownish rufous; a beautiful yellow in form of a bracelet on the feathers of lower part of leg; feet brown; length, 6 inches, 3 lines." VIEILLOT, Ois. Am. Sept. I, 69.

8. MYADESTES ELISABETH (LEMB.).

1850.—*Muscicapa elisabeth* LEMBEYE, Aves de Cuba, p. 39, tab. 5, fig. 3.
1856.—*Myiadestes elisabeth* CABANIS, Jour. f. Ornith. 1856, p. 2.—GUNDLACH, ibid.
    1861, p. 328; 1872, p. 428.—ID. Ann. Lyc. N. Y. 1858 (p. 271). Extr. p. 5.—
    ID. Repert. Fis.-Nat. de Cuba, I, 1865-66, p. 240.—ID. Ornith. Cuban.
    Anales 1873, p. 79.—BAIRD, Rev. Amer. Birds, I (1866), p. 425.
1859.—*Myiadestes elisabethæ* NEWTON, Ibis, 1859, p. 110.—ALBRECHT, Journ. f. Ornith.
    1861, p. 209.—SCL. and SALV. Exot. Ornith. (1867) p. 55, pl. xxviii.
1873.—*Myiadestes elisabethæ* SCL. and SALV. Nomencl. Neotr. p. 4.

The adult bird has so often been described (see the above references), that I shall give here only a short description of the young.

COLL. LAWRENCE. (♂ juv., Cuba. GUNDLACH.)

General color that of the adult. Upper parts more rusty, with a subterminal yellowish spot and terminal blackish edge on each feather, except on the rump, which is uniform; spots very obsolete on the upper tail-coverts, where the darker edges are scarcely perceptible; the upper wing-coverts, except the primary coverts, marked like the back. Underparts whitish, with a faint ochraceous tinge and very obsolete dark edgings; mustachial stripe hardly recognizable.

---

* Perhaps *M. dominicanus.*

EXPLANATION OF PLATE II.

FIG. 1. *Myadestes montanus* Cory.
FIG. 2. *Myadestes solitarius* Baird.
FIG. 3. *Myadestes genibarbis* Swainson.
FIG. 4. *Myadestes sanctæ-luciæ* Stejneger.
FIG. 5. *Myadestes dominicanus* Stejneger.
FIG. 6. *Myadestes sibilans* Lawrence.

PLATE 2

WEST INDIAN MYADESTES.

HAB.—*Cuba.* Mr. GUNDLACH informs us that this species lives "in the rocky mountains of Western Cuba. After the breeding season it frequents the woods at the foot of the mountains" (J. f. Orn. 1856, p. 2), and that he also has observed it in the eastern, but neither in the middle part of the island nor in the Isla de Pinos, as he has previously indicated (J. f. Orn. 1872, p. 429).

Table of dimensions.

Collection.	Museum number.	Collector's number.	Collector.	Locality.	Sex and age.	When collected.	Total length.*	Wing.	Tail-feathers.	Tarsus.	Middle toe, with claw.	Exposed culmen.
						1861.	mm.	mm.	mm	mm.	mm.	mm.
U. S. Nat. M.	21645	.....	Wright.	Donna del Gate, Cuba.	♀ ad..	Dec. 11	200	89	81	23	21	12
Do......	23542	...·.	...do	Mt. Libano..	.. ad..	Sept. 23	190	84	77	22	21	12
Do......	23543	....	...do	... do	.. ad..	Sept. 24	190	88	84	23	21	12
Coll. Lawr..	....	.....	Gundlach	Cuba	♂ ad..	....	.....	90	88	23	21	13
Do......	....	.....	...do	... do	♂ juv.	....	.....	88	84	23	21	12
Average measurements of the above four adult specimens............							193	88	82	23	21	12

* Fresh.

Table of comparative measurements.

Name of species.	Average of—	Total length (fresh).	Wing.	Tail-feathers.	Tarsus.	Middle toe, with claws.	Exposed culmen.
		mm.	mm.	mm.	mm.	mm.	mm.
M. sibilans	5 specimens.......	185	87	74	24	21	11
M. genibarbis	3 specimens.......	192	86	82	21	19	11
M. sanctæ-luciæ............................	7 specimens.......	.....	88	91	22	21	11
M. dominicanus..............................	5 specimens.......	.....	90	88	23	20	11
M. montanus....................................	1 specimen........	177	85	85	23	20	10
M. solitarius ....,..	7 specimens.......	.....	91	91	22	20	10
M. elisabeth....................................	4 specimens.......	193	88	82	23	21	12

Table of geographical distribution.

Name of species.	St. Vincent.	Santa Lucia.	Martinique.	Dominica.	Haiti.	Jamaica.	Cuba.	
M. sibilans	×	—	—	—	—	—	—	—
M. genibarbis	—	—	×	—	—	—	—	—
M. sanctæ-luciæ	—	×	—	—	—	—	—	—
M. dominicanus	—	—	—	×	—	—	—	—
M. montanus	—	—	—	—	×	—	—	—
M. solitarius..................................	—	—	—	—	—	×	—	—
? M. armilatus...............................	—	—	—	—	—	—	×	?
M. elisabeth..................................	—	—	—	—	—	—	×	—

ON SOME GENERIC AND SPECIFIC APPELLATIONS OF NORTH AMERICAN AND EUROPEAN BIRDS.

By LEONHARD STEJNEGER.

Looking at the ornithological nomenclature and the changes which it daily undergoes, in consequence of the radical introduction of the oldest generic or specific appellations, without considering that many commonly adopted names thus become expelled, we feel it to be our duty to make this transitional state as short as possible, by presenting the evidences we may possess, and by proposing those changes which appear necessary.

The following remarks are chiefly proposed in reference to the names of North American and European birds, as they are given in the latest catalogues of birds from those regions, viz: ROBERT RIDGWAY'S "Nomenclature of North American Birds" (Washington, 1881), and H. E. DRESSER'S "List of European Birds" (London, 1881).

As to the rules of the nomenclature, it seems to me that the best are those which present the smallest number of exceptions, and which, once adopted, give the least occasion for disputes. I therefore propose to use the oldest available name in every case, where it can be proved, and to spell it exactly as it was spelled when published for the first time, notwithstanding incorrect derivation, barbarous offspring, error facti, &c.

The significance of a name, by means of the sound and the appearance, is to give a conception of the named object as being different from all other objects. If it, at the same time, can be formed so that it indicates one or another chief property of the object, then it is the better. The main point is, however, that we, by hearing or seeing the name, will get an idea of the object as being different from any other.

That names which do not signify anything cause no inconvenience worth mentioning is evident from the numberless specific names, indicating a quality common to all the species within the same genus, e. g. cinereus, fuscus, etc. It may be rather tedious that the names are incorrect, but the simply endless number of incorrect names with which we daily work, without feeling especially troubled, and which probably no one intends to change or to correct, shows better than anything else how unimportant the corrections and improvements are for facilitating the work. I think that we may very soon agree that many corrections have caused more trouble than relief, as for instance such improvements as *Heniconetta* for *Eniconetta*, and the like, and that they only have succeeded in swelling our lists of synonyms.

The only rule which can be carried out with safety, is the use of the oldest name, without regard to its appearance, derivation, or signification. If this be adopted, most differences would disappear from the nomenclature, and it is in fact the only rule which is able to establish a passable uniformity in place of the present variety. Once universally accepted and put in practice, it would save much time, labor, and dispute; disputes concerning year and date may easily be settled, while all philological and linguistic disagreements may be thereby avoided. The question as to which species one or another name is to be referred has nothing to do with the rules of nomenclature, and are therefore liable to come up at any time.

As to the following remarks, it will be seen that some of the pages quoted are given in brackets. This signifies that the author has not had opportunity of verifying them, and that he therefore does not answer for their correctness. All the other citations have been carefully gone over, and are thought to be quite correct. He has followed this method in his later papers, and intends to proceed so in all his works in the future.

In order to show how far carelessness in quotation and excessive zeal concerning philological correctness may bring it, I select from the synonymy of *Phoenicurus titys*, p. 30, the following bouquet: *titys, tithys, thytis, tythis, tithis, thitis, tites, tethys;* and many other modes of spelling this word are surely to be found by a scrupulous investigation through the whole literature. Now please, dear reader, if you are as learned a philologist as an ornithologist, choose the right one!

WASHINGTON, D. C., *February* 21, 1882.

RUTICILLA

is untenable as a generic name for the European Redstart and its allies, as the group had already, in 1817, received the name

Phoenicurus FORST.

SYN: = 1817—*Phoenicurus* FORST. Syn. Cat. Brit. Birds, p. 53.
      1822—*Ruticilla* NAUMANN, Naturg. Vög. Deutschl. I, p. iii.
      1831—*Phoenicura* SWAINS. Fauna Br.-Amer. II, Append. p. 489.

The synonymy of the European species is as follows:

1. *Phoenicurus erithacus* (LIN.) 1758.

1758.—*Motacilla phoenicurus* LIN., Syst. Nat. ed. 10, I, p. 187.
1758.—*Motacilla erithacus* LIN., *ut supra.*
1817.—*Phoenicurus ruticilla* FORSTER, Syn. Cat. Brit. Birds, p. 16.
1817.—*Phoenicurus muralis* FORSTER, *op. cit.* p. 53.
1831.—*Ruticilla sylvestris* BRM., Handb. Vög. Deutschl. p. 363.
1831.—*Ruticilla arborea* BRM., *ut supra.*
1831.—*Ruticilla hortensis* BRM., *tom. cit.* p. 364.
1831.—*Phoenicura muraria* SWAINS., Fauna Bor.-Amer. II, App. p. 489.
1836.—*Phoenicura rutacilla* SWAINS., Classif. Birds, II, p. 240 (*nec Motacilla ruticilla* LIN.).
1854.—*Sylvia phenicurus* MACHADO, Aves Andal. p. 8.
1863.—*Ruticilla pectoralis* v. Hengl. Journ. f. Orn. 1863 (p. 165).

2. *Phoenicurus titys* (LIN.).

1758.—*Motacilla titys* LIN., Syst. Nat. ed. 10, I, p. 187.
1766.—*Motacilla phoenicurus* LIN., Syst. Nat. ed. 12, I, p. 335 (*passim*).
1769.—*Sylvia tithys* SCOP. Ann. I, Hist. Nat. p. 157.
1788.—*Motacilla gibraltariensis* GRU., Syst. Nat. I, p. 987.
1788.—*Motacilla atrata* GRU., tom. cit. p. 988 (*nec* JARD. & SELB.).
1792.—*Motacilla erithacus* BECHST., Gemeinn. Naturg. I, p. 538 (*nec* LIN.).
1803.—*Sylvia tythus* BECHST., Taschb. Vög. Deutschl. p. 179.
1810.—*Motacilla erythrourus* RAFINESQUE, Caratt. (p. 6).
1829.—*Sylvia tites* EHRB., Symb. Phys. (fol. dd.).
1831.—*Ruticilla atra* BRM., Handb. Vög. Deutschl. p. 365.
1840.—*Sylvia tithis* SCHINZ, Eur. Fauna I, p. 190.
1840.—*Phœnicura tethys* JARD. & SELB., Ill. Orn. (pl. 86).
1845.—*Ruticilla thitis* RÜPP., Syst. Ueb. (p. 57).
1848.—*Ruticilla cairii* GERBE, Dict. Univ. d'Hist. Nat. XI (p. 259).
1854.—*Lusciola thytis* SCHLEG., Vog. v. Nederl. (p. 156).
1855.—*Ruticilla montana* BRM., Naumannia 1855, (p. 281).

The other species given in DRESSER'S List of Eur. Birds should stand as

3. *Phoenicurus mesoleucus* (EHR.).—Ehrenberg's Redstart.
4. *Phoenicurus rufiventris* (VIEILL.).—Indian Redstart.
5. *Phoenicurus moussieri* (OLPH-GALL.).—Moussier's Redstart.
6. *Phoenicurus erythrogaster* (GULD.).—Güdenstädt's Redstart.

CINCLUS AQUATICUS BECHST.

is the usually adopted name of the red-bellied Water Ouzel of Central Europe. The oldest name is, however,

Cinclus merula SCHÄFF.

SYN:=1789.—*Tringa merula* SCHÄFFER, Mus. Orn. p. 52.

REGULUS CRISTATUS VIEILL. 1807,

is a name which antedates LICHTENSTEIN'S *Regulus satrapa* (1823). As it is not preoccupied, there is no reason for rejecting it. VIEILLOT, indeed, states that his bird is identical with the European species, which he erroneously calls *Motacilla regulus* LIN., but he gives a description and plate, which represent the American bird better than the *Regulus ignicapillus* BRM. The following is thought to be a tolerably exhaustive synonymy of these species:

1. *Regulus cristatus* VIEILL. 1807.

1807.—*Regulus cristatus* VIEILL., Ois. Amer. Sept. II, p. 50, pl. 106 (*nec* KOCH 1816).[*]
1808.—*Sylvia regulus* WILS., Am. Orn. I (p. 126, pl. 8, fig. 2), (*nec Motacilla regulus* LIN.).
1823.—*Regulus satrapa* LICHT., Doublettenverz., p. 35.
1832.—*Regulus tricolor* NUTT., Man. Orn. I, p. 420.
1864.—*Regulus satrapa olivaceus* BAIRD, Rev. Am. Birds I, p. 65.
1866.—*Regulus satrapus* COUES, Pr. Phil. Acad. 1866 (p. 66).

---

[*] It may, perhaps, be to this species that BARTRAM, Trav. Flor. (1791) p. 291, refers the same name. *Cfr.* E. COUES, Pr. Phil. Acad. 1875, p. 351.

2. *Regulus vulgaris* LEACH.

1758.—*Motacilla regulus* LIN., Syst. Nat. ed. 10, I, p. 188.
1816.—*Regulus vulgaris* LEACH, Cat. M. B. Brit. Mus. p. ——.
1816.—*Regulus cristatus* KOCH, Bair. Zool. I (p. 199), (*nec* VIEILL. 1807).
1822.—*Regulus aureocapillus* MEY., Tasch. Vög. Deutschl. III, p. 108.
1822.—*Regulus crococephalus* BRM., Beitr. Vogelk. II (p. 120).
1823.—*Regulus flavicapillus* NAUM., Vög. Deutschl. III (p. 968).
1831.—*Regulus septentrionalis* BRM., Handb. Vög. Deutschl. p. 479.
1831.—*Regulus chrysocephalus* BRM., op. cit. p. 481.
1833.—*Regulus auricapillus* SELBY, Brit. Orn. I (p. 229).
1877.—*Regulus linnei* MAHN, Göteb. och Bohusl. Fauna, p. 170.

3. *Regulus ignicapillus* (TENM.).

1815.—*Motacilla regulus* TEMM., Man. d'Orn. I ed. p. ——.
1820.—*Sylvia ignicapilla* TEMM., Man. d'Orn. I, p. 231.
1822.—*Regulus mystaceus* VIEILL, Faun. Franc., p. 231 (part).
1822.—*Regulus pyrocephalus* BRM., Orn. Bectr. II (p. 130); Handb. Vög. Deutschl., p. 482 (1831).
1831.—*Regulus nilssonii* BRM. Handb. Vög. Deutschl., p. 482; Naumannia 1855, p. 285.
1831.—*Regulus brachyrhynchos* BRM., op. cit., p. 483.

HIRUNDO, CHELIDON, and COTILE.

It has almost unanimously been considered, that BOIE (Isis, 1822), was the first who subdivided the genus *Hirundo* after the species belonging to *Cypselus* had been removed, and consequently his names *Hirundo* (type *rustica* LIN.), *Chelidon* (type *urbica* LIN.), and *Cotile* (type *riparia* LIN.) have been generally adopted. The same species, however, had already five years earlier been made types of three different genera, by THOMAS FORSTER, who, in his "Synoptical Catalogue of British Birds" (London, 1817), establishes the genera *Chelidon*, *Hirundo* and *Clivicola*, having as types respectively *rustica*, *urbica*, and *riparia*.

These names, which are as well founded as the later names of BOIE, cannot, so far as I can see, be rejected. Mr. FORSTER himself states, p. 40, that he has "in the following catalogue attended to generic and specific differences, and thereon founded a nomenclature, regardless of the modern names, wherever they appeared to disagree with facts." ·

I suppose the following will stand as the correct synonymy:

Chelidon FORSTER, 1817.

< 1758.—*Hirundo* LIN. S. N. ed. 10, I, p. 191.
= 1817.—*Chelidon* FORSTER, Syn. Cat. Brit. B. p. 55 (*nec* BOIE, 1822), (type *H. rustica* LIN.).
= 1822.—*Hirundo* BOIE, Isis, 1822, p. 550 (*nec* FORSTER, 1817), (same type).

Of this genus we only have one species in North America, viz:
1. *Chelidon erythrogastra* (BODD.), Barn Swallow.
The European species are:
Chelidon rustica (LIN.), and
Chelidon savignii (STEPH.).

Hirundo LIN. 1758.

< 1758.—*Hirundo* LIN. S. N. ed. 10, I, p. 191.
= 1817.—*Hirundo* FORSTER, Syn. Cat. Brit. B. p. 55 (*nec* BOIE, 1822), (type *H. urbica* LIN.).
= 1822.—*Chelidon* BOIE, Isis, 1822 p. 550 (*nec* FORSTER, 1817) (same type).

This genus has no American representative. The European species is *Hirundo urbica* LIN.

Clivicola FORSTER 1817.

< 1758.—*Hirundo* LIN. S. N. ed. 10, I, p. 191.
= 1817.—*Clivicola* Forster, Syn. Cat. Br. B. p. 55 (type *H. riparia* LIN.).
= 1817.—*Riparia* FORSTER, t. c. p. 17 (same type).*
= 1822.—*Cotile* BOIE, Isis, 1822, p. 550 (same type).
= 1826.—*Cotyle* BOIE, Isis, 1826, p. 971 (same type).

In North America only occurs—
1. *Clivicola riparia* (LIN.).

PLECTROPHANES and CENTROPHANES.

In his " Ornithologisches Taschenbuch von und für Deutschland oder kurze Beschreibung aller Vögel Deutschlands" (Leipzic, 1803), BECHSTEIN separates the *Fringilla lapponica* from the other Fringillæ, and gives to this group, which he characterizes "by having an acute pointed bill with considerably in .ected tomia, and a long straight claw on the hind toe", the name *Calcarius*. This is, as far as I know, not preoccupied, and must therefore necessarily stand as the name for the genus, which has *F. lapponica* for its type. The Snow Bunting he left in the genus *Emberiza*. In 1815 Dr. BERNHARD MEYER, in his "Kurze Beschreibung der Vögel Liv- und Esthlands" created the genus *Plectrophanes* for the same type in the following words : " Fringilla calcarata *Pall.* (this bird does not at all belong to the genus Fringilla, but forms a separate genus, which I call Plectrophanes, Longspur)." He also did not include the Snow Bunting in this genus, but treated it under the head of *Emberiza*, following the example of BECHSTEIN.† In the third volume of the "Taschenbuch" (1822) Mr. MEYER first unites the two species under the same genus, *Plectrophanes*. In 1829 JAKOB KAUP, in his " Skizzirte Entwickelungo-Geschichte und Natürliches System der Europäischen Thierwelt," again separates the two species, selecting

---

* FORSTER uses this name a few pages earlier than *Clivicola*. As, however, the adoption of *Riparia* would necessitate the change of the specific name of *H. riparia* into *europœa* FORST. 1817 (which would be inconvenient, because the species is by no means limited to Europe), or into *cinerea* VIEILL. 1817 (which has only been used for the supposed American form), I have preferred to accept the name *Clivicola*.

† Mr. DRESSER in his " Birds of Europe " erroneously cites *Plectrophanes lapponica* Mey. & Wolf, Tasch. Vög. Deutschl. I (1810), p. 187, and *P. nivalis* Mey. & Wolf, op. cit. p. 176; but these authors l. c. only give the names *Fringilla calcarata* and *Emberiza nivalis*, and the word *Plectrophanes* is not to be found either in the first or the second volume of their work. Consequently, the statement of TEMMINCK, Man. d'Orn. 2 ed. I (1820), p. 318, is also false, viz, that " Mr. Meyer has made of this species [*E. nivalis*] and of the following [*E. calcarata*] the genus *Plectrophanes*."

each as type for different genera, *nivalis* for *Plectrophanes* and *lapponica* for *Centrophanes*, and herein he has been followed by later writers. But from the foregoing statement it is evident that—

(1) BECHSTEIN'S *Calcarius* is the oldest name;

(2) the type of this is *Fr. lapponica* LIN.;

(3) MEYER'S *Plectrophanes* is merely a synonym of *Calcarius*, having the same type;

(4) the same is the case with KAUP'S *Centrophanes;*

(5) his *Plectrophanes* cannot be used for the genus having *E. nivalis* for type, because preoccupied as synonymous with *Calcarius;*

(6) the genus which has *E. nivalis* for its type should be supplied with a new name, as no later name has been given. In order to make as little change as possible, I propose for it the name *Plectrophenax.**

The synonymy of the two genera is then the following:

Calcarius BECHST. 1803.

=1803.—*Calcarius* BECHST.   Taschb. Vög. Deutschl. p. 130. (Type *Fringilla lapponica* LIN.)

=1815.—*Plectrophanes* MEYER, Vög. Liv- & Estl. p. xii (*nec* KAUP, 1829) (same type).

=1829.—*Centrophanes* KAUP, Entw. Eur. Thierw. p. 158 (same type).

=1850.—*Leptoplectron* REICHL. Av. Syst. pl. LXXV. (Type *Emberiza picta* SWAINS.)

To this genus belong the North American species:

1. *Calcarius lapponicus* (LIN.).—Lapland Longspur; [a] u
2. *Calcarius pictus* (SWAINS.).—Smith's Longspur;
3. *Calcarius ornatus* (TOWNS.).—Chestnut-collared Longspur.

Plectrophenax STEJNEGER, 1882.

<1817.—*Passerina* VIEILL. Analyse Ornith. p. 30 (type *Tanagra cyanea* LIN.) (preoccupied in Botany).

=1829.—*Plectrophanes* KAUP, Entw. Eur. Thierw. p. 138 (*nec* MEYER, 1815). (Type *Emberiza nivalis* LIN.)

The North-American and only species of this genus is—

1. *Plectrophenax nivalis* (LIN.).—Snow Bunting.

EREMOPHILA BOIE, 1828,

cannot be used in Ornithology, because already preoccupied in Ichthyology and Botany. As the following synonymy shows, the proper name of the genus will be—

Otocoris BONAP. 1839.

= 1828.—*Eremophila* BOIE, Isis, 1828, p. 322 (preoccupied in Botany; *nec Eremophilus* HUMB. 1805).

= 1831.—*Phileremos* BREHM, Handb. Vög. Deutschl. p. 312 (*nec* LATR. 1809).

= 1837.—*Brachonyx* LESSON, Compl. de Buffon, VIII, p. 126 (*nec* SWAINS. 1827, *nec* SCHÖNHEN, 1826).

= 1839.—*Otocoris* BP. Faun. Ital. Ucc. Introd.

= 1840.—*Philammus* G. R. GRAY, List Gen. Birds (p. 47.)

= 1845.—*Otocornis* RÜPP. Syst. Uebers. (p. 78).

= 1851.—*Otocorys* CAB. Mus. Hein. I, p. 121.

= 1854.—*Otocoryx* LICHT. Nomencl. p. 38.

---

† $\pi\lambda\tilde{\eta}\kappa\tau\rho\sigma\nu$ = spur; $\varphi\varepsilon\nu\alpha\xi$ = impostor.

June 5, 1882.

The names of the North-American species and races will then be:

1. *Otocoris alpestris* (LIN.)—Shore Lark.
2. *Otocoris alpestris leucolæma* (COUES).—White-throated Shore Lark.
3. *Otocoris alpestris chrysolæma* (WAGL.).—Mexican Shore Lark.

ARCHIBUTEO LAGOPUS. (BRÜNN.), 1764.

Authors who reject names given before 1766, usually cite *Archibuteo lagopus* (GM.), 1788. As GUNNERUS, in 1767, has described the species very distinctly in LEEM'S Finm. Beskr.* p. 237, as *Falco norvegicus*, they will have to adopt the name *Archibuteo norvegicus* (GUNN.), 1767, being the first name applied to the bird after the 12th edition of LINNÆI Syst. Nat., in which the species is not included. The description of GUNNERUS is as follows:

"Falconis hujus * * * pullum vivum * * * accepi in nido captum, jam 8 menses natum : magnitudo est gallinacei. In dorso, alis et subtus fuscus est, maculis canis sublematis supra, præsertim in alis adspersis. Caput, collum & pectus ad medium usque dilute brunnea sunt, maculis longitudinalibus fuscis; color tamen capitis dilutior est, & maculæ longitudinales colli pectus adtrigentes, reliquis latiores & longiores sunt. In infima parte frontis supra ceram, nec non sub oculis s. in superiore regione genarum, color est dilute canus. Iris dilute cana, & membrana nictitans cærulea. Remigeo saturate fuscæ non ad extremam caudam pertingunt, alis scilicet complicatis. Rectrices supra & infra albæ extremitatibus latitudine trium digitorum fuscis. Rostrum, ad instar ungvium, lividum, breve & inde a radice curvum, cera autem cum digitis flava. Pedes ad talos usque lanati & sordide albi, femoribus extrorsum parvis maculis fuscis adspersis. * * * Character ejus pro præsenti ita formari potest : Falco *Norvegicus* dorso alis, sterno subtus & abdomine fuscis maculis sublemulatis canis supra, præsertim in alis, adspersis, rectricibus albis, extremitatibus late nigris."

CICONIA ALBA.

The oldest author for this name has been stated to be BECHSTEIN, in his Naturgesch. Vög. Deutschl. III (1793), p. 48. It is, however, antedated by SCHÄFFER, who in 1789, Mus. Orn., p. 52, gave the name *Ciconia alba*.

STREPSILAS ILLIG. 1811,

is untenable, being one year younger than *Morinella* MEY. & WOLF, Taschb. Vög. Deutschl. II, p. 383 (1810).

The two North-American species are:

1. *Morinella interpres* (LIN.).—Turnstone.
2. *Morinella melanocephala* (VIG.).—Black Turnstone.

---

* As to this work see p. 37, footnote under *Totanus glottis.*

VANELLUS CRISTATUS M. & W.

for a long time was considered to be the oldest name as given in 1805, in their " Hist. Nat. Ois. de l'Allem." (p. 110). DRESSER has shown that BECHSTEIN'S *Vanellus vulgaris* of 1803 (Orn. Taschb. Vög. Deutschl., p. 313)·is older, and substitutes this latter name for *cristatus*. The oldest name, however, is

Vanellus capella SCHÄFFER. Mus. Orn., p. 49 (1789).

AEGIALITIS CANTIANUS (LATH.)

had already, in the tenth edition of LINNÆI Syst. Nat. I, p. 150 (1758), received the name *Charadrius alexandrinus*.* Hence

Aegialitis alexandrinus (LIN.), 1758; and for the form occurring in North America.

Aegialitis alexandrinus nivosus (CASS).—Snowy Plover.

GALLINAGO MEDIA LEACH, 1816,

is antedated by *Scolopax media* BOCK, Naturforscher, XIII (1779), p. 211, which belongs to the bird subsequently called *Scolopax major* by GMELIN in 1788, and must therefore give place to *Gallinago coelestis* FREUZEL.† The North-American form will then stand as

Gallinago coelestis wilsoni (TEMM.).—Wilson's Snipe.

TOTANUS GLOTTIS (LIN.) BECHST.

is the name usually adopted for the Greenshank, and for this is quoted either Syst. Nat. ed. 10, i, p. 146 (1758), Fauna Svec., ed. 2, p. 61 (1761), or Syst. Nat., ed. 12, i, p. 245 (1766). Any one who will take the trouble to compare these three quotations will soon find that they refer to a bird totally different from the *Totanus glottis* of BECHST. The fact that the three descriptions of LINNÆUS do not fully agree, will be mentioned later; for the present we will only consider those characters which occur in all the three editions, or which occur only in the one without being contradictory to any character given in the others.

The following phrase of the diagnosis is the same in all the editions: "*Rostro recto basi inferiori rubro*"; and the same phrase is repeated in the description in the Fauna, thus: "*Rostrum nigrum basi inferioris matilla rubra.*" BECHSTEIN'S *glottis* has the bill "gray at the base" (under Wurzel grau), and never red or reddish at any age or season.

After the diagnosis follows a reprint of the diagnosis of the first edition of the Fauna, viz, "*Remigibus lineis albis piscisque undulatis.*" In BECHSTEIN'S *glottis* the primaries, however, are black, and the

---

* (Cf. R. COLLETT, in Christiania Vidensk. Forh. 1881, No. 10, p. 4.—R. R.)

† *Scolopax cœlestis* FREUZEL, Beschreibung der Vögel und ihrer Eier in der Gegend um Wittenberg. 1801. (p. 58).

secondaries grayish, with white edges.   In the 12th edit. is said: *"Alba immaculata sunt .   .. tectrices alarum,"* while those (upper wing-coverts) in the summer plumage of *glottis* BECHST. are dark grayish-brown with a black stripe along the shafts, in the autumnal plumage brownish-gray with such a stripe, and in the young blackish-brown with rusty-gray edges.   Further in the same edition LINNÆUS says : *"Remiges primores scapo albo";* in the *glottis* BECHST. only the shaft of the first quill is white, while the shafts of the remaining primaries are black. From these quotations it is evident that the *glottis* of LINNÆUS is a bird totally different from the species so named by BECHSTEIN, while most authors since BECHSTEIN'S time, however, mean the bird of the latter when they are speaking about *Totanus glottis* (LIN.).

It remains to determine to which species the descriptions of LINNÆUS really belong.   In order to clear up this question it will be necessary to compare those characters which in the above-mentioned three editions do not agree.   It will thus be seen that while in the diagnoses the legs are said to be greenish (*"pedibus virescentibus"*), they are given as plumbeous (*"pedeo plumbei"*) in the description of the Fauna.   In the same work is said: *"pectus griseum,"* but in the 12th edition, *"Alba immaculata sunt pectus. . . ."*   From these disagreements of the descriptions it seems to be very probable that LINNÆUS in this case did not give his diagnosis and descriptions from the specimens themselves, but only from the statements of earlier writers.   The phrase *"pedes plumbei"* may thus have been taken from STRÖM, who, in his Söndmörs Beskrivelse, I, p. 235, quotes the Linnæan diagnosis in the following manner: *"Numenius pedibus virescentibus* (more correctly *plumbei*). . . ."*   This opinion seems also to be well founded when one compares the very meager description of the Fauna (*l. c.*) with the much fuller descriptions of other species, of which LINNÆUS had specimens before him when describing.   It will therefore be very useful to know which species the authors cited by LINNÆUS may have meant.   In the 12th edit. LINNÆUS quotes as synonymous *Limosa grisea major* BRISS., ed. 5, p. 272, t. 24, f. 2.   To this species BRISSON himself cites the same authors, which are given by LINNÆUS, and besides, the diagnosis of LINNÆI Systema, 10th edit., and Fauna, 1st ed.   From the excellent description of BRISSON it is unquestionable that his species is *Limosa lapponica* in winter-plumage. The description of STRÖM (l. c.) also shows that the bird in question belongs to this species.   It then only remains to determine whether the characters given by LINNÆUS agree with those of *L. lapponica*.   This species has in fact the base of the lower mandible reddish, as described above, and also the shafts of the first quills white.   The two other marks, "quills varied with white and black lines," and "wing-coverts white, unspotted," do not agree so well, but the difference is not great, and is easily understood when one attends to the manner in which the description has been made ; the inner webs of the first quills are variegated as above described, and although the wing-coverts are not unspotted white, this color, however, occupies a larger space on the wing-coverts of this species than in any other which here reasonably can be in question. The other characters agree as well with both species, and consequently they speak in favor of the opinion here expressed.

Having thus proved that the *Scolopax glottis* LINNÆUS is merely a synonym of *Limosa lapponica*, we proceed to select a new name for it, the first binominal one given to the bird in question, the Green-shank.

In 1767 KNUD .LEEM published his "Beskrivelse over Finmarkens Lapper," with a Latin translation following the Danish text, in which work the Norwegian bishop GUNNERUS, a very distinguished and, at that time, everywhere in Europe, highly esteemed naturalist, and one of the more prominent of the pupils of LINNÆUS, gives a tolerably complete account of the natural history of northern Norway in form of foot-notes. In these we find many good descriptions, and several species named for the first time, and there is not the slightest reason for rejecting his names, as he was a strict binominalist, whose descriptions are very clear, and published in a proper manner after 1766.

As the work is perhaps somewhat scarce, I think it proper to give its title in full below.*

From his diagnosis of *Scolopax nebularia* (p. 251), "*Rostro levi, acuto, sub-recurvato, collo pectoraque albido, maculis fuscis ; rachi prima remigis nivea*" it is beyond doubt that he means the species here in question. The description of the bill is sufficient to separate this bird from the other species of *Totanus* BECHST., which have the beak straight, and it cannot be confounded with any *Limosa*, having the shaft of the first quill white. The comparison with other species which he gives corroborates this opinion, as also does his quotation of STRÖM's Söndm. Beskr. I, p. 252.

As GMELIN's name *canescens* was bestowed 21 years later, and his description is by far not so precise as that of GUNNERUS, the name of the latter must be adopted, and the species for the future stand as *Totanus nebularius* (GUNN.).

The synonymy may be given as follows :

1766.—*Scolopax totanus* LIN., Syst. Nat. 12 ed. I, p. 245 (*nec* 1758 *quæ T. calidris*).
1767.—*Scolopax nebularius* GUNNERUS, in Leem, Lapp. Beskr. p. 251.

---

* Knud Leems, | Profe sor i det Lappiske Sprog | Beskrivelse | over | Finmarkens Lapper, | deres Tungemaal, Levemaade og forrige Afgudsdyrkelse | oplyst ved mange Kaabberstykker : | med | J. E. Gunneri, | Biskop over Trondhjems Stift, og S. S. Theologiæ Doctor, | Anmærkninger ; | og | E. J. Jessens, | Justitz-Raad, General-Kirke-Inspector og Cancellie-Secretaire, | Afhandling om de 'Norske Finners og Lappers Hedenske Religion. | ——— Canuti Leemii, Professoris Linguæ Lapponicæ. | De | Lapponibus Finmarchiæ, | eorumque lingua, | vita et religione pristina | commentatio, | multis tabulis æneis illustrata : | una cum | J. E. Gunneri, | Episcopi | Diœces. Nidros. & S. S. Theologiæ Doctoris | Notis ; | & | E. J. Jessen-s |.Conciliar Justit., Rer. Ecclesiast. p. utr. regn. Inspector, General. & Secret. Cancellar | Tractatu Singulari de Finnorum Lapporumque Norvegic religione pagana. | ——— Kiöbenhavn, 1767. Trykt udi det Kongel. Wäysenhuses Bogtrykkerie af | G. G. Salikath.

1787.—*Scolopax glottis* LATH., Synops. Suppl. p. 292 (*nec* LIN. *quæ Limosa lapponica*).

1788.—*Scolopax canescens* GMEL., Syst. Nat. I, p. 668.

1803.—*Totanus glottis* BECHST., Ornith. Taschenb. p. 287.

1809.—*Totanus griseus* BECHST., Gem. Naturg. Deutschl. 2 ed. IV, p. 231 (*nec Scolop. grisea* GMEL. *quæ Macrorhamphus gr.*).

1809.—*Totanus fistulans* BECHST., tom. cit. p. 241.

1810.—*Totanus chloropus* MEY. & WOLF, Taschb. Vög. Deutschl. II, p. 371.

1816.—*Glottis natans* KOCH, Syst. Pair. Zool. II (p. 305) (*nec Scol. natans* OTTO 1797 *quæ T. fuscus*).

1831.—*Glottis nivigula* HODGS. in Gray, Zool. Misc. II (p. 36).

1831.—*Totanus glottoides* VIGORS, Proc. Zool. Soc. 1831 (p. 173).

1838.—*Glottis floridanus* BP. Comp. List. (p. 51).

1844.—*Glottis vigorsii* GRAY, Cat. Brit. Mus. III, Grallæ (p. 99).

1844.—*Glottis horsfieldii* GRAY, *ut supra.*

1877.—*Glottis linnei* MALM, Göteb. och Bohurl. Fauna pp. 81 & 278.

MACHETES CUV. 1817,

must give place to the one year older *Pavoncella* LEACH, Cat. M. B. Brit. Mus. (1816), which is not, so far as I can detect, preoccupied. Then *Pavoncella pugnax* (LIN.).

TADORNA CORNUTA (GMEL.)

is not the oldest name given to that species. It is recognizably described as *Anas damiatica* in HASSELQUIST'S Palest. Reise, Deutsche Ausgabe (1762), p. 318, and should therefore stand as *Tadorna damiatica* (HASSELQU.), 1762.

HARELDA GLACIALIS (LIN.), 1766,

should be changed into *Harelda hyemalis* (LIN.), 1758, the name *Anas glacialis* not being found in his 10th edition. The three earliest names of this species are:

1758.—*Anas hyemalis* LINN., Syst. Nat. ed. 10, I, p. 126 (*nec* PALL).

1764.—*Anas hiemalis* BRUNN., Ornith. bor. p. 17.

1766.—*Anas glacialis* LINN., Syst. Nat. ed. 12, I., p. 203.

POLYSTICTA EYT. 1836,

is preoccupied by *Polysticte* SMITH, 1835. The next name in date is *Stellaria* BP. 1838, which is preoccupied in botany. The name given by GRAY in 1840 is not occupied, and the genus should therefore in the future bear the name

Eniconetta GRAY.

Syn : =1834.—*Macropus* NUTTALL, Man. II, p. 450 (*nec* SPIX, 1824).

=1836.—*Polysticta* EYTON, Brit. Birds, p. — (*nec Polysticte* SMITH, 1835, *nec Polystictus* REICH, 1850.

=1838.—*Stellaria* BONAP., Comp. List, p. 57 (preoccupied in Botany).

=1840.—*Eniconetta* G. R. GRAY, List Gen. Birds (p. 95).

=1840.—"*Stelleria* BP." GRAY, *ut supra.*

=1846.—*Heniconetta* AGASS., Ind. Univers. p. 178.

=1872.—"*Polysticte* EYT." SUNDEV., Tentam. Av. Disp. p. 148.

Species : *Eniconetta stelleri* (PALL.).—Steller's Duck.

PAGOPHILA EBURNEA (Phipps).

As to this bird, both the generic and the specific appellations are to be changed. *Pagophila* is antedated by *Gavia* BOIE, 1822, to whom the first use of this name, as a generic one, is to be referred. MÖHRING, it is true, had already used the same word in another sense, but as his genera are not recognized there cannot be any obstacles for adopting them by later authors. KAUP's name *Pagophila* is 7 years younger and based upon the same type as BOIE's genus.

Synonymy of the genus:

Gavia-BOIE 1822.

<1822.—*Gavia* BOIE, Isis, 1822, p. 563 (*nec* KAUP, 1829, *nec* BRUCH, 1853).
=1829.—*Pagophila* KAUP, Entwg. Eur. Thierw. p. 69 (*nec Pagophilus* ——).
=1842.—*Cetosparactes* MACGILL, Man. Brit. Orn. II, p. 251.
=1845.—*Catosparactes* G. R. GRAY, Gen. of Birds III, p. 655 (*err. typogr.*)

The Ivory Gull had already, in 1767, received a binominal appellation by GUNNERUS, who applied to it the name *Larus albus* in the following words: " Larus *albus* . . . . qui toto interdum corpore albus esse & Laro cano vel & fusco magnitudine convenire perhibetur . . . . Ni valde fallor, Larus hic habendus est idem ac *Senator Martensii*, qui toto corpore albus, nostro & pedibus nigris describitur esse." The restoration of this name, which is 7 years older than the *eburneus* of PHIPPS, cannot meet with any objection, as the later use of the same name by SCOPOLI and STAT. MÜLLER never has been adopted.

Gavia alba (GUNN.).

SYN: 1767.—*Larus albus* GUNNERUS in Leem, Beskr. Finm. Lapp., p. 265 (*nec* SCOP.
  1769, *nec* S. MÜLL. 1776).
  1774.—*Larus eburneus* PHIPPS, Voy. N. Pole, App. (p. 187).
  1876.—*Larus candidus* O. F. MULLER, Prodr. Zool. Dan. p. VIII.
  1783.—*Larus niveus* BODD., Tabl. Pl. Enl. (p. 58), (*nec* PALL.).
  1846.—*Larus brachytarsus* HOLBÖLL, Fauna Grönl. (p. 52).

LARUS GLAUCUS (BRÜNN.).

The appellation of this species exhibits a close analogy to the facts referred to under *Archibuteo lagopus* (BRÜNN.), and writers, who choose the 12th edit. of LINNÆI Systema as the starting point for specific names, will have no alternative but to adopt the name of GUNNERUS, given in 1767, as the bird has not received any name by LINNÆUS. There can be no doubt as to which species the following description of GUNNERUS belongs: " Larus *hyperboreus* dorso dilute ciuereo, extremitatibus remigum albis. A Martensio in itril. Spitzberg. dicitur Germanice *Burgemeister*. A Brünn. in Ornith. p. 44, n. 148 evocatur Larus *glaucus* totus albus, dorso & alis canis, remigum extremitatibus albis." GUNNERUS in LEEM's Beskr. Finm. Lapp. p. 283 (1767).

HYDROCHELIDON LARIFORMIS (LIN.).

I see no reason whatever for rejecting the name *Sterna nigra* for this species, because there can be no doubt about which species LINNÆUS has described under this name.   That GRAY and others have used the names *fissipes* for this species, and erroneously given the appellation *nigra* to another, cannot be a hinderance to the restoring of the proper name.   The following citations from LINNÆI's description in Fauna Suec., 2d ed. p. 56, will be sufficient to show that he means the common black Tern : " *Color totius avis supra canus*" and " *remiges & rectrices omnes unicolores & subtus albescentes*."   From these it is evident that the rump and upper surface of the tail are gray, which are the very characteristic features of the bird occurring in Scandinavia, in opposition to *leucoptera* MEISN., which has the rump and the upper surface of the tail white, being one of the rarest stragglers in the country of LINNÆUS. His statement " *Habitat Ultunæ prope Upsaliam*" therefore corroborates the opinion here expressed.

The synonymy of the European form is the following:

Hydrochelidon nigra (LIN.).

1758.—*Sterna nigra* LIN., Syst. Nat. ed. 10, I, p. 137 (*nec Gray*).
1758.—*Rallus lariformis* LIN , tom. cit. p. 153.
1766.—*Sterna nævia* LIN., Syst. Nat. ed. 12, I, p. 228 (*nec Pall.*).
1766.—*Sterna fissipes* LIN., ut supra (*nec Pall.*).
1769.—*Sterna merulinus* SCOP.. Ann. I, Hist. Nat. p. 81.
1788.—*Sterna obscura* GMEL., Syst. Nat. I, p. 608 (*nec LATH.*).
1831.—*Hydrochelidon nigricans* BRM., Handb. Vög. Deutschl. p. 794.

The American form has the following synonymy :

Hydrochelidon nigra surinamensis (GMEL.).

1788.—*Sterna surinamensis* GM., Syst. Nat. I, p. 604.
1813.—*Sterna plumbea* WILS. Am. Orn. VII (p. 83, pl. 60).
1828.—*Sterna nigra* BP., Syn. (p. 355).
1860.—*Sterna frenata* SALVIN, Ibis, 1860, p. 278 (*nec GAMBEL*).
1862.—*Hydrochelidon fissipes* COUES, Pr. Phil. Ac. 1862 (p. 554).
1874.—*Hydrochelidon lariformis* COUES, Birds of N. W. p. 704.
1880.—*Hydrochelidon lariformis surinamensis* RIDGW. Pr. U. S. Nat. Mus. 1880, p. 208.

STERCORARIUS CREPIDATUS and PARASITICUS.

Since the first separation of the two species, which are called "Richardson's Jaeger" and "Long-tailed Jaeger," much dispute among authors has arisen from the question, to which of these species the *parasiticus* of LINNÆUS really belongs.*   At one time the one opinion has been the prevailing one, and at other times the opposite belief.   For a long time the "Richardson's Jaeger" held the name in unquestioned possession,

---

* All the authors before 1800 (except BRISSON and GUNNERUS, who in 1767 give the name *parasiticus* to Richardson's "Jaeger;" see LEEM, Beskr. Finm. Lapp. pp. 239 and 287) confound the two species.

until in the later years especially English ornithologists (SAUNDERS, DRESSER) have tried to vindicate the name *parasiticus* for the long-tailed species.

In the 10th edition of LINNÆI Systema we find nothing which justifies the change of the generally adopted appellation. Mr. SAUNDERS (Pr. Zool. Soc. Lond. 1876, p. 327) thinks, however, that the short diagnosis, viz, "*rectricibus duabus intermediis longissimis,*" is sufficient to prove the contrary, and exclaims, "Nothing could well be clearer!" This might perhaps have been right, if LINNÆUS had had before him more than one species, of which one or more were supplied with *rectricibus longis*, and the *parasiticus* then was given as having *rectrices longissimi;* but as he knew but one species, his expression would be quite correct if the middle pair of rectrices in his *parasiticus* had been still shorter than they are in "Richardson's Jaeger." The phrase "*Rectrices intermedii longissimi*" indicates only that the middle pair was longer than the other, or it may also signify that it in this "*Larus*" was very long compared with the other *Lari* described by him, which all had a square tail. Mr. SUNDEVALL (Tent. Meth. Av. Disp. p. 136) uses the same phrase exactly in the same meaning, when he characterizes the whole genus in the words, "*Cauda subaquali, pennis 2 mediis longissimis,*" and nobody will consider this to be incorrect, although he, in the genus thus characterized, includes the *Cataracta skua* BRÜNN. Besides, the quotations and the habits of the bird, as they are given in the 10th edition, agree better with the common Jaeger than with the long-tailed species. The authors who take this edition for their starting point in nomenclature have not the slightest reason for changing the name.

SAUNDERS, DRESSER, and most English writers, however, found their nomenclature on the 12th edition of 1766. In this the diagnosis from the 10th edition is reprinted verbatim. Besides, there is nothing new which can justify the change; it might then be that LINNÆUS here cites as synonymous the *longicaudatus* of BRISSON, but as he does not adopt the other species of the same author, the "*Stercorrarius*," the erroneous citation is of no importance, as it only shows that he did not recognize more than one species. Thus the 12th edition does not furnish any reason supporting the change. But—and this is the main point—this edition contains a phrase which corroborates the opinion here defended, and which appears to have been overlooked by Mr. SAUNDERS, viz, "*Rachis remigum rectricumque, imprimis subtus, nivea.*" From this quotation it seems to be evident that LINNÆUS means the bird which has the shafts of all the primaries white, and that his *parasiticus* of 1766 belongs less to the long-tailed species than even the *parasiticus* of 1758. If it is so that both editions of the Systema Naturalis entitle the common Jaeger to bear the name *parasiticus* LIN., it will be of no importance that the description in Fauna Svecica (1761)—the diagnosis is the same as in S. N.—is perhaps made from a specimen of the long-tailed species which LINNAEUS himself confounded with the common Jaeger. It will have no influence on those authors who start from 1758, because the description of the Fauna is of later date, and it must have less influence on those who date their specific appellations from 1766, because the said description is older, and those ornithologists pay no attention to names given before that date. It will be the less justifiable for them to use the name of 1761, as this is opposite to the appellation of 1766.

Mr. SAUNDERS himself states (tom. cit. p. 651) that "these violent transfers must always be productive of confusion *even when justifiable.*" Where the case is clear and allows no doubt, we will have, however, to accept the oldest name, even if the restoration for a short time would produce some confusion, but it should never be performed where the case is doubtful, or, like the present, more than doubtful.

In order to show the proper names of the two species, I give the following synonymy :

1. *Stercorarius parasiticus* (LIN.)

1758.—*Larus parasiticus* LIN. Syst. Nat. ed. 10, I, p. 136 (*nec descr. Faun.* 1761 (?), *nec* LESS.).
1764.—*Catharacta cepphus* BRUNN. Orn. Bor. p. 36 (*nec* STEPH.).
1764.—*Catharacta coprotheres* BRÜNN. op. cit. p. 38 (?).
1773.—*Larus crepidatus* BANKS in Hawkesw. Voy. II (p. 15) (*nec* BRM. & SCHILL.).
1824.—*Lestris boji* BRM., Lehrb. Eur. Vög. II, p. 991.
1824.—*Lestris schleepii* BRM., tom. cit. p. 993.
1824.—*Lestris benickii* BRM., tom. cit. p. 996.
1811–31.—*Catarractes parasita* PALL., Zoogr. Ross.-As. II, p. 310.
1831.—*Lestris richardsoni* SWAINS., Faun. Bor.-Amer. p. 433.
1862.—*Lestris thuliaca* PREYER, Reise n. Isl. (p. 418).
1865.—*Stercorarius tephras* MALINGR., Journ. f. Orn., 1865, p. 392.
1873.—*Stercorarius asiaticus* HUME, Stray Feathers (p. 269).

2. *Stercorarius longicaudus* VIEILL.

1819.—*Stercorarius longicaudus* VIEILL., Nouv. Dict. d'Hist. Nat. xxxii, p. 157.
1822.—"*Lestris buffonii* H. BOIE," F. BOIE, Isis, 1822, pp. 562 and 874.
1822.—*Lestris crepidata* BRM. and SCHILL., Britr. Vogelk. (p. 861) (*nec* BANKS).
1826.—*Stercorarius cepphus* STEPH. in SHAW, Gen. Zool., XIII, I, p. 211.
1828.—*Lestris parasitica* LESS., Man. d'Orn. II, p. 288 (*nec* LINN. 1758).
1831.—*Lestris microrynchus* BRM., Handb. Vög. Deutschl. p. 725.
1838.—*Lestris lessoni* DEGL., Mem. Acad. Roy. de Lille, 1838 (p. 108).
1842.—*Stercorarius longicaudatus* DE SELYS, Faune Belg. (p. 156).
1855.—*Lestris brachyrhynchus* BRM., Vogelf. (p. 337).
1867·—"*Lestris brissoni* BOIE," DEGL. and GERBE, Ornith. Europ. II, p. 400.

PODICEPS and COLYMBUS.

LINNÆUS united the Grebes and the Loons or Divers in the same genus, *Colymbus*, but in 1760 BRISSON had already separated the Loons from the Grebes, retaining the name *Colymbus* for the latter. In 1777 SCOPOLI followed his example. Ten years later LATHAM applied the name *Podiceps* to the same group, this consequently being a mere synonym of *Colymbus* as restricted by BRISSON. As the name given by the latter author to the Loons was preoccupied, the next name, which is CUVIER's *Urinator*, is to be used. The name *Eudytes* ILLIGER, although 12 years younger, has been generally adopted, but it must give way to the older name, for the suppression of which I see no reason.

The following is the synonymy of the genus:

Urinator CUV.

< 1758.—Colymbus LIN., Syst. Nat. ed. 10, I, p. 135.
= 1760.—<i>Mergus</i> BRISS., Ornith. VI, p. 104 (<i>nec</i> LINN. 1758.)
< 1777.—Uria SCOPOLI, Introd. p. 473 (nec BRISS. 1760).
= 1799.—<i>Urinator</i> CUV., Anat. Comp. I, Tabl. II.
= 1811.—<i>Eudytes</i> ILLIG., Prodr. Syst. p. 282.
< 1811-31.—Cepphus TALL., Zoogr. Ross.-As.
> 1829.—<i>Eudites</i> KAUP, Entwg. Eur. Thierw. p. 144.

All the species belonging to this genus occur in North America, viz:

1. Urinator immer (BRÜNN.),

usually known among North American ornithologists as <i>C. torquatus</i> BRÜNN.; but as the former name also is acceptable to those writers who follow the Stricklandian code of nomenclature, and who reject specific names older than 1766, I have found the name <i>C. immer</i> preferable, because it also occurs in the 12th edition of LINNÆI Syst. Nat.

The synonymy of the species is given as follows:

1764.—<i>Colymbus immer</i> BRÜNN., Ornith. Bor. p. 38.
1764.—<i>Colymbus torquatus</i> BRUNN., tom. cit. p. 41.
1765.—<i>Colymbus maximus</i> GUNN., Tr. Selsk. Skr. III, p. 125.
1766.—<i>Colymbus glacialis</i> LIN., Syst. Nat. ed. 12, I, p. 221.
1790.—<i>Mergus nævius</i> BONNAT., Enc. Meth. Orn. I, p. 73.
1810.—<i>Colymbus atrogularis</i> MEYER & WOLF, Taschb. Vög. Deutschl. II, p. 449 (<i>partim</i>).
1824.—<i>Colymbus hyemalis</i> BRM., Lehrb. Eur. Vög. II, p. 883.
1831.—<i>Colymbus hiemalis</i> BRM., Handb. Vög. Deutschl. p. 972.

2. Urinator adamsii (G. R. GRAY), 1859.
3. Urinator arcticus (LINN.), 1758.
4. Urinator pacificus (LAWR.).
5. Urinator lumme (BRÜNN.), 1764.

This is the <i>C. septentrionalis</i> LIN. 1766. But as LINNÆUS does not give the name either in the 10th edition of Syst. Nat. or in 2d ed. of Fauna Svecica (1761), the name of BRÜNNICH is to be used.

ON THE GENERA HARPORHYNCHUS, CABANIS, AND METHRIOPTERUS, REICHENBACH, WITH A DESCRIPTION OF A NEW GENUS OF MIMINÆ.

By ROBERT RIDGWAY.

In treating of the genus <i>Harporhynchus</i>, in its comprehensive sense, most authors have alluded more or less strongly to the great difference in form presented by the type of the genus (<i>H. redivivus</i>) on the one hand and certain species associated with it under the same generic name (notably *"H."* *rufus*). The various attempts at subdivision, however, have either proven failures, on account of the gradual transition between the two extremes in certain characters, or unsatisfactory, by reason of the line having been variously drawn by different authors.* As long ago as 1858, however, the line separating *Methriopterus* from *Harporhynchus* appears to have been properly drawn, by Professor Baird in a "Synopsis of the species" under the heading of the latter, on p. 348 of "Birds of North America." While, however, arranging the species in a table under the two separate headings (*Harporhynchus* including *redivivus*, *lecontei*, and *crissalis*, *Methriopterus* comprising *curvirostris*, *longirostris*, and *rufus*), Professor Baird hesitated to separate the two groups generically, but remarked as follows concerning the matter: "The transition from the one extreme in structure in *H. redivivus* to the other in *T. rufus* is so gradual as to render it very difficult to separate them; *T. curvirostris* has a shorter tarsus (about equal to the middle toe) than the others, and the graduation of the tail is less. It is very difficult to say whether it should more properly be assigned to the first section or the second. In the character of the bill there is the most gradual transition from its very long greatly curved shape in *H. redivivus* to the straight and short one of *H. rufus*." It appears, however, that other characters of more importance than the mere size and shape of the bill, serve, when taken in connection with the latter, to very readily distinguish two groups which it seems to me are of generic rank. These distinctions I have been able to verify in the case of all the species known to date, including two (*M. palmeri* and *M. bendirei* unknown when Professor Baird's "Review" was published, besides two others *M. cinereus* (Xantus), and *M. ocellatus* (Scl.) not taken into consideration in the "Birds of North America." These characters are as follows:

1. HARPORHYNCHUS. Tarsus much shorter than culmen; gonys equal to or longer than middle toe, without claw; tail exceeding the wing by much more than the length of the tarsus. Lower parts wholly immaculate.

2. METHRIOPTERUS. Tarsus longer than the culmen; gonys much shorter than the middle toe, without claw; tail exceeding the wing by much less than the length of the tarsus. Lower parts more or less distinctly spotted or streaked (markings nearly obsolete in *M. palmeri*).

To *Harporhynchus*, as thus restricted, belong only *H. redivivus*, *H. lecontei*, and *H. crissalis*, while to *Methriopterus* may be referred the

---

* Thus, Dr. Sclater, in his "Synopsis of the Thrushes (*Turdidæ*) of the New World," (P. Z. S. 1859, pp. 338–40), includes *Orpheus curvirostris* Swains. under *Harporhynchus*, along with *H. redivivus*, *H. lecontei*, and *H. crissalis*, "*Methriopterus*" embracing only *O. longirostris* Lafr. and *Turdus rufus* Linn. In his "Catalogue of a Collection of American Birds," published two years later (1862; pp. 7–8), Dr. Sclater adopts essentially the same arrangement, *Harporhynchus* being represented by *"H."* *curvirostris*, and *Methriopterus* by *"H."* *longirostris*, *"H."* *rufus*, and *"H."* *cinereus*. It is proper to state, however, that the term *Methriopterus* is not used in a generic sense, but merely as a convenient subgeneric heading.

following: *M. rufus*, *M. longirostris*, *M. ocellatus*, *M. cinereus*, *M. bendirei*, *M. palmeri*, and *M. curvirostris*. This arrangement, I am aware, removes *M. palmeri* much further from *H. lecontei* than Mr. Brewster (*cf*. Bull. Nutt. Orn. Club, vi, Apr. 1881, p. 67) has suggested should be its position, but after a very careful comparison with all the species, made in connection with Mr. Brewster's remarks upon the subject, I am convinced that the two birds have in fact nothing in common beyond a general superficial resemblance in coloration. In fact, these two species, which exhibit the nearest approach in the two genera, may be as readily distinguished by the characters given above as may *H. redivivus* and *H. rufus*, although the difference is of course far greater between the two latter. With but a single specimen of *H. lecontei* for comparison, I cannot verify a single one of the characters adduced by Mr. Brewster as distinguishing this species from *H. redivivus*, although I am obliged to indorse his view of their specific distinctness, since very positive specific characters distinguish them, the most important of which, it appears to me, are the following:

1. H. REDIVIVUS. Tail slightly darker and somewhat browner than the back; lower parts chiefly ochraceous-buff, the crissum more fulvous; auriculars dusky, with distinct whitish shaft-streaks; no distinct dusky "bridle" or paler malar stripe. Wing 3.90–4.30, tail 4.90–5.80, culmen (to exposed base) 1.35–1.75; width of maxilla at nostrils .25–.30, *the lateral outlines gradually but decidedly divergent toward the base;* tarsus 1.45–1.60, middle toe .95–1.12.

2. H. LECONTEI. Tail very decidedly darker but scarcely browner than the back; lower parts (especially abdomen) chiefly dull white, the crissum ochraceous, in marked contrast; auriculars light brownish gray, like occiput, without distinct paler streaks; throat bordered on each side by a distinct dusky "bridle," and a distinct malar stripe of whitish, narrowly barred with dusky. Wing 3.70–3.90, tail 4.57–5.20, culmen (from exposed base) 1.25–1.35; width of maxilla at nostrils .20–.21, *the lateral outlines parallel from near the tip back nearly or quite to nostrils;* tarsus 1.25, middle toe .85.

It will thus be seen that aside from positive characters afforded by the plumage, the proportions of the two species are radically distinct. While the wing and tail average slightly less in *H. lecontei*, the tarsus and middle toe are disproportionately shorter. In fact, both the feet and bill are altogether slenderer, and much more like those of *H. crissalis*, to which there is also a much nearer resemblance in the dusky "bridle" and whitish malar stripe.

The "*Harporhynchus*" *graysoni*, from the island of Socorro, I propose to make the type of a new genus, as follows:

Genus MIMODES, Ridgway.

CH.—Somewhat like *Mimus*, but with the bill decidedly stouter the wing much more rounded, and the colors much more uniform.

Tail much longer than the wing, rounded, but with the four middle rectrices of equal length. Fourth, fifth, and sixth quills longest, the third about equal to the seventh; second not longer than the tenth. Depth of the bill through the base decidedly more than half the length of the gonys, or of the maxilla from the nostril to the tip; gonys less than half the total length of the mandible. Colors plain brown, paler below, without distinct white markings on wings or tail.

Type, *Harporhynchus graysoni* Baird.

In general appearance, the type and only known species of this genus is somewhat intermediate between the species of the genera *Methriopterus* and *Harporhynchus*, having the straighter bill of the former and the uniform brownish coloration of the latter. A close comparison, however, reveals the fact that the species in question is very much more nearly related to the genus *Mimus* than to either of those named above, while at the same time it becomes obvious that it cannot be included in the latter genus, by reason of the very marked distinctive characters pointed out above, in which it differs from every species of *Mimus* with which I have been able to compare it.* The distinctive characters of the two genera may be contrasted as follows:

MIMUS. Depth of bill through base decidedly less than half the length of the maxilla from nostril to tip, and not more than half the length of the gonys; the latter decidedly more than the distance from its base to the malar apex; third, fourth, and fifth quills longest, second longer, equal to or longer than eighth. Tail with more or less of white.

MIMODES. Depth of bill through base decidedly more than half the length of the maxilla from nostril to tip, and also decidedly more than half the length of the gonys; the latter decidedly less than the distance from its base to the malar apex. Fourth, fifth, and sixth quills longest, the second equal to the tenth. Tail without white.

ON A PHOSPHATIC SANDSTONE FROM HAWTHORNE, IN FLORIDA.

BY GEORGE W. HAWES, PH. D.,

Curator in the National Museum.

In connection with the work, upon the products of quarries which is being performed under the auspices of the Tenth Census at the National Museum in Washington, analyses have been made of a stone that is

---

*Including the following: *M. polyglottus* (including its West Indian races), *M. triurus*, *M. gilvus*, *M. saturninus*, *M. calandria*, *M. thenca*, *M. longicaudatus*, *M. "nigriloris,"* *M. hilli*, and *M. melanotis*. I have not seen specimens of *M. dorsalis*, *M. patachonicus*, *M. trifasciatus*, or *M. parvulus*, but these species (except possibly the two latter) appear to be congeneric with those named above. It may prove advisable, however, to separate the three species of the Galapagoes (*M. melanotis*, *M. trifasciatus*, and *M. parvulus*) on account of their very lengthened and slender bill, but I am not prepared to say that this should be done.

quarried in Florida, which has proved to contain ingredients that make it valuable for other than building purposes. To render this information available to those interested in agricultural resources, the analyses that have been made upon this material are now published.

There are very few stone quarries in the State of Florida—in fact almost the only one in actual operation is that at Hawthorne, in Alachua County, which is operated by Mr. C. A. Simmons.

When saturated with its quarry water this stone is quite soft and can be cut with an axe or sawn with much facility, and bricks of any desired shape can be very easily cut from it. The chimneys of the region, and the walls and houses, so far as stone has been used in their construction, are made from blocks that have been taken from this quarry. The material rapidly hardens when exposed to the air and sun, and some structures that were made of it thirty years ago are said to be still in good condition. Cubes 34 inches upon their edges have been extracted, and it is stated that a cube two or three times as large might be obtained. The cubic contents of the excavated space is 800 yards, but the space occupied by the deposit covers a large area and the material is said to be practically inexhaustible. The marl beds which are associated with this rock contain sharks' teeth and bones which mark the Tertiary age of the formation. Professor Smith, who has so recently written upon the geology of Florida, in the American Journal of Science, April, 1881, page 292, states that this bed belongs with the Vicksburg beds which cover so large a portion of the interior of Florida.

This stone possesses properties which evidently render it valuable as a material of construction, especially in the southern latitudes, where frost does not act as a disintegrating agent. It was examined by one of the southern chemists, who stated that it consisted almost entirely of silica and would be good for glass making. The examination of a thin section of this stone, however, indicated that it possesses such a peculiar structure, foreign to a quartz rock, that the necessity of analyses was suggested. These analyses were performed by Dr. A. B. Howe, upon two specimens taken from different portions of the quarry. The first specimen gave the following results:

	1.	II.	Mean.
SiO_2	46.70	46.83	46.765
Al_2O_3	19.53	19.61	19.57
Fe_2O_3	1.79	1.64	1.715
CaO	2.91	2.75	2.83
MgO	.16	.27	.215
P_2O_5	16.12	16.02	16.07
H_2O	14.28	(14.28)	14.28
	101.49	101.40	101.445

The second specimen was like the first, porous, and slightly yellowish in color, but it was softer—a circumstance due to the fact that it had been lately quarried. Its composition was as follows:

	I.	II.	Mean.
SiO$_2$	50.70	50.76	50.73
Al$_2$O$_3$	12.84	12 86	12.85
Fe$_2$O$_3$	1.81	1.85	1.83
CaO	12.07	11.96	12.015
MgO.....	.36	.33	.345
Na$_2$O	.32	.32	.32
K$_2$O	.33	.33	.33
P$_2$O$_5$	12.97	13.12	13.045
H$_2$O	8.39	8.39	8.39
CO$_2$	.86	.86	.86
	100 65	100 78	100.715

The composition of this rock indicates therefore that it might be advantageously employed as a fertilizing material. Although the percentage of phosphoric acid is less than in the best Carolina phosphate, there is no lime to be neutralized by sulphuric acid before liberating the phosphoric acid. I am informed that the extent of the deposit which is represented by these analyses is very large. But the investigation of the value of this material as a fertilizer would of necessity involve further analyses and a more extensive investigation of this aspect of the question than interests us in our consideration of the substance as a building material.

The microscopic structure of this rock indicates that it is composed largely of angular grains of sand which are cemented together by a fibrous material which is probably the phosphate, and by a simple refracting substance which appears to be a mixture of kaolin and hydrous silica. By treating the rock with caustic potash, Dr. Howe dissolved over 7 per cent. of silica from it. The solution used contained 50 per cent. of caustic potash (K O H.); in the first experiment 8.71 per cent. of silica was dissolved, and in the second 7.93 per cent. of silica. This determination is an indication that the hardening of the rock on exposure is due to the presence of this hydrous silica, which might be, in part at least, in a gelatinous condition in the rock, when soaked with its quarry water. Owing to the nature of the components it is not easy to calculate the mineral nature of the phosphate, which is apparently different in the two specimens analyzed. In the first case the acid is apparently combined with alumina and in the second case with lime.

NATIONAL MUSEUM, *June* 29, 1881.

NOTES ON THE NATIVE TREES OF THE LOWER WABASH AND WHITE RIVER VALLEYS, IN ILLINOIS AND INDIANA.

By ROBERT RIDGWAY.

[The accompanying notes on the forest-growth of the Lower Wabash Valley were prepared originally for the use of Professor Sargent in his report upon the forest trees of North America for the Tenth Census. It being impossible, however, for him to utilize more than occasional extracts, he suggested to the writer their publication *"in toto* in some convenient form," so that all interested in this important subject might have the benefit of these observations. It is, therefore, in deference to Professor Sargent's advice that the present paper is herewith presented.]

INTRODUCTION.

Although the field of this paper ostensibly includes the valley of the Wabash and that of its main tributary, White River, from the mouth of the former stream north to where the Ohio and Mississippi Railroad crosses them both (or from latitude 37° 50′ to 38° 50′, approximately), it is proper to state that actual investigations have been made at very few points within the district named, and chiefly in the immediate vicinity of Mount Carmel, Wabash County, Illinois, which alone has been carefully explored. In the limited area comprised within five miles' radius from Mount Carmel, 86 species of trees have been found growing wild, including several which are commonly classed as shrubs, but which there grow to a height of 30 feet or more. Rather protracted observations in Knox County, Indiana, some twenty-five miles to the northeast of Mount Carmel, and in Posey County, 20 miles or more southward, did not increase the list, but extremely desultory observations, made by Dr. J. Schneck, of Mount Carmel, in Gallatin County, Illinois, near the mouth of the Wabash River, where the country is very broken, resulted in the addition of *Juniperus virginiana, Chamæcyparis sphæroidea,* and a *Pinus;** while White County, the next one south of Wabash, adds one more (*Aralia spinosa*). *Robinia pseudacacia* occurs plentifully in the hilly districts in the southern part of both Illinois and Indiana, but has not been met with in the wild state by the writer.

Halesia tetraptera is quoted from Evansville, Ind. (only forty miles south of Mount Carmel), and from "Southern Illinois"; while the Prince Maximilian von Wied, who passed one winter (October 19, 1832, to March 16, 1833) at New Harmony, Posey County, gives, in his *Reise in das innere Nord-America,*† vol. i, p. 209, a list of about 60 species of trees which came under his observation in that vicinity, and among which are included several which have not been found by Dr. Schneck or myself, though it should be stated that our observations in Posey County have been confined to a very limited field. These species are, *"Juglans"*

---

* It is as yet undetermined whether the species is *P. mitis* or *P. inops.*

† Published in Coblenz, 1839.

June 12, 1882.

[= *Carya*] *aquatica*, "*J.*" [= *C.*] *myristicæformis*, *Acer* "*striatum*" [= *A. pennsylvanicum*], *Robinia pseudacacia*, "*Cerasus*" [= *Prunus*] *virginianus*, and *Nyssa sylvatica*. It is, therefore, very likely that several species are to be added to those given in the appended list, thus making an actual total of nearly 100 species of trees which are native to the valley of the lower Wabash.

The most marked features of the woods in the region under consideration, as compared with those of more eastern districts, are, (1) the entire absence of coniferous trees, except in special and usually very restricted localities, and (2) the great variety of species growing together. They are emphatically "mixed woods," it being very rare indeed to find a single species predominating over all others, though in limited sections or particular localities one or another of the oaks (most frequently *Q. alba*), the Sugar Maple or Sweet Gum, may largely prevail; indeed, even the Honey Locust and Catalpa have been noticed, in a single instance each, to form the prevailing growth on a restricted area. Usually, however, from 40 to 50 species of trees are mixed together indiscriminately upon an area approximating, say, 50 to 75 acres, the relative abundance of the component species varying with the location, character of soil, geological formation, and other local causes. The two following lists, made on the spot, are given as typical:

(1) *Area, about 50 acres; situation, about 1½ miles west of Mount Carmel, Wabash County, Illinois, in bottoms of Greathouse Creek; date, September 16, 1876.*

1. Pawpaw, *Asimina triloba.*
2. Silver Maple, *Acer dasycarpum.*
3. Red Maple, *Acer rubrum.*
4. Sugar Maple, *Acer saccharinum.*
5. Honey Locust, *Gleditschia triacanthos.*
6. Coffee-bean, *Gymnocladus canadensis.*
7. Red-bud, *Cercis canadensis.*
8. Wild Plum, *Prunus virginiana.*
9. Wild Cherry, *Prunus serotina.*
10. Crab Apple, *Pirus coronaria.*
11. Cock-spur Thorn, *Cratægus crus-galli.*
12. "Red Haw", *Cratægus* (species undetermined).
13. Sweet Gum, *Liquidambar styraciflua.*
14. Flowering Dogwood, *Cornus florida.*
15. "Black Gum", *Nyssa* (*sylvatica?*).
16. Persimmon, *Diospyros virginiana.*
17. White Ash, *Fraxinus americana.*
18. Blue Ash, *Fraxinus quadrangulata.*
19. Red Ash, *Fraxinus pubescens.*
20. Sassafras, *Sassafras officinale.*
21. White Elm, *Ulmus americana.*

22. Slippery Elm, *Ulmus fulva.*
23. Hackberry, *Celtis occidentalis.*
24. Mulberry, *Morus rubra.*
25. Sycamore, *Platanus occidentalis.*
26. Black Walnut, *Juglans nigra.*
27. Butternut, *Juglans cinerea.*
28. Shell-bark Hickory, *Carya alba.*
29. "Big Shellbark", *Carya sulcata.*
30. "Little Shellbark", *Carya microcarpa.*
31. Black Hickory, *Carya tomentosa.*
32. Broom Hickory, *Carya amara.*
33. Pig-nut Hickory, *Carya porcina.*
34. White Oak, *Quercus alba.*
35. Swamp White Oak, *Quercus bicolor.*
36. "Chinquapin" Oak, *Quercus muhlenbergi.*
37. Michaux's Oak, *Quercus michauxi.*
38. Scarlet Oak, *Quercus coccinea.*
39. Laurel Oak, *Quercus imbricaria.*
40. Water Oak, *Quercus palustris.*
41. Red Oak, *Quercus rubra.*
42. Black Oak, *Quercus tinctoria.*
43. Red Birch, *Betula nigra.*
44. Shining Willow, *Salix lucida.*
45. Cottonwood, *Populus monilifera.*
46. Swamp Cottonwood, *Populus heterophylla.*

The following additional species grew within half a mile of the woods in question, some of them just beyond its borders:

1. Tulip Poplar, *Liriodendron tulipifera.*
2. Box Elder, *Negundo aceroides.*
3. Stag-horn Sumac, *Rhus typhina.*
4. Black Haw, *Viburnum prunifolium.*
5. Winged Elm, *Ulmus alata.*
6. Pecan, *Carya olivæformis.*
7. Spanish Oak, *Quercus falcata.*
8. Black-jack Oak, *Quercus nigra.*
9. Post Oak, *Quercus stellata.*
10. Hornbeam, *Carpinus caroliniana.*
11. Black Willow, *Salix nigra.*
12. Aspen, *Populus tremuloides ?*

Making a total of 58 species of trees, all "hard woods," actually found growing on an area of less than one mile square. In addition to these there would be added in certain portions of the river bottoms the following, so that it is possible to find as many as 75 species on the same area in the vicinity of Mount Carmel:

1. Linden, *Tilia americana.*

2. Large-leafed Linden, *Tilia heterophylla.*
3. Buckeye, *Æsculus glabra?*
4. Water Locust, *Gleditschia monosperma.*
5. Narrow-leafed Crab Apple, *Pirus angustifolia.*
6. Scarlet-fruited Thorn, *Cratægus coccinea.*
7. "Red Haw," *Cratægus subvillosa.*
8. Service Tree, *Amelanchier canadensis.*
9. Green Ash, *Fraxinus viridis.*
10. Black Ash, *Fraxinus sambucifolia.*
11. Catalpa, *Catalpa speciosa.*
12. Mississippi Hackberry, *Celtis mississippiensis.*
13. Overcup Oak, *Quercus lyrata.*
14. Hop Hornbeam, *Ostrya virginica.*
15. Beech, *Fagus ferruginea.*
16. Black Birch, *Betula lenta.*
17. Bald Cypress, *Taxodium distichum.*

The larger number of the species in the last list are, of course, more or less local, but it is believed that every one of them, and also those of the two preceding lists (excepting, perhaps, *Ulmus alata, Quercus falcata, Q. nigra,* and *Q. stellata,* which prefer poorer soils), could be found on an area of less than a square mile in extent, commencing at the bank of the Wabash River, immediately above the mouth of White River, and extending back through the cypress swamp to the bluffs which border the bottom lands. This gives for one square mile of woods, a grand total of more than 70 species of trees, not including several of the larger shrubs (as *Amorpha fruticosa* and *Ilex verticillata*), which here attain almost the stature of trees.

(2) *Area, about 75 acres; location, about 2 miles west of Wheatland, Knox County, Indiana, adjoining the western border of Monteur's Pond; date, May,* 1881.

1. Tulip Poplar, *Liriodendron tulipifera.*
2. Pawpaw, *Asimina triloba.*
3. Silver Maple, *Acer dasycarpum.*
4. Red Maple, *Acer rubrum.*
5. Sugar Maple, *Acer saccharinum.*
6. Box Elder, *Negundo aceroides.*
7. "Dwarf" Sumac, *Rhus copallina.*
8. Smooth Sumac, *Rhus glabra.*
9. Honey Locust, *Gleditschia triacanthos.*
10. Coffee-bean, *Gymnocladus canadensis.*
11. Red-bud, *Cercis canadensis.*
12. Wild Plum, *Prunus americana.*
13. Wild Cherry, *Prunus serotina.*
14. Crab Apple, *Pirus coronaria.*
15. Black Thorn, *Cratægus tomentosa.*

16. "Haw," *Cratægus* (species undetermined.)
17. Sweet Gum, *Liquidambar styraciflua.*
18. Flowering Dogwood, *Cornus florida.*
19. "Black Gum," *Nyssa sylvatica ?*
20. Black Haw, *Viburnum prunifolium.*
21. Persimmon, *Diospyros virginiana.*
22. White Ash, *Fraxinus americana.*
23. Black Ash, *Fraxinus sambucifolia.*
24. Red Ash, *Fraxinus pubescens.*
25. Catalpa, *Catalpa speciosa.*
26. Sassafras, *Sassafras officinale.*
27. White Elm, *Ulmus americana.*
28. Slippery Elm, *Ulmus fulva.*
29. Hackberry, *Celtis occidentalis..*
30. Mulberry, *Morus rubra.*
31. Sycamore, *Platanus occidentalis.*
32. Black Walnut, *Juglans nigra.*
33. Shell-bark Hickory, *Carya alba.*
34. Big Shellbark, *Carya sulcata.*
35. Black Hickory, *Carya tomentosa.*
36. Pig-nut Hickory, *Carya porcina.*
37. Broom Hickory, *Carya amara.*
38. White Oak, *Quercus alba.*
39. Swamp White Oak, *Quercus bicolor.*
40. Bur Oak, *Quercus macrocarpa.*
41. Scarlet Oak, *Quercus coccinea.*
42. Laurel Oak, *Quercus imbricaria.*
43. Water Oak, *Quercus palustris.*
44. Red Oak, *Quercus rubra.*
45. Black Oak, *Quercus tinctoria.*
46. Beech, *Fagus ferruginea.*
47. Hornbeam, *Carpinus caroliniana.*
48. Black Willow, *Salix nigra.*
49. Shining Willow, *Salix lucida.*
50. Swamp Cottonwood, *Populus heterophylla.*
51. Common Cottonwood, *Populus monilifera.*
52. Aspen, *Populus tremuloides.*

Originally, much the larger part of the district under consideration was heavily timbered, and at present the nearest actual prairies to Mount Carmel are distant about 20 to 30 miles in Lawrence and Richmond Counties, Illinois. Since the first settlement of the country,[*] however, the distribution of the timber has very materially changed, much of the original forest having been cleared for cultivation, while on the other hand nearly all the smaller prairies have become trans-

---

[*] Mount Carmel was laid out as a town in 1818, but the surrounding country had already become sparsely settled.

formed into woodland. It is difficult to now estimate what proportion of the orignal growth (considered as to area, little if any being now in its primitive condition) is now standing, but it is stated by those most competent to judge, that on account of this encroachment of the woods upon the former prairies, there is now a greater extent of woodland in Wabash and adjoining counties (in Illinois) then there was fifty years ago. The growth of this new forest is so rapid that extensive woods near Mount Carmel (consisting chiefly of Oaks and Hickories, averaging more than 80 feet high, and 1 to nearly 2 feet in diameter), were open prairie within the memory of some of the present owners of the land!

The original growth of the richer bottom lands and slopes of the bluffs was probably equal in magnitude to that of any other hard-wood forest in Eastern North America; at least the taller trees even now standing considerably exceed in height the dimensions given in standard text-books, and evidently based on the growth of other sections of the country. That this discrepancy of size indicates actual superiority I am, however, loth to believe, but am rather inclined to attribute it to a paucity of measurements of trees in other sections, a view of the case which is considerably strengthened by the fact that the diameter of the larger trees does not greatly exceed that attained in the original forest along the Atlantic seaboard, except, perhaps, in the case of particular species. Certain it is, that the virgin forests of the western slope of the Alleghanies, in West Virginia, and, possibly, that of some portions of Southern Ohio, appear to compare very favorably with those of the lower Wabash region; at least that is the impression which I have received from passing through them repeatedly by rail; while I am confident that in Jackson County, Indiana, near the line of the Ohio and Mississippi Railroad even a larger growth exists at the present time than in most parts of the Lower Wabash Valley, but I have no measurements wherewith to substantiate this impression.

The investigations upon which my knowledge of the timber of the Lower Wabash region is based extend over many years, during which time an opportunity for taking a desirable measurement was never neglected. I have furthermore received much assistance from friends and correspondents interested in the subject, among whom I may especially mention Dr. J. Schneck, of Mount Carmel; his brother, Charles Schneck, of Posey County, Indiana; and Mr. Thos. J. Johnston, county surveyor of Posey County. Dr. Schneck has already published, in Professor Cox's *Geological Survey of Indiana* (volume for 1875, pp. 504–579), a "Catalogue of the Flora of the Wabash Valley, below the mouth of White River," in which may be found most important information respecting the subject in hand; and in reply to letters asking for measurements of the timber of their localities, both of the other gentlemen named above responded with the desired information. The measurements taken by Messrs. Johnston and Schneck are herewith given.

(1.) *Measurements of trees in New Harmony Township, Posey County, Indiana, by Thos. J. Johnston, county surveyor.*

Name of tree.	Circumference at 3 feet from ground.	Distance to first large limb.	Total height.	Remarks.
	Ft. In.	*Feet.*	*Feet.*	
Yellow Poplar. (*Liriodendron tulipifera*)	21	80	145	Hollow base.*
Do	19	60	130	Sound.
White Poplar. (*Liriodendron tulipifera*)	16	70	125	Do.
Do	15° 6'	50	110	Do.
Yellow Poplar. (*Liriodendron tulipifera*)	14° 9'	55	120	Do.
White Oak. (*Quercus alba*)	15	60	115	Do.
Do	15° 4'	54	110	Do.
Do	13° 6'	45	97	Do.
Do	13° 4'	48	107	Partially hollow.
Do	13	43	95	Sound.
Do	12° 5'	35	87	Do.
Black Oak. (*Quercus tinctoria?*)	18	75	128	Do.
Do	17° 6'	60	118	Do.
Do	20	50	102	"Swell but."
Do	14	49	100	Sound.
Do	12° 6'	43	96	Do.
Bur Oak. (*Quercus macrocarpa*)	18° 3'	35	75	Do.
Do	17° 2'	37	80	Do.
Do	14° 7'	31	77	Do.
Do	12° 9'	32	76	Do.
Sweet Gum. (*Liquidambar styraciflua*)	13° 6'	70	115	Hollow
Do	12	60	100	Sound.
Do	11° 8'	62	104	Do.
Do	11° 2'	58	98	Do.
Mulberry. (*Morus rubra*)	10	20	60	Do.
Sassafras. (*Sassafras officinale*)	7° 6'	75	95	Do.
Sugar-tree. (*Acer saccharinum*)	12	48	90	Do.
Maple. (*Acer rubrum?*)	11° 7'	70	108	Do.

* This tree and the next growing near together.

(2.) *Measurements of trees in vicinity of Big Creek, Posey County, Indiana, by Mr. Charles Schneck.*

Name of tree.	Circumference in feet, 3 feet from ground.	Distance to first large limb.	Total height.	Remarks.
Cotton. (*Populus monilifera*)	18	70	165	Bottoms; sound.
Ash. (*Fraxinus americana?*)	13	65	137	Hills; sound.
Oak. (*Quercus alba?*)	14			Hills.
Poplar (*Liriodendron tulipifera*)	15	78	140	Hills; sound.
Do	17½	81	142	Do.
Do	20	91	155	Do.
Do	19½			Hill.
Cotton. (*Populus monilifera*)	19	74	135	Bottoms; sound.
Walnut (*Juglans nigra*)	17½	60	130	Bottoms, a small hollow.
Bur Oak (*Quercus macrocarpa*)	21			Bottoms; sound.
Sycamore. (*Platanus occidentalis*)	22			Do.
Bur Oak (*Quercus macrocarpa*)	18	60	130	Do
Gum (*Liquidambar styraciflua*)	17			Bottoms.
Sycamore (*Platanus occidentalis*)	24			Bottoms: sound.
Lin. (*Tilia americana*)	17			Bottoms

The following extracts from Mr. Johnston's letter accompanying the measurements sent by him may also be of interest:

" The decayed stump of a poplar [*i. e.*, Tulip Tree] is now partly standing near here (New Harmony) that is said by good citizens to have been, when standing, about 37 feet in circumference. . . . There are some Cottonwoods here that I have not mentioned [in the list], some 5 to 6 feet diameter. Some large Sycamores, 'swell-buts,' reach even 37 to 40 feet circumference, but they are hollow."

The heaviest timber in Posey County is said to be in Point Township, in the lower end of the county.

In Dr. Schneck's "Catalogue of the Flora of the Lower Wabash Valley," already referred to, the author gives (on p. 512) a table of measurements, which are said to show the maximum size attained by 23 species of trees, "the measurements in each case being those of one individual."

Name.	Circumference 3 ft. from ground, or above roots and swell.	Height of trunk from root, to first branch.	Total height.
	Feet.	Feet.	Feet.
Pecan (Carya olivæformis)	16	90	175
Black Oak (Quercus coccinea var tinctoria)	20	75	160
Bur Oak (Quercus macrocarpa)	22	72	165
White Oak (Quercus alba)	18	60	150
Persimmon (Diospyros virginiana)	5½	80	115
Black Walnut (Juglans nigra)	22	74	155
Honey Locust (Gleditschia triacanthos)	18	61	129
Catalpa (Catalpa bignonoides). [= C. speciosa!]	6	48	101
Mulberry (Morus rubra)	10½	20	62
Scarlet Oak (Quercus coccinea)	20½	94	181
Sassafras (Sassafras officinale)	7¾	75	95
Bass-wood (Tilia americana)	17½	50	109
Bald Cypress (Taxodium distichum)	18¾	74	146
Red Maple (Acer rubrum)	13	60	108
Sycamore (Platanus occidentalis)	33½	68	176
Tulip Tree (Liriodendron tulipifera)	25	91	190
White Ash (Fraxinus americana)	17½	90	144
Cottonwood (Populus monilifera)	22	75	170
Sweet Gum (Liquidambar styraciflua)	17	80	164
Black Hickory (Carya tomentosa)	10½	55	112
Sugar Maple (Acer saccharinum)	12½	60	118
Water Oak (Quercus palustris)	12	23	120
Beech (Fagus ferruginea)	11	10	122

It may be remarked that the size indicated by the above figures is, in the case of some species, highly exceptional, and that I have measured none so large. Not that a single one of the three measurements given is so very unusual (though this is in some cases true as regards height), but that it is exceedingly uncommon to find such extreme measurements of girth, length of trunk, and total height combined in a single tree.

According to measurements thus far made it has been determined beyond doubt that at least thirty-four species of trees reach or exceed a height of 100 feet, and it is all but certain that some ten or a dozen more, of which no measurements have been taken, also reach this height. No less than eleven reach, occasionally, at least, a height of 150 feet, the greatest height of any tree, so far as determined by accurate measurements, being 190 feet (*Liriodendron*); two (*Liriodendron* and *Quercus coccinea, fide* Dr. Schneck) reach a height of 180 feet; four reach 170 feet;

eight attain 160 feet; eleven grow 150 feet high; thirteen 140; sixteen reach 130; twenty-three reach 120 feet; twenty-seven 115 feet; twenty-nine grow to 110 feet; and thirty-two exceed 105 feet.

The following list of the species determined as growing to 100 feet elevation or more shows the maximum height according to the independent measurements of Dr. Schneck, Mr. Charles Schneck, Mr. Thomas J. Johnston, and myself:

List of trees attaining a height of 100 feet or more in the Lower Wabash Valley.

No.	Name.	Maximum height.
1	Liriodendron tulipifera	+150, R. R.; 155, C. S ; 145, T. J. J.; 190, Dr. S.
2	Tilia americana	130, R. R.; 109, Dr. S.
3	Acer dasycarpum	118, R. R.
4	Acer rubrum	108, R. R , Dr. S , T. J. J.
5	Acer saccharinum	115, R. R, 118, Dr. S.; 90, T. J. J.
6	Gymnocladus canadensis	109, R. R.
7	Gleditschia triacanthos	137, R. R ; 129, Dr. S
8	Liquidambar styraciflua	144, R. R , 115, T. J. J.; 164, Dr. S.
9	Nyssa (sylvatica?)	125, R. R.
10	Diospyros virginiana	115, Dr. S.
11	Fraxinus americana	144, R R., Dr. S.; 137, C. S.
12	Fraxinus quadrangulata	124, R. R.
13	Catalpa speciosa	101, Dr S.
14	Ulmus americana	119, R R.
15	Celtis occidentalis	134, R. R.
16	Platanus occidentalis	168, R. R , Dr. S.
17	Juglans cinerea	117, R. R.
18	Juglans nigra	156, R R , 155, Dr. S.; 130, C. S.
19	Carya alba	129, R. R.
20	Carya amara	113, R. R.
21	Carya olivæformis	175. Dr. S.
22	Carya tomentosa	+107, R R.; 112, Dr. S.
23	Quercus alba	142, R. R.; 150, Dr. S., 115, T. J. J.
24	Quercus bicolor	+100, R. R.
25	Quercus coccinea	181, Dr. Schneck.
26	Quercus imbricaria	100, R R
27	Quercus macrocarpa	162, R. R.; 165, Dr. S., 130, C. S.; 80, T. J. J.
28	Quercus muhlenbergi	122½, R. R.
29	Quercus palustris	119, R R ; 120, Dr. S.
30	Quercus rubra	150, R R.
31	Quercus tinctoria	128, T. J. J.; 160, Dr. S : +100, R. R.
32	Fagus ferruginea	122, Dr. S
33	Populus monilifera	140, R. R , 165, C. S , 170, Dr. S.
34	Taxodium distichum	147, R. R.; 146, Dr. S.

In addition to the above there are several other trees large specimens of which have not been measured, but which, with scarce a doubt, occasionally, at least, reach 100 feet in height, thus rendering it very probable that in reality about fifty species attain this elevation. These species are the following:

* 1. *Magnolia acuminata.*
2. *Tilia heterophylla.*
* 3. *Robinia pseudacacia.*
4. *Prunus serotina.*
5. *Fraxinus pubescens.*
6. *Fraxinus sambucifolia.*
7. *Fraxinus viridis.*
8. *Celtis mississippiensis.*
9. *Carya porcina.*
10. *Carya sulcata.*
11. *Quercus michauxi.*
12. *Quercus falcata.*
13. *Quercus lyrata.*
14. *Quercus stellata.*
* 15. *Castanea vulgaris americana.*
* 16. *Chamæcypharis sphæroidea.*
* 17. *Pinus (mitis?).*

* These trees, though growing within the field of this paper, have not been met with by the writer.

The measurements given under the head of the species enumerated in the following list include all the reliable ones which 1 have made up to date, or which I have been able to get upon unimpeachable authority, and, it should be understood, cancel all measurements or estimates previously published by me *when in excess of those here given.* They include *no estimates of height,* but only actual tape-line measurements of prostrate trees or else very careful measurements of isolated standing trees with a thoroughly-tested "dendrometer," although the specimens measured by the latter method are very few indeed.

The following species, usually classed as shrubs, are not included, though some of them may occasionally reach 30 feet in height. No measurements, however, have been taken of any of them :

1. *Xanthoxylum americanum.* Prickly Ash.
2. *Ptelea trifoliata.* Hop Tree; Wafer Ash.
3. *Euonymus atropurpureus.* Burning Bush; Waahoo.
4. *Hydrangea arborescens.* Wild Hydrangea.
5. *Hamamelis virginica.* Witch Hazel.
6. *Ilex decidua.* Deciduous Holly.
7. *Forestiera acuminata.* Forestieria.
8. *Lindera benzoin.* Spice Bush.
9. *Alnus serrulata.* Smooth Alder.
10. *Aralia spinosa.* Hercules' Club; "Devil's walking-stick."

On the other hand, a small number which are not usually classed as trees are so considered here, having been found to attain, occasionally, at least, a height of 30 feet or more. They are the following :

No.	Name.	Maximum height as measured.
1	Ilex verticillata	28 feet, but taller ones seen.
2	Rhus glabra	30 feet
3	Rhus copallina	33½ feet.
4	Amorpha fruticosa	35 feet.

SMITHSONIAN INSTITUTION, *July* 20, 1881.

CATALOGUE.

1. (1.) *Magnolia acuminata.* Cucumber Tree.

I have never seen a tree of this species growing in any part of the district under consideration. I have *heard,* however, that a few grow on Sugar Creek, in the southern part of Wabash County, but have been unable to verify the rumor. It grows quite abundantly in the extreme southern portion of Illinois (Johnson and Union Counties), where the

---

* The number in parenthesis prefixed to the name of a species corresponds in each case with that given in Professor Sargent's *Catalogue of the Forest Trees of North America,* published by the Census Bureau (Washington, 1881). When no second number is given, the species is one not included in the catalogue in question.

country is very hilly, and therefore adapted to it. The nearest point in Indiana where I can find a record of its occurrence is Orange County, the third county east from Knox.

2. (8.) *Liriodendron tulipifera.* Tulip Tree; "Poplar."

Formerly very abundant, and still common in some localities. The great demand for poplar lumber for weatherboarding, etc., has greatly depleted the supply, however. Although growing both on the hills and in the river bottoms, the growth of the former will probably average larger than the latter. The larger trees of this species now standing will average about 5 feet diameter and 140 feet high, though specimens of much larger size may still be found, and formerly were numerous. A few yet exist, having a diameter of 7 or even 8 feet, but they are exceedingly rare. Straight trunks of 50 to 70 feet clear are occasionally found, and twenty years ago trunks 100 feet long were not so very unfrequent.

Lumbermen recognize three varieties of the "poplar"—the "yellow," "white," and "blue," distinguished, however, only by the color of the wood. The first is the most abundant, and produces the best lumber.

This species flowers during the first half of May, leafing the first half of April.

List of specimens measured. [*]

Specimen.	Girth above swell at base.	Distance from ground to first large limb.	Total height.	Locality.	Authority.
a	15	78	140	Hills, Posey County, Indiana	Charles Schneck.
b	17½	81	142	......do	Do.
c	20	91	155	......do	Do.
d	19½			......do	Do.
e	21	80	145	Posey County, Indiana	Thos. J. Johnston. ("Yellow.")
f	19	60	130	......do	Do.
g	16	70	125	......do	Thos. J. Johnston. ("White.")
h	15½	50	110	......do	Do.
i	14¾	55	120	......do	Thos. J. Johnston. ("Yellow.")
j	37			......do	Thos. J. Johnston. (Stump.)
k	12		143	Wabash County, Illinois	Dr. J. Schneck.
l	20	70	153	......do	Do.
m	23½	50	139	......do	Do.
n	20	60	168	......do	Do.
o	19½	82	145	......do	Do.
p	12	88	120	......do	Do.
q	23	74	158	......do	Do.
r	19½	61	142	......do	Do.
s	19	70	140	......do	Do.
t	23	72	158	......do	Do.
u		120		......do	Do.
v	20	100		......do	R. Ridgway.
w	19	58		......do	Do.
x	17	70		......do	Do.
y	19	64		......do	Do.
z	+ 26	+ 50		......do	Do.
a′	15¾		145	Knox County, Indiana	Do.
b′	15¼	47	145½	......do	Do.
c′	14	84		Wabash County, Illinois	Thos. Hoskinson.
d′	23	62	158	......do	Dr. J. Schneck.
e′	22			......do	R. Ridgway.

[*] The measurements are in feet.

With the exception of the last two, the trees of the above list were all felled, and the total length measured with a 100-foot tape-line. The two exceptions were fine, vigorous, standing trees, and their height measured with a "dendrometer." Standing isolated, this was easily done, and the measurements are no doubt perfectly accurate.

The finest tree of all those given above was example q, which at 74 feet measured 6 feet in diameter, the trunk being perfectly sound even at the extreme base, and straight as a column.

The longest trunk (example u) was cut into ten 12-foot logs. It was not very large, however, measuring, if I remember rightly, about 4 feet in diameter at the butt and less than three feet through at the top of the last cut. A trunk measuring 84 feet in length (sawed into seven 12-foot logs), measured 54 inches in diameter at the butt and 42 inches at the small end of the last cut. This is the tree marked c' in the list.

At the "Timber Settlement" in Wabash County, I measured, in May, 1881, a solid stump of this tree, which, although entirely denuded of bark and with a considerable portion hewn off for firewood, was still 26 feet in circumference at 4 feet from the ground. A portion of the trunk still lying on the ground was 50 feet or more in length, and had apparently supplied the occupants of a deserted cabin near by with firewood for many years.

The example marked v was 35 feet in circumference at the ground, and at 150 feet from the base the several branches were 1 to $1\frac{1}{2}$ feet in diameter. The top branches, broken off and scattered by the falling of the tree, had been collected for firewood, so that its total height could not be measured, but could not have been much less than 190 feet, which is the maximum height as given by Dr. Schneck in his "Flora of the Wabash Valley" (Cox's *Geological Survey of Indiana*, 1875, p. 512).

3. (10.) *Asimina triloba.* Pawpaw.

The Pawpaw is a very abundant underwood in all bottom lands and other damp woods, growing usually to a height of 20 to 30 feet, and 2 or 3 inches in diameter, but not unfrequently 40 feet or more in height, and, in exceptional cases, nearly a foot in diameter. The two largest specimens measured (both in the bottoms below Mount Carmel) were 46 and 43 feet, respectively, in height, the larger being 32 inches in circumference, the smaller only 10 inches around.

Two well-marked varieties are distinguished by the fruit, which in one has the pulp a rich golden yellow, very aromatic, and exceedingly sweet, and much liked by most people, though too rich for many. This variety is known as the "Yellow Pawpaw"; the other, called "White Pawpaw," has a whitish or very faintly yellow, insipid, or disagreeable tasting fruit, and is seldom eaten. I am unable to state whether any peculiarity of flower or foliage distinguishes the two varieties.

4. (14.) *Tilia americana.* American Linden. "Lin."

A very common tree, growing chiefly near the river banks, but occurring in all rich woods. The average height of the larger trees is about 100 feet, but an elevation of 125 or even 130 feet is sometimes reached, the diameter of large trees averaging about 3 feet. In the Wabash bottoms single trunks of the "Lin" are exceedingly rare, fully 80 per cent. of the trees consisting of compound trunks, as if several trees had grown up close together and become more or less completely coalesced at the base.

The following measurements are of trees of rather exceptional size:

Specimen.	Girth above swell at base.	Distance from ground to first large limb.	Total height.	Locality.	Authority.
a	8	53	110	Wabash County, Illinois...............	R. R. (Hills.)
b	22⅓	....	....	Gibson County, Indiana................	R. R. (Bottoms.)
c	17	....	....	Wabash County, Illinois..............	Do.
d	....	....	130	......do	Do.
e	17½	40	....	......do	Do.
f	13	....	125	......do	Do.
g	....	62	....	......do	Do.
h	17½	50	109	......do	Dr. J. Schneck.

Example *b* was the largest I have seen, but was divided into three trunks a short distance from the ground.

5. (15.) *Tilia heterophylla.* White Basswood.

This tree has been found near Mount Carmel by Dr. Schneck, but I am not autoptically acquainted with it. Possibly some of the measurements given under the head of *T. americana* belong to this species.

6. (—.) *Ilex verticillata.* Black Alder.

Very abundant about the borders of ponds and swamps, and the mouths of the creeks, forming dense almost impenetrable thickets. In some localities it grows to a height of 20 feet and upwards, with a stem 2 to nearly 3 feet in circumference. The two largest measured were 2 feet 11 inches, and 2½ feet, respectively, in girth, and the tallest (cut down especially for measurement) 28 feet high. Taller specimens, which were apparently about 35 feet high, were seen in the Cypress swamp, in the lower part of Knox County, Indiana.

7. (40?) *Æsculus glabra?* Smooth Buckeye?

Although I give the species as *Æ. glabra*, on the strength of Dr. Schneck's identification, I am not sure but that we have the *Æ. flava* also. The specimens examined by me (a considerable number, in the bottoms nearly opposite the village of Rochester, Wabash County), were 70 or 80 feet high, and some of them 2 feet or more in diameter, thus appearing too large for _Æ. glabra._ Whichever it may be, however, the Buckeye is a very local tree in the Wabash Valley, and I have only seen it in the locality mentioned, where it appears to be confined wholly to a belt of only a few hundred yards width, a few trees only being found on the opposite side of the river. I am unable to ascribe any reason for this restriction of its range, since the same trees, and other vegetation associated with it, occur throughout the bottoms on either side. It is said to be common among the hills of Gibson County, several miles back from the river, and there to attain a height of 100 feet or more, and a diameter of 3 feet.

8. (47.) _Acer dasycarpum._ Silver Maple.

A very abundant tree along the banks of rivers and large streams, attaining an average elevation of 90 to 100 feet, and a diameter of 2 to 3 feet. Unlike the Red Maple (_A. rubrum_) the trunk usually divides low down, usually at about 8 to 15 feet from the ground; the three or more secondary trunks, however, extending upward for a considerable distance before branching.

Of four trees measured, the extremes were: height, 90 and 118 feet; circumference, $12\frac{1}{2}$ to 14 feet; trunk, 20 feet (only one measured).

Flowers early in April, leafing from March 31 to April 12.

9. (51.) _Acer rubrum._ Red Maple.

A very common tree, but much more local than _A. dasycarpum._ Is almost wholly confined to swamps or very wet bottoms, where it grows tall, straight, and slender. In size it is about equal to _A. dasycarpum_ and _A. saccharinum_, but is much more slender than either, with a less spreading top. Three specimens measured 70 to 108 feet in height, the average being $95\frac{1}{3}$ feet; $10\frac{1}{2}$ to 15 feet in circumference (average 12.83), clear trunk, 49 to 60 feet. Decidedly taller trees occur, however, those measured being prostrate ones, of by no means the largest size.

Flowers middle of February to March 20, according to the season; leafs out last of March to April 12.—(SCHNECK.)

10. (52.) _Acer saccharinum._ Sugar Maple.

A very abundant tree in some localities, rare or wholly wanting in other portions. Occasional "sugar groves" occur where, over a space of several acres, scarcely a single tree of any other species can be found.

The larger trees of this species average about 100 feet high (the average of the five specimens measured being $108\frac{2}{3}$ feet, the extremes 90 and 118), and $2\frac{1}{2}$ to 3, occasionally over 4, feet in diameter. The trunk, like that of _A. rubrum_, is frequently tall and straight, four specimens measured being, respectively, 47, 48, 60, and 70 feet to the first limb. The var. _nigrum_ and the common form appear to be about equally numerous, each, however, predominating, or even wholly replacing the other, in particular localities.

Flowers as early as March 10 (SCHNECK), leafing April 15 to 20.

11. (53.) *Negundo aceroides.* Box Elder.

A very common, and in some localities abundant, underwood in rich bottoms. The larger trees of this species are 2½ to 3—rarely 4—feet in diameter, and 50 to 60, possibly 70, feet high. No measurements for height have been made, but the tallest specimens do not approach the elevation of the oaks and other trees with which they are associated.

Flowers March 20, leafing the last week in April.

12. (56.) *Rhus typhina.* Stag-horn Sumac.

In most localities less common than *R. glabra.* The largest speci-mens observed were about 30 to 35 feet high, and 4 inches in diameter.

13. (——.) *Rhus glabra.* Smooth Sumac.

Much the commonest species, and when growing in woods or thickets attaining a height of 30 to 35 feet. Near Monteur's Pond, in Knox County, Indiana, I found this species and *R. copallina* growing together, and to about an equal size.

14. (——.) *Rhus copallina.* "Dwarf Sumac." (!)

A very common species in some localities. Near the northwestern border of Monteur's Pond, in Knox County, Indiana, it is an abundant underwood, growing frequently to a height of 25 to 30 feet, and 4 inches or over in diameter. Three specimens (the only ones measured) were, respectively, 25½, 31¼, and 32½ feet in height (all being cut down for measurement), 6, 7, and 1½ feet trunk, and 14, 8, and 29 inches in circumference. The last consisted properly three stems united at the base, though near the ground the coalescence of the wood was almost complete, while externally there was no evidence of the triple nature of the trunk. A section of this trunk, also leaves and fruit of the same tree, has been deposited in the museum of the Agricultural Department.

15. (——.) *Amorpha fruticosa.* False Indigo.

In the cypress swamps of Knox County, Indiana, I found this shrub growing to a very unusual size, many specimens being 20 feet and up-wards in height. The largest one seen was cut down for measurement, and found to be 35 feet high; it was 17 inches in circumference at the base, and contained eighteen annual rings.

16. (65.) *Gymnocladus canadensis.* Coffee-bean; Coffee-nut.

Scarcely one of our native trees is more local in its distribution than the present species, and there are few localities indeed where it can be said to be abundant. It is usually found scattered through the richer bottoms.

It is never a large tree, but grows tall and slender, frequently reaching 100 feet in height, though seldom over 2 feet in diameter, and with a rather scant top. One tree, cut expressly for measurement, was 109 feet in length, 76 feet to the first limb, and only 20 inches in diameter across the stump. The largest trunk was that of a tree growing in a door-yard, and possibly a cultivated specimen. It was 8 feet in circumference, but ramified at about 4 feet from the ground into several upright branches. The top was dense and symmetrical, the summit elevated about 80 feet.

17. (66.) *Gleditschia monosperma.* Water Locust.

An abundant species in the cypress swamps in the lower part of Knox County, Indiana, where it grows along with the Large-leafed Cottonwood (*Populus heterophylla*), White Ash (*Fraxinus americana*), Black Willow (*Salix nigra*), and other swamp trees. It is a very much smaller tree than *G. triacanthos* and of quite different appearance, having a smoothish, dull-gray bark (much like that of the Hackberry, *Celtis*), and very crooked, scraggy growth. The largest specimen measured was 7 feet in circumference and 65 feet in height.

18. (67.) *Gleditschia triacanthos.* Honey Locust.

When growing to its full perfection, the Honey Locust is one of the most majestic trees of the forest in which it is native. Many trees occur which are 120 to nearly 140 feet high, with straight trunks of 50 to 70 feet clear, and 4 to 5, occasionally even 6, feet in diameter. There are none of our trees, excepting only the Bald Cypress and Catalpa, which have a more thoroughly characteristic appearance, its tall, straight, but usually inclined trunk of a dark iron-gray or nearly black color being much darker than any other species, and thus easily identified at a considerable distance, while the extremely delicate foliage renders its top equally conspicuous by its contrast with the adjacent tree tops. The Honey Locust usually, like very many other trees, occurs singly throughout the richer woods, but it is occasionally multiplied so as to form the prevailing growth. It was found thus multiplied over an area of a hundred acres or more in the White River bottoms of Gibson County, Indiana, where the trees of this species constituted more than half the forest, and averaged 2 to 3 feet in diameter and 100 feet high, with occasional specimens of considerably larger size.

The finest tree of this species which I have ever seen was an isolated one standing near the roadside in Posey County, Indiana. It was tall and straight, with a widely-spread, symmetrical top, the trunk measuring 18 feet in circumference at a yard from the ground, and about 60 feet to the first limbs. It was apparently sound throughout, and was not less than 120 feet high.

The following measurements are of rather unusually large specimens:

Specimen.	Girth above swell at base.	Distance from ground to first large limb.	Total height.	Locality.	Authority and remarks.
a	17	50	....	Posey County, Indiana	R. R. About 130 feet high.
b	15	(70?)	....	Gibson County, Indiana	Do.
c	14	63	137	Wabash County, Illinois...............	R. R.
d	13	70	130	. do	R. R. Ambitus, 50 feet
e	18	61	129	Posey County, Indiana	Dr. J Schneck and R R.

19. (58.) *Robinia pseudacacia.* Black Locust.

Not observed in a native state by Dr. Schneck or myself in Wabash or adjoining counties in Illinois, or in Knox, Gibson, and Posey Counties, Indiana. Given by Maximilian, however, in his list of the trees found in the latter county, where, probably, found only in hilly localities.

20. (70.) *Cercis canadensis.* Red-bud.

A very abundant underwood in all rich woods, but attaining its greatest development in the bottom lands, where specimens 40 to 50 feet high and 1 foot in diameter are not uncommon. The following measurements have been taken:

Specimen.	Girth above swell at base.	Distance from ground to first large limb.	Total height.	Locality.	Authority.
a	3⅜	15	50	....	Wabash County, Illinois...............	R.. R.
b	5¼	19	46	41	......do	R. R.
c	2⁷⁄₄	23½	54	....	Knox County, Indiana...............	R. R.
d	1₁⁷₂	16⅔	41	....	......do	

Flowers April 10 to 15, leafing from the 15th to the 20th of the same month.

21. (76.) *Prunus americana.* Wild Plum.

22. (78.) *Prunus chicasa.* Chickasaw Plum.

Wild Plums are very abundant, but whether the *P. chicasa* is common in the wild state I do not know. I have seen cultivated trees, however, which were about 20 feet high and nearly a foot in diameter. *P. americana* is usually 15 to 20, sometimes 30 feet high, and flowers April 10.

June 12, 1882.

23. (81.) *Prunus serotina.* "Wild Cherry."

Once very common, the wild cherry is now rare in most portions of the Wabash Valley. It is partial to the hilly country back from the river, and it is there that the trees of this species attain the largest size. They were formerly found 100 or more feet high and 3 to 4 feet in diameter, and a few may perhaps still be found having this stature. I have measured but a single tree of this species, however, the one in question being 7¼ feet in girth, 31 feet to the first branch, and 94 feet high, being by no means so large as some that might be found.

Flowers about the middle of April, leafing a little later.

24. (86.) *Pirus angustifolia.* Narrow-leaved Crab Apple.

This species has been found in Wabash County by Dr. J. Schneck (see Cox's *Geological Survey of Indiana*, 1875, p. 528). It is perhaps not so common as *P. coronaria*, but blooms at about the same time. According to Dr. S. it is "usually taller than *P. coronaria*."

25. (87.) *Pirus coronaria.* Crab Apple.

Common in rich woods, sometimes forming extensive thickets. I have made no measurements, but would say that trees 25 to 30 feet high and nearly a foot in diameter are occasionally found; trunks 6 to 8 inches through, being, however, more common. It blooms in April and May, leafing about the middle of the former month.

26. (94.) *Cratægus coccinea.* Scarlet-fruited Thorn.

"Open upland woods; not rare; April, May." (SCHNECK.) No measurements taken.

27. (95.) *Cratægus cordata.* Washington Thorn.

Given in Patterson's catalogue of the plants of Illinois on Dr. Schneck's authority.

28. (96.) *Cratægus crus-galli.* Cockspur Thorn.

"Low moist thickets; common; March to May." (SCHNECK.) No measurements.

29. (101.) *Cratægus subvillosa.* "Red Haw."

River banks chiefly; common; blossoms in April and May. A specimen (cut down) measured 37 feet in height, 2¼ feet in circumference.

30. (102.) *Cratægus tomentosa.* Black Thorn.

"Thickets; rare; March, April." (SCHNECK.) No measurements; begins to leaf April 22 to 25. (SCHNECK.)

31. (105.) *Amelanchier canadensis.* June Berry.

Found by Dr. Schneck, but not recognized by the writer, and probably rare.

32. (106.) *Liquidambar styraciflua.* Sweet Gum.

One of the most abundant trees in the river bottoms, where in some places it constitutes the prevailing growth. It is one of the tallest and stateliest of forest trees, frequently attaining an elevation of 130 feet, and occasionally of 150 feet or more, with straight trunks 60 to 80 feet clear and 4 feet in diameter. Only the Tulip Tree (*Liriodendron*) rivals it in altitude of the trunk, but in symmetry cannot be compared to it, except in occasional instances. As frequently seen, it has by far the tallest and straightest shaft of any tree in the forest. One trunk 71 feet long measured only 8 inches less in diameter at the small end than at the lower, where the diameter was a little less than 3 feet. Another trunk 94 feet long was only 11½ feet in girth at the large end. The two largest specimens seen each measured 17 feet in circumference, one of them having a trunk of 80 feet clear. The tallest tree measured was one cut for lumber, and was 164 feet in total length.

Blossoms in May.

Specimen.	Girth above swell at base.	Distance from ground to first large limb.	Total height.	Locality.	Authority.
a	17	....	....	Posey County, Indiana..........	Schneck.
b	13½	70	115	..... do	Thomas J. Johnston.
c	12	60	100	.....do	Do.
d	11⅜	62	104	.....do	Do.
e	11½	58	98	..... do	Do.
f	....	....	144	Wabash County, Illinois	R. R.
g	9¼	....	140	.....do	Do.
h	7	....	120	.....do	Do.
i	13	....	....	.....do	Do.
j	....	83	....	..... do	Do.
k	13	81	....	.....do	Do.
l	....	76	....	.....do	Do.
m	12½	78	....	..... do	Do.
n	11½	94	137	..... do	Do.
o	9	71	....	.....do	Do.
p	17	80	164	......do	Dr. J. Schneck.
q	11	(70 ?)	127	Knox County, Indiana	R. R.
r	10¾	....	128	..... do	Do.
s	12	(90 ?)	129	..... do	Do.
t	13	....	128½	......do	Do.

The tree marked *m* was straight as an arrow, and not less than 135 feet high; the top spread 85 feet. No. *o* was 2 feet 2 inches in diameter at the upper end.

33. (114.) *Aralia spinosa.* Angelica Tree; "Devil's Walking Stick."

Not seen in Wabash County, but grows in White, the next county south.

34. (115.) *Cornus florida.* "Dogwood."

A very abundant tree in upland woods. Occasionally reaches 50 feet or more in height, and a foot or more in diameter, but is usually much smaller. The only trees measured, two of rather exceptional size, were 3½ and 4⅛ feet in circumference, with trunks 30 feet clear. The total height of the first (a standing tree) was estimated at 60 feet; the latter (prostrate, and measured with tape-line) was 50 feet long.

Blossoms in April or May, and commences to leaf about April 20.

35. (119?) *Nyssa multiflora?* " Black Gum."

A very abundant tree both on uplands and in the bottoms. Grows tall and slender, with few large branches except at the extreme summit, but the trunk frequently thickly set with small horizontal branches to near the ground, thus closely approximating the "excurrent" growth characteristic of many *Coniferæ*. Growing on thin or dry soils, its height does not usually much exceed 70 or 80 feet, but on rich lands an elevation of 100 to 120 feet or more is sometimes reached, one specimen being 125 feet long, 13 feet in circumference, and the trunk entirely free from branches for 64 feet. An exceptionally large specimen, which may possibly have been *N. uniflora*, growing in the bottoms of Posey County, Indiana (but not in water), was 18 feet in circumference, and proportionately tall.

It may be that some of our so-called "Black Gums" may be *N. sylvatica*, but of this I am not certain.

Begins to leaf May 1.

36. (123.) *Viburnum lentago.* Sweet Viburnum; Sheep Berry.

"Dry, open wood, scarce." (Dr. SCHNECK.)   No measurements.

37. (124.) *Viburnum prunifolium.* Black Haw.

Very abundant on rich lands. Blossoms in April or May. No measurements have been taken, but no specimens exceeding 25 feet in height have been observed.

38. (———.) *Viburnum dentatum.* Arrow-wood.

"I have seen but one tree. May, June." (SCHNECK.)

Begins to leaf the last week in March, and blooms about the 10th of April.

39. (143.) *Diospyros virginiana.* Persimmon.

Common everywhere. When growing in the thick bottom-forest is frequently 100 feet or more in height, the tallest specimen measured being 115 feet high, 80 feet to the first limb, but only 5½ feet in girth at the base, or less than 2 feet in diameter! When growing in open fields or along roadsides (where it is most frequently seen), it forms a more spreading tree, usually 30 to 40, and rarely more than 50, feet high.

40. (148.) *Fraxinus americana.* White Ash.

Very common in the bottom lands, where it becomes one of the very tallest trees, an altitude of 140 feet being not uncommon, while clear trunks of 60 to 90 feet are occasionally met with. When growing in very wet lands it becomes greatly enlarged at the base, some such trees measuring 30 feet in girth at the ground, but rapidly contracting, so that at 20 feet they diminish one-half to two-thirds in bulk. These "swell-butt ashes" are said to decay first at the top, and to be sometimes solid at the base. Following is a list of measurements of large trees of this species:

Specimen.	Girth above swell at base.	Distance from ground to first large limb.	Total height.	Locality.	Authority.
a	16½	....	....	Wabash County, Illinois..............	R. R.
b	17½	....	....	......do	Do.
c	13	83	144	......do	Do.
d	10	90	....	......do	Do.
e	13	65	137	Posey County, Indiana	Charles Schneck.
f	27	....	....	Wabash County, Illinois..............	R. R.
g	20	....	....	......do	Do.
h	29	....	....	......do	Do.
i	17½	90	144	......do	Do.
j	15½	45½	....	......do	Do.
k	11	50	143	Knox County, Indiana.................	Do.
l	12	47	105	......do	Do.

Tree marked c was 9 feet in circumference at the small end of the trunk, which was perfectly solid throughout; d was 7½ feet in girth at the small end; j was a prostrate tree with the top totally destroyed, but at 100 feet from the base were six branches averaging nearly 1 foot in diameter, so that it could not have been much less than 140 feet long.

41. (154.) *Fraxinus pubescens.* Red Ash.

Rather rare. No measurements.

42. (155.) *Fraxinus sambucifolia.* "Black Ash"; "Hoop Ash."

"Swamps and wet places; not rare." (SCHNECK.) Abundant in the northern portion of Mouteur's Pond, Knox County, Indiana, where it grows tall and slender, frequently 80 and occasionally nearly or quite 100 feet high, the only specimen measured being 83 feet long, trunk 57 feet, diameter (at 5 feet from the ground—the base being considerably swollen), 1½ feet. This tree presents so very close a resemblance in bark, foliage, and general aspect to young Pecan trees (*Carya olivæformis*), as to be not readily distinguished, except by experts.

43. (156.) *Fraxinus quadrangulata.* Blue Ash.

Common in rich hilly woods; resembles in general appearance *F. americana*, but is smaller and more slender. Four freshly cut trees, felled on a space including not more than two acres, were 2 to 2½ feet in diameter (across top of stump), 51 to 76 feet clear trunk, and 116 to 124 feet long. A fine tree still standing on the same piece of ground was 13 feet in girth, and at least 50 feet to the first limb.

44. (157.) *Fraxinus viridis.*   Green Ash.

Not uncommon in wet woods; no measurements.

45. (165.) *Catalpa speciosa.*   Catalpa; "Patalpha"; "Wahoo."

Formerly abundant in rich bottom lands, but now nearly exterminated in many localities. Trees of 100 feet or more in height were formerly not uncommon, while a diameter of 4½ feet has been reported (see Cox's *Geological Survey of Indiana*, 1873, p. 417). The usual dimensions, however, are, for the larger trees, 70 to 90 feet high, and 2½ to 3 feet diameter. It is usually, however, decidedly smaller, and when growing in open situations forms a low spreading tree, seldom more than 50 feet in height, and frequently much less. Trees of this character were formerly very abundant in the bottoms about a mile above Mount Carmel, but they have nearly all been cut for fence-posts.

In Posey County, Indiana, while making inquiries of an intelligent gentleman regarding the timber of his neighborhood, I was informed that the day before he had cut a Catalpa, the trunk of which produced eight 7-foot post-cuts, the diameter at the base being 4 feet, while the total length of the tree he estimated at about 130 feet.

Blossoms late in May or early in June (seen in full bloom near the O. and M. R. R., between Shoals and Huron, Indiana, May 30, 1881.)

Specimen.	Girth above swell at base.	Distance from ground to first large limb.	Total height.	Locality.	Authority.
a	8	....	90	Wabash County, Illinois..............	R. R.
b	10	60	... .	....do	R. R.
c	6	48	101	Posey County, Indiana................	Dr. J. Schneck.

46. (171.) *Sassafras officinale.*   Sassafras.

Very common, and in rich woods growing to a large size. The lumber of this tree is more highly prized than any other for skiffs, being light, strong, and durable. It is also much used for fence-posts and rails. Although averaging perhaps not more than 50 feet in height and a foot in diameter it is occasionally much larger, reaching in rare instances a diameter of 4 feet. The largest trees measured by me, however, were much less, being respectively, 7, 7½, and 7¾ feet in girth; the last 60 feet high, with a clear trunk of 30 feet; the second 95 feet high, with a trunk 75 feet long.

47. (176.) *Ulmus alata.*   Winged Elm.

A rather rare tree, chiefly in river bottoms and along banks of streams; no measurements.

48. (177.) *Ulmus americana.* White Elm; "Red Elm."

A very common tree, most abundant in rich bottoms, where it attains a large size. Trees fully equaling the finest New England specimens are not uncommon, many being 5 feet in diameter and 120 feet or more in height. A very remarkable specimen was seen in the bottoms below Mount Carmel. It had grown in a thick wood, but the surrounding trees having been cleared away, was thus exposed to full view. The trunk, $3\frac{1}{2}$ feet in diameter, extended straight upward like a shaft or column for about 40 feet, and then gradually enlarged, and subdivided, the subdivisions coalescing in places, but finally taking the character of distinct branches, of which about 13 could be counted; these main upright branches gradually diverged, now and then dividing, to near the top, which was gracefully inclined outwards all round, and with an extremely regular outline. This bouquet-shaped top had an ambitus of about 50 feet, while its summit was elevated about 120 feet above the ground. In the immediate vicinity of Mount Carmel are several very beautiful elm trees of the dome-shaped type, one having an ambitus of about 90 feet, the ends of the branches nearly touching the ground, and the total height about 70 or 75 feet. Another one expands 91 feet, though the total height of the tree is scarcely 60 feet, and the diameter of its trunk only a little over 3 feet. It is needless to remark that both these trees are completely isolated. The largest specimen which I have measured was 16 feet in circumference (above the spurs), the trunk undivided for about 50 feet, and the total height more than 120 feet. The ambitus of this tree was 105 feet, but another, also a very large tree, expanded 111 feet.

A conspicuous peculiarity of this tree, when growing in wet situations, consists in the very prominent spurs or buttresses thrown out from the base. These thin walls extend sometimes many feet from the body of the tree, some specimens with a trunk 3 feet or less in diameter above the spurs being 12 to 15 feet in diameter at the ground. The only other tree exhibiting this feature to a marked degree is the Red Oak (*Quercus rubra*), in which, however, the spurs are thicker and do not project so far as they do with the present species in extreme cases.

The White Elm is the tree to which the mistletoe (*Phoradendron flavescens*) is most partial, fully 90 per cent. of the trees affected by this parasite in the White River and Wabash bottoms being elms; in fact, I have never seen it except on this tree and the Honey Locust (*Gleditschia triacanthos*). In the vicinity of Evansville, however, only 40 miles southeast from Mount Carmel, the case is said to be quite different, according to Professor John Collett, who gives a list of thirteen species of trees upon which this parasite was found growing, the Black Gum being first, the "Red Elm" (*i. e., Ulmus americana*) second, and the Honey Locust fifth, in the order of numbers upon which it grows. (See Cox's *Geological Survey of Indiana*, 1875, p. 242.)

The following specimens of *Ulmus americana* have been measured by
me :

Specimen.	Girth above swell at base.	Distance from ground to first large limb.	Total height.	Ambitus.	Locality.	Authority.
a	15½	(50?)	....	85	Wabash County, Illinois......	R. R.
b	....	59	....	....	......do	R. R.
c	16	(55?)	....	105	......do	R. R.
d	10	....	(60?)	91	......do	R. R.
e	11	50	....	80	...... do	R. R.
f	15	35	....	.. ....	......do	R. R.
g	15½	40	....	85	......do	R. R.
h	....	....	....	111	Knox County, Indiana	R. R.
i	10	(50?)	119	50	Wabash County, Illinois......	R. R.

Flowers March 10 to 20, and begins to leaf the last week in April.

49. (179.) *Ulmus fulva.* "Slippery Elm."

A common tree in rich woods, but much less abundant than *U. americana.* Grows commonly from 50 to 70 feet high, and 1½ to 2 feet in
diameter, although much larger specimens undoubtedly occur. No
measurements, however, have been taken.

50. (183.) *Celtis mississippiensis.* Mississippi Hackberry.

A very common tree, though less numerous than *C. occidentalis*, with
which it is found associated in very rich bottoms. It is usually a
smaller tree than that species, commonly 60 to 80 feet high, the branches
growing lower down, the bark of the trunk covered with prominent
warty excrescences, and the leaves smaller, more coriaceous, and entire.
The only specimen measured was 60 feet high and 11 in circumference.

51. (184.) *Celtis occidentalis.* Hackberry.

A very tall and beautiful tree in rich bottoms, growing frequently 120
to 130 feet high and 3 feet in diameter, with a tall, straight trunk of 60
to 70, or even 80, feet to the first limb. When growing to its full perfection in a dense forest, there is an individuality in the aspect of this
tree which it is difficult to describe. It does not excel either in height
or girth, yet it has the appearance of being one of the very tallest trees
in this lofty forest, this illusion being doubtless due to the extreme slen-
derness and great length of the trunk, which not unfrequently comprises
three-fourths of the total height of the tree, the smooth, gray bark conspicuously clouded on the north side, with blackish moss or lichen for
the entire length. This striking appearance is sometimes still further
increased by vines of the Virginia Creeper ascending to the topmost
branches, which are wreathed and matted with its foliage. Although
83 feet is the greatest length of the tape-line actually stretched along a

trunk of this species, one tree was seen whose silvery shaft gleamed among the surrounding tree-tops in a wood where the summit level was considerably more than 100 feet aloft, and though only ten feet in circumference must have been upwards of 90 feet to the first limb, which grew not more than 25 feet from the extreme summit of the tree.

The following tape-line measurements of prostrate specimens have been made in the vicinity of Mount Carmel:

Specimen.	Girth above swell at base.	Distance from ground to first large limb.	Total height.	Locality.	Authority.
a	9	70	....	Gibson County, Indiana	R. R.
b	10	75	....	......do	Do.
c	13	46	....	Knox County, Indiana	Do.
d	9	46	....	Gibson County, Indiana...............	Do.
e	11	83	134	......do	Do.

52. (189.) *Morus rubra.* Mulberry.

Very common on rich lands. The largest specimens measured were the following:

Specimen.	Girth above swell at base.	Distance from ground to first large limb.	Total height.	Locality.	Authority.
a	10	20	60	Posey County, Indiana	Thomas J. Johnston.
b	10½	20	62	Wabash County, Illinois (?)	Dr. J. Schneck.
c	4¾	19½	68	......do	R. R.

53. (191.) *Platanus occidentalis.* Sycamore.

This very abundant tree is unquestionably the largest hard-wood of North America, though there are several which it does not excel in height. The largest specimens are 140 to 160 feet high, with an ambitus of 100 to 130 feet, the diameter of single trunks averaging 5 to 7 feet, but of compound trunks (*i. e.*, those which fork comparatively near the ground), 8 to 10 feet. The chief superiority of the Sycamore over other trees, in point of size, consists, however, in the massiveness of the branches, each of the principal limbs of a very large tree of this species fully equaling an average forest tree in bulk. Twelve trees measured the same day in the bottoms of Gibson County, Indiana, below the mouth of Patoka Creek, averaged 127 feet spread of top and 23½ feet in circumference, the extremes being 100 to 135 and 14 to 30 feet;

two other trees had an ambitus of 108 and 97 feet, respectively, while another was 33 feet in girth. These being all standing trees, their height could not be measured accurately, but not one of them was less than 100 feet high. The average height of eight trees, which are all that have been actually measured, was 145¼ feet, the extremes being 129 and 168 feet.

Begins to leaf May 1.

The following detailed list of all the specimens measured may be of interest, as showing the great amount of variation in proportions in this tree:

Specimen.	Girth above swell at base.	Distance to first large limb or fork.	Total height.	Ambitus.	Locality.	Authority.
a	30	7	160	134×112	Gibson County, Indiana..............	R. R. (Photographed.)
b	30	18 (?)	(160 ?)	126	......do	Do.
c	31	12	145	105	......do	Do.
d	24	....	140	....	......do	R. R
e	....	....	....	108	......do	Do.
f	....	....	....	97	......do	Do.
g	14	....	....	100	.... do	Do.
h	19	....	....	100	......do	Do.
i	18½	....	....	128	......do	Do.
j	22	....	....	135	......do	Do.
k	33	....	....	....	......do	Do.
l	28½	....	....	129	......do	Do.
m	22½	....	....	110	.... do	Do.
n	29½	....	....	134	......do	Do.
o	23	....	....	100	......do	Do.
p	25	....	....	130	... do	Do
q	27	50	....	....	Wabash County, Illinois	Do.
r	25	40	....	....	......do	R. R. (Photographed.(
s	30	60	....	....	......do	R. R
t	25	68	168	....	......do	Do.
u	33½	13	....	....	......do	Do
v	18	74	....	....	......do	Do.
w	9	83½	....	....	......do	Do.
x	22	....	140	....	......do	Do.
y	15	61	129	....	......do	Do.
z	14½	63	141	....	......do	Do.
a′	13	55	139	....	......do	Do.
b′	22	....	....	....	Posey County, Indiana..............	Charles Schneck.
c′	24	....	....	....	......do	Do.

a. This is probably the largest tree of any kind which I have seen anywhere in the Wabash Valley, or any other part of the Eastern Province of North America. It is of very vigorous growth, and apparently perfectly sound. Circumference at the ground, 42 feet; round smallest part of the trunk, 30 feet; greatest diameter, 15 feet, least diameter, 10 feet, the average diameter being about 11 feet. Ambitus, 134 feet in one direction, the least spread of top being 112 feet. Total height, as determind by several measurements with "dendrometer," and by shadow, about 160 feet. The trunk first divides at about 7 feet from the ground, but above this division the main stem is still 8 feet in diameter; this extends upward, *gradually enlarging*, to about 15 feet from the ground, where the next division takes place, the next fork being nearly 30 feet up. No horizontal branches are thrown out until a height of 70 or 80 feet is reached (or about half the total height of the tree), the great bulk of the broadly spreading top being elevated above 90 feet from the ground.

b. This tree, though slightly less in diameter and spread of top, is a more symmetrical, and in this respect a decidedly finer tree than the preceding. Although the trunk first ramifies at a distance of about 18 feet from the ground, both forks extend straight upward, the larger straight as a column, and averaging about 6 feet in diameter, for 50 feet, the smaller 70 feet or more (but the upper portion curving gracefully outward). The top constitutes, when in full leaf, a compact dome of foliage, the great bulk of which constitutes the upper third of the total height. The tree is in perfect vigor, without a single dead branch, and showing no signs of decay about the base. Its trunk is wreathed with vines of the Virginia Creeper, which, extending upward for more than 100 feet, show in beautiful contrast to the smooth snow white bark of the larger branches and upper portion of the trunk.

c. Trunk divides at about 10 or 12 feet from the ground, where the circumference is much greater than at the ground.

r. Probably the handsomest trunk of any sycamore which I have ever seen. It rises like a huge column, 8 feet in average diameter, without any perceptible diminution for at least 40 feet, from a widely expanded base, measuring 17 feet in diameter and more than 50 feet in circumference, from which spring four "sprouts," the largest of which is nearly three feet in diameter, and all extending nearly straight upward, to almost the height of the main tree. The base is covered with dark green moss, and the trunk ornamented with the Virginia creeper. The trunk is hollow, and has recently been disfigured on one side by the axe of some vandal.

u. Not a handsome tree, the three main forks widely diverging.

y and *z.* Solid trees, newly felled, growing only 12 feet apart!

a′. 84 feet to second limb.

The decaying prostrate remains were found in the bottoms of Gibson County, Indiana, a short distance below Mount Carmel, of a huge sycamore, which must have been much larger than any tree that I have measured. The space covered by the crumbled base was 66 feet in circumference. The three upright forks, found lying near together, two of them still united, the other broken off, were each 5 feet in diameter, and careful measurements of them indicated a circumference of about 62 feet, below their ramification, which took place some 20 feet from the ground, and the base of the tree. Each of the three trunks, which were still intact, though much decayed exteriorly, was 70 feet long, but the branches were, of course, entirely decayed. When standing in its full vigor, this tree must have been a grand one, indeed. There is said to be still standing, near Worthington, Greene County, Indiana, a tree of this species which has a solid trunk measuring 48 feet in circumference, and dividing at 25 feet into three or four main branches, the largest of which is more than 5 feet in diameter.—(See *Case's Botanical Index*, April, 1880, and *Botanical Gazette*, June, 1880, p. 70.)

54. (195.) *Juglans cinerea.* Butternut; White Walnut.

By no means a common tree, except in certain restricted localities. Though very much inferior to *J. nigra* in stature, it sometimes attains a considerable size, two felled trees, in the "Timber Settlement," Wabash County, measuring 97 and 117 feet in length, and each 1 foot 10 inches in diameter, with clear trunks 50 and 32 feet long. These trees grew within a few rods of one another, the species being very common in that locality.

55. (196.) *Juglans nigra.* Black Walnut; "Walnut."

The Black Walnut was, originally, a very abundant tree throughout the rich bottom lands of the Wabash and White Rivers, but is now rapidly becoming scarce. Trees of this species, 5 or 6 feet in diameter, with straight, solid trunks 40 to 60 feet in the clear, were formerly common, but the finest trees have long been destroyed. Eight walnut trees, of less than medium size, were found freshly felled, in the bottoms of Greathouse Creek, about two miles west of Mount Carmel, and carefully measured, with the following result: Average length, 106¼ feet; average length of trunk, 47¼ feet; average circumference, 9⅛ feet. Extreme measurements: length, 97½ to 119½; trunk, 35½ to 60; circumference, 8 to 10½. In the river bottoms the growth is much larger. One very fine tree measured 5½ feet across the top of the stump, 42½ feet to the first limb, 75 feet to the second limb, and 131 feet to the extreme top. A perfectly sound and very symmetrical standing tree, of which photographs were taken, measured 18 feet in girth at a yard from the ground, had an ambitus of 97 feet, and was little, if any, less than 150 feet high, the trunk alone being over 70 feet to the first limb, on main fork.

The following measurements represent, very fairly, the size of Black Walnut trees which have been cut for lumber in the vicinity of Mount Carmel:

	Girth.	Trunk.	Height.	Ambitus.	Locality.	Authority.
a	15¼	40	....	....	Wabash County, Illinois............	R. R.
b	17½	60	130	....	Posey County, Ind ana	Charles Schneck.
c	20	64	150	....	Gibson County, Indiana	R. R.
d	18½	....	....	....	..... do	R. R. (Photographed.)
e	18	(75?)	(+150?)	97	.....do . ... :	Do.
f	15	70	....	....	Wabash County, Illinois	R. R.
g	15	71	144	....	.......do	Do.
h	13	94	156	....	.......do	Do.
i	15	67	144	....	.......do	Do.
j	17	43½	131	....	.......do	Do.
k	9	44½	97½	....	..... do	R. R. (Greathouse Creek.)
l	10½	54½	119½	....	.......do	Do.
m	8	54	103	....	.......do	Do.
n	8¼	38¼	106¼	....	.......do	Do.
o	9⅜	60	113	....	.......do	Do.
p	8½	35½	101	....	.......do	Do.
q	9	45½	107½	....	.......do	Do.
r	9⅜	45½	102	....	.......do	Do.
s	22	74	155	....	Wabash County, Illinois (?).........	Dr. J. Schneck.

Remarks.—*f*, trunk 3 feet diameter at upper end; *g*, ditto.

56. (198.) *Carya alba.* "Shell-bark."

Very common, attaining its greatest size on rich sand ridges in the bottom-lands, where specimens 3 to 4 feet in diameter and 130 feet or more high are not rare. The maximum height attained by this species has not been ascertained, but it is one of the very tallest trees of the forest, the tough and elastic top branches not being liable to be broken by the wind, as is so often the case with tall "Poplars" and "Sycamores." Some tall shell-barks are certainly 150 feet high, and probably more, many trunks, apparently constituting less than half the total height, being 70 or 80 feet to the first limb. The following measurements may in part refer to *C. sulcata*, it being impossible to distinguish this species from *C. alba*, except by the fruit and foliage, and some of the measurements were taken in winter.

Flowers April 15 to 20, leafing from the 10th to the 13th of the same month.

Specimen.	Girth.	Trunk.	Total height.	Locality.	Authority.
a	11	70	....	Wabash County, Illinois	R. R.
b	13	(80?)	... .	........do	Do.
c	14⅝	78	....	...... do	Do.
d	11	75	....	...... do	Do.
e	10½	51	129	........do	Do.
f	5	39	101	Knox County, Indiana, *young tree!* ...	Do.
g	4½	51	88	........do	Do.

The so-called *C. microcarpa*, which may be a distinct species, is also found. Dr. Schneck, in his catalogue (p. 560), says: "Heavy damp soil, scarce. Has very little loose bark, one of our smallest hickories." One specimen, however, of what was apparently this form, measured 14 feet in girth and was considerably over 100 feet high.

57. (199.) *Carya amara.* Swamp Hickory; White Hickory.

Not uncommon in the bottoms, growing tall and slender, being occasionally 100 feet or more high and 3 feet in diameter. The largest measured was 11 feet in circumference; another was 113 feet high and 6¼ in circumference, the trunk 64 feet.

58. (——.) *Carya olivæformis.* Pecan (pronounced *Pe-cawn'*).

Common in rich bottom lands. This is by far the largest of the hickories, being, in truth, one of the very largest trees of the forest. With the single exception of the White Elm the Pecan tree has, in proportion to its size, the most widely-expanded head of any tree, while in altitude and majestic appearance the largest and finest elms bear no comparison. The dome-like head may occasionally be seen reared conspicuously above the surrounding tree-tops, even in a very lofty forest, some trees being as much as 175 feet high (by actual measurement) and with an ambitus of 100 feet or more. The trunk, like that of the shell-bark hick-ories (*C. alba* and *C. sulcata*), is very long, often measuring more than 50 feet, and occasionally 80 or even 90 feet, to the first limb. A very large tree of this species, cut down in the "Timber Settlement," Wabash County, and measured by Dr. Schneck, was found to be 175 feet high, with a clear trunk 90 feet long and 16 in circumference. Another still standing, only fifteen yards distant, had exactly the same circumference, and apparently agreed very closely in other measurements. A very fine tree in the White River bottoms of Gibson County, Indiana, was 30 feet in circumference at the ground and 18½ feet around above the swollen base; the column-like trunk was more than 50 feet to the first limb, while the lofty top spread 100 feet. Near Sandborn, in Knox County, Indiana, according to Professor Collett (Cox's *Geological Survey of Indiana*, 1873, p. 364), there is a tree of this species measuring 8 feet in diameter, but its height is not stated.

59. (202.) *Carya porcina.* "Pig-nut"; "Broom Hickory."

Common, usually in upland woods. No measurements.

60. (203.) *Carya sulcata.* "Big Shell-bark"; "Bottoms Shell-bark."

A very common tree in rich bottom lands, where, growing to a large size, and in the character of its bark, as well as in general appearance, exactly resembling *C. alba*. For this reason it is possible that some of the measurements given under *C. alba* may be intended for the present species.

61. (204.) *Carya tomentosa.* "Black Hickory"; "White-heart Hickory"; "Bull-nut."

A very common tree in upland woods, growing frequently more than 100 feet high and 3 feet or more in diameter, one specimen measuring 112 feet in length, 10½ in circumference, the trunk 55 feet.

62. (207.) *Quercus alba.* White Oak.

Perhaps the most abundant and generally distributed of all our trees, growing to a large size, especially in the bottoms, where trees of this species 130 feet or more in height and 3 feet in diameter are not uncom-mon. Indeed, even in upland woods, the average height of the larger White Oaks is 100 feet or more. Ten trees, cut for rails, on one piece of ground, averaged as follows: Total length, 100.05 feet; trunk, 40.1 feet; diameter (across top of stump), 2¾ feet. All but one grew on high ground. The extremes of size were: height, 87 to 111 feet; trunk, 26 to 54 feet; diameter, 2 feet 3 inches to 3 feet. One, measuring 2 feet 4 inches in diameter and 98 feet in height, exhibited 190 annual rings of growth. All but one were perfectly solid, and the one exception was hollow only in the stump, the first cut being sound. The tallest and largest tree grew at the edge of the creek bottoms, its height being 111, trunk 54, and diameter 3 feet. In rich bottom lands the size averages considerably greater, or about 120 feet in height by 3½ to 4 in diameter, very large trees having an ambitus of 75 to 95 feet. The following measurements show pretty well the difference in size between trees growing in rich bottoms and those growing in the drier upland woods:

Size of White Oak trees growing in bottom lands, as measured.

	Diameter	Trunk.	Height.	Ambitus.	Locality.	Authority.
a	3.39	65	128	....	Knox County, Indiana	R. R.
b	3.55	....	111	....	......do	Do.
c	4.53	40	123	92	......do	Do.
d	3.66	....	121	76	......do	Do.
e	3.00	54	111	....	......do	Do.
f	5.00	60	115	....	Posey County, Indiana	Thos. J. Johnston.
g	5,10	54	110	....	......do	Do.
h	5.50	30	142	....	Wabash County, Illinois	R. R.
i	5.83	....		....	......do	Do.
j	... .	68	125	....	.... do	Do.
k	6.00	60	150	....	......do	Dr. J. Schneck.
Av.	4.59	52	123.60			

Size of White Oak trees growing on uplands.

	Diameter	Trunk.	Height.	Ambitus.	Locality.	Authority.
l	2.50	56	104.50	....	Knox County, Indiana............	R. R.
m	2.83	39	99	....	......do	Do.
n	2.33	36	98	....	......do	Do.
o	2.25	38	99	..,.......	......do	Do.
p	2.33	43.50	103	....	......do	Do.
q	2.25	41.50	109	....	......do	Do.
r	2.25	30	93	....	......do	Do.
s	2.50	35	87	....	......do	Do.
t	2.33	38	97	....	......do	Do.
Av.	2.40	40	99.82			

The following measurements are given in Mr. Johnston's list, but it is not stated whether the trees grew in uplands or in the bottoms; most probably the former, however:

	Diameter.	Trunk.	Height.	Locality.	Authority.
u	4.50	45	97	Posey County, Indiana	Thos. J. Johnston.
v	4.40	48	107	......do	Do.
w	4.33	43	95	......do	Do.
x	4.12	35	87	......do	Do.
Av.	4.34	43	94		

The White Oak begins to leaf, near Mount Carmel, about the 12th of April.

63. (209.) *Quercus bicolor.* Swamp White Oak.

A very common, or in some places abundant, tree, fully equal to *Q. alba* in size, but more resembling in form *Q. macrocarpa.* Only two specimens have been measured; one of these, a somewhat decayed prostrate one, measured 4 feet 8 inches across the top of the stump (not including the bark), the trunk 67 feet to the first limb; the topmost branches were gone, but at 100 feet from the base the five limbs were 10 inches to 1 foot in diameter, so that the tree when standing must have been 130 feet or

more high. The extreme base was hollow. The other was a standing
tree, measuring 15⅓ feet girth at four feet from the ground, the trunk
about 20 feet, and the total height 100 feet or more. The top was widely
spreading, probably measuring nearly or quite 100 feet ambitus.

64. (213.) *Quercus coccinea.* Scarlet Oak; "Black Oak" (?).

This tree is apparently not popularly distinguished from *Q. tinctoria.*
Dr. Schneck, in his catalogue, gives the maximum measurements of this
species as 20¼ feet girth, 94 feet trunk, and 181 feet total height. I am
unable to give measurements of my own, however. It is apparently our
tallest oak, though I had supposed *Q. rubra* to be entitled to this distinction.

65. (218.) *Quercus falcata.* Spanish Oak.

Common, along with *Q. nigra* and *Q. imbricaria,* in poor soils. Very
rare in rich grounds, only one tree being seen in the bottoms; this a very
large one near White River, in Gibson County. It measured 14 feet in
circumference, and was estimated to be 130 feet high, with a crooked
trunk of 60 to 70 feet clear. The bark was remarkably light colored,
appearing almost as pale as some of the white oak section, but the
leaves, a number of which were obtained (the date being November 2,
and the ground beneath the tree covered with them, while many, still
adhering to the branches, afforded proof that those on the ground were
from the same tree), were unquestionably those of *Q. falcata.* A photograph of this tree is in my possession, and specimens of the leaves
were deposited in the herbarium of the Agricultural Department. As
usually found growing, however, in drier and poorer soils, this oak is
by no means a large tree, seldom exceeding 80 feet in height, and probably not averaging over 50 or 60 feet, with a diameter of 1 to 2 feet.

66. (222.) *Quercus imbricaria.* Laurel Oak; Shingle Oak.

With possibly the exception of *Q. alba,* this is the most abundant and
generally distributed species, at least in Wabash County. It is the
most slender of all the oaks, and in some rich bottoms trees 100 feet in
height and 50 feet to the limbs are only 6 to 7 feet in girth; one tree,
however, measuring nearly 4 feet in diameter (11 feet in circumference)
and over 100 feet high, has been measured. The largest prostrate tree
measured was 100 feet long, 50 feet to the first limb, and 6½ feet in
girth. It is only in very rich lands, however, that this species attains
such large dimensions, and on poorer soils, where it is more abundant,
it does not usually much exceed half this size.

Flowers May 9 to 12, leafing about the 2d or 3d of the same month.

67. (226). *Quercus lyrata.* Overcup Oak; Swamp Post Oak.

Not uncommon in some places, but very local—more so, indeed, than
any other of our oaks. It is confined almost entirely to the low "swales"
or depressions in the bottom lands, where the ground is either often over-

flowed or very wet for the greater part of the year, and in such places is found along with the "swell-butt" ashes (*Fraxinus americana*) and other swamp trees. In general appearance it very closely resembles the Swamp White Oak (*Q. bicolor*), branching, like that species, comparatively near the ground, the lower branches drooping so as to often touch the ground at their extremities. It is a smaller tree, however, no specimens exceeding 80 feet high and 2½ in diameter having been noticed, though, like other species, it may occasionally much exceed its usual size.

68. (227.) *Quercus macrocarpa.* Bur Oak.

Very common in rich bottom lands. Much the largest, though not the tallest of all our oaks, being frequently 5 to 6, sometimes 7, feet in diameter, and 130 feet or more high, with an ambitus of 100 feet or more. Dr. Schneck gives the maximum dimensions of the Bur Oak as follows: Circumference, 22; clear trunk, 72; total height, 165. Trees of this size are exceedingly rare, however, if not wholly exceptional. The largest that I have measured was 124 feet long to where the top branches had been broken off, the trunk 63 feet in the clear and 21 feet in circumference, the measurements, in the same order, of the next largest being 162, 30, and 20 feet. A standing tree more than 5 feet in diameter (16 feet circumference) had an ambitus of 130 feet in one direction and 134 feet the opposite way.

Following are the measurements that I have taken of this tree, including several by Mr. Thos. J. Johnston and Dr. Schneck:

Specimen.	Girth	Trunk.	Height.	Ambitus.	Locality.	Authority
a	18. 25	35	75	....	Posey County, Indiana ..	Thomas J Johnston.
b	17. 20	37	80	....	...... do	Do.
c	14. 65	31	77	....	......do	Do.
d	12. 75	32	. 76	....	......do	Do.
e	19. 50	70	149	....	Wabash County, Illinois..	Dr J Schneck.
f	20	30	162	....	......do	R R
g	21	....	....	....	Posey County, Indiana...	Charles Schneck
h	18	40	130	....	......do	Do
i	18. 50	66	... .	100	Wabash County, Illinois..	R R
j	15	60	140	....	......do	Do.
k	21	63	+124	....	......do	Do.
l	16	....	....	130×134	......do	Do
m	22	72	165	....	Wabash County, Illinois (?)	Dr J Schneck
Average ..	17. 95	48. 73	+115 80	....		

REMARKS.—*e*, trunk perfectly solid throughout; *g*, "trunk apparently sound"; *h*, trunk sound.

69. (——.) *Quercus michauxi ?*

To this species I refer provisionally an oak which is not a common species in the vicinity of Mount Carmel (the only place I have seen it), but which grows sparingly in rich alluvial soils. So far as I have observed, it is rather a small species, resembling in general appearance the *Q. muhlenbergi* more than any other of our oaks, but having very different fruit and foliage. The leaves, 3.25 to 7.00 inches long and 1 50 to 3.50

Proc. Nat. Mus 82——6   **June 12, 1882.**

wide, are obovate, acute at each end, *long petioled* (petiole .70 to 1.50 long), *coriaceous, very glossy above, pale and very velvety beneath*, the margin deeply cuspidate-toothed. The acorn is very large (.90 to 1.00 inch long by the same in breadth), broadest at the base, the summit somewhat depressed, the color a rich leather-brown; cup saucer-shaped, flattish beneath, *very thick*, velvety inside, roughly clad exteriorly with very distinct and prominent claw-like, somewhat carinate scales, the margin thin, and turned slightly outward; peduncle very short (.30 or less) or wanting, the acorn being usually sessile. This tree can hardly be a form of *Q. bicolor* (to which *Q. michauxi* is referred by Dr. Englemann), its principal characters being directly the reverse of those of that species. Thus, the leaves of *Q. bicolor* are very short-petioled or almost sessile, while those of the present species have the petiole an inch or more, frequently an inch and a half in length ; in *Q. bicolor* the acorn is attached to a longer peduncle than any other of our oaks (usually 2 inches or more in length!), while in this species, if present at all, it does not exceed .30 of an inch! The acorn of *Q. bicolor* is also very much smaller, and of a *totally* different character.

Whatever this species may be, I leave it for botanists to decide.*

70. (228.) *Quercus muhlenbergi.* "Yellow Oak"; "Chinquapin" (!).

This fine tree is a very common species in the bottom lands as well as on rich hillsides. The trunk may be recognized at a distance by its thin-scaled, very light-colored bark, and tall slender growth, this oak being probably the tallest in proportion to its diameter of any of the white-barked species. One felled tree measured 130 feet in length, the trunk 40 feet, and the circumference 13 feet; another (a photograph of which, taken before the tree was cut, is in my possession) was 122½ feet long, 73 feet to the first limb and 84 feet to the main fork, the diameter across the top of the stump being only 3½ feet! A standing tree, whose height could not be ascertained, was 14 feet in circumference above the spurred base, which, at the ground, measured 10 feet in diameter.

The acorns of this tree are very small and sweet, much resembling in both appearance and taste, and certainly not inferior to, the nuts of the Chinquapin (*Castanea pumila*), whence the popular name. The wood is said to be tougher than that of *Q. alba*, and is much used by wagon-makers.

71. (229.) *Quercus nigra.* "Black Jack"; "Jack Oak."

A very abundant species in poor, sandy soils, growing 30 to 50 feet high and 8 inches to 1½ feet diameter, being, perhaps, the smallest of all our oaks. No actual measurements having been made, it may be that the dimensions given above are sometimes exceeded.

---

* Since the above was written, Professor Sargent writes me as follows: "This is, no doubt, *Q. michauxi*, and it must now be considered a good species. It is one of the most beautiful and useful of the American oaks."

72. (231.) *Quercus palustris.* "Water Oak"; "Turkey Oak."

A very common species in wet bottoms, distinguished by its comparatively smooth, grayish bark, and usually by the numerous small drooping branches which grow from the trunk, sometimes to quite near the ground. In close woods, however, it frequently has a clean straight stem of 50 feet or more, one of 73 feet having been measured. The Water Oak is usually 100 to 120 feet high, and 2 to 3 feet in diameter, but much larger specimens sometimes occur, trunks even 4 and 5 feet through being occasionally met with. But few specimens have been measured, as follows:

	Girth	Trunk.	Height.	Locality.	Authority.
a	6	73	116	Wabash County, Illinois	R. R.
b	12	23	120	Wabash County, Illinois (?)...........	Dr. J. Schneck.
c	9¼	61	119	Knox County, Indiana	R. R.
d	9	55	117	......do	Do.

This species blossoms about the middle of April.

(?) 73. (232.) *Quercus phellos.* Willow Oak.

This species I give with some doubt, not being quite positive that it occurs. I have seen, however, along the road between Mount Carmel and Olney (Richland County) several trees which, at the time of inspection, I unhesitatingly decided to be *Q. phellos* (a tree with which, as growing in Maryland and Virginia, I was perfectly familiar), but not having seen it since, while Dr. Schneck has not recorded it, I place the interrogation mark as above.

74. (234.) *Quercus rubra.* Red Oak; "Spanish Oak"; "Turkey Oak."

With the possible exception of *Q. coccinea*, this is the tallest oak growing in the district under consideration, and, excepting *Q. macrocarpa*, is the largest also. Trunks, straight as an arrow, of 5 or even 6 feet diameter (above the spurs), and 50 to more than 70 feet clear, were formerly not at all rare, but at the present time most of them have been cut for barrel-staves or clap-boards. The largest Red Oak which I have measured was 23 feet in girth (round the top of the stump), the trunk 76 feet long and 3 feet in diameter at the small end. The top branches beyond 120 feet from the base were destroyed, but at this point the several main limbs were a foot in thickness. Another tree, measuring 19 feet in girth and 71 feet to the first limb, was 150 feet long. At the ground these large Red Oaks measure much more than they do a few feet up, on account of the projecting spurs, or buttresses, which, as in the White Elm (*Ulmus americana*), are a very characteristic feature of the species. Thus, a Red Oak measuring 6 feet through at two yards from the ground may be 12 feet or more in diameter at the base.

Flowers April 18 to 20, and leafs out a few days later.

The extent to which this tree is cut for barrel-staves and clapboards has afforded the opportunity of taking several measurements, which are herewith appended:

	Girth	Trunk.	Height.	Locality.	Authority.
a	23	76	....	Gibson County, Indiana	R. R.
b	13	60	150	Wabash County, Illinois...............	R. R.
c	12	....	125	......do	R. R.
d	19	71	150	......do	R. R.
e	12	63	130	......do	R. R.
f	14	65	....	......do	R. R.
g	11. 50	40	132	......do	R. R.
h	17	62	....	Gibson County, Indiana	R. R. (Photographed.)
i	16	60	....	Wabash County, Illinois...............	R. R.
j	15	75	....	.. do	R. R.
k	15	72	....	Gibson County, Indiana	R. R.
l	14. 50	54	134	Knox County, Indiana	R. R
m	9	57	115	......do	R. R.
n	11	62	115	.... do	R. R.
o	14	55	143	.... do	R. R.
p	9	65	127	.... do	R. R.
Av.	14 00	62. 50	132. 10		

REMARKS.—*a*, trunk 3 feet in diameter at upper end; at 120 feet branches 1 foot thick; *h*, circumference at ground, 36 feet; *i*, circumference at ground, 28 feet; *j*, diameter at ground, 11 feet; *k*, diameter across stump, over spurs, 6 feet; through upper end of trunk, 3 feet; *l*, 181 annual rings to central hollow, 15 inches across; *n*, 242 annual rings.

75. (235.) *Quercus stellata.* Post Oak.

A very common tree in clay soils. No measurements have been taken, but the usual size of the heavier growth is about 50 to 80 feet high, and 2 to 3 feet in diameter. Larger trees, however, no doubt occur.

76. (236.) *Quercus tinctoria.* Black Oak.

A very common, large tree, chiefly in upland woods. Frequently 100 feet or more in height, and 3 feet in diameter. It is occasionally larger, however, as may be seen from the annexed measurements.

	Girth.	Trunk	Height.	Locality.	Authority.
a	9	39	100	Knox County, Indiana	R. R.
b	18	75	128	Posey County, Indiana.................	Thomas J. Johnson.
c	17½	60	118	......do	Do.
d	20	50	102	......do	Do.
e	14	49	100	......do	Do.
f	12½	43	96	......do	Do.
g	20	75	160	Wabash County, Illinois (?)	Dr. J. Schneck.

REMARKS.—*a*, 179 annual rings.

Flowers April 17th to 20th, and begins to leaf about a week later.

77. (242.) *Castanea vulgaris americana.* American Chestnut.

The chestnut does not properly belong to the district under consideration, but in Indiana extends westward very nearly to the junction of the two forks of White River, having been noticed from the railroad, growing wild between Loogootee and Shoals, in Martin County, the second county east of Knox. In Jackson and other counties in the southern and southeastern part of the State it is abundant, and grows to a large size, a specimen near Seymour, being mentioned in *Case's Botanical Index*, which measured 22 feet in circumference 2 feet from the ground, and 70 feet to the first limb.

A few trees, raised from imported seed, are to be found in various parts of Wabash County, where they grow finely, and under proper conditions, fruit plentifully. Trees near Mount Carmel flower about March 20, and begin to leaf about the middle of April.

78. (243.) *Fagus ferruginea.* Beech.

I have never seen, nor, indeed, heard of a single beech tree growing on the Illinois side of the Wabash; but immediately across the river, in Knox County, Indiana, a few large trees begin to occur, while back on the hills of both that county and Gibson it is a very common tree. Trees of 3 to 4 feet diameter are not uncommon, while Dr. Schneck records one which measured 122 feet in height. Ordinarily, however, the finest beech trees are decidedly inferior in altitude to the surrounding oak, gum, and other tall forest trees, and I should estimate their average height at not more than 90 feet.

79. (244.) *Ostrya virginica.* Hop Hornbeam.

By no means a common tree, but occasionally found, and possibly more numerous in some localities not visited. No measurements.

80. (245.) *Carpinus caroliniana.* "Blue Beech"; "Water Beech."

Very common in rich bottom lands. The largest trees measured were 30 to 32 feet high, and 1 to 1½ feet in diameter, but larger ones may occur. Only four trees were measured, their dimensions being as follows:

	Girth.	Trunk.	Height.	Locality.	Authority.
a	4½	....	30	Knox County, Indiana	R. R.
b	3½	10	30	......do	Do.
c	....	....	31	Wabash County, Illinois	Do
d	3½	7½	32	Knox County, Indiana	R. R. Ambitus 35 feet.

81. (247.) *Betula lenta.* Cherry Birch; "Black Birch"; "Mahogany Birch."

Not uncommon along banks of streams. One tree, forking several feet from the ground, measured 17½ feet in circumference, and was about 80 feet high.

82. (249.) *Betula nigra.* Red Birch; River Birch.

Commoner than the last in similar situations. Young trees, as well as some old ones, with very scaly bark, the projecting laminæ very thin, paper-like. Grows commonly 70 to 80 feet high, and occasionally 3 or even 4 feet in diameter. The only one actually measured was 84 feet in length.

83. (260.) *Salix lucida.* Shining Willow.

"Moist banks of streams; common." (SCHNECK.) No measurements.

84. (———.) *Salix discolor.* Glaucous Willow.

"Moist banks and along stream; rare." (SCHNECK.) No measurements.

85. (261.) *Salix nigra.* Black Willow.

Much the most abundant and also by far the largest of our native willows. In some swamps the trees of this species average 60 to 70 feet high and more than a foot in diameter, while trees considerably larger are occasionally met with. Two trees growing on the border of Monteur's Pond, in Knox County, Indiana, measured, respectively, 80 and 87½ feet in length, the latter being more than 3 feet in diameter (10 in girth), the former 7¼ feet in circumference, and 18⅓ feet to the first limb. One cut expressly for measurement, near the mouth of Crawfish Creek (Wabash County, Illinois), was 77 feet long, 55 feet to the first limb, and only 2½ feet around! Two other trees, measuring respectively 8½ and 9 feet in girth, were also measured, the former being 30 feet to the first limb.

86. (266.) *Populus heterophylla.* "River Cottonwood"; "Swamp Cottonwood"; "Stupy Gum" (Knox County, vern.).

Very common about the borders of swamps, usually associated with the Black Willow (*Salix nigra*). Much inferior in size to *P. monilifera*, the largest trees scarcely exceeding 90 feet in height and 2 to 2½ in diameter. The trunk, however, is usually very long in proportion, frequently occupying two-thirds or more of the total length. Only three trees of this species have been actually measured, the following being their dimensions:

	Girth.	Trunk.	Height.	Locality.	Authority.
a	7½	34	88	Knox County, Indiana................	R. R.
b	7½	51	92	......do	R. R.
c	7¼	38	80	......do	R. R.

87. (267.) *Populus monilifera.* Cottonwood; "Big Cottonwood."

A very common tree in rich bottom lands and along the alluvial banks of streams, where it occasionally attains an immense size and altitude. Trees of 5 to 6 feet diameter are not uncommon, while trunks of 7 or even 8 feet are occasionally to be met with; the stem being usually more than 50 feet clear. The total height of the tallest cottonwoods is generally more than 130 feet, as may be seen from the following measurements:

	Girth.	Trunk.	Height.	Locality.	Authority.
α	9	40	140	Wabash County, Illinois	R. R. (Coffee Creek bottoms.)
b	16			......do	Do.
c	16			......do	Do.
d	18	70	165	Posey County, Indiana	Charles Schneck.
e	19	75	134	......do	Do.
f	14	15	130	Wabash County, Illinois...........	Dr. J. Schneck.
g	24			Gibson County, Indiana	R. R. (Photographs.)
h	20			......do	Do
i	20			......do	Do
j	18¾	75	170	Wabash County, Illinois (?).......	Dr. J. Schneck.
k	11	58	114	Knox County, Indiana.............	R. R.

REMARKS.—*f*, a very fine tree, formerly standing on the commons within the corporation limits of Mount Carmel, but destroyed by the tornado of June 4, 1877; height measured by its shadow, the result verified by subsequent tape-line measurement; *g, i*, three majestic trees standing near together on the bank of a bayou opposite Rochester, the gradually tapering trunks estimated to be 70 to 80 feet clear, the total height of the tree is nearly 150 feet. In the immediate vicinity many others nearly as large (5 to 6 feet through).

88. (268.) *Populus tremuloides.* Aspen; "Quaking Asp."

A very rare tree in upland woods of Wabash County, but co mmon in both uplands and bottoms near Monteur's Pond, in Knox County, Indiana, where it forms a small slender tree, 50 to 70 feet high and 6 inches to a little over a foot in diameter. Only two trees were measured, one, blown over by the wind, but still growing, being 71 feet long and 1 foot 2 inches in diameter; the other, cut for measurement, being 51½ feet long, though only 14 inches in circumference at the base, and measuring 24 feet to the first limb.

89. (277.) *Juniperus virginiana.* Red Cedar.

Not native, so far as known, in any part of Wabash County, nor adjoining counties in Indiana, the soil being everywhere far too rich for it. It is abundant, however, on the hills of Gallatin County, near the mouth of the Wabash. The miniature *J. communis* is found sparingly in Wabash and adjoining counties, but becomes only a small bush in stature.

90. (283.) *Chamæcyparis sphæroidea.* White Cedar. "Wet places near the mouth of the Wabash River." (SCHNECK.) Not seen by me; no measurements.

91. (287.) *Taxodium distichum.* Bald Cypress; "Cypress."

I have never heard of any cypress growing anywhere on the Illinois side of the Wabash, but in the lower part of Knox County, Indiana, or that portion embraced between the Wabash and White Rivers, and known as "The Neck," it is very abundant, the area embraced by the cypress swamps of that district, and largely timbered with cypress, being estimated at 20,000 acres (see Cox's *Geological Survey of Indiana*, 1873, p. 338). The cypress swamps of this region comprise two quite distinct tracts, of which the northern is very much the larger, its natural outlet being the river Deshee, which empties into the Wabash between Mount Carmel and Vincennes. The "Little Cypress Swamp" is situated immediately above the mouth of White River, into which it empties through what is termed the "White River Slough." Although known as the "Cypress Swamp," it consists of a series of beautiful, secluded ponds, hidden in the dense forest, and difficult of access by any one not familiar with the locality. The principal ponds are the Cypress, Beaver-dam, Washburne's, and Forked Ponds, of which Washburne's is perhaps the largest. The cypress trees here grow chiefly around the borders of these ponds and along the sloughs connecting them, as well as the one which empties into the river. Being so near the river, into which the logs are floated at "high water," the finest trees have long since been destroyed, and there are very few left whose symmetry is not marred by low-growing branches or knots upon the trunks. The largest standing tree observed by me was a very old and exceedingly rough specimen, entirely unfit for lumber or shingles. The swollen base measured 45 feet in circumference at the ground, the girth immediately above the conical portion being 21 feet; the trunk consisted of several upright stems grown together for the greater part of their length, but in places distinct, with one very conspicuous transverse growth joining the two main stems, at a height of about 50 feet from the ground. The top expanded 94 feet, the greater part of it elevated over 100 feet from the ground. A solid stump, measuring 38 feet around at the ground, was 22 feet in girth at 8 feet; at about 15 feet it divided into two main trunks of equal size, which were cut off immediately above the fork, a scaffold being necessary for the purpose. Another stump was 13 feet in diameter across the top, but was hollow, and from its decaying wood grew several tall, but slender, birch trees, some of which were 50 feet high. Several other stumps of 9 and 10 feet in diameter (across the top) were measured. Several single, solid trunks of 50 to 92 feet in the clear were measured, their diameter at the base being 3 to 5 feet, while the largest one measured, a standing tree, was 27 feet in girth above the swollen base. The tallest of these trees did not, however, much exceed 140 feet (the two tallest measured being 146 and 147 feet), their average height being little, if any, over 100 feet; and even the finest of them would not compare for symmetry and length with the Sweet Gums and Ashes with which they were associated.

92. (324.) *Pinus mitis?* Yellow Pine.

For obvious reasons there are no pines growing native in Wabash or adjoining counties of Illinois or Indiana; but, according to Dr. Schneck (catalogue, p. 562), the Yellow Pine occurs on the "hills near the mouth of the Wabash River, in Gallatin County, Illinois." Professor Sargent, however, suggests that the pine of Southern Illinois may be *P. inops*, which "is common and reaches its best development on the 'Knobs' of Southeastern Indiana."

NOTES ON FISHES COLLECTED BY CAPT. CHAS. BENDIRE, U. S. A., IN WASHINGTON TERRITORY AND OREGON, MAY TO OCTOBER, 1881.

By TARLETON H. BEAN.

The United States National Museum has again received from Captain Bendire a consignment of alcoholic fishes secured by him last summer and fall. A large collection previously sent by the captain was only partially examined and reported on in the summer of 1881[*]; the greater portion of the fishes are yet to be studied.

The lot just received includes eleven species, nearly all of which are well represented by individuals, giving opportunity for comparison of forms which are mostly rare in museums. Captain Bendire's field notes are included in the remarks upon the species to which they apply.

I think there is no reasonable doubt that the material thus brought together will enable us to prove the identity of *Coregonus Couesii* with *C. Williamsonii*, and to make, eventually, a consolidation of several species of *Apocope*.

The following is a list of the species:

1. *Uranidea marginata.*
2. *Coregonus Williamsonii.*
3. *Oncorhyncus chouicha.*
4. *Oncorhynchus nerka.*
5. *Acrochilus alutaceus.*
6. *Rhinichthys transmontanus.*
7. *Apocope nubila.*
8. *Mylochilus caurinus.*
9. *Richardsonius balteatus.*
10. *Lampetra tridentata.*
11. *Ammocœtes plumbea ?*

1. Uranidea marginata Bean.

U. marginata Bean, Proc. U. S. N. M., iv, p. 26.

30324 (383) 5 specimens. Garrison Creek, Wash. Ter., July 1, 1881. Length of specimens, $2\frac{3}{5}$ to 3 inches.

A.—D. VIII, 18; A. 14$\frac{1}{4}$; V. I, 3-4; origin of anal vertically under second ray of dorsal; pectoral reaches to origin of anal; lateral line 24-25, ending under the 14th ray of dorsal; head $3\frac{1}{3}$ in length; depth 5.

B.—D. VIII, 19; A. 14 $\frac{1}{4}$; V. I, 3; origin of anal vertically under second ray of dorsal; pectoral reaches to origin of anal; lateral line ends under the 13th dorsal ray of left side and the 16th of right side, containing 22 to 25 short tubes; head $3\frac{1}{4}$; depth 5.

C.—D. VIII, 20; A. 14$\frac{1}{4}$; V. I, 3; 13th anal ray divided at tip; last anal ray deeply divided; origin of anal under third ray of dorsal; pectoral reaches to vent; lateral line 27, ending under 17th dorsal ray; head $3\frac{1}{3}$; depth 5.

D.—D. VII, 18; A. 14$\frac{1}{4}$; V. I, 3; the left ventral has, however, a fourth ray which is quite rudimentary; lateral line 20, ending under 10th dorsal ray.

---

[*] Bendire, Notes on Salmonidæ, Proc. U. S. N. M. iv, pp. 81-87, June 2, 1881.

2. Coregonus Williamsonii Girard.

30301 (344–345) ♀ 2 specimens. Mill Creek, tributary of Walla Walla R., May 1, 1881.
30302 (352) 1 spec.   Garrison "   " 9, "
30303 (353) 1 "   "   "   " " "
30304 (354) 1 "   "   ‹ " ‹
30305 (355) 1 "
30306 (356) 1 "   "   "   " " "

Numbers 344 and 345 are the "fresh-water herring" of Mill Creek, "caught with hook and line."

Numbers 352 to 356, inclusive, are the "small-mouthed whitefish caught in Garrison Creek, Walla Walla, by turning the water off. The fish takes a hook occasionally."

Number 344 is a female with the following characters:

Head a little greater than depth of body, 4 in length to end of anal when this is extended backward, slightly more than twice dorsal base. Eye $4\frac{2}{3}$ in head. Maxilla $3\frac{3}{4}$ in head, mandible $3\frac{1}{4}$. 13 or 14 gill-rakers below angle. 13 rows of scales under dorsal base. Scales 10–90–8.

Compare with this the type of *Coregonus Couesii* Milner, from Chief Mountain Lake. This type, number 14146, has: Head a little less than depth of body, $4\frac{1}{5}$ in length to end of extended anal, $1\frac{2}{3}$ times dorsal base. Eye $4\frac{2}{3}$ in head. Maxilla $3\frac{3}{4}$, mandible 3 in head. 14 gill-rakers below angle. 13 rows of scales under dorsal base. Scales 9–88–8.

Number 345 is a female with the following characters: Head a little less than depth of body, $4\frac{1}{3}$ in length to end of extended anal, $1\frac{2}{3}$ times dorsal base. Eye $4\frac{1}{2}$, maxilla $4\frac{1}{4}$, mandible 3 in head. 15 gill-rakers below angle. 13 rows of scales under dorsal base. Scales 10–87–8.

Number 354 shows the following: Head $\frac{7}{8}$ of depth of body, $4\frac{1}{3}$ in length to end of extended anal, $1\frac{1}{2}$ times dorsal base. Eye $4\frac{1}{2}$, maxilla $4\frac{1}{2}$, mandible 3 in head. 14 gill-rakers below angle. 15 rows of scales under dorsal base. Scales 10–90–8.

3. Oncorhynchus chouicha (Walb.) Jor. & Gilb.

30290 (363) Grilse. Walla Walla R. May 18, 1881.
30326 (383)   Garrison Creek. July,   "

In determining the species of *Oncorhynchus*, to which the small example number 383 belongs, I have relied upon the numerous anal rays and branchiostegals as a guide.

Number 363, the "salmon grilse" of this invoice, is a handsomely spotted young male $16\frac{1}{2}$ inches long, with the following characters: Gill-rakers 22; branchiostegals 17; a few weak teeth on head of vomer only; teeth in jaws all small, trout-like; dorsal with 11, anal with 16 divided rays; scales from end of dorsal to lateral line 26, from dorsal line midway between dorsal and snout to lateral line 33; lateral line 145; from ventral origin to lateral line 28; pyloric cæca very small and numerous.

4. Oncorhynchus nerka (Walb.) Gill & Jor.

30291 (359) ♂ head. Celilo, 10 miles above the Dalles, Oregon, May 15, 1881.
30292 (360) ♂ "   "   "   " " "
30293 (361) ♀ "   "   "   " " "
30294 (362) ♀ "   "   "   " " "

Numbers 359–362 are "heads of *Oncorhynchus nerka* caught at Celilo, 10 miles above the Dalles, Oregon, May 15, 1881. Color of fish, as appearing then: Back, steel blue with greenish reflections; sides and belly, pure silvery white. In a number of specimens I examined about that time the vomerine teeth were not perceptible to the touch, but the two rows where they are located can be seen plainly in nearly all the specimens."

Gill-rakers in number 361, 40; branchiostegals 14.

5. Acrochilus alutaceus Ag. & Pick. Hard mouth.

30297 (368) 1 specimen, John Day River, Oregon, Aug. 15, 1881.
30298 (369) 1   "   "   "   "   "

Number 368 has: scales 22–89–16; persistent teeth on left side 5, and one deciduous; greatest depth equals head, $4\frac{1}{2}$ in length to end of scales; least depth of caudal peduncle 3 in head; eye $1\frac{1}{3}$ in snout, $4\frac{1}{2}$ in head; greatest width of cartilaginous plate of lower lip equals lower jaw, $3\frac{1}{4}$ in head; longest anal ray nearly $1\frac{1}{2}$ times anal base; pectoral 5, ventral $6\frac{1}{4}$ in length to end of scales; dorsal origin midway between snout and end of scales; D. 10; A. 9; V. 9; length of fish 10 inches.

Number 369 has: scales 22–87–16; persistent teeth 4–5, one deciduous tooth on one side and two on the other; greatest depth of body equal to head, $4\frac{1}{2}$ in length to end of scales; least depth of caudal peduncle 3 in head; eye 5 in head, $1\frac{1}{3}$ in snout; width of cartilaginous plate on lower lip equals lower jaw and 3 in head; longest anal ray $1\frac{1}{3}$ times anal base; pectoral $5\frac{1}{3}$ in length to end of scales; dorsal origin midway between snout and end of scales; ventral 7 in length to end of scales; D. 10; A. 9; V. 9; length of fish 11 inches.

6. Rhinichthys transmontanus Cope.

30332 (383) 4 specimens, Garrison Creek, Wash. Ter., July, 1881.

Teeth 2, 4–4, 2; scales in three individuals examined were as follows: 14–77–14, 14–72–14, 14–68 to 70–14. In one of these I counted 68 scales in the lateral line of one side and 70 on the other side.

Dorsal midway between anterior nostril and end of scales, its base equals $\frac{2}{3}$ of its longest ray, which is 6 in length to end of scales; head $4\frac{1}{3}$, depth 5. pectoral 5 in length to end of scales; ventrals reach to vent; pectorals do not extend to ventral origin; D. 8; A. 7; length of specimens $3\frac{1}{2}$ to $4\frac{1}{3}$ inches.

7. Apocope nubila (Grd.) Jor. & Gilb.

30323 (383) 5 specimens, Garrison Creek, Wash. Ter., July 1881.

The larger of the two types of *Argyreus nubilus* Grd. has the following characters: Greatest height of body very little more than length of head, $4\frac{1}{4}$ in length to end of scales; upper jaw reaching to vertical through hind margin of posterior nostril; eye 5 in head; snout 3 in head; pectoral $5\frac{1}{2}$, ventral $6\frac{1}{2}$ in length to end of scales; longest dorsal ray equal to longest anal, which equals head without snout; D. 8; A. 7; V. 7; scales 12–60–10; length 4 inches; teeth 2, 4–4, 2, slightly hooked, and with a very narrow groove beneath the hook.

The examples sent by Capt. Bendire show the following characters: Greatest height of body slightly exceeds length of head, 4⅕ in length to end of scales; upper jaw as in the above; eye 4½ in head; snout 3¼ in head; pectoral 5, ventral 6 in length to end of scales; longest dorsal and anal rays as in last; ventral reaches to anal; D. 9; A. 7; V. 7; scales 12 to 13–55 to 50–10 to 12; length 3 to 3¼ inches; teeth 2, 4–4, 2.

These specimens show considerable variation in the number of scales in the lateral line, and there is constantly one more dorsal ray than in the types of *A. nubila;* they are, however, certainly not specifically distinct from Girard's form.

I have examined a fish collected by Prof. Jordan in Utah Lake and correctly identified by him with *Apocope vulnerata* Cope. While the teeth of one side of the specimen identified by Prof. Jordan are 1, 4, as he states, on the other side of the same fish I find 2, 4. If this condition occurs frequently the margin of separation between *A. vulnerata* and *A. nubila* will become uncomfortably small, as there will be little left besides the slightly greater number of scales.

Description of a female specimen of *A. nubila*, number 24195, collected by Capt. Bendire at Walla Walla.

D. ii, 7+; A. ii, 6+; V. 8; P. 15; scales 13–53–10; teeth hooked, slightly grooved, 1, 4–4, 1.

Barbels minute. The end of the maxilla reaches the vertical through the anterior margin of the nostril; snout contained 3 times, eye 5 times in length of head. Eye 1½ times in width of interorbital area. Length of head nearly 4 times in total length caudal excluded, 4½ times caudal included. Greatest depth 5 times. Longest dorsal and anal rays equal and contained 5½ times in total length without caudal; pectoral contained 5 times in the same length. Ventral equal to length of head without postorbital part. The origin of the dorsal is a little behind that of the ventrals, about midway between the tip of the snout and the end of the middle caudal rays. Length 81 millimeters.

Color of the alcoholic specimen grayish olive. There is a faint indication of a dark stripe on the nose.

8. Mylochilus caurinus (Rich.) Girard.

30299 (342) ♀ 1 specimen, Mill Creek, trib. of Walla Walla R., Apr. 26, 1881.

"Chub, taken Apr. 26, 1881, in Mill Creek, tributary of Walla Walla River, Washington Terr'y."

"Above bluish brown; sides paler. A carmine red stripe along the sides. Belly silvery white. Nose steel blue. Stripe below the eye brick red. Called Red Horse occasionally."

Eye equal to preorbital, 1½ in snout, 5 in head. Maxilla reaching vertical through hind margin of posterior nostril. Head 1⅕ in depth, 4⅔ in length to end of scales. Depth 4¼ in length to end of scales. Pectoral equals longest dorsal ray, 3 in distance from snout to dorsal. Ventral is under 3rd ray of dorsal, does not reach vent, equals head without snout. D. 8; A. 8; V. 9; scales 14–74–9; teeth 1, 5–5, 1; length 11⅘ inches.

While it is certain that the persistent pharyngeal teeth are as stated, I must note that a small tooth was found loose in the tissues covering the dentigerous bones. It may be that this fish had the normal number and two of them were displaced by accident. Four of the teeth of each side are molar-like.

9. Richardsonius balteatus (Rich.) Grd.

30322 (383) ♀ 1 specimen, Garrison Creek, Wash. Ter., July, 1881.

Length of example 4 inches. Teeth 2, 5–5, 2, hooked, without grinding surface. Body compressed, resembling *Notemigonus*. Snout ⅔ as long as eye, 4 in head. Eye 3 in head. Head ¼ of greatest height of body, almost 4½ in length to end of scales. Maxilla 3 in head, mandible 2½. .Dorsal behind ventrals, much nearer caudal than end of snout, its base equal to ¼ of its distance from snout. Longest dorsal ray equals length of pectoral, 5½ in length to end of scales. Anal basis nearly equals head, 4⅔ in length to end of scales. Ventral nearly equi-distant from snout and end of scales. D. 10; A. 13; V. 9;. scales 12–63–8.

10. Lampetra tridentata (Gairdner) Jor. & Gilb.

30295 (347) 1 specimen, Walla Walla R., Wash. Ter., May 6, 1881.
30296 (351) 1 specimen, Garrison Creek, Wash Ter., May 9, 1881.

"Lamprey eel." Number 347 is 19 inches long; number 351 is almost exactly as long. The teeth are as in Richardson's description in Fauna Boreali-Americana; the dorsals, however, are separated simply by a deep emargination; the base of the first is from one-half to two-thirds as long as that of the second; the second dorsal is higher than the first, and is subcontinuous with the caudal. The length of the space occupied by the gill-openings is contained 8½ times in total length, and is a little more than the length of the head from end of snout to first gill-opening. Greatest height of body 6 in distance from snout to first dorsal.

I have compared the type of *Petromyzon astori* Grd. with *Lampetra tridentata* and find that they are certainly identical, as already pointed out by Professor Jordan. The types of *P. ciliatus* and *P. lividus* have the dorsals separated by a space nearly or quite half as long as the first dorsal, but otherwise they have the characters of *tridentata*.

11. ?Ammocœtes plumbea (Ayres).

30321 (383) juv., 1 specimen, Garrison Creek, Wash. Ter., July, 1881.

1 am in doubt whether or not this small lamprey, 4⅘ inches long, is the larval form of the above-named species or not. The maxillary plate is bicuspid, the cusps well separated; the mandibulary plate has 7 teeth of uniform size. I am unable to determine the structure of the other teeth. The lips are fringed. Head 8½ in length, equal to space occupied by gill-openings. Dorsals subcontinuous. Height of body almost equal to head. Perhaps this is *Ammocœtes cibarius* Girard, and may be distinct from *A. plumbea*.

3.—NEW MOLLUSCAN FORMS FROM THE LARAMIE AND GREEN RIVER GROUPS, WITH DISCUSSION OF SOME ASSOCIATED FORMS HERETOFORE KNOWN.

By C. A. WHITE.

[Extract from the Annual Report of the United States Geological Survey for 1882, by permission of the Director]

Notwithstanding the large number of specific and generic forms of fossil mollusca that have been obtained from the Laramie and fresh-water Eocene groups of Western North America, every fresh examination of those deposits in any region in which they occur is sure to add something to our knowledge of the faunæ which respectively characterize them. While studying the Laramie Group in Northeastern Colorado during the season of 1881, I obtained no less than four new species, and extended the known geographical range of several others. Besides the new forms just mentioned I have recognized two others among collections made by other persons that have been in the National Museum for several years past. All of these new forms are described in the following paragraphs ; and remarks are made upon other forms concerning which new facts have been discovered. These descriptions are also to appear in the Annual Report of the United States Geological Survey for 1882, in a "Review of the Non-Marine Fossil Mollusca of North America."

Genus UNIO Retzius.

Unio clinopisthus (sp. nov.), Plate III, figs. 1 and 2.

Shell transversely elongate, short in front of the beaks, elongate and narrowing behind them to the posterior end ; basal margin having a gentle sinuosity, there being a slight emargination just behind the mid-length ; front margin regularly rounded ; dorsal margin proper rather short ; postero-dorsal margin forming a long, convex, downward slope from the dorsal to the postero-basal margin, which latter margin is narrowly rounded ; beaks depressed and placed near the front of the shell. A somewhat prominent, but not sharply defined, umbonal ridge extends from the beak of each valve to the postero-basal margin, giving a flattened space at the postero-dorsal portion of each valve. Surface marked only by concentric lines of growth.

Length, 63 millimeters ; height, 30 millimeters ; thickness, both valves together, 23 millimeters. (Museum No. 8359.)

Position and locality.—Strata of the Green River Eocene group near Washakie Station, in Southern Wyoming, where it was collected by Dr. Hayden.

Genus CORBICULA Mühlfeldt.

Corbicula berthoudi (sp. nov.), Plate IV, figs. 1, 2, and 3.

Shell very large, subtrigonal in marginal outline, moderately gibbous; front concave immediately in front of the beaks; front margin regularly rounded; basal margin broadly rounded; postero-basal margin abruptly rounded up to the postero-dorsal margin, which latter margin slopes obliquely downward with a gentle convexity from between the beaks; hinge strong; all the teeth well developed, the lateral ones especially being long and large and crenulated upon their edges as is usual with all the known species of *Corbicula* of the Laramie Group; muscular and pallial impressions having the usual characteristics; surface marked with the usual concentric lines.

Length of one of the largest examples in the collections, 62 millimeters; height from base to umbo, 54 millimeters; thickness, both valves together, 44 millimeters.

This fine large species, the largest yet known in North America, has been found only in the Laramie strata east of the Rocky Mountains in Colorado. It is named in honor of Capt. E. L. Berthoud, the first discoverer of the rich shell deposits of the Laramie Group in that region. (Museum No. 11556.)

Position and locality.—Laramie Group; valley of South Platte River; Northeastern Colorado.

Corbicula augheyi (sp. nov.), Plate IV, figs. 4, 5, and 6.

Shell moderately large, subtetrahedral in marginal outline, posterodorsal region not flattened, as in *C. berthondi;* umbones full, rounded, considerably elevated above the hinge line, front regularly rounded; basal margin broadly convex; posterior end truncated, the direction of the truncated margin usually a little backward of a line drawn perpendicularly with the base of the shell; postero-dorsal margin a little convex; hinge well developed; muscular and pallial markings of the usual character; surface marked by the usual concentric line of growth, and usually by very faint umbonal ridges extending from the umbo to the postero-dorsal and postero-basal margins respectively.

Length of an adult example, 46 millimeters; height from base to umbones, 38 millimeters; thickness, both valves together, 30 millimeters.

This species has yet been foun‸ only in the valley of South Platte River, in Northern Colorado, east of the Rocky Mountains. It is named in honor of Prof. Samuel Aughey, of Nebraska State University, who assisted me in the collection of the type specimens. (Museum No. 11557.)

Position and locality.—Laramie Group; valley of South Platte River; Northeastern Colorado; associated with the preceding.

Genus NERITINA Lamarck.

Neritina bruneri (sp. nov.), Plate IV, figs. 7 and 8.

Shell subglobose; volutions about four; spire much depressed; suture moderately distinct; inner lip broad, its inner edge a little irregular. Surface of adult examples marked by numerous raised revolving lines, which are crossed by strong, dark, zigzag color-markings. Upon young examples the revolving lines are absent, or nearly so, and the color-markings are less distinctly zigzag in their direction.

Axial length, 10 millimeters; transverse diameter, 13 millimeters.

The specific name is given in honor of Mr. Lawrence Bruner, who first discovered the species. It differs from *N. volvilineata* White, in being somewhat more globose, having a less elevated spire, and the inner lips broader and less retreating. It is marked by revolving lines, somewhat like that species, but they are sometimes obsolete. It is also ornamented by zigzag color-markings. The type specimen is represented by figs. 7 and 8 on Plate IV.

Associated with the foregoing is still another form, much smaller, which seems to be the young of *N. volvilineata*. It is without color-markings, and the inner border of the inner lip is dentate.

Position and locality.—Laramie Group; valley of South Platte River, Northeastern Colorado, where it is associated with the two last-described species.

Genus MELANOPSIS Lamarck.

Melanopsis americana (sp. nov.), Plate IV, figs. 9 and 10.

Shell very small, sides straight, and meeting at the apex at an acute angle; volutions six or seven, those of the spire not convex, but so flattened as to show only a linear suture between them, which is somewhat irregular; proximal portion of the last volution gently convex, its length being more than half the entire length of the shell; outer lip thin, not expanded, its margin not distinctly sinuous; inner lip having a very strong callus nearly filling the distal end of the aperture, leaving a narrow groove between it and the margin of the outer lip, and gradually diminishing in thickness towards the proximal end of the aperture; aperture, as bounded by the outer lip and callous inner lip, rudely subelliptical, angular at its distal end, rounded at its proximal end, and terminating at the end of the columella in a distinct, narrow canal, which is slightly bent to the left. Surface marked only by faint lines of growth.

Length, 7 millimeters; diameter of last volution, 3½ millimeters. (Museum No. 11559.)

If we except the species which were published by Conrad under the generic name of *Bulliopsis*, but which probably belong to the genus *Melanopsis*, no species of the latter genus have hitherto been known in North America, either fossil or living. The species which is here described is plainly congeneric with the living *Melanopsis costellata* Ferussac, and with the Eocene *M. buccinoidea* Ferussac, both of Western Europe.

Position and locality.—Laramie Group, Valley of South Platte River, Northeastern Colorado, where it is associated with the three last described forms, and also with *Corbula*, *Melania*, *Anomia*, and *Ostrea*.

Genus CAMPELOMA Rafinesque.

Campeloma producta (sp. nov.), Plate III, figs. 7, 8, and 9.

Shell, elongate-ovate; test, moderately thick; spire, more than usually produced for a species of this genus; volutions, six or seven, usually slightly flattened, or having a faint revolving depression upon the distal side near the suture, which is more apparent upon the larger than the smaller volutions; suture, deep and abrupt upon the proximal side; aperture and lips having the usual characteristics of *Campeloma*; surface marked by the usual lines of growth, and by somewhat numerous revolving striæ which are often obscure. Among these examples are others which possess the general characteristics of those which are regarded as the types; but two or three of the revolving striæ upon the smaller volutions of these examples are much more prominent than in the case of typical examples. I at present, however, regard these as only varieties of a very variable species.

Length of an example regarded as typical, 32 millimeters; breadth of the last volution, 14 millimeters; but some examples, evidently referable to the typical forms, are proportionally less elongate. (Museum No. 8140.)

Position and locality.—Laramie strata in the Valley of Yellowstone River, Montana, where they were collected several years ago by Mr. J. A. Allen.

The under valve of ANOMIA MICRONEMA *Meek.*

It has been the subject of frequent remark that not a single example of the under valve of either of the two species of *Anomia*, *A. micronema* and *A. gryphorhynchus* Meek, both of the Laramie Group, has ever been discovered, although hundreds of examples of the upper valves of both of these species have been obtained, at many different localities, in a good state of preservation. I was lately so fortunate, however, as to find in the Laramie strata of Northeastern Colorado several examples of the under valve of *A. micronema*, one of which is illustrated by fig. 3, Plate III. That the under, or byssus-bearing, valves of *A. micronema* at least have been so generally destroyed is due to the fact, first, of their extreme thinness, and, secondly, to the fact that, with the exception of a thin, porcelanous layer in the middle portion, the whole valve is composed of a prismatic layer, like the shell of *Pinna;* the pearly layer, which gives such strength to the upper valve, being apparently entirely wanting in the lower. This prismatic layer breaks up into its component prisms with great facility. The characteristics of the under valve of *A. micronema*, as well as those of the upper valve, show it to be a true *Anomia;* thus presenting evidence of the great antiquity of the genus essentially as it exists to-day.

Both valves of recent species of *Anomia* have, as a covering to the pearly layer, a very thin prismatic layer, which is often obsolete. This layer is also sometimes distinguishable upon the upper valves of these

June 24, 1882.

fossil species. The latter seem to differ from the shells of living species of *Anomia* only in the lack of development in the under valve of the pearly layer, and the excessive development of the prismatic layer.

PYRGULIFERA *Meek and* PARAMELANIA *Smith.*

There occurs somewhat abundantly in the Bear River Laramie beds of Southwestern Wyoming and the adjacent parts of Utah a shell which Mr. Meek first referred to *Melania*, but to which he afterward gave the new generic name of *Pyrgulifera*, describing it under the name of *Pyrgulifera humrosa*.* It is illustrated on Plate III, figs. 10, 11, and 12. Meek placed this shell among the Ceriphasiidæ or American Melanians, but as it seems to differ quite as widely from the typical forms of that family as it does from the true Melanians, I have placed it provisionally with the latter family. It is the only known species of the genus which has been proposed to receive it, either fossil or living, if we except the two living forms which were described by Mr. Edgar A. Smith from Lake Tanganyiki, in Africa,† under the new sub-generic name *Paramelania*. Mr. Smith gave these two forms the names *P. damoni* and *P. crassigranulata*, respectively. Copies of his figures of both these forms are given on Plate III for comparison.

Paramelania, as represented both by these figures and Mr. Smith's description, seems to be exactly equivalent with *Pyrgulifera* Meek. It is true that we can never know whether the animal of the latter was generically the same as that of the former, and the wide chronological and geographical separation of the fossil and living forms is presumptive evidence against their generic identity. But if we are justified in establishing genera upon shells alone, as we must do in paleontology, we are entitled to hold them as against anything except direct proof of error.

EXPLANATION OF PLATE III

UNIO CLINOPISTHUS (sp. nov.).

Fig. 1.—Left side view; natural size.
Fig. 2.—Dorsal view of the same example.

ANOMIA MICRONEMA Meek.

Fig. 3.—View of the under valve, showing the byssal plug.
Fig. 4.—Exterior view of an upper valve.
Fig. 5.—Similar view of another example, showing coarser radiating lines.
Fig. 6.—Interior view of a very large upper valve, showing muscular scars and process beneath the umbo. All of natural size.

CAMPELOMA PRODUCTA (sp. nov.).

Fig. 7.—Lateral view of type specimen; natural size.
Fig. 8.—Opposite view of the same.
Fig. 9.—Lateral view of a more robust example.

---

* For diagnosis of this genus, and description and figures of the species, see U. S. Geol. Sur. 40th Parallel, vol. iv, p. 146, pl. 17, fig. 19.
† See Proc. Zool. Soc. Lond. for May, 1881, pp. 558–561.

PLATE III.

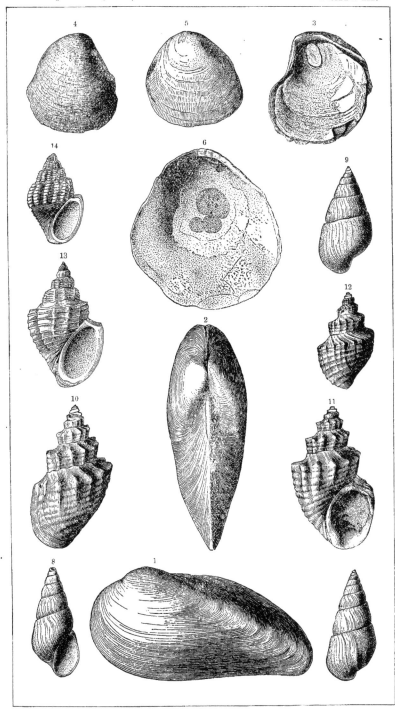

PLATE IV.

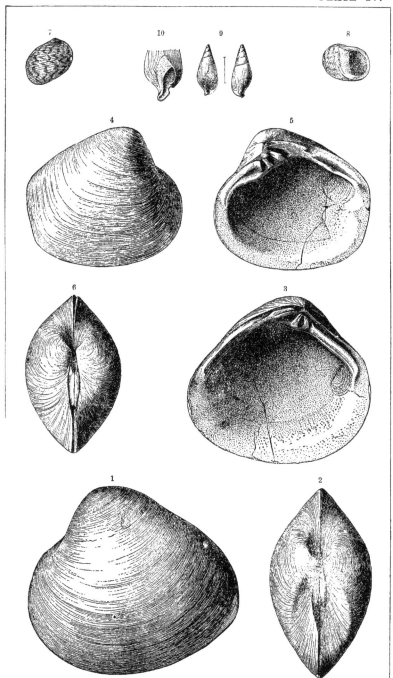

PYRGULIFERA HUMEROSA Meek.

Fig. 10.—Lateral view of type specimen; natural size.
Fig. 11.—Opposite view of the same.
Fig. 12.—Similar view of a smaller example.

PYRGULIFERA (PARAMELANIA) DAMONI Smith.

Fig. 13.—Copy of Mr. Smith's original figure.

PYRGULIFERA (PARAMELANIA) CRASSIGRANULATA Smith.

Fig. 14.—Copy of Mr. Smith's original figure.

EXPLANATION OF PLATE IV.

CORBICULA BERTHOUDI (sp. nov.).

Fig. 1.—Left side view; natural size.
Fig. 2.—Dorsal view of another example.
Fig. 3.—Interior of left valve of another example.

CORBICULA AUGHEYI (sp. nov.).

Fig. 4.—Right side view; natural size.
Fig. 5.—Interior view of the same example.
Fig. 6.—Dorsal view of another example.

NERITINA BRUNERI (sp. nov.).

Fig. 7.—Lateral view; natural size.
Fig. 8.—Apertural view of the same example.

MELANOPSIS AMERICANA (sp. nov.).

Fig. 9.—Two different lateral views; enlarged.
Fig. 10.—Another view of the lower part of the same example, showing the beak and the callus of the inner lip.

4.—THE MOLLUSCAN FAUNA OF THE TRUCKEE GROUP, INCLUDING A NEW FORM.

By C. A. WHITE.

[Extracted from the Annual Report of the United States Geological Survey for 1882, by permission of the Director]

In volume II, Paleontology of California, Mr. Gabb described and figured two species of fresh-water fossil mollusca from the valley of Snake River, Idaho, which he stated to be of Tertiary age. In volume IV, United States Geological Survey of the Fortieth Parallel, Mr. Meek described and figured seven other species, one from Southwestern Idaho and the others from the Kawsoh Mountains, in Northern Nevada. He referred these to the Tertiary period, and they evidently came from strata that are geologically equivalent with those which furnished Mr. Gabb's specimens. Mr. King, in volume 1 of the last named survey, referred these strata to the Miocene epoch of the Tertiary period, and gave them the name of Truckee Group.

While arranging the fossils of the National Museum, I lately found among other undistributed material a small mass of fossiliferous rock, which bore the label "50 miles below Salmon Falls, Snake River."

Upon breaking up this mass it was found to contain examples of both the species that were described by Gabb, and also another form that has not hitherto been described. This makes ten species of mollusks that are now known to exist in that formation.

Although this fauna, so far as it is now known, is a very meager one, it is, nevertheless, very interesting, because it differs so much from any other fresh-water fauna, either fossil or now living, in North America. This difference is all the more remarkable because the fresh-water faunæ of the Laramie, and the several Eocene groups, all of which are older than the Truckee Group, consist largely of types that are now living in the Mississippi drainage system.

Illustrations of all the molluscan species that are yet known to exist in the Truckee Group are brought together on Plate V for the purpose of presenting them all at a single view. All of them, except the four figures of *Latia dallii*, are copies of the original figures published by Meek and Gabb respectively.

Not deeming it necessary to repeat the descriptions of the species that have already been published, the new form only will be herein described.

The names of the others, however, are given in the following paragraphs, together with references to the respective works in which the species were originally described.

Melania sculptilis Meek, U. S. Geol. Sur. 40th Parallel. Vol. IV, p. 195.

Melania subsculptilis Meek, Ib., p. 196.

Melania taylori Gabb. Paleoutology of California. Vol II, p. 13.

Lithasia antiqua Gabb, Ib., p. 13.

Carinifex (Vorticifex) tryoni Meek, U. S. Geol. Sur. 40th Parallel. Vol. IV, p. 188.

Carinifex (Vorticifex) binneyi Meek, Ib., p. 187.

Ancylus undulatus Meek, Ib., p. 186.

Sphærium rugosum Meek, Ib., p. 182.

Sphærium idahoense Meek, Ib., p. 183.

Latia dallii (sp. nov.), plate V, figures 17, 18, 19, and 20. Shell subovate in marginal outline; irregularly convex above; the posterior portion narrowing rapidly to a small prominent umbo which ends in a very small closely incurved beak which is turned a little to the right side, and makes about one full volution. Semilunar shelf or septum comparatively large. Surface marked by many strong irregular concentric undulations; but otherwise it is comparatively smooth, being marked only by fine lines of growth.

Length, 16 millimeters; breadth, 10 millimeters; height, 7 millimeters. (Museum No. 11547.)

The specific name is given in honor of Mr. W. H. Dall, whose important works upon the mollusca are well known.

This interesting shell seems to agree with the genus *Latia* Gray in all essential characteristics, so far as they are observable upon the specimens that have yet been discovered. The form is *Crepidula*-like, the test thin, and the semilunar septum, well developed; but the "projecting free lamina" upon the right side of the septum, described by Dr. Gray, has not been observed upon our examples.

Although in form and structure this shell is so much like a *Crepidula*, its fresh-water associations forbid its reference to the Calyptriidæ. I am not entirely satisfied that it ought to be referred to the Ancylidæ, but for the present I place it provisionally in that family.

The fact that this shell is entirely unlike any form that is now known in North America, either living or fossil, gives it peculiar interest. This interest is also largely increased by the fact that the genus to which it is here referred has hitherto been known only in New Zealand or other parts of Oceanica, and only in the living state.

The molluscan fauna, to which this shell belongs is, as a whole, quite unlike any other fresh-water fauna of North America, either living or fossil. The reason of this difference between the Truckee molluscan fauna and that which now characterizes the Mississippi drainage system is doubtless that the outlet of the Truckee lake has had no continuous connection or identity with the streams that, persisting from Tertiary time and earlier, have become portions of that system.

The forms among the Truckee fauna that are most nearly like species now living in North America are the *Ancylus* and the two species of *Sphærium* just mentioned; and yet the latter present some noteworthy differences from any North American congeneric form either living or fossil. It is true there is a species of *Carinifex* in the Pacific drainage waters of California, but its difference from those of the Truckee fauna was regarded by Mr. Meek as of subgeneric importance. The three forms of *Melania* and the *Lithasia* of the foregoing list have no true type-representatives, either living or fossil, in North America; and the newly discovered form herein described differs still more widely from any member of any North American fauna.

The Truckee Group is understood to have quite a large geographical extent in northern Nevada, southwestern Idaho and southeastern Oregon, but it has yet received very little investigation as regards its molluscan fauna. The presence in that group of a molluscan fauna so widely differentiated as it is indicated to be by the few species that have hitherto been discovered encourages the hope that large additions to it will hereafter be made.

EXPLANATION OF PLATE V.

MELANIA SCULPTILIS.

Fig. 1.—Copy of Meek's original figure.

MELANIA SUBSCULPTILIS.

Fig. 2.—Copy of Meek's original figure.

MELANIA TAYLORI.

Fig. 3.—Copy of Gabb's original figure.

LITHASIA ANTIQUA.

Fig. 4.—Copy of Gabb's original figure.

CARINIFEX (VORTICIFEX) TRYONI.

Figs. 5, 6, and 7.—Different views of the type specimen. After Meek.

CARINIFEX (VORTICIFEX) BINNEYI.

Figs. 8 and 9.—Different views of the type specimen. After Meek.

ANCYLUS UNDULATUS.

Fig. 10.—Dorsal view of type specimen. After Meek.
Fig. 11.—Lateral outline of the same.

SPHÆRIUM? IDAHOENSE.

Figs. 12 and 13.—Copies of Meek's original figures.

SPHÆRIUM RUGOSUM.

Figs. 14, 15, and 16.—Copies of Meek's original figures.

LATIA DALLII (sp. nov.).

Fig. 17.—Dorsal view of the largest known example.
Fig. 18.—Lateral view of the same.
Fig. 19.—Dorsal view of another example.
Fig. 20—Dorsal view of another example which has been cut away so as to reveal the transverse semilunar septum.

All the figures on this plate are of natural size except Figs. 14, 15, and 16, which are a little enlarged.

DESCRIPTION OF FOUR NEW SPECIES OF SHARKS, FROM MAZATLAN, MEXICO.

By DAVID S. JORDAN and CHARLES H. GILBERT.

Carcharias fronto, sp. nov. (28167.)

Allied to *Carcharias amblyrhynchus* Bleeker, but with much larger second dorsal.

a. Description of No. 28167, a young (female) example, 36 inches in length:

Body comparatively short and stout. Head very broad, depressed, broadly rounded anteriorly, the outline of the snout nearly parallel with that of the broad V-shaped mouth. Length of snout from mouth equal to half the distance between the angles of the mouth, or to the distance from the line connecting these angles to the chin, about six-sevenths the distance between the nostrils. Eye a little nearer nostril than angle of

PLATE V.

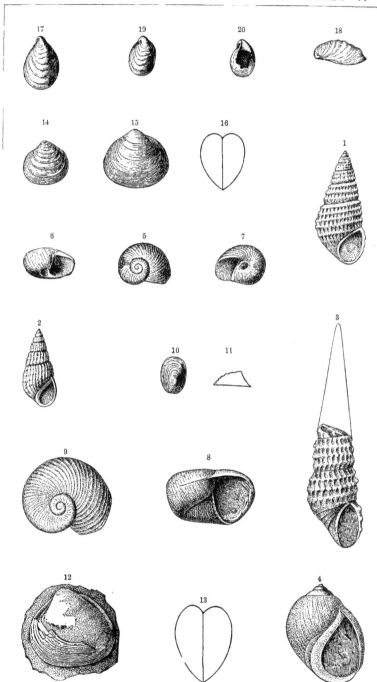

mouth. Nostril a little nearer eye than tip of snout. Interorbital width a trifle more than half distance from snout to base of pectoral, about twice length of snout, measured from eye. Angle of mouth with a deep pit which scarcely extends as a furrow on either lip. Nostrils near margin of head, their length half their distance from eye, and about the same as length of eye, the anterior margin with a moderate flap. Teeth ot both jaws narrowly triangular, more than twice as high as broad, those of the upper jaw rather broadest, all of them nearly erect and not evidently notched on the outer margin. Edges of teeth appearing minutely serrulate under a lens. Teeth about $\frac{20}{20}$.

Gill-openings rather deep, the last two over the base of the pectorals, the depth of them about equal to the distance from nostril to middle of eye, the branchial area scarcely longer than broad.

Free margins of all the fins concave. Insertion of first dorsal rather nearer pectorals than ventrals, its distance behind pectoral being nearly equal to the length of its anterior margin. Anterior lobe, when depressed, reaching past the base of the fin, but not to the end of the posterior lobe, which reaches nearly to the base of the ventrals. Length of base of first dorsal slightly more than its (vertical) height, and notably less than interorbital width. Distance between dorsals slightly more than twice the base of first dorsal, a little more than three times base of second.

Second dorsal similar in form to the first dorsal, its base one-fifth shorter, its posterior lobe reaching to within an eye's diameter of the pit at root of caudal.

Lower lobe of caudal half length of upper, both of the ordinary form in this genus; length of the upper lobe a little less than the distance from snout to posterior part of root of pectorals, a little less than one-fourth the total length.

Anal smaller than second dorsal and proportionately higher, its margin deeply concave, its anterior lobe reaching the tip of posterior when reflexed; length of anterior margin about equal to base of second dorsal. Distance of anal from caudal a little more than its base, and a little less than distance of front of anal from base of ventrals. Middle of anal under middle of second dorsal.

Ventrals moderate, their anterior margins about equal to the distance between the angles of the mouth.

Pectorals rather large, their angles not acute; their tips reaching a little past front of dorsal; their anterior margins half longer than interorbital width, and $2\frac{1}{2}$ times the free margin.

Color uniform slaty-gray; all the fins edged with darker brownish, darkest on the lower lobe of the caudal, but nowhere fully black.

b. Description of a large specimen:

A large example of this species, some 10 feet in length, was speared from the wharf at Mazatlan. The jaws of this specimen (collector's number, 997) were taken.

In this specimen the teeth of the upper jaw are broadly triangular, the breadth of the base being about equal to the vertical height, the inner margin nearly straight or slightly convex, the outer margin a little concave, but not distinctly notched. Edges of upper teeth conspicuously serrate, especially towards the base. Median tooth very broad and short, serrate, with concave margins.

Teeth of lower jaw narrowly triangular, with broad bases, which spread out abruptly. Edges of lower teeth weakly serrate; median teeth small, triangular, entire; middle teeth of sides of each jaw largest. Number of teeth about $\frac{2}{2}\frac{3}{6}$.

The following notes on this specimen were taken by Mr. Gilbert. The measurements were taken without instruments as the shark was lying on the beach, and are not all entirely accurate:

Head very heavy and short, the snout exceedingly broad and short; its preoral portion much longer, shorter than distance between nostrils, and nearly equal to the length between the inner margins of the pectorals. Eyes on the depressed margin of the head, the nostrils very close to the margin. A short deep fold at angle of mouth, extending a very short distance on each lip.

Gill openings wide, the last two above base of pectorals, the longest slit about equal to half base of pectoral.

Origin of first dorsal behind vertical from posterior base of pectoral, a distance about equal to a gill-slit, the fin considerably nearer pectorals than ventrals. Anterior margin of dorsal two-thirds anterior margin of pectorals. Anal inserted slightly behind front of second dorsal and somewhat smaller than the latter; its length a little more than that of branchial area.　　　　　　　　　　　　　　　　　　·

Pectorals long, not half longer than broad, their anterior margins convex; inner margin of pectoral about half longer than a gill-slit; about one-third the length of the free edge, which is six-sevenths the anterior edge.

Ventrals short; length of anterior margin less than one-third the length of pectorals, nearer second dorsal than first. Distance from anterior base of ventrals to vertical from first dorsal equal to distance of the posterior base from anal. A pit above and below root of tail. Caudal with lower lobe half length of upper, the lower lobe broadly scythe-shaped.

This species is rather common about Mazatlan, where it is known as *Tiburon*. Its liver is valued for the oil it produces. It was not seen elsewhere.

The fresh-water shark of Lake Nicaragua (*Eulamia nicaraguensis* Gill) is closely allied to this species, but apparently distinct.

Carcharias æthalorus, sp. nov. (28202, 29549).

Allied to *Carcharias lamia* (Risso), but with smaller dorsal and pectoral fins and longer and more pointed snout.

Body rather robust, the back somewhat elevated. Head depressed, but rather long and pointed, the snout low and flat, somewhat acute in outline. Length of snout from mouth just equal to the distance between the angles of the mouth, about half greater than the distance from the line connecting the angles of the mouth, forward to the chin. Eye moderate, a little nearer nostril than angle of mouth; distance from eye to nostril nearly two-thirds distance from nostril to tip of snout. Interorbital width less than half distance from snout to first gill-opening, slightly more than length of snout measured from eye.

Angle of mouth with a pit from which radiate three very short furrows· Nostrils not far from margin of head, their length equal to diameter of eye and rather less than half their distance from eye, the flap on the anterior margin nearly obsolete.

Teeth of both jaws narrowly triangular, nearly erect and not evidently notched on the outer margin; all the teeth distinctly though finely serrated on both margins. Lower teeth considerably narrower than upper and somewhat flexuous, more faintly serrate; their roots broad. Number of teeth about $\frac{24}{24}$.

Head without conspicuous pores. Gill-openings rather large, the last one shorter than the others, above base of pectoral, the depth of the middle one about equal to distance from nostril to middle of eye; the branchial area about half longer than broad.

Free margins of all the fins concave. Insertion of first dorsal close behind base of pectorals, its distance from the latter being not more than the diameter of the eye.

Anterior lobe when depressed extending beyond posterior lobe; distance from base of posterior lobe to ventrals somewhat more than length of snout from mouth. Length of base of first dorsal somewhat less than vertical height of the fin and equal to interorbital width. Distance between dorsals 2 to $2\frac{1}{8}$ times base of first dorsal, and about 4 times base of second.

Second dorsal much smaller than first; its posterior lobe longer than anterior and somewhat longer than base of fin, reaching to within $1\frac{1}{2}$ eyes' breadth of the large pit at root of caudal. Base of second dorsal nearly half length of first dorsal.

Lower lobe of caudal subfalcate, moderately pointed, two-fifths length of upper lobe, which is a little more than one-fourth the total length, and a little less than distance from snout to posterior part of root of pectorals. Anal a little longer than second dorsal and inserted nearly opposite its front; its posterior lobe extending considerably beyond the anterior when reflexed; length of its base nearly equal to its distance from caudal, a little less than distance from its front to ventral. Ventrals moderate, their anterior margins about three-fourths distance between angles of mouth.

Pectorals rather large, somewhat falcate, twice as long as broad, reaching to about opposite posterior part of base of dorsal, their tips somewhat pointed, the length of the anterior margin 1⅔ times interorbital width, and nearly 1½ times the free margin; length of pectoral about one-sixth the total length of the fish.

Color light slaty-gray; belly white; middle line of back posteriorly and upper edge of tail blackish; tips of all the fins distinctly blackish, especially the pectorals and lower lobe of caudal₍

This species is rather common at Mazatlan, where several young specimens were taken, the largest (29549), a young male, being 30 inches in length. A species thought to be the same was also seen at Panama, but no specimens were brought to the museum.

Carcharias longurio, sp. nov.   (28306, 28330, 28331, 29541, 29551.)

  ? *Squalus* (*Scoliodon*) *porosus* Poey, Memorias Cuba, II, 339 tab. 19, f. 11, 12, 1861 (Cuba.)   (Not *Carcharias porosus* Ranzani, 1839.)
  ? *Scoliodon porosus* Poey, Synops. Pisc. Cubens, 1868, 452.   (Cuba.)

Subgenus *Scoliodon* Müller and Henle.

Body rather slender and elongate, the back little elevated. Head depressed, long and narrow, rather pointed anteriorly. Length of snout from mouth greater by about the diameter of the eye than the distance between the angles of the mouth and a little more than half greater than the distance from a line connecting the angles of the mouth to the tip of the chin.

Eye rather large, a little nearer nostril than angle of mouth; distance from eye to nostril scarcely half the distance from nostril to tip of snout.

Interorbital width two-fifths distance from snout to first gill-opening, somewhat less than length of snout measured from eye.

Angle of mouth with a pit from which a furrow extends on the lower lip a distance about equal to the length of a nostril, and on the upper lip considerably farther. Length of nostril about two-thirds diameter of eye, and about half its distance from eye, the anterior margin with a narrow projecting flap. Distance between outer angles of nostrils slightly more than their distance from tip of snout.

Teeth of the upper jaw subtriangular, very oblique, deeply notched on the outer margin, those of the lower jaw similar, narrower and rather less oblique. Bases of upper teeth with a few weak serrations. No evident serrations on enameled parts of any of the teeth. Number of teeth about ²⁵⁄₂₅. Teeth all somewhat narrower and less oblique than in *Carcharias* (*Scoliodon*) *terræ-novæ*.

Gill-openings narrow, the last two over base of pectoral, the depth of one about ⅘ the distance from snout to mouth, the middle or largest about half the length of the branchial area; first and last gill-openings about equal.

Top of head with numerous mucous pores. A series of numerous large pores in a line above angle of mouth, and a band of them behind eye, extending upward on the nape. Under side of snout with many minute pores, these forming an elliptical area on the lower side of snout, outside of which on each side is a crescent-shaped area of pores.

Insertion of first dorsal well behind pectoral, but much nearer to pectoral than ventral; its distance behind posterior base of pectoral $1\frac{2}{5}$ in preoral part of snout.

Anterior lobe of dorsal, when depressed, not reaching the tip of the posterior lobe; the distance to the base of posterior lobe from ventrals nearly half preoral part of snout.

Length of base of first dorsal about equal to the height of the fin, or to the interorbital width, about one-third the distance between dorsals, which is 10 times base of second dorsal.

Second dorsal very small, its free margin scarcely concave; the pointed posterior lobe nearly twice the anterior.

Caudal, $4\frac{1}{4}$ in total length; its lower lobe bluntish, about $\frac{2}{5}$ the upper; length of caudal equal to distance from snout to base of pectoral.

Anal fin small, but considerably larger than second dorsal, inserted in front of the latter, so that its posterior part is below the insertion of the dorsal. Length of anal about equal to its elongate posterior lobe, and less than half its distance from the ventrals. Ventrals moderate, their anterior margins two-thirds distance between angles of mouth.

Pectorals rather small, not quite reaching middle of first dorsal, their tips bluntish, the length of the anterior margin $1\frac{2}{5}$ times interorbital width and $1\frac{1}{4}$ times the free margin. Pectoral scarcely more than one-eighth the total length of the fish.

Color slaty-gray with a distinct bluish tinge; white below; upper edge of tail and tip of caudal dusky; vertical fins faintly margined with dark.

This species is common in the harbor of Mazatlan, where five specimens were obtained; the largest of these (28330), a male, 32 inches long, is apparently nearly mature, the claspers reaching the anal. It has especially served for the basis of the above description.

Our specimens agree in most respects with Professor Poey's accounts of his *Scoliodon porosus* from Cuba. If the generic value of the group called *Scoliodon* be not admitted, the latter species must receive a new name, as there is already a *Carcharias porosus* Ranzani.

Carcharias, sp. incog.

The jaws of a large shark were obtained at Mazatlan, the species of which we have not been able to ascertain. The following is a description of the teeth:

Teeth $\frac{2\cdot2}{3\cdot2}$. Teeth of the upper jaw rather narrowly triangular, the tip rather strongly curved outward; the inner margin rather strongly convex, the outer rather strongly concave; the outer margin with a broad, shallow basal angle, a continuation of the curve of the outer edge of the tooth. Both margins of the teeth strongly serrate, the serræ stronger on the outer margin towards the base of the tooth. Upper jaw with a moderate, triangular median tooth, denticulated like the others.

Lower teeth very different in form, long, slender, sharp, straight, sub-terete, rising from broad roots, somewhat turned outward, but more erect than the upper teeth. No distinct notch on outer margin, where a slightly obtuse angle is formed. Edges of lower teeth everywhere strictly entire.

Its relations seem to be, so far as can be judged by the teeth, with such species as *Carcharias tjutjot* Bleeker and *C. menisorrah* Val., members of the "genus" *Platypodon* Gill.

Mustelus lunulatus, sp. nov. (29211.)

Allied to *Mustelus canis* (Mitch.) Dek.

Body elongate; the back little elevated. Head depressed, rather long and pointed; length of snout from mouth about one-sixth more than the distance between angles of mouth, and nearly twice the length of the mouth, from a line connecting the angles, to the chin. Eye oblong, large, a little nearer angle of mouth than nostril; distance from eye to nostril three-fifths distance from nostril to tip of snout. Interorbital space narrow, $2\frac{1}{3}$ times in distance from snout to first gill-opening, some-what less than length of snout as measured from eye. Angle of mouth with a pit from which furrows extend on each lip for a short distance, about equal on each lip and considerably less than length of nostril. Nostrils inferior, large, their length a little more than half eye and nearly half their distance from eye; the anterior flap large; posterior flap turned inward, half as long as eye. Distance between nostrils 3 in snout as measured from eye.

Teeth, as in *M. canis*, paved; some of the inner teeth somewhat pointed; spiracles small, but evident; head without conspicuous mucous pores. Gill-openings small, the last two above base of pectoral, the depth of the middle one about equal to the length of the eye, and less than the length of the branchial area.

All the fins with their free margin notably lunate or concave. Inser-tion of first dorsal well behind pectorals, and nearly opposite the tip of the inner lobe of the latter, the outer lobe extending about to the mid-dle of the fin; distance of insertion of first dorsal from anterior root of pectorals about $\frac{1}{4}$ its distance from tip of snout. First dorsal high, its anterior lobe when depressed reaching a little past tip of posterior lobe. Length of base of first dorsal about equal to its vertical height, and nearly half greater than interorbital width. Distance between dorsals $2\frac{2}{3}$ times base of first and a little more than 3 times base of second. Posterior angle of dorsal produced but not reaching to vertical from front of ventrals. Middle of dorsal nearer pectoral base than ventral base by a distance nearly equal to the diameter of the eye.

Second dorsal not very small, proportionately lower than first, its posterior lobe extending farther than the anterior when depressed, its distance from base of caudal more than the length of its base.

Lower lobe of caudal short but pointed; tail forming a little more than one fifth the total length, its length about equal to distance from snout to front of pectorals. Terminal lobe about two-fifths length of tail.

Anal much smaller than second dorsal, its posterior margin a little behind posterior margin of the latter. Ventral moderate, its distal margin concave, the angles sharp. Pectorals comparatively sharp, half longer than broad, three fourths to four-fifths length of head (to first gill-slit), their length 7 to 7½ in total.

Color very light gray above, pale below; fins all pale.

Two half-grown specimens about 20 inches long, both numbered 29,211, were obtained at Mazatlan, where it is known to the fishermen as "Gato."

The following analysis of American species of *Mustelus* shows the relation of *M. lunulatus* to the other species of this genus:

a. Middle of first dorsal evidently nearer root of pectoral (posteriorly) than root of ventral (anteriorly); snout long, its length from mouth more than width of mouth; teeth bluntish; free margins of fins concave; first dorsal high, its narrow anterior lobe reaching tip of the slender posterior lobe when reflexed; the fin about as high as long. Interval between dorsals 2¾ times base of first; lower lobe of caudal pointed; tail 5 in body, its terminal lobe more than one-third its length; pectorals rather sharp, their free margin incised, their tips about reaching middle of dorsal; embryo unknown (probably without placenta); color pale..LUNULATUS.

aa. Middle of first dorsal about midway between pectorals and ventrals; snout shorter, its length from mouth about equal to width of mouth.

   b. First dorsal higher than long, the tip of anterior lobe usually reaching, when depressed, beyond tip of posterior lobe, its free margin deeply incised, its base 2¾ times in interval between dorsals; teeth bluntish; lower lobe of caudal blunt; tail more than one-fifth body, its terminal lobe more than one-third its length; pectorals rather obtuse, their free margin little incised, their tips reaching first third of dorsal; inner lobe of ventrals produced; embryo not attached to uterus by a placenta; colors rather pale................CANIS.*

   bb. First dorsal longer than high, its tip not reaching tip of posterior lobe, its free margin scarcely incised, its base about half the interval between dorsals; teeth rather sharp (in adults?); lower lobe of caudal not acute; tail less than one-fifth total length, its terminal lobe less than one-third its length; pectorals obtuse, their free edges almost straight, their tips reaching first fourth of dorsal; inner lobe of ventrals not produced, the free edge of the fin straight; embryo unknown (probably with placenta); color rather dark, axils of pectorals and ventrals dusky.................................DORSALIS.†

---

* *Squalus canis* Mitchill, Trans. Lit. and Phil. Soc. N. Y. i, 1815, 486 := *Mustelus asterias* Cloquet, Dict. Sci. Nat. xiv, 407, 1820: = *Squalus hinnulus* Blainv. Faune Française, 1820-'30, 83 := *Mustelus vulgaris* Müller & Henle, Plagiost. 1839, 64, and of many authors. Cape Cod to Cuba and on all coasts of Southern Europe. We are thus far unable to detect any permanent difference between European and American specimens. The American name has clear priority.

† *Mustelus dorsalis* Gill, Proc. Ac. Nat. Sci. Phila. 1864, 149. Panama (Gill; Gilbert).

aaa. Middle of first dorsal much nearer root of ventrals than pectorals; snout rather short, its width a little greater than distance between angles of mouth. First dorsal longer than high, its blunt tip when depressed not reaching tip of posterior lobe, its margin deeply incised, its base 2¼ times in the interval between dorsals; teeth rather sharp; lower lobe of caudal blunt; tail less than one-fifth length of body, its terminal lobe more than one-third its length; pectorals rather obtuse, their free margin little concave, their tips reaching little past front of dorsal; inner lobe of ventrals somewhat produced; embryo attached to uterus by a placenta; color rather dark; axils of pectorals and ventrals dusky..CALIFORNICUS.*

DESCRIPTION OF A NEW SHARK (CARCHARIAS LAMIELLA) FROM SAN DIEGO, CALIFORNIA.

By DAVID S. JORDAN and CHARLES H. GILBERT.

Allied to *Carcharias lamia* (Risso).

Body comparatively robust, the back elevated, the greatest depth half more than the height of the dorsal fin and equal to the distance from the nostril to the gill-openings.

Head broad and flat, the snout long, but wide and rounded. Length of snout from mouth greater than the distance between outer angles of nostrils, a little more than width of mouth. Nostrils considerably nearer the eye than tip of snout, but nearer snout than angle of mouth. Eyes moderate.

Teeth about $\frac{30}{30}$, not very large, the upper rather narrowly triangular, nearly erect, slightly concave on the outer margin, but not notched, rather finely serrated everywhere. Lower teeth similar, but considerably narrower, finely serrated. Middle teeth in both jaws smaller than the others.

A pair of jaws taken from a much larger specimen have, as usual, the teeth considerably broader than in the young and more distinctly serrate. They are quite similar to the teeth of *C. lamia.*

First dorsal beginning at a distance four-fifths the length of its own base behind the root of pectorals, and ending at a point somewhat more than its own base before the ventrals, its height slightly more than the distance from the snout to the posterior margin of the eye, slightly more than its base, and considerably less than greatest height of body. Space between dorsals equal to the distance from snout to first gill-opening, 2⅗ times base of first dorsal, 7 times base of second.

Second dorsal very small, not one-sixth the size of the first and considerably smaller than the anal, which is deeply emarginate, the two fins nearly opposite each other. Ventrals small, nearly midway between the two dorsals. Tail long, forming nearly two-sevenths of the total length. Pectorals broad and long, not pointed, their tips reach-

---

Mustelus californicus Gill, Proc. Ac. Nat. Sci. Phila. 1864, 148: = *Mustelus hinnulus* Jor. & Gilb. Proc. U. S. Nat. Mus. 1881, 31 (not of Blainville). Coast of Southern California, San Francisco, Monterey, Santa Barbara, San Pedro, San Diego (Jor. & Gilb).

ing somewhat past the end of the base of the dorsal, the inner margin a little less than one third the outer, their length 5⅔ in total.

Color, plain light gray, white below; edge of pectorals and caudal narrowly dusky.

A young male specimen of this species (27,366), two feet in length, was obtained by us in San Diego Bay, California. It is said to be not uncommon along the coast of Lower California and it is known at San Diego as "Bay Shark." The jaws of an adult example taken on the coast of Lower California were also procured.

It is evidently closely related to *C. lamia*, but the smaller dorsal and pectorals and the more backward position of the dorsal seem to distinguish it sufficiently. The fins seem to be less falcate than in *C. lamia*. * In the Proc. U. S. Nat. Mus. 1881, p. 32, this species is mentioned by us under the name of *Eulamia lamia*.

CRITICAL REMARKS ON THE TREE-CREEPERS (CERTHIA) OF EUROPE AND NORTH AMERICA.

By ROBERT RIDGWAY.

The question of whether the American tree-creeper is separable from the European as a distinct race or species has long been a mooted point, and one in regard to which there is great difference of opinion among writers. Several eminent authorities, both in Europe and America, consider the European and North American birds of this genus as identical, or not separable even as races; but not a few authors, who base their conclusions on ample material, and are not influenced by ultra-conservative views regarding geographical variations, agree in recognizing two European races or species (according to the individual views of the author), one being the true *C. familiaris* Linn. of northern Europe, the other of more southern range, and variously designated as *C. costæ* Bailly or *C. trachydactyla* Brehm;* and in considering the common American bird as distinct from both the European forms, though some of them have referred it to *C. costæ*.

The North American creeper was first separated, as *C. americana* (by which name it has been known by American ornithologists up to a comparatively recent date), by Bonaparte in 1838; but having been already named many years previously (by Bartram, in 1791, as *C. rufa*, and Barton, in 1799, as *C. fusca*), Bonaparte's name cannot be used. The Mexican creeper was also separated in 1834, by Gloger, as *C. mexicana*. Thus two European and two American races or species of *Certhia* have been recognized by many ornithologists of standing. Others, however, pro-

---

* It is unnecessary for me to discuss here the question of which of these names should be adopted; therefore, without inquiring particularly into the case, I adopt provisionally the former.

fess an inability to distinguish between specimens from the two continents, and therefore insist upon their identity, although some of the
best authorities rank *C. mexicana* as a distinct *species*.

For the purpose of carefully reviewing the subject in all its bearings,
I have brought together a considerable number of specimens, and after
a very deliberate comparison of this material (embracing many skins
not included in the following tables of measurements), and an equally
careful consideration of all that has been written on the subject, I am
forced to the conclusion that the *C. mexicana* itself cannot stand even
as a race, or else it becomes necessary to recognize a larger number
of races than have usually been claimed for the species. In other
words, it is simply a question of whether geographical variations of
form and colors are to be completely ignored as a factor in the genesis
of species, or whether they should receive due consideration in connection with this important subject. Believing the latter view to be the
more scientific one, and since they are each "associated with definite
geographical areas," I find the following races susceptible of definition.

A.—*Primary coverts distinctly tipped with whitish.*

1. familiaris Linn. (based upon Scandinavian specimens).

Of this form I have three examples before me from Bergen, Norway
(coll. L. Stejneger). These agree in having the lower parts of a brilliant
silvery white, never seen in American specimens, though this pure white
color is somewhat obscured by a grayish tinge undoubtedly caused by
contact with carbonaceous substance upon burnt trees. The crissum is
very faintly tinged with buff; the lores are either wholly white or else
merely tinged with dusky in front of the eye; the dark ground color of
the upper parts is much tinged with yellowish tawny (which prevails on
the rump), and the maxilla is either very dark brown or black. These
specimens measure as follows:

Catalogue number.	Locality.	Sex and age.	Wing from carpal joint.	Tail to basal end of feathers.	Tarsal joint.	Hind toe.	Hind claw.	Culmen to extreme base.	Bill from nostril.
296 L S.	Bergen, Norway	♂ ad	2.60	2.75	.62	.30	.40	.72	.45
301 L S.	......do	....do	2.60	2.60	.60	.28	.35	.70	.40
224 L S.	......do	....do	2.50	2 50	.60	.30	.38	.60	.35
	Average		2 57	2.62	.61	.29	.38	.67	.40

2. ?costæ Bailly (described from Savoy).

I have five examples from central Europe, which may be readily distinguished from the Scandinavian specimens described above. The

lower parts are of a yellowish rather than silvery white, the crissum and flanks are more decidedly tinged with buff, the lores are distinctly dusky, and the upper parts are decidedly more tawny. Two of the five specimens have the maxilla a clear light-brown color, *which I have never seen in an American specimen.* The measurements are as follows:

Catalogue number.	Locality.	Sex and age.	Wing.	Tail.	Tarsus.	Hind toe.	Hind claw.	Culmen from extreme base.	Bill from nostril.
23416	Hungary	♂ ad	2.60	2.50	.60	.30	.40	.70	.45
56747	Saxony	...do	2.65	2.80	.60	.30	.38	.70	.42
56751	Silesia	...do	2.65	......	......	.30	.32	.82	.55
18947	France	...do	2.60	2.70	.63	.32	.40	.80	.52
17006	France (?)	...do	2.50	2.50	.60	.35	.40	......	......
	Average		2.60	2.62	.61	.31	.38	.75	.48

3. brittanica Subsp. nov.

Two examples from England differ from all continental specimens which I have seen very nearly as much as *C. mexicana* does from the ordinary North American bird; and since it would appear from descriptions that these specimens represent the normal style of coloration of specimens from the British Islands, I see no alternative but to characterize the British specimens as a race always distinguishable from the two continental forms. These British examples are very much browner above than those from the continent (closely resembling, in this respect, Californian specimens hereinafter described as *occidentalis*), the rump is more deeply tawny, and the lower parts appear to be of a much duller white, though this may be owing to a soiling of the plumage. These are the specimens which in *History of North American Birds* (i, pp. 124, 125) were supposed to be the true *C. familiaris*, thus leading to the erroneous views of their relationships therein given. The measurements are as follows, the tail being in both examples much worn at the tip, and therefore not included:

Catalogue number.	Locality.	Sex and age.	Wing.	Tail.	Tarsus.	Hind toe.	Hind claw.	Culmen from extreme base.	Bill from nostril.
18760	England	♂ ad	2.50	......	.62	.30	.38	.70	.45
18761	...do	♀ ad	2.50	......	.60	.30	.35	.70	.40
	Average		2.50	......	.61	.30	.36½	.70	42½

4. rufa Bartr. (Pennsylvania.)

Creepers from eastern North America have almost invariably a decidedly shorter bill and hind claw than European specimens, while

Proc. Nat. Mus. 82——8   **July 8, 1882.**

other measurements are on the average quite different. In coloration, they most resemble *C. costœ*, but as a rule have the the crissum more decidedly buff, and the rump brighter tawny, while the maxilla is never light brown, as often occurs in the South-European form. The following measurements are from fully adult birds, in perfect plumage:

Catalogue number.	Locality.	Sex and age.	Wing from carpal joint.	Tail to basal end of feathers.	Tarsal joint.	Hind toe.	Hind claw.	Culmen to extreme base.	Bill from nostril.
82701.........	Massachusetts	♂ ad........	2. 65	2. 90	. 60	. 30	. 30	. 65	. 38
827.........	Carlisle, Pa.................	♂ ad........	2. 60	2. 75	. 60	. 32	. 32	. 65	. 40
—— H.W H.	District of Columbia	♂ ad........	2. 70	(2. 60)	. 60	. 30	. 32	. 70	. 40
82707.........	Wabash County, Ill	♂ ad........	2. 60	(2. 60)	. 60	. 30	. 35	. 70	. 42
82706.........	......do.................	♂ ad........	2. 70	2 75	. 60	. 30	. 30	. 70	. 47
	Average of males	. ...	2. 65	2 72	. 60	. 30	. 32	. 68	. 41
577 H.W.H.	Watertown, Mass	♀ ad	2 50	2. 70	. 55	. 28	. 30	. 65	. 40
63288.........	......do....	♀ ad	2. 50	2 50	. 55	. 30	. 32	. 65	. 40
578 H.W.H.	Concord, Mass	♀ ad	2 40	2 50	. 58	. 28	. 30	. 65	. 40
11724.........	Pennsylvania	♀ ad	2. 55	2 50	. 5ⁿ	. 28	. 28	. 65	. 40
82705.........	District of Columbia.......	♀ ad	2. 50	2 50	. 55	. 30	. 32	. 65	. 35
82704.........	......do..................	♀ ad	2 60	(2. 55)	. 58	. 27	. 30	. 67	. 38
—— H.W. H.	......do	♀ ad	2. 50	2. 60	. 60	. 30	. 32	. 60	. 35
82708.........	Wabash County, Ill	♀ ad........	2 45	2 50	. 60	. 30	. 32	. 68	. 40
	Average of females ...	....	2. 42	2. 54	. 57	. 29	. 31	. 65	. 38
.	Average of both sexes of *C. rufa*.........	....	2. 53	2. 63	. 58	. 29	. 31	. 66	. 39

5. montana Subsp. vov.

   Middle Province of North America; (north to Kadiak, Alaska) breeding south to New Mexico and Arizona, in wooded mountains.

While I have been able to examine a smaller series of this form than any other except *familiaris* proper and *brittanica*, the six examples inspected show such well-marked peculiarities of form and coloration as to leave no doubt of the propriety of separating the Rocky Mountain bird as a geographical race. The general tone of coloration is decidedly grayer above than in any other form of the species, the flanks are decidedly grayish, the crissum more pronounced buff than in either of the three European races, and the tawny of the rump in more abrupt contrast with the grayish of the back. The most decided differences, however, are in proportions: thus, while the wing averages shorter than in either *familiaris* or *costœ*, the tail is decidedly longer; the bill also averages much longer than in *familiaris* or *costœ*, but is altogether more slender, both the vertical height and the transverse thickness being much less. As is the case with *all* the American races, the hallux and hind claw—the latter especially—are almost constantly shorter than in the European forms.

Catalogue number.	Locality.	Sex and age.	Wing from carpal joint.	Tail to basal end of feathers.	Tarsal joint.	Hind toe.	Hind claw.	Culmen to extreme base.	Bill from nostril.
68793	Colorado	♂ ad	2.60		.60	.30	.35	.77	.50
66704	Arizona	♂ ad	2.65	2.75	.60	.28	.30	.82	.50
79550	......do	♂ ad	2.55	2.70				.80	.52
	Average of males		2.60	2.72	.60	.29	.32	.80	.51
53443	Nevada	♀ ad	2.50	2.65	.58	.30	.30	.70	.42
13114	New Mexico	♀ ad	2.55	2.75	.60	.30	.35	.70	.40
	Average of females		2.52	2.70	.59	.30	.32	.70	.41
7154	New Mexico	— ad	2.55	2.65	.60	.28	.30	.82	.52
	Average of both sexes		2.56	2.69	.60	.29	.32	.77	.48

6. occidentalis Subsp. nov.

Pacific coast of North America, breeding from mountains of southern California to British Columbia.

Next to *mexicana*, this is the darkest colored of all the races of this species. In extremely slender bill it agrees with *montana*, but, apparently, has a shorter tail (although this apparent difference may be due to an insufficient number of specimens compared—one specimen having the tail .15 of an inch longer than the longest-tailed specimen of *montana*), but the colors are strikingly different. Instead of being grayer than *rufa*, *occidentalis* is much browner, extreme examples having the light patches of the remiges a bright ochreous-buff and the general cast of the upper parts a decidedly rusty brown, such specimens coming chiefly from the coast of Washington Territory and British Columbia. The rump is a bright *rusty* fulvous, and the crissum always a deep ochreous-buff. Of the European races, this most resembles *brittanica* in the color of the upper parts, some specimens being very similar indeed; but the crissum is constantly much more deeply buff. In the darker-colored examples there is some resemblance to *mexicana*, in fact some of them have been labeled as such; but the rump is much less chestnut, the primary coverts are always tipped with whitish, and the lower parts more whitish. Specimens measure as follows:

Catalogue number.	Locality.	Sex and age.	Wing from carpal joint.	Tail to basal end of feathers.	Tarsal joint.	Hind toe.	Hind claw.	Culmen to extreme base.	Bill from nostril.
11810	Hiniahmoo, Wash	♂ ad	2.50	2.60	.60	.32	.35	.70	.50
17433	Simiahmoo, Wash	♂ ad	2.50	2.70	.55	.30	.32	.68	.42
13743	Fort Tejon, Cal	♂ ad	2.50		.58	.30	.30	.75	.45
16175	Fort Crook, Cal	♂ ad	2.50	2.60	.60	.28	.35	.80	.50
82709	Nicasio, Cal	♂ ad	2.50	2.50	.65	.32	.32	.80	.50
71950	Kern River, Cal	♂ ad	2.50	2.62	.60	.32	.32	.70	.47
	Average of adult males		2.50	2.60	.60	.31	.33	.74	.47

Catalogue number.	Locality.	Sex and age.	Wing from carpal joint.	Tail to basal end of feathers.	Tarsal joint.	Hind toe.	Hind claw.	Culmen to extreme base.	Bill from nostril.
22092	Fort Crook, Cal	♀ ad........	2.50	......	.58	.30	.32	.70	.42
82710	Nicasio, Cal	♀ ad........	2.40	2 30	.58	.30	.30	.70	.42
	Average of adult females.....	....	2.45	2 30	.58	.30	.31	.70	.42
45951	British Columbia	— ad........	2.35	2.50	.60	.30	.30	.75	.45
9520	Puget Sound	— ad........	2.75	......	.60	.32	.32	.68	.43
7125	Fort Steilacoom, Wash	— ad	2.45	2.50	.60	.32	.35	.72	.40
76686	California	— ad........	2.75	2.90	.60	.30	.32	.72	.45
76687	......do	— ad........	2.35	2.40	.55	.30	.30	.67	.42
73900	Calaveras County, Cal..............	— ad........	2.60	2.50	.60	.30	.32	.76	.52
73899	......do	— ad........	2.60	2.60	.60	.30	.32	.80	.50
	Average sex undetermined ...	....	2.55	2.57	.59	.31	.32	.73	.45
	Average of both sexes	....	2.50	2.49	.59	.31	.32	.72	.45

b.—*Primary coverts without whitish tips.*

7. mexicana Gloger.

Guatemala and southern Mexico.

This form differs conspicuously from all the others in the total absence of light tips to the primary coverts. The lower parts are also decidedly grayish, with only the throat and chin white, the rump a bright chestnut-rusty, and the ground-color of the anterior upper parts a blackish-brown, with the lighter streaks of a rather grayish tone. In slender bill and other features of form it scarcely differs from the more northern races, *montana* and *occidentalis*, and is by no means smaller, thus affording another of the very numerous "exceptions" to the supposed law of smaller size to the southward of resident species.* The three specimens which I have been able to examine measure as follows:

Catalogue number.	Locality.	Sex and age.	Wing from carpal joint.	Tail to basal end of feathers.	Tarsal joint.	Hind toe.	Hind claw.	Culmen to extreme base.	Bill from nostril.
13388	Mexico...............................	— ad........	2.60	2.65	.58	.30	.32	.70	.46
8176	......do.............................	— ad........	2.55	2.65	.60	.30	.30	.73	.48
69835	V. de Fuego, Guatemala.............	— ad........	2 60	......	.60	.35	.35	.70	.42
			2 58	2 65	.59	.32	.32	.71	.44

SMITHSONIAN INSTITUTION, *March 23, 1882.*

---

* In perhaps a majority of cases where I have recently tested the matter by measurements of large series of many Passeres I have been unable to verify this supposed law of latitudinal variation in size.

NOTE ON THE OCCURRENCE OF A SILVER LAMPREY, ICHTHY-OMYZON CASTANEUS, GIRARD, IN LOUISIANA.

By TARLETON H. BEAN.

Nearly two years ago Mr. N. B. Moore, of Forlorn Hope, Louisiana, sent to the National Museum, at the request of the Director, a lamprey which he perceived to be different from the sea-lamprey (*Petromyzon americanus*, = *marinus*), by comparing it with a description in Zell's Encyclopedia. About that time the writer was on his way to Alaska, and had no opportunity to examine the specimen forwarded by Mr. Moore. It was placed in storage and came to my notice again only a few days ago. As this individual shows some unusual characters, it is worth while to call attention to them. It agrees pretty well with the account of the species described by Girard from Galena, Minn., under the name *Ichthyomyzon castaneus*, but shows a variation from the ascribed characters of the genus in having three close-set maxillary teeth, while some of the lateral teeth are bicuspid.

I insert here Mr. Moore's description of the recent specimen:

"I have a lamprey—first ever seen by me, and identified by a description in Zell's Cyclopædia. Thinking it quite far south for one of this species—*Petromyzon americanus*, if it be this one—I put it in whisky, and, as I found it to differ from that given in Zell in one particular, the dorsal fin being 4¼ (inches) in length, continuous, not separated, I thought you would like to examine it. If so, I will send it to you. Total length, 9¾ inches; anus to tip of tail fin, 2; between anterior and posterior gills, 1⅛; tip of snout to anterior gill, 1¼; commissure of mouth, ¾; greatest depth of fish at interior part of dorsal, ⅞; greatest width 2 (inches) from tip of snout, ⅝; thence a true taper to tip of tail in lateral outline; body not cylindrical. Color ochraceous about head, then yellowish gray; small blue dots from head to tail and on under side of neck."

For convenience of comparison, I copy here Girard's description of *I. castaneus*[*]:

"SPEC. CHAR.—Head depressed, constituting the ninth of the total length; body and tail compressed. Buccal disk sub-elliptical, provided with a double series of short, tentacular fringes upon its periphery. Posterior margin of buccal aperture exhibiting a series of nine teeth, disposed upon an arc of a circle. Eyes small and inconspicuous; spiracle sub-tubular, raised above the surface of the head. Origin of the dorsal fin equidistant between the anterior margin of the buccal disk and the apex of the tail. Vent situated immediately in advance of the most elevated portion of the dorsal fin. Chestnut-colored, of a darker tint above than beneath.

---

[*] < Rep. U. S. Pacific R. R. Surv., Fishes, 1858, pp. 381-2.

" What we have termed the head is measured from the anterior extremity of the buccal disk to the first branchial orifice, the chest being the region occupied by the entire series, seven in number, of the same branchial orifices.

" The length of the head is equal to that of the chest. The tentacles, at the periphery of the buccal disk, are inserted into a shallow groove, formed exteriorly by the thickened edge of the disk, and interiorly by a soft and flexible membranous ridge. The fringes themselves are more developed posteriorly than anteriorly. The branchial orifices are sub-circular, provided with two semi-circular lips, an anterior and a posterior one, fringed upon their edge, and somewhat raised above the surface of the chest.

" The dorsal fin exhibits two convex elevations, one anterior to the vent, the other posterior to it. Its continuity with the caudal is marked by a gradual shallow depression. The lower lobe of the caudal is rather more developed than the upper lobe. The tail itself is bluntly spear-shaped.

" The color is of a uniform chestnut tint, somewhat lighter along the abdominal region than over the sides and back, which is much darker."

The single typical specimen was catalogued at number 979. It was collected by Dr. George Suckley at Galena, Minn.

In the example received from Mr. Moore the following characters are observed:

Head $7\frac{3}{4}$ in total length; body and tail compressed. Labial fringes short everywhere, but more developed posteriorly than anteriorly. Maxillary teeth pointed, close-set, three in number; mandibulary plate crescent-shaped, with nine pointed teeth very gradually diminishing in size from the middle tooth to each end. Two of the lateral teeth on each side of the oral aperture bicuspid, the rest unicuspid. Both series of lingual teeth finely pectinate. Eyes very small, obscure. Spiracle about once its own length in front of eyes. Origin of dorsal fin midway between spiracle and end of tail. The dorsal fin is continuous, low in the first half of its length (about one-sixth the height of the part of the body under it), thence gradually rising to its greatest height a little behind the vent and again gradually diminishing to the emargination which separates it from the tail. The greatest height of the dorsal is less than one-half that of the body at the same point. The distance of the vent from end of tail equals 3 times height of body at vent. The space occupied by the gill-openings is about equal to length of head. Greatest height of body equals head to hind margin of eye. The tail small, no part of the fin surrounding it being much higher than the anterior half of dorsal.

The alcoholic specimen now is almost uniformly light brown replaced by chestnut in one small area on the belly a little in front of vent. The spots on head, chest, and back, which Mr. Moore described as blue, are now dark brown or nearly black, resembling fly-specks.

The length of the specimen (numbered 30334) is now 9.3 inches; head 1.1; chest very nearly the same; greatest height of body, $\frac{17}{20}$; vent to tip of tail, $1\frac{18}{20}$; dorsal from end of head, 5.

I have thought it worth while to describe this lamprey in some detail because of the interest which attaches to the locality and on account of the slightness of our knowledge of *I. castaneus*, to which our present example is most closely related.

UNITED STATES NATIONAL MUSEUM,
Washington, March 24, 1882.

Since the above was written I have found and examined the types of *Ichthyomyzon hirudo* and *I. castaneus* Girard. The first is 5 inches long and is certainly congeneric with *castaneus*, from which it differs in the number of mandibulary cusps. The maxillary tooth is tricuspid and a few of the lateral teeth are bicuspid. *Ichthyomyzon hirudo* and *I. castaneus*, therefore, show a departure from the type of the genus, which is *Petromyzon argenteus* Kirtland; they have the dorsals continuous as in *argenteus*, but the dentition is different. The maxillary cusps in *hirudo* and *castaneus* are placed close together. The lingual teeth are pectinate throughout, as in *Lampetra tridentata*. We are called upon now to decide whether *Petromyzon argenteus* Kirtland and Girard's two species of *Ichthyomyzon* (*hirudo* and *castaneus*) are all members of the genus *Ichthyomyzon*. In my opinion they are, but I should refer the three species of *Ichthyomyzon* to *Petromyzon*.

UNITED STATES NATIONAL MUSEUM,
June 14, 1882.

NOTES ON A COLLECTION OF FISHES FROM JOHNSTON'S ISLAND, INCLUDING DESCRIPTIONS OF FIVE NEW SPECIES.

By ROSA SMITH and JOSEPH SWAIN.

The specimens which form the subject of the present paper were obtained in the spring of 1880 at Johnston's Island, by the captain of a vessel belonging to the North Pacific Guano Company. A can of alcohol was sent out on this vessel by Professors David S. Jordan and Charles H. Gilbert during their stay on the Pacific Coast of the United States in the interest of the United States Fish Commission. Johnston's Island is located about 700 miles southwest of the Hawaiian Islands, and approximates 17° north latitude, 170° west longitude. This collection, containing five new and many little known species, Professor Jordan has turned over to the writers for study. The specimens are now in the United States National Museum.

We are greatly indebted to Professor Jordan for the use of his library and for many valuable suggestions.

The following is a list of the species obtained:

1. *Ophichthys (Pisodontophis) stypurus* Smith & Swain.
2. *Gymnomuræna tigrina* (Less.) Blkr.
3. *Aulostomus chinensis* (L.) Lacépède.
4. *Polynemus kuru* Bleeker.
5. *Scombroides sancti petri* (C. & V.) S. & S.
6. *Caranx gymnostethoides* (Blkr.) Gthr.
7. *Holocentrus leo* Cuvier & Valenciennes.
8. *Holocentrus erythræus* Günther.
9. *Kuhlia tæniura* (Cuv. & Val.) S. & S.
10. *Upeneus crassilabris* Cuv. & Val.
11. *Upeneus velifer* Smith & Swain.
12. *Upeneus (Mulloides) vanicolensis* (C. & V.) S. & S.
13. *Upeneus (Mulloides) preorbitalis* Smith & Swain.
14. *Chilinus digrammus* (Lac.) C. & V.
15. *Scarus perspicillatus* Steindachner.
16. *Julis verticalis* Smith & Swain.
17. *Julis clepsydralis* Smith & Swain.
18. *Harpe bilunulata* (Lac.) Smith & Swain.
19. *Chætodon setifer* Bloch.
20. *Acanthurus triostegus* (Linn.) Bloch. & Schn.
21. *Naseus lituratus* (Forst.) C. & V.
22. *Balistes aculeatus* Linn.
23. *Balistes buniva* Lac.
24. *Ostracion punctatum* Bl. & Schn.
25. *Tetrodon meleagris* Lacép.
26. *Diodon hystrix* Linn.
27. *Platophrys mancus* (Brouss.) S. & S.

1.—OPHICHTHYS STYPURUS sp. nov.

Subgenus *Pisodontophis* Kaup.

Head 5¼ in trunk. Head and trunk together slightly longer than tail, exceeding the latter by the length of the snout. Snout blunt, 5½ in head. Eye 2½ in snout, 3 in interorbital space. Gape of mouth moderate, extending beyond eye, 3½ in head. Anterior nasal tubes turned downward, conspicuous; posterior nostrils large. Teeth in lower jaw less blunt than in *O. xysturus* J. & G., in two series in front, becoming three posteriorly; two rows (the outer row being larger) of bluntish, conical teeth on each side of upper jaw, preceded by a patch of eight on extremity of nasal bone; smaller teeth on vomer in a band of two series.

Dorsal and anal fins rather high, the highest part of dorsal exceeding length of snout; dorsal beginning at the nape, at a distance from the snout equal to half the length of the head. Pectoral short, 1½ in its base, 1¾ in snout; its free margin lunate. Gill-opening oblique, its width equaling base of pectoral and 1¾ in isthmus. End of tail rather blunt and little compressed.

Ground color, in spirits, light olivaceous; round brown spots in four series on the sides, extending on the dorsal but becoming fainter on the fin; second series on lateral line, the spots of third mostly smaller; the spots of the different series sometimes alternating regularly, sometimes without definite order; the diameter of most of the spots in upper two series exceeding the snout; a fourth series of much smaller spots (not half the diameter of the largest ones) along sides of belly, almost disappearing on tail; small, irregular, more or less confluent spots on upper half of dorsal, the fin narrowly margined with whitish. Anal plain, light olivaceous. Pectorals with one or two small, obscure brown spots.

One fine specimen (26817 U. S. Nat. Mus.), 24¼ inches in length.

2.—GYMNOMURÆNA TIGRINA (Lesson) Bleeker.

Ichthyophis tigrinus "Lesson, Mem. Soc. d'Hist. Nat. Paris, iv, 399, and Voy. Coq. Zool. ii, 129, Atl. Pois. pl. 12; Richard's Voy. Ereb. and Terr. Fish, p. 96; Bleek. Versl. Ak. Wet. Natuurk. xv, 463."

Murænoblenna tigrina Kaup, Cat. Apod. Fish. Brit. Mus. 98, 1856 (Strong Island, Havre, Carteret, Moluccas, Celebes); Blkr. 8th Bijdrage der Vischfauna van Amboina, p. 93, 1857 (Java); Blkr. Index Pisc. Actorum Sci. Soc. Indo-Mer, 5 (name only).

Gymnomuræna tigrina "Blkr. Atl. Ichth. Mur. p. 113, pl. 21, fig. 3; Kner, Novara Fisch, p. 387"; Gthr. Cat. Fish. Brit. Mus. viii, 133, 1870 (Mauritius, Zanzibar. East Indian Archipelago, India).

Habitat.—Java, Mauritius, Zanzibar, East Indian Archipelago, India, Strong Island, Havre, Carteret, Moluccas, Celebes, Johnston's Island.

Head 4⅘ in trunk. Head and trunk together exceed the tail by the width of the gape, the length from the tip of snout to vent being greater than that from vent to extremity of tail. Snout 8 in head. Eye 1½ in snout, its position being over the middle of the gape. Gape 3¼ in head.

Teeth pointed, recurved, some of them depressible; in two series on upper jaw; anteriorly in two series in lower jaw, only one at the corner of the mouth; a few teeth on front of vomer. Posterior nostrils with tubes more conspicuous than anterior.

Color, in spirits, brownish, everywhere irregularly covered with nearly round blackish spots, varying in size from the orbit to ⅔ of the gape; in front of the occiput the head is thickly covered with very small, dark spots; the spots on the body run together in places, forming blotches.

One specimen (26823), 17½ inches long.

3.—AULOSTOMUS CHINENSIS (L.) Lacép.

Fistularia chinensis (in part) L. Syst. Nat. i, 515, 1766.
Aulostomus chinensis (in part) Lacép. v, 357, 1803.

Aulostoma chinense "Schleg. Faun. Japon. Poiss. 320"; "Richards, Ichth. Chin. 247";
"Peters in Wiegm. Arch. 258, 1855"; Gthr. Cat. Fish. Brit. Mus. iii, 538,
1861 (Amboyna); "Playfair in Fish. Zanz. 79"; Blkr. Quat. Mem. Ichth. N.
Guinée, 6 (name only); Street Bull. Nat. Mus. No. 774, 1877 (Honolulu);
Blkr. Enum. Poiss. Japan 14, 1879 (name only); Gthr. Jour. Mus. Godeff.
Fische der Südsee, 221, taf. 123, figs. B and C, 1881 (Indian Ocean and Archi-
pelago, Sandwich, Society, and Paumatu Islands, Aneiteum).
Polypterichthys valentini "Bleck. Ternate, ii, 608."

Habitat.—China, New Guinea, Honolulu, Amboyna, Indian Ocean,
Indian Archip., Society and Paumatu Islands, Aneiteum, Ternate, Johnston's Island.

Head 3 ($3\frac{1}{5}$); depth 12 ($12\frac{5}{6}$); D. X-3, 26; A. 3, 27.

Snout nearly $1\frac{1}{2}$ in head; its profile somewhat concave from end of
snout to occiput. Eye 9 in snout, its diameter a little greater than
interorbital width.

Spines of first dorsal free, remote, equal in length to orbital diameter; each spine attached by a broad membrane, and received into a
narrow groove. Soft dorsal with a somewhat irregular margin, first
branched rays $3\frac{4}{5}$ in snout, last ones about 6 in snout. Anal similar to
soft dorsal, its base slightly longer than the dorsal; base of soft dorsal
slightly longer than caudal peduncle. Caudal fin $3\frac{1}{2}$ in snout. Ventrals subtruncate, $5\frac{1}{4}$ in snout, about as long as pectorals. Peduncle
nearly straight, narrow, its depth being $\frac{1}{4}$ of greatest depth of body.

Color, in spirits, brownish olive above, light olive below; four lighter
horizontal olive bands on body, the two below lateral line not distinct
except posteriorly—where the ground color is brownish below as well
as above the lateral line—and on head, behind eye, three of these light
bands being very distinct across the opercles; three light oblique bands
across snout, with many other irregular light markings; a black band
across maxillary, horizontal with snout; a small black spot in front of
each pectoral and on a level with its lower edge; a larger one behind
each pectoral in a line with the first. On median line of belly are six
black spots about the size of the pupil; between vent and anal fin are
four more similar ones. A black streak from the pupil through the nostrils. Peritoneum reddish brown.

One fine specimen (26819), $26\frac{1}{2}$ inches in total length.

4.—POLYNEMUS KURU Bleeker.

Polynemus kuru Blkr. Nat. T. Ned. Ind. iv, 600, 1853 (Ternate); Blkr. Conspect. Spec.
Pisc. p. 6 (name only); Blkr. Enum. Spec. Pisc. Archip. Indic. 40, 1859 (Halmaheira, Ternate, Amboyna, Saparua; name only); Gunther, Cat. Fishes
Brit. Mus. ii, 325, 1860 (taken from Bleeker); Blkr. Conspec. Moluce. Cognit.
p. 5 (Ternate, Amboyna, Saparua; name only); Blkr. Beschrij. Visch. Amboina, p. 4 (name only); Blkr. Achtste Bijd. Visch. Amboina, pp. 3, 14
(name only); Blkr. Tweede Bijd. Schth. Fauna, Halmaheira, pp. 2, 4 (name
only).
Trichidion kuru Gill, Proc. Ac. Nat. Sci. Phila. 275, 1861 (name only).

Habitat.—Halmaheira, Ternate, Amboyna, Saparua, Batjan, Archip. Molucca, Johnston's Island.

Head $3\frac{2}{7}$ ($4\frac{2}{3}$); depth $3\frac{1}{2}$ ($4\frac{2}{3}$); length (26837) $18\frac{1}{2}$ inches. D. VIII–I, 13; A. II, 12; scales, 8–66–12.

Allied to *P. approximans*, Lay & Bennett, from which it chiefly differs in coloration, number of anal rays, and in its higher dorsal and anal fins.

Body robust, compressed. Snout comparatively blunt; the profile from snout to dorsal gently and regularly curved. Mouth moderate, horizontal; tip of mandible behind posterior nostril. Maxillary 2 in head. Teeth in a continuous villiform band on palatines and pterygoids, the patch broader, especially on the pterygoids, than in *P. approximans*, and much broader than in *P. opercularis;* the premaxillary band also broad. Eye in head about $3\frac{1}{2}$ times, interorbital space slightly convex, 4 in head. Preopercle with few and irregular serrations. Tooth above the lobe of preopercle well developed. Gill-rakers one-fourth length of maxillary; about 20 on lower limb.

Scales large, almost smooth. Small scales nearly covering the soft dorsal, anal, and ventral fins; the membraneous flap of the spines of first dorsal scaled, but the connecting membrane entirely naked. Upper pectoral rays scaled nearly to their tips; all the rays covered with scales at their base; upper rays also closely scaled on posterior surface.

Dorsal fins well developed; second and third spines longest, $1\frac{2}{3}$ in head, about three-fourths length of longest soft ray. The first two rays of second dorsal longest, about $1\frac{2}{5}$ in head. Caudal large, widely forked, the upper lobe slightly the longer, and one-fourth longer than head. Anal shorter than soft dorsal, the base of the fin three-fourths of its greatest height; when depressed the tips of the two anterior rays extend very nearly as far as the tip of the last ray; the free margin of the fin lunate, similar to soft dorsal; first rays four-fifths length of first rays of soft dorsal, and $1\frac{3}{4}$ in head. First anal spine very small; second $4\frac{1}{4}$ in head. Ventral fins in head $1\frac{4}{5}$ times. Pectorals $1\frac{2}{3}$ in head; pectoral filaments 6, the superior one longest, $1\frac{1}{5}$ the length of head, reaching tip of ventrals.

Air-bladder large.

Adipose eyelid well developed.

Color olivaceous, the scales finely punctulate with brown; these punctulations most numerous on the upper and lower margins of the scales, forming longitudinal streaks along the rows of scales. The scales from the snout to the first dorsal, on the belly, and the space between the anal and caudal fins smaller, and on these regions the brown points are aggregated on the margins of the scales, making their outline conspicuous. Vertical fins dark; margin of dorsals blackish. Pectoral dark, purplish underneath. Ventrals light, but with some brown punctulations. Preopercle plain except its flap, which, together with the other opercles, is rather dark.

One fine specimen (26837), $18\frac{1}{2}$ inches in length.

5.—SCOMBROIDES SANCTI PETRI (C. & V.) Smith & Swain.

? *Chorinemus toloo* Cuv. & Val. Hist. Nat. Poiss. viii, 377, 1831; Blkr. Spec. Pisc. Archip. Indic. 61, 1859 (Java, Sumatra, Nias); Blkr. Makr. Visch. 45 (Celebes, Ternate, Amboyna, Ceram).

Chorinemus sancti petri Cuv. & Val. Hist. Nat. Poiss. viii, 379, 1831; Blkr. Enum. Spec. Pisc. Archip. Iudic. 61, 1859 (Java, Bali, Sumatra, Singapura, Bintang, Banka, Celebes, Ternate, Halmaheira, Amboyna, Saparua, Ceram); Blkr. Makreelachtige Visschen p. 45 (Batavia, Pasuruan); Gthr. Cat. Fishes Brit. Mus. ii, 473, 1860 (Madagascar, Ceylon, China, Amboyna, Malayan Peninsula, Moluccas); Gthr. Jour. Mus. Godeff. Fische der Südsee, 138, 1873 (Kingsmill, Schiffer, Society and Sandwich Islands); Streets, Bulletin U. S. Nat. Mus. No. 7, 89, 1877 (Christmas Island); Blkr. Conspec. Moluc. Cognit. 11 (Halmaheira, Amboyna, Ternate, Ceram, Archip. Molucca, Saparua, Banda; name only); Blkr. Vier. Bijd. Ichth. Fauna Japan, 5 (name only); Blkr. Zes. Bijd. Visch. Fauna Sumatra, 20 (Priamam; name only); Blkr. Beschrij. Visch. Amboina, 15 (name only); Blkr. Beschrij. Visch. Manado Makassar, 4, 18 (Macassar, Manado, Kema; name only); Blkr. Achtste Bijd. Visch. Amboina, 5, 18 (name only); Blkr. Bijd. Ichth. Fauna von Midden en oost Java, 8 (Pasuruan; name only); Lütk. Spolia Atlantica, 508, 511, 1878 (name only); Blkr. Tweede Bijd. Ichth. Fauna Halmaheira, 4 (name only); Blkr. Nieuwe Verzam. Visschen, Batjan, 4 (name only); Day's Fish. Malabar, 95 (E. Coast Africa, Sea of India, Malasia, China, Malabar; name only).

? *Chorinemus mauritianus* C. & V. Hist. Nat. Poiss. viii, 382, pl. 286, 1831.

Head $4\frac{5}{7}$ ($5\frac{5}{7}$); depth $4\frac{3}{8}$ ($5\frac{2}{5}$); D. VI–I, 20; A. II–I, 18.

Body elongate; profile straight to occiput, thence gently curved. Snout bluntish, $3\frac{1}{2}$ in head; mouth oblique, lower jaw longest; maxillary terminating beneath posterior margin of eye, the supplemental bone well developed. Villiform teeth on jaws, tongue, vomer, palatines, and a *broad patch on pterygoids*. Eye $4\frac{5}{6}$ in head, $1\frac{1}{3}$ in interorbital space. Edge of upright limb of preopercle somewhat convex, slanting backward slightly. Gill-rakers strong, the longest $1\frac{1}{6}$ in eye, 19 on lower part of arch. Scales elongate-rhombic.

Margin of soft dorsal less concave than anal, second articulate ray highest, $2\frac{1}{3}$ in head, the tenth ray $2\frac{1}{5}$ in second. Caudal deeply forked, lower lobe longer, equal to length of head. First ray of anal highest, equaling highest dorsal ray. Ventrals 2 in head; pectorals 2 in head.

Color, in spirits, bluish above, silvery below; upper part of dorsal black, its base yellowish, the semi-detached dorsal finlets dusky; caudal irregularly washed with dark blue, middle rays yellowish; anal with a central black blotch, its semi-detached finlets yellowish-white; pectoral dusky, black at base posteriorly, a dark spot in the axil; ventrals yellowish. Top of head dark blue with metallic luster, below light silvery; an irregular band of very dark blue from occiput to caudal, making the dorsal outline dark; the greatest width of this band is one-half the ocular diameter; the two series of spots on the side are more or less indistinct, on one side seven above lateral line and four below; on the other side five above and five below lateral line, the lower anterior two larger and more distinct than the others.

One fine specimen (26825), $20\frac{1}{4}$ inches long.

6.—CARANX GYMNOSTETHOIDES (Bleeker) Günther.

Carangoides gymnostethoides Blkr. Makreelachtige Visschen, p. 61; "Blkr. Nat. T. Ned.
Ind. i, p. 364"; Blkr. Spec. Pisc. Archip. Indic. p. 41, 1859 (Java).
Caranx gymnostethoides, Gthr. Cat. Fish. Brit. Mus. ii, 431 (Sea of Batavia).

Habitat.—Java, Sea of Batavia, Johnston's Island.

Head $3\frac{2}{7}$ (4); depth 3 ($3\frac{2}{3}$); D. VII-I, 31; A. II-I, 26.

Body elliptical, compressed; profile convex from snout to nostril, thence regularly arched to caudal; the ventral outline less convex, being almost straight from head to anal. Head longer than deep; snout rather blunt, 3 in head. Mouth low, somewhat oblique; maxillary $2\frac{3}{10}$ in head, extending opposite front of pupil; lower jaw little produced.

Teeth in villiform bands on jaws, vomer, palatines, and a patch on the tongue. Eye large, $1\frac{1}{2}$ in snout, $4\frac{2}{3}$ in head. Adipose eyelid little developed. Cheeks and temporal regions with fine scales; head otherwise naked. Scales rather small, those below pectoral smaller. A naked area on breast not widening forward from base of ventrals as much as in *C. orthogrammus* J. & G.

Lateral line but little curved, arched above pectorals, and gradually becoming straight at their tips; greatest depth of the arch about equal to pupil, the arched part of the line longer than the straight. Plates developed only in the posterior half of the straight part; the plates small with low keels, their spines little prominent; 25 developed plates, including small ones.

Spinous dorsal rather weak, the highest spine $1\frac{3}{4}$ in snout (these spines probably varying according to the age). Soft dorsal long and low, with slender rays; a well-developed scaly basal sheath anteriorly; the first articulate ray is $1\frac{3}{4}$ in base of fin and $1\frac{1}{4}$ in head. Anal similar to soft dorsal. First free anal spine nearly obsolete, second small.

Caudal lobes moderate, equal, $1\frac{1}{5}$ in head; their length is much less than the depth from tip to tip.

Pectorals falcate, their tips slender, reaching tenth ray of anal; their length $2\frac{1}{2}$ in body (from snout to base of caudal fin); ventrals $2\frac{3}{5}$ in head.

Color, in spirits, about as in *C. orthogrammus*.

One fine specimen (26839), $15\frac{1}{2}$ inches in length.

7.—HOLOCENTRUS LEO Cuvier & Valenciennes.

Holocentrum leo C. & V. Hist. Nat. Poiss. iii, 204, 1829 (Society and Waigiou Islands);
"Less. Voy. Coquille, ii, 222"; "Cuv. Règne Anim. Ill. pl. 14, f. 1."; "Blkr. Kokos
Islands, iii, 355, 1855"; Blkr. Enum. Spec. Pisc. Archip. Indic. 2, 1859 (Cocos,
Batu, Celebes, Batjan, Amboyna); Blkr. Nat. T. Ned. Ind. vii, 355; "Blkr.
Voy. Astrol. Poiss. p. 678, pl. 14, f. 3"; Blkr. Conspec. Molucc. Cognit. p. 5
(Batjan, Amboyna, Archip. Molucca; name only); Blkr. Beschrij. Visch.
Manado Makassar, pp. 3, 13 (Manado; name only); Blkr. Achtste Bijd. Visch.
Amboina, pp. 3, 13; Blkr. Nieuwe Verzam. Visschen Batjan, p. 3.

Holocentrum spiniferum Gthr. Cat. Fishes Brit. Mus. i, 39, 1859. (In part; the specimens from the Pacific.)

Habitat.—Society, Waigiou, Kokos, Batjan, Amboyna, Batu, Celebes, Manado, and Johnston's Islands.

Head 3 (3½); depth 2½ (3); D. XI, 15–16; A. IV, 10; scales 3½–45–8.

Body ovate, compressed, elevated. Profile rather steep; from snout to occiput slightly concave, thence evenly curved. Mouth oblique, maxillary terminating opposite anterior half of pupil; lower jaw produced; snout pointed, 3⅕ in head; eye 4 in head; interorbital space 7 in head; intermaxillary groove as long as snout. The lower of the two opercular spines smaller than the upper. The prominent striæ of opercle and suprascapula end in points, producing sharply serrate margins; all the other bones of shoulder-girdle smooth; subopercle scarcely striate, rather reticulate, its margin nearly smooth; posterior half of interopercle serrate; preopercle with a strong spine at the angle, which varies in length from 1½ to 2⅓ in the height of the straight upright limb of preopercle; posterior edge of preopercle coarsely serrate and slightly slanted forwards. Nasal bones prominent. Fan-like striations on occiput, and all the occipital bones coarsely serrate on their margins. The orbital rim much narrower than in *H. erythræus*, also less deeply lobed and more finely denticulate. Supraocular region rough with minute spines. As in *H. erythræus*, the infraorbital bone has a blunt tooth in front of the supplemental maxillary bone, and another beneath front part of eye, leaving the intervening space lunate and more or less serrate.

First dorsal spine is 1⅔ in the third, which is the highest, and 1⅚ in depth of body; the fourth is a trifle lower than the third, and thence the spines decrease regularly in height to the eleventh, which is 3⅔ in the highest spine. In soft part of dorsal the third, fourth, fifth, and sixth rays are highest and equal the highest spine; the last ray less than a third of the highest; first ray unbranched, 1½ in greatest depth of the fin, the margin rounded.

Caudal not deeply forked, its lobes rounded and about equal. Anal similar to soft dorsal, its third spine strong, 1⅕ in third dorsal spine. Ventrals 1⅗ in head. Pectorals scarcely longer.

Color, in spirits, nearly uniform. Cheeks and dorsal region somewhat darker than elsewhere, there being dark punctulations on the scales. Faint whitish lines follow the rows of scales along the sides, and are most noticeable on the caudal peduncle. The "halved" scales at base of spinous dorsal are of a bluish white superiorly. Fins plain, except that in one specimen the pectoral shows on the base of the rays on its posterior side a small gray spot formed by very minute punctulations. Peritoneum light.

Two fine specimens (29180), 11¼ inches in length.

8.—HOLOCENTRUS ERYTHRÆUS Günther.

Holocentrum erythræum Gthr. Cat. Fishes Brit. Mus. i, 32, 1859 (Sea of S. Christoval); Gthr. Jour. Mus. Godeff. Fische der Südsee, 99, 1873 (Soliman, New Hebrides, Hervey, Kingsmill, Society, Paumatu, and Sandwich Islands).

Habitat.—Soliman, New Hebrides, Hervey, Kingsmill, Society, Paumatu, Sandwich, and Johnston's Islands.

Head $2\frac{3}{5}$ ($3\frac{3}{5}$); depth $2\frac{5}{6}$ ($3\frac{1}{2}$); D. XI, 14; A. IV, 9; scales 3–50-6.

Body more elongate than in *H. leo.* Profile gently curved. Snout rather pointed, slightly shorter than eye, 4 in head. Mouth somewhat oblique; jaws about equal; maxillary extends to posterior margin of pupil, and is $2\frac{1}{3}$ in head; eye large, $3\frac{2}{3}$ in head; interorbital space $5\frac{1}{2}$ in head; intermaxillary groove slightly longer than diameter of eye. "The infraorbital bone has a rather short tooth in front of the supplementary bone of the maxillary, and another rather smaller one beneath anterior half of orbit; between them are five or seven small ones" (*Günther*). The remainder of the orbital rim is broad, with four or five serrate lobes. Occipital region with fan-like striations which end in points. Posterior part of supraorbital with small, distinct spines. Suprascapula striate, each stria ending in a point; otherwise the shoulder girdle is smooth. Preopercular spine variable, $1\frac{3}{5}$ to 2 in posterior edge of preopercle; upright limb of preopercle serrate, slightly convex, and slants backward a very little. Opercle striate, dentate, and having two spines, the upper one larger. Sub- and interopercle serrate only on posterior half, occasionally smooth.

Spinous part of dorsal fin rather low, first spine $1\frac{1}{2}$ in highest, which is $2\frac{3}{4}$ in depth of body; third, fourth, and fifth are highest, the spines evenly decreasing to the last, which is five-sevenths of first; soft part higher than spinous, its margin describing a nearly perpendicular line, the first ray unbranched, second ray highest, $1\frac{4}{5}$ in depth of body, thence regularly decreasing in height to the last, which is $3\frac{3}{4}$ in first.

Caudal well forked, upper lobe longer, $1\frac{2}{5}$ in head. Anal similar to soft dorsal; third anal spine large, $1\frac{4}{5}$ in head; fourth slender; $1\frac{1}{5}$ in third. Ventrals $1\frac{2}{3}$ in head; pectorals $1\frac{2}{3}$.

Color, in spirits, light olivaceous with indistinct bands along the sides; superiorly these bands are dark, caused by punctulations beneath the scales, the scales themselves evenly and very finely punctulate; above anal a few narrow, silvery bands, the scales not punctate in this region. The spinous dorsal is marked by a series of roundish, white spots along middle of fin, and a triangular white spot behind tip of each spine; the fins otherwise uniform. Peritoneum light.

"This species appears to be near *H. pœcilopterum*, from which, however, it differs in several respects * * *; from *H. tiere* it may be distinguished by eleven dorsal spines, and from *H. tieroides* by a greater number of scales in the lateral line." (*Günther.*)

Two fine specimens (26813), 11 inches long.

9.—KUHLIA TÆNIURA (C. & V.) Smith & Swain.

Dules tæniurus C. & V. Hist. Nat. Poiss. iii, 114, 1829 (Java); "Blkr. Perc. 49";
    Gthr. Cat. Fish. Brit. Mus. i, 267, 1859 (Chinese Sea); Blkr. Enum. Spec.
    Pisc. Archip. Indic. 4, 1859 (Java).
Moronopsis tæniurus Blkr. Sur Genre Moronopsis, p. 2 (Java, Sumatra, Buro).

Habitat.—Java, Chinese Sea, Sumatra, Buro, Johnston's Island.

Head $3\frac{1}{3}$ ($4\frac{1}{4}$); depth $2\frac{5}{6}$ ($3\frac{2}{3}$); D. X, 11; A. III, 11; scales, 8–54–13; Br. 6.

Greatest width on head behind eye, $2\frac{1}{3}$ in greatest depth. Ventral outline well arched to beginning of caudal peduncle, thence slightly concave to caudal fin, somewhat more curved than dorsal; profile nearly straight from end of snout to occiput, thence gently curved to caudal peduncle, which is little concave.

Snout rather pointed, very short, not two-thirds of orbital diameter. Eye large, $2\frac{4}{5}$ in head; interorbital width slightly less than diameter of eye. Mouth moderate, maxillary reaching almost to pupil, $2\frac{2}{5}$ in head. Teeth in villiform bands, the teeth on upper jaw higher and the band wider in front than on the sides; the band on mandible similar but narrower; minute teeth on palatines and pterygoids, and in a $\wedge$-shaped band on vomer.

Preopercle finely pectinate on whole length of lower limb, becoming coarser at the angle, extending only on lower third of upright limb, which is scarcely oblique. The inferior of the two opercular spines longer and narrower than the superior one. Preorbital narrow, notched, the lobe in front of the notch serrate. Gill-rakers slender, long, 2 in eye, about 28 on lower part of gill-arch.

Scales moderate, minute ones extending upon caudal; a well developed basal sheath of small scales on dorsal and anal pectorals, with small scales on their base.

First dorsal spine shortest, $2\frac{1}{3}$ in eye, second $1\frac{2}{3}$ in eye, third spinal three times height of first, fourth and fifth highest, $1\frac{5}{6}$ in head, thence decreasing to ninth, which is 4 in head, the tenth spine 3 in head; soft part of dorsal obliquely truncate, its first ray $2\frac{1}{2}$ in head, the last $4\frac{1}{3}$ in head. Anal somewhat concave, its soft portion longer than articulate part of dorsal, and the median rays lower than those of soft dorsal; first anal spine $1\frac{3}{4}$ in third, second stronger and little shorter than third, which is 3 in head. Caudal deeply forked, upper lobe slightly longer, $1\frac{1}{5}$ in head. Ventral not reaching vent, 2 in head. Pectoral $1\frac{2}{3}$ in head.

Color, in spirits, bluish, with metallic luster above, bright silvery below; fins light yellowish, thickly dotted with brown, except ventrals and lower part of pectorals; an obscure light band conforms to the caudal outline near its margin, and the fin is narrowly edged with dusky. Lining of mouth bluish. Peritoneum brown.

Our alcoholic specimens do not show the markings on the caudal fin which previous writers have described.

"Elle est caractérisée * * * par les cinq bandes brunes de la caudale." (*Bleeker.*)

Two fine specimens (26814), 11¾ inches in length.

10.—UPENEUS CRASSILABRIS Cuvier & Valenciennes.

Upeneus crassilabris Cuv. & Val. Hist. Nat. Poiss. vii, 523, 1831 (New Guinea); Blkr. Enum. Spec. Pisc. Archip. Indic. 38, 1859 (name only); Gthr. Cat. Fishes Brit. Mus. i, 411, 1859 (taken from Cuv. & Val.); Blkr. Vischfauna Nieuw-Guinea, p. 8 (name only).

Parupeneus crassilabris Blkr. Quatrième Mem. Faune Ichthyologique Nouvelle Guinée; Blkr. Révision Mulloïdes, 33, 1874? (from Cuv. & Val.) (New Guinea).

Habitat.—Indian Archipelago, New Guinea, Johnston's Island.

Head 3 (3⅔); depth 3 (3⅔); D. VIII, 9; A. I, 7; scales 2–31–6.

Body oblong, compressed, robust. Head and anterior part of body heavy. Profile concave from snout to a point midway between the nostrils, thence regularly curved to first dorsal; snout long, blunt, 1¾ in length of head. Ventral outline little curved; caudal peduncle two-thirds length of head, its least depth almost twice in its length. Mouth moderate, little oblique, the lower jaw included; maxillary 2⅓ in head, terminating between the nostrils.

Strong, blunt, but conical, teeth in one series in each jaw, the teeth more or less widely separated. Eye 2¾ in snout and 5 in head; interorbital space very convex, 3⅔ in head.

Preopercle with upright limb slanting obliquely forwards; opercular spine strong. Gill-rakers 2 in eye, 4½ in maxillary, about 28 on lower limb of arch.

Barbels two-thirds length of head, reaching posterior margin of subopercle. Scales large, ctenoid.

Dorsal fins well developed; third and fourth spines longest, 1⅕ in head, twice the height of soft dorsal, the last of which are scarcely longer than the first; soft dorsal two-thirds as high as long, its length 2 in head. Caudal moderate, well forked, upper lobe more rounding, 1¾ in head. Anal differs from soft dorsal in having its first branched rays one-fourth longer than the last, the unbranched ray also slightly exceeding the last ray; the greatest height of the fin 2¾ in head. The membrane of the first soft ray envelopes a small spine, which, being thus covered, might easily be overlooked; the first articulate ray of anal, as in soft dorsal, not branched. Ventral fins large, 1⅓ in head, about reaching tips of pectorals.

Air-bladder large.

Color, in spirits, olivaceous, lighter below, the fish, as a whole, having a smutty appearance; exposed part of each scale punctulate with brown; first dorsal and caudal dusky; base of soft dorsal dusky, upper half irregularly light and dark; anal with irregular dusky bars; soft dorsal,

caudal, and anal tipped with black; ventrals yellowish, spine and connecting membrane smutty; pectorals yellow, their base, the preorbital,
and upper part of head purple. Barbels very dark; branchiostegal membrane purple. Peritoneum light.

According to Cuvier and Valenciennes *Upeneus crassilabris* is "voisin
du *cyclostome*. * * Ce poisson parait avoir été jaune, avec des points
ou des lignes peu marquées sur les côtés. La première dorsale est vio-
lette; la seconde n'a que la base de cette couleur; la moitié supérieure
est rayée de cinq à six raies parallèles longitudinales, alternativement
blanches et violettes. L'anale, beaucoup plus pâle, a des points violets
et un plus grand nombre de raies obliques. La caudale est plus foncée
que la dorsale, et elle a des points blancs plus ou moins effacés. Les
pectorales sont jaunes, plus ou moins olivâtres. Les ventrales ont les
trois rayons externes colorés en violet, et les internes jaunâtres. La
membrane branchiostège et les barbillons sont d'un brun violet plus ou
moins foncé."

11.—UPENEUS VELIFER sp. nov.

Head $3\frac{1}{5}$ ($3\frac{5}{6}$); depth $3\frac{1}{5}$ ($3\frac{5}{6}$); D. VIII, 9; A. I, 7; scales 2-29-6.

Allied to *Upeneus trifasciatus* (Lac.) Cuv. & Val.

Body less robust than in *U. crassilabris*. Profile from beginning of
dorsal fin to a point above center of orbit, a gentle and regular curve;
thence to the snout a straight oblique line. Snout rather pointed, lower
jaw included; lips thin; maxillary $2\frac{3}{5}$ in head; eye rather small, high,
and far back, its diameter contained $3\frac{1}{5}$ times in the snout (measured
obliquely from eye) and $5\frac{1}{2}$ times in the head.

Teeth blunt, isolated, in a single series in each jaw; the overlapping
upper jaw shows all the teeth of the premaxillary in the closed mouth.
No teeth on vomer or palatines. A strong, blunt spine on opercle.
Gill-rakers 28 on lower limb.

Barbels slender and long, extending slightly beyond base of ventral,
nearly as long as head. Scales ctenoid, moderate.

Dorsal fins well developed; third spine highest, equaling two-thirds of
the greatest depth, the spines following about evenly decreasing in
height to the eighth, which is two-sevenths of the third. In the second
dorsal the first ray is shortest, $3\frac{3}{5}$ in depth of body, all between the first
and ninth about equal and slightly higher than the first; the posterior
half of the ninth, or split ray, is $2\frac{1}{2}$ times as high as the intermediate
rays, and exceeds the highest dorsal spine; when depressed it extends
onto the caudal one-fifth of the greatest length of the caudal; outline
of soft dorsal slightly concave. Anal about one-fifth higher than soft
dorsal; its last ray, however, is less produced, being six-sevenths of the
last ray of the dorsal, and, depressed, does not reach base of caudal;
anal outline somewhat convex from first ray to the split ray, which,
similarly to that ray in the soft dorsal, is produced beyond rest of fin.
The first ray of second dorsal and anal unbranched but plainly articu-

late. Caudal well developed, forked, its lobes rounded. Pectoral 1⅓ in head. Ventrals not quite reaching the vent, 1⅛ in head. Air-bladder large, lined with black.

Color, in spirits, yellowish, with dark markings; head gray; a black spot on the median line between occiput and first dorsal; a blackish band across the back between the dorsal fins, another at base of soft dorsal, and a third on the caudal peduncle; all these bands becoming lighter below the lateral line; a gray blotch below lateral line, nearer second band than the one on the tail. First and second dorsal spines black, the whole upper portion of spinous dorsal dark, the membrane connecting the fourth, fifth, sixth, and seventh spines nearly black, forming a blotch; second dorsal very dark, with about three narrow whiteish stripes, and inconspicuous white spots on its base; caudal plain, yellow, narrowly margined with brown above, less narrowly margined with black below; anal dark, its marginal third with three light lines disposed horizontally; nearer the base the membrane is crossed by white bars at right angles with the rays, almost forming a network of lines; pectorals dusky at base, otherwise plain yellow; ventral spine nearly black, the rest of the fin 'dusky, with about nine slightly waving lines of white across the rays. Lower lip and barbels gray, tips of the barbels fading into yellowish. Branchiostegal membrane dusky. Peritoneum light.

This species is closely allied to *U. trifasciatus* (Lac.) C. & V., from which its most conspicuous differences (according to a figure in the Jour. Mus. Godeffroy) are the extreme elongation of the last ray in both the soft dorsal and the anal fin, and the greater length of the barbels, which pass beyond the base of the ventrals, while in this figure the barbels only reach the posterior margin of the subopercle. Lacépède's figures, copied from Commerson, represent the specimen which he calls *Mullus bifasciatus* as having the barbels shorter than the head, and *M. trifasciatus* with the barbels nearly attaining the extremity of the ventral fins. Of the barbels Cuvier & Valenciennes say: "ils ne dépassent pas l'angle du préopercule, et ceux du *mulle trois-bandes* dépassent même l'opercule. Capendant la figure, qui est de Sonnerat, les exagère, en les faisant aller jusques sous les ventrales."—(Hist. Nat. Poiss. iii, 468.) In the figures by Lacépède the dorsal and anal fins are low in both *M. bifasciatus* and *M. trifasciatus*.

One specimen (26822), 10¾ inches in length.

12.—UPENEUS VANICOLENSIS (C. & V.) Smith & Swain.

Upeneus vanicolensis C. & V. Hist. Nat. Poiss. vii, 521, 1831 (Vanicolo).
Upeneus vanicolensis Blkr. Nat. T. Ned .Ind. iv, 601, 1853.
Mulloides vanicolensis Blkr. Ternate ii, 601, 1853; Gthr. Cat. Fishes Brit. Mus. i, 404,
　　1859 (seas of Ternate and Vanicolo); Bleeker, Enum. Spec. Pisc. Archip.
　　Indic. 39, 1859 (name only); Blkr. Conspect. Spec. Pisc. p. 6 (Halmaheira,
　　Ternate, Archip. Molucca; name only); Blkr. Révision Mulloides, p. 14,
　　1874? (Ternate, Sangir).

Habitat.—Vanicolo, Ternate, Sangir, Halmaheira, Archip. Molucca, Johnston's Island.

Head $3\frac{4}{5}$ ($4\frac{3}{4}$); depth $4\frac{1}{3}$ ($5\frac{1}{4}$); D. VIII–I, 8; A. II, 6; scales, $2\frac{1}{2}$–36–6.

Body rather slender; ventral outline almost as much curved as dorsal; profile gently, and nearly evenly, curved from snout to first dorsal fin. Caudal peduncle tapers evenly from dorsal and anal to the caudal fin, and nearly equals length of head; its least depth $2\frac{1}{3}$ in its length.

Snout short, bluntish, $2\frac{3}{4}$ in head; mouth small, maxillary reaching posterior nostril, $2\frac{4}{5}$ in head. The bands of villiform teeth very narrow; in front two series in each jaw, on the sides only one. Eye large, $1\frac{1}{4}$ in snout, $3\frac{1}{2}$ in head; interorbital space moderately convex, 3 in head. Upright limb of preopercle straight; opercular spine small.

Gill-rakers slender, $2\frac{1}{2}$ in maxillary, 7 in head, about 25 on lower limb of arch.

Barbels $1\frac{2}{3}$ in head, extending beyond posterior margin of eye.

Scales, moderate, ctenoid; preorbital smooth.

Dorsal fins moderate; spines of first dorsal rather weak, depressible into a groove; first spine *very minute*, second and third longest, $1\frac{1}{4}$ in head, those following evenly decreasing in height to the eighth spine, which is 3 in head. The first ray of the second dorsal fin is unbranched, and, showing no articulations, has the appearance of a true spine, slightly shorter than the last soft ray, and 4 in head; the second articulate ray is contained $1\frac{4}{5}$ times in the highest dorsal spine. Caudal well forked, its longest rays $1\frac{1}{4}$ in head. Anal with two spines, the first of which is *very minute*; otherwise, similar to soft dorsal, though a very little higher. Pectorals $1\frac{2}{3}$ in head. Ventrals $1\frac{1}{3}$ in head. Air-bladder moderate. Peritoneum, black.

Color, in spirits, grayish-green above lateral line, lighter below, with yellow metallic luster; minute black punctulations on scales above lateral line; none below.

One specimen (30,873), $6\frac{1}{8}$ inches long.

13.—UPENEUS PREORBITALIS sp. nov.

Head $3\frac{2}{5}$ ($4\frac{1}{6}$); depth $4\frac{1}{6}$ ($5\frac{1}{3}$). D. VIII–9; A. I, 7; scales 2–37–5.

Body more slender than in U. *vanicolensis* C. & V.; ventral outline almost straight, dorsal outline well curved; profile from snout to dorsal regularly curved; caudal peduncle $1\frac{1}{3}$ in head, its least height $3\frac{1}{4}$ in head.

Mouth nearly horizontal, maxillary 3 in head, terminating behind anterior nostril; lower jaw produced. The band of villiform teeth moderate in both jaws, in a patch in front, narrowing posteriorly. Eye moderate, $2\frac{1}{2}$ in snout, $4\frac{1}{3}$ in head; interorbital space slightly concave, $4\frac{1}{4}$ in head; preorbital very deep, $2\frac{1}{4}$ in head.

Gill-rakers short and rather slender, $4\frac{1}{2}$ in maxillary; 19 on lower limb of arch.

Barbels 1½ in head, reaching posterior margin of preopercle. Scales large, ctenoid.

Dorsal fins medium; spinous dorsal depressible into a groove; first spine rudimentary, scarcely perceptible, second and third spines longest, 1⅔ in head, eighth spine 4 in second. The first ray of soft dorsal is unbranched but evidently articulate, shorter than the first branched ray, which is 2⅓ in head, thence about regularly decreasing in height. Caudal well forked, its longest rays 1¼ in head; anal similar to soft dorsal, its spine *very minute* and first ray unbranched but plainly articulate. Ventrals 2 in head. Pectorals 1⅔ in head. Air-bladder moderate.

Color, in spirits, nearly uniform yellowish; snout dusky, fins plain. Peritoneum dark.

One fine specimen (29,662), 14¾ inches long.

14.—CHILINUS DIGRAMMA (Lacép.) Cuv. & Val.

Sparus radiatus, Bl. Schn. p. 270, tab. 56 (not of Linn.).
Labrus digramma Lacép. Hist. Nat. Poiss. iii, pp. 448, 517, 1802.
Cheilinus commersonii, "Benn. Proc. Comm. Zool. Soc. i, 167."
Cheilinus coccineus, "Rüpp. Atl. Fische, p. 23."
Cheilinus diagrammus, Cuv. & Val. Hist. Nat. Poiss. xiv, 98, 1839 (Isle of France,
    Séchélles, Madagascar, New Guinea); Blkr. Overzigt Labroieden, 4 (name
    only), 38 (descriptions); Blkr. Tweede Bijd. Ichth Fauna Halmaheira, 2, 4
    (name only).
Cheilinus radiatus, Bleek. Atl. Ichth. i, p. 68, tab. 26, fig. 1, 1862; Gthr. Cat. Fishes
    Brit. Mus. iv, 131 (Moluccas, Amboyna, Louisiade Archipelago, Cape Flattery,
    N. E. Australia); Gthr. Fish. Zanz. p. 89; "Klunz. Fisch. d. Roth. Meer. p. 556";
    Blkr. Visschsooten van Amboina, p. 21; Blkr. 8th Bijdrage Amboina, 7, 25
    (name only); Blkr. Conspec. Moluc. Cognit. 18 (Halmaheira, Batjau, Am-
    boyna, Ternate, Archip. Moluce.; name only); Blkr. Bijd. Visch. Nieuw Guinea'
    3, 11 (name only); Blkr. Beschrij. Visch. Manado Makassar, 5, 25 (Manado,
    Macassar); Blkr. Nieuwe Verzam Visschen Batjau, 5 (name only); Blkr. Quat.
    Mem. Ichth. Nouv. Guinée, 6 (name only).

Habitat.—Isle of France, Séchélle, Madagascar, New Guinea, Moluc-
cas, Amboyna, Louisiade Archipelago, Cape Flattery, N. E. Australia,
Halmaheira, Batjau, Ternate, Archip. Moluce., Manado, Macassar,
Johnston's Island.

Head 2¼ (3⅙); depth 3 (3⅔); length (26815) 12½ inches; D. IX, 10;
A. III, 8; scales, 1½–21–5½; Br. 5.

Body rather slender; profile not steep; snout rather pointed, 2⅔ in
head; lower jaw produced; mouth little oblique, maxillary not reaching
eye; anterior canine teeth strong; eye 7 in head; interorbital space 4¾
in head; nostrils very small; gill-rakers placed wide apart, 10 on lower
part of arch, the longest ones 3 in eye; slit behind last gill obsolete;
gill-membranes not joined to the isthmus.

First dorsal spine as high as orbital diameter, the spines increasing
slightly in height to the last, which is one-half higher than first; soft
part of dorsal higher than spinous, its highest rays 3 in head, the outline rounded; caudal subtruncate; first anal spine equals diameter of

eye, second and third increasing in height, as usual; soft portion little higher than that of dorsal. Ventrals short, $2\frac{4}{5}$ in head; pectorals fan-shaped, $2\frac{2}{5}$ in head.

Most of the tubes of lateral line simple.

Color in spirits, chocolate above, lighter below; a narrow light band crosses the back immediately behind soft-dorsal fin, fading out just before it reaches the space close behind anal, thus outlining the base of caudal peduncle; this light mark extending up on the dorsal fin, almost meeting the central light line of the dorsal, the pedunculate band seeming to be a continuation of the middle dorsal line. A light streak outlines the cheek superiorly, becoming fainter where it extends across opercles to the upper base of pectoral. A series of similar light lines extends obliquely downward from eye across cheeks and opercles; these lines somewhat waving, and coalescing more or less; two or three light streaks on preorbital, running from eye toward snout; nasal region vermiculate, with similar light markings. Dorsal fin chocolate, with two light lines running lengthwise of the fin; soft part brownish only at base, transparent superiorly, yet somewhat dusky. Caudal copper-green, its outer rays of a brown, like the back; anal light, tinged with green; a very dark brown spot on base of ventrals, covering half the fin; pectorals transparent, uniform yellowish; teeth greenish.

Three fine specimens in the collection.

The name *radiatus*, having been given to this species through an erroneous identification, cannot be retained.

15.—SCARUS PERSPICILLATUS Steindachner.

Scarus (*Scarus*) *perspicillatus* Steind. Neue Seltene Fisch-Arten aus. k. k. Museen Wien, etc., p. 16, taf. iv, f. 1, 1879 (Sandwich Islands).

Habitat.—Sandwich Islands. Johnston's Island.

Head $3\frac{1}{3}$ (4); depth $2\frac{3}{5}$ ($3\frac{1}{6}$); D. IX, 10; A. III, 9; Br. 5; scales, $1\frac{1}{2}$–23–6.

Body oblong, compressed. Ventral outline well and regularly curved, exceeding dorsal. Head longer than deep; snout rather blunt, $2\frac{1}{3}$ in head; eye moderate, 3 in snout; interorbital space equals snout. Lips thin, covering half the dental plate; upper lip double only at the corner of the mouth. Dental plates crenulate; no posterior canines. Gill-rakers numerous, short, and very slender.

Scales large; one series of six scales on cheek, and an extra scale below this series. Tubes of lateral line irregularly branched, and the branches more or less waved; not very prominent.

Dorsal and anal spines rather flexible, not pungent; dorsal rather low and of nearly equal height throughout, $3\frac{1}{3}$ in head. Anal similar to soft part of dorsal, its height $2\frac{4}{5}$ in head. Caudal emarginate, $1\frac{3}{5}$ in head. Pectorals $1\frac{1}{3}$ in head. Ventrals $1\frac{4}{5}$ in head.

In spirits olivaceous brown, little lighter below. Dorsal and anal smutty, both lighter anteriorly and on marginal half of the fins; a well defined dusky line running horizontally near the margins of both, leaving the edges of the fins plain light colored. Pectorals and caudal dusky; ventrals plain, light colored. Head marked with yellow as follows: a line in front of the eyes outlines a brownish oblong figure, which extends vertically across the snout, not quite equal to the eye in width, becoming narrower on median line of snout, its length six times that of its greatest width; an indistinct line passes over the interorbital space, curves around anterior margin of eye, and ends before reaching the scales on the cheek; a more distinct line runs nearly parallel with preopercular membrane; a horizontal, waving band on the middle of the cheek is continuous with a wider band on chin, this having its lower edge evenly waved; a narrow, nearly lunate, band on each side of the lower jaw; round and oblong spots on cheek and jaws. Obscure dots on some of the scales of ventral region, above anal, and more noticeable ones behind pectoral fins. All the foregoing markings yellow. Teeth rosy; peritoneum dark.

One fine specimen (26833), 15 inches long.

16.—JULIS VERTICALIS sp. nov.

Head $3\frac{1}{5}$ ($3\frac{2}{3}$); depth $3\frac{1}{5}$ ($3\frac{2}{3}$); length (26829) 11 inches. D. VIII, 13; A. III, 11; Br. 6; scales $2\frac{1}{2}$–27–$8\frac{1}{2}$. (These measurements were taken to end of middle rays of caudal.)

Body oblong, compressed, rather robust; dorsal and anal outlines about equally curved; head longer than deep; snout somewhat blunt, $2\frac{2}{7}$ in head; eye moderate, $6\frac{2}{3}$ in head. Branchiostegal membranes forming a fold across the isthmus.

Scales moderate, becoming smaller in front of ventrals; small scales on base of dorsal, anal, and caudal fins. Gill-rakers short, 16 on lower part of arch; slit behind last gill wholly obsolete. Tubes of lateral line forked once on most of the scales anterior to caudal peduncle.

Spinous dorsal low; the anterior spines about $5\frac{1}{2}$ in head; the spines increasing slightly in height posteriorly; height of soft portion of dorsal fin $3\frac{1}{3}$ in head. Caudal with outer rays produced, but not greatly so; the greatest length of the fin $1\frac{2}{3}$ in head. Anal similar to soft dorsal. Pectorals $1\frac{3}{5}$ in head. Ventrals short, $2\frac{2}{5}$ in head.

In spirits olivaceous brown; each scale with a conspicuous dark, vertical streak, forming two vertical lines on the body for each scale of the lateral line. Head, dorsal, and produced rays of caudal purplish-brown, caudal otherwise olivaceous; anal with a light purplish basal band, otherwise brownish; ventrals purplish dusky; pectorals olivaceous, washed in part with purple. Peritoneum light.

One fine specimen.

17.—JULIS CLEPSYDRALIS sp. nov.

Head $3\frac{1}{4}$ ($3\frac{5}{6}$); depth $3\frac{1}{4}$ ($3\frac{5}{6}$); length (26826) $7\frac{1}{2}$ inches. D. VIII, 13; A. II, 11; Br. 6; scales 2–27–8. (These measurements taken to end of middle rays of caudal.)

Body rather stout; head longer than high; jaws about equal; snout little pointed, $2\frac{4}{5}$ in head; eye $5\frac{1}{2}$ in head, $1\frac{1}{2}$ in interorbital space. Branchiostegal membrane forming a fold across the isthmus. Gill-rakers short and weak, about 12 on lower part of arch. No slit behind last gill.

Pores of lateral line mostly branched, forming three or four spreading tubes on the scale. The scales extend up on base of dorsal and anal fins, smaller ones on base of caudal. Scales on breast reduced in size.

Dorsal low, its first spine equaling orbital diameter; spines gradually increasing, the last being twice the height of first; soft portion slightly higher than spinous. Anal similar.

Outer caudal rays greatly produced, the filamentous part nearly as long as the head. Ventrals $1\frac{2}{3}$ in head; although their first ray is produced the ventrals do not quite reach the vent. Pectorals $1\frac{1}{5}$ in head.

Color, in spirits, blackish. Head, dorsal and anal fins black; posterior portion of body blackish olivaceous; an hour-glass-shaped lighter area on shoulders, extending across thorax, this area bounded in front by the outline of the black head. Pectoral blue-black in the axil; base of fin yellowish, followed by a black blotch that covers the upper rays to their extremities, descending obliquely forward leaves the lower rays plain yellowish and transparent at their tips. Caudal blackish. Ventrals light, transparent, the spine and first ray dusky. Peritoneum gray, with a pinkish shade.

One fine example.

18.—HARPE BILUNULATA (Lacép.) Smith & Swian.

Labrus bilunulatus Lacép. Hist. Nat. Poiss. iii, 454, 526, pl. 31, 1802.

Cossyphus bilunulatus Cuv. & Val. Hist. Nat. Poiss. xiii, 121, 1839 (Seas of India, Isle de France); Gthr. Cat. Fish. Brit. Mus. iv, 105, 1862 (Isle de France, Amboyna, Mauritius); "Gthr. Fish. Zanz. p. 87"; Blkr. Atl. Ichth. i, 160, tab. 38, fig. 3; Blkr. Neg. Bijd. Visch. Amboina, 4 (Amboyna); Blkr. Index Pisc. Actorum Sci. Soc. Indo-Neer. 4 (name only); Gthr. Jour. Mus. Godeff. Fische der Südsee, 240, pl. 130, 1881 (Mauritius, Zanzibar, Amboyna, Misol, Sandwich Islands).

Cossyphus albotæniatus C. & V. Hist. Nat. Poiss. xiii, 141, 1839 (Sandwich Islands); Gthr. Cat. Fish. Brit. Mus. iv, 105 (taken from Cuv. & Val.).

Gymnopropoma (*bilunulatum*) Gill, Proc. Phil. Acad. Nat. Sci. 1863 (generic diagnosis only).

Habitat.—Seas of India, Isle of France, Amboyna, Mauritius, Zanzibar, Misol, Sandwich Islands, Johnston's Island.

Head $2\frac{7}{8}$ ($3\frac{4}{7}$); depth $2\frac{2}{3}$ ($3\frac{2}{5}$); D. XII, 10; A. III, 12; scales $5\frac{1}{2}$–35–13

Head longer than deep; upper lip thin, lower lip narrow. Gill-rakers short, 13 on lower arch. Ventrals reach vent, nearly as long as head. The outer ray of ventral and outer rays of caudal produced. Scales on cheeks and occiput much smaller than elsewhere, in about 12 rows on the cheeks.

The fatty hump on forehead, which is usually seen on mature individuals in this genus, is wholly undeveloped.

Color, in spirts, yellow olivaceous, with darker olivaceous waving-streaks running horizontally between the rows of scales; these lines becoming mere brown spots above and below, but forming two bands behind the eye, which converge into one in front of the eye; under jaw scarcely spotted except near the gape of the mouth. A dark brown blotch between the soft dorsal and the lateral line extending around the posterior part of the soft dorsal, Λ-shaped, as seen from above, close up under the fin, but not extending on it; the dorsal fin with a dark spot anteriorly, the margin of connecting membrane brownish along whole of spinous portion. Teeth white.

A larger specimen (17 inches in length) is darker everywhere, with the dorsal, caudal and ventral fins dusky, and the blotch between soft dorsal and lateral line less prominent. The ventrals are longer, reaching third anal spine.

Two fine specimens, numbered 26830, 15$\frac{1}{4}$ inches in length.

19.—CHÆTODON SETIFER Bloch.

Chætodon setifer Bl. Naturg. ausländ. Fische, t. 426, f. 1, 1797; Bloch & Schn. Systema Ichthyologiæ, 225, 1801 (Tranquebar); Cuv. & Val. Hist. Nat. Poiss. vii, 76, 1831 (Bolabold); "Guérin, Iconogr. Poiss. pl. 22, f. 1"; "Less. Voy. Cog. Zool. ii, 175, Poiss. pl. 29, f. 2"; "Richards, Ichth. China, p. 246"; Cuv. Règne Anim. Ill. Poiss. pl. 38, f. 1; "Jenyns, Zool. Beagle, Fishes, p. 61"; Gthr. Cat. Fishes Brit. Mus. ii, 6, 1860 (Isle de France, Amboyna, China, Aneiteum); Gthr. Jour. Mus. Godeff. Fish. Süd. 36, taf. 26, B. 1873.

Pomacentrus filamentosus, Lacép. iv, pp. 506, 511, 1802.

Chætodon sebanus C. & V. vii, 74, 1831 (Timor, Guam, Tongatabou, Isle of France, Batavia).

" *Chætodon auriga,* var. Rüpp. N. W. Fische, p. 28."

Chætodon lunaris " Gronov. Syst. ed. Gray, p. 70."

Chætodon auriga Bleek. Celebes, iv, 164.

Habitat.—Tranquebar, Bolabold, Isle of France, Amboyna, China, Aneiteum, Timor, Guam, Tongatabou, Celebes, Johnston's Island.

Head 3$\frac{1}{10}$ (3$\frac{2}{3}$); depth 1$\frac{2}{3}$ (2). D. XII, 24; A. III, 20; scales 6–44–13. (In a straight horizontal series from head to caudal 15, about 44 in first row above lateral line.)

Body ovate. Profile steep; from dorsal to occiput convex, thence concave to snout. Snout pointed, conical; mouth nearly horizontal, maxillary 4 in head; teeth ordinary; eye 1$\frac{3}{4}$ in snout, 4 in head; interorbital space 1$\frac{1}{2}$ in snout.

Gill-rakers very short and slender; suprascapula striate and dentate.

First dosal spine 1¾ in snout, thence almost evenly increasing in height to the sixth, which is 2⅖ in head, seventh to eleventh about equal, twelfth 2 in head; the soft portion of dorsal higher than spinous, its highest rays 1⅛ in head, its margin rounded. The filiform elongation of the fifth ray is broken off. Caudal truncate. Middle rays of anal longest, giving a bluntly pointed outline to the fin; second anal spine not as long as third and scarcely stronger, 2⅕ in head. Ventrals 1⅔ in head. Pectorals 1⅛ in head. Scales finely ctenoid. Fins scaled as usual. Preopercle crenulate at the angle and on lower margin.

Color, in spirits, pale olivaceous with dark and black markings; five dark lines ascend obliquely from shoulder-girdle to dorsal, seven similar ones ascend obliquely from anal meeting the anterior lines at a right angle; above the seven lines and running parallel with them are about four wider greenish-brown bands, which anteriorly describe a right angle ascending to the dorsal, this part of the wide bands thus running parallel with the lines on front part of body, the ground color between these bands being of a sulphur yellow. A black band descends perpendicularly across the eye, narrower and fading out above, a third wider than orbital diameter below the eye, the bands of the two sides, extending across the interopercle, meet at the isthmus. A black oval spot near the margin on the sixth, seventh, eighth, ninth, and tenth soft rays of dorsal. Between the posterior spines of the dorsal the connecting membrane is narrowly margined with black, the soft dorsal edged with black to the twenty-third ray. The extreme edge of caudal fin is white, anterior to which is a narrow space of gray margined with black, then a dull-yellow lunate space equal in width to both the other marginings; the base of the caudal being grayish leaves this lunate space outlined by a gray line, the dusky ground color covering the anterior two-thirds of the length of the fin. And with a thread-like line of black near its margin. Pectorals and ventrals plain. Peritoneum dusky.

One fine specimen (26831), 7½ inches long.

20.—ACANTHURUS TRIOSTEGUS (L.) Bloch. & Schn.

Chætodon triostegus Linn. Syst. Nat. i, 463; Brouss. Ichthy. fig. and description, 1782; Gmel. Syst. Nat. 1246, 1788 (Pacific & Indian Oceans).

Chætodon couagga Lac. iv, 727, 1802.

Acanthurus zebra Lac. Hist. Nat. Poiss. iii, pl. 25, fig. 3, 1802, iv, 160, pl. 6, fig. 3, 1802; C. & V. Hist. Poiss. x, 197, 1835 (Isle de France, Séchellés, Marianna, N. Zealand, Oulan, Society and Sandwich Islands).

Acanthurus triostegus Bl. & Schn. Syst. Ichthy. 215, 1801 (Pacific and Indian Oceans); Blkr. Enum. Spec. Pisc. Archip. Indic. 75, 1859 (name only); Blkr. Conspec. Moluc. Cognit. 12 (Buro, Amboyna, Ternate, Ceram, Archip. Moluc.; name only); Blkr. Bijd. Visch. N. Guinée, 4, 10 (name only); Blkr. Zes. Bijd. Visch. Sumatra, 5, 21 (Kauer, Trussan, Padang, Ulakan, Sibogha, Priamam); Blkr. Twaalfde Bijd. Visch. Celebes, 2 (name only); Blkr. Beschrij. Visch. Amboina, 6 (name only); Blkr. Beschrij. Manado Makassar, 4, 20 (Manado; name only); Blkr. Achtste Bijd. Visch. Amboina, 5, 20 (name only); "Blkr. Verhand. Batav. Genootsch, xxiii, Teuth. 13; Jenyns, Voy. Beagle, Fishes, 75"; Gthr. Cat.

Fish. Brit. Mus. iii, 327, 1861 (Amboyna, Celebes, Malayan Archipelago, Sandwich Islands, Aneiteum, W. Coast of Australia, Mauritius, New Zealand); Gthr. Jour. Mus. Godeff. 108, 1873 (Polynesia, Sandwich Islands, Indian Ocean).

Harpurus fasciatus "Forst. Descr. Anim. ed. Licht. 216."
Acanthurus hirudo "Benn. Ceyl. Fishes, 11, pl. 11; Blkr. Bat. Gen. xxiii, Teuth. 13."
Acanthurus subarmatus "Benn. Whal. Voy. ii, 278."
Teuthis australis "Gray in King's Survey of the coasts of Austr. ii, 435."

Habitat.—Isle of France, Séchellés, Marianna, New Zealand, Oulan, Society Islands, Sandwich Islands, Buro, Amboyna, Ternate, Ceram, Archip. Molucc., New Guinea, Sumatra, Kauer, Trussan, Padang, Ulakan, Sibogha, Priamam, Celebes, Manado, Macassar, Malayan Archip., Aneiteum, W. Coast of Australia, Mauritius, Polynesia, Indian Ocean, Johnston's Island.

Two fine specimens, numbered 26820, 7 inches in length.

21.—NASEUS LITURATUS (Forst.) Cuv. & Val.

Harpurus lituratus "Forst. Descr. Anim. ed. Licht. 218."
Acanthurus lituratus Bl. & Schn. 216, 1801.
Acanthurus harpurus "Shaw, Zool. iv, 381."
Aspisurus elegans "Rüpp. Atl. Fische, 61, taf. 16, fig. 2."
Aspisurus carolinarum "Quoy & Gaim. Voy. Uran. Zool. 375, pl. 63, fig. 1 (New Ireland)."
Prionurus eoume "Less. Voy. Cog. Zool. ii, 151 (Otaïti, Matavaï)."
Naseus lituratus C. & V. x, 282, 1835; "Blkr. Celebes, in, 763"; Gthr. iii, 353, 1861 (Olaheiti, Aneiteum, Malayan Archipelago, Red Sea, Polynesia); Gthr. Jour. Mus. Godeff. 124, taf. 82, 1873 (Society Islands, Red Sea, East Coast of Africa, Sandwich Islands); Blkr. Conspec. Molucc. Cognit. 12 (Amboina, Archip. Molucc; name only); Blkr. Zes. Bijd. Visch. Sumatra, 11 (Batœ; name only); Blkr. Beschrij. Visch. Amboina, 17 (name only); Blkr. Beschirj. Visch. Monado, Makassar, 20 (Macassar); Blkr. Achtste Bijd. Visch. Amboina, 20 (name only); Blkr. Visschen Van Diemensland, 11 (name only).
Acanthurus lituratus Blkr. Bijd. Visch. N. Guinea, 3, 11 (name only); "Blkr. Nat. T. Ned. Ind. iii, 763."
Aspisurus lituratus "Rüpp, N. W. Fische, 130."

Habitat.—New Ireland, Uran, Otaïti, Matavaï, Celebes, Aneiteum, Malayan Archip., Red Sea, Polynesia, Society Islands, East Coast of Africa, Sandwich Islands, Amboyna, Archip. Molucc., Sumatra, Batoe, Manado, Macassar, Van Dieman's Land, New Guinea, Ulea, Gulf of Suez, Isle of France, Johnston's Island.

Two fine specimens, numbered 26812, measuring 10¼ inches. ·

22.—BALISTES ACULEATUS Linnæus.

Balistes aculeatus "L. Syst. Nat. i, 406, 1766"; Bloch, Naturgeschichte ausländ. Fische, i, 183, 194, 1786. pl, 149 (Red Sea); Gmel. Syst. Nat. 1466, 1788 (Indian Ocean, Red Sea); Bloch & Schn. Syst. Ichth. 465, 1801 (Indian Ocean, Red Sea); Lay & Benn. in Beechey's Voy. Zool. 69, pl. 22, fig. 2, 1839 (Loo-Choo Islands); "Jenyns, Zool. Beagle Fish. 155, 1842; Blkr. Vehr. Bat. Gen. xxiv, Balist. 15, 1852; Hollard, Ann. Sc. Nat. i, 333, 1854"; Gthr. Cat. Fish. Brit. Mus. viii, 223, 1870 (Isle of France, Island of Johanna, Zanzibar, W. Doast of Africa, Moluccas, Amboyna, China Seas, Fiji Islands, Micronesia, Seychelles,

Mauritius); Blkr. Conspec. Moluec. Cognit. 21 (Amboyna, Archip. Moluec., Banda; name only); Blkr. Zes. Bijd. Visch. Sumatra, 13 (Nias; name only); Blkr. Ichth. Fauna Borneo, 12 (Ignata; name only); Blkr. Twaalfde Bijd. Visch. Celebes, 2 (name only); Blkr. Beschrij. Visch. Amboina, 823 (name only); Blkr. Beschrij. Visch. Manado Makassar, 6, 29 (Manado; name only); Blkr. Achtste Bijd. Visch. Amboina, 8, 28 (name only); Blkr. Einige Visschen Van Diemensland (name only); Streets, Bull. U. S. Nat. Mus. No. 7, 79, 1877 (Fanning Islands).

Balistes ornatissimus "Less. Voy. Cog. Zool. Pois. i, 119, pl. 10, fig. 1, 1830."

Balistes armatus "Cuv. R. An. Ill. Poiss, pl. 112, f. 2, 1829-'30."

Balistes striatus "Gronov. Syst. ed. Gray, 32."

? *Balistes assasi* "Forsk. 75, n. 112"; Gmel. 1471, n. 12, 1788 (Red Sea).

Balistes (Balistapus) aculeatus "Blkr. Atl. Ichth. Balist. 120, pl. 2, f. 3."

Baliste epineux Lac. Hist. Nat. Poiss. i, 367, pl. 17, f. 1, 1798.

Habitat.—Indian Ocean, Red Sea, Loo-Choo Islands, Isle of France, Island of Johanna, Zanzibar, W. Coast of Africa, Moluccas, Amboyna, China Seas, Fiji Islands, Micronesia, Seychelles, Mauritius, Archip. Molucca, Banda, Nias, Ignata, Manado, Van Diemen's Land, Sumatra, Fanning Islands, Johnston's Island.

Two fine examples, numbered 26316 and 29760; length 11 inches.

23.—BALISTES BUNIVA Lacépède.

Balistes ringens "Osbeck, Voy. Chin. ii, 93, 1771, not of L.;" Bloch, Ausl. Fisch. 183 (footnote), pl. 152, fig. 2, 1786; Bloch & Schn. Syst. Ichth. 472, 1801 (Indian and Chinese Seas); "Rich. Voy. Samar. Fish. 21, pl. 16, f. 1-4, 1848; Rep. Ichth. Chin. Rep. 15th Meet. Brit. Assoc. 201, 1846; Hollard, Ann. Sc. Nat. 4th series, 1854, Zool. i, 317; Bleek. Act. Soc. Sc. Ind. Néere viii, 1860; Sumatra viii, 69" (not of Linn.).

Baliste sillonné Lac. Hist. Nat. Poiss. i, 370, pl. 18, fig. 1, 1798 (China Sea, E. Coast of Africa).

Balistes buniva Lac. Hist. Nat. Poiss. v, 669, pl. 21, f. 1, 1803; Gthr. Cat. Fish. Brit. Mus. viii, 227, 1870 (Jamaica, St. Croix, St. Helena, Zanzibar, China Seas, Sandwich Islands); Streets, Bulletin U. S. Nat. Mus. No. 7, 56, 1877 (Honolulu).

? *Balistes piceus* Poey, Proc. Acad. Nat. Sci. Phila. 180, 1863 (Cuba); Poey, Proc. Acad. Nat. Sci. Phil. 177, 1863 (name only); "Repert. Fis. Nat. Cuba, ii, 435, 1868."

Melichthys ringens Bleek. Act. Soc. Sc. Indo-Néere. vi, Sumatra viii, 69; "Blkr. Atl. Ichth. v, 108, pl. 220, f. 2, 1865"; "Blkr. Balist. pl. vi, f. 2."

Balistes niger "Gthr. Fish. Zanz. 135, pl. 19, f. 1, 1866."

Habitat.—Indian Ocean, China Sea, Sumatra, E. Coast of Africa, Cuba, Zanzibar, Jamaica, St. Croix, St. Helena, Sandwich Islands, Johnston's Island.

Three fine specimens, numbered 26818, 8 inches long.

24.—OSTRACION PUNCTATUM Bl. & Schn.

Ostracion pointillé 'Lacép. i, 442, 445, pl. 21, f. 1, 1798 (Isle de France).

Ostracion punctatus Bl. & Schn. 501, 1801; "Cuv. Règne An."; "Jenyns, Zool. Beagle, Fish. 158; Blkr. Nat. Tyds. Ned. Ind. xi, 108, and Atl. Ichth. Ostræ, 39, pl. 2, fig. 4; Hollard, Ann. Sc. Nat. vii, 165, 1857"; Gthr. Cat. Fishes, viii, 261, 1870 (Ind. Ocean and Archip.; Pacific); Blkr. Conspec. Moluec. Cognit. 22 (Archip. Moluce, Banda; name only); Blkr. Quat. Mem. Ichth. N. Guinée, 4, 22 (name only).

Ostracion lentiginosus Bl. & Schn. 501 (India).
Ostracion meleagris, "Shaw, Zool. v, 428, pl. 172, and Zool. Misc. pl. 253."

Habitat.—Indian Ocean and Archipelago, Zanzibar, India, Archip. Molucc., Banda, New Guinea, Isle of France, Johnston's Island. One fine specimen (26821).

25.—TETRODON MELEAGRIS Lacép.

Tetrodon meleagris Lac. i, 476, 505, 1798 (Seas of Asia); Bl. & Schn. Syst. Ichthy. 507, 1801 (Asia); "Richards, Voy. Sulphur, Fish. 122, pl. 57, figs. 1-3"; Gthr. Cat. Fish. Brit. Mus. viii, 299, 1870 (Polynesia).
Tetrodon lacrymatus "(Cuv.) Quoy & Galm. Voy. Uran. Poiss. 204."

Habitat.—Asia, Polynesia, Johnston's Island.
Three fine specimens (26811).

26.—DIODON HYSTRIX L.

Diodon hystrix, L. Syst. Nat. i, 413; Bl. Naturgeschichte ausländ. Fische. i, 91, 1786 (name only); Bris. Barnev. Rev. Zool. 141, 1846"; Gthr. Cat. Fish. Brit. Mus. viii, 306, 1870 (Gaboon, Fernando Po, Calabar, West Indies, Jamaica, Cape Seas, Amboyna, Indian Ocean, Society Islands); Bean (No. 23779), Proc. U. S. Nat. Mus. iii, 75, 1880 (Bermuda ; name only); Poey, Proc. Acad. Nat. Sci. Phil. 179, 1863 (name only).
Diodon atinga Bl. tab. 125, i, 91 (name only); Bl. & Schn. Syst. Ichth. 511, 1801 (American Seas, Cape of Good Hope); "Kaup Wiegm. Arch. 227, 1855 (not L.)"; Poey. Proc. Acad. Nat. Sci. Phil. 179, 1863 (name only).
Diodon plumieri, Lacép, ii, pp. 2, 10 ; i, pl. 3, fig. 3.
Diodon brachiatus Bl. & Schn. p. 513, 1801.
Diodon punctatus, "Cuv. l. c. 132; Blkr. Conspec. Molucc. Cognit. 21 (Amboyna, Ternate, Ceram, Archip. Molucc., Banda ; name only); Bleeker, Verh. Bat. Gen. xxiv, Blootk. p. 19"; Blkr. Elfde. Bijd. Visch. Celebes, 4 (name only); Blkr. Zez. Bijd. Visch. Sumatra (Lampong); Blkr. Beschrij. Visch. Amboina, pp. 8, 23 (name only); Blkr. Achtste Bijd. Visch. Amboina, 28 (name only); Blkr. Tweede Bijd. Ichth. Fauna Ba toë, 4 (name only).
Holocanthus hystrix "Gronov. Syst. ed. Gray, 27."
Paradiodon hystrix "Bleek. Atl. Ichth. Gymnod, 66, pl. 3, fig. 2."

Habitat.—Gaboon, Fernando Po, Calabar, West Indies, Jamaica, Cape Seas, Amboyna, Indian Ocean, Society Islands, Cape of Good Hope, Burmudas, Ternate, Ceram, Archip. Molucc. Banda, Celebes, Sumatra, Lampong, Batoë, Johnston's Island. (Much of the above synonymy is doubtful.)

Length (26842) 24½ inches.

Spines all more or less flattened except behind the pectorals, where they are round ; in about 18 series between nostrils and dorsal fin. First spine behind nostril, 2¼ in pectoral ; shorter and stronger spines in front of dorsal, becoming longer again on tail; spines behind pectoral about as long as that fin.

Color, in spirits, purplish dusky above and on sides; belly light; back, sides, and fins with small dark spots, much more numerous than the spines; lips purplish brown with small dark spots.

A specimen (28267) from Mazatlan, $10\frac{1}{2}$ inches in length, differs in the following particulars: spines behind pectorals somewhat shorter, and all the spines more keeled; more spines on interorbital space; color darker above and the spots larger, scarcely more numerous than the spines.

One large specimen.

27.—PLATOPHRYS MANCUS (Broussonet) Smith & Swain.

Pleuronectes mancus Brouss. Ichth. description on figure, 1782 (Pacific). (Not *Rhomboidichthys mancus* Gthr.)

Habitat.—Pacific Ocean.

Head $3\frac{2}{7}$ ($4\frac{1}{4}$); depth 2 ($2\frac{1}{4}$); length (26838) 16 inches. D. 98; A. 78; scales about 95; Br. 6.

Body elliptical, the profile continuous with the dorsal curve, the snout projecting and the nasal bones forming a prominent knob; ventral outline a regular and gentle curve from gill-opening to caudal peduncle; lower jaw produced beyond upper, a pointed knob below and behind symphysis.

Head not much higher than long; mouth moderately oblique, small for a large mouthed species, the maxillary reaching little beyond anterior rim of eye, $2\frac{2}{3}$ in head. Pointed teeth in two series in each jaw, those of the inner and larger series becoming somewhat smaller posteriorly, the teeth on maxillary not extending as far back on the blind side; the outer series of few small teeth. Eyes small, the lower orbit 7 in head, the upper one slightly smaller; the lower orbit wholly in advance of the upper; the concave interorbital space $2\frac{5}{6}$ in head; the orbital rim a sharp ridge without distinct knobs.

Nostrils apparently wanting. Cheeks and opercles more or less scaly. Gill-rakers rather long, the length of longest 2 in upper orbit; 10 on lower part of arch, none above.

Scales cycloid, not deciduous, similar on both sides, but without accessory scales on the blind side.

Dorsal fin beginning on the snout, the first ray on the blind side, about as long as superior orbit, the rays gradually increasing in height to the posterior third of the fin, where they are $2\frac{2}{3}$ in head; thence rapidly decreasing to end of fin. Anal similar, its highest rays not opposite the highest part of dorsal, but a little farther back. Pectoral of eyed side falcate, the second ray one-fourth longer than head, produced into a filament; pectoral of blind side $1\frac{5}{6}$ in head. Ventrals moderate, when depressed reaching past front of anal. Caudal bluntly pointed, $1\frac{3}{5}$ in head.

Coloration in spirits: everywhere mottled with gray and brown; the fins (except pectoral on blind side) marked with same colors, but the spots more nearly round and less complicated. On the colored side there is a large, irregular blackish blotch behind pectoral, a round black spot on the lateral line half way between head and caudal fin. About twelve blackish spots at regular intervals on dorsal fin, six or seven similar ones on anal. The ventral on the eyed side is marked like the anal. The colors and spots extend over on the blind side on the nasal bones, premaxillary, chin, and interopercle. The skinny flap in the mouth between the teeth and vomer is also spotted.

One specimen (26838), 16 inches in length.

This species, well described and figured by Broussonet in 1782, seems not to have been seen by any succeeding author. The specific name *"mancus"* has been wrongly transferred by Dr. Günther to the very different *Platophrys heterophthalmus* of the Mediterranean.

INDIANA UNIVERSITY, *April* 4, 1882.

DESCRIPTION OF A NEW CYPRINODONT (ZYGONECTES INURUS), FROM SOUTHERN ILLINOIS.

By DAVID S. JORDAN and CHARLES H. GILBERT.

Zygonectes inurus sp. nov. (29666.)
> *Zygonectes melanops* Jordan, Bull. Ills. Lab. Nat. Hist. No. 2, 52: not *Haplochilus melanops* Cope,=*Gambusia holbrooki* (Agassiz).
> *Zygonectes melanops* Jord. & Gilb. Syn. Fish. N. A. 340.

Closely allied to *Zygonectes dispar* Agassiz.

Body rather short and high, compressed, the back considerably arched; caudal peduncle deep and compressed; head small, much narrowed forwards; interorbital space twice as wide as diameter of orbit; eye rather large, $3\frac{1}{2}$ times in head—as long as snout, which equals mandible; teeth small, in villiform bands, the outer series not at all enlarged; height of caudal peduncle at vertical behind anal fin $1\frac{1}{4}$ in head; at base of caudal $1\frac{1}{2}$ in head.

Dorsal small, posteriorly inserted; distance from its origin to snout twice that to base of caudal; length of base of fin $2\frac{3}{4}$ in head; the vertical from origin of dorsal passes through middle of anal base. Distance from origin of anal to base of caudal $1\frac{1}{2}$ times in that to tip of snout; length of anal base equalling one-half that of head; its longest ray two-thirds head; pectoral broad, reaching beyond base of ventrals, $1\frac{1}{4}$ in head.

Scales in regular series, the humeral scale not enlarged; 29 oblique series of scales from scapula to base of caudal fin; 9 in an oblique series from vent to middle of back.

Head $3\frac{1}{6}$ in length; depth $3\frac{1}{2}$. D. 6; A. 9; scales 29.9.

Color: Brownish, light on belly and sides of head; sides and back with a few scattered dark brown specks, these forming inconspicuous series behind pectorals; caudal peduncle punctate with brown specks below; opercles silvery; a very decided, well-defined, brownish-black bar through eyes and across cheeks, the bar $\frac{2}{3}$ as wide as orbit; vertical fins with irregular cross series of brown dots.

Several specimens of this species were collected by Prof. S. A. Forbes in the streams of Southern Illinois. A single specimen, 2 inches long (No. 29066, U. S. Nat. Mus.), from Cache River, serves as the type of the species.

NOTES ON BIRDS COLLECTED DURING THE SUMMER OF 1880 IN ALASKA AND SIBERIA.

By TARLETON H. BEAN.

The collection which is the subject of the present paper was made by the writer while investigating the fish and fisheries of Alaska for the United States Fish Commissioner, in company with the Coast Survey party commanded by Mr. W. H. Dall.

Owing to the engrossing nature of the primary object of my inquiry and the limited number of days passed in port, there was little opportunity for collecting and observing birds. However, as fish were not plentiful north of the Arctic circle and birds were comparatively abundant and desirable for the Museum, much of my time was devoted to making bird skins while in that region. Especial effort was made, also, to procure a good series of skins of the species of *Melospiza*, inhabiting the mainland and islands of the Territory. The region in which *M. fasciata rufina* occurs is so interesting and rich ichthyologically that birds were necessarily neglected, and this sparrow is represented by only a few skins. The insular species (*cinerea*), on the other hand, came in for a larger share of attention.

In order to give an idea of the time which one may ordinarily devote to bird-collecting who is not sent upon that special duty, I will state here the number of days which were devoted mainly or partly to that work during the six months of our cruise: May, 2 days; June, 11; July, 11; August, 9; September, 9; October, 7; in all 49 days.

Although the number of species secured is small (less than a third of the whole number known to occur in Alaska), there are some interesting features about the collection. Many of the species here mentioned are from localities north of the Arctic circle, and some of them were not previously recorded from the Territory. The specimen of *Eurinorhynchus pygmæus* obtained by our party in Plover Bay was the first secured for an American museum, and is in a plumage which has not yet been illustrated. Six young individuals of *Saxicola œnanthe* were found between Port Clarence and Cape Lisburne. *Empidonax difficilis* and *Buteo borealis calurus* were obtained for the first time in the Territory. The range of *Actodromas acuminata* was extended northward to Port Clarence.

Larus marinus was found to be common on Unalashka Island, though previously unknown in Alaska.

Diomedea melanophrys was seen within 1,060 miles west of Cape Mendocino, California, and may safely be claimed for the fauna of the United States. The nests of *Hirundo erythrogastra* and *Melospiza fasciata rufina* are worthy of more extended examination than I have been able to give them.

On the 23d of October, when about 700 miles south of Unalashka Island, a small flock of geese was seen flying towards the southeast, and sometimes resting on the water. We did not sail close enough to make them out, but there is little doubt that they belonged to the genus *Anser*. Mr. Dall, Mr. Baker, Captain Herendeen, and I looked at them with a glass, and all agreed as to the identification. Owing to the distance, we could not determine the species.

So far as most of the species are concerned I have simply transcribed my hastily made field notes. In a few cases, however, I have supplemented these fragments by subsequent investigations in the Museum. Even in this small collection there still remains some material that is worthy of the attention and will, doubtless, eventually receive the notice of an ornithologist.

To Mr. Dall I owe the opportunity of joining his party, and he, as well as his assistants, contributed as much as possible to the collection; the probability of finding *Eurinorhynchus* at Plover Bay was first suggested by Mr. Dall.

I am much indebted to Mr. Robert Ridgway for aid in determining the species collected by me and for advice in the preparation of these notes.

1. HYLOCICHLA ALICIÆ Baird.

81334 (3493) ♂. St. Paul, Kodiak, July 13, 1880.

Found in the timber, in the vicinity of the potato fields of the St. Paul people.

2. HYLOCICHLA UNALASCÆ (Gmel.) Ridgw.

81333 (3236) ♀. Sitka, June 15, 1880.
81331 (3340) ♂. Chugachik Bay, Cook's Inlet, July 1, 1880.
    (3341) ♂ ad.   "   Cook's Inlet,   "   "   "
81332 (3342) ♀.   "   "   "   "   "
81330 (3492) ♂. St. Paul, Kodiak,   " 12,   "
81692 (3428) alcoholic, Wooded Id., Kodiak, " 13,   "

At Chugachik Bay this species was not uncommon in the little grove of Sitka spruce on the spit near our anchorage, associated with *Ægiothus linaria*.

3. MERULA MIGRATORIA (Linn.) Sw. & Rich.

I saw one of this species at Sitka, May 29, 1880, in the vicinity of Piseco Lake.

4. HESPEROCICHLA NÆVIA (Gmel.) Baird.

(3235) ad. ♂. Sitka, Alaska, June 15, 1880.

Found near the mouth of Indian River. The only one seen, although,

July 8, 1882.

it is common. Its mouth was filled with insects and an earthworm. Measurements from the fresh bird: Length, 9.87; extent, 15.50; tarsus, 1.31; middle toe and claw, .94.

5. SAXICOLA ŒNANTHE (Linn.) Bechst.

81336 (3639) ♂ juv. Cape Lisburne, Alaska, Arctic O., Aug. 21, 1880.
81337 (3640) ♂   "   "   "   "   "   "   "   "
      (3641) ♂   "   "   "   "   "   "   "   "
81338 (3743). Port Clarence,   "   Sept. 6,   "
      (3787). Chamisso Id., Kotzebue Sd., Alaska, Aug. 31,   "
      (3788).   "   "   "   "   "   "   "   "

At Cape Lisburne this bird was found with *Anthus ludovicianus*, but was not nearly so abundant as the titlark. I saw it also 10 miles to the eastward of Cape Lisburne. Its movements are similar to those of *Anthus*. It was feeding here on grass seeds and fruit of *Saxifraga*, and was, in consequence, excessively fat.

On Chamisso Island the only two seen were secured; they were on a sand and gravel beach and the low bluffs adjacent.

Capt. E. P. Herendeen went ashore, September 1, on the east side of Choris Peninsula and reported having seen stonechats, one of which he wounded but could not get. Owing to the rain he found it useless to attempt to collect small birds.

At Port Clarence only one was seen; this was near the beach on the west side of the spit. The day was cold and windy, with a little rain and some snow. *Plectrophanes nivalis* was of more frequent occurrence, though not plentiful; *Centrophanes lapponicus* was rather common.

6. PARUS ATRICAPILLUS SEPTENTRIONALIS (Harris) Allen.

81679 (3352) alcoholic. Port Chatham, Cook's Inlet, July 6, 1880.
81680 (3428) alcoholic. St. Paul, Kodiak, July 13, 1880.

The Port Chatham specimen was shot with a rifle by Mr. Baker in the timber near Refuge Cove.

On Kodiak Island we found the species in the timber near the potato ground of the St. Paul people.

7. ANORTHURA ALASCENSIS (Baird) Coues.

81339 (3896) ♂ ad. Iliuliuk, Unalashka, Oct. 13, 1880.
81340 (3897) ♀.   "   "   "   "   "

This was one of only eight species of land birds seen by me October 6 to 18, 1880. The other birds were *Passerculus sandwichensis, Melospiza cinerea, Corvus corax carnivorus, Leucosticte griseinucha, Arquatella Couesii, Heteroscelus incanus,* and *Haliæetus leucocephalus. Lagopus rupestris* was, of course, present, but I did not find it.

Anorthura was more abundant at Chernoffsky than at Iliuliuk. At both places I found it frequenting the rocks near the water's edge and the faces of cliffs looking seaward.

8. MOTACILLA OCULARIS Swinhoe.

81341 (3595) ♀ (?). Port Providence, Plover Bay, Siberia, Aug. 14, 1880.

Only one individual certainly seen; occurring with *Budytes flava* in the vicinity of the native summer tents; very hard to approach. Judging from the uniformly small number of specimens of this bird secured by collectors in Plover Bay, the species seems to be rare in that locality.

9. BUDYTES FLAVA (Linn.) Gray.

81342 (3594). Port Providence, Plover Bay, Siberia, Aug. 13, 1880.
81343 (3596) ♂. "   "   "   "   "   " 14, "

These wagtails were present in small numbers on the spit at Port Providence, and they were exceedingly shy. They were seldom seen at the tents, but usually in the grass. They were sometimes heard in the air, chirping while in flight.

A single wagtail was seen flying towards the point of the spit at Port Providence, September 13, but I could not determine the species.

It may not be out of place here to remark that, on the 15th of September, Mr. Baker and Captain Herendeen, of the coast-survey party, walked through a divide leading from Port Providence to Moore Lake and saw no birds except ravens. The few land birds still remaining at this port were near the sea-shore.

10. ANTHUS LUDOVICIANUS (Gm.) Licht.

81665. Little Koniushi Island, Shumagins, July 16, 1880.
81344 (3776). Cape Lisburne, Alaska, Arctic O., Aug. 21, 1880.
81673 (alcoholic)."   "   "   "   "   "   "   "
81674   "   "   "   "   "   "   "   "   "
81682 (3647) "   "   "   "   "   "   "   "   "

The Little Koniushi Island example was found at the top of the ridge overlooking Northwest Harbor, at least 1,000 feet above the sea level.

At Cape Lisburne most of the specimens seen of this wagtail were in a little valley through which runs a small stream, and on the low plateau east of this stream. They were feeding on seeds of *Saxifraga* mainly.

11. DENDRŒCA ÆSTIVA (Gmel.) Baird.

81675 (1539) alcoholic. St. Paul, Kodiak Island, July 9, 1880.

Found in the Sitka spruce near the village of St. Paul.

12. MYIODIOCTES PUSILLUS PILEOLATUS (Pall.) Ridgw.

81345 (3295) ♂. Port Mulgrave, Yakutat Bay, June 24, 1880.
81676 (3432) alcoholic. St. Paul, Kodiak, July 13, 1880.

Common at Port Mulgrave around the head of the harbor; found in Sitka spruce near St. Paul.

13. HIRUNDO ERYTHROGASTRA Bodd.

(3533) ♂ & ♀, alcoholic. Cave Rock, Unalashka, July 28, 1880.
(3510) nest of above, with 4 young, July 28, 1880.

A pair of the above species of swallow was observed for some time circling around Cave Rock, on Amaknak Island, near Iliuliuk. In the mouth of the cave was the nest here to be described. The swallows were not seen on the nest, but there is no reasonable doubt that the pair obtained were the owners.

The nest in its present condition is 5½ inches long; the greatest depth of the front wall is 3 inches; the grass lining, on which is placed an additional cushion of feathers, is 3½ inches long and nearly 3 inches wide on top; the back wall of the nest contains only grasses and sea-weeds; the mud wall separates readily into only four layers. The mud in its dry state crumbles very readily, and could not have had great cohesive power originally. To remedy this defect, the pellets were intermingled with a long, narrow, red sea-weed which has considerable strength and furnishes a sticky secretion well adapted for holding them together, and the same sea-weed was employed between the layers. The mud was evidently found on the shore close to high-water mark, as it contains numerous small shells which may be always seen in such location. The grass seems to be mainly the common wild rye of the vicinity. The inner lining is ample and prettily arranged. It consists of soft feathers of young bald eagle, raven, and gull, tastefully inter-twined, and forming a shallow, but luxurious, cushion.

The structure and situation of this nest are similar to what Mr. Ridg-way observed at Pyramid Lake and the Ruby Mountains;* but the limited number of mud layers of the Unalashka nest and the introduction of a glutinous sea weed to supply the defective cohesive power of the pellets forming the wall, afford a new illustration of the faculty which this swallow possesses of adapting itself to the conditions of its environment.

14. LEUCOSTICTE GRISEINUCHA (Brandt) Baird.

81349 (3487) ♀. Little Koniushi Id., Shumagins, July 16, 1880.
81348 (3889) ♀. Iliuliuk, Unalashka, Oct. 7, 1880.
     (3890) ♀. "   "   " " "

Found on the low ground on Little Koniushi, near Northwest Harbor. Not common.

Abundant at Iliuliuk late in July and also in October; frequents the hillocks and cliffs, and comes into the village yards.

---

* Orn. 40th Parallel Surv., p. 441.

15. ÆGIOTHUS CANESCENS EXILIPES (Coues) Ridgw.

81678 (3679) 5 juv., in alcohol. Chamisso, Id., Aug. 31, 1880.
81362 (3752) ♂ ad. Chamisso Id., Kotzebue Sound, Aug. 31, 1880.
    (3753).   "   "   "   "   "   "   "
    (3755) ♂ juv.,   "   "   "   "   "   "   "
    (3756) ♂.   "   "   "   "   "   "   "
81365 (3757) ♂ juv.   "   "   "   "   "   "   "
    (3758) ♀ (?)   "   "   "   "   "   "   "
    (3759) ♀ (?) juv., first plumage. Chamisso, Id., Kotzebue Sound, Aug. 31, 1880.
81363 (3760)   "   "   "   "   "   "   "
    (3761)   "   "   "   "   "   "   "

From the above list of skins, secured on Chamisso Island, it will be observed that these red polls were quite abundant there; indeed it was the only land bird found in numbers. Some grouse were seen, but not by me. A single *Passerculus* was observed and secured. Two young stone chats (*Saxicola œnanthe*) were obtained; no others were seen. On this island the land rises gradually from the sides and ends, so that a very regular curve is shown. With the exception of numerous hummocks, which greatly impede walking, there are no serious hinderances to collecting. The island is covered with grass, alder, and willow, and there is also a dwarf birch. Wild rye is present in considerable patches in some places. Salmon berries, whortleberries, empetrum, and another berry which is not edible, were all abundant. There are some little rills of tolerably good water. We found *Ægiothus* most abundant, swaying on the stalks of wild rye and in the small trees lining the rivulet banks. There are some cliffs in a tumble-down condition, and occasional deep cuts between and small stretches of sand beach. On these cliffs were immense numbers of puffins.

Many of these skins of the white-rumped redpoll show a great amount of wearing of the feathers, particularly of the tail, and in one bird the tail is finely graduated.

16. ÆGIOTHUS LINARIA (Linn.) Caban.

    (3338) ♂ ad. breeding plumage. Chugachik Bay, Cook's Inlet, July 1, 1880.
81367 (3339) ♀ ad. Chugachik Bay, Cook's Inlet, July 1, 1880.
81366 (3754) ♂ ad. Chamisso Id., August 31, 1880.

On the spit adjoining that portion of Chugachik Bay which is known as Ugolnoi Bay, this bird was observed sparingly in a small patch of Sitka spruce.

Out of 15 *Ægiothi* secured on Chamisso Island, only one proved to be *linaria*; *Æ. linaria exilipes* was the common form.

17. PLECTROPHANES NIVALIS (Linn.) Meyer.

81347 (3483) juv. Little Koniushi Id., Shumagins, July 16, 1880.
81346 (3484) ♂.   "   "   "   "   "   "   "
181666·   "   "   "   "   "   "   "

(3570). Port Providence, Plover Bay, Siberia, Aug. 14, 1880.
(3749) ♂. Port Clarence, Alaska,   Sept. 6,   "
(3751).   "   "   "   "   "   "
81352 (3790) ♂.   "   "   "   "   "   .
(3784) ♂ (?). Point Belcher, "Arctic O.,   Aug. 27,   "
(3770) ♂. Point Belcher, Alaska, Arctic O., Aug. 27, 1880.
81351 (3771) ♂.   "   "   "   "   "   "   "
81354 (3773) ♀ (?). Icy Cape,   "   "   "   " 25,   "
(3592). Port Providence, Plover Bay, Siberia, Aug. 13, 1880.
81353 (3593) ♂.   "   "   "   "   "   "   "   "
(3570)   "   "   "   "   "   Aug. 14,   "
81355 (3826) ♀.   "   "   "   "   "   Sept. 12,   "

On Little Koniushi Island I obtained one adult male and several young birds—all of them from the top of the ridge several times referred to. The young were able to make short flights only.

At Plover Bay *Plectrophanes nivalis* was found in small numbers, generally feeding on refuse near the summer tents; difficult to shoot on that account, and when it flew away it was hard to overtake, since it remained only a few seconds in once place.

Near Icy Cape, Alaska, *Plectrophanes* was again scarce, shy, and hard to shoot. One of these buntings, which was followed for a long time, but not secured, showed a nearer approach to the winter plumage than the individual brought down.

At Point Belcher *P. nivalis* was more abundant than at any of the other localities where we obtained it, although even here there were comparatively few, *Centrophanes lapponicus* being much more common.

At Port Clarence few of the species were seen; they were usually found not far from the beach, not going inland on the spit like *Centrophanes*.

As we approached Cape Upright, Saint Mathew Island, September 22, 1880, small flocks of the snow bunting from the land flew around the vessel; they were in winter plumage, or nearly so.

18. CENTROPHANES LAPPONICUS (Linn.) Caban.

81361 (3588) juv. ♂ (?). Belkoffsky, Aliaska, July 23, 1880.
81360 (3589) juv.   "   "   "   "   "   "
(3774) ♂.   Cape Lisburne, Alaska, Aug. 22   "
(3775) ♀ (?).   "   Alaska, Aug.   "   "
81356 (3777) ♂.   "   Alaska, Aug. 21   "
81683 (3647)   "   Alaska, Aug 22   "
81358 (3785) ♂. Point Belcher, Alaska, Aug. 27, 1880.
(3786) ♂.   "   "   "   "   "   "
(3767) ♂. Point Belcher, Arctic O., Aug. 27, 1880.
81359 (3768) ♀ (?). "   "   "   "   "   "
81677 (3671) alcoholic. Point Belcher, Alaska, Aug. 27, 1880.
81357 (3824) ♂. Port Clarence, Sept. 9, 1880.

The specimens obtained at Belkoffsky, which is on the peninsula of Aliaska, were young; the species was by no means common at the date of my collecting, but it was more abundant than any other land bird except *Passerculus sandwichensis*. The birds were on the low plateau bordering the sea-shore near the village.

At Port Clarence the bird was common in small flocks, feeding on seeds, usually near the small lagoons which are present on the spit.

At Cape Lisburne and 10 miles to the eastward I observed numerous examples on the 21st and 22d of August, feeding, as usual, on seeds of species of *Saxifraga*, and congregating in small flocks.

At Point Belcher, August 27, there were more of this species than at any other place visited by us. They were, as elsewhere, feeding on grass seeds and the seeds of flowering plants, among which *Saxifraga* was most common. Number 81358 of this lot is worthy of mention on account of the deformity of its bill; the gonys is nearly twice as long as the culmen and decidedly hooked.

19. PASSERCULUS SANDWICHENSIS (Gmel.) Baird.

81687 (3501) alcoholic. Belkoffsky, Aliaska, July 23, 1880.
81371 (3590) ♀ ad. Belkoffsky, Aliaska, July 23, 1880.
81370 (3881) ♀ ad. Chernoffsky, Unalashka, Oct. 1, 1880.

Moderately common at Belkoffsky as well as at Chernoffsky, on the island of Unalashka. Also common during our stay at Iliuliuk, on the same island, but no skins of it were made there.

20. PASSERCULUS SANDWICHENSIS ALAUDINUS (Bp.) Ridgw.

(3353) ♂ ad. Chugachik Bay, Cook's Inlet, July 1, 1880.
(3354) ♀ "   "   "   "   "   "   " "
(3355) ♀ "   "   "   "   "   "   " "
81372 (3356) ♀ ad.   "   "   "   "   "   " "
(3496) ♂ ad. St. Paul, Kodiak, July 13, 1880.
81369 (3497) ♂ ad. St. Paul, Kodiak, July 13, 1880.
81368 (3762) ♂ ad. Chamisso Id., Kotzebue Sound, Aug. 31, 1880.

The spit in Chugachik Bay, on which I collected birds July 1, 1880, is low and level, its beaches higher than the interior. At some high tides the sea breaks over and carries with it immense numbers of fish, which are left stranded when the waters recede. This occurred a few days before our visit, and we saw thousands upon thousands of fishes lying uncovered on the ground. Great quantities of drift-wood are found here. Wild wheat abounds, and there are many pretty flowering plants, among which are serrana, violets, chickweed, vetch, and Jacob's ladder. There is also a little grove of Sitka spruces, in which I found the redpolls (*Ægiothus*) and thrushes. *Passerculus* was quite abundant in the wild wheat. On this spit was found the young eider which I have numbered in my catalogue.

On the 2d of July we visited Glacier spit, distant 9 miles from our anchorage. Here a pair of eagles had a nest on one of the tall pines. A small plover, resembling the killdeer and with similar actions, was shot but badly mutilated and finally lost.

The specimen of *Passerculus* obtained on Chamisso Island was the only one seen there.

21. ZONOTRICHIA CORONATA (Pall.) Baird.

81373 (3490) ♀ ad. Popoff Id., Shumagins, July 18, 1880.
81690 (3428) (alcoholic). St. Paul, Kodiak Id., July 13, 1880.
81693 (3429)   "   "   "   " 9, "
81714 (3429 bis.) "   "   "   " 9, "
81686 (3430)   "   "   "   " 12, "

Common on the island of Kodiak.

22. JUNCO OREGONUS (Towns.) Scl.

81350 (3238) ♂. Sitka, June 15, 1880.
81681 (1404) alcoholic. Sitka, June 15, 1880.

23. MELOSPIZA FASCIATA RUFINA (Brandt) Baird.

(3299) (1451, alc.) ♂. Port Althorp, June 19, 1880.
81386 (3300) ad. ♀. Port Althorp, George island, June 19, 1880.
(3251) nest containing 4 young, the young preserved in alcohol. Port Althorp, June 19, 1880.
81380 (3358) ♀ ad. Graham Harbor, Cook's Inlet, July 4, 1880.
81385 (3357) ♂ juv.   "   "   "   "   "   "   "

Common at Graham Harbor; frequently seen feeding on the beach.

The nest found on George Island (Port Althorp) is made of coarse grasses, loosely laid together below, and interlaced with strips of what appears to be the leaf of *Panax horridum*, and with the light inner bark of the same. The superstructure is of fine grasses more intimately woven. The greatest depth of the nest is 4 inches, and its diameter is from 5 to 7 inches. The inside lining is 2½ inches across the top and 2 inches deep. The nest was supported by a dead stalk of *Panax* and concealed in the tall, coarse grass which is abundant in that locality.

Number 81385 bears a wonderfully close resemblance in coloration and general appearance to number 81384 from Kodiak, which is supposed to be *cinerea;* it will be observed, however, that there is considerable difference in the measurements.

As nearly as I can determine from the material in the collection, the conclusions expressed in the History of North American birds by Baird, Brewer, and Ridgway are fully justified. There is a large series of skins of *cinerea* from Kodiak and Unalashka, but the representation of *fasciata rufina* is still unsatisfactory, and the song sparrow of the western islands of the Aleutian chain has a meager showing. A study of the collection in its present state, as already remarked, will lead us to the adoption of the views advanced in the History of North American Birds: *Melospiza fasciata rufina* is notably smaller than *M. cinerea* in its wing, tail, tarsus, middle toe, and all measurements of the bill; in coloration, also, adult birds of the two species differ greatly.

The following table of measurements deals with all the skins now accessible in the collection which have been referred to *fasciata rufina*. For convenience of reference, the average measurements of the large series of skins of *M. cinerea* are brought on the same sheet:

Measurements.

Species, *Melospiza fasciata rufina.*

Catalogue number.	Locality.	Sex and age.	Depth of bill through base.	Wing from carpal joint.	Tail to basal end of feathers.	Tarsal joint.	Middle toe.	Gonys.	Culmen.	Bill from nostril.	Nature of specimens measured.	Date.
F1385	Graham Harbor, Cook's Inlet	juv. ♂	.25	3.10	2.50	1.00	.70	.31	.54	.35	Skin	July 4, 1880.
81380	" "	ad. ♀	.30	3.10	2.80	1.00	.72	.41	.66	.43	"	" 4, "
68287	Lituya Bay		.25	2.70	2.50	.93	.66		.60	.39	"	May 16, 1874.
46006	Sitka		.25	2.70	2.60	.98	.70	.38	.60	.40	"	Oct., 1865.
46004	"		.25	2.80	2.78	.90	.70	.41	.55	.35	"	" "
46011	"		.24	2.60	2.50	.95	.60	.35	.57	.38	"	Sept., 1865.
	"		.28	2.95	2.98	.95	.70	.34	.63	.45	"	" "
46014	"		.21	2.90	2.85	1.00	.76	.40	.55	.31		Dec., "
46009	"		.21			.95		.31				Sept., "
	Average of the specimens		.25	2.83	2.68	.98	.70	.36	.59	.38		
	Average of *M. cinerea*		.30	3.25	3.10	1.07	.79	.44	.67	.47		

24. MELOSPIZA CINERA (Gm.) Ridgw.

(3494) ♂.	St. Paul, Kodiak,				July 9, 1880.		
83184 (3495) ♂ juv.	"	"	"		"	13,	"
81382 (3488) ♂ ad.	Little Koniushi Id., Shumagins,				"	16,	"
81381 (3489) ♀ ad.	"	"	"	"	"	"	"
81383 (3491).	Popoff	"	"		"	18,	"
81377 (3882) ♀ ad.	Chernoffsky, Unalashka Id.,				Oct.	1,	'
(3883) ♀.	"	"	"		"	"	'
(3884) ♀.	"	"	"		"	"	.
81376 (3885) ♂ ad.	"					2,	'
(3886)	"					"	'
81378 (3887) ♀ ad.	Iliuliuk,	..	..			7,	'
(3888) ♀.	"	..	..			"	'
(3892) ♂.	"				.	12,	'
(3898) ♂.	..				'	13,	'
(3899) ♂ ad.	"	..			:	"	'
81375 (3900) ♂ ad.	"	"				"	.
(3901) ♀.	..	"					
81379 (3902) ♀ ad.	"						
(3903) ♀.	..						
(3904) ♀.	..				"	.	
(3905) ♂.	..				16,	'	
(3906) ♂.	.:	..			"	'	

A nest of this sparrow containing 4 eggs was sent over from Wooded Island, Kodiak, July 11, 1880, by Nicolas Pavloff.

On little Koniushi Island it was not uncommon on the low ground bordering Northwest Harbor. Not seen on the beach, because there is none, properly speaking, and small crustacea would scarcely occur in sufficient numbers to attract these sparrows.

It frequents the sea-shore at Chernoffsky and Iliuliuk, feeding among sea-weeds.

Upon examination of the measurement tables of *M. cinerea* the peculiarities of birds from Kyska and Attu will appear; the small bill, even of the adult bird, is noteworthy, and it is to be hoped that sufficient material will soon be obtained to determine the extent and value of this divergence.

Measurements.

Species, *Melospiza cinerea.*

Catalogue number.	Locality.	Sex and age.	Depth of bill through base.	Wing from carpal joint.	Tail to basal end of feathers.	Tarsal joint.	Middle toe.	Gonys.	Culmen.	Bill from nostril.	Nature of specimens measured.	Date.
54539	Kodiak		.28	3.30	3.10	1.00	.80	.41	.64	.43	Skin	Aug. 12, 1868.
70097	''		.26	3.16	3.40	1.09	.78	.50	.70	.48	''	Sept. 25, 1868.
54536	''	♂ ♀	.30	3.35	3.12	1.10	.80	.41	.63	.43	''	May 24, ''
52477	''		.30	3.36	3.20	1.05	.73	.45	.72	.50	''	Aug. 27, 1867.
52476	''			3.00	2.91	1.03	.80				''	Aug. 5, 1842.
60161	''	juv.	.28	3.10	3.10	1.10	.80	.46	.65	.48	''	June 10, 1868.
52479	''	♂	.30	3.13	3.10	1.05	.79	.51	.74	.51	''	July 13, 1880.
81384	''	♂	.28	3.20	3.10	1.00	.83	.38	.62	.40	''	'' 9,
(3494)	''		.30	3.20	3.35	1.05	.80	.39	.61	.41	''	——, 1844.
60162	''		.31	3.20		1.05	.86	.42	.70	.50	''	
	Average of the specimens		.29	3.20	3.13	1.05	.80	.44	.67	.46		
81381	Little Koninshi Id	ad. ♀	.28	3.20	3.20	1.08	.78	.48	.70	.50	Skin	July 16.
81383	Popoff Id	juv.	.30	3.20	2.95	1.10	.80	.40	.70	.47	''	'' 18.
62717	do	ad. ♂	.30	3.23	3.10	1.08	.72	.46	.70	.48		June 22.
	Average of the specimens		.29	3.21	3.08	1.09	.77	.45	.70	.48		

Measurements.

Species, *Melospiza cinerea.*

Catalogue number.	Locality.	Sex and age.	Depth of bill through base.	Wing from carpal joint.	Tail to basal end of feathers.	Tarsal joint.	Middle toe.	Gonys.	Culmen.	Bill from nostril.	Nature of specimens measured.	Date.
67801	Iliuliuk	ad. ♂	.31	3.30	3.20	1.10	.80	.50	.70	.56	Skin	May 13, 1874.
73497	"	ad. ♂	.29	3.40	3.30	1.05	.80	.45	.70	.48	"	June 5, 1877.
(3906)	"	ad. ♂	.30	3.40	3.20	1.10	.75	.43	.65	.44	"	Oct. 16, 1880.
(3802)	"	ad. ♂	.30	3.45	3.30	1.10	.85	.37	.64	.46	"	" 12, "
81374	"	ad. ♂	.30	3.40	3.30	1.15	.82	.46	.62	.43	"	" 13, "
79012	"	ad. ♂	.30	3.35	3.00	1.00	.80	.46	.68	.50	"	
61320	"	ad. ♂	.30	3.40	3.30	1.10	.80	.45	.71	.50	"	Dec. 19, 1871.
67799	"	ad. ♂	.30	3.20	3.10	1.08	.83	.40	.80	.50	Mounted	May 14, 1874.
81375	"	ad. ♂	.31	3.30	3.40	1.05	.80	.44	.79	.46	Skin	Oct. 13, 1880.
78905	"	ad. ♀	.30	3.30	3.00	1.10	.80	.50	.70	.48	"	
78910	"	ad. ♀	.30	3.10	3.10	1.04	.80	.47	.63	.50	"	
78898	"	ad. ♀	.29	3.20	2.90	1.07	.80	.40	.71	.50	"	
78914	"	ad. ♀	.31	3.25	2.47	1.10	.83	.41	.68	.42	"	
78906	"	ad. ♀	.28	3.20	2.80	1.05	.80	.48	.70	.46	"	May 11, 1874.
67797	"	ad. ♀	.30	3.10	3.40	1.12	.76	.48	.70	.50	"	" 1877.
73495	"	ad. ♀	.30	3.20	3.20	1.05	.80	.45	.70	.50	"	Oct. 13, 1880.
(3063)	"	ad. ♀	.30	3.20	3.10	1.00	.80	.44	.70	.50	"	" 7,
81379	"	ad. ♀	.30	3.18	3.10	1.10	.76	.40		.50	"	
(3904)	"	ad. ♀	.30	3.20	3.08	1.08		.41			"	
81378	"	ad. ♀	.30		3.20							
	Average of all		.30	3.26	3.17	1.07	.80	.44	.65	.46		
	Average of 9 males		.30	3.34	3.23	1.08	.81	.44	.70	.48		
	Average of 11 females		.30	3.20	3.11	1.07	.80	.44	.71	.48		

Measurements.

Species, *Melospiza cinerea.*

Catalogue number.	Locality.	Sex and age.	Depth of bill through base.	Wing from carpal joint.	Tail to basal end of feathers.	Tarsal joint.	Middle toe.	Gonys.	Culmen.	Bill from nostril.	Nature of specimens measured.	Date.
86	Chermofsky	ad. ♂	.30	3.40	3.20	1.08	.80	.43	.66	.49	Skin	Oct. 2, 1880.
87	"	♀	.29	3.25	3.10	1.08	.75	.40	.60	.48	"	Oct. 1, 1880.
88	"	♂	.30	3.20	3.00	1.08	.80	.41	.62	.43	"	" " 2
89	"	♀	.28	3.10	3.00	1.10	.80	.44	.60	.44	"	
90	"	♀	.30	3.25	3.10	1.08	.78	.41	.62	.44	"	
	Average of the specimens		.29	3.24	3.05	1.08	.79	.42	.62	.45		
85591	Atkha	ad. ♂	.30	3.35	3.20	1.10	.80	.38	.66	.50	Skin	May 9,
85588	"	♂	.31	3.28	3.28	1.08	.84	.48	.70	.50	"	
85590	"	♀	.31	3.34	3.10	1.18	.82	.46	.63	.50	"	
85593	"	♀	.35	3.15	2.70	1.10	.75	.43	.68	.50	"	May 9.
85594	"	ad.	.32	3.40	3.00	1.00	.80		.65	.48	"	
	Average of the specimens		.32	3.30	3.06	1.09	.80	.44	.66	.50		
65474	Kyska Harbor	♂ juv.	.30	3.32	3.20	1.10	.78	.40	.58	.40	Skin	July 7,
65476	"	♂ juv.	.28	3.32	2.20	1.00	.80	.34	.55	.37	"	" 15.
	Average of the specimens		.29	3.32		1.05	.79	.37	.57	.38		
65478	Attu	♂ juv.	.26	3.20	2.60	1.05	.80	.36	.59	.39	Skin	June 20.
	General average, excluding the Attu bird		.30	3.25	3.10*	1.07	.79*	.44*	.67*	.47*		

*Kyska and Attu both omitted.

25. PASSERELLA ILIACA UNALASCENSIS (Gm.) Ridgw.

81389 (3319) ♂ ad.   George Island, Port Althorp,   June 19, 1880.
    (3359) ♂.   Graham Harbor, Cook's Inlet,   July   4,   "
81689 (3428) alcoholic.   Wooded Id., Kodiak,   " 13,   "
81716 (3429)   "   St. Paul,   "   " 9,   '
81387 (3498) ♂ ad.   St. Paul, Kodiak,   " 12,   '
    (3499) ♂.   "   "   "   " 13,   '
81388 (3500) ♀ ad.   "   "   "   "   "   '
81688 (3458) alcoholic.   Popoff Id., Shumagins,   " 18,   "
81390 (3485) ♂ ad.   Little Koniushi Id., Shumagins,   " 16,   "
81391 (3486) ♀ ad.   "   "   "   "   "   "   "

The afternoon of July 4, 1880, was decidedly warm for bird-collecting at Graham Harbor, and I have a lively recollection of the difficulties encountered in the timber at that place. The sound of a woodpecker tapping on a dead tree allured me to the chase. There was a lavish display of flowering plants—American cowslip, salmon berry, anemones, and a beautiful blue cranesbill. Frost had nipped the detestable wild ginseng (*Panax horridum*), but unfortunately had not destroyed the entire crop. Mosquitos were at the climax of their capacity for making life wretched. The only bird that could be approached with a degree of comfort was the song sparrow (*Melospiza fasciata rufina*), which frequented the beach and its immediate vicinity. *Passerella* hid in the recesses of the timber, and the way to him led through stinging acres of *Panax*, over legions of briar-beset, snaggy fallen trees, into numberless concealed pitfalls, and within the jurisdiction of the most relentless mosquitos known to man. Bird-collecting here was simply a painful duty, and the reward of honest labor was inadequate, because one was almost sure to lose a bird after killing it in that maze of undergrowth.

On Little Koniushi Island I found this *Passerella* associated with *Plectrophanes nivalis* and *Anthus ludovicianus* on the top of the ridge overlooking Northwest Harbor, about 1,200 feet above the sea level. Walking on this island is simply torture, especially for one who is intent on birds and takes no heed to his steps. The soil is soft and yielding, and in most places thickly covered with loose rocks, scrub alder, and a kind of wild apple—all mingled in such a way as to impede one's progress and multiply his toil. Sitka and Port Mulgrave are little better for comfortable walking than the localities just described; indeed, most of the timbered region, so far as I have observed, is a most discouraging field for pedestrianism.

26. CORVUS CORAX CARNIVORUS (Bartr.) Ridgw.

    (3075) ad.   Sitka, Alaska.
81394 (3076) ad.   Sitka, Alaska.
    (3291) sternum.   Port Mulgrave, Yakutat Bay, Alaska, June 24, 1880.
81667 (3292) head.   "   "   "   "   "   "   "   "

Extremely abundant at Sitka. Mr. A. T. Whitford informed me that he has seen ravens catch rats in a very expert manner; swooping swiftly upon the victims, they carry them up into the air and let them fall from a great height. If the first fall does not kill the rat, he is captured again and carried higher. The rats are eaten by ravens.

I shot at a raven at Cape Lisburne, August 21, but failed to kill it. On the following day I heard one at a distance on one of the hills, 10 miles to the eastward of this cape.

Again, at Chamisso Island, Eschscholtz Bay, I attempted to kill a raven with small shot and failed.

I saw, but could not obtain, a fine bird of this species at Elephant Point, Eschscholtz Bay, September 2, 1880.

At Port Providence, Plover Bay, Siberia, ravens were extremely abundant September 14, and so gorged with blubber and overrun with parasites that it was too disgusting to prepare skins of them. At the head of the spit I watched their movements for some time, as they did not fear me while I sat still and made no sudden movement. They would alight close to my head and look at me with apparent curiosity, uttering now and then a hoarse call to other ravens flying near at hand. Hopping forward a step or two, they would pull off pieces of moss from the stones and jump slightly into the air in an affected sort of way, sometimes taking a good-sized stone in the beak, perhaps to see if any food might be concealed underneath. Occasionally, one would find a morsel, and then another would try to take it out of his bill, the lucky one seeming to hold out the prize temptingly, but firmly, to tantalize his covetous neighbor. In starting to fly they would strike the ground with their feet several times to gain an impetus.

27. Corvus caurinus, Baird.

81396 (3239) ♀ ad. Sitka, June 16, 1880.
81395 (3240) ♂ "   "   "   "   "

Abundant at Sitka, May 28 to June 16, 1880, associated with the preceding. Voice variable, usually less ringing and hoarser than that of *C. frugivorus*, but sometimes an exact counterpart of it. Without some definite and reliable mode of recording the notes of this fish crow for comparison with those of the common eastern species, there must be doubt as to the relation between the voices of the two birds.

A small flock was seen at George Island, Port Althorp, June 19, 1880, perched on the rocks, and feeding on a gravel beach at low tide.

Measurements.

Catalogue number.	Locality.	Sex and age.	Depth of bill through base.	Wing from carpal joint.	Tail to basal end of feathers.	Tarsal joint.	Middle toe.	Gonys.	Culmen.	Bill from nostril.	Nature of specimens measured.	Date.
81395	Sitka, Alaska	♂♀	+	11.	6.50	1.87			1.83		Fresh	June 16.
81396	" "			10.75	6.12	1.75			1.75		"	"
(*)	Sitka	♂♀	.70	11.	6.50	1.85	1.30	.83	1.83	1.17	Dried skin	May 31.
81396	" "		.75	10.75	6.10	1.75	1.25	.83	1.75	1.18	"	June 16.

* Bird not kept. The two sets of measurements are from the same birds—first in the fresh state and afterward from the dried skins.

28. CYANOCITTA STELLERI (Gm.) Caban.

(3037) ♂. Old Sitka (mouth of river), June 2, 1880.
81392 (3063) ♂. Near Hot Springs Bay, Baranoff Id., June 5, 1880.

Measurements of number 3037 in the fresh state: Length, 13; extent, 18.37; wing, 6; tail, 6; crest, 2; testis, .37. When shot, this bird had its mouth and crop crammed full of insects.

A bird of this species was shot at Port Althorp, June 19, but was lost in a dense thicket.

29. EMPIDONAX DIFFICILIS Baird.

81393 (3067) ad. ♂. Near Hot Springs Bay, (Sitka), Alaska, June 5, 1880.

This individual was one of a few examples seen at the place noted. It is the first speccimen of the species recorded from Alaska, and, so far as I know, the only one. From the size of the testes (.25) it is probable that this date represents very nearly the breeding time of this fly-catcher in the locality named.

The fresh bird furnished the following measurements: Length, 5.25; extent, 7.50; wing, 2.62.

30. SELASPHORUS RUFUS (Gmel.) Aud.

(3097) juv. Sitka.
(3098) " "

A live humming bird, with its nest and eggs, was brought into Mr. Whitford's store at Sitka, June 9, 1880, but none of our party were present at the time, and we did not get them.

31. NYCTEA SCANDIACA Linn.

(3681) sternum of 81397.
81397 (3689) ♂ ad. Point Belcher, Alaska, Arctic O. Aug. 27, 1880.

Common on the gently rising ground inland from the small lake near our anchorage. I saw as many as six at one time on small grassy mounds. They were uniformly hard to approach, never allowing me to come within gun-shot, except in the one instance when I crept along under cover of the low bluff forming one of the lake borders, and rose suddenly within easy range.

32. HIEROFALCO GYRFALCO SACER (Forst.) Ridgw.

81398 (3838) ♀. Bering Sea, 60 miles E.S.E. from St. George Island. Sept. 24, 1880.
(3838) sternum of above.

This individual was shot while trying to alight on the vessel; it dropped into the leach of the mainsail, and from thence into the cockpit, where it was secured. Two examples of this species, according to my belief, were around the vessel between St. Mathew and St. Lawrence Islands, a few days previous to this date. One of them was shot, but lost.

July 25, 1882.

The following color notes and measurements were taken from the bird : Iris brown; tarsus and toes bluish gray; bill the same at base, but black at tip; eyebrows bluish gray.

Ovaries little developed; eggs not distinguishable to the unaided eye.

Length, 21; extent, 44; wing, 14; tail, 9; tarsus, 2.37; bill, 1.12; head, 2.50; middle toe, 1.94; middle toe claw, .81.

33. PANDION HALIAËTUS CAROLINENSIS (Gm.) Ridgw.

81668 (3150) head. Hot Springs, Baranoff Island, Alaska, June 9, 1880.
  (3151) sternum.   "   "   "   "   "   "   "

This specimen of the osprey was shot by Capt. E. P. Herendeen near Hot Springs.

34. CIRCUS HUDSONIUS (Linn.) Viell.

81401 (3720) ♀. Elephant Point, Eschscholtz B., Alaska, Sept. 2, 1880.

Several individuals of this hawk were seen flying over the marshes in the vicinity of Elephant Point. The following color notes and measurements were taken from the recently-killed bird:

Length, 21.50; extent, 47.50; wing, 15.50; tail, 10.50; bill, 1.19; head, 2; tarsus, 3.37; middle toe and claw, 2.37; middle toe claw, .75.

Iris brown. The upper tail coverts are not white, as is usually recorded of this species, but whitish, with many blotches of rufous.

35. BUTEO BOREALIS CALURUS (Cass.) Ridgw.

81399 (3060) ♀ (?) juv. Baranoff Id., near Sitka, Alaska, June 5, 1880.

This young hawk was shot by Lieutenant Rockwell, U. S. N., near Hot Springs Bay. I have the following notes from the recently-killed bird: Iris very light hazel; length, 21.50; extent, 47.50; wing, 14.50; tail, 9. This species has not been previously recorded from Alaska.

36. ARCHIBUTEO LAGOPUS SANCTI-JOHANNIS (Gmel.) Ridgw.

81400 (3466) ♀. Popoff Island, Shumagins, July 18, 1880.

Iris hazel. Cere yellow, with a greenish tinge. Lips and feet lemon yellow. Eggs very small.

Measurements from the fresh bird: Length, 23; extent, 56.50; wing, 18; tail, 10.06; bill, 1.37; head, 2.25; tarsus, 2.94; middle toe and claw, 2.19; middle toe claw, .81.

37. HALIÆETUS LEUCOCEPHALUS (Linn.) Savig.

(3293) sternum. Port Althorp, Alaska, June 19, 1880.

Very abundant in the vicinity of Sitka, May 28 to June 16, 1880, usually around shallow coves in the neighborhood of the mouths of fresh water streams.

A pair of young birds of this species was seen at Iliuliuk, Unalashka, October 13, 1880.

38. LAGOPUS ALBUS (Gm.) Aud.

81402 (3482) ♀. Unga Id., Shumagins, July 21, 1880.

The crop was filled with leaves of a species of willow. Several of the birds were seen on low ground not far away from the ocean beach, in the vicinity of a small trout stream.

This specimen corresponds very closely in most respects with number 33548, a female from Norway, collected July 2, 1862; the claws, however, are considerably shorter than in the Norway example, and in all other specimens of *albus* in the Museum.

39. HÆMATOPUS NIGER Pall.

(3096). Old Sitka, Alaska, June 1, 1880.
81669 (3122) head. Sitka Bay, Alaska, June 8, 1880.
(3124) sternum of 3122.

A pair were seen at Port Althorp, June 21; they passed and repassed the vessel at anchor, drawing near when their peculiar whistle was imitated, and circling around us several times.

40. STREPSILAS INTERPRES (Linn.) Illig.

81709 (3543) alcoholic, St. Paul. Id., Bering Sea, Aug. 6, 1880.
81403 (3764) ♂. Point Belcher, Alaska, Arctic O., Aug. 27, 1880.
81404 (3602) ♂. Port Providence, Plover Bay, Siberia, Aug. 14, 1880.

No. 3602 was shot on the end of the spit. Toes semipalmate, though when the skin dries this may not be evident. Legs and feet yellow and olive brown. Bill nearly black at base and tip, the remaining portion greenish gray.

Measurements from the fresh bird: Length, 9; extent, 19; wing, 6; tail, 2.37; bill, .81; tarsus, 1.12; middle toe and claw, 1.12. Testes elongate, minute.

41. STREPSILAS MELANOCEPHALA Vig.

81405 (3789) ♂. Elephant Point, Eschscholtz Bay, Alaska, Sept. 2, 1880.

Only a few of these turnstones were seen here.

42. SQUATAROLA HELVETICA (Linn.) Cuv.

(3115) ♂ ad. Sitka, Alaska, June 8, 1880.
81406 (3828) ♂ juv. Port Providence, Plover Bay, Siberia, Sept. 12, 1880.
(3829) ♀ juv. Port Providence, Plover Bay, Sept. 13, 1880.

The single example secured in Alaska was in adult male summer plumage. It was found on a small rock in the cove near the old fish-house at Sitka. The following measurements were taken from the fresh bird: Length, 12.50; extent, 24.87; wing, 7.69.

The individuals obtained at Plover Bay were the only two of the species seen there. They were found on the spit which forms the harbor of Port Providence. Land birds were very scarce here during the time of our second visit, September 12 to 17. Besides the *Squatarola* I saw only *Stercorarius crepidatus, Heteroscelus incanus, Corvus corax carnivorus, Plectrophanes nivalis,* and one wagtail in flight.

43. CHARADRIUS DOMINICUS Müll.

81407 (3772) ♂. Icy Cape, Alaska, Arctic O., Aug. 25, 1880.

Only one small flock of this plover was definitely seen, containing perhaps not more than a half dozen individuals.

44. ARQUATELLA COUESII Ridgw.

(3879) ♂.	Chernoffsky, Unalashka, Oct. 1, 1880.					
81409 (3880) ♀.	"	"	"	"	"	
(3891) ♀.	Iliuliuk,	"	"	11,	"	
81408 (3893) ♂.	"	"	"	13,	"	
(3894) ♂.	"	"	"	"	"	
(3895) ♀.	"	"	"	"	"	

Not uncommon on small rocks in Chernoffsky Harbor, near its head and around the shores. At Iliuliuk, also, I found it feeding on sea-washed shores, usually on small islets.

45. ACTODROMAS ACUMINATA (Horsf.) Ridgw.

81410 (3825) ♂. Port Clarence, Alaska, Sept. 9, 1880.

Found near the margin of one of the small fresh-water lagoons. Rare. This species has not previously been obtained north of St. Michael's.

46. ACTODROMAS MACULATA (Viell.) Coues.

(3765) ♂. Point Belcher, Alaska, Arctic O., August 27, 1880.
81411 (3782). " " " " "

Quite common, with *Pelidna alpina americana,* at small fresh-water ponds, and sometimes near tide-pools.

47. ACTODROMAS MINUTILLA (Viell.) Bp.

81715 (3501) alcoholic. Belkoffsky, Aliaska, July 23, 1880.
81412 (3591) ♀ (?). Belkoffsky, Aliaska, July 23, 1880.
81413 (3597) ♂. Port Providence, Plover Bay, Siberia, August 13, 1880.

It was a real pleasure to collect land birds at Belkoffsky, although few species were found—only *Centrophanes lapponicus* and *Passerculus sandwichensis* besides the small sand piper. Walking was comfortable and there were many small, rapid streams of delightfully cool water rushing down from the steep hill behind the village. The valley between this hill and the sea is undulating, free from alder and other impediments to travel, rich in grasses and flowers, and abounding in patches of exceedingly hard stones covered with lichens. Iris, geranium, aster, *Pinguicula,* azaleas, Jacob's ladder, painted cups, yarrow, and water willow were in bloom. A fine salmon river falls into Belkoffsky Bay, and salmon were beginning to ascend. On the low ground birds were not abundant, but I heard more up the hillside. The volcano, Pavloff, is visible from the village, and was sending up columns of smoke during our stay.

48. PELIDNA ALPINA AMERICANA. Cass.

    (3598) ♀. Port Providence, Plover Bay, Siberia, Aug. 13, 1880.
    (3599).   "   "   "   "   "   "   "   "
81417 (3600) ♀.   "   "   .   "   "   "   "   "   "
81415 (3601) ♂.   "   "   "   "   "   " 14,   "
81416 (3778) ♂. Cape Lisburne, Alaska, Arctic O., "   " 21,   "
    (3779) ♂ juv."   "   "   "   "   "   "   "   ..
81414 (3780) ♂. Cape Lisburne, Alaska, Arctic O., Aug. 21, 1880.
81418 (3766) ♀. Icy Cape, Alaska, Arctic O.,   " 25,   "
81419 (3783) ♀. Point Belcher, "   "   "   " 27,   "

A very common species at all of the places named above. Feeding on the beach or at tide-pools and fresh-water ponds.

49. EURINORHYNCHUS PYGMÆUS (Linn.) Pearson.

81434 (3795) juv. Port Providence, Plover Bay, Siberia, 1880.

Shot on the end of the spit by a native, most probably late in August. Ammunition was left with this boy on the 13th of August for the express purpose of getting this sand piper, and one month later we were rewarded by receiving from him the only specimen we saw of the species.

50. HETEROSCELUS INCANUS (Gmel.) Coues.

    (3831) ♂. Port Providence, Plover Bay, Siberia, Sept. 14, 1880.
81421 (3832) ♂.   "   "   "   "   "   "   "   "
81420 (3907) ♂ ad. Iliuliuk, Unalashka, Oct. 16, 1880.
    (3822) 2 sternums, of 3831 and 3832.

The Unalashka specimen was in winter plumage, the only one seen there. Its call drew me towards it.

At Port Providence no others were observed except the two here recorded. They were found standing on the rocks near the eastern border of the harbor, teetering like some of the small species of *Actodromas*.

51. PHALAROPUS FULICARIUS (Linn.) Bp.

    (3603) ♂ (?). Off Cape Tchaplin, Siberia, Aug. 15, 1880.
81422 (3604) ♂ (?). Off Cape Tchaplin, Siberia, Aug. 15, 1880.
81423 (3781) ♀ (?). Point Belcher, Alaska, Arctic O., Aug. 27, 1880.

Immense flocks of this phalarope were heard and seen off Cape Tchaplin. Their twittering was a very pleasant sound. The day was calm, clear, and pleasant, so that a fine opportunity was afforded for the use of the dredge and the pursuit of phalaropes. At Point Belcher again we saw large flocks of the same species, feeding in the swash of the tide along the beach, and drifting shoreward with the incoming current from short distances at sea. The northern phalarope, on the other hand, was observed at the margins of fresh-water lagoons.

52. LOBIPES HYPERBOREUS (Linn.) Cuv.

81424 (3791). Port Clarence, Alaska, Sept. 6, 1880.

In small flocks, feeding at the margins of fresh-water lagoons. Four individuals were shot.

The spit at Port Clarence, where I collected birds, is long, narrow, and curved. The width at the astronomical station of the "Yukon" party must have been about three-fourths of a mile. The ground is level, and walking good; there are numerous fresh lagoons of very palatable water, around which birds collect. There are no trees except the very scrubby dwarf willows. We found a few flowering plants, and many exquisite lichens. The shallow lagoons are well stocked with sticklebacks. *Centrophanes lapponicus* was common; a few *Plectrophanes nivalis* were seen and only one *Saxicola œnanthe*. A wagtail was observed on the 6th and again on the 8th of September, but too far off for identification. *Larus glaucescens* was abundant, associated with kittiwakes.

I saw here a bird which I supposed to be a small wren; it appeared unexpectedly, when my attention was fixed on other species, alighted not very far off, was marked down and diligently searched for in a place where there was no apparent chance of escape, but unfortunately could not be found.

53. GRUS CANADENSIS (Linn.) Temm.

On the 18th of August, in the vicinity of the Diomede Islands, sand-hill cranes were seen flying towards the American shore.

On the 1st of September, Capt. E. P. Herendeen went ashore on the east side of Choris Peninsula, and here he saw a sand-hill crane.

54. BERNICLA NIGRICANS (Lawr.) Cass.

81425 (3667) ad. ♂. Near Icy Cape (Lat. 70° 13′ N.), Arctic O., Aug. 25, 1880.

On the 22d of August, while at anchor 10 miles to the eastward of Cape Lisburne, we first observed brant migrating southward; great numbers of them passed us during the day. On the 25th of August we found them very abundant on the brackish-water lagoons of the spit near Icy Cape.

Measurements taken from number 81425 in the fresh state are the following: Length, 23.50; extent, 46.62; wing, 12.87; tail, 4.19; bill, 1.31; head (from base of bill), 2.50; tarsus, 2.31; middle toe and claw, 2.25.

55. MARECA AMERICANA (Gmel.) Steph.

81710 (3678) heads in alcohol. Eschscholtz Bay, Aug. 31, 1880.

Two individuals were shot at Elephant Point, Eschscholtz Bay, September 2, 1880.

56. FULIX sp.

81717 (3481) embryo. Unga Id., Shumagins, July 18, 1880.
    (3481) 3 eggs. Unga Id., Shumagins, July 18, 1880.

This nest was obtained by Mr. Marcus Baker; it contained, when found, 7 eggs.

57. CLANGULA ALBEOLA (Linn.) Steph.

Many small flocks were seen at Chernoffsky, Unalashka, October 1 to 4, 1880, and again at Iliuliuk, on the same island, October 5 to 18, 1880.

58. SOMATERIA V-NIGRA Gray.

81426 (3337) jnv. Chugachik Bay, Cook's Inlet, July 1, 1880.
    (3320) 4 eggs.   "   "   "   " June 30, "

The young, not able to fly, but wonderfully expert in diving, were abundant near the head of Plover Bay, Siberia, August 12, 1880; their disappearance under water was so sudden that I failed to secure even a single specimen. One of the adult females feigned to be crippled and labored off through the water with much make-believe effort, to draw us away from the young.

59. SOMATERIA SPECTABILIS (Linn.) Boie.

(3793) heads in alcohol. Port Clarence, Sept. 9, 1880.

Eight of these ducks were brought to us by an Eskimo as we were leaving Port Clarence. They were moulting, and the native speared them.

60. PELIONETTA PERSPICILLATA (Linn.) Kaup.

    (3123) sternum. Near Hot Springs, Baranoff Id., June 5, 1880.
81712 (3125) head. Sitka Bay, Alaska, June 9, 1880.
81711 (3126) "   "   "   "   "   "
81713 (3127) "   "   "   "   "   "

These were heads of ♂, ♀ and young.

61. MERGUS MERGANSER AMERICANUS (Cass.) Ridgw.

I shot a female of the above species, July 13, 1880, near the margin of a small fresh-water lake not far from the village of St. Paul, Kodiak Island.

An egg (number 3389), said to be of this merganser, was obtained from Nicolas Pavloff, at Wooded Island, Kodiak, about the same time.

62. RISSA TRIDACTYLA KOTZBUEI (Bp.) Coues.

(3685) 2 sternums. Cape Lisburne, Arctic O., Aug. 21, 1880.
(3605) ♀. Port Providence, Plover B., Siberia, Aug. 14, 1880.
(3673) feet of two. Cape Lisburne, Arctic O., " 21, 1880.
(3836) ♀ jnv. St. Mathew Id., Bering Sea, Sept. 22, 1880.

The species was abundant in Plover Bay, August 11 to 14 and September 12 to 17. I have the following notes of colors from number 3605:

Bill light greenish yellow; eyelids, commissure, and inside of mouth deep orange red; legs and feet black.

The feet of two individuals shot at Cape Lisburne are preserved in alcohol. There is a well-developed nail on the hind toe of one of these feet, while the rest of the nails are quite rudimentary. The pair of feet having the best developed nails had, when fresh, a mere trace of yellowish on the skin of the under surface of the toes, while the other pair had bright yellow areas on the corresponding parts.

The young female shot at St. Mathew Island was in the nest. The nests were built of sea-weeds on high, narrow ledges of the inaccessible cliffs. Abundant in this locality with *Fratercula corniculata* and *Fulmarus glacialis Rodgersi*. At St. Mathew Island we saw a great many beautiful young kittiwakes flying near Cape Upright, the black collars and wing patches making them attractive objects of pursuit.

63. LARUS GLAUCUS Brunn.

81696 (3668) head in alcohol. Cape Lisburne, Arctic O., Aug. 21, 1880
    (3669) sternum.   "   "   "   "   "   "

The species was abundant at Cape Lisburne.

64. LARUS GLAUCESCENS Licht.

81695 (3729) alcoholic head. Port Clarence, Alaska, Sept. 6, 1880.
    (3729) sternum.   "   "   "   "   "   "

This gull was shot on the western side of the spit, near the point. Common.

65. LARUS MARINUS Linn.

81694 (3841) juv. head. Chernoffsky, Unalashka, Oct. 1, 1880.

Abundant, feeding at the mouth of the river falling into the head of Chernoffsky Bay. The first recorded instance of its occurrence on the west coast of America.

66. STERCORARIUS POMATORHINUS (Temm.) Viell.

    (3738) sternum. Point Belcher, Aug. 27, 1880.
    (3670) sternum of 81427.
81427 (3690) ♀ ad. Point Belcher, Alaska, Arctic O., Aug. 27, 1880.
    (3686) 2 sternums. "   "   "   "   "   "   "
81702 (3672) head in alcohol. Point Belcher, Aug. 27, 1880.

The eggs of this bird (81427) were very small. The species was very common, with *Nyctea scandiaca*, on the rising ground, industriously feeding upon something which I could not make out because of the difficulty of approaching the birds. The flights of this jaeger from seaward to the land and back again were frequent. We found *Stercorarius* very abundant in the vicinity of the whaling ships, where it fared sumptuously.

67. STERCORARIUS CREPIDATUS (Banks) Viell.

81428 (3830) ♀. Port Providence, Plover Bay, Siberia, Sept. 12, 1880.
   (3818) 2 sternums. "   "   "   "   "   "   "
81701 (3818) head in alcohol.   "   "   "   "   "   "

Common. Two examples were shot near the head of the spit.

68. DIOMEDEA NIGRIPES Aud.

(3009) ad. ♀. Pacific Ocean, Lat. 36° 32′ N., Long. 126° 13′ W., May 15, 1880.

The "Yukon" sailed from San Francisco May 13, proceeding to the westward several hundred miles and then laying her course for Sitka. From the time we left the bar until we neared north latitude 52 degrees on this voyage *D. nigripes* was with us every day, soaring around us when we had a good breeze and leisurely following in our wake or floating astern when the wind was light or wanting. On the above date (May 15) we caught three of these birds with a fish-hook baited with pork. Soon after taking them on deck one of them became seasick, and ejected a piece of pumice. One of the calls of this albatross is similar to the peeping sound of very young chickens. It makes a peculiar sound, too, by striking its jaws together when approached on deck, and can inflict a painful wound with its sharp hook. These gonies pick up whatever floating food is cast from the vessel, and it is surprising to see how soon after anything is cast overboard a flock of the birds will approach, although none may be in sight at the time. In alighting after rapid flight they back air with their wings, drop their legs and thrust their feet forward to back water, making a light splashing. As soon as a small flock has gathered the gonies begin to fight and scream over their floating food, watching one another to see when anything turns up, the quickest and strongest getting the most. One of the most laughable things we saw was the chase of an overloaded *Fratercula* by one of these gonies; the *Fratercula* skimmed along close to the surface of the water and sometimes apparently floundering through it, as if its body were too heavy for its wings; the gony followed in hot haste but was soon foiled, astonished, and apparently much disgusted by the unexpected diving of the little struggler.

As we proceeded northward we observed a larger proportion of old birds with the upper and under tail coverts and part of the belly white. We saw no individuals of *D. nigripes* as far north as Sitka on the northward voyage. On our homeward way late in October, we saw the species frequently when about 700 miles south of Unalashka; a few were reported within 300 miles south of this island.

Measurements of number 3009 from the fresh specimen: Length, 28.50; extent, 79.50; wing, 19.50; tail, 6; bill, 3.75; head, 2.87; tarsus, 3.50; middle toe, 4; middle toe claw, .62; iris umber; tarsus, foot, base and tip of bill black; remainder of bill plumbeous.

A second living one, caught with the last, had the bill 4.

A very large one caught May 16 gave the following record: Length, 32.50; wing, 21.50; bill, 4.31; upper and under tail coverts white; crissum with some white; iris umber or golden brown.

69. DIOMEDA BRACHYURA Temm.

(3331 alc. 1474) 4 heads. Alexandrovsk, Cook's Inlet, July 4, 1880.
(3301) ad. ♀. Cook's Inlet, June 29, 1880.
(3333) sternum of 3301.

The specimen here mentioned was shot by Capt. E. P. Herendeen near the mouth of the inlet, not far from Fort Alexander. The species was abundant. This example was moulting; some of the primaries are rudimentary. It has been extremely difficult to kill these birds because they never come near the vessel nor allow it to approach them closely. Unlike *D. nigripes*, it is extremely shy.

Measurements from the recently-killed bird: Extent, 88; wing, 21; tail, 6.75; head, 3.75; bill, 5.19; tarsus, 3.87; middle toe and claw, 5.12. Bill flesh color, with a faint purplish tinge; hook light horn color; iris dark brown.

The Kodiak native name for this gony is *Kay-măh-rye erk'*.

In about north latitude 51 degrees we begun to lose sight of *D. nigripes*, and *D. brachyura* took its place. From latitude 52 degrees north the latter species increased in numbers. We found it at various points around the Gulf of Alaska, but the mouth of Cook's Inlet, and the vicinity of the Barren Islands, seemed to be its favorite summer resort. Natives of the trading village Alexandrovsk frequently spear this bird from their bidarkas. I picked up four skins of this species from a pile of refuse at this village.

We saw *D. brachyura* in Unimak Pass July 25, and in Bering Sea, off Makushin, on the following day. A single individual was seen August 10 about 40 miles to the westward of the entrance to Plover Bay. Another individual was seen September 18 to the northward of St. Lawrence Island. On the 22d of September we saw a few of these birds in the vicinity of St. Mathew Island. On the 5th of October we saw a few individuals, in beautiful plumage, while under sail from Chernoffsky, along the west coast of Unalashka, to Iliuliuk.

70. DIOMEDEA MELANOPHRYS Temm.

On the 31st of October a single *Diomedea* was seen on the Pacific not far from the following position: North latitude 40° 30′, west longitude 142° 23′. Observing that it differed greatly from the common *D. nigripes*, I made these notes concerning it: Head, neck, lower parts, and rump white; the under surface of the wings, too, shows considerable light color; elsewhere the bird is dark gray like *nigripes;* in size it is slightly less than the *nigripes* around it; the dark part of the wing of this bird is very different from the black of *D. brachyura* so far as observed; the bill is light; a dark streak runs from the bill behind the eye; the bird could not be secured.

This description, taken while the bird was flying near the vessel, evidently indicates *D. melanophrys*, as suggested to me by Mr. Ridgway recently, and, if so, the range of that species will be extended to within about 1,060 miles west of Cape Mendocino, California, thus coming well within the limits of *D. nigripes.*

71. FRATERCULA CORNICULATA (Naum.) Gray.

81429 (3837) ad. ♀. St. Mathew Island, Bering Sea, Sept. 22, 1880.

Abundant on the cliffs near Cape Upright, where they were inaccessible except by shooting. They fairly cover the narrow ledges in company with fulmars and kittiwakes. Colors of the fresh bird: Bill red and pale lemon, with narrow stripes of black in the grooves; corners of mouth yellow; iris white; eyelids red; palpebral appendages black; feet and legs orange; lighter on the upper surface of the toes and front of the feet; worn and soiled so as to appear grayish on the under surface of feet and toes.

The corners of the mouth are *soft* and *not callous.* The palpebral appendages are also soft.

72. PHALERIS PSITTACULA (Pall.) Temm.

(3465) sternum. Little Koniushi Id., Shumagins, July 16, 1880.

Abundant. The bird whose sternum was prepared flew into a crevice in the rocks, and was caught without being injured.

73. SIMORHYNCHUS CRISTATELLUS (Pall.) Merrem.

81430 (3827) juv. Big Diomede Island, Bering Strait, Sept. 10, 1880.

Changing to first plumage.

Great bunches of these little auks were brought to us by Eskimo at Big Diomede. Mr. Baker secured six of the young also on the island.

74. SYNTHLIBORHAMPHUS ANTIQUUS (Gm.) Coues.

81706 (3116) alcoholic. Sitka Bay, Alaska, June 9, 1880.
81708 (3117)   "   "   "   "   "   "   "
81707 (3118)   "   "   "   "   "   "   "

These specimens were shot by Lieut. Com. Chas. H. Rockwell, U. S. N.; small flocks were occasionally met with in the bay.

75. BRACHYRAMPHUS MARMORATUS (Gm.) Brandt.

81431 (3069) ad. ♂. Sitka Bay, Alaska, June 5, 1880.
      (3070) sternum of last.
81705 (3119) alcoholic. Sitka Bay, Alaska, June 9, 1880.
81703 (3120)   "   "   "   "   "   "   "
81704 (3121)   "   "   "   "   "   "   "

Not abundant; found in small flocks.

76. URIA COLUMBA (Pall.) Cass.

81700 (3128) head. Sitka Bay. Alaska, June 9, 1880.
81698 (3129) " " " " " " "
81699 (3289) heads. Port Althorp, " " 19, "
   (3294) sternum. " " " " " "

At Port Althorp, on the 19th of June, 1880, I saw a dozen or more of these birds feeding in a small cove on George Island. They are very graceful in their movements. While feeding they put the head under the water and paddle along with it in that position—moving rather quickly. When one sees something in the water at a little distance he makes a rush for it, and others follow to get the prize. In alighting, after a short, rapid flight, they come down on the water with a tumble. One of their calls resembles the chipping of a sparrow, and I mistook it for that several times. They have a low whistle also.

We found it very abundant in the harbor of St. Paul, Kodiak, July 9 to 14, 1880.

77. LOMVIA TROILE (Linn.) Brandt.

81697 (3303) head in alcohol. Chugachik Bay, Cook's Inlet, June 30, 1880.
   (3305) sternum of last. " " "
   (3321) 7 eggs. Chugachick Bay, " " "

Abundant in the inlet.

Distribution of species.

		Sitka.	Port Althorp.	Port Mulgrave.	Cook's Inlet.	Kodiak Island.	Little Koniushi Island.	Popoff Island.	Belkoffsky.	Unalashka Island.	St. Paul Island.	Port Clarence.	Eschscholtz Bay.	Cape Lisburne.	Icy Cape.	Belcher Point.	Plover Bay.
1	*Hylocichla aliciæ*	×			×	×											
2	*Hylocichla unalascæ*	×			×	×			—								
3	*Merula migratoria*	×							—								
4	*Hesperocichla nævia*	×															
5	*Saxicola œnanthe*									×	×	×					
6	*Parus atricapillus septentrionalis*				×	×											
7	*Anorthura alascensis*									×							
8	*Motacilla ocularis*																×
9	*Budytes flava*																×
10	*Anthus ludovicianus*						×						×				
11	*Dendrœca æstiva*					×											
12	*Myiodioctes pusillus pileolatus*			×		×											
13	*Hirundo erythrogastra*									×							
14	*Leucosticte griseinucha*						×			×							
15	*Ægiothus canescens exilipes*										×						
16	*Ægiothus linaria*				×						×						
17	*Plectrophanes nivalis*						×				×		×	×	×	×	×
18	*Centrophanes lapponicus*							×		×		×		×			
19	*Passerculus sandwichensis*							×	×						×		
20	*Passerculus sandwichensis alaudinus*										×						
21	*Zonotrichia coronata*				×	×			×								
22	*Junco oregonus*	×				×											
23	*Melospiza fasciata rufina*		×		×												
24	*Melospiza cinerea*					×	×	×		×							

Distribution of species—Continued.

		Sitka.	Port Althorp.	Port Mulgrave.	Cook's Inlet.	Kodiak Island.	Little Koniushi Island.	Popoff Island.	Belkoffsky.	Unalashka Island.	St. Paul Island.	Port Clarence.	Eschscholtz Bay.	Cape Lisburne.	Icy Cape.	Belcher Point.	Plover Bay.
25	Passerella iliaca unalascensis	×	×		×	×	×	×									
26	Corvus corax carnivorus...	×		×	×									×	×		×
27	Corvus caurinus...........	×															
28	Cyanocitta stelleri	×															
29	Empidonax difficilis	×															
30	Selasphorus rufus	×															
31	Nyctea scandiaca															×	
32	Hierofalco gyrfalco sacer*																
33	Pandion haliaetus carolinensis	×															
34	Circus hudsonius..........												×				
35	Buteo borealis calurus	×															
36	Archibuteo lagopus sancti-johannis							×									
37	Haliæetus leucocephalus ...		×							×							
38	Lagopus albus**																
39	Hæmatopus niger	×															
40	Strepsilas interpres........									×						×	×
41	Strepsilas melanocephala ..												×				
42	Squatarola helvetica	×															×
43	Charadrius dominicus......														×		
44	Arquatella couesi									×							
45	Actodromas acuminata.....										×						
46	Actodromas maculata														×		
47	Actodromas minutilla								×								
48	Pelidna alpina americana.													×	×	×	×
49	Eurinorhynchus pygmæus..																×
50	Heteroscelus incanus									×							×
51	Phalaropus fulicarius																×
52	Lobipes hyperboreus										×						
53	Grus canadensis.												×				
54	Bernicla nigricans.........												×			×	
55	Mareca americana												×				
56	Fulix sp**																
57	Clangula albeola																
58	Somateria v-nigra				×												×
59	Somateria spectabilis										×						
60	Pelionetta perspicillata	×															
61	Mergus merganser americanus					×											
62	Rissa tridactyla kotzbuei...														×		×
63	Larus glaucus.............														×		
64	Larus glaucescens											×					
65	Larus marinus									×							
66	Stercorarius pomatorhinus.															×	
67	Stercorarius crepidatus....																×
68	Diomedea nigripes †																
69	Diomedea brachyura					×											
70	Diomedea melanophrys ‡...																
71	Fratercula corniculata § ...																
72	Phaleris psittacula							×									
73	Simorhynchus cristatellus||																
74	Synthliborhamphus antiquus	×		,													
75	Brachyrhamphus marmoratus	×															
76	Uria columba	×	×			×											
77	Lomvia troile	×			×												

* Bering Sea, 60 miles ESE. from St. George Island.
† Lat. 36° 32′ N., long. 126° 13′ W.
‡ Lat. 40° 32′ N., long. 142° 23′ W.
§ St. Mathew Island, Bering Sea.
|| Big Diomede Island, Bering Strait.
** Unga Island.

U. S. NATIONAL MUSEUM, *May 22, 1882.*

OUTLINES OF A MONOGRAPH OF THE CYGNINÆ.[*]

By LEONHARD STEJNEGER.

["It is better to err on the side of minuteness than of vagueness."—GOSSE.]

CYGNINÆ BONAP.

1838.—*Cygninæ* BP. Comp. List. p. 55.
1850.—*Cygnidæ* KAUP (fide GRAY).
1852.—*Olorinæ* REICHB. Syst. Av. p. x.
1860.—*Cycnidæ* DES MURS, Tr. Ool. Ornith. p. 537.

DIAGN.—*Anatidæ having the hind toe without web and the lores naked, coincident with reticulate tarsi, the latter shorter than the middle toe with claw.*

The preceding marks combined appear to express the essential characters of the *Cygninæ*. By this diagnosis I follow Mr. SUNDEVALL[†] in excluding the genus *Coscoroba* REICHB., which has the lores feathered at all ages. As early even as RÜPPELL'S monograph of the genus *Cygnus*, (Mus. Senkenb. III), it was separated from the Swans. Here, however, it may be remarked, that this diagnosis refers only to the adult birds, because the young have the lores more or less downy or feathered, except in the genus *Chenopis*, which has the loral space naked at all ages. The removal of *Coscoroba* to the *Anatinæ* will be discussed more explicitly below. The criterion "tarsi reticulate" further excludes the genera *Cairina* FLEM. and *Plectropterus* LEACH, which, it is true, have the lores naked, but the tarsi of which are scutellate instead of reticulate. *Anseranas* LESS. has certainly both naked lores and reticulate tarsi, but differs in having the tarsus longer than the middle toe with claw.

Anatidæ which do not at once unite all the above characters consequently belong to one of the other subfamilies.

The whole family *Anatidæ* forms, as to structural features, a very homogeneous group, and intermediate links are everywhere to be found. Thus it is very difficult to define the subfamilies anatomically, and to

---

[*] The present treatise comprises merely the outlines of a monograph of the Swans, intended by the author to be much more complete, but which his departure for the Commander Islands prevented him from finishing according to the original plan. The paper contains so many valuable hints and so much important information upon this interesting group of birds, that it has been thought advisable to publish in it its present form, as preliminary to the more elaborate monograph contemplated by the author after his return.—R. R.

[†] Tent. Meth. Av. Disp. p. 147.

give the structural differences by which they are to be separated, so that I find it not improbable that an exact investigation, based on a more abundant material than I can at present procure, will reduce the subfamilies to groups of lower rank.

CHARACTERISTICS OF THE SUBFAMILY.

External characters.

Neck very long, as long as, or longer than, the body. Bill longer than the head, broad, and of nearly equal breadth for the whole length, rounded at the end, culmen high, depressed at the tip; nail rather large, only slightly arched; lamellæ of upper mandible vertical, in one row; nostrils situated nearly at the middle of the bill, in the fore part of the oblong nasal sinus. Lores naked in the adults; in all species, except one, thinly covered with small down or feathers in the young. Legs short, stout; lower part of tibia naked; tarsi compressed, much shorter than the middle toe with the claw, and covered with small hex-agonal plates, the size of which diminishes laterally and posteriorly; the anterior toes reticulate as far as the second joint, then scutellate; mid-dle toe longest, longer than the tarsus, the outer longer than the inner, which has a broad margin; hind toe short, elevated, and without web,* the claws strong, arched, compressed except the middle, which is only compressed on the one side, the claw of the inner toe in old birds the largest and most arched. Wings long, ample, the inner remiges highly developed, with about 32 quills. Tail composed of 20–24 rectrices, short, rounded, or cuneate.

Sexes similar.

Osteological characters.

The Swans, restricted as above, have a rather elongated skull, the intermaxillar portion being especially lengthened, but their cranium does not otherwise differ materially from that of the other *Anatidæ.* As a rule, however, the *Cygninæ* lack the two apertures on the occiput just above the *foramen magnum,* which always are to be found in the other members of the family† as well circumscribed and often large foramina. The *glandular depressions* along the roof of the orbits are more or less well marked. They are rather distinct in the genus *Cygnus,* whereas they seem to be wanting in most of the other *Anatidæ.*

The neck is extremely long, longer than the body, and is composed of the greatest number of *vertebræ* yet discovered in any recent bird, viz,

---

* This expression is not quite correct, for I have, in the freshly-killed bird, always found a narrow, very slightly developed lobe.

† One specimen of *Cairina moschata* (LIN.), which I have examined, had no fontanelles. I have seen two skulls of *Olor columbianus* (ORD.) which presented corresponding open-ings, their limits, however, being lacerated and in a state indicating that the ossifi-cation was not yet finished. The other crania of the same species show no trace of these fontanelles.

from twenty-two to twenty-six. (The next in order are *Coscoroba candida* having 21, and *Branta canadensis* with 20, and of birds belonging to other families, the long necked *Plotus anhinga* with 20 vertebræ colli, and *Phœnicopterus*, in which I have found only 18.)

The number of the *dorsal vertebræ* amounts to eight, and consequently there are eight pairs of *dorsal pleurapophyses*, the first five usually supporting *epi-pleural appendages*. The three last have no *uncinate processes* as do likewise neither the two cervical ribs nor the sacral one.

The body of the *sternum* is square, with the lateral margins quite parallel, and not narrower at the hind termination of the costal border, where the last dorsal rib articulates, as in the other *Anatidæ*. (See figs. 1 and 2.) The hind border, with two proportionally shallow notches, their length making as a rule about one-sixth of the greatest length of the sternum. The middle portion of the end of the sternum usually slightly sinuated. The *crista sterni* is rather high, but the *carinal angle* does not protrude forward longer than the short *manubrium*, the fore border of the

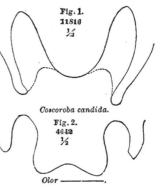

Fig. 1.
11810
¼

Coscoroba candida.

Fig. 2.
4642
¼

Olor ————.

crista being more or less arched. In the one genus (*Olor*), the *carina* of which is shallow for the reception of a long fold of the windpipe, the anterior margin consequently is double; in the other genera only a little concave. The lower limit of the *crista* is slightly curved. The greater portion of the lateral margin of the *corpus sterni* is occupied by the *costal border*, from which eight *hæmapophyses* ascend to meet the dorsal ribs, the free border behind being proportionately very short. The *pectoral ridge* on the body of the *sternum*, defining the origin of *musculus pectoralis secundus*, does not run parallel with the external margin or to the end of the keel, as is the case in the other *Anatidæ* (figs. 3–5), but passes obliquely towards the middle, which it reaches before the termination of the *crista*. This feature, however, is not always equally marked. In one of the skeletons of the *Olor columbianus* which I have examined, the course has some resemblance to that of *Coscoroba*, not dismissing, however, its peculiar swanlike character.

The *clavicles* form a broad, rather robust, U-shaped arch, except in the genus *Olor*, where the lower end is bent upwards and backwards to admit the fold of the *trachea* to enter the hollow keel of the breast bone.

The *coracoids* are rather short and very stout bones. The *scapula* is proportionally short.

The most marked feature in the osteology of the Swans, wherein they differ from the other members of the family, and which characterizes them as powerful flyers, seems to be the considerable length of the *humerus* and *antibrachium*, these being almost of equal length·

When folded and lying close to the body their elbow-joint reaches far beyond the *acetabulum*. Their length is greater than that of the hand, and considerably more than twice the *tarsus*.

The *pelvis* presents only few differences from that of the other *Anatidæ*. It is, however, proportionally longer and narrower, the breadth between the *acetabula* making only about one-fifth of the total length of the *ilia*. The very prominent ridges, forming the internal borders of the post-acetabular parts of the *ilia*, run from the *acetabula* backward nearly parallel, the hinder sacral roof being rather narrow and of equal

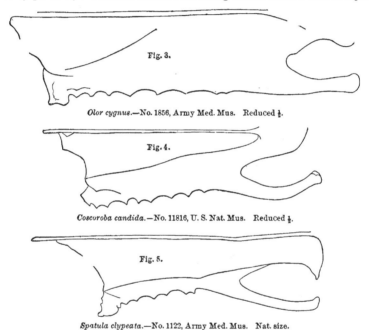

Fig. 3.

Olor cygnus.—No. 1856, Army Med. Mus.   Reduced ¼.

Fig. 4.

Coscoroba candida.—No. 11816, U. S. Nat. Mus.   Reduced ¼.

Fig. 5.

Spatula clypeata.—No. 1122, Army Med. Mus.   Nat. size.

breadth, whereas in the other *Anatidæ*, the ridges converge backwards from the more distant *acetabula*, forming a wedge-shaped and rather flat and broad roof. The *foramen ischiadicum* is rather large.

The *pelvic limbs* agree in every respect with those of the typical members of the family, except in the proportional length of the single bones mutually.

The number of free *caudal vertebræ* is seven, to which is to be added the *pygostyle*.

The *Cygninæ* are more nearly related to the *Anatinæ* than to the *Anserinæ*, the *Coscoroba* REICHB. being among Ducks the genus most closely allied to the Swans. It has also, as stated above, usually been referred to the latter group, but an accurate examination undoubtedly shows that *Coscoroba* belongs to the *Cygninæ* as little as does *Cairina* to the *Anserinæ*, and that SUNDEVALL was right in removing it from the Swans.

Jul 25 1882.

Though both the exterior proportions and the color of the *Coscoroba* are much like those of the Swans, there are, however, considerable differences.

In the first place, the bill is not of equal breadth for the whole length as in the Swans, but broadens, comparatively, considerably towards the tip, being, besides, more depressed in front of the nostrils, so that, on the whole, it is a perfect duck-bill. Add to this that the lores, at all ages, are completely feathered. The relation of the wing-feathers is nearly identical, but there is, however, a difference, the inner web of the third primary of *Coscoroba* not being sinuated, as is the case in all species of *Cygninæ*. The relatively much longer hind toe of *Coscoroba* is another not unimportant difference, the whole nail touching the ground when the bird walks.

The interior differences are even more essential.

As I have just above given a short characteristic of the most interesting and peculiar facts in the osteology of the Swans, I here only intend to enumerate the more essential osteological features wherein *Coscoroba* differs from the *Cygninæ*, mostly leaving to the reader himself to draw the comparison.

The *skull* shows only few differences besides the above-mentioned peculiar shape of the bill. The *os lacrymale*, however, is more duck-like than in the Swans, the fore *processes* being more elongated. On the oc-

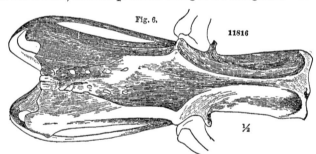

Pelvis of *Coscoroba candida.*—No. 11816, U. S. Nat. Mus.

ciput the two fontanelles, above the *foramen magnum*, are to be found as well circumscribed, long, and rather narrow apertures (4 by 1.5^{mm}). The *vertebræ colli* amount to 21, the last supporting a free rib without *epipleural appendage*. Seven *vertebræ of the dorsal section* with their *pleurapophyses* and *hæmopophyses*, the five first having *uncinate processes*. One sacral rib. The number of free *coccygeal vertebræ* is only five plus the *pygostyle*. The *sternum* is quite duck-like in its outlines, the lateral margins converging to the articulations of the last *dorsal hæmapophyses*, and from this point again diverging. The *costal border* is comparatively short. The notches of the hind margin very deep, making about ⅓ of the whole length of the *sternum*. The *pectoral ridge* for the origin of *musc. pectoralis secundus* runs backwards to the end of the *crista*, the lower border of which forms an undulating line, being higher on the fore portion; the *carinal angle* overhangs the *manubrium* considerably; the fore border of the keel is sharp and rather straight. Most of these features will be well seen in fig. 4. The *brachium* and *antibrachium* are of the same length, proportionally much shorter than in the Swans, their length being shorter than twice the tarsus, and only equal to the distance between the shoulder and hip joints. The *pelvis* is, as fig. 6 shows, quite typically duck-like. Compared with fig. 7, the *pelvis* of a swan, and with fig. 22, in OWENS Anat. Vertebr., II, p. 32, representing a typical *pelvis* of a duck, the differences from the former and the identity with the latter are easily perceptible, the greater breadth and wedge-shaped form of the post-acetabular sacral roof being the most essential characters.

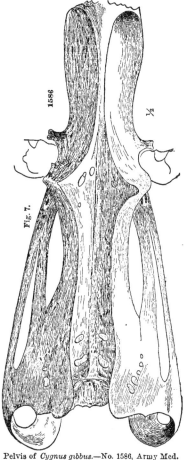

In nearly all the above-mentioned features *Coscoroba* differs from the Swans, while the same characters draw it near to the true Ducks; or, in other words, in nearly all the points wherein the *Cygninæ* differ from the *Anatinæ*, the *Coscoroba* agrees with the latter.

There can, after this, be no doubt where it, for the future, should be placed in the system.[*]

It is, however, unquestionable that the Swans, through *Coscoroba*, are more nearly allied to the *Anatinæ* than they are to the *Anserinæ*. But

Pelvis of *Cygnus gibbus.*—No. 1586, Army Med. Mus.

---

[*] The genus *Coscoroba* only comprises two species, of which one is known merely from a single specimen. The following is a short synopsis:

Coscoroba REICHB.

= 1852.—*Coscoroba* REICHB. Syst. Avium, p. x.

= 1855.—"*Pseudolor* G. R. Gray, MSS." Catal. Gen. Subgen. Bird's Brit. Mus., p. 122.

= 1872.—*Pseudocycnus* SUNDEV., Tent. Meth. Av. Disp. p. 147.

Key to the species:

a^1 Primaries with black tips; the nail of the bill flesh-colored; feet red.

1. *candida* (Vieill.) 1816.

a^2 Primaries entirely white; the nail of the bill black; feet orange-colored.

2. *davidi* SWINH. 1870.

this statement refers only to the recent forms, because we, in reality, have a fossil species, which seems to be an intermediate link between the swans and geese. This form is the *Cygnus falconeri* PARKER,[*] a gigantic swan from the Zebbug-Cave, Malta, nearly one-third larger than average individuals of the Mute Swan. It stood on longer legs, and had the comparatively short toes of a goose. In fact, the tarsi were considerably longer in proportion than those of the recent swans, the toes being very short, so that, whilst the proximal joint of the middle toe is one-fourth thicker than that of the Mute Swan, it is only three fourths the length. As this species evidently is generically quite distinct from any of the recent genera, I propose for it the name

PALÆOCYCNUS [*] Stejneger gen. nov.

Type *Palæocycnus falconeri.*

Fossil Swans have not been found longer back than the diluvium. Mr. R. OWEN indicating the existence of a Swan from the diluvious strata of Essex, alongside of the bones of *Elephas primigenius* and *Rhinoceros tichorhinus.* In the caves of France and of Malta, in the so-

1. Coscoroba candida (VIEILL.)

1782.—*Anas coscoroba* MOLIN., Stor. Nat. Chili (p. 207).
1818.—*Anser candidus* VIEILL., Nouv. Dict. d'Hist. Nat. xxiii, p. 331.
1831.—*Cygnus anatoides* KING, Pr. Zool. Soc. Lond. 1830–31, p. 15.
1837.—*Cygnus hyperboreus* D'ORBIGNY, Mag. Zool. p. —.
1854.—*Cygnus chionis* LICHT., Nomencl. p. 101.

Hab.—South America, from Chili and Buenos Ayres southward to the Falkland Islands.

List of specimens and dimensions.

Museum.	Catalogue number.	Locality.	Sex and age.	From the tip of the bill to—			The breadth of the bill at the nostrils.	Length of toes, with claws.				Tarsus.	Tail.	Wing.
				The mouth.	The nostrils.	The eye.		Outer toe.	Middle toe.	Inner toe.	Hind toe.			
				mm	mm	mm	mm	mm	mm	mm	mm	mm	mm	mm
Leyden	Cy. c. 1	Rio de la Plata.	♂...	75	42	98	32	112	125	93	28	87	161	455
Gotheburg	1072...	Montevideo...	♂ad.	82	44	106	30	11	128	96	32	89	138	480
Gotheburg	1071...	....do	♀ad.	71	38	92	22	906	102	71	24	76	120	405

2. Coscoroba davidi SWINHOE.

1870.—*Cygnus (Coscoroba) davidi* SWINH., Proc. Zool. Soc. Lond. 1870, p. 430.
Hab.—China. Only one specimen known.
Both species are white with red bill.

[*] Proc. Zool. Soc. Lond. 1865, p. 752.
[*] παλαιὸς = antiquus, κυκνος = cygnus.

called "Kjökkenmöddings" of Denmark, and in the leavings of the lake-dwellings of Switzerland the bones of *Olor cygnus* are found tolerably common, and likewise from the peat-bogs of England. From Belgium, Prof. P. I. VAN BLUEDEN has described a separate species as *Cygnus herenthalsii*,* from "une phalange du pied."

The subfamily *Cygninæ* is at the present time considered to embrace nine recent species, distributed in four different genera.

This number is only one more than RÜPPELL already indicated in his monograph, but amongst the eight considered by him are enumerated *Cairina moschata* (LIN.) and *Plectropterus rüppellii* SCLAT. (= *gambensis* RÜPP. nec LINN.). BLAINVILLE gave, in Compt. Rend. VII, 1838, pp. 1022–1026, and LESSON, in Rev. Zool. 1839, pp. 321–324, an enumeration of the species belonging to the genus, also comprising eight, having, instead of the two last, adopted *C. coscoroba* (MOL.) and *immutabilis* YARR., whilst, however, they did not distinguish between *bewickii* YARR. and *columbianus* (ORD). EYTON having published in 1838 his "Monograph of the Anatidæ" separates these, and thus makes 9 species. SCHLEGEL, in his synopsis of the genus (Mus. P.-B. 1866, VI, *Anseres*, pp. 78–83) enumerates eight species. He, it is true, adopts also *coscoroba* (MOL.) as belonging to this genus; on the other hand, however, he does not recognize *C. immutabilis* YARR. as a distinct species.

LINNÆUS only described one species of Swan under the name of *Anas cygnus*, enumerating, however, the tame Swan among the synonyms as var. *mansuetus;* but in 1779 PALLAS had already separated the latter specifically, and retained for it the title of *cygnus*, whilst the Hooper received the name *olor*.

MOLINA, in 1782, described the Chilian Swan as *Anas melancorypha*.

In 1788 GMELIN described *nigricollis* after BOUGAINVILLE and *melanocephala* after MOLINA. In the mean time, as these two are synonymous with MOLINA's *melancorypha*, the number of species known at that time amounted still to only three.

In 1790 LATHAM described *atratus*. In the same year it was described by BONNATERRE under the name of *Anser Novæ-Hollandiæ*.

LEWIS and CLARKE, in their "Travels" (1814), separated the American Swan, which ORD, in the second American edition of GUTHRIE'S Geography in the following year (1815), gave the systematic name *Anas columbianus*, thus making the fifth species.

In 1830 YARRELL described *Cygnus bewickii* as new, by which the number of species was increased to six. In the same year BREHM indicated *C. islandicus* as a supposed new species, which, however, is only a synonym of *Olor cygnus*.

The seventh species was added in the following year (1831) by RICHARDSON, viz: *C. buccinator* from North America.

The eighth dates from the year 1838, when YARRELL introduced *C. immutabilis* into the system.

---

* Jour. de Zool. I, 1872, p. 288. (*C. herrenthalsi* Ibis, 1873, p. 434.)

· Three Swans, which von PELZELN in 1862 described as belonging to *immutabilis*, are in the present work introduced under *C. unwini*, a species which HUME founded in 1871 on two immature specimens, and which has usually been regarded as the young of *C. gibbus* BECHST. Thus we at present allow nine species of Swans.

The *C. passmori*, described by HINCKS in 1865, seems only to be a young *buccinator*. *Cygnus davidi*, which was described by SWINHOE in 1870, does not belong to the Swans, but to the genus *Coscoroba* REICHB. amongst the *Anatinæ*. (See p. 180.)

· Until BECHSTEIN in 1803 indicated the genus *Cygnus*, the species belonging to this group were referred to the great LINNÆAN genus *Anas*. The new genus was soon commonly adopted, and remained undivided until 1832, when WAGLER* divided it into three, viz: *Chenopis, Olor,* and *Cygnus*.

At first I was inclined to regard all the Swans as belonging to only one genus. But since Prof. THEO. GILL has drawn my attention to several differences in the structure I have convinced myself that the genera in question are as well founded as a greater part of genera among the *Anatidæ*, which I never hesitated to admit. If one would adopt the view of Mr. SEEBOHM,† that the color is the most important generic criterion, only two genera ought to be established, the one white and the other black; but the greatest differences are even to be found between the white species, this fact, for one, showing the untenability of Mr. SEEBOHM's standpoint. The color can indicate where the limits of a genus are to be drawn, and may in many cases be of great value as instruction when the matter is doubtful, or may also add an important character to the other ones, but it ought not to be the only or even the main character of a genus, which should merely be based upon structural marks.

In the matter now before us it will, however, be seen that if we admit any subdivision of the genus, the black-necked Swan must be separated from the Palæarctic knob-billed Swan (*Cygnus gibbus*) and its congeners to obtain equivalency with the different groups.‡ I therefore propose the new genus *Sthenelus*, the number of recent genera thus being four. For the fossil *C. falconeri* I have introduced a fifth genus, *Palaeocycnus*.

SYNOPSIS OF THE GENERA.

a^1. Predominant color of the adults white; young with downy or feathered lores; tertiaries and scapulars normal, not crisp; tail longer than the middle toe with claw.

---

* Eearlier than this BOIE had asserted the necessity of this divison (OKEN'S Isis 1822, p. 564, nat).

† Cat. Birds Brit. Mus., vol. v, p. viii: "These so-called structural characters have no generic value at all." (!)

‡ REICHENBACH, in his Naturg. Vög. Neuholl, p. 343, expresses the same opinion, nowhere, however, as far as I can detect, giving a name.

b^1. Tail cuneate; the young with the down on the sides of the bill not forming loral antiæ.*

c^1. Inner webs of outer four primaries and outer webs of the second, third, fourth, and fifth sinuated; the young with the down on the sides of the bill reaching almost to the nostrils; webs of the feet scalloped.

1. *Sthenelus* STEJNEGER 1882.

c^2. Inner webs of outer three primaries and outer webs of the second, third, and fourth sinuated; the young with the down on the sides of the bill terminating far back of the nostrils; webs of the feet straight, not scalloped.

2. *Cygnus* BECHST. 1803.

b^2. Tail rounded; the young with the down on the sides of the bill forming very distinct loral antiæ.

3. *Olor* WAGL. 1832.

a^2. Predominant color of the adults blackish; the young with naked lores; tertiaries and scapulars crisp; tail shorter than the middle toe with claw.

4. *Chenopis* WAGL. 1832.

Geographical distribution.

The *Cygninæ* appear both in the northern and the southern hemispheres as extra-tropical birds, no representatives of these large *Lamellirostris* being found within the tropics. They are consequently wanting both in the Indo-African Tropical—they do not at all breed in Africa—and in the American Tropical Region, only one species being met with in the South American Temperate and one in the Australian Region. The remaining seven species occur in the Arctic and the North Temperate Regions, the greatest number, viz, five, being found in the Old World, and here they only extend their winter migrations to the two southern provinces, the Mediterranean and the Manchurian, without breeding there. The two North American species only breed within the American division of the Arctic Region.

The following table gives a synopsis of their distribution:

TABLE I.

Name of species.	Arctic reg.		North temp. reg.		Amer. trop. reg.	Indo-Afr. trop. reg.	South Amer. temp. reg.	African temp. reg.	Antarctic reg.	Australian reg.
	Old world.	New world.	Old world.	New world.						
Sthenelus melancorypha	—	—	×	—	—	—	+	—	—	—
Cygnus gibbus	—	—	×	—	—	—	—	—	—	—
immutabilis	—	—	××	—	—	—	—	—	—	—
unwini	—	—	××	—	—	—	—	—	—	—
Olor cygnus	×	—	×	—	—	—	—	—	—	—
bewickii	×	—	—	—	—	—	—	—	—	—
columbianus	—	××	—	×	—	—	—	—	—	—
buccinator	—	—	—	×	—	—	—	—	—	—
Chenopis atratus	—	—	—	—	—	—	—	—	—	×

* This term denotes the projecting angle of the loral feathering at the base of the bill.

TABLE II.—Table of average comparative measurements.

Name of species.	Number of specimens.	Length of bill from tip to anterior border of—				Breadth of bill at nostrils.	Height of the knob.	Length of toes with claw.				Tarsus.	Tail.	Wing.
		The mouth.	The knob.	Nostrils.	The eye.			Outer.	Middle.	Inner.	Hind.			
	mm.	*mm.*	*mm.*	*mm.*	*mm.*	*mm.*	*mm.*	*mm.*	*mm.*	*mm.*	*mm.*	*mm.*	*mm.*	*mm.*
Sthenelus melancorypha (MOL.)	13	68	41	39	91	24	11	101	107	85	20	81	136	425
Cygnus gibbus BECHST	4	104	77	59	133	35	17	152	157	121	31	106	222	591
Cygnus immutabilis YARR	1	102	82	60	131	25	6	139	145	112	30	95	158	565
Cygnus unwini HUME	1	91		51	113	22		138	148	108	30	96	193	533
Olor cygnus (LIN.)	12	97		44	124	30		138	148	114	24	110	168	596
Olor bewickii (YARR.)	19	80		38	110	29		112	121	95	22	94	162	516
Olor columbianus (ORD.)	23	90		42	117	33		130	138	108	23	106	169	545
Olor buccinator (RICH.)	10	110		53	139	35		154	162	126	26	115	185	603
Chenopis atratus (LATH.)	8	73		40	97	25		114	123	96	26	91	103	475

STHENELUS * Stejneger gen. nov.

DIAGN.—*Predominant color of the adults, white; young with downy or feathered lores, the down on the sides of the bill reaching almost to the nostrils, but not forming distinct loral antiæ; tertiaries and scapulars normal, not crisp; tail longer than the middle toe with claw, cuneate; inner webs of outer four primaries and outer webs of the second, third, fourth, and fifth sinuated; webs of the feet scalloped.* (See fig. 8, and compare with fig. 10.)

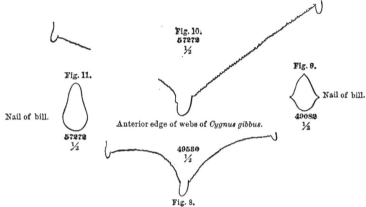

Fig. 10.
57272
½

Fig. 11.

Fig. 9.

Nail of bill.

Nail of bill.

49082
½

Anterior edge of webs of *Cygnus gibbus.*

57272
½

49580
½

Fig. 8.
Anterior edge of webs of *Sthenelus melancorypha*

Sthenelus melancorypha (MOL.).

Black-necked Swan.

DIAGN.—*Head and neck brownish black; body white; young in the down pure white; legs pale flesh-colored.*

SYN.—1782—*Anas melancorypha* MOLINA, Stor. Nat., Chili, p. 207.
    1786—*Anas melancoripha* BRANDIS, Uebers MOLIN. Naturg. Chili, p. 207.
    1788—*Anas nigricollis* GMEL., Syst. Nat. I, p. 502.
    1788—*Anas melanocephala* GMEL., ut supra.
    1810—*Anas melancorypha* MOLINA, Stor. Nat. Chil., 2 ed. (p. 199).
    1837—*"Anas melanocorphynphus* MOL." Less. Compl. Buff. IX, p. 528.
    1839—*"Anas melanocoryphea* MOL." Less. Rev. Zool. 1839, p. 322.
    1839—*"Anser melanocoryphus* BONN." Less., ut supra.

COLL. STEJNEGER No. 716, (♂ ad. *South America*).
Length of bill from tip to mouth 70ᵐᵐ, to anterior border of the nostrils 40ᵐᵐ, to the forward angle of the eye 95ᵐᵐ. Breadth of bill at the nostrils, 26ᵐᵐ. Length of toes with claw: outer toe 111, middle toe 118, inner toe 95, and hind toe 21ᵐᵐ. Tarsus 87, longest tail feathers 140, and wings 450ᵐᵐ. From tip of bill to the base of the frontal knob 43ᵐᵐ; the height of which amounts to 16ᵐᵐ.
The bill, in front of the tubercle and a point under the middle of the nostrils, is dark plumbeous, while the tubercle, the base of the bill behind the point mentioned and the naked lores, are yellowish brown. In

---

* Σένελος, nom. myth., father of CYCNUS, Ov. Met. 2, 367.

the live bird, these parts are stated to be, respectively, light plumbeous, with white nail, and intense rose-red. Iris is said to be brown or almost black. The legs are, in the skin, light brownish; in the live bird, pale flesh color.

The head and the upper two-thirds of the neck are of a beautiful blackish-brown color, with velvet gloss; a narrow white stripe surrounds the eye, from the hind angle of which it extends backward into the nape, but without meeting the stripe from the other side. On the chin a large white spot. The whole remaining plumage is pure white.

♀ differs from the ♂ only in being smaller.

COLL. STEJNEGER No. 711, (♂ jun. South America).

Length of bill along gape 69ᵐᵐ, from the tip to the front of the nostrils 39ᵐᵐ, to the fore border of the eye 98ᵐᵐ, breadth at the nostrils 25ᵐᵐ. Length of toes with claw: outer toe 103, middle toe 110, inner toe 91, and hind toe 22ᵐᵐ. Length of tarsus 87, tailfeathers 110, and wing 400ᵐᵐ.

The bill, which does not show the slightest trace of a frontal knob, is, in the dried condition, dark yellowish-red at the base, gradually changing into dark brownish towards the tip; the nail light yellowish. Legs light yellowish gray, with darker webs.

The plumage is white with pale rusty edges on each feather, this tinge being most intense on the upper parts. The head and the upper two-thirds of the neck, as in the adult described above; the brown, however, being considerably lighter. The limit of the feathering round the base of the bill very light, becoming almost white round the eye and on the chin, on which the light color forms a rather large spot; from the hind angle of each eye the white stripe extends backward, nearly meeting its fellow on the median line of the nape. The primaries are white, the tips broadly edged with dark chocolate-brown on the outer five, becoming narrower and fainter on the following quills; in the former, the colored edge is about 20ᵐᵐ broad at the tips, tapering towards the base on both webs, and becoming first obsolete on the outer web; the shafts of the outer quills are brown for the most part, gradually decreasing towards the innermost, the shafts of which are almost white to the very tip. The primary coverts are also more or less marked with brown shadings on the tips of the webs and shafts.

Another young specimen, U. S. Nat. Mus., No. 49530 (♀ jun.—*Conchitas, Buenos Ayres, June,* 1866), shows the following dimensions: Length of bill from the tip to the mouth 61ᵐᵐ, to the fore border of the nostrils 34ᵐᵐ, to the front of the eye 83ᵐᵐ, breadth 21ᵐᵐ. Length of toes with claw: outer toe 87, middle toe 95, inner toe 74, and hind toe, 20ᵐᵐ. Tarsus 88, tailfeathers 104, and wings 395ᵐᵐ.

No trace of frontal knob, the culmen only slightly rising above the nostrils.

Color as in the foregoing specimen, with the exception that the white behind the eyes is almost wanting, and the edges of the feathers of the middle part of the neck are conspicuously lighter brown, becoming almost white above towards the limit of the white part of the neck. Besides, the middle tail feathers are brownish gray on the inner web towards the tip; this color on the outer ones also extending into the outer web, which, however, is edged with white to the very tip; the shafts are brown.

In specimen *No. 66605, U. S. Nat. Mus.*, which has the knob still very small, viz, only 3^{mm}, the plumage has already become pure white, without any trace of brown shadings or spots, the same being the case in *No. 2, Mus. Leid.*, in which the height of the tubercle only amounts to 5^{mm}.

Fig. 12.

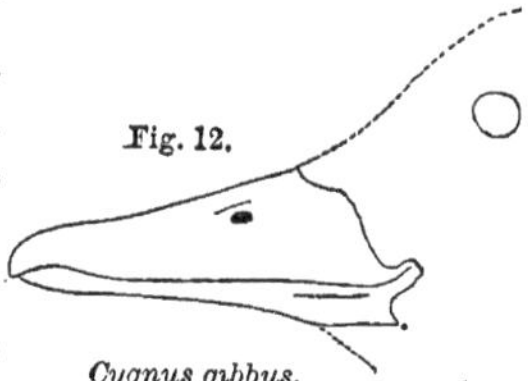

Cygnus gibbus.

Fig. 13.

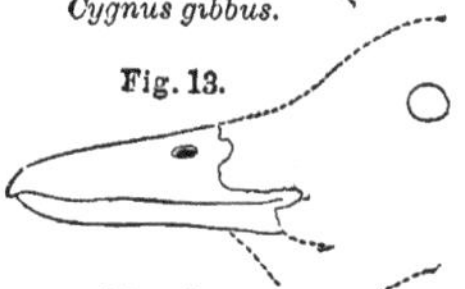

The *downy plumage* is white. The following dimensions and descriptions are from two cygnets hatched in the Zoological Garden in Rotterdam:

MUS. LEIDEN. (♀ *Pullus*, 34 *days old.*)

Sthenelus melancorypha.

Fig. 14.

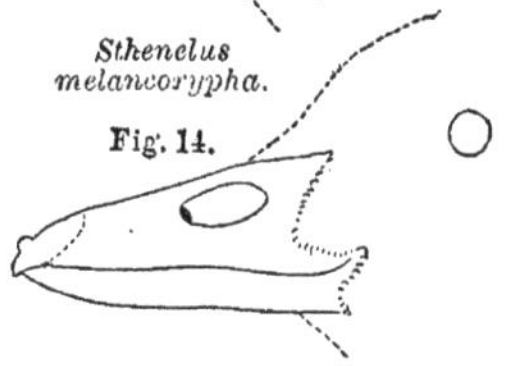

Length of bill along gape 36^{mm}, from tip to fore border of the nostrils 20^{mm}, to the eye 55^{mm}, breadth 12^{mm}. Length of toes with claw: outer toe 43, middle toe 44, inner toe 34, and hind toe 10^{mm}. Tarsus, 35^{mm}.

Pure white; the down of the upper parts is gray at the base, giving the upper surface a faint grayish tinge. Bill lead-black, with the nail light. Legs yellowish gray, with the webs grayish yellow.

Olor columbianus.

MUS. LEIDEN. (♂ *Pullus*, 1 *day old.*)

Length of commissure 21^{mm}, bill from tip to the nostrils 12, to the eye 30^{mm}, breadth of bill 8^{mm}. Length of toes with claw: outer toe 29, middle toe 31, inner toe 24, hind toe 6, and tarsus 23^{mm}.

Pure white; the grayish tinge on the upper surface almost imperceptible.

As may be seen by reference to fig. 13, almost the whole of the base of the bill in this species is covered with down, which reaches much more than half way to the nostrils, both above and below, and having a very different anterior outline from the same stage of species of *Cygnus* and *Olor*, as shown in figs. 12 and 14.

TABLE III.—*Sthenelus melancorypha* (Mol.).

TABLE OF DIMENSIONS.

Collection	Mus. No.	Locality	Date	Sex	Age	Bill from tip to anterior border of—				Breadth at the nostrils	Height of the Knob	Toes with claw				Tarsus	Tail	Wing
						Mouth	Knob	Nostrils	Eye			Outer	Middle	Inner	Third			
						mm.	*mm.*	*mm.*	*mm.*	*mm.*	*mm.*	*mm.*	*mm.*	*mm.*	*mm.*	*mm.*	*mm.*	*mm.*
Mus. Sth.	716	Kili		♂	ad	72	43	39	99	26	16	113	122	93	24	88	135	450
Coll. Steineger	15653	South America			ad	70	39	40	95	26	14	111	118	95	21	87	140	420
U. S. Nat. Mus	66607	Chili			ad	70	41	38	91	24	3	98	105	85	22	83	140	420
"		Buenos Ayres			ad	70	49	49	98	25	9	98	102	85	21	84	122	410
Mus. Ad., Phil		Kili			ad	73	45	42	98	26	12	108	114	89	24	90	143	450
"		South America			ad	69	42	37	98	25		103	111	87	22	80	132	427
Mus. Leiden	C. nigric. No. 3	...o, Chili	June, 1863	♂	ad	68	41	39	94	25	11	105	113	88	21	77	145	435
"	C. nigric. No. 3			♂	ad	66	39	40	91	25	14	99	107	83	20	78	147	420
U. S. Nat. Mus	49082	Ch..	June, 1864	♂	ad	63	34	36	86	25	9	95	100	80	18	78	127	420
"		Chili	July, 1855	♂	ad	62	42	36	85	23	5	97	105	85	17	80	137	
Mus. Leiden	C. nigric. No. 2	Santa Fé, Rio Salad		♀	ad	62	40	35	87	23	11	96	102	77	18	76	132	415
Mr. Mus., N. Y		Kili		♀	ad	61	33	37	84	23	11	95	94	78	20	76	127	400
U. S. Nat. Mus	49083	City of Santiago. Chili	July, 1865	♀	ad				80	23		91		78	17			

CYGNUS BECHST. 1803.

DIAGN.—*Predominant color of the adults white; young with downy or feathered lores, the down on the sides of the bill terminating far back of the nostrils, and not forming distinct loral antiæ; tertiaries and scapulars normal, not crisp; tail longer than the middle toe with claw, cuneate; inner webs of outer three primaries and outer webs of the second, third, and fourth sinuated; webs of the feet straight, not scalloped.*

SYN.— < 1803.—*Cygnus* BECHST., Taschb. Vög. Deutschl. p. 404.
    < 1840.—*Cycnus* TEMM., Man. d'Orn. 2 ed. IV, p, 526.
    = 1842.—*Olor* BP., Catal. Meth. Uccell. Europ. (gen. 206). (Nec WAGL.)

Synopsis of the species.

a^1. Culmen with a knob at the base.
    b^1. Frontal knob larger; legs in the adults black; young gray or brownish gray, with the bill lead-color.

1. *gibbus* BECHST. 1809.

    b^2. Frontal knob smaller; legs in the adults gray or yellowish gray; young white, with the bill light pinkish red.

2. *immutabilis* YARR. 1838.

a^2. Culmen without knob.

3. *unwini* HUME 1871.

Cygnus gibbus BECHST.

Mute Swan.

DIAGN.—*Culmen with a large knob at the base; legs in the adults black; young gray or brownish gray, with the bill lead-color.*

SYN.—1758.—*Anas cygnus* LINN., Syst. Nat. x, ed. I, p. 122 (*part*).
    1783.—*Anas cygnus* BODD., Tabl. Pl. Enl. p. 54 (ed. TEGETM.).
    1788.—*Anas olor* GMEL., Syst. Nat. I, p. 501 (nec PALL. 1779 quæ *Olor cygnus* (L.).).
    1809.—*Cygnus gibbus* BECHST., Gemeinn. Naturg. Deutschl. IV, p. 815.
    1811.—*Cygnus sibilus* PALL., Zoogr. Rosso-As. II, p. 215.
    1817.—*Cygnus mutus* FORSTER, Synopt. Cat. Br. Birds, p. 64.
    1820.—"*Cygnus gibbosus* MEYER," KUHL, Buff. Fig. Av. Nom. Syst. pp. 16 and 26.
    1828.—*Cygnus mansuetus* FLEM., Brit. Anim. (p. 126).
    1858.—"*Cygnus sibilans* PALL.," NILSS. Skand. Fauna, Fogl. 3 ed. II, p. 386.

This species has usually been called *Cygnus olor* (GMEL. nec PALL. 1811). But, as will be shown below, PALLAS has given the name *Anas olor* to the Hooper long before GMELIN compiled his *Systema Naturalis*, for which reason the name of the latter must be suppressed for the present species. The matter stands as follows: PALLAS, in 1779, in the introduction to a treatise on *Anas glocitans* (Sv. Vetensk. Acad. Handl. XL, p. 26–27), says as follows: . . . "Duck-genus (*Anas*) . . . most kinds occur very generally both in Europe, Asia, and America, but not in the Tropics. Thus also . . . ·the Swan is to be found, not only that which is rightly called *Cygnus*, which has a hoarse and hissing voice, but also the *Olor*, which the newest zoologists hesitate over, and which has a clear and pleasant voice, which can be heard far; it ought thus to belong to a different species." In a foot-note he adds: "I mean here the so-called *Cygnus ferus*, . . . which . . . ' really is a distinct species from the so-called *Cygnus mansuetus*." He gives the name clearly, in a scientific and highly distinguished journal, which, at that time, was widely spread over everywhere where the science of natural history was cultivated, and four years after the above-quoted remarks were translated into German (Schwed. Abhandl. Uebersetzt von A. G. Kästner, 41 vol. p. 23, Leipzic, 1783). He urges repeatedly that the same species (in opposition to LINNÆUS, who had only distinguished between the two Swans as the wild and the tame state of the same species) must be separated, and his indication of which kind he means is fully unmistakable. By this he has fulfilled all demands by adopting a name given by an author. In this case it is not less than nine years older than GMELIN'S. But of his errors, the same applies to this as to many others—that science must not allow itself to be bound by them, even if it should cause the greatest difficulties to rectify the mistake.

Among the synonyms of this species Mr. DRESSER (Birds of Eur.) cites "*Anas* (*Cygnus*) *mansuetus*, LATH. Gen. Synopsis, Suppl. p. 297 (1787)." This quotation is not correct, and can be misapprehended, as if LATHAM had given the name *mansuetus* as a specific one, but he only writes—

GENUS XCII.
Anas.

. . . *Cygnus* (*ferus*) . . .
. . . —— (*mansuetus*) . . .

TABLE IV.—*Cygnus gibbus* (BECHST.).

TABLE OF DIMENSIONS.

Collection.	Mus. No.	Locality.	Date.	Sex.	Age.	Bill from tip to fore border of—				Breadth of bill at nostrils.	Height of knob.	Toes with claw.				Tarsus.	Tail.	Wing.
						Mouth.	Knob.	Nostrils.	Eye.			Outer.	Middle.	Inner.	Hind.			
						mm.	*mm.*	*mm.*	*mm.*	*mm.*	*mm.*	*mm.*	*mm.*	*mm.*	*mm.*	*mm.*	*mm.*	*mm.*
Coll. Stejneger	441	Copenhagen, Denmark	Feb. 22, 1881	♂	ad	106	80	62	136	..	..	160	162	125	31	117	245	615
Mus. Copenh.	Eur. birds, 129	Sealand, Denmark			ad	103	75	56	133	34	22	154	165	126	32	108	218	595
Museum Leiden	57171	Harderwijk, Holland	Feb. 17, 1871	♂	ad	105	79	60	134	34	16	148	152	118	32	101	239	580
				♂	ad	101	73	58	127	36	13	144	150	116	30	97	187	575

Cygnus immutabilis YARR.

Polish Swan.*

DIAGN.—*Culmen with a smaller knob at the base; legs in the adults slate-gray or yellowish gray; young white, with the bill light pinkish red.*

SYN.—1838.—*Cygnus immutabilis* YARR., Proc. Zool. Soc. Lond. 1838, p. 19 (nec v. PELZ. 1862, quæ *C. unwini* HUME).

Since YARRELL, in 1838, described this species, but few contributions to the elucidation of the questions concerning its habitat and its relation to *C. gibbus* have been made. The time was when its right as a species was generally denied, essentially for the reason that a few instances of mixed broods with both white and gray cygnets were stated to have occurred. But at present, the opinion being inclined to regard such a case as "the result of an alliance between a Mute and a Polish Swan," the distinction of these two species seems to be generally admitted—at least in England. The various investigations about this question are described at great length, and important new observations given, by Mr. DRESSER in his Birds of Europe, Parts lxxvii, lxxviii, and lxxix, April, 1880, but not even he has answered the inquiry as to the true habitat of the *immutabilis*. At first I regarded it as an eastern form, confounding it with *C. unwini;* but I have now convinced myself that the latter constitutes a different species, and I am inclined to believe that the English *immutabilis* will show itself to be a western bird. Specimens can easily be overlooked, and a few may, perhaps, be found in one or another of the European museums (as, for instance, the example in Mus. Leiden.), but I see no reason why it should be supposed that the ornithologists of the continent have been less exact in this case than those of England. I therefore regard the species as being very scarce on the European continent; the only specimen from there was killed in Holland, just opposite to England, in which latter country it seems to be not even rare.

BLAINVILLE has already questioned whether the *immutabilis* is not the wild form of the Tame Swan, and we see that Mr. DRESSER for a long time also was inclined to indorse the same view, which, however, my investigations most positively contradict. It appears to me that the question, with more right, could be asked conversely, viz, whether the Polish Swan is not a race originated by domestication; but even this seems not to be the case, as it appears from the quotation in Mr. DRESSER'S Birds of Eur. (l. c.) of the experiences of Mr. SIMPSON, "who had from seventy to a hundred cygnets through his hands yearly for the past thirty years, and who never saw a white one," and from the statement of Mr. DRESSER himself, that the Changeless Swan, "so far

---

*Not "Polar Swan (Cygne du Pôle)" as BLAINVILLE, Compt. Rend. VII, 1838, p. 1024, and after him DEGL. & GERBE, Ornith. Eur. II, p. 477, indicates.

as he can ascertain, has only been recorded in a wild state from the shores of Great Britain."* Should it, after all, be an absurd supposition that *immutabilis* is the indigenous wild English Swan, while *gibbus* is indigenous only to the continent, but introduced, in a half domesticated state, to England during the time of Richard I?

This Swan presents the peculiar fact that the young of it are better distinguishable from its nearest allies than the adults of both species in their perfect plumage. This is, however, no objection to its right to be considered a species any more than in the case of two other species, the plumages of whose young are quite alike.

The most conspicuous distinctive mark of the two species is that the young (in down and in the first plumage) of *immutabilis* are white, and not gray or brownish, as in *gibbus*. They are, however, not pure white, at least not always, as they were described as being on the back more or less tinged with warm buff.

They differ also in the color of the bill, this being pale pinkish red in the young *immutabilis* and plumbeous in *gibbus*. It cannot here be objected that the Mute Swan in the later youth also has the bill of a similar color, as it, during the transition to the white plumage, begins to take a reddish tone, because the mentioned red color on the bill of the young Polish Swan is to be found already in the first summer simultaneous with the first feathers, as is evident from Mr. SOUTWELL'S (DRESSER l. c.) description of the plumage of three young the 20th of August: "They had then assumed nearly all their feathers and were more than half grown; the color was white, apparently stained or sullied by a yellowish tint, which was strongest on the wing-coverts; feet pale ash-color, and beak a purplish flesh-color, differing entirely from the lead-color of the bill in the young Mute Swan of the same age." Also the color of the bill of the adult birds is different, the Polish Swan having it rather redder than the continental species.

The frontal knob is said to be smaller in *immutabilis* at all ages. It is, however, present also in the quite young, as is evident from Mr. DRESSER'S plate, fig. 2. The eye and the lamella, too, are said to be smaller.

The character now to be mentioned belongs only to the adult birds. In the adult *gibbus* the legs are jet-black, sometimes with a shade of red shining through the black color; in *immutabilis* their color is variously stated to be from pale plumbeous or slate-gray to a light drab color. This latter color they had in the specimen examined by me. In the young the color of the feet is nearly the same in the two species, and it

---

* Is the statement, p. 4, about the captures of *immutabilis* in Norfolk, enumerated by Mr. STEVENSON, contrary to this? He says: "Some, at least, if not most of these, however, were undoubtedly birds which had straggled from other waters, and not genuine wild birds." I cannot plainly see if these words are the reflections of Mr. DRESSER himself or only a quotation of Mr. STEVENSON.

July 25, 1882.

is expressly stated "that at no stage of growth is this a character to be depended upon."

In their size they seem not to differ. Mr. DRESSER gives the total length of the adult male of *gibbus* at about four and a half to five feet, the gape 3.55 inches, and the tarsus 4.5 inches; and of the adult *immutabilis*, respectively at about five feet, 3.6, and 4.25 inches. The differences in the length of the wing, 27 inches as against 23.5, and yet more in the length of the tail, 10 to 6.8, are certainly quite considerable, but not more shan sometimes occurs in the same species, especially as it is probable that the feathers of the specimen from North-Repps are not fully developed.

Finally, there are the osteological differences described by Mr. PELERIN (Mag. Nat. Hist., 1839, p. 178), which I have had no occasion to verify, and which I cannot remember to have seen confirmed or denied by any other thar Mr. YARRELL himself.

The English ornithologists may after this be right when they urge the independence of *C. immutabilis*, and it should be a great offense against the science if one would unite these two forms and hereby cut off, or at least trouble, the study of this particular phenomenon.

As far as my investigations go, they also agree with the results of the English authors. In SCHLEGEL's Catal. Mus. P.-B., VI, *Anseres*, p. 79, a male "de l'année" is enumerated under *Cygnus olor* (GMEL.) as killed on the Lake of Haarlem in the month of December, 1840. The description of this interesting specimen, which certainly belongs to *C. immutabilis* YARR., is as follows:

MUS. LEIDEN, *C. olor No. 3* (♂, *Lake of Haarlem, Holland, December, 1840*).

Length of the bill along the gape, 102^{mm}; from the tip to the fore border of the nostrils 60, and to the eye 131^{mm}. Length of toes with claws: Outer toe 139, middle toe 145, inner toe 112, and hind toe 30^{mm}. Tarsus 95, tail 158, and wing 565^{mm}. The distance from the tip of the bill to the fore border of the knob 82^{mm}, the knob itself being 6^{mm} high.

The whole plumage pure white, with a faint rose-colored shade on the wing-coverts, and a rust-colored tinge on the crown and chin. The tarsus and toes yellowish-gray, the webs grayish-yellow. The original color of the bill cannot be recognized in the dried specimen.

If one compare the above dimensions with those given on Table IV, it will be seen that they agree quite well with the smallest specimen. The small size of the frontal knob, and the remarkably light feet, are very characteristic features, combined with the white plumage. I therefore regard the identification of this specimen with *C. immutabilis* to be unquestionable.

As to the colors of the young, I refer to the descriptions given above.

Cygnus unwini HUME.

Knobless Swan.

DIAGN.—*Culmen without knob; legs in the adults slate-colored; young gray or brownish gray.*

SYN.— ? 1804.—*Anas dircœa* HERMANN, Observ. Zool. I, p. 139.
 1862.—*Cygnus immutabilis* V. PELZELN, Schr. Zool. Bot. Ver. Wien, xii, p. 785 (nec YARR. 1838).
 1871.—*Cygnus unwini* A. O. HUME, Ibis 1871, p. 413.
 1871.—*Cygnus olor* SALVIN, Ibis 1871, p. 413 (nec PALL., nec GM.).
 1872.—*Cygnus urwini* GIEBEL, Thes. Orn. I, p. 857.

Note 1 to the synonymy.—DRESSER cites HERMANN'S *Anas dircœa* as belonging to *immutabilis* YARR. with a query. Because the description of the said author contains the phrase "*corpore cinereo*" I regard this reference unadvisable. The resemblance of the title *Cygnus polonicus*, cited by HERMANN, and the English name, "Polish Swan," is of no consequence for the reason that such a title is not to be found in GESNER, in spite of the quotation.* It belongs rather to the species here in question, but the phrase "*rostro rubro*" makes me hesitate, because I am not satisfied whether the young of this species has a red bill or not. From the description of HUME it seems that it should not be the case. HERMANN does not speak about the knob, it is true, but if it had been completely absent he should not have failed to mention it. I have therefore introduced it into the above synonymy with some doubt.

Note 2.—The museum at Vienna received in the year 1857 three adult swans which Mr. ZELEBOR had captured in the month of March the foregoing year, and which had been deposited in the imperial menagerie at Schönbrunn, near Vienna, where they died in the beginning of the said year. Misled by the statement that white and gray cygnets had been found in the same brood, Mr. A. V. PELZELN, in a short article (l. c.), identified the specimens with the *C. immutabilis* YARR.

Mr. A. V. PELZELN has had the great courtesy to send one of the specimens a great distance for my inspection, and I am thus enabled to make up my own opinion.

As far as I can judge the specimens in question are distinct from both the *gibbus* and the *immutabilis*. From both these species they are separated by the complete absence of even the slightest trace of a tubercle or knob, by their inferior size, and by the different form of the bill. From the former, with which they agree in having the plumage of the young brownish, they are further distinguishable by the legs and webs, which are "slate-colored, changing into olive," and from the latter by having a brownish and not white plumage of the young.

---

* HERMANN quotes: GESNER, Edit. Francof. 1604, p. 273 B; but on p. 373 B (on p. 273 he treats of *Ciconia*) he only says: "*In Polonia cygni sunt diversi generis; sunt enim alij feri, pari magnitudine, alij domestici, quorum vox suavis est, & tuba refert.*"

In 1871 Mr. A. O. HUME described two young swans from India under the name *C. unwini*, which I consider not to be identical with the *gibbus*, as is the general opinion, for the following reason:

They are said to have not the faintest trace of a tubercle. The young of the common Mute Swan get their knob very early, as soon as they have been full feathered. The fact that the female has a smaller and more indistinct knob is of no consequence, while HUME had before him both male and female. They were, too, full grown, ready to take the white plumage in the first spring, and in this age the young *gibbus* has a very distinct knob, even in the female sex.

Nor can these birds be identified with the *immutabilis* YARR., the total absence of the knob and the brownish plumage being invincible obstacles.

On the other hand, the description in these points agrees very well with the above-mentioned birds from Egypt. The following description is by HUME: "From the frontal feathers to beyond the end of the nasal fossa, a distance of very nearly 1½ inch, the culmen is a perfectly straight line. Beyond this there is a very shallow concavity to the posterior margin of the nail."

To the foregoing five specimens I add a sixth, which agrees in the eastern habitat, the absence of the tubercle, and the apparently dark young plumage, viz: the bird, which C. A. WRIGHT mentions (Ibis, 1874, p. 241), as follows: "There is an example of *C. olor* in the Malta University Museum nearly pure white, but with scarcely any appearance of the frontal knob."

The Polish Swans, indicated to have been found in Corfu and Epirus (Ibis, 1860, p. 351, and 1870, p. 338), probably may also belong to the species here in question.

K. K. HOF.-NAT. CAB. VIENNA. (*Taken alive in March, 1856, on Lake Menzaleh, Egypt; died in confinement at Vienna, 1857. By H. ZELEBOR.*)

Total length 1,300^{mm} (V. PELZELN *in litt.*).*

Length of the bill along gape 91^{mm}, from the tip to the front of the nostrils 51^{mm}, to the fore border of the eye 113^{mm}; breadth of bill at nostrils 32^{mm}. Length of toes with claws: outer toe 138, middle toe 148, inner toe 108, and hind toe 30^{mm}. Tarsus 96, wing 535, and tail 193^{mm}.

Mr. A. V. PELZELN† describes the color of the bill on the newly dead bird as orange changing into crimson, with the same black markings as the Mute Swan. On the stuffed bird the black color has the follow-

---

* The dimensions of the two other specimens were: Total length 1,360 and 1,330^{mm}; wing 540 and 550; bill along gape 105 and 85^{mm}. (v. PELZ. *in litt.*)

† Sehr. Zool. Bot. Ver. Wien, Bd. XII. 1862, p. 785: Notiz. über *Cyqnus immutabilis* YARRELL. Von A. v. PELZELN.

ing extension: The naked skin between the base of the bill and the eye; further, a large spot 8ᵐᵐ long, on the culmen adjacent to the frontal feathers, and connected with the black loral space by a small black stripe; besides, the whole skin covering the nasal cavity is black, surrounded by the red-bill color; finally, the nail both on the upper and lower mandible with the edges of both jaws. "The legs and webs are not black, but slate-colored, changing into olive" (v. PELZELN).

Plumage pure white with a few brownish feathers here and there, the remains of the young plumage.

For the sake of completeness, I here give the main points of Mr. HUME's description of the coloration of the young.

(♂ and ♀ jun. Jubbee stream, on the borders of the Hazara and Rawulpindes districts, India.—17th January, 1871.—By Capt. UNWIN.)

"If from each side of the frontal tongue of feathers, about half an inch from its point, a slightly curving line be drawn to a point on the edge of the upper mandible about a quarter of an inch from the gape, the whole of the space inclosed by such line between it and the eye is perfectly black. At the extreme point of the frontal feathers, again, is a black band about a quarter of an inch wide, which extends right and left over the whole narial space. The nail is black; the rest of the bill was light gray. The legs and feet, I may add, were grayish black.

"The general color of the lower surface is a dull white; of the upper whitey-brown. The crown and occiput wood-brown; the greater portion of the wing, the scapulars, and rump are wood or sandy brown. There is nowhere any trace of a 'sooty gray.' The brown is essentially a buffy or sandy brown, though here and there, as in the feathers of the base of the neck, a faint grayish shade is intermingled.

"Both male and female, though differing somewhat in size, are precisely similar both as regards plumage and coloration of the bill."

I am aware that of late there have been published two or three papers about the Indian Swans in "Stray Feathers," and in the "Journal of the Asiatic Society of Bengal," but as I have not been able to procure any of them, I cannot say whether they have any influence on the question discussed above. If they really prove that Capt. UNWIN's young birds belong to Cygnus gibbus, I would propose that the present species, which certainly at all events is distinct from the Knob-Swan, should be called Cygnus pelzelni.

OLOR WAGL. 1832.

DIAGN.—Predominant color of the adults white; the young with downy or feathered lores, the down on the sides of the bill terminating far back of the nostrils, and forming very distinct loral anticæ; tertiaries and scapulars normal, not crisp; tail longer than the middle toe with claw, rounded; in-

ner webs of outer three primaries, and outer webs of the second, third, and fourth, sinuated; webs of the feet not scalloped.

SYN.—= 1832.—*Olor* WAGLER, Isis 1832, p. 1234 (nec BP. 1842, quæ *Cygnus*).

1845.—*Cygnus* GERBE, Rev. Zool., 1845, p. 244.

Synopsis of the species.

a¹. The distance from the anterior angle of the eye to the hind border of the nostrils much longer than the distance from the latter to the tip of the bill.

b¹. The yellow color at the base of the bill extending beyond the nostrils.

1. *cygnus* (LIN.), 1758.

b². The yellow color at the base of the bill not extending to the nostrils.

c¹. Smaller: Total length about 1,150^{mm}; middle toe with claw about 125^{mm}; the yellow spot at the base of the bill making at least ⅓ of the surface of the bill and lores.

2. *bewickii* (YARR.), 1830.

c². Larger: Total length about 1,400^{mm}; middle toe with claw about 140^{mm}; the yellow spot at the base of the bill making, at most, $\frac{1}{15}$ of the surface of the bill and lores.

3. *columbianus* (ORD.), 1815.

a². The distance from the anterior angle of the eye to the hind border of the nostrils equal to the distance from the latter to the tip of the bill.

4. *buccinator* (RICH.), 1831.

Olor cygnus (LIN.).

Hooper Swan.

DIAGN.—*The distance from the anterior angle of the eye to the hind border of the nostrils is much longer than the distance from the latter to the tip of the bill; the yellow color at the base of the bill extending beyond the nostrils, making ⅔ of the surface of the bill and lore.*

SYN.—1758.—*Anas cygnus* LINN., Syst. Nat. x ed. I, p. 122 (*part*).

1779.—*Anas olor* PALL., Sv. Vet. Acad. Handl. XL, p. 27 (nec GMEL. 1788 quæ *Cygnus gibbus* BECHST.)

1809.—*Cygnus musicus* BECHST., Gemein. Naturg. Deutschl. IV, p. 830 (nec BP. 1826, quæ *O. columbianus* (ORD)).

1810.—*Cygnus melanorhynchus* MEY. & WOLF, Taschb. Vög. Deutschl. II, p. 498.

1816.—*Cygnus ferus* LEACH, Syst. Cat. Mam. & Birds, Br. Mus. (p. 37) (nec BARTR. 1791, quæ ? *O. columbianus* (ORD)).

1830.—*Cygnus islandicus* BREHM, Isis, 1830, p. 1135 (nec NAUM. 1838, quæ *O. bewickii* (YARR)).

1842.—*Cygnus xanthorhinus* NAUM., Vög. Deutschl. XI, p. 478.

1877.—*Cygnus linnei* MALM., Götebs. och Bohusl. Fauna, pp. 90 and 343.

Since BECHSTEIN raised the specific name *cygnus*, given by LINNÆUS, to the rank of a generic name, the species has been called *musicus* or *ferus*. As synonymous, *Cygnus olor* PALL., Zoogr. Rosso-Asiat. II, p. 211 (nec

GMEL. 1783) has been thus quoted. But the name of PALLAS has, without doubt, the priority in this species, as I have shown above (p. 189), for which both BECHSTEIN'S name for this and GMELIN'S *olor* must give way, and I cannot see but that the authors, who only admit one genus of Swans, viz, *Cygnus*, must adopt the name of PALLAS as the oldest for the present species. It is certainly a serious matter to transfer the name, which the Mute Swan has borne so long, to the Hooper, but if we consider the right of priority, there is nothing else to be done. In this case it does not depend upon a question which can be disputed, how the old authors' descriptions can be interpreted (as, for example, with *Sterna hirundo* and *Stercorarius parasiticus*). [See the note under *Cygnus gibbus*.]

TABLE V.—*Olor cygnus* (LIN).

TABLE OF DIMENSIONS.

Collection.	Mus. No.	Locality.	Date.	Sex.	Age.	Bill from tip to anterior border of— Mouth.	Bill from tip to anterior border of— Nostrils.	Bill from tip to anterior border of— Eye.	Breadth of bill at nostrils.	Toes with claw. Outer.	Toes with claw. Middle.	Toes with claw. Inner.	Toes with claw. Hind.	Tarsus.	Tail.	Wing.
						mm.	*mm.*	*mm.*	*mm.*	*mm.*	*mm.*	*mm.*	*mm.*	*m.*	*mm.*	*mm.*
Mus. Acad., Phil	17–No. 70	Europe			ad	108	52	136	32	155	167	130	25	114	171	615
"		"			ad	103	46	133	29	147	156	121	27	111	162	590
Mus. Copenh		Ringkjöbing, Jutland, Denmark	Janr., 1863	♂	ad	101	45	127	34	147	159	118	25	121	183	628
Mus. Leiden	C. Mus. No 5	New York		♂	ad	110	44	B1	30	143	154	118	30	112	163	585
Mus. Bergen		Bergen, Norway			ad	99	45	128	30	139	146	115	24	108	164	630
"		"	22 Febr., 1877		ad	93	46	128	31	142	148	118	29	115	163	505
"		"	29 Novbr.,1877		ad	93	42	122	31	137	145	111	23	102	180	620
Mus. Acad., Phil	17–No. 28	Lappland, Sweden			ad	94	41	119	29	133	140	106	28	101	175	625
U. S. Nat. Mus	57173			♂(?)	ad	95	42	119	28	132	141	111	26	98	160	558
Coll. Stejneger	436	Stavanger, Norway	Febr., 1881	♀	ad	97	41	122	28	132	140	111	23	98	163	575
Mus. Copenh		Greenland	13 Novbr, 1861		ad	93	42	119	32	129	146	113	23	99	163	560
					ad	90	40	114	30	124	132	107	26	119	174	570

Olor bewickii (YARR.).

Bewick's Swan.

DIAGN.—*The distance from the anterior angle of the eye to the hind border of the nostrils is much longer than the distance from the latter to the tip of the bill; the yellow color at the base of the bill does not extend to the nostrils, making at least ⅓ of the surface of the bill and lores. Smaller: Total length about* 1150^{mm}; *middle toe with claw about* 125^{mm}.

SYN.—1830.—*Cygnus bewickii* YARRELL, Trans. Lin. Soc. XVI, p. 453 (nec RICH. 1831 quæ *O. columbianus* (ORD)).

1838.—*Cygnus islandicus* NAUM., WIEGM. Archiv, IV, 1838, p. 364 (nec BREHM, 1830, quæ *Olor cygnus* (LIN.)).

1838.—*Cygnus berwickii* EYTON, Monogr. Anat. Pl. 18 (*err. typ.*).

1840.—*Cygnus minor* KEYS. & BLAS., Wirbelth. Europ. p. LXXXII.

1842.—*Cygnus melanorhinus* NAUM., Vög. Deutschl. XI, p. 497.*

1851.—*Cygnus musicus* KJÆRBÖLL., Orn. Dan., Pl. XLIV (nec BECHST. quæ *O. cygnus* (LIN.)).

1854.—*Cygnus americanus* HARTL., Naumannia, 1854, p. 327 (nec SHARPL. quæ *columbianus* (ORD)).

1856.—"*Cygnus altumi* HOMEYER" BP., Cat. PARZUD., p. 15.

1866.—"*Cygnus altumii* BÄDEKER," SCHLEGEL, Mus. P. B., VI, *Anseres*, p. 82.

1880.—*Cygnus bewicki* DRESSER, Birds of Eur. pt. lxxvii-lxxix.

Note 1.—In PALLAS'S Zoographia Rosso-Asiatica I, p. 214, this species is found to be separated from the common Hooper, but only as variety "*β minor*" under *Cygnus olor*.† In 1840 KEYSERLING and BLASIUS altered the name given by PALLAS to a binominal, and called the species *minor*; but already, 10 years earlier,‡ YARRELL had described it under the name *bewickii*. The same year BREHM, in *Isis*, had named a little Swan as *islandicus*, but after what I have tried to show, in my second note on the synonymy, it does not belong here, but to the Hooper, whilst the species described and well drawn by NAUMANN in WIEGMANN'S Archiv. 1838, under BREHM'S name *islandicus* certainly belongs to *bewickii*; he altered the name 4 years later to *melanorhinus*. In Naumannia for the year 1854, p. 145, Taf. I and II, Professor ALTUM described and delineated a little Swan after specimens killed in North-western Germany, and which he considered to be a new species, different from *bewickii*, but without giving it a name, and whose principal

---

*GIEBEL, Thesaur. Ornith. I, cites, "WIEGM. Arch. IV, 1838, p. 361, Taf. 9," and DRESSER, B. of Eur., has the same quotation.

†Although PALLAS'S description in the above-named place only partly refers to *C. bewickii*, it will not do to place his name as unconditionally synonymous with the Hooper, as Mr. FINSCH does (Verh. Zool.-Bot. Ver. Wien XXXIX, 1879, p. 255).

‡A specimen was killed in France in 1807, and was deposited in the museum at Leiden under the name *Cygnus musicus*, until I, the last summer, identified it with *bewickii*.

character consisted in the unmixed black color of the whole culmen. During the discussion which followed, the name ALTUM'S Swan was occasionally employed to indicate the specimens described by him, and hence probably "*Cygnus altumii*" originated.

Note 2.—Mr. DRESSER indicates the year of publication of YARRELL'S name to be 1833, probably because the volume of the "Transactions" in question bears that date on the title-page. But the part in which YARRELL'S treatise was printed was published in 1830. Besides, Mr. DRESSER quotes "P. 445," which also is that on which the treatise begins, but the name and the diagnosis first occur on p. 453.

Note 3.—Prof. J. REINHARDT has already made a statement* which, strange to say, has generally been overlooked, to the effect that C. L. BREHM'S *Cygnus islandicus* is not synonymous with the species in question. His description in OKEN'S Isis, 1830, p. 1135, and in Handb. Vög. Deutschl. 1831, p. 832, contains nothing on which the identity can be founded, with the exception that the Iceland Swan was smaller, "frequently 6 inches shorter," than the Hooper. Besides, the shape of the bill of the two supposed species was indicated to be different, but not in such a manner that anything about the present question is to be concluded from this. It is highly improbable that BREHM could really have a *C. bewickii* before him without taking notice of the difference between the extension of the yellow on the beak. The matter will be found to be quite certain when we look at the drawing in his Handb. Vög. Deutschl., pl. xli, fig. 1, which, according to p. 1035, is meant to represent *C. islandicus*. Though drawn by GOETZ, and belonging to the class of unlucky representations, it still unmistakably shows the extension of the yellow color, both in the upper and the lower mandible, precisely as in the Hooper, viz, the yellow color is carried to a point under the nostrils, and BREHM expressly assures us that all the figures are drawn from nature. Neither can it be pleaded as a proof against the opinion here expressed that Iceland is stated as the habitat after it is known that *C. bewickii* has never been seen there. Neither do BREHM'S small specimens allow themselves to be referred to as any pigmy variety of the Hooper. Prof. J. REINHARDT, in Copenhagen, has, at all events, kindly informed me that those Swans occurring in Iceland cannot be separated from those of the continent on account of smaller size.† Here it must be remarked that the so-called considerable difference in size, viz, "6 inches," is not especially extraordinary. The difference between the largest and smallest individuals of the latter species which I have measured (except an unusually small specimen from Greenland) amounts to 5 inches.

Note 4.—The uppermost figure to the right on Plate xliv in KJÆR-

---

* Natuhistorisk Tidsskrift, II (p. 532).

† Personally I have had no opportunity of examining skins of specimens from Iceland. In the mean time this affair ought to be very closely examined. I refer here to the remarkably small specimen of the Hooper from Greenland, included in the table of dimensions on p. 202.

BÖLLING'S "Ornithologia Danica" represents undoubtedly a *C. bewickii*, although that on the plate is called *musicus*, and although the author in the text under the latter species refers to the same representation. The yellow color on the beak has in fact precisely the peculiar limit of that in *bewickii*.

Note 5.—In SCHLEGEL'S Mus. P.-B. *Anseres*, p. 82, and in DEGLAND and GERBE'S Ornith. Europ. II, p. 474 (probably on his authority), as synonymous with the supposed species, *Cygnus altumii*,* BÄDEKER is stated, without date and without naming the place from which the quo; tation is taken. In BONAPARTE'S Cat. Ois. Eur. PARZUDAKI, 1858, in SEVERZOW'S "Turkistanskie Jevotnie," 1873, and again in Mr. DRESSER'S translation of the same in the Ibis, 1876, p. 416, also in CAB. Journ. Ornith. 1875, p. 184, the name *Cygnus altumi* occurs, but with the author's name, HOMEYER, added. Mr. E. V. HOMEYER has in the meantime had the kindness to inform me as follows: "Neither I nor any of my friends in Berlin have any knowledge that BÄDEKER has anywhere spoken of a *C. altumii*. Neither have I ever done so. . . . I repeat that I have never spoken about *C. altumii*, and do not know how SEVERZOW can have quoted me." Prof. B. ALTUM writes to me that the Swan described by him in Naumannia IV, p. 145, BÄDEKER has had figured with the name in question. He can, however, neither give place nor date.

COLL. STEJNEGER no. 394.   (♀ *ad. Sömme, Jadderen, Norway, 58° 53″ N. lat., 22d January,* 1880.   *By Mr.* SOPHUS A. BUCH.)

Total length of the newly-killed bird, 1,135^{mm}; length of the bill along gape, 89^{mm}; from the hip to the front of the nostrils, 39^{mm}; to the fore border of the eye, 108^{mm}; length of toes with claws: outer toe 116, middle toe 124, inner toe 99, and hind toe 19^{mm}. Tarsus 90, wing 530, and tail 163^{mm}.

Bill, black on the whole surface from the tip to the front, and on the sides to a point about 15^{mm} behind the nostrils; the remainder and the naked lores intense reddish-yellow, about of the same color as the pulp of the blood-orange; the border of the black color forms a very jagged line; on that part of the culmen which lies between the lateral yellow spots the yellow color shines through the black, like the shadings in marble; along the forehead towards the eye both the beak and the lores are black; the lower jaw black, the margins with the lamellæ dark flesh-colored; the naked skin of the chin grayish-black, with transparent faint yellowish marbled shadings. Feet, grayish-black.

The plumage pure white, with a fine ashy-gray tinge on the sides of the head, and edged with pale rust-color on the feathers of the forehead, crown, and cheeks.

---

* Or *altumi*, as DEGL. and GERBE, 1867, write it.

Mus. Bergen (*Balestrand, Sogn, Norway,* 61° 8″ *N. Lat.,* 19th *January,*
1880. *By Mr.* Svφrdrup.)

Length of the bill from the tip to the mouth 89ᵐᵐ, to the front of the
nostrils 39ᵐᵐ, and to the fore border of the eye 113ᵘᵐ. Length of toes
with claws: outer toe 118, middle toe 127, inner toe 96, and hind toe
21ᵐᵐ. Tarsus 92, wing 550, and tail 166ᵐᵐ.

In this specimen, when fresh, the lateral spots on the bill are of a
purer yellow color than the preceding, without red; also the whole
culmen is jet black, without the shaded yellow transverse stripe towards
the forehead; the margin of the lower mandible with the lamella quite
light flesh-colored.

On the whole like the former, although without the grayish tinge on
the sides of the head.

Coll. of Norway Scientific Soc. Trondhjem. (♀ ? ad., *Stjord-
alen, Norway,* 63° 25″ *N. Lat.—The first half of February,* 1880.)

Length of bill along gape, 87ᵐᵐ; from the tip to the front of the
nostrils, 39ᵐᵐ ; to the fore border of the eye, 116ᵐᵐ. Length of toes
with claws: Outer toe 117, middle toe 125, inner toe 95, and hind toe
22ᵐᵐ. Tarsus 92, wing 560, and tail 163ᵐᵐ.

The shortest distance from the openings of the nostrils to the yellow
lateral spot on the base of the bill amounts to 18ᵘᵐ. The lateral spots
are in connection with each other over the culmen by a narrow yellow
stripe, which forms an angle in the middle of the culmen with the point
turned towards the front; the margin nearest the feathers of the forehead, blackish. From the upper and hinder border of the skin of the
nostrils, but not in immediate connection with the yellow at the base
of the bill, a yellowish brown spot almost 8ᵐᵐ long extends towards the
tip to cr. 10ᵐᵐ from the hind border of the openings of the nostrils. Skin
of the chin brownish.

The whole plumage dazzlingly white, with faint yellowish edges on
the feathers of the fore part of the head.

Coll. of Norway Scientific Soc. Trondhjem (*Jun.; Hitren, Norway,* 63° 30″ *N. Lat. Last of December,* 1879. *By Mr.* Arnet).

Mr. Storm* gives the total length at 1,040ᵐᵐ. The distance from the
tip of the bill to the mouth I found in the stuffed specimen to be 83ᵐᵐ,
to the fore border of the nostrils 37ᵐᵐ, and to the anterior angle of the
eye 109ᵐᵐ. Length of the toes with claws: Outer toe 114, middle toe
120, inner toe 91, and hind toe 18ᵐᵐ. Tarsus 92, wing 500ᵐᵐ, and tail
134ᵐᵐ. The slight differences in the dimensions given by Mr. Storm
(*l. c.*) of the same specimen probably arise from some difference in our
respective modes of measurement. The dimensions given here are carried out in the same manner as all those undertaken and introduced by
me into this treatise.

---

*Kgl. N. Vidmsk. Selsk. Skr. 1879, p. 129.

From the base forward to between the nostrils and the nail, the bill on the unskinned bird was flesh-colored (Mr. STORM, *l. c.* and *in litt.*). The red color in the dried condition has now exactly the same extent, but has changed to a dull yellowish-red in the hinder part and dark crimson in the front part; border, tip, and a spot round the opening of the nostrils, black. Mr. STORM describes the feet in the freshly-killed specimen as grayish, lighter than in the adult, and the iris as light grayish.

The upper part of the head and neck dull bluish-gray, with the edges of the feathers on the head lighter; chin and throat dirty-white; forehead partly with rather strong rusty-yellow tinge; round the eyes a sharply-defined, downy, white ring. Rest of the surface of the body light violet-gray, with the edges of the feathers tawny yellow; on the back, shoulders, wing-coverts, sides, and the rather purer light bluish-gray rump, the shafts are blackish, forming very distinct dark streaks; on each shoulder a pure white feather protrudes, with a few gray rays. The underside whitish, with the edges of the feathers rust-colored, especially on the middle of the belly; crissum shaded with dull grayish. The primaries a trifle darker than the back, the first with a white stripe in the outer web, along the shaft; the primary coverts rather light. Rectrices gray, lighter along the edge of the inner web; a cluster of the outer tail-coverts on each side pure white.

The tail consists of 20 rectrices.*

MUS. UNIVERSITY COPENHAGEN (♀ *jun. Velling, Jutland, Denmark,* 6th March, 1859.)

Length of the bill along gape, 82^{mm}; to the fore border of the nostrils, 36^{mm}; and to the fore border of the bill, 104^{mm}. Breadth at the nostrils, 28^{mm}. Length of toes with claws: Outer toe 114, middle toe 119, inner toe 97, and hind toe 24^{mm}. Tarsus 90, wing 475, and tail 138^{mm}.

Lores almost bare, and the light color on those and the bill yellow. This color extends along the edge of the upper mandible not farther than is usual in the adult birds, whilst that on the culmen reaches as far as the fore border of the nostrils; likewise the hinder part of the skin of the nostrils is yellow. On the culmen, straight up from the upper posterior point of the skin of the nostrils, a large horseshoe-shaped black spot, with the opening towards the back. The limits between black and yellow less distinct than in the adult.

The color of the plumage about the same as that of the young specimen in the collection at Trondhjem, described above, although not so bluish; the tint on the back, wings, and tail feathers being, on the contrary, brownish. Also, the shafts are light, except on the remiges and rectrices, the shafts of which are brownish. The forehead and the abdomen with rusty-yellow tinge.

---

* Mr. DRESSER, Birds of Eur., part for April, 1880, says: "The young bird is said to have only eighteen or nineteen tail-feathers."

? MUS. ACAD. NAT. SCI. PHILADELPHIA, No. 1794.   (*Pull. Eurcpe.*)

Length of bill along gape, 26ᵐᵐ; from the tip to the fore border of the nostrils 14, and to the anterior angle of the eye 36ᵐᵐ. Breadth of the bill at the nostrils, 10ᵐᵐ. Toes with claws: Outer toe 32, middle toe 33, inner toe 27, and hind toe 8ᵐᵐ. Tarsus 33ᵐᵐ.

Color of the bill, brownish, with whitish nail. Legs yellowish-gray.

The down on the upper surface has a distinct tinge of brownish on the white ground, this tinge changing into a lighter tone on the under-side of the neck, while it forms a very well defined limit against the white on the rest of the under surface.

This specimen is admitted to the present species with doubt. Perhaps it may belong to *O. cygnus;* but the proportionately great height of the bill at the base, and the position of the nostrils parallel to the commissure and not to the culmen, seem to indicate it to be a true *bewickii.*

BEWICK'S Swan has often been confounded with both the Hooper and *O. columbianus,* and even quite recently doubts about the difference from the first mentioned have been stated; whilst the erroneous identification of *columbianus* with *bewickii* has caused the impression that the latter is to be found in the Nearctic region.

When once attention has been drawn to the difference between the adult Hooper and the adult *bewickii,* it is almost impossible afterwards to make a mistake between them, as one, only from the color of the beak, will be able to distinguish them from each other, apart from the size and structural differences, which will be spoken of later. In the Hooper, the yellow color on the bill and lores embraces really a larger surface than the black, and reaches, even on the jaw, in a pointed angle to under and in front of the nostrils, whilst that in the *bewickii* only embraces about one-third of the surface of the bill and lores, as also that in the latter ends in a curved line behind the nostrils, without reaching them. On some individuals one sees a very little portion or spot of the yellow, stretching itself on to the skin of the nostrils, where it occasionally is said to extend in a narrow stripe to the hind border of the opening of the nostrils, but on the jaw itself the yellow color does not reach by a long way near the opening of the nostrils. The mentioned relation concerning the extent of the yellow on the skin of the nostrils I have most frequently observed on specimens from Denmark.

As is clearly shown by a comparison of the measurements given in Tables V and VI, the difference in size alone is sufficient to separate the adult birds of both species from each other.

With regard to the adults there is thus no difficulty. On the contrary, it is not always so easy to distinguish the young birds of the two species from each other, as in them the given distinctions in the color of the bill and the size do not always hold good. Dr. O. FINSCH thinks, after having spoken about the mentioned difficulties, which for him even appear to raise doubts about the specific value of *bewickii** (Verh. Zool. Bot. Ver. Wien, 1879, p. 256), that "only the shorter tarsus and middle toe can be given as distinctions"; but even this mark cannot be employed with individuals that have not yet reached their full size. I have therefore looked about after another distinctive mark, and believe I have found one, which is characteristic in all ages.

Fig. 15.

Fig. 16.

Olor cygnus, juv.

Olor bewickii, juv.

What there is most peculiar in *O. bewickii*, when compared with the Hooper, is without doubt the higher and rather shorter form of the bill, and on the whole the bill is that part in which we can expect to find the most essential characters in these birds.

I have had the beaks of two full-grown young birds, in gray plumage, photographed, the one of BEWICK'S, the other of the Hooper, so that the former, in order to be more easily compared, is so much enlarged that it has obtained exactly the same size as the latter. Figs. 15 and 16 are taken very carefully after these photographs.

If one takes the distances from the tip of the bill to the hind border of the nostrils, and from this point again to the mouth, in the one figure, between the feet of the dividers, and places these measurements on the other figure, it will be very easy to convince oneself that the nostrils in *bewickii* lie nearer the tip of the bill than in the Hooper, which can also be expressed thus, viz, that in the Hooper the distance from the mouth to the hind border of the nostrils is equal to the distance between this and the hind border of the nail of the bill, whilst in *bewickii* the former distance is equal to that between the hind border of the nostrils and a point on the middle

---

*When Dr. FINSCH, l. c., in his comparative table of the dimensions, quotes the measurements of Professor SCHLEGEL, and thereby makes out that the difference in size between *cygnus* and *bewickii* is only slight, it should not be forgotten that one of the specimens which Professor SCHLEGEL measures as *cygnus* is only a female of *bewickii*, and, moreover, a very small one, too, as is fully evident from my table.

The mistake of Professor SCHLEGEL is the more strange from the fact that he in his catalogue (p. 81) expressly adverts to the peculiar extension of the yellow color on the bill being exactly that of the typical *bewickii*.

of the nail. It will further be easily seen in the same figure that a straight line laid along the upper border of the nostrils in the Hooper will go almost parallel with the culmen, whilst this in BEWICK's Swan will form a much more obtuse angle with the same.*

It will not be difficult in general in these birds to notice through the skin of the bill the outlines of the bones which lie underneath. Especially easy will one be able to discern the outlines of the *processus maxillaris* of the nasal bone with the open angle lying back and below the same, between the named *processus* and the *arcus zygomaticus* (x on figs. 15, 16), together with the angle lying above and to the front (y on the same figures), formed by the *processus maxillaris* and *intermaxillaris* of the *os nasale*.

In all the specimens which I have examined it has shown itself that the *processus maxillaris* in *O. cygnus* is much more inclined than in *O. bewickii*, in which it is more perpendicular, so that perpendicular lines through the upper points of the angles x and y in the figures, descending to a line parallel with the commissure, have a not inconsiderable distance from each other in *bewickii*, whilst they come together, or almost so, in the Hooper; or, in other words, in the latter the point of the angle x extends so far forward that it comes almost under the point of the angle y, which is far from being the case in BEWICK's Swan. The relation can be very clearly seen in the sketches.

I have thus always found the formation of the bill, in old as well as in young specimens; and I have but little doubt that this relation, which agrees with the greater height of the bill in *bewickii*, will show itself to be an excellent, easily perceived, and constant mark, and that by this the difficulty of distinguishing the young birds of both species by the assistance of outward marks is satisfactorily settled.

Besides, if one compare the two above-mentioned young birds, separately described (see pp. 206 and 207), which would have taken, the ensuing spring, the white plumage of the old birds, the color does not show any particular difference. Exactly the contrary to what Mr. DRESSER (Birds of Eur., April, 1880) describes,† the young *bewickii* now before me is considerably lighter than my specimen of the Hooper. Besides, the former has on the back numerous blackish hairlike stripes, formed by the dark-colored shafts, whilst they in the other are not darker than the *radii*. Another young specimen of the Hooper, belonging to the Bergen Museum, and which I have described in Nyt Mag. for Naturv., xxv, p. 145, is similar to the one in my collection.

---

* NAUMANN has already drawn attention to this feature.

† Said to resemble the young of *C. musicus*, but is, of course, much smaller, and the coloration of the plumage is rather darker.

TABLE VI.—*Olor bewickii* (YARR.).

TABLE OF DIMENSIONS.

Collection.	Mus. No.	Locality.	Date.	Sex.	Age.	Bill from tip to anterior border of— Mouth.	Nostrils.	Eye.	Breadth of bill at nostrils.	Outer.	Middle.	Inner.	Hind.	Tarsus.	Tail.	Wing.
						mm.	*mm.*	*mm.*	*mm.*	*mm.*	*mm.*	*mm.*	*mm.*	*mm.*	*mm.*	*mm.*
Mus. Trondhjem		Stjordalen, Norway	? Febr., 1880	♀(?)	ad	87	39	116	33	117	125	95	22	92	163	560
Mus. Bergen		Hareide, Söndmor, Norw	27 Febr., 1880	♂	ad	88	39	113	32	118	125	102	21	91	162	550
Mus. "		Balestrand, Sogn, Norw	19 Janr., 1880		ad	89	39	113	30	118	127	96	21	92	116	550
Mus. Leiden	C. minor, No.1	Holland	1 July, 1855	♂	ad	92	38	116	29	114	133	93	21	94	140	505
		Örlandet, Trondhjem, Norw	? Janr., 1880		ad	90	39	109	31	118	126	98	21	93	176	530
Mus. Stavanger		Sole, Jædderen, Norw	16 Janr., 1879	♂	ad	84	38	112	26	110	125	99	20	95	173	555
Mus. "		Inderöen, Indherred, Norw	? Mar., 1880	♀♂	ad	88	41	111	31	117	120	99	19	98	161	540
Coll Steenegger	394	Sömme, Jæderen, Norw	22 Janr., 1880		ad	89	39	108	31	116	124	103	22	90	163	530
Mus. Copenhagen		Rinkobing, Jutland, Denmark	26 Octbr., 1881	♀♂	ad	85	39	116	30	116	128	94	22	110	159	488
"		Stauning, Jutland, Denn	10 March, 1859		ad	94	35	106	27	115	123	99	25	99	169	
Amer. Mus., N Y		Europe			ad	87	40	110	28	115	123	99	26	89	152	
Mus. Copenhagen		Bulow	30 March, 1863	♂(?)	ad	84	35	108	29	110	122	92	18	100	172	495
"		Ringkobingsfjord, Jutland, Denn	March, 1857		ad	85	35	105	28	108	118	91	22	99	163	
Coll. Gram		Bolgen, Kristiansund, Norw	Decbr., 1879	♂(?)	ad	86	39	105	29	106	117	85	18	94	163	480
		Reinsvandet, Stenkjær, Norw	? Janr., 1880		ad	83	39	111	29	110	116	93	22	89	161	477
Mus. Acad., Philad'a.		Europe			ad	86	36	109	30	112	120	92	23	91	156	497
Mus. Leiden	C. musicus, No.7	France	1807		ad	81	36	113	29	110	110	93	21	87	160	510
Mus. "	C. minor, No. 2.	Holland	15 March, 1856	♀♂	ad	81	38	103	29	102	110	85	26	87	158	478
Ms. Bergen		Hamburg, Germany	20 Octbr., 1880		ad	81	36	102	29	106	115		20	87	163	510

July 25, 1882.

Olor columbianus (ORD.)

Whistling Swan.

DIAGN.—*The distance from the anterior angle of the eye to the hind border of the nostrils much longer than the distance from the latter to the tip of the bill; the yellow color at the base of the bill does not extend to the nostrils, making at most $\frac{1}{15}$ of the surface of the bill and lores; larger. Total length about 1,400^{mm}; middle toe with claw about 140^{mm}.*

SYN.— ? 1791.—*Cygnus ferus* BARTRAM, Travels (p. 294) (nec LEACH, 1816 quæ *O. cygnus* (L.)).

1815.—*Anas columbianus* ORD, GUTHRIE'S Geogr. 2d Amer. ed. (p. 319).

1826.—*Anas cygnus* BP., Obs. Nomencl. Wils. p. — (nec LIN. 1758).

1826.—*Cygnus musicus* BP., ut supra (nec BECHST. 1809, quæ *O. cygnus* (L.)).

1831.—*Cygnus ferus* SHARPLESS, DOUGHTY'S Cab. Nat. Hist. I. No. 8, p. 181 (nec LEACH, 1816).

1831.—*Cygnus americanus* SHARPL., op. cit. p. 185.*

1831.—*Cygnus bewickii* RICH., in SW. & RICH., Fauna Bor., Amer. II, p. 465 (nec YARR. 1830).

Note 1 to the Synonymy.—As it seems impossible to decide whether BARTRAM has met with the Trumpeter or the Whistling Swan, I have admitted it to the latter species with query. Probably it may belong to this, but on the probability alone I should not like to transfer to any species a name which another bird has borne during a long time.

Note 2.—In order to justify the change of the name given by SHARPLESS, and the reinstatement of ORD'S title, I quote below Dr. ELLIOTT COUES'S investigation in this matter —:" By their size and the difference in the voice, the two American species are correctly discriminated by LEWIS and CLARKE;† unfortunately, however, they blunder in the matter by saying that the large species (i. e., the one subsequently called *Cygnus buccinator* by Sir JOHN RICHARDSON) is the same as that common on the Atlantic coast; whereas, it is their other species, here called by them the Trumpeter, that is found also in the Atlantic States. But this confusion must not be allowed to stand in the light of the main point of this case, which is that in 1815, ORD based his *Anas columbianus* exclusively upon the Whistling Swan of LEWIS and CLARKE, i. e., upon the smaller of the two species, subsequently named *Cygnus americanus* by SHARPLESS. The blunder of the original authors does not extend to ORD, to whose name *columbianus* should be restored its rightful priority." (Bull. U. S. Geol. and Geogr. Surv. Terr. 2d ser. No. 6, p. 444.)

Note 3.—In opposition to those American ornithologists who have regarded the specimen from Igloolik (in 66° N. Lat.), described by RICH-

---

* Only the word "*Americana*" occurs, the whole name, *Cygnus americanus* first being found in SHARPLESS'S paper in the Americ. Journ. Sc. Art. xxii, 1832, p. 83. The date of number 8 of DOUGHTY'S Cabinet is 1831 and not 1830, as generally quoted.

† History of the Expedition under the command of Captains LEWIS and CLARKE. By PAUL ALLEN, Philad., 1814, II, (p. 192).

ARDSON in Faun. Bor.-Amer. II. p. 465, as belonging to *O. columbianus*, Professor SCHLEGEL (Mus. T. B. VI, *Anseres*, p. 82) refers it to *bewickii*, under which name RICHARDSON also described it. In the mean time, after it has been shown that this species does not at all occur in the New World, it appears to be certain that the American ornithologists are right. The description contains, besides, nothing that speaks in favor of SCHLEGEL'S opinion. "Cere orange (that color entirely behind the nostrils)" agrees fully as well to *O. columbianus*. It is not so remarkable that RICHARDSON himself identifies it with YARRELL'S *bewickii*, for this was first described the previous year, and that without special details, concerning the color of the bill. Besides, it would almost seem as if the specimen had not been preserved, and the description compiled from memory, or from a short notice in his journal. When the specimen was killed at Igloolik the *O. bewickii* was not yet described. It is therefore most probable that RICHARDSON at the time overlooked the species, and then, when first informed of the description of YARRELL, has remembered that he had killed a Swan on which the yellow color did not extend to the nostrils. In confirmation of this, it may be stated that the dimensions given are not of the specimen described, but copied from YARRELL, and that it is not indicated where it was deposited, as is the case with the other specimens collected by him.

U. S. NAT. MUS. No. 85578. (♂ *ad. Currituck, North Carolina*, 1st *December*, 1881. By Mr. ISAAC HINCKLEY.)

Total length of the newly killed bird 1,355^{mm}*, between the tips of the outstretched wings 2,180^{mm}. Length of the bill along gape 100^{mm}, from the tip to the front of the nostrils 46^{mm}, to the fore border of the eye 121^{mm}. Distance from the anterior angle of the eye to the hind border of the nostrils 64, and from that point to the tip of the bill 56^{mm}. Breadth of bill at the nostrils 33^{mm}. Length of toes with claws: outer toe 143, middle toe 154, inner toe 121, and hind toe 30^{mm}. Tarsus 117, wing 575, and tail 192^{mm}.

Tip of tail beyond folded wings 92^{mm}. Outstretched legs reach 50^{mm} beyond the tip of tail. Length of cubitus, measured inside of the wing, 290^{mm}.

Largest secondaries 23^{mm} longer than the longest primary. 2nd primary longest, 8^{mm} longer than the 1st, which is equal to the 3d. The inner web of the the three first primaries and the outer web of the second, third, and fourth, sinuated. Number of primaries 10, of secondaries 25.

Number of tail-feathers 20, one in the sheath.

Color of the bill black, with a 15·^{mm} long, pale greenish yellow spot in front of the eye. Tomium of the lower mandible, with the lamella, dark pinkish red, changing into plumbeous black at the base. Naked skin

---

*Another freshly killed bird, which I measured 15th December, 1881, was 1,390^{mm} long. The spot before the eye was 12^{mm} long, and intense orange colored.

of the *angulus mentalis* black, with higher shadings of pinkish lead-color.

Legs brownish black.

Iris dark brown.

Plumage pure white, with a faint tinge on rusty of the forehead and crown.

COLL. STEJNEGER, NO. 437. (♂ ad. *Koshkonong Lake, Jefferson County, Wisconsin, 9th November*, 1880. *By Prof.* THURE KUMLIEN.)

In the freshly killed bird the length from the tip of the bill to the end of the tail amounted, according to Mr. KUMLIEN'S kind information, to 52.72 inches, *i. e.*, 1,333^{mm}, and the tail reached 2 inches, *i. e.*, 51^{mm}, beyond the tips of the folded wings. The remaining dimensions are as follows: Length of bill from tip to mouth, 99^{mm}; to the fore border of the nostrils, 44^{mm}; to the front of the eye, 117^{mm}; the breadth of the bill at the nostrils, 32^{mm}. Length of toes with claws: Outer toe 141, middle toe 151, inner toe 121, and hind toe 30^{mm}. Length of tarsus 123, wing 547, and tail 152^{mm}.

The whole of the bill and lores black, with exception of a spot about 20^{mm} long and 8^{mm} broad (now of a yellowish-gray color), which extends from the eye forward and downward, along the borders of the plumage of the cheeks, and which in the fresh condition, according to Mr. KUMLIEN'S statement, was "very conspicuously orange-yellow; feet and tarsi black, the naked portion of tibia a little lighter; iris brownish black."

The whole plumage pure white, with exception of a great many small, narrow, but regularly spread, rusty yellow longitudinal spots on the crown, the points of many of the feathers being of this color. Besides, the points and the edge of the outer web of some of the first primaries, and the large upper coverts of these, are shaded with brownish gray.

U. S. NAT. MUS. NO. 85579. (—jun. *Currituck, North Carolina, 28th November*, 1881. *By Mr.* ISAAC HINCKLEY).

Total length of the bird in flesh 1,183^{mm}. L. of bill along gape 84^{mm}, from the tip to the front of the nostrils 39^{mm}, to the fore border of the eye 108^{mm}. Breadth of bill at the middle of the nostrils 31^{mm}. Distance from the anterior angle of the eye to the hind border of the nostrils 58, and from this point to the tip of the bill 48^{mm}. Length of toes with claws: outer toe 130, middle toe 138, inner toe 112, and hind toe 26^{mm}. Length of tarsus 110, wing 510, and tail 137^{mm}.

The tip of the tail reaches 63^{mm} beyond the folded wings, outstretched legs 100^{mm} beyond the tip of tail. Length of cubitus, measured inside of the wing, 258^{mm}.

The longest secondaries are equal to the longest primary. 2nd and 3d primaries equal and largest, the first considerably shorter. The sinuation of the four first quills as in the adult bird, with the exception that the inner web of the fourth primary also is slightly sinuated. The number of the primaries is 10, and of the secondaries 23.

Tail feathers 20.

The middle portion of the bill (in the newly killed bird) is of a dull purplish lead-color, lighter and changing into pale pinkish red on the hind part of the skin, covering the nasal fossæ and the *processus maxillaris* of the nasal bone, becoming plumbeous at the borders of this area; the remaining portion of the bill and the nearly naked lores, is plumbeous black, a small stripe of which also is to be found behind the openings of the nostrils. The borders of the dark color are very indistinct, forming numerous more or less perceptible islets within the light area. The tomium of the lower mandible dark purplish plumbeous, becoming almost black at the base.

Legs light plumbeous-gray, dark, almost black in the midst of the web. The underside of the feet blackish with a stripe mostly of bluish white on the webs along the toes.

Iris dark.

The color of the plumage is dull ash-gray, tinged with lavender, and on neck, shoulders, and middle wing-coverts each feather bordered with light yellowish gray. The head is much darker, the crown being especially dull brownish, while the chin is much lighter, and a grayish white spot is to be found right under the eye. The hinder back, and upper part of the rump are quite white, the rest of the rump and the tail-coverts the same as the shoulders; one of the tail-coverts was quite white, and as its base was still in the sheath,* showing itself to be a feather of the coming white plumage. The remiges are white, with broad pure gray tips, this color reaching back as far as the sinuation; on the first and second still longer. The tail feathers are darker ashy gray, the basal half of the shafts being white. Whole of the under surface light grayish with a slight tinge of yellowish. Under wing-coverts and axillaries pure white. The shafts of the upper surface are somewhat darker than the webs, but not very perceptibly so, and do not form any distinct dark stripes.

Another young specimen in flesh (for the examination of which I am indebted to the kindness of Professor S. F. Baird), killed 14th December, 1881, measured from tip of the bill to the end of the tail 1,225^{mm}. The bill had the same color as the foregoing specimen, but the light portion was somewhat more pinkish red. The plumage was also similar, with the exception that the whole underside behind the neck was white with faint rusty tinge on the border of each feather.

In addition to the statements above about the color of the bill of the young bird, I give the following note, kindly given me by Mr. E. W. Nelson, showing the color of several freshly killed specimens, shot at St.

---

* "The outer follicle," Nitsch, Pterylographie.

Michael's, Alaska, September 19, 1879: "Bill purplish flesh-color, rather light, and bordered along gape by black. Iris hazel."

March 15, 1882, I had the opportunity of examining a living young specimen. The bill was black, except the portion between the nostrils, the posterior half of the upper tomium, and the whole margin of the lower mandible, which were of a vivid pinkish flesh-color. The yellow spot in front of the eye was very perceptible, of common length, but still narrower and duller than in the quite adult bird. Iris hazel. The plumage white, except head and neck, which were gray, somewhat lighter than in the specimen described above.

TABLE VII.—*Olor columbianus* (ORD.).

TABLE OF DIMENSIONS.

Collection.	Mus. No.	Locality.	Date.	Sex.	Age.	Bill from tip to anterior border of—			Breadth of bill at nostrils.	Toes with claw.				Tarsus.	Tail.	Wing.
						Mouth.	Nostrils.	Eye.		Outer.	Middle.	Inner.	Hind.			
						mm.	*mm.*	*m.*	*mm.*	*mm.*	*mm.*	*mm.*	*mm.*	*mm.*	*mm.*	*mm.*
U.S. Nat Mus	85578	Currituck, North Carol	1 Decbr, 1881	♂	ad	100	46	121	33	143	154	121	30	117	192	575
Mus Acad, Phil.	437	Lake Koshkonong, Wisc.	9 Novbr., 1880	♂	ad	98	47	120	33	141	151	121	30	107	186	570
Cl. Steqneger.		Oregon			ad	98	48	117	33	140	148	116	26	123	152	547
U.S. Nat. Ms					ad	97	41	121	33	137	150	116	25	107	173	570
Mus. Copenhag	C. Americ. No. 1	Washington, D. C	Decbr., 1865		ad	95	46	118	33	134	139	110	30	102	165	630
Ms. Leiden	78352	California		♂	ad	93	45	117	34	131	138	110	22	110	178	543
U.S. M. Mus		Oregon			ad	90	41	120	31	129	137	107	25	110	168	550
Mus. Acad., Phil.		Roanoke River, Virginia.	March, 1867		ad	93	40	115	32	128	138	110	26	105	178	655
U.S. Nat Mus	80057	Columbia River.	10 Febr, 1836	♀	ad	94	45	114	29	124	133	104	26	102	162	643
"	1197	Yankton, Dakota		♂	ad	83	45	122	31	125	133	98	26	D7	170	512
"	40249	Potomac River	Decbr., 1843	♀	ad	96	41	120	30	121	127	97	25	103	153	648
"	12740	Washington, D. G	Decbr., 1865	♂	ad	85	42	119	33					99		530
"	44061	Franklin Bay	8 July, 1864	♀	ad	94	40	122	29							
"	54620	Koyoutuk Isl., W. T	21 May, 1808	♂	ad	92	41	113	32							
"	9981	Vancouver Isl., W. T	Decbr., 1833	♀	ad	91	44	117	33							
"	44063	Anderson River, N. of Bear L	June, 1865	♂	ad	83	41	115	32							
"	44064	Franklin Bay	5 July, 1864	♀	ad	92	41	113		124	131	105	25	101	164	
Mus. Leiden	C. Americ. No. 2	Maumee Bay	19 May, 1877	♀	ad	88	35	109	33	120	129	102	23	109	165	530
Mus. Acad., Phil.		California		♀	ad	86	40	111	29	113	123	94	20	93	173	520

Olor buccinator (RICH.).

Trumpeter Swan.

DIAGN.—*The distance from the anterior angle of the eye to the hind border of the nostrils equal to the distance from the latter to the tip of the bill; color of the bill and lores entirely black.*

SYN.—1831.—*Cygnus buccinator* RICH. in SW. & RICH. Fauna Bor.-Amer. I, p. 464.
      1844.—*Cygnus bucinator* GIRAUD, Birds of Long Island, p. 299.
      1865.—*Cygnus passmori* HINCKS, Journ. Linn. Soc. Zool. VIII, p. 1.
      1872.—" *Cygnus passmorei* HINCKS," COUES, Key, N. A. B. p. 281.
      1876.—*Olor passmorii* BOUCARD, Catal. Av. p. 57.

Note to the Synonymy.—Prof. W. HINCKS laid before the Linnean Society, on January 21, 1864, the description of a supposed new species of Swan, *Cygnus passmori*, from Canada, which could be distinguished from *O. buccinator* RICH., by several anatomical differences, also, amongst other things, by a smaller size (the whole length from the bill to the end of the tail being 1,295^{mm}, in opposition to 1,524^{mm}, the distance between the tip of the bill and the hind border of the nostrils 51^{mm}, in opposition to 76^{mm}); also, by faint dirty gray tinge in opposition to *buccinator's* generally more or less rust-colored tinge on the head and neck; by the same gray tinge on the inner web and points of the remiges, and by the naked black skin of the lores only reaching to the eyes and not surrounding them. There is, however, reason to suppose, and Mr. HINCKS himself expresses strong doubts, that these differences only arise from age. The smaller size, gray tinge on head and wing-feathers, feathering of the skin surrounding the eye, are all features which prove the young age of the bird, and *C. passmori* may therefore be regarded as a young *buccinator* until the reverse has been demonstrated.

Though the present species is a very distinct one, and the most remote of the genus to which it belongs, it has been very difficult to point out a character which will hold good in birds of all ages. I am not at all acquainted with the quite young bird, but think, however, that the above diagnosis will be sufficient even for identification of the younger specimens.

As both *Olor cygnus* and *bewickii* are easily recognizable by the yellow color of the base of their bills, a nearer comparison only is needed with the *O. columbianus*, of which specimens are said to be found which want entirely the yellow spot. I may here remark that I myself never met, amongst the numerous birds of this species which I have examined, a specimen on which I could not detect distinct traces of the spot by a careful inspection.

Besides the larger size, which is not always sufficient to distinguish the two species, as a comparison of the Tables VII and VIII will show, it has often been stated as a good criterion that *buccinator* has twenty-four tail feathers in contradistinction to *columbianus*, which only has twenty. Independent of the inconvenience of this character, when the birds moult their rectrices, I may confess that I only in a few cases have been able to count twenty-four tail feathers; and the inconstancy of the number of these feathers I have found pervading the whole group, this character changing individually, so that it is not at all to be depended upon.

As a rule, the frontal apex of the ptilosis forms a sharp angle in *buccinator*, whilst it always is rounded in *columbianus;* but I have also seen specimens of the former which had the limit of the feathering rounded as in the latter. In *buccinator* I also usually found the distance from the eye to the point of the mentioned frontal apex to be larger than from the same point to the hind border of the nasal fossæ, whilst the relation is quite the reverse in *columbianus;* but I have also met specimens of both species in which this character was only very slightly expressed, the young *columbianus* especially having the culmen feathered longer forward than the older birds.

The position of the nostrils, those being situated more backwards in the Trumpeter than in the Whistling Swan, is thus the only mark which it is possible to express in a short diagnosis, and which I have found constant and easily perceptible.

TABLE VIII.—*Olor buccinator* (RICH.).

TABLE OF DIMENSIONS.

Collection.	Mus. No.	Locality.	Date.	Sex.	Age.	Bill from tip to anterior border of—			Breadth of bill at nastrils.	Toes with claw.				Tarsus.	Tail.	Wing.
						Mouth.	Nostrils.	Eye.		Outer.	Middle.	Inner.	Hind.			
						mm.	*mm.*	*mm.*	*mm.*	*mm.*	*mm.*	*mm.*	*mm.*	*mm.*	*mm.*	*mm.*
U. S. Nat. Mus	81290	Lake Koshkonong, Wisc.	20 Apr., 1880	♂	ad	113	55	142	34	163	170	133	27	122	191	652
Mus. Leiden	C. bucc. No. I	Missouri			ad	112	54	140	36	163	173	128	23	108	204	620
U. S. Nat. Mus	5470	Oregon			ad	115	57	142	34	158	165	128	..	127	179	630
"		Yellow Stone River	27 Aug., 1856	♂	ad	109	50	142	35	14	168	132	26	126	173	545
"	70317	St. Claire Flats, Mich	20 Novbr., 1875	♂	ad	111	55	141	37	153	164	129	27	123	175	600
Mus. Leiden		North America	21 Febr., 1877	♀	ad	11 0	53	140	35	150	164	122	30	113	196	630
Mus. Acad. Phil		Oregon			ad	106	51	133	36	156	165	127	27	113	189	
U. S. Nat. Mus		Ft. Resolution	May ....,		ad	108	53	136	34	141	151	118	22	103	178	605
"	46224	Mouth of Fraser River	8 Janr, 1806		ad	110	53	137	33	142	147	118	27	111	173	550
"					ad	105	51	133	35	150	156	124	25	108	195	593

CHENOPIS WAGL. 1832.

DIAGN.—*Predominant color of the adults blackish; the young with naked lores; tertiaries and scapulars crisp; tail shorter than the middle toe with claw, rounded; inner webs of outer three primaries and outer webs of the second, third, and fourth sinuated; webs of the feet not scalloped.*

SYN.—=1832.—*Chenopis*, WAGLER, Isis, 1832, p. 1234.
=1852.—*Chenopsis* REICHENB., Syst. Av., p. X.*
=1864.—*Chenopis* JERDON, Birds of India, III, p. 777.

Chenopis atrata (LATH.).

Black Swan.

DIAGN.—*Plumage of the adults blackish, with white wing feathers; bill red, with a white band behind the nail; legs black.*

SYN.—1790.—*Anas atrata* LATH., Ind. Ornith., II, p. 834.
1790.—*Anser Novæ-Hollandiæ* BONNAT., Encycl. Méth. Ornith. I, p. 108.
1791.—*Anas plutonia* SHAW, Natur. Miscell. III (tab. 108).

COLL. STEJNEGER, NO. 710. (*Pullus, Victoria, Australia.*)

Length of bill along gape 24^{mm}, from the tip to the fore border of the nostrils 14^{mm}, to the front of the eye 34^{mm}. Length of toes with claws: Outer toe 34, middle toe 36, inner toe 29, and hind toe 7^{mm}. Tarsus 29^{mm}.

The bill and an entirely naked 2-^{mm} broad stripe from that to the eye dark horn colored, or brownish black; the nail of the upper mandible as well as the lower is white at the tip. The feet dull grayish brown.

The faintly glossy plumage, is on the whole of the upper surface, the cheeks the tibia, and the crissum, light brownish gray, which color, especially behind the feet, is tolerably distinctly marked against the white color of the undersurface; this on the throat is shaded with the same tinge as the back; the white color of the chin and throat goes imperceptibly over into the grayish tinge on the cheeks.

---

* Usually is quoted "*Chenopsis* AGASSIZ," and GIEBEL, in his Thes. Ornith., adds "Nomencl. univers"; but I have not been able to find it in his Nomenclator Zoologicus, Aves, nor in either of the two editions of his Index Universalis.

TABLE IX.—*Chenopis atrata* (LATH).

TABLE OF DIMENSIONS.

Collection.	Mus. No.	Locality.	Sex.	Age.	Bill from tip to fore border of—			Breadth of bill at nostrils.	Toes with claw.				Tarsus.	Tail.	Wing.
					Mouth.	Nostrils.	Eye.		Outer.	Middle.	Inner.	Hind.			
					mm.	*mm.*	*mm.*	*mm.*	*mm.*	*mm.*	*mm.*	*mm.*	*mm.*	*mm.*	*mm.*
U. S. Nat. Mus	15652	Australia		ad	81	44	109	26	125	136	107	29	97	116	508
Mus. Acad., Philada	1443*	N. South Wales, Australia	♀	ad	76	41	101	25	121	129	99	25	89	100	470
"	1442*	N. South Wales, Australia		ad	73	40	100	22	117	123	94	28	95	98	495
U. S. Nat. Mus	17–No. 8	Australia	♂	ad	75	41	102	26	118	122	100	28	93	95	470
"	71664	Victoria, Australia		ad	70	39	90	26	112	118	92	26	92		490
Mus. Leiden	71656	Victoria, Australia		ad	68	36	90	23							470
Mus. Copenhagen	C. atratus, No. 1	Australia		ad	70	40	98	26	116	118	90	19	83	107	428
				ad		36	89	25	110	117	90	25	88	102	465

* "Type of Gould's B. of Austr."

NOTE ON THE HABITS AND THE REARING OF THE AXOLOTL, AMBLYSTOMA MEXICANUM.[*]

By M. CARBONNIER.

These amphibians live very well in an aquarium of suitable capacity—30 to 40 liters of water for each pair. This water should be renewed about once a fortnight. Some clusters of aquatic plants (*Elodea canadensis*) will assist in maintaining the purity of the water, and their topmost branches will serve, at the same time, as a support for the eggs deposited by the female. These eggs resemble frogs' eggs; they are covered with a similar viscous material and are deposited in strings instead of being agglomerated; they hatch in from 15 days to 3 weeks, depending upon the temperature of the water.

---

[*] Translated from the French by Tarleton H. Bean.

According to my observations, several days before the spawning the male spermatizes all the water in the aquarium and the zoosperms (spermatozoa) penetrate directly into the oviduct of the female, thus fecundating the eggs. The axolotl is capable of spawning five or six times a year, and produces each time from 150 to 200 young. A dim light is better for the maintaining of axolotls than a bright light, which they dislike.

I have some individuals in which the branchiæ are altogether absorbed; they have thus passed into the *Amblystoma* state and respire entirely by the lungs (*poumons*). I have never been able to secure reproduction under this last condition.

I feed my axolotls with earth-worms; they are fond of tadpoles also; in the absence of these things I frequently give them calf liver, presenting it to them in small morsels by the aid of a piece of wood.

The axolotl in its normal state is black; the albino is a variety which I have obtained among the spawnings of the former, and which became permanent and fertile like the black form.

DESCRIPTION OF A NEW SPECIES OF URANIDEA (URANIDEA POLLICARIS) FROM LAKE MICHIGAN.

By DAVID S. JORDAN and CHARLES H. GILBERT.

Uranidea pollicaris sp. nov. (29663.)

Body robust; nape prominent, the profile of head steeply declined, thence to tip of snout in a straight or slightly concave line; head much depressed, broad and flat above, evenly narrowed forwards to the broad, much depressed, bluntly-rounded snout; eyes small, with extensive vertical range, their diameter less than snout or than the flat interorbital width; mouth rather small, anterior, with but little lateral cleft; maxillary reaching vertical from front of orbit; teeth villiform on jaws and vomer, none on palatines; preopercular spine large and strong, spirally curved upwards and inwards, wholly invested with membrane; a single, sharp, concealed spinous point below angle of preopercle; isthmus broad, without fold, its width equaling distance from snout to middle of pupil.

Spinous dorsal rather low, nearly uniform in height, connected with second dorsal by a low membrane; longest spine equaling length of snout; soft dorsal long, and its longest ray $2\frac{1}{5}$ in head; origin of anal fin under third dorsal ray, its last ray under sixteenth of dorsal; highest anal ray $2\frac{1}{2}$ in head; ventrals I, 4, reaching two-thirds distance to vent; pectoral rays all simple, unbranched, the longest reaching vertical from vent, and contained $1\frac{1}{6}$ times in head. Vent equidistant between tip of snout and base of caudal fin.

Skin everywhere smooth.

Head $3\frac{2}{3}$ in length to base of caudal; depth $4\frac{3}{4}$; eye $5\frac{1}{3}$ in head.

D. VII—19; A. 13; V. I, 4; P. 17. Lat. l. complete.

Color olivaceous above, little punctulated; lower two-fifths of sides and whole under side of head and body uniform whitish; above, head and body with irregular spots and blotches of black; these in finer pattern on head, and not forming bands on back; dorsals, caudal, and pectorals with black spots arranged in more or less distinct series; anal, ventrals, and lower rays of pectorals translucent, unmarked.

A single specimen (No. 29663) 4½ inches in length, was taken in Lake Michigan, off Racine Wis., by Dr. P. R. Hoy, and presented to the National Museum.

OBSERVATIONS ON FOUR MULES IN MILK.*

By Professor ALFRED DUGÈS.

[Translation of a note contained in "El Repertorio" of Guanajuato, Mex., No. XVII, 1876.]

Although observations relative to the milk given by animals which have not passed through the state of gestation are few, still a number have been recorded, including some regarding the human species. Frèmy has given an analysis of the milk of a sheep and Schlossberger of that of a goat. Facts of this nature being so uncommon, I believe that the note which, conjointly with my learned friend Prof. Vicente Fernandez, I now publish, will prove of considerable interest.

On the 11th of May, 1876, having learned that there was a mule in milk at the Hacienda d'Argent de San Pedro de Rocha, on the Marfil road, a quarter league from Guanajuato, I went to the place, accompanied by my friend Fernandez. Through the kindness of the employés of Mr. Bernardo Lopez, proprietor of the farm, we were permitted to examine the phenomenal animal, which was then working in an ore mill.

The mule is of a chestnut color, with the nose, lower parts of the limbs, belly, tail, and mane white. Its height is about 1½ meters; its proportions are perfect, without fullness of the abdomen; the breast is also larger than those of hybrids of the same kind ordinarily; the back is quite concave. Except in these particulars, however, there is not the least doubt but that we had before our eyes an ordinary mule. We were told that it had been bought five years before, and, according to the workmen, it was at least seven years old. On examination, however, I discovered that the teeth resembled those of a horse four and a half or five years old. It is possible that there is an anomaly here co-ordinate with the peculiar appearance presented by the mammæ. The latter are shaped like the alligator pear (*Persea grattissima*), black, and without nipples. Their length is 12 centimeters, exclusive of the base, which is

---

*Translated by Frederick W. True, from Professor Dugès' French version of his original Spanish.

buried, as it were, in the skin of the abdomen; as a whole the organs somewhat resemble testicles. According to the information given us, the animal had never given birth to offspring, nor had ever been served by an ass or horse. It appeared that two years before a work-man in the establishment, seeing that the mammæ were a little large, attempted to milk the animal, and that the repetition of this act had brought about the condition in which we found the animal. In a moment, and before us, more than four hundred grams of milk were drawn, which issued with much force and fell foaming into the vessel prepared to receive it. When it had remained undisturbed for a little time it appeared of a dead white color, resembling that of milk of almonds. Its odor was slight, not at all resembling that of the mule. Its taste seemed to me oily and a little sweet. but as I tasted of it with repugnance I cannot describe the flavor accurately. Regarding its other peculiarities I refer to the note of my friend, Professor Fernandez. The microscopic characters were those of ordinary milk.

Such are the more important facts which I learned regarding the hybrid in question. The matter is known to a large number of the cit-izens of Guanajuato.

The following note on the nature of the mule's milk is extracted from the report of Prof. Vincente Fernandez, which appeared in the same number of the "Repertorio" in which my own observations were first printed:

"The liquid obtained from the mule has the appearance of whey, is without sensible odor, and has a sweet taste. Its reaction is slightly alkaline. Density, 1.0270. Heat alone does not coagulate it. Acetic and hydrochloric acids coagulate it, however, and leave oil globules upon the filter. Sulphuric acid coagulates the milk also, and gives a white precipitate by forming an insoluble compound with the casein. It contains, therefore, two of the principal constituents of cow's milk— fats and casein.

" By pouring into a test-tube 80 drops of pure sulphuric acid, 5 cen-tigrams of ox-besoar,* and a drop of milk, and heating to 60° or 80° F., I obtained a reddish purple color similar to that of a solution of permanganate of potash. This demonstrates the presence of glucose, which is formed by the sulphuric acid at the expense of the lactose— another principle of cow's milk.

"In order to prove the existence of butter and of casein, I mixed 20 centiliters of milk with an equal volume of a saturated solution of sul-phate of soda and one gram of carbonate of soda.

"Filtration gave a clear liquid, and butter remained on the filter. The liquid, neutralized by acetic acid, gave a precipitate of casein, which the carbonate held in solution.

---

* This reagent, very delicate for use in recognizing the presence of glucose, is a dis-covery of Vicente Fernandez, and has always been of great service to me in testing diabetic urine.—A. D.

"A quantitative analysis gave the following figures, the process being carried on with the greatest care:

	Liter.	Hundredths.
Water	908.50	90.850
Casein	19.45	1.945
Butter	17.00	1.700
Sugar of milk	51.30	5.130
Fixed salts	3.75	0.375
	1,000.00	100,000

"The result proves that the liquid in question is a true milk, and that this milk does not differ from that of horses in general, except by the presence of a little more fat, which diminishes its density. Possibly the predominance of fat is due to the fact that the milk remained a long time in the mammæ, and that the casein underwent a regressive change. Otherwise it is a liquid almost entirely composed of olein."

Subsequent to the time of this observation my friend, Mr. Epifanio Jimenez, brought to Guanajuato a mule five years old, which gave about a liter of milk daily for four months. The animal was taken away again, however, so that I was unable to examine it.

I have been made aware of an additional fact. I received milk from two mules of the Hacienda de Luna, near Guanajuato, in February, 1880. It is salt, very fat, and whiter than that of which an analysis has been given. The facts which I obtained are as follows: One mule is fifteen years old, the other eighteen. The first furnishes milk at all times of the year, and has done so from the time it was purchased. The second mule has been under observation only a month. Neither has given birth to young. The quantity of liquid given by the first animal is 250 grams per day; by the second, a liter or a liter and a quarter.

GUANAJUATO, *November* 24, 1880.

ON LAGOPUS MUTUS, LEACH, AND ITS ALLIES.

By LUCIEN M. TURNER.

The following paper is based upon an examination of the specimens contained in the National Museum collection, to which I have been kindly allowed access by Professor Baird. A sufficiency of material alone can demonstrate to a certainty the relationship of birds subject to almost daily mutations of plumage as are exhibited in the various species of the genus *Lagopus*.

It is well known that individual birds of this genus differ greatly, though they inhabit a restricted locality, such as a single mountain. The birds from the lowlands are larger and have a looser plumage,

July 29, 1882.

while those from the more elevated localities are perceptibly smaller and have a denser, closer fitting plumage.

During the winter season the entire plumage is white with the exception of the tail, and in some of the males of *L. mutus*, also the greater number of the females, a black stripe from the base of the side of the bill produced through the eye to the auricular region. This black stripe, however, varies in position and distribution. When nearly obsolete it occupies the auricular region, and when greatly developed is continuous across the forehead of the bird, and is even present in the summer plumage of some females. This feature is specially characteristic of the winter plumage alone, however, and at this season it is almost impossible and even hazardous to assert that this or the other example is to a certainty this or that race. The table of measurements proves only such variability of size as may be met with in individuals of any other series of birds belonging to the same species.

The summer plumage is assumed at variable periods of the months of April, May, or even in early June, according to the locality. The moult for the summer is usually shown first on the head and neck, followed by the lower back, sides, breast, middle back, flanks, and abdomen, in the order named. The abdomen and chin are the last areas to show the complete moult. The parts named are also the first to assume, in the order given, the white winter plumage.

During the time of the summer plumage scarcely a single day passes but that the general color of the feathers is not modified by the appearance or loss of some feather. How, then, is it possible to state just where the plumage of an individual shall constitute the summer stage when it is scarcely possible to find two birds of the same sex, age, and locality which do not differ in an appreciable degree of coloration, and where there are no other characters on which to base a comparison? In the examples just compared I find the plumage of birds from Norway, France, Switzerland, and two localities in the "Barren Grounds" of Arctic America which do not vary in an essential color, and the pattern of coloration scarcely more divergent than will be found in birds of the same sex from the same locality of either region mentioned.

The birds from the western coast of Arctic America and the easternmost Aleutian Islands do not, so far as I can see, differ appreciably from the European specimens in point of plumage during the breeding season. The males perhaps show a slight variation in shade of the ground color, but not in an essential degree. Hence the American and the European bird should be separated only as races, if at all, although most authors who have separated the American bird have distinguished it as a species by a binomial appellation—*Lagopus rupestris* (Gm.) Leach.

It seems to me, however, that the European birds *mutus* and *alpinus* should constitute, as is held by many authors, but a single species having the name *Lagopus mutus* LEACH, while the American bird may be recognized as a fairly definable race to be called *Lagopus mutus rupestris* (Gm.) Ridgw.*

The most striking variation of coloration is to be found in the examples from Greenland and Cumberland Gulf. If the summer plumage is to be taken as the consideration which shall constitute a species or race in this genus, then the birds from Greenland and Cumberland Gulf should be recognized as a definable form, for which the name *Lagopus mutus reinhardti* (BREHM) should be used, unless the Iceland bird should prove to be identical (and this I have had no opportunity of verifying), in which case the birds of all the localities named above should then receive the name *Lagopus mutus islandorum* FABER.

The birds procured by me at Atkha Islands (Aleutian chain) present still greater variations of coloration, and appear to represent a well-marked local race, for which I propose the name *Lagopus mutus atkhensis*.

The following descriptions of summer specimens, together with the table of comparative measurements, will help to establish the relationship of the four races recognized in this paper:

I. **Lagopus mutus** (*typicus*).

No. 34120, Lapland (67° N.), ♂, ad., July 17, 1855.

Head and neck dusky, with light gray tips to many of the feathers, and others having an obscure yellowish-brown spotting near, but anterior, to the gray. The back, rump, tail-coverts, and scapulars very dusky, much vermiculated with grayish and fulvous, the rump having a tendency to zigzag, fine markings almost approaching bars on the lower portion. Jugulum and breast having few light yellowish-brown spots, especially on upper breast and sides of the neck. The sides and flanks are strongly but sparsely barred with dusky and light buff. Tail entirely black. This example is identical in plumage with No. 33546, ♂, marked "T. *lagopus*," from Norway, summer; and with 43686, ♂, marked "*L. rupestris*," from the Barren Grounds of Arctic America, late spring.

No. 33547 ♂, ad., labeled "*L. alpinus*," Norway, July 9, 1862.

Head black, feathers narrowly tipped with brownish-yellow; entire neck black, the feathers tipped with pure gray; upper back black with narrow bars of light fulvous; back and rump black with fine dots of gray and fulvous, which latter disappear on the lower rump and upper tail-coverts, where replaced with small gray dots, and each feather tipped with a narrow crescentic band of grayish white. Jugulum and sides black with fine dots of white and buff, inclined to spotting. The tendency to produce bars is in this example nearly obsolete. The tail with a rather broader tip of white than in other specimens.

No. 34119, ♀, ad., "*L. alpinus*," Lapland, July, 1855, and No. 18897, ♀, ad., "*L. mutus*," France, late spring.

---

* See "Hist. N. Am. Birds," vol. iii, pp. 456, 462.

These two females are only distinguishable, the one from the other, by a slight variation in the shade of the yellowish-brown. The bird from France is a little lighter in color than the other; the tendency to produce distinct bars of black, alternating with yellowish-brown, is very well marked, while on the inferior surface there is a somewhat distinct tendency to broader gray tips to the feathers. These markings are so little different from the pattern of coloration of the other specimens that it is not easy to exactly define the points of discrepancy.

No. 56825, ♀, ad., "*L. mutus*," Switzerland, summer.

General color above similar to No. 44582, "*L. rupestris*," Barren Grounds of Arctic America. The yellowish-brown is lighter and the bars narrower. The black bars also narrower and somewhat broken into dots or spots. The ends of most of the feathers of the upper parts, jugulum, breast, sides, and flanks, broadly tipped with white. The best expression to define the coloration of this example in contradistinction to No. 44582, is to state that it (the Switzerland bird) is paler.

No. 33549, ♀, ad., "*T. lagopus*," Norway, June 11.

No. 856, ♀, yng., "*T. lagopus*," Norway (nearly two-thirds grown).

These two birds are conspicuous for the finer, narrower bars of yellowish-brown and black. The back, rump, tail-coverts, shoulders, sides, and upper part of the flanks distinctly tipped with white on the greater number of the feathers. The jugulum and upper breast less marked with the white tips of the feathers, but more distinctly barred with black and the yellowish-brown.

2. Lagopus mutus rupestris (Gm.) Ridgw.

No. 2855, Barren Grounds of Arctic America, ♂, ad., summer. Crown blackish, with white tips to some of the feathers, others very narrowly tipped with faint yellowish-brown. Neck and sides of head with greater area of white on tips of feathers. Back, rump, and tail-coverts very dusky with fine vermiculations of fulvous and gray, having but little tendency to barring. The upper breast, sides, and jugulum barred with black and very light fulvous, some of the feathers broadly tipped with gray.

No. 43675, ♀, ad., Fort Yukon, Alaska, June, 1864.

Head, entire neck, sides, breast, flanks, and abdomen light yellowish-brown, distinctly barred with black. Back, rump, and upper tail coverts very distinctly barred with bright yellowish-brown, each feather of the upper parts broadly tipped with a crescentic margin of grayish. The tail merely tipped with whitish.

No. 80100, ♀, ad., Gens de Large Mountains, Arctic America. This example presents a lighter yellowish-brown coloration, occupying a slightly greater area than No. 43675, and the black bars being more restricted in width are not less conspicuous and the tips of the feathers more grayish. No other essential differences can be distinguished.

Catalogue No. 73488, Unalashka, May 18, 1877.

♂ ad. The ground color of back, scapulars, rump, and upper tail-coverts dark liver-brown, the nape and crown light reddish brown barred with black, and on the back and other posterior parts very finely and densely vermiculated with black, producing the dark liver-brown general aspect. The jugulum similar to the crown and nape, but with the black bars broader and more distinct, but becoming finer and less distinct on the upper breast. The wing, including primaries, secondaries, and some of the tertiaries white, with few scattered feathers of same pattern of coloration as the upper back. The longer upper tail-coverts are some-what darker than the color of the back, owing to the finer vermiculation of the black and brown colors. Chin and lower sides of head white. The black stripe from base of side of bill is much spotted with white. The lower breast, abdomen, and under tail-coverts white. Tail black, with very narrow tip of white, and decidedly rounded in outline.

Catalogue No. 73489. Unalashka, May 18, 1877.

♀, ad. Upper parts, including head, neck, and upper tail-coverts bright brown-ochre, the tips of each feather either brighter or else white, coarsely barred, having a tendency to spotting with black, which, on elevating the superincumbent feathers, is greater in area on each side of the shaft. The lower parts, including foreneck, breast, and sides, bright yellow-ochre with sparser, but more regular bars of black. The wings, including primaries and secondaries, white. The wing-coverts similar to the coloration of the hind neck. The flanks and sides broadly barred with black and light yellowish-ochre. The lower tail-coverts very distinctly barred with black and yellowish-ochre, the latter color finely dotted with black and narrowly tipped with white. Abdomen white. The claws black with light edges and tip. Tarsus and toes of both sexes covered with fine white downy feathers containing few bristles.

No. 43682, ♀, ad., Arctic coast, east of Fort Anderson, H. B. 7, July 25, 1867.

This example is in full breeding plumage and scarcely differs in any regard from No. 43675 and No. 80100 from near the same region.

3. Lagopus mutus reinhardti (Brehm) Turner.

No. 20346, ♂, ad., Sukkertoppen, in lat. 65° 22′ N. and long. 53° 05′ W. on the West coast of Greenland, July 24, 1860; marked *L. reinhardti.* Ground color grayish-fulvous, minutely dotted with black and fulvous-brown, nowhere producing bars, except on jugulum, upper breast, and sides of neck, where these bars are very narrow, and of black and yel-lowish-brown color.

No. 20347, ♂ ad, (from the same locality as the preceding example) marked *L. reinhardti.* Is similar to the preceding, but has a more grayish ground color and greater tendency to barring on the rump, some of the tail-coverts, upper breast, sides of neck and jugulum. The tendency to produce bars is scarcely evident in No. 20346.

No. 70997, ♀, yng., Niantalik, Cumberland Gulf.

No. 7,0998, ♀, ad., Niantalik, Cumberland Gulf, August 10, 1876.

No. 20345, ♀, ad., Sukkertoppen, West coast of Greenland.

All of these birds are labeled *L. rupestris*, but are so entirely different in plumage that they should be referred to *L. reinhardti* BREHM, or else to *L. islandorum* FABER, should these two prove to be the same bird, a statement which I am not prepared to make, as there are no accessible specimens of the Iceland bird with which to compare them.

The birds from Niantalik and Sukkertoppen present such great distinctions from the corresponding plumage of *rupestris*, that they should be recognized as distinct from *rupestris*. The pattern of coloration in these three birds is not appreciably different in the adult birds from the two localities separated by an expanse of water, which would hardly admit them being considered as a rare bird in those respective localities.

The crown, hind neck, back, rump, and upper tail-coverts black, each feather distinctly edged with white, many of the feathers obscurely marked with short bars of light fulvous-gray, most conspicuous on wing-coverts and sides of neck The entire lower parts black, with buffy bars distinctly alternating with the black bars, each feather tipped with gray. The under tail-coverts show the bars very plainly. The breeding plumage of this bird is very similar to the corresponding plumage of the female of *Canace canadensis*.

4. Lagopus mutus atkhensis Turner.

Catalogue No. 85597. ♂. May, 29, 1879. Ground color of upper parts light olive-brown, altogether lighter than in the corresponding plumage of *rupestris*. The whole surface very finely and densely vermiculated with black. The tips of many of the feathers lighter and more grayish, with very narrow crescentic terminal bar of whitish. The ground color of head and nape above is more yellowish than that of the back. The crown spotted with black. Ground color of foreneck, jugulum, and upper breast light fulvous or yellowish-brown, distinctly and somewhat regularly barred with black. The upper breast, sides, and flanks similar, but more finely and distinctly barred with dusky. The wings, lower breast, abdomen, and under tail-coverts pure white. The inferior upper tail-coverts in this example are little lighter than the rump, simply the obliteration of the prevailing ground color of the back. Tail black and decidedly truncate (not rounded as in *rupestris*), and narrowly tipped with white.

No. 85598. ♂. Same locality, June 7, 1879.

This example of few days later plumage presents no appreciable difference from the one of May 29, 1879. The extent of the white on the upper breast is little greater. The dusky shaft of the wing quills is quite conspicuous in both examples. The black patch from base of bill is continuous around the eye, and embraces the auricular region. The tarsus and toes are only moderately feathered, and have but few bristly terminating feathers. The claws are long and narrow, black at their insertion, and white tipped and edged. The bill is pure black, as is also the iris.

Catalogue No. 85600. May 29, 1879, ♀ adult.

Ground color of head, neck, breast, sides, flanks, and upper tail-coverts light brown-ochre, paler and much less rusty than in corresponding plumage of *rupestris*. The upper parts irregularly barred with black. The most of the feathers tipped with a crescentric bar of white, the black bar immediately preceding which is much broader than the others. The fore part of the back is irregularly spotted with black. Crown spotted with black, the feathers tipped with yellowish-white. Jugulum and breast more sparsely but regularly barred with black. The sides and abdomen similarly, but more broadly, barred with black and light yellowish-brown. But few feathers of white occur on the breast and abdomen. The under tail-coverts are very distinctly barred with black and light yellowish-brown, the tips of the upper tail-coverts and tail have a narrow band of pure white. The wings white, the dusky shaft extending not quite to the tips. The tarsus and toes are but slightly feathered. The claws black, with white edge and tips. The bill and iris black.

Example No. 85599 is similar.

When I first obtained these birds I was struck with the apparent greater size and also the difference in the shape of the bill and claws. These birds frequent the low lands, where, amongst the rank grasses and weeds, a nest, composed of grasses and other plants, is loosely arranged. The number of eggs reaches as high as seventeen, though I never found more than fifteen in a single nest. The eggs are much darker in color than those of *L. albus* and but little inferior in size. I had a number of eggs of this bird, but they were broken *in transitu*.

The following tables of measurements of specimens in the National Museum collection will serve to show the differences of size and proportions which, to a certain degree, distinguish the several races of this species:

a.—MUTUS.

Catalogue number.	Locality.	Sex and age.	Gape.	Nostril to tip of maxilla.	Culmen.	Gonys.	Height of maxilla at nostril.	Tail feathers.	Tarsus.	Middle toe.	Middle toe claw.	Wing.	Remarks.
33547	Norway	♂ ad	.80	.45	.80	.38	.19	4.10	1.29	1.14	.46	8.00	July 9, 1862.
33545	...do	♂ ad	.85	.41	.70	.30	.80	4.30	1.23	1.10	.55	7.50	Apr. 25, 1862.
33546	...do	♂ ad	.80	.40	.75	.30	.19	4.30	1.22	1.00	.65	7.60	July ——
34114	Norway, 63° N	♂ ad	.80	.41	.80	.30	.20	4.00	1.30	.90	.65	7.80	Jan, 1832.
34113	...do	♀ ad	.70	.38	.71	.29	.17	3.50	1.11	.85	.58	7.70	Jan, 1832.
34120	Lapland, 67° N	♀ ad	.83	.40	.72	.30	.18	4.30	1.20	1.00	.40	7.50	July 17, 1855.
34119	...do	♂+♀ ad	.80	.40	.73	.30	.20	3.90	1.10	.98	.40	7.00	July 17, 1855.
33549	Norway	♂+♀ ad	.80	.40	.80	.30	.19	3.70	1.15	1.00	.40	7.30	June 11 ——
5682 0	Switzerland	♂+♀ ad	.80	.36	.72	.30	.21	4.00	1.20	.85	.60	7.60	Winter.
18897	France	♀ ad	.80	.34	.71	.30	.19	4.00	1.20	1.08	.45	7.40	Late summer.
	Average ♂		.81	.40	.76	.31	.19	4.10	1.23	1.00	.53	7.50	
	Average ♀		.77	.38	.72	.30	.19	3.80	1.14	.98	.46	7.30	

b.—REINHARDTI.

Catalogue number.	Locality.	Sex and age.	Gape.	Nostril to tip of maxilla.	Culmen.	Gonys.	Height of maxilla at nostril.	Tail feathers.	Tarsus.	Middle toe.	Middle toe claw.	Wing.	Remarks.
20346	Sukkertoppen, Greenland	♂ ad	.80	.40	.71	.28	.17	4.40	1.10	.98	.39	7.70	July 24, 1860.
79081	Greenland	♂ ad	.80	.43	.78	.33	.20	4.00	1.23	.88	.62	7.80	Winter.
76127	Goothaab, Greenland	♂ ad	.80	.40	.71	.30	.20	4.10	1.16	.93	.44	8.00	May, 1878.
30370	Rigolet, Labrador	♂ ad	.80	.38	.70	.30	.21	4.00	1.20	.83	.50	7.50	Winter.
20347	Sukkertoppen	♂ (?) ad	.87	.43	.73	.35	.19	3.75	1.10	1.10	.40	7.40	July (?).
26345	Sukkertoppen	♂+♀ ad	.80	.38	.80	.31	.19	3.60	1.20	.90	.40	7.30	On nest, July 24, 1860.
70998	Niantalik, Cumberland Gulf	♀ ad	.79	.38	.75	.31	.20	3.90	1.15	1.10	.40	7.30	Aug. 10, 1876.
30371	Rigolet, Labrador	♂ ad	.78	.38	.75	.31	.20	3.90	1.15	.80	.40	7.30	Winter.
	Average ♂		.82	.40	.73	.31	.19	4.10	1.16	.91	.49	7.70	
	Average ♀		.79	.37	.76	.31	.19	3.75	1.15	.90	.45	7.35	

c.—RUPESTRIS.

No.	Locality	Sex/age											Date
2585	Barren grounds	♂ ad	.80	.38	.80	.31	.19	...	1.10	.90	.52	7.20	Summer.
3486	Barren grounds	♂ ad	.76	.38	.80	.30	.20	4.20	1.20	1.00	.70	7.30	Late spring.
31641	Fort Rae	♂ ad	.70	.40	.70	.31	.17	3.80	1.28	1.00	.70	7.50	Feb., 68.
1813	Fort Rae	♂ ad	.80	.37	.75	.31	.19	3.90	1.10	.98	.68	7.50	Jan. 28, 68.
3675	Gens de Large Mountains, Arctic America	♂ ad	.70	.40	.75	.37	.19	4.20	1.18	.92	.50	7.80	Jan., 1864.
5056	St. Michael's, Alaska	♂ ad	.80	.36	.75	.30	.19	3.90	1.20	.95	.70	7.20	Mar.
221	Alaska	♂ ad	.85	.38	.75	.30	.20	4.00	1.28	1.00	.60	7.60	Mar. 15, 1876.
3488	Alaska	♂ ad	.77	.37	.80	.34	.19	4.60	1.31	.98	.56	7.75	May 18, 1877.
44582	Alaska	♂ ad	.80	.38	.80	.30	.20	3.60	1.12	.90	.68	7.50	June 23, 1864.
9876	Fort Rae	♀ ad	.80	.35	.71	.30	.18	4.00	1.10	1.00	.45	7.20	
3482	Arctic coast, E. of Fort	♀ ad	.80	.36	.71	.30	.17	3.70	1.10	.90	.45	6.80	July, 1865.
8000	...miles NE. of Fort Yukon	♀ ad	.85	.38	.70	.32	.17	3.90	1.16	.92	.62	7.10	Smer.
3675	Alaska	♀ ad	.80	.26	.75	.31	.20	3.90	1.20	1.00	11	7.00	June, 68.
1626	Alaska	♀ ad	.80	.37	.71	.31	.18	4.20	1.00	1.00	.58	7.20	Dec. 14, 1871.
73489	Unalashka	♀ ad			.71	.31	.18	4.50	1.00	1.00	.45	7.10	May 18, 187.
	Average ♂		.77	.38	.76	.32	.19	4.10	1.21	.97	.62	7.50	
	Average ♀		.80	.37	.71	.30	.18	3.90	1.10	.94	.55	7.10	

d.—ATKHENSIS.

No.	Locality	Sex/age											Date
85597	Atkha Island	♂ ad	.94	.44	.87	.35	.23	4.25	1.37	1.06	.75	7.82	May 29, 1879.
85598	do	♂ ad	.89	.44	.87	.37	.25	4.25	1.32	1.06	.57	7.82	June 7, 1879.
85599	do	♀ ad	.89	.44	.82	.37	.23	4.25	1.19	1.12	.57	7.87	June 7, 1879.
85600	do	♀ ad	.88	.44	.83	.36	.25	4.00	1.37	1.09	.56	7.50	May 29, 1879.
	Average ♂		.91	.44	.87	.36	.24	4.25	1.34	1.06	.65	7.82	
	Average ♀		.89	.44	.83	.36	.24	4.00	1.28	1.10	.57	7.69	

WASHINGTON, D. C., *April 9, 1882.*

GENERA OF THE SCOLOPENDRELLIDÆ.

By J. A. RYDER.

There are two forms of this group; the first has the body very slender, tapering anteriorly, with the eyes or stemmata placed on the upper surface of the narrow, elongate head; the second form has a broader, more robust body of nearly uniform width anteriorly and posteriorly, with the eyes or stemmata at the sides of the head and not visible from above, the head itself being nearly circular or subquadrate in outline from above. The first is the type to which we may assign the old designation of *Scolopendrella* originally proposed for it by Gervais; the second, of which Newport's species becomes the type, may be distinguished generically from the first as pointed out above, under the name of *Scutigerella*. The latter form is also distinguished from the first by the much greater development of the basal appendages of the legs.

Scolopendrella comprehends:

 S. notocantha Gervais.

 S. microcolpa Muhr.

Scutigerella comprehends:

 S. immaculata Newport.

 S. gratiæ Ryder.

The literature of the subject has been fully cited by me in a paper entitled *The Structure, Affinities, and Species of Scolopendrella*, in Proc. Acad. Nat. Sci. Philad., 79–86, 1881.

A LIST OF THE SPECIES OF FISHES RECORDED AS OCCURRING IN THE GULF OF MEXICO.

By G. BROWN GOODE and TARLETON H. BEAN.

For the convenience of collectors in the Gulf of Mexico we have prepared the following list of fishes recorded as occurring in its waters. Of the species marked by an asterisk there are specimens in the National Museum from this region. We have not inquired into the validity of the other species, but have given them under the names by which they are cited in the works of Günther, Cuvier & Valenciennes, Girard, and other authorities, preferring to leave their nomenclature unchanged until studies have been made upon specimens.

Nearly 300 species are enumerated in this list and several undescribed forms remain to be added. The list was prepared in January, 1881, but has since been somewhat enlarged.

MALTHEIDÆ.

* Malthe cubifrons Richardson.
* Malthe vespertilio (L.) Cuv.
* Halieutichthys aculeatus (Mitch-ill) Goode.

ANTENNARIIDÆ.

* Antennarius annulatus Gill.
* Antennarius pleurophthalmus Gill.
* Pterophrynoides histrio (L.) Gill.

CERATIIDÆ.

Ceratias, sp.

DIODONTIDÆ.

* Chilomycterus geometricus (Linn.) Kaup.
* Chilomycterus reticulatus (L.) Günther.
* Diodon hystrix L.
* Diodon novemmaculatus Cuv.

TETRODONTIDÆ.

* Tetrodon nephelus G. & B. MSS.
* Tetrodon testudineus Linn.
* Tetrodon Spengleri Bl.
* Lagocephalus lævigatus (L.) Gill.

OSTRACIONTIDÆ.

* Ostracion quadricorne Linn.
* Ostracion triquetrum Linn.
* Ostracion trigonum Linn.

BALISTIDÆ.

* Alutera Schoepfii (Walb.) Goode & Bean.
* Aiutera scripta (Osbeck) Blkr.
* Monacanthus occidentalis Gün-ther.
* Monacanthus pullus Ranz.
Monacanthus Davidsonii Cope.
Monacanthus spilonotus Cope.
* Balistes capriscus Linn.

HIPPOCAMPIDÆ.

* Hippocampus zosteræ J. & G. MSS.
* Hippocampus stylifer J. & G. MSS.
Hippocampus puncticulatus Guich.

SYNGNATHIDÆ.

* Siphostoma louisianæ (Linn.) Jor. & Gilb.
* Siphostoma zatropis J. & G. MSS.
* Siphostoma affine (Gthr.) Jor. & Gilb.
* Siphostoma floridæ J. & G. MSS.

FISTULARIIDÆ.

* Aulostoma maculatum Val.

SOLEIDÆ.

Achirus Brownii Günther.
* Achirus brachialis Bean MSS.
* Aphoristia plagiusa (L.) Jor. & Gilb.
* Etropus crossotus Jor. & Gilb.

PLEURONECTIDÆ.

* Hemirhombus aramaca (Cuv.) Gthr.
* Citharichthys spilopterus Gün-ther.
* Paralichthys dentatus (Linn.) Jor. & Gilb.
* Paralichthys ommatus Jor. & Gilb.
* Paralichthys squamilentus J. & G. MSS.

OPHIDIIDÆ.

* Ophidium Josephi Girard.
Ophidium Holbrookii Putnam.
* Ophidium Graëllsi Poey.
* Leptophidium profundorum Gill.
* Genypterus omostigma J. & G. MSS.

FIERASFERIDÆ.

Fierasfer dubius Putnam.

BLENNIIDÆ.

* Blennius Stearnsii J. & G. MSS.
* Chasmodes Boscianus (Lac.) C. & V.
* Chasmodes saburræ Jor. & Gilb. MSS.
* Isesthes punctatus (Wood) Jor. & Gilb.
* Isesthes scrutator J. & G. MSS.
* Isesthes ionthas Jor. & Gilb. MSS.
Hypleurochilus multifilis (Girard) Gill.
* Labrosomus nuchipinnis (Q. & G.) Poey.
* Cremnobates marmoratus Steind.

OPISTHOGNATHIDÆ.

Opisthognathus maxillosus Poey.
* Opisthognathus . lonchurus Jor. & Gilb. MSS.

LEPTOSCOPIDÆ.

* Dactyloscopus tridigitatus Gill.

URANOSCOPIDÆ.

Astroscopus anoplus (C. & V.) Brev.
Astroscopus y-græcum (C. & V.) Gill.

BATRACHIDÆ.

* Batrachus tau Linn., subsp. beta Günther.
* Batrachus pardus Goode & Bean.
* Porichthys plectrodon J. & G. MSS.

GOBIESOCIDÆ.

* Gobiesox virgatulus J. & G. MSS.

GOBIIDÆ.

* Gobiosoma molestum Girard.
* Gobionellus oceanicus (Pall.) Jor. & Gilb.

* Lepidogobius gulosus (Girard) J. & G.
* Gobius soporator Cuv. & Val.
* Gobius lyricus Girard.
* Gobius boleosoma J. & G. MSS.
* Eleotris gyrinus Cuv. & Val.
* Dormitator maculatus (Bloch) Jor. & Gilb.
* Philypnus dormitator Cuv. & Val.
* Culius amblyopsis Cope.
* Ioglossus calliurus Bean MSS.

TRIGLIDÆ.

* Cephalacanthus volitans (Linn.) J. & G.
* Prionotus tribulus Cuv. & Val.
Prionotus carolinus (L.) Cuv.
* Prionotus punctatus (Bloch) C. & V.
* Prionotus scitulus J. & G. MSS.

SCORPÆNIDÆ.

* Scorpæna Stearnsii.
* Scorpæna Plumieri Bloch.

SCARIDÆ.

* Scarus radians C. & V.
* Hemistoma croicense (Bloch) G. & B.
* Hemistoma guacamaia (C. & V.) G. & B.

LABRIDÆ.

* Platyglossus humeralis Poey.
* Platyglossus florealis J. & G. MSS.
* Platyglossus radiatus (L.) J. & G.
* Platyglossus caudalis Poey.
* Platyglossus bivittatus (Bl.) Gthr.
* Harpe rufa (L.) Gill.
* Xyrichthys vermiculatus Poey.
* Lachnolæmus falcatus (L.) Val.

POMACENTRIDÆ.

* Pomacentrus leucostictus M. & T.
* Glyphidodon declivifrons (Gill) Gthr.

*Chromis insolatus (C. & V.) J. & G.

*Chromis enchrysurus J. & G. MSS.

POLYNEMIDÆ.

*Polynemus octonemus Girard.

ACANTHURIDÆ.

* Acanthurus nigricans (Linn.) Gill.

*Acanthurus chirurgus Bl. & Schn.

CHÆTODONTIDÆ.

*Chætodon capistratus L.

*Pomacanthus arcuatus (L.) Cuv.

Holacanthus tricolor (Bloch) Lac.

* Holacanthus ciliaris (L.) Lac.

TRICHIURIDÆ.

*Trichiurus lepturus Linn.

SCOMBRIDÆ.

*Orcynus alliteratus (Raf.) Gill.

*Scomberomorus maculatus (Mitch.) Jor. & Gilb.

Scomberomorus regalis (Bloch) Jor. & Gilb.

Scomberomorus caballa (C. & V.) J. & G.

CARANGIDÆ.

*Decapterus punctatus (Mitch.) Gill.

* Caranx pisquetus Cuv. & Val.

* Caranx hippus (Linn.) Gill.

* Caranx fallax Cuv. & Val.

* Caranx trachurus (L.) Lac.

* Caranx amblyrhynchus Cuv. & Val.

*Selene argentea (Lac.) Brevoort.

Selene capillaris (Mitch.) G. & B.

Vomer setipinnis (Mitch.) C. & V.

*Blepharis crinitus (Akerly) De Kay.

*Trachynotus carolinus (Linn.) Gill.

*Trachynotus ovatus (L.) Gthr.

*Trachynotus goreënsis C. & V.

*Trachynotus glaucus (L.) C. & V.

*Seriola Stearnsii Goode & Bean.

Seriola Lalandii C. & V.

*Seriola falcata Cuv. & Val.

Seriola Rivoliana C. & V.

*Oligoplites occidentalis (Linn.) Gill.

*Elagatis pinnulatus Poey.

*Chloroscombrus chrysurus (Linn.) Gill.

Naucrates ductor (L.) Raf.

CORYPHÆNIDÆ.

*Coryphæna punctulata (Cuv. & Val.) Gthr.

STROMATEIDÆ.

*Stromateus alepidotus (Linn.)

LATILIDÆ.

*Caulolatilus microps G. & B.

BERYCIDÆ.

*Holocentrum sogo (Bloch).

SCIÆNIDÆ.

*Eques acuminatus Schn.

*Eques lanceolatus Gmel.

*Cynoscion maculatum (Mitch.) Gill.

*Cynoscion nothum (Holbrook) Gill.

* Pogonias chromis (Linn.) Cuv.

*Micropogon undulatus (L.) C. & V.

* Liostomus xanthurus Lac.

*Sciæna lanceolata (Holbrook) Gthr.

*Sciæna punctataa (L.) J. & G.

* Sciæna ocellata (Linn.) Gthr.

*Menticirrus alburnus (Linn.) Gill.

*Menticirrus nebulosus (Mitch.) Gill.

*Menticirrus littoralis (Holbr.) Gill.

GERRIDÆ.

*Gerres harengulus (G. & B.) J. & G.

* Gerres homonymus (G. & B.) J. & G.

PIMŒLEPTERIDÆ.

Pimelepterus Boscii Lac.

SPARIDÆ.

* Stenotomus caprinus Bean, MSS.
* Lagodon rhomboides (Linn.) Holbrook.
* Sparus Milneri (G. & B.) J. & G.
* Sparus pagrus L.
* Sparus macrops (Poey).
* Sparus bajonado Bloch.
* Diplodus probatocephalus (Walb.) J. & G.
* Diplodus Holbrookii Bean.
Diplodus caribbæus (Poey) Jor. & Gilb.

PRISTIPOMATIDÆ.

* Pomadasys fulvomaculatus (Mitch.) Jor. & Gilb.
* Pomadasys bilineatus (C. & V.) Jor. & Gilb.
* Conodon nobilis (L.) Jor. & Gilb.
* Rhomboplites aurorubens (Cuv. & Val.) Gill.
* Lutjanus synagris (L.) J. & G.
* Lutjanus caxis (Schneider) Poey.
* Lutjanus Stearnsii G. & B.
* Lutjanus Blackfordii G. & B.
Lutjanus campeachianus Poey.
Lutjanus caballerote Poey.
* Diabasis fremebundus (G. and B.) J. & G.
* Diabasis chrysopterus (L.) J. & G.
* Diabasis formosus (L.) Jor. & Gilb.
* Diabasis aurolineatus (Cuv. & Val.) Jor. & Gilb.
* Diabasis elegans (C. & V.) Jor. & Gilb.
* Diabasis chromis (Brouss.) Jor. & Gilb.
Diabasis albus (C. & V.) Jor. & Gilb.
* Diabasis jeniguano (Poey) G. & B.
* Pomadasys virginicus (L.) J. & G.

* Lutjanus chrysurus (Bl.) Vaill.

CENTRARCHIDÆ.

* Micropterus salmoides (Lac.) Henshall.
* Lepomis pallidus (Mitch.) Gill & & Jor.
* Lepomis Holbrooki (C. & V.) McKay.
* Lepomis punctatus (C. & V.) J.

SERRANIDÆ.

* Rhypticus pituitosus G. & B.
* Epinephelus morio (C. & V.) Gill.
* Epinephelus Drummond-hayi Goode and Bean.
* Epinephelus nigritus (Holbrook) Gill.
* Epinephelus lunulatus Poey.
Epinephelus striatus (Bloch) Gill.
* Epinephelus atlanticus (Lac.) J. & G.
Epinephelus punctatus (L.) J. & G.
Epinephelus tæniops (C. & V.) J. & G.
* Epinephelus guasa (Poey) J. & G.
* Hypoplectrus nigricans (Poey) Gill.
* Trisotropis falcatus Poey.
* Trisotropis microlepis G. & B.
* Trisotropis stomias G. & B. MSS.
Trisotropis petrosus Poey.
* Serranus atrarius (J. & G.)
* Serranus subligarius (Cope) J. & G.
* Serranus trifurcus (L.) J. & G.
* Diplectrum fasciculare (C. & V.) Holbr.

LABRACIDÆ.

* Roccus lineatus (Bl.) Gill.

CENTROPOMIDÆ.

* Centropomus undecimalis (Bloch) C. & V.

EPHIPPIIDÆ.

* Chætodipterus faber (Brouss.) Jor. & Gilb.

POMATOMIDÆ.

* Pomatomus saltatrix (Linn.) Gill.

ELACATIDÆ.

* Elacate canada (L.) Holbr.

LOBOTIDÆ.

* Lobotes surinamensis (Bloch)Cuv.

CHILODIPTERIDÆ.

* Apogon alutus (Poey) J. & G. MSS.
* Apogon maculatus (Poey) J. & G.

ECHENEIDIDÆ.

* Echeneis naucrates L.

SPHYRÆNIDÆ.

* Sphyræna picuda Schneider.
* Sphyræna guaguancho Poey.

MUGILIDÆ.

* Mugil albula Linn.
* Mugil brasiliensis Agassiz.

ATHERINIDÆ.

* Atherina Velieana G. & B.
* Menidia peninsulæ (G. & B.) J. & G.
* Menidia vagrans (G. & B.) J. & G.

BELONIDÆ.

* Tylosurus longirostris (Mitch.) J. & G.
Tylosurus caribbæus (Les.) J. & G.
* Tylosurus notatus (Poey) J. & G.
* Tylosurus gladius Bean MSS.

SCOMBRESOCIDÆ.

* Hemirhamphus unifasciatus Ranz.
* Exocœtus noveboracensis Mitchill.
* Exocœtus Hillianus Gosse.

CYPRINODONTIDÆ.

* Cyprinodon variegatus Lacép.
* Cyprinodon gibbosus B. & G.
* Cyprinodon elegans B. & G.

* Cyprinodon bovinus B. & G.
* Mollienesia latipinna Le Sueur.
* Mollienesia lineolata Grd.
* Fundulus grandis B. & G.
* Fundulus similis (B. & G.) Gthr.
* Fundulus ocellaris J. & G. MSS.
* Fundulus xenicus J. & G. MSS.
* Fundulus adinia Jor. & Gilb. MSS.
* Gambusia patruelis (B. & G.) Grd.
* Lucania venusta Grd.

STOMIATIDÆ.

Astronesthes niger Rich.

SYNODONTIDÆ.

* Synodus fœtens (Linn.) Gill.
* Trachinocephalus myops (Frost.) Gill.
* Synodus intermedius (Spix) Poey.

ELOPIDÆ.

* Megalops atlanticus C. & V.
* Elops saurus L.

ALBULIDÆ.

* Albula vulpes (L.) Goode.

CLUPEIDÆ.

* Brevoortia patronus Goode.
* Opisthonema thrissa (Osbeck) Gill.
Harengula clupeola C. & V.
* Harengula pensacolæ G. & B.
* Clupea chrysochloris (Raf.) J. & G.
* Culpea sapidissima Wilson.
* Clupea pseudohispanica (Poey) Gthr.

DOROSOMIDÆ.

Dorosoma mexicanum (Gthr.) J. & G.
* Dorosoma Cepedianum (Les.) Gill.

ENGRAULIDIDÆ.

* Stolephorus Brownii (Gmel.) Jor. & Gilb.
* Stolephorus Mitchilli (C. & V.) Jor. & Gilb.

SILURIDÆ.

⁂ Arius felis (Linn.) Jor. & Gilb.
* Ælurichthys marinus (Mitch.) B. & G.

ANGUILLIDÆ.

* Anguilla vulgaris Raf.
* Anguilla tyrannus Girard.
* Conger caudicula Bean MSS.

MURÆNIDÆ.

* Ophichthys macrurus Poey.
* Ophichthys chrysops Poey.
* Pisodontophis longus Poey.
* Crotalopsis mordax (Poey) G. & B.
* Sphagebranchus scuticaris G. & B.
* Sphagebranchus teres G. & B. MSS.
* Letharchus velifer G. & B. MSS.
* Herpetichthys ocellatus (Le Sueur) Goode & Bean.
* Neoconger mucronatus Girard.
* Myrophis lumbricus J. & G. MSS.
* Myrophis microstigmius Poey.
* Gymnothorax ocellatus Agassiz.
* Gymnothorax afer Bl.
* Gymnothorax moringa (Cuv.).

AMIIDÆ.

* Amia calva L.

LEPIDOSTEIDÆ.

* Lepidosteus osseus (L.) Ag.
* Lepidosteus platystomus Raf.
Lepidosteus tristœchus (Bl. & Schn.) Jor. & Gilb.

ACIPENSERIDÆ.

Acipenser sp.

CEPHALOPTERIDÆ.

Ceratoptera birostris (Walb.) Goode.

TORPEDINIDÆ.

* Narcine brasiliensis (Olfers) M. & H.

MYLIOBATIDÆ.

* Rhinoptera quadriloba (Lac.) Cuv.

RHINOBATIDÆ.

* Rhinobatus lentiginosus Garman.

TRYGONIDÆ.

Pteroplatea maclura (Les.) M. & H.
* Trygon sabina Le Sueur.

PRISTIDÆ.

* Pristis pectinatus Lath.

SPHYRNIDÆ.

* Sphyrna tiburo (Linn.) Raf.

GALEORHINIDÆ.

* Hypoprion brevirostris Poey.
* Scoliodon terraenovæ Rich.
Carcharinus platyodon (Poey) J. & G.

LAMNIDÆ.

* Isurus DeKayi (Gill) Jor. & Gilb.

GINGLYMOSTOMIDÆ.

* Ginglymostoma cirratum (Gmel.) M. & H.

PETROMYZONTIDÆ.

* Petromyzon castaneus (Grd.) Bean.

UNITED STATES NATIONAL MUSEUM,
Washington, March 31, 1882.

NOTES ON FISHES OBSERVED ABOUT PENSACOLA, FLORIDA, AND GALVESTON, TEXAS, WITH DESCRIPTION OF NEW SPECIES.

By DAVID S. JORDAN and CHARLES H. GILBERT.

The greater part of the month of March, 1882, was spent by Professor Jordan at Galveston and Pensacola, in the collection and study of fishes, in the interest of the United States National Museum. Fifty-one species of marine fishes were observed at Galveston and about 110 at Pensacola; making a total of 129. These are enumerated below. The "common names" here given are, in all cases, those in use among the Gulf fishermen. The letters P. or G. after the name of a species indicate that it was observed at Pensacola or Galveston, respectively. The specimens obtained are mostly in the United States National Museum.

Professor Jordan wishes to make especial acknowledgment of his indebtedness to Mr. Silas Stearns, of Pensacola, for enthusiastic and intelligent assistance. Mr. Stearns is a member of the firm of Warren & Co., wholesale fish-dealers at Pensacola, and the resources of this firm were in the most generous way placed at our disposal. The most valuable portions of the present collection were obtained from the vessels sent out for Red Snappers, the captains of these vessels being directed by Mr. Stearns to save for the Museum all small fishes taken from the mouths or stomachs of these fishes. Several interesting species were also obtained by Mr. Stearns and Professor Jordan, with a fine-meshed seine in the shallow waters of the Laguna Grande at Pensacola.

It will be observed that the shore-fishes, even as far westward as Galveston, are essentially the same as those found along the Carolina coast. The forms found in deeper water have a close relation with the West Indian fauna.

LAMNIDÆ.

1. Isurus dekayi (Gill) J. & G. P.

> *Lamna punctata* Dekay, New York Fauna Fish. 352, pl. 63, f. 205 (not *Squalus punctatus* Mitch.; not *Lamna punctata* Storer Hist. Fish. Mass., which seems to be *Lamna cornubica*.)
>
> *Isuropsis dekayi* Gill, Ann. Lyc. Nat. Hist. N. Y. vii, 409. (After Dekay.)
>
> *Isuropsis glaucus* Poey, Synops. Pisc. Cubens. 1868, 446. (Probably not *Oxyrhina glauca* Müller & Henle.)

The synonomy of the American species of *Isurus* has been much confused, as appears from the above account of it. It is certain that two species of this type, viz, *Lamna cornubica* and *Isurus dekayi*, occur on our Atlantic coast. We find no evidence of the existence of *Isurus spallanzani* Raf. in our waters, although Dr. Günther has referred the descriptions both of Storer and Dekay to the latter species. We recognize the American *I. dekayi*, provisionally, as a species distinct from *I. glauca*, which inhabits the coasts of Asia and Africa, as, in the speci-

Aug. 8, 1882.

men examined by us, the pectoral is much larger than in the description and figure of the latter, published by Müller & Henle.

A single individual of *Isurus dekayi*, a female ten feet in length, was found on the beach of Santa Rosa Island, near Pensacola. It showed the following characters:

Color dark sooty-gray above, white below, the color abruptly changing on the tail. The whole of the caudal, the dorsal and upper edge of pectoral, dark. Anal and under side of pectoral white.

Head 5 in total length with caudal, the upper lobe of caudal $5\frac{1}{3}$ in the same. Pectoral fin falcate, as long as head; front of dorsal inserted well behind axil of pectoral, at a distance equal to $\frac{1}{4}$ the head or a little more than half the dorsal base, which is $2\frac{1}{3}$ in head. Height of dorsal, $1\frac{3}{5}$ in head. Distance from posterior edge of base of dorsal to front of ventral, $1\frac{3}{4}$ in head. Dorsal and pectoral somewhat falcate.

Second dorsal very small, in front of the slightly larger anal, and not twice as large as eye. Interspace between dorsals, $2\frac{3}{4}$ times base of first dorsal.

Gill area deeper than long; its depth $2\frac{3}{5}$ in length of head. Snout sharp, conical. Eye large, $4\frac{3}{4}$ in snout, which measured from eye, is 3 in head. Nostril half nearer eye than snout; eye slightly nearer tip of snout than angle of mouth. Labial fold very short. Caudal keel strong, a pit above and below it.

Greatest depth of body, three-fourths length of head. Teeth about $\frac{28}{28}$, none of them with basal cusps; those of the middle of each jaw much longer and narrower than the others, as in other species of the genus.

CARCHARIIDÆ.

2. Carcharias,* *sp. incert.* *Sharp-nosed Shark.* G.

The jaws of an unknown species of shark were obtained at Galveston. The teeth in the upper jaw are narrowly triangular, little oblique, and slightly notched on the inner side. Median teeth smaller and narrower than those on the sides. Bases of the teeth coarsely serrate, especially on the inner edge; crown of the teeth finely serrate. Lower teeth very narrow, nearly erect, their edges very minutely serrulate, appearing entire, except under a lens. Teeth about $\frac{32}{32}$.

* The name *Carcharias* first appears in Rafinesque's Caratteri di Alcuni nuovi Generi, etc., 1810, 10. Only *new* species are noticed in this paper, and but one is mentioned. *Carcharias taurus* Raf., a species of *Odontaspis* Ag., which does not agree with the original diagnosis of *Carcharias*. In Rafinesque's Indice d' Ittiologia Siciliana, 1810, p. 44, a work which appeared almost simultaneously with the preceding, we find three species mentioned under the head of *Carcharias*, viz, *lamia*, *glaucus*, *taurus*. It seems evident from the context that the former species was intended by Rafinesque as the type of the genus *C rcharias*. It is, however, not described and not identifiable, although the species called later "*Carcharias lamia*" by Risso, was probably intended. In view of the fact that nearly all modern writers have adopted the name *Carcharias* for the genus, to which *glaucus* and "*lamia*" belong, it seems to us that Cuvier's restriction of the name *Carcharias* may be retained, in spite of the evident objection to it. If *Carcharias* be retained, *C. glaucus* should be considered its type, being a species certainly identified and agreeing with the original diagnosis of the genus, with which *C. taurus* was associated by error, an error several times since repeated. The two papers of Rafinesque may well be considered as parts of the same memoir, the "Caratteri" containing an account of "new" species, "the Indice" an enumeration of known species.

3. Carcharias platyodon (Poey.) J. & G.   *Shovel-nosed Shark (Galveston).*

?*Squalus platyodon* Poey, Memorias, Cuba, II, 331..

?*Squalus obtusus* Poey, Memorias, Cuba, II, 337.

?*Eulamia obtusa* Poey, Rep. Fis. Nat. Cuba, 1868, 447.

This is the commonest of the large sharks found on the coast of Texas in the summer. A young male specimen 32 inches long was obtained at Galveston, and the jaws of a very large example, in the possession of Mr. E. Gabriel, of Galveston, were also examined.

The following is a description of the specimen obtained

Color slaty, with a distinctly bluish tinge above, whiter below, the white extending higher posteriorly, and forming a faint lateral stripe. Caudal fin all blackish; second dorsal and anal tipped with dusky.

Body comparatively short and stout. Head very short, broad, bluntly rounded anteriorly, and much depressed. Mouth very broad and short. Length of snout from mouth $1\frac{2}{3}$ in distance between angles of mouth. Breadth of mouth between angles twice length of mouth. Angle of mouth with a pit from which radiate two very short furrows.

Inner edge of nostril with a very blunt lobe. Distance between nostrils but a trifle less than length of snout from mouth. Length of nostril greater than eye and half its distance from eye. Eye slightly nearer nostril than angle of mouth. Nostril a little nearer eye than tip of snout. Distance from eye to snout $1\frac{1}{4}$ times in interorbital width, which is $1\frac{2}{3}$ in length of head to first gill opening. Gill openings short, the height of one a little more than half length of gill area. Top of head with numerous mucons pores.

First dorsal beginning close behind pectoral, at a distance from the posterior root of the latter equal to about $1\frac{1}{2}$ diameter of the eye; the fin moderate in size, its anterior lobe rather obtuse, the posterior little produced; the free edge of the fin little concave. Anterior lobe extending when depressed a little beyond posterior lobe; the fin a little higher than long, its base $2\frac{1}{3}$ times in the interspace between dorsals, and about equal to the distance from the posterior base of the first dorsal and the vertical from the insertion of the ventrals. Length of posterior lobe two-fifths base of the fin.

Second dorsal very small, its base 5 times in the interspace between dorsals, less than half base of first dorsal; the fin scarcely as long as high; its posterior lobe moderately produced.

Caudal moderate, the lower lobe not falcate, $2\frac{1}{3}$ times in the length of the upper lobe; the latter $3\frac{2}{3}$ in the total length, about equal to the distance from the snout to the base of the dorsal.

Anal a little larger than second dorsal and placed a little further back; its lobes more falcate, its distance from base of caudal $1\frac{1}{2}$ its base.

Ventrals moderate, their lobes bluntish, the anterior margin scarcely more than half the length of the base. Pectorals rather small, their tips not falcate, reaching slightly past posterior part of dorsal; their free margins a little concave, the anterior margin a little shorter than head, 6 times in total length of body. Width of pectoral a little less than than two-thirds its length; the posterior lobe contained $3\frac{2}{3}$ times in its anterior lobe.

Claspers, in specimen described, not reaching nearly to edge of ventral.

Teeth of upper jaw broadly triangular, nearly erect, not notched on the outer margin, the edges distinctly and rather coarsely serrate. Lower teeth narrowly triangular, with broad base, the edges finely serrate. Teeth in the young scarcely narrower than in the adult.

The specimen here described was not preserved, it having spoiled before the arrival of alcohol.

Among the described species of this genus *Carcharinus platyodon* (Poey) (=*obtusus* Poey) seems to be most nearly related to the species examined by us. The pectoral in *C. platyodon* is larger, the teeth somewhat different, and the second dorsal is said to be "assez grande," whereas in *C. cœruleus* the latter fin is very small. *C. fronto*, lately described by us from Mazatlan, is also very similar, but has a much larger second dorsal.

Another species, similar, but with longer snout, has been described by Dekay under the name of *Carcharias cœruleus*. This description has been referred by Professor Gill to the synonymy of the very different species, *Carcharias plumbeus (Nardo)* = *Carcharias milberti* M. & H., and has been called "*Eulamia milberti*".

There is, however, no good evidence that *C. milberti* (*plumbeus*) has ever been taken in our waters. The only record is that of Müller & Henle, who mention " ein Exemplar in Paris, von New York durch Milbert". This specimen is apparently not the type of the original description; it belonged to a collection in which there were several confusions of localities, and if really from New York it may have belonged to some species different from the type in the museum at Berlin—perhaps to *C. obscurus* or *cœruleus*.

There are apparently seven species of *Carcharias* (in the broad sense in which the genus is understood by Müller & Henle, Gunther, etc.,) now known to inhabit the waters of the Atlantic and Gulf coasts of the United States. If others exist, their occurrence is yet to be verified.

These are,

1. *C. glaucus* (L.) Cuv.
2. * *C. obscurus* (Le S.) M. & H. (*Platypodon*.)

---

*The first four of these species may be readily recognized by the following characters:

   a. First dorsal inserted nearer ventrals than pectorals. (*Carcharias*).

                                                     GLAUCUS.

  aa. First dorsal inserted close behind pectorals.

   b. Upper teeth oblique, very deeply notched on the outer margin; pectorals. long. (*Platypodon* Gill)..OBSCURUS.

  bb. Upper teeth suberect, triangular, scarcely notched. (*Eulamia* Gill.)

   c. Snout moderate; its length from mouth not less than width of mouth.

                                                CŒRULEUS.

  cc. Snout very short; its length from mouth much less than width of mouth .... PLATYODON.

3. *C. cœruleus* (Dek.) J. & G. (*Eulamia.*)

4. *C. platyodon* (Poey) J. & G. (*Eulamia.*)

5. *C. limbatus* M. & H. (*Isogomphodon maculipinnis* (Poey) Gill).

6. *C. brevirostris* (Poey) G'thr (*Hypoprion*).

7. *C. terræ-novæ* Rich. (*Scoliodon.*)

The *Squalus punctatus* Mitch. (Trans. Lit. and Phil. Soc. 1, 484), agrees well enough with the common *Scoliodon terrænovæ*, and was probably founded on that species. It has, however, been identified by Gill with *Carcharias isodon* M. & H., a species of *Aprionodon*. This species is known only from a specimen collected by Milbert—the locality not stated; but as some other collections of Milbert were made at New York, this type of *C. isodon* has been assumed to be from that locality. So far as we know, no American collector has ever obtained a specimen of the species, and *Carcharias isodon*, or *Aprionodon punctatus*, should be erased from our lists.

It is not likely that the type of "*Scoliodon terrænovæ*" really came from Newfoundland. It is a southern species, and is very abundant along our South Atlantic and Gulf coasts.

4. Scoliodon terræ-novæ (Rich.) Gill. P.

Two young specimens obtained at Pensacola, where the species are said to be common.

SPHYRNIDÆ.

5. Sphyrna tiburo (L.) Raf.—*Shovel-nosed shark* (Pensacola). P.

Abundant at Pensacola.

PRISTIDIDÆ.

6. Pristis pectinatus Latham.—*Saw-fish.* G.

Common. There is thus far no evidence of the occurrence of *Pristis antiquorum* in American waters, although the name occurs in several lists of species.

TRYGONIDÆ.

7. Trygon sabina Le Sueur.—*Sting-ray; Sting-a-ree.* G. (31045).

Generally common. Also seen in the markets of New Orleans, being obtained in Lake Pontchartrain and Lake Borgne.

SILURIDÆ.

8. Arius felis (L.) J. & G.—*Sea cat-fish; Blue cat.* G.

Very common on the sandy beaches. It is seldom brought into the markets, and is eaten chiefly by the negroes. The specimens seen belong to the form described by Baird & Girard as *Arius equestris.* This form agrees in dentition, character of bony plates, etc., fully with the *Arius felis* of the Atlantic coast. The barbels in specimens of *equestris* examined are, however, somewhat longer, the maxillary barbel extending to about the end of the first fourth of the base of the pectoral; the others lengthened in proportion. In *felis* the barbel does not usually reach the gill opening. The pectoral in *equestris* extends slightly beyond last ray of dorsal. These peculiarities are not likely to be constant. There is probably no permanent difference on which to base a subspecies *equestris*.

9. Ælurichthys marinus (Mitch.) B. & G.—*Sea kitten; Sea cat-fish; Gaff-top-sail cat.* G.

Generally abundant.

ELOPIDÆ.

10. Megalops atlanticus C. & V.—*Grande Écaille; "Grandacoy"; Tarpun; Silver fish.* G.

This species is generally common along the Gulf coast, but only scales were obtained. It reaches a length of some 6 feet. Its habit of leaping out of water like the mullet causes it to be dreaded by fishermen. It is said that several persons have been killed or injured when in small boats by the "Grande Ecaille" leaping into the boat.

It seems to us that the specific name *atlanticus* should be adopted as the name of this species, being the oldest name ever really conferred on it. The earlier names "*cyprinoides*" Bloch, "*thrissoides*" Bloch & Schn., and "*giganteus*" Shaw, were alike based on a figure and description of Broussonet, as *Clupea cyprinoides*. Broussonet had evidently a specimen of the Indian species, *Megalops cyprinoides* (Brouss.) Bleeker, and for this species the name *cyprinoides* should be retained. Bloch took his name "*cyprinoides*" and his description from Broussonet, but added a figure from Plumier, of the American species. The names "*thrissoides*" and "*giganteus*" were given as substitutes for "*cyprinoides*," and were likewise based primarily on Broussonet's description. The earliest name intended for our species is *Megalops atlanticus* C. & V. The reference to *Clupea apalike* Lac., given by Günther, is fallacious. Lacépède describes *Clupea cyprinoides*, "la clupe apalike," after Broussonet, his synonymy, like that of all writers before Cuvier and Valenciennes, including references both to *M. cyprinoides* and *M. atlanticus*.

11. Elops saurus L.—*Lady-fish.* P.

Very abundant in summer; at Pensacola, largely salted as bait for the Red Snapper. Not used as food.

CLUPEIDÆ.

12. Brevoortia patronus Goode.—*Alewife.* G. P. (31046, 30907).

Generally common; reaching a length of about 13 inches; no use is made of it.

In life this species is bluish above, silvery below; a faint narrow dark stripe along the middle of each row of scales on the back. Caudal fin bright yellow, its posterior margin blackish; dorsal and anal dull yellowish; paired fins, pale; opercle, yellowish; a blackish blotch on its upper edge; a round blackish humeral spot.

13. Opisthonema thrissa (Osbeck) Gill. G. P.

Two specimens obtained at Pensacola, where it does not appear to be very abundant.

14. Clupea sapidissima Wils.—*Shad, Alewife.* P. (30809.)

Head, 3⅝ (4¼ in total); depth, 3⅝ (4⅗). D. I, 16. A. I, 20. Scutes, 21+15.

This species is not uncommon about Pensacola, where numerous young specimens were obtained. It is known to the fishermen as "alewife" or "shad," and is used only for bait. The specimens seen were 8 to 9 inches in length. They are somewhat more elongate than the young of the northern shad, and the number of gill-rakers is pretty constantly smaller (about 38 below the angle of the arch, instead of 45 to 50).

15. Clupea chrysochloris (Raf.) J. & G.—*Blue herring.* P. G. (30809.)
(*Meletta suœii* Cuv. & Val. xx, 375.)

Not rare on the Gulf coast. Known to the fishermen only as a marine species. One specimen obtained at Galveston and one at Pensacola.

The following is a description of the Galveston specimen:

Color in life deep bluish-green above, the color abruptly ceasing on level of upper edge of gill opening; sides white, with a strong tinge of golden, especially on head. Dorsal yellowish, more or less dusky at base and in front. Caudal soiled yellowish, dusky at tip. Ventrals and anal pale; pectorals pale, a dusky streak on the inner side, behind first ray; tips of jaws blackish; mouth yellowish within; tongue bluish; lining of opercle mostly pale; peritoneum white.

Body comparatively long and slender; head not very deep; lower jaw strongly projecting, its tip fitting into an emargination of the upper jaw and entering the profile; tip of lower jaw with a few slender deciduous teeth. Premaxillaries with a narrow band of rather strong *permanent teeth;* those of the outer series strongest. Tongue with feeble teeth; vomer toothless. Gill-rakers numerous, but not long, not so long as eye; about 5½ in head (about 22 below angle). Eye not large, 4¼ in head. Maxillary reaching past middle of pupil, a little less than half head. Cheeks longer than deep; their depth below eye 4 in head; lower limb of preopercle 2¼ in upper. Longest ray of dorsal 1¼ in head. Ventrals small, nearer snout than base of caudal. Pectorals 1½ in head.

Head 4 (5 in total); depth 3½ (4¾). D. 2, 17; A. 1, 18. Lat. l. 48. Scutes 16 + 15 (19 + 15 in the Pensacola specimen, 20 + 15 in a specimen from White River, Indiana).

The Pensacola specimen is remarkable for its extraordinary fatness, the body being very plump and full of oil. It is very greasy to the touch, even after having been for some time in alcohol.

16. Clupea pseudohispanica (Poey) Gthr. P. (30820.)

Four specimens of this species, each 6½ inches long, were obtained at Pensacola. Its resemblance to the European sardine (*Clupea pilchardus*

Walb.) is very great; hence its name of "Sardina de España," among the Cuban fishermen.

Head $4\frac{1}{6}$ to $4\frac{1}{3}$ in length; depth 5 to $5\frac{1}{3}$; D. 16, A. 16; lat. l., about 45.

Body slender, little compressed, the belly scarcely carinated, its scutes not prominent; mouth small, the maxillary not extending quite to front of pupil, its length $2\frac{3}{5}$ in head; gill-rakers long, very slender and numerous, about two-thirds diameter eye, between 30 and 40 on arch below angle. Lower jaw with a few feeble teeth, visible with lens; tongue with some asperities; cheeks much longer than deep, the vertical depth below eye about two-thirds diameter of eye; eye $3\frac{3}{4}$ in head. Opercle without distinct striæ; interopercle with very few. Caudal well forked; the lower lobe a little the longer as long as head. Ventrals inserted nearly below middle of dorsal, a little nearer base of caudal than tip of snout; pectorals $1\frac{1}{3}$ in head; a conspicuous sheath of scales at base of pectorals.

About 45 scales in a longitudinal series; the scales being thin and deciduous, their number cannot be exactly ascertained.

Color bluish above, becoming golden and silvery below, with no distinct markings anywhere.

Peritoneum pale; lining of opercle somewhat dusky. Intestinal canal somewhat elongate, about $1\frac{1}{2}$ times length of body.

This species is mostly readily distinguished from *C. pilchardus* by the absence of radiating striæ on the opercles, these being very conspicuous in the sardine.

DOROSOMATIDÆ.

17. Dorosoma cepedianum (Le S.) Gill.—*Shad.* G. (30913.)

Generally abundant, especially along the coast of Texas. The specimens all differ somewhat from the usual form of this species, and apparently constitute a local variety or subspecies, perhaps worthy of a distinctive name. Compared with specimens from White River, Indiana, the Galveston form has a slenderer body (depth $3\frac{1}{3}$ to 3 in length, instead of $2\frac{1}{2}$ to $2\frac{2}{3}$), and larger head (4 in length, instead of $4\frac{1}{3}$). The dorsal filament is in all specimens shorter than the head. There seem to be no other permanent differences. D. 12; A 1, 32. Scales 56 to 20. Scutes 18 + 12.

This species is not used for food. It must spawn in or near the sea at Galveston, as individuals of all sizes are abundant in the bay

ENGRAULIDIDÆ.

18. Stolephorus mitchilli (C. & V.) J. & G. G. P. (30892 Galv.); (30857 Pens.).

Engraulis mitchilli, C. & V., Hist. Nat. Poiss. xxi, 50, 1848 (not *Engraulis mitchilli* Günther vii, 391; not *Clupea vittata* Mitch).
Engraulis vittata Storer, Hist. Fish. Mass. pl. xxvii, f. 3 (not description).
? *Engraulis duodecim* Cope, Trans. Am. Philos. Soc. 1866, 405.

Head $3\frac{4}{5}$ in length ($4\frac{2}{3}$ in total); depth 4 (5) in adults, the young more slender; D. 14; A. 25 to 26; lat. l. 37.

Body rather short and deep, strongly compressed; the belly compressed and slightly serrated. Head short, compressed, bluntish. Snout extremely short, not longer than the pupil of the very large eye. Eye about 3 in head. Mouth somewhat oblique; mandible extending farther forward than eye. Maxillary extending beyond root of mandible about to margin of opercle. Both jaws well provided with teeth. Cheeks broadly triangular, almost equilateral, smaller than eye. Opercle short, little oblique. Gill-rakers rather long, about two-thirds diameter of eye.

Insertion of dorsal about midway between base of caudal and middle of eye. Caudal deeply forked, the lower lobe slightly the longer, about as long as head. Anal long and high, its base $3\frac{2}{3}$ in body, considerably longer than head. Pectorals long, $1\frac{1}{4}$ in head, reaching about to the front of the small ventrals, which do not reach the vent and are about $2\frac{1}{4}$ times in head.

Scales thin, caducous.

Color in life translucent, very pale, with bluish reflections. Sides with a narrow and not sharply defined but bright silvery shade, scarcely wider than the pupil, distinct for the whole length of the body. Snout yellowish; top of head dusty; the occiput nearly black; sides of head lustrous silvery. Middle line of head blackish; a series of dark points along the base of the dorsal, becoming a well-defined dark streak behind the fin. Dark points along base of anal, these also becoming a dark stripe behind the fin. Caudal distinctly yellowish, with many dark points; its tip dusky; other fins pale; the dorsal slightly yellowish.

This species is very common in the Bay of Galveston, where many specimens were obtained. The longest about $2\frac{1}{2}$ inches in length. One specimen was obtained at Pensacola; another is in our collection from Wood's Holl, Mass., where it is the commonest species of *Stolephorus*. From most of the North American species of this genus, *S. mitchilli* is distinguished by the length of the anal and by the less sharply-defined lateral stripe.

SCOPELIDÆ.

19 **Synodus intermedius** (Spix) Poey.—*Sand Diver, Sand Launce.* P. (30877.)

    ? *Saurus intermedius* Spix. Pisc. Bras. 81. Günther v, 396.
    Saurus anolis C. & V., xxii, 483.
    Synodus intermedius Poey, Syn. Pisc. Cub. 414 (No. 68).

Numerous specimens, most of them badly mutilated, were obtained from the stomachs of Red Snappers at Pensacola. Many of these were full of spawn. The most perfect specimens, about a foot in length, shows the following characters:

Color grayish-white above, becoming abruptly paler on the level of the upper margin of the pectorals; back and sides with eight broad dark cross-bands, which are broadest near the lateral line; lower part of sides with a pinkish tint. A jet-black blotch on shoulder girdle above, hidden by upper part of opercle: some irregular dark blotches on cheeks and opercles; opercle with some yellow; membrane joining maxillary to head black. Dorsal with about 6 narrow dark bars formed by series of dark spots; caudal yellowish, margined posteriorly with black; a dark blotch at its base; pectoral faintly barred with dusky and light yellow; ventrals, anal, and gill membranes sulphur yellow. Tip of snout not black; jaws mottled with dark; top of head with dark cross-line; axil blackish.

Head 4 (4⅔); depth 8 (9); D. I, 10, A. I, 10. Scales about 4-50-7.

Body fusiform, somewhat depressed, especially posteriorily. Head bluntish, rather large; snout short, broader at base than long, scarcely longer than eye, 4½ in head. Bones of top of head weakly striate; region behind eyes with strong radiating ridges; interorbital space deeply concave, its width 6 in head, superciliary bone prominent, scale-like, with radiating striæ.

Jaws subequal in front, the lower scarcely included. Maxillary 1⅔ in head, considerably longer than pectoral. Teeth not very large, those on palatines and tongue rather small.

Scales on cheeks large, in 4 or 5 rows. Scales on body everywhere large, those on breast not reduced; three series between adipose fin and lateral line; lateral line conspicuous, slightly keeled on the tail.

Origin of dorsal midway between adipose fin and nostrils, the fin high, as high as long, the longest rays 1⅔ in head. Caudal 1⅔ in head; pectoral 1⅞, reaching about to seventh scale of lateral line; ventral 1⅙; insertion of ventrals under second third of pectoral, the fin extending to slightly beyond base of last ray of dorsal; base of anal as long as max·illary.

Our specimens are evidently identical with Poey's " Species dubia, an *Synodus intermedia*. No. 68."

CYPRINODONTIDÆ.

20. Cyprinodon variegatus Lac. (30829.)
Cyprinodon gibbosus Baird & Girard, Proc. Acad. Nat. Sci. Phila. 1853, 390.

Body very short and robust, in adults high and much compressed, the females abruptly constricted at base of caudal peduncle; caudal peduncle rather short and high, rapidly narrowed backwards to tail, its greatest height nearly equal to length of head, its least height one-half head; head short, little depressed, narrowed upwards and forwards, with sharp snout and small mouth; width of mouth rather less than length of snout; teeth large, in a single series, consisting of wedge-shaped incisors, much widened towards tips, the cutting edge tricuspid; no villiform teeth; eye moderate, its diameter longer than mandible, slightly less than interorbital width, about equal to length of snout, and contained 3½ times in length of head; interorbital width 3 in head. Opercle joined by membrane to shoulder-girdle from a point slightly above base of pectoral.

Intestinal canal long, but not much convoluted, $2\frac{2}{3}$ times length of body.

Dorsal moderate, in females as high as the length of its base, in males much higher; origin of dorsal midway between base of caudal and end of snout; base of fin $1\frac{1}{3}$ to $1\frac{2}{5}$ in length of head; longest ray (in ♂ $2'$ long) reaching half way from base of fin to base of caudal; the anterior rays equaling length of head and extending beyond tips of posterior rays where the fin is depressed; in females, the longest ray about $1\frac{1}{2}$ in head. Origin of anal under eighth or ninth ray of dorsal; the fin very small, and much higher than long; length of base about equaling snout; longest ray half length of head (less in females). No external oviduct. Caudal truncate or slightly emarginate, $1\frac{1}{4}$ in head. Ventrals, in adult males, reaching front of anal, $2\frac{1}{3}$ in head; in females, reaching vent. Pectorals long, reaching middle of ventrals, $1\frac{1}{6}$ in head.

Scales large, tuberculate in males, arranged in regular series; humeral scale much enlarged, its height nearly half length of head; 26 or 27 oblique series of scales from opercle to base of tail; 13 scales in an oblique series from vent to middle of back.

Head, $3\frac{2}{5}$ to $3\frac{2}{3}$ in length; depth, 2 to $2\frac{2}{3}$; D. 11; A. 10. Scales, 26–13.

Color: ♂, olivaceous; from dorsal forward above pectoral to head deep lustrous steel-blue, the color very intense and conspicuous in life; rest of upper parts with rather greenish luster, becoming dull slaty blue; and on cheeks, opercles, sides anteriorly, and belly, deep salmon-color; lower lip and preopercle, violet. Dorsal blackish, the anterior margin of fin orange; caudal dusky olive, with a jet-black bar at tip, and a narrow black cross-streak at base. Anal dusky at base, bordered entirely around with bright orange. Ventrals dusky, bordered with orange. Pectorals dusky-orange, darker below. Smaller specimens show some orange shading on the sides, and sometimes also traces of the cross-bands of the female.

♀, very light olive; lower half of sides with about 14, alternately wide and narrow, vertical, dark bars, those anteriorly narrower and closer together; usually 7 or 8 dark cross-bars on the back, alternating with the wide bars below; these bars are of various degrees of distinctness, sometimes almost obsolete; a dusky area below eye; young with broad greenish cross-shades wider than the interspaces. Belly pale or yellowish; lower jaw largely blue; cheeks brassy. Dorsal dusky, with an intense black, faintly ocellated spot near tip of last rays. Caudal faintly reddish, with a black bar towards base. Other fins pale orange, with some dark points.

Found very abundant at Galveston and still more so at Pensacola. Specimens from the Gulf ("*gibbosus*") are larger and somewhat brighter colored than those from the Atlantic farther north, but a careful comparison with specimens from Beaufort, N. C., and Wood's Holl, Mass., failed to show any differences of even varietal value. It is possible that this species is identical also with *C. bovinus* (Baird & Girard, Proc. Acad. Nat. Sci. Phil. 1853, 389), and with *C. eximius* (Grd. Proc. Acad. Nat. Sci., Phil. 1859, 158). But as *bovinus* is described as having head 3 in length, eye 4 in head, ventrals under anterior margin of dorsal, fin rays fewer in number, D. 9, A. 8, and with somewhat different coloration, and *C. eximius* with head about $3\frac{1}{3}$ in length, eye 4 in head, D. 12, A. 12, and different coloration, it is not advisable to include them, for the present, in the synonomy of *variegatus*.

21. Fundulus similis (Girard) Jor. P. G. (30812 Pens.; 30920 Galv.)

Body very long and slender, the outlines scarcely arched; adults much deeper than young; head narrow, very long, and regularly narrowed forwards; preorbital exceedingly wide, as wide as eye, $4\frac{1}{2}$ to 5 in length of head; eye small, 5 to $5\frac{1}{2}$ in head, $1\frac{1}{2}$ to $1\frac{3}{4}$ in interorbital width; posterior margin of orbit slightly behind middle of head; eye $1\frac{3}{4}$ in length of mandible; mouth small, maxillary not nearly reaching vertical from anterior nostril; teeth very small, in broad villiform bands, the outer series not at all enlarged; interorbital width $3\frac{1}{5}$ in head.

Dorsal fin long and rather low, the height less than length of base in adult males, $1\frac{1}{5}$ in length of base in females; in males the last rays are but little higher than some of those preceding, in females the last are the lowest; longest ray (in ♂) $2\frac{1}{5}$ in head; origin of dorsal midway between middle of eye and tip of caudal. Origin of anal under third dorsal ray, the fin much higher than dorsal, the longest ray $1\frac{1}{4}$ in head; the rays regularly increase in length to the sixth; the seventh, eighth, and ninth then rapidly shortened, the last again somewhat longer; thus the anterior outline of the fin is convex, and the posterior deeply emarginate or falcate, or in females nearly vertically truncate; posterior margins of oviduct adnate along either side of third anal ray, forming a pouch at base of first and second rays, covering one-fourth length of first ray. Pectorals reaching origin of ventrals, $1\frac{3}{5}$ to $1\frac{3}{4}$ length of head; ventrals not reaching vent, $2\frac{1}{5}$ in head; their base midway between pectorals and origin of anal; caudal subtruncate, $1\frac{2}{3}$ in head.

Scales large, in regular series; 33 oblique series from opercle to base of tail; 11 in an oblique series upwards from vent to middle of back; humeral scales not enlarged.

Head $3\frac{1}{4}$ in length; depth $3\frac{3}{4}$ to $4\frac{2}{5}$; D. 11 to 13.

A. 10; scales 33—11.

Color: ♂, olivaceous, bronze below; lower parts of head strongly orange; sides with 10 to 15 narrow dark bars, one-third to two-thirds as wide as the interspaces, and not very dark, although distinct; a large, diffuse, dark humeral blotch, extending from above opercle to about base of pectoral; each scale with a distinct >-shaped intermarginal series of dots, forming conspicuous reticulations. Dorsal dusky with black specks, mostly black at base; a small ocellated black spot behind, disappearing in adults; caudal faintly clouded with dusky, especially about the middle; ventrals pale, somewhat soiled.

♀, olivaceous, sides paler olive, with metallic lustre; belly white; 7 to 15 very narrow sharply-defined black bars on sides, not extending on the back, scarcely broader than the pupil; scales marked as in the males, but much more faintly. Fins pale, almost immaculate.

This species is very abundant at Pensacola, where many specimens were collected; it was also obtained at New Orleans. The Galveston specimens show quite constantly: D. 11, A. 9, head 3½ in length; eye smaller, 1¾ in interorbital width, and width of preorbital 5½ in head; and may represent a tangible variety.

22. Fundulus grandis Grd. G. P. (30836.)

Fundulus floridensis Grd. Proc. Acad. Nat. Sci. Phil. 1859, 157.

Body stout, robust; adult females much compressed and elevated; caudal peduncle short and rather deep, its greatest depth (in ♀, 5 inches long) equaling its length, which equals one-half length of head; head long, broad, and heavy, the lower jaw conspicuously longer than the upper, and very strong; teeth in a narrow villiform patch, the outer series in each jaw enlarged; preorbital narrow, about one-half diameter of orbit; eye large, slightly less than length of snout or mandible, 4 to 4½ in head, slightly more than one-half interorbital space.

Dorsal fin small and low, but little elevated, in males 4½ inches long, where the tips reach scarcely more than half way from base of fin to root of caudal; the rays still shorter in adult females; origin of dorsal usually slightly nearer tip of caudal than tip of snout; base of dorsal contained from 2½ to 3 times in head; longest rays in male about one-half head, somewhat less in females. Origin of anal under fourth or fifth ray of dorsal, its base equalling length of snout; longest ray in males 1¾ in head, in females 2 to 2½ times; oviduct attached to first anal ray for a distance more than one-third length of ray; ventrals barely reaching vent in males, about 2½ in head; pectorals large, reaching to or beyond base of ventrals, and half or more than half length of head; caudal about 1½ in head.

Scales in 35 to 38 oblique rows; 15 in an oblique series from vent forwards to middle of back.

Head 3 to 3⅕ in length; depth 3⅗ to 3⅝; D. 11; A. 10 or 11; scales 35 to 38–15.

Color: ♂, very dark green above, paler posteriorly; sides with numerous small, round, pearly-white spots, occasionally some of them arranged in vertical series; posteriorly with traces of 8 to 10 very narrow, pale, vertical bars, alternating with broader, faint, dusky ones; belly yellowish; sides of head dusky. Caudal greenish, almost black behind, its edge translucent; the basal part with numerous small white spots. Dorsal olive, anteriorly orange, blackish on basal half, and marked with numerous small white spots. Anal and ventrals bright orange, the former sometimes dusky, and frequently with several white specks at base. Pectorals light yellow.

♀, sometimes nearly plain silvery, dusky olive above, and with much minute dark specking on lower half of sides; sides usually showing traces of from 12 to 15 narrow, silvery, vertical bars, less than one half as wide as the dusky interspaces; no white spotting on body or fins; fins all nearly plain dusky olive, with some yellow; top of head blackish.

This species is very closely allied to *F. heteroclitus*, but differs constantly in the much lower fins; the interorbital width is slightly less, and the fins show some slight differences in coloration. *F. grandis* was found very abundant in the Laguna Grande at Pensacola, and was also found at Galveston, Tex.

23. Fundulus ocellaris sp. nov. (29667, 29667, 30853.) P.

Head comparatively small and narrow, with short depressed snout, and weak jaws; body rather slender; lower jaw but little longer than upper; eye small, 4 in head, 1⅔ in interorbital width, equaling snout, which equals length of mandible; teeth all villiform, in narrow bands in each jaw, the outer series but little enlarged, but projecting appreciably beyond the others; preorbital narrow, less than half diameter of orbit.

Dorsal fin (in ♂ 3 inches long) much elevated, reaching, when depressed, beyond base of rudimentary rays of caudal; much shorter than this in females and young males. Origin of dorsal midway between tip of caudal and tip of snout, or slightly nearer snout; the base of the fin 1½ in height of longest ray, which is contained 1¼ in head; outline of fin rhomboid, the upper edge straight, the last rays highest. Anal fin similar to dorsal, but narrower and slightly lower, not reaching caudal when depressed; its origin under second ray of dorsal and distant from caudal half as far as from tip of snout; base half height of longest ray; greatest height of caudal peduncle two-thirds its length and half length of head; oviduct not attached to first anal ray, but forming a low sheath along base of first six rays. Caudal short, rounded, 1¼ in head; pectorals slender, reaching base of ventrals, 1⅔ in head; ventrals (in adult ♂) extending beyond front of anal, half length of head.

Scales moderate, in somewhat irregular oblique series, of which there are 35 between gill opening and base of caudal; 15 scales in an oblique series from vent forwards to middle of back; about 18 cross series between nape and front of dorsal; humeral scale not enlarged.

Female with somewhat deeper body, and different coloration; the fins smaller, the last ray of dorsal shorter than those preceding, and not reaching half way from its base to rudimentary caudal rays; length of longest ray greater than base of fin; ventrals not nearly reaching vent; front of dorsal nearer tip of caudal than end of snout.

Head 3 to 3⅓ in length; depth 4. D. 11; A. 10; V. 6; P. 13; scales 35–15.

Color: ♂, dark olive brown above, golden on sides and below, the golden tint extending farther up on caudal peduncle than on trunk;

scales margined with darker; sides with 13 to 15 dark cross-bands of the color of the back, not extending on the belly, but almost reaching lower median line behind ventrals; these bands usually approximately parallel, and the anterior ones, at least, narrower than the interspaces, the widest of which is about two-thirds diameter of orbit; sides posteriorly to origin of dorsal finely speckled with small pearly spots which cover both bands and interspaces. Dorsal and anal margined with orange anteriorly, the color deeper on front of anal; the two fins tinged with orange and checked with black and pearl color; caudal light orange, indistinctly barred at base with series of linear blotches; pectorals and ventrals plain orange, the former slightly dusky.

♀ dark above, sides finely dusted with dark points, pale below, tinged with yellowish; middle of sides with about 13 very narrow, short, dark half bars; back sometimes with small dark blotches; dorsal dusky with a very distinct black spot ocellated with white, on its posterior rays; caudal and anal plain dusky; ventrals light yellowish.

About 15 specimens, the longest about 3 inches long, obtained in Laguna Grande, at Pensacola.

24 Fundulus xenicus nom. sp. nov. P. (29668; 30821; 30841.)

> *Adinia multifasciata* Girard, Proc. Acad. Nat. Sci. Phil. 1859, 117 (not *Hydrargyra multifasciata* Le Sueur, nor *Fundulus adinia* Jor. and Gilb. Synopsis Fishes N. A. 334).

Body very deep and much compressed, with very high caudal peduncle, rapidly tapering head, and very slender, sharp, conical snout; tip of snout on axis with body, the ventral outline somewhat more arched than the dorsal; profile rising rapidly from tip of snout to origin of dorsal, slightly depressed at nape; body highest at origin of dorsal fin, where the profile is angulated; depth much greater in adults than in the young; in a male specimen, 2¼′ long, the depth equals one half the length; in younger males the depth is contained 2⅓ to 2½ times in length; greatest depth of caudal peduncle 3¼ in length. Head high and narrow; snout conical, pointed; jaws equal, the gape horizontal in closed mouth; mouth protractile downwards and forwards; teeth very small, in a villiform band, the outer series in each jaw enlarged and conical. Eye large, 3 in head, 1⅓ in the narrow interorbital space, equal to length of snout, rather more than length of mandible. Branchiostegal membranes broadly joined across throat, united as far back as vertical from preopercular margin. Branchiostegal 5. Opercle joined by membrane to shoulder-girdle, down to a point just above base of pectoral.

Intestinal canal equaling length of body.

Dorsal in advance of anal, its origin midway between base of caudal and middle of orbit; the fin much higher than long, the longest rays reaching, in adult males, beyond rudimentary caudal rays; highest dorsal ray 1⅔ in head. Anal beginning opposite middle of dorsal base, similar to dorsal, but lower, scarcely reaching base of caudal; the base

of the fin is very oblique and is about equal to length of caudal peduncle; distance from origin of anal to base of caudal, 2 in distance to tip
of snout; longest anal ray $1\frac{1}{2}$ in head. Caudal broad, $1\frac{1}{4}$ in head. Ventrals (in $\male$) reaching anal, $2\frac{1}{3}$ in head. Pectorals long, reaching middle of ventrals, $1\frac{1}{2}$ in head. Oviduct not adnate to first anal ray.

Female specimens have body less deep, fins much lower, and differ-
ent coloration; the depth is $2\frac{2}{3}$ to $2\frac{4}{5}$ in length, and the longest dorsal
ray $1\frac{2}{3}$ in head.

Head $2\frac{9}{10}$ in length; depth 2; D. 9 or 10; A. 11 or 12; V. 6; P. 14;
B. 5; scales 25–10.

Color: $\male$, dark green, sides with 10 to 14 narrow bands of bluish-
silvery, the first of which is somewhat in front of dorsal; these bands
are slightly oblique below, and are a little narrower than the interspaces;
they become wider and farther apart behind; the interspaces are frequently divided by fainter silvery bands; a diffuse, broad, dusky blotch
below and behind eye. Lower jaw bright orange; lower side of head
and belly yellow.

Dorsal blackish, with very numerous round blue spots, the lower
spots, and sometimes most of them, orange; anal similarly colored;
caudal with irregular alternately dark and light bars, and a few white
basal spots; ventrals dusky, tipped with sulphur-yellow; pectoral translucent.

$\female$ greenish, with a faint trace of a dusky lateral stripe, and with
about 8 obscure pale cross-bands; dorsal, caudal, and pectorals plain
dusky, the lower edge of caudal tipped with orange; anal and ventrals orange-yellow; lower jaws yellow; a dusky shade below and
behind eye.

Very numerous specimens, the largest about 2 inches long, were
obtained from the Laguna Grande, at Pensacola, in salt-water.

25. Lucania venusta Girard. P. (30819.)

Lucania affinis Grd. Proc. Acad. Nat. Sci. Phila. 1859, 118.

Body fusiform, rather strongly compressed, the dorsal and ventral
outlines about equally arched; head narrow, compressed, flattened
above the eyes, the upper profile of snout both longitudinally and transversely convex; snout compressed, conspicuously shortened and vertically rounded, its height greater than its width; caudal peduncle long
and rather slender, its greatest height $1\frac{2}{3}$ in head, its length slightly
less than head; mouth very small, protractile forwards, the lower jaw
very much projecting in open mouth; mandible heavy, short, and
strongly convex, less than diameter of orbit; teeth small, but firm and
strong, conical, in a single series in each jaw, or forming an irregular
double series anteriorly; no villiform teeth behind this outer series;
eye large, 3 in head, slightly shorter than interorbital width, and greater
than length of snout.

Intestinal canal rather less than length of body.

Origin of dorsal fin midway between tip of snout and base of caudal, or very slightly nearer the latter; the length of its base contained 1⅔ in head; the upper margin of the fin rounded, the longest ray (in ♂) equalling the length of its base.

Origin of anal fin under middle of dorsal; distance from its origin to base of caudal from four-sevenths (♀) to five-sevenths (♂) of distance to top of snout; oviduct not attached to first anal ray, but produced backwards, forming a low sheath on both sides at base of first 6 rays; length of anal base, two-fifths head; longest ray (♂), one-half head; caudal 1⅔ in head. Pectorals long, reaching beyond base of ventrals; 1¾ in head. Ventrals to slightly beyond vent; 1⅛ in head.

Head 3½ in length; depth 3½. D. 11 or 12; A. 9 or 10; Scales 26–8.

Color ♂ light olive, pale on belly, sides with some silvery lustre and with indistinct trace of an obsolete dusky lateral stripe; scales conspicuously dark-edged; opercles and cheeks bright silvery; dorsal and caudal light yellow, and, as well as the anal, narrowly margined with black; dorsal with an elongate, vertical, black blotch at anterior margin, a yellow spot behind it; a vertical dusky streak behind each dorsal ray, composed of fine black points. Anal orange or translucent, white at base; ventrals similar to anal. Pectorals pale yellowish. A dark vertical streak through iris.

♀ similar, fins all plain.

Exceedingly abundant in the lagoons at Pensacola.

26. Gambusia patruelis Girard. N. O. G. 30922.

Heterandria affinis Baird & Girard, Proc. Acad. Nat. Sci. Phil. 1853, 390.
Gambusia gracilis Girard, Proc. Acad. Nat. Sci. Phila. 1859, 121.
Gambusia humilis Günther, vi, 335.

The specimens described are all femâles.

Body rather slender, compressed, the belly much distended with ova, projecting much beyond normal outline of body, and abruptly constricted at the vent; greatest height of caudal peduncle one-third greater than its least height, and three-fourths length of head; head small, very broad, and much depressed; teeth strong, in a broad villiform band in each jaw, the outer series much enlarged, the teeth not movable, straight; eye small, 1¾ in interorbital width, slightly greater than length of snout, and 3⅛ to 3½ in length of head; interorbital width 1⅔ in head.

Intestinal canal short, about equal to length of body.

Dorsal small, inserted far back, its base scarcely greater than diameter of orbit; distance from its origin to base of caudal equaling one-half the distance to tip of snout; the origin of fin over middle of anal; highest ray 1¾ in head. Anal larger than dorsal, with longer base and higher rays; the longest anal ray slightly less than length of head; origin of anal about midway between rudimentary caudal rays, and gill opening. Caudal acutely rounded, slightly less than length of head

Ventrals short, not nearly reaching front of anal, 2 in head. Pectorals nearly as long as head, reaching to beyond base of ventrals.

Head 4 in length; depth 3 to 4; D. 7; A. 8 or 9. Scales 30 or 31–10.

Color, light olive with some bluish reflections; each scale edged with dark; a very narrow dark line along median row of scales on sides; top of head and upper part of opercle, dusky; an oblique, narrow and rather obscure, dark blue-black band below eye; a black spot on each side of belly, a dark median line on caudal peduncle below. Fins dusky.

Exceedingly abundant in the marshes about Lake Pontchartrain. A few specimens were also obtained at Galveston. This species is most closely allied to *Gambusia holbrooki* (Agassiz); a comparison with specimens of the latter from Indian River, Florida, show certain constant differences. Thus, in *holbrooki* the eye is larger, more than one-third length of head, and is contained 1½ in interorbital width; and the head is larger, 3⅜ in body. These slight differences may disappear on the examination of an extended series, but with our present material no variation is apparent. In the synonymy of *holbrooki* must be placed *Haplochilus melanops* Cope. Proc. Amer. Philos. Soc. 1870, 457 (*nec Zygonectes melanops* Jordan. Bull. Ill. Lab. Nat. Hist. No. 2, 52); and *Zygonectes atrilatus* Jordan & Brayton, Bull. U. S. Nat. Mus. xii, 1878, 84.

27. Mollienesia latipinna Le Sueur. P. (30823, 30870.)

> *Pœcilia multilineata* Le Sueur, Journ. Acad. Nat. Sci. Philad. 1823, ii, 4.
> ? *Limia matamorensis* Grd. Proc. Acad. Nat. Sci. Phila. 1859, 116.

Body oblong, much compressed in males, of nearly equal height from dorsal backwards, the greatest height of body but one-third greater than that of caudal peduncle; females, with gibbous belly and narrower caudal peduncle; head very small, depressed, not narrowed forwards; mouth very small, vertical, and without lateral cleft; length of mandible about two-thirds diameter of orbit; teeth all very small, movable, in a rather narrow band; the outer series much larger than the others, but still very small, composed of slender pointed teeth, strongly curved inwards; eye moderate, 1½ to 1⅝ in interorbital width, equal to or slightly greater than snout, and 3¼ to 3½ in head.

Dorsal very long, in adult males enormously elevated, exceeding height of body; the fin is almost square, the base slightly longer than the height, the upper margin nearly straight; longest ray 2½ in length of body, the last ray reaching beyond base of caudal; base of fin 2¼ in body; origin of dorsal distant from base of caudal, 2⅕ times its distance from the tip of snout. In females, the dorsal is low, the longest ray equaling two-thirds length of head, the last ray reaching but half way to base of caudal; the base of the fin 3⅔ times in length of body, its origin distant from base of caudal 1¼ times the distance from tip of snout.

Anal very small; in the male, modified into an intromittent organ, and inserted in advance of middle of dorsal, its origin about half way between snout and base of caudal, the fourth ray longest and thickest, 1¼ in head ; in females the origin is under twelfth ray of dorsal, and about midway between tip of caudal and tip of snout. Caudal rounded, about equaling length of head in females, one-fourth greater than head in males. Ventrals inserted behind vertical from origin of dorsal, reaching beyond vent in females ; in males the first and second rays are thickened, the second filamentous, 1⅛ in head. Pectoral long, longer in males, where it reaches beyond middle of ventrals, and is very slightly less than length of head.

Scales in very regular rows, 26 in a longitudinal series, 9 or 10 in an oblique series forward from vent to middle of back; humeral scale not enlarged. Intestinal canal about 2½ times total length of fish (with caudal).

♂. Head 4 in length; depth 2¾ to 3. ♀. Head 3½ to 3¾ in length; depth 2½ to 2⅔.

D. 15 or 16; A. 8; scales, 26 — 9 or 10.

Color: ♂. Light olive-green, marbled with darker and spotted with pale green ; each scale on back and sides with an oblong, blackish spot, these forming continuous lengthwise stripes; head dusky above, opercle and cheek minutely speckled; an orange stripe above opercle; lower parts of head mostly orange; some orange tinge on breast. Dorsal translucent, its basal half with about five series of linear blackish horizontal spots, forming interrupted lines; above middle of fin, on membrane between each pair of rays, is a large, roundish dark spot. Between these spots and above them are many small, round bronze spots. Membrane between second and·third rays red at base; all of these markings irregular on first and last rays; caudal narrowly margined all around with black, its base lavender; its lower parts mostly whitish; the middle orange; the upper parts pale, with round orange spots; other fins pale orange. Females have dorsal and caudal olivaceous, with indistinct, narrow cross-bands, formed by series of small dark spots on the rays.

Very abundant at Pensacola, where numerous specimens were procured from the Laguna Grande. It is also very common about the wharves, the gorgeous dorsal fin of the male being conspicuous in the shallow water.

28. Mollienesia lineolata (Grd.) J. & G. G.; N. O. (30891.)

   ? *Mollienesia pœcilioides* (Girard).

Four female specimens and one male, from Galveston, Tex. (the largest 2′ long), and two females from Lake Pontchartrain, are referred by us to this species. They show the following differences from *M. latipinna:*

Eye small, the iris jet black ; diameter of orbit 3⅓ to 3½ times in head, and 1¾ to 2 times in interorbital width (the eye 2⅞ in head, and 1⅜ in interorbital width, in *latipinna* of same size); dorsal fin smaller, its base 3⅞ in body in females, 3 in males, the rays constantly 13 or 14 in number (usually 13); origin of dorsal equidistant from tip of snout and ru dimentary caudal rays in females; in males, distant from snout by length of base of fin; ventrals inserted in advance of vertical, from origin of dorsal, or, in male, opposite origin of dorsal; color the same as in *latipinna*, except that all the specimens show the 5 faint, dark, vertical-half bars on the sides.

This species can hardly be regarded as more than a representative form of *M. latipinna*, and, on the examination of a sufficient number of specimens of the various localities, may be found to vary into the typical form. The differences pointed out above are, however, constant in the specimens in our possession, and seem to warrant the retention of the name for the present.

The male fish described by Girard as *Limia poeciloides*, is probably referable to this species. *Limia matamorensis*, on the contrary, seems to be a typical *latipinna*.

MURÆNIDÆ.

29. Muræna ocellata (Ag.) Jen. P.

One small specimen in good condition, together with the remains of several larger ones, were taken from the stomachs of Red Snappers at Pensacola. Color light olive green, darker above, becoming light yellowish on the belly, the dark color forming reticulations around whitish spots of various sizes; most of them round, some oblong and some confluent, the largest not quite as large as eye; spots becoming smaller toward head and largest toward the tip of the tail. ·Dorsal with dark marginal blotches; anal black edged; a small jet-black spot at angle of mouth; no black around gill-opening.

Teeth uniserial, the larger ones distinctly serrated on the posterior margin, rather strong and turned backward, those in front little larger than the others. Vomer, in all specimens examined, without trace of teeth; gape in head; dorsal beginning a little in advance of gill-opening. Head 2⅓ in trunk; head and trunk a little shorter than tail; eye 3 in gape, half broader than gill-slit, equal to interorbital space.

ANGUILLIDÆ.

30. Ophichthys mordax (Poey) J. & G. P.

One specimen, nearly digested, from the stomach of a Red Snapper, at Pensacola. The dentition agrees better with Poey's account of his "*Macrodonophis mordax*," than with Günther's description of *Crotalopsis punctifer* Kaup. Dr. Günther considers the two identical.

31. Ophichthys macrurus Poey. P. (30895.)

A single specimen, in good condition 11 inches long, was presented to the National Museum by Dr. August Galny, of Galveston.

Color light olive, the back closely punctulate but pale, the belly whitish; fins all pale; dorsal and pectoral without darker margin.

Head 2⅔ in distance from snout to vent; the distance from snout to vent 2⅛ to 2⅔ in total length; gape 2⅔ in head, a little less than length of pectoral, which is about equal to greatest depth of body; teeth all distinctly biserial. Dorsal beginning a little in front of tip of pectoral. Fins all edged with black.

Body not very slender. Head narrow and pointed, the upper jaw projecting beyond lower. Eye large, more than half length of snout, its position over the middle of the gape, its diameter more than the interorbital width; gape 2⅔ in length of head; teeth biserial on jaws and vomer, subequal, short, slender, and sharp, all of them more or less directed backward; no large canines; some of the vomerine teeth larger than the others; nasal tubes short and inconspicuous; gill-openings small, their height about ⅔ eye.

Tail almost exactly twice length rest of body. Head slightly more than half trunk, nearly 9 in total length. Distance from snout to front of dorsal 2⅓ in distance from snout to vent. Dorsal beginning opposite anterior fourth of pectoral, rather low. Pectorals long and narrow, about 2⅔ in head. Free tip of tail sharp. This species is allied to *Ophichthys parilis* (Rich.), but seems to be well distinguished by the short tubes of the nostrils.

32. Ophichthys chrysops Poey. P.

? *Ophisurus gomesii* Castelnau, Anim. Amér. Sud., Poiss. p. 84.

Two specimens, one male and one female, the male about 20 inches long, in poor condition, were taken from the stomach of a Red Snapper at Pensacola. The male with the testes well developed; the female with two large ovaries extending for the entire length of abdominal cavity.

33. Myrophis lumbricus sp. nov. (G.) 30896.

A single specimen, 9 inches in length, obtained at Galveston.

Color light olivaceous, scarcely translucent in life, with a slight bluish luster towards the head. Everywhere, except on belly, finely and densely punctulate with black, besides which are small faint spots of greenish yellow. Eyes bright green. Underside of belly and head with steel-blue luster.

Body subterete, worm-like, tapering backward almost to a point, even the tail scarcely compressed. Diameter of head much less than of body. Head extremely small, slender, and pointed, the narrow upper jaw projecting well beyond lower. Eye small, considerably nearer angle of mouth than tip of snout, its length about half snout. Gape short, about 4 in head. Teeth all strong, slender, sharp, directed backward, apparently in single series, some of the anterior in the upper jaw canine-like, a single series of teeth on the vomer rather stronger than the teeth in the jaws.

Gill openings small, oblique, rather close together, subinferior, just below the minute rounded pectorals, which are narrower than the gill openings and not much larger than the eye. Opercular region long, with very conspicuous concentric striæ.

Head 10⅔ in total length; greatest depth of body 33. Length of head and trunk 2⅔ in total. Dorsal very low, beginning at a point nearer gill opening than vent, at a distance behind gill opening about equal to length of head. Lateral line distinct.

This species is evidently distinct from the *Myrophis* found at Panama, which Dr. Günther calls *Myrophis punctatus*. This species has larger head, larger mouth, longer pectorals, and the body more compressed, etc. *Myrophis microstigmius* Poey, from Cuba, is said to have the dorsal inserted farther back. Kaup's description of *M. longicollis* (= *M. punctatus*), from Surinam, also indicates a species with a longer head; but too little is known of that species to afford a comparison with *M. lumbricus*, *M. microstigmius*, or the Panama species, if that be really different from *M. punctatus* Lütken.

Three other eels, two of them *Ophichthys*, and the other perhaps an *Ophiosoma*, and all new to our coast, were obtained from stomachs of Red Snappers at Pensacola, but in such bad condition that they cannot be identified.

34. Anguilla rostrata (Le S.) DeKay.—*"Fresh-water eel."* N. O.

Seen only in the New Orleans markets.

35. Conger caudicula Bean, MSS. P.

A species of *Conger* with the skin entirely digested was taken from the stomach of a Red Snapper. We were unable to distinguish its remains from the common species.

SCOMBERESOCIDÆ.

36. Tylosurus longirostris (Mitch.) J. & G.—*Needle-fish.* G.; P. (31010, G.)

(*Belone scrutator* Girard, U. S. Mex. Bound. Surv. 1859, 30, pl. xiii.)

Generally common; rarely brought into the markets, although considered good eating. It is not tangibly different from the northern form.

37. Hemirhamphus unifasciatus Ranzani. G. (31027.)

(*Hemirhamphus roberti* and *H. richardi* C. & V. xix, 24, 26.)

Generally common.

38. Exocœtus hillianus Gosse. P. (30866.)

One fine specimen, 5½ inches long, from the "Snapper Banks" at Pensacola.

Color, back and sides to middle of base of pectoral dark green, thence abruptly bright silvery, this shade covering the lower two-thirds of the sides, belly, and sides of head. A rather faint purplish band from upper edge of pectoral base backward, parallel with back; occiput, snout, sides of head and silvery area on sides more or less flushed with pinkish purple. Dorsal fin translucent, with a large black blotch covering upper part of first 6 rays; the fin with narrow white edging posteriorly;

caudal bright brick-red, speckled with dark points and edged posteriorly with translucent. Pectorals dusky translucent, with reddish tinge on basal two-thirds of upper rays. Ventrals translucent, with some reddish on base of central rays and with a distinct small dusky spot at base of outer ray, externally visible through the covering scale. Anal translucent, somewhat white anteriorly.

Head 4⅜; depth 5. D. 12; A. 14; scales 38-5.

Body moderately compressed. Head rather short, the short snout 4¾ times in its length; the large eye 3 times, interorbital space flat, 3 in head. Gill rakers rather long.

Pectoral fin reaching about to middle of anal, its length 1¾ in body, its second ray scarcely shorter than third, not forked. Ventral fin inserted slightly nearer root of caudal than tip of snout, its tip extending very slightly past front of anal, its length 1¼ in head. Dorsal much higher than long, its longest rays slightly longer than head, reaching caudal. Lower lobe of caudal slightly longer than head.

This rare and beautiful species has not been hitherto recorded from our coast.

The species of the restricted genus *Exocœtus* (exclusive of *Halocypselus* and *Cypselurus*) represented in the National Museum from our Atlantic coast, may be recognized in the following analysis:

a. Ventrals moderate, shorter than head, reaching little past front of anal; second ray of pectoral simple. (*Parexocœtus* Bleeker.)

 b. Dorsal higher than long, with a black blotch in front; ventrals plain; anal about as long as dorsal; D. 12, A. 14 .... HILLIANUS.

aa. Ventrals long, longer than head, reaching usually past anal fin; second ray of pectoral forked. (*Exocœtus.*)

 c. Ventrals pale; snout not very blunt.

  d. Anal rather long, its base about three-fourths that of dorsal; its insertion nearly opposite front of dorsal; lower caudal lobe shorter than head; D. 11, A. 12 .... EXILIENS.*

  dd. Anal short, its base less than half that of dorsal; its insertion behind that of dorsal; lower caudal lobe nearly one-third longer than head; D. 14, A. 9. .... NOVEBORACENSIS. †

 cc. Ventrals black, with white edgings; snout very blunt; anal rather long, its base more than ⅔ dorsal; its insertion slightly behind front of dorsal; lower caudal lobe half longer than head; D. 12, A. 12.

RONDELETII. ‡

SYNGNATHIDÆ.

39. Siphostoma floridæ sp. nov. P. (30826.)

Body comparatively slender, the belly scarcely keeled, even in the females. Head slender, the snout long, from one-third to one-half longer

---

* *Exocœtus exiliens* Gmelin, Syst. Nat. i, 1400, 1788; Günther vi, 291; Goode, Bull. U. S. Nat. Mus. v, 64.

 † *Exocœtus noveboracensis* Mitchill, Amer. Monthl. Mag. ii, 233, 1817: *Exócœtus melanurus* C. & V. xix, 101.

 ‡ *Exocœtus rondeletii* Cuv. & Val. xix, 115. A specimen we examined (21870) from open sea, lat. 46°; long. 61°.

than the rest of the head, its upper edge with a low sharp keel; top of head without keel; supraocular ridge a little elevated, the region between eyes concave; opercle striate, without median keel. Lateral line not continuous with upper edge of tail. Dorsal fin on one body-ring and 6 or 7 caudal rings, the distance from its insertion to the tip of the snout $1\frac{1}{5}$ to $1\frac{2}{3}$ in total length. Head $5\frac{1}{2}$ to $6\frac{1}{2}$ in length. Dorsal rays 27. Rings 17 or 18 + 31 or 32. Caudal pouch in the male, covering about 18 rings. Tail longer than trunk, $1\frac{5}{8}$ in total.

Color in life, dark green ; tail with faint darker bars broader than the interspaces; sides of body with horizontal pale streaks or vermiculations; sides of tail with some round pale spots, snout dusky, marbled or barred on side with paler; lower part of opercle nearly plain. Dorsal translucent, yellowish at base ; caudal yellow, dusky at tip.

Many specimens, the longest about seven inches in length, were taken with the seine in sea-wrack and algæ in Pensacola Bay, especially in the Laguna Grande. In our paper on the Fishes of Beaufort Harbor (Proc. U. S. Nat. Mus. 1878, 368) we have recorded a " *Siphonostoma fuscum* " from that locality. The specimens referred to under that name belong to *Siphostoma louisianæ* chiefly ; among them are examples of the present species.

40. Siphostoma affine (Gthr.) J. & G. P. (30827.)

(*Siphostoma* sp. Jordan, Proc. Ac. Nat. Sci. Phila. 1880, 22 ; Saint John's River.)

Abundant in Pensacola Bay in the same localities as the preceding, from which it is readily distinguished by the much shorter snout and the peculiar coloration.

Color in life: Females deep olive-green, varying to brown, blackish, or slightly reddish, according to the character of the surroundings; females with a black keel on the belly, which is obsolete in the male. Dark color of the back forming about 15 dark cross-bars, very faint and much wider than the interspaces. Plates of anterior parts of body, each with two narrow vertical stripes of shining silvery, very conspicuous in life. Sides of head mottled, especially on lower half of opercle. Snout dark above, abruptly paler below. Dorsal dark, like the body, with narrow dark oblique paler streaks formed of small pale spots. Caudal and anal dusky. Males olivaceous, mottled with darker, the vertical silvery streaks absent. Dorsal rays 28 to 31. Rings 16 + 32.

Specimens of this species from Saint John's River, Florida, are in our collection.

41. Siphostoma zatropis sp. nov. P. (30865.)

A single specimen, $5\frac{3}{8}$ inches long, obtained from the mouth of a Red Snapper.

Color brown, marbled with darker and with reddish. Back and sides with ten broad dark bands, the anterior portion of each band paler than the posterior; all the bands broader than the whitish interspaces.

Snout whitish, with two narrow dark bands; opercle and lower part of head with white vertical streaks. Behind the vent the dark bands encircle the body; before the vent the belly is immaculate. Caudal tipped with black. Dorsal rays 20; rings 18+30. Dorsal much shorter than head, on 1+4 rings. Head 9 in length; snout short, $2\frac{2}{3}$ in head; tail longer than rest of body, $1\frac{3}{4}$ in total length.

Occiput crested; opercle with a conspicuous keel. This species is very different from any other thus far found in our waters, and is an interesting addition to our fauna.

Our specimen is doubtless identical with *Syngnathus albirostris* Günther (viii, 170) from "Mexico." The original *Corythroichthys albirostris* of Kaup from Bahia and Mexico is uncertain, and must apparently have been some other fish. It is said to have D. 27; rings 12+29. *Syngnathus elucens* Poey is closely related, but apparently different.

42. Hippocampus zosteræ sp. nov. P. (30852.)

Two specimens, each less than two inches long; a male with distended egg-sac, and a female were taken with seine in the Laguna Grande. They were found in the sea-wrack (*Zostera*) in water about 3 feet deep.

Snout very short, about $2\frac{2}{3}$ in head; supraorbital spines moderate, diverging, each with a smaller spine in front of it. Coronet stoutish, high, fully two-thirds as long as snout, ending in five small, bluntish spines, besides which are a few filaments, which are about as long as snout; some filaments on the back of the neck; temporal spines sharp, pointing nearly straight out. Spine on side of throat rather short. Spines on body small, subequal, sharp, straight. A spine at base of pectoral, and one below it. Length of head about equal to greatest depth of body. Dorsal fin covering most of two body rings and one caudal ring; the fin rather high and very short, the number of rays but 12. Rings 11+26 to 30.

Color olive-green, the sides of the head mottled and with some paler spots, especially about the eye; dorsal mottled with dusky, in the male with a broad conspicuous red margin, in life.

The smaller number of dorsal rays seems to fully distinguish this species from all others found in the Atlantic.

43. Hippocampus stylifer sp. nov. P. (30876.)

One specimen (♀) about three inches long, "spewed up" by a Red Snapper at Pensacola.

Snout not very short, but little shorter than rest of head, equal to distance from middle of eye to gill-opening; a small tubercle on the median line at base of snout above; supraocular and temporal spines long, simple; a long spine on the median line in front of coronet, its length scarcely less than diameter of eye; coronet stoutish, high, its five spines broadly spreading, slender; the three posterior spines shortest and less divergent; extent of coronet greater than its height; spines of head with dermal tentacles.

Each alternate plate on the neck, armed on each of the dorsal ridges, with a long slender spine, which is as long as the eye, and scarcely tapering toward the tip; each provided with a filament about as long as the spine; upper lateral ridges of each plate similarly armed, the spines shorter; lower lateral and ventral ridge on two plates, likewise armed. Each fourth plate on the tail similarly armed with a long, slender spine on its upper and lower ridges. A strong spine in front of pectoral, and one below it. About fifty well-developed spinous processes on the body, besides numerous smaller spinous points. Base of dorsal elevated, the fin covering about 4 body rings, its number of rays 16. Rings about 12 + 31.

Color brownish, crossed at intervals by darker bars, which have a grayish center. These bars cross the plates which have the largest spines. Snout blackish, with two or three oblique whitish streaks, one of them forming a ring.

Two other specimens of this species, taken in gulf-weed off the east coast of Florida, by Prof. J. H. Comstock, are in the museum of Cornell University.

Another specimen of *Hippocampus*, in bad condition, was taken from the stomach of a Red Snapper.

Snout rather longer than postorbital part of head; coronet and supraorbital spines high; spines on body and tail large and sharp. Dorsalrays apparently 16. Whether this specimen belongs to *H. stylifer* or not, we are unable to say.

MUGILIDÆ.

44. Mugil albula L.—*Mullet, Molly, Meuille.* G. (30912, 30915, 30923, 31039,31050.)

Mugil berlandieri Girard, U. S. Mex. Bound. Surv. Ichth. 20.

Mugil mexicanus Steindachner, Ichthyol. Beiträge, iii, 58, 1875.

Excessively abundant, particularly about Galveston, where they are found even in the gutters along the streets. Held in low esteem as a food-fish, and largely used for bait.

We do not believe that the mullet of Southern California and the west coast of Mexico, *Mugil mexicanus* Steind. can be distinguished as a species from the Atlantic fish. Both *Mugil albula* and *M. brasiliensis* appear to be equally abundant on both coasts, and their range on both sides is similar, *M. albula* reaching to Cape Cod, and Monterey *M. brasiliensis* to Virginia and Lower California.

ATHERINIDÆ.

45. Menidia peninsulæ (Goode & Bean) J. & G. P. (30918.)

Very abundant about Pensacola, in schools along the sandy beaches.

Light green; edges of scales with dark dots; lips and top of head dusky; a dusky streak along base of anal; eye silvery; lateral streak narrow, tapering behind; bases of pectoral and caudal bright yellow; fins otherwise nearly plain; D. IV–I, 8; A. I, 16. Scales 40–9. Scales thin and smooth, their edges entire, as in *M. notata*, from which closely allied species it differs mainly in the shorter anal. Vertical fins scaleless. Length 4 inches.

46. Menidia vagrans (Goode & Bean) J. & G. G. (30893.)

Very abundant about Galveston, in schools along the sandy beaches; originally described from Pensacola, but not obtained there by us.

Color in life, light greenish above, the lateral band broad, covering two half-rows of scales, becoming narrow posteriorly; sides and belly silvery. Tip of snout and of lower jaw yellow, soiled with blackish. Each scale of back with one to three dark points, these forming about 5 conspicuous streaks as seen from above; caudal yellow, with dark punctulations, its margin dusky; dorsal and pectorals somewhat dusky, lower fins white, the anal with dark points at base.

Head $4\frac{2}{5}$ in length ($5\frac{1}{4}$ with caudal); depth $5\frac{1}{4}$ ($6\frac{1}{4}$). D. IV–I, 8; A. I, 15 to I, 17. Scales 43–6.

First dorsal very small, its insertion over front of anal, midway between base of caudal and posterior angle of opercle; distance from its front to front of second dorsal $\frac{2}{3}$ head. Pectorals slightly shorter than head. Vertical fins with large scales.

Scales firm, adherent, their edges crenate or laciniate, feeling very rough to the touch. Scales of head large. Length 4 inches.

This species appears to represent in the Gulf the allied *Menidia bosci* (*Atherinia menidia* L.) of the South Atlantic coasts. *M. vagrans* differs from the latter chiefly in the shorter anal (A. I, 20 to I, 22 in *M. bosci*.).

ECHENEIDIDÆ.

47. Echeneis naucrates L. P.

One specimen 25 inches long and another 8 inches long were taken at Pensacola. The larger example shows the following characters: Color nearly uniform dusky, the black lateral band little marked, the tips of dorsal and caudal lobes little paler than the rest of the fin. D. XXII–34; A. 35. Caudal lunate, the lobes pointed.

The small specimen has but 20 laminæ; the lobes of the dorsal and anal are yellowish white, as are the upper and lower rays of the caudal; the median (black) rays of the caudal being abruptly produced.

TRICHIURIDÆ.

48. Trichiurus lepturus L.—*Sabre-fish; Silver Eel.* G. (30983.)

Rather common about Galveston.

SCOMBRIDÆ.

49. Scomber ?grex Mitchill. P. (30825.)

The anterior half of the body of a small mackerel was obtained at Pensacola, the posterior part having been cut off for bait. This specimen differs from others of this species examined by us in having the body very slender, the depth 1⅜ in length of head. The coloration is peculiar, the back and sides being reticulated with black in fine pattern, on an olivaceous ground, there being about 12 cross streaks of black between the occiput and the dorsal fin. In *S. grex* these streaks are not usually half so numerous. The lower part of the sides is plain silvery. The air-bladder is developed, and the ovaries in this specimen, which was about a foot long, are full of eggs.

Scomber grex Mitchill (Trans. Lit. & Phil. Soc. N. Y. 1815, 422) of the Atlantic and *Scomber diego* Ayres (Proc. Cal. Ac. Sci. 92, 1855) of the coast of Southern California are apparently identical. The Mediterranean species, *Scomber colias* Gmel. (= *S. pneumatophorus* Delaroche), seems to differ in some particulars, slight, but constant in the specimens examined. These are shown in the following analysis:

a. Air-bladder present.
 b. Dark bands on back broad, as broad as interspaces, usually confluent below with a wavy dark, lateral streak on the level of upper edge of pectoral; sides and belly below the streak immaculate; head 3¼ to 3¾ in length; longest dorsal spine not more than half head............................GREX.
 bb. Dark dorsal bands narrow, more wavy, not so broad as interspaces; lateral streak obsolete or represented by a line of dots; lower part of sides with numerous irregular, wavy vertical streaks and reticulations of a dull gray color, which are usually broader than the interspaces; head 4 in length; longest dorsal spine a little more than half head....................COLIAS.

50. Scomberomorus maculatus (Mitch.) J. & G.—*Spanish mackerel.* P.

Abundant in spring and summer; one of the most important food-fishes.

51. Scomberomorus caballa (C. & V.) J. & G.—*King-fish.* P.

A specimen 4 feet in length was taken at Pensacola. Color in life steel-blue, paler below, slightly clouded, but without spots; upper fins dusky; lower fins whitish. Head 5 in length; depth 6. Maxillary 1⅕ in head, reaching posterior margin of eye. Eye 6 in head; snout pointed, 2⅖ in head. Teeth broad, triangular, smallest in front, those in lower jaw largest, their breadth at base ¾ their height. Gill-rakers very short, scarcely higher than broad. Pectorals 1¾ in head; ventrals 3¼. Dorsal lobe 3; anal lobe 2¾. Interspace between dorsals a little longer than eye. D. (spines injured) I, 14–9; A. III, 12–10.

CARANGIDÆ.

52. Decapterus punctatus (Agass.) Gill.—*Cigar-fish.* P.

Rather common at Pensacola, where several specimens were obtained.

53. Caranx trachurus (L.) Lac. P. (30833.)

Two specimens, one of them in fair condition, the other partly digested, taken from the stomach of a Red Snapper at Pensacola.

We identify the Gulf species with the *Caranx trachurus* proper, our specimen agreeing well with the detailed accounts of Cuvier & Valenciennes (ix, 11) and of Day (Fishes of Gt. Brit. 1881, 124, pl. xliv). There are at least three well-defined species or varieties of the type called *Trachurus* represented in our collections. These appear to correspond to the three species described, but not named, by Cuvier & Valenciennes (ix, p. 17), and all three are, if descriptions are to be trusted, found in the Mediterranean, and pretty widely distributed over the globe.

The following characters are shown by our specimens:

a. Body comparatively deep and compressed, the depth 4 to $4\frac{1}{6}$ in length; scutes 34 to 36 + 36 to 38 in number, the anterior scutes scarcely lower than the posterior, their height about $\frac{3}{4}$ diameter of eye; length of curve of lateral line $1\frac{2}{3}$ to $1\frac{3}{5}$ in the straight part; maxillary reaching past front of pupil $2\frac{1}{4}$ to $2\frac{1}{3}$ in head; lining of opercle blackish ....TRACHURUS.*

aa. Body moderately compressed, the depth $4\frac{1}{4}$ to $4\frac{3}{4}$ in length; scutes 38 to 40+38 to 42 in number, the anterior little lower than the posterior, their height about three-fifths diameter of eye; curve of lateral line $1\frac{1}{4}$ to $1\frac{1}{3}$ in straight part; maxillary reaching to front of pupil, $2\frac{3}{4}$ in head; lining of opercle scarcely blackish.

DECLIVIS.†

aaa. Body elongate, little compressed, the depth 5 in length; scutes 50+46 to 48 in number, the anterior one-third lower than the posterior, their height $2\frac{1}{4}$ in diameter of eye; curve of lateral line scarcely shorter than straight part; maxillary reaching to just beyond front of eye, $2\frac{3}{4}$ in head; lining of opercle scarcely blackish...PICTURATUS,†

54. Caranx hippus (L.) J. & G.—*Jack-fish; Crevallé.* N. O.

(*Carangus hippos* and *Carangus chrysos* Gill, Proc. Ac. Nat. Sci. Phila. 1862, 434. *Caranx carangus* Günther, ii, 448. *Caranx esculentus* Gerard, U. S. Mex. Bound. Surv. Ichth. 23. *Caranx defensor* Holbr. Ichth. S. C. 1860, 87.)

Specimens of enormous size, weighing more than 25 pounds, were seen in the markets of New Orleans, having been taken in Lake Borgne.

---

* ? *Scomber trachurus* L. Syst. Nat. 298. *Scomber trachurus* Gmelin, Syst. Nat. 1335. *Caranx trachurus* Cuv. & Val. ix, 11. *Caranx trachurus* Risso, Ichth. Nice, 1810, 173. *Trachurus trachurus* Day, Fishes G't Brit. 124. ? *Caranxomorus plumierianus* Lacép. Hist. Nat. Poiss. iii, 84, pl. 11. *Trachurus saurus* Rafinesque, Indice d'Ittiol. Sicil. 1810, 20. Specimens examined from Pensacola and from Newport, Rhode Island.

† *Caranx trachurus* "première subdivision" C. and V. ix, 17 (specimens from various points in the Mediterranean). *Caranx declivis* Jenyns, Voyage Beagle, Fish. 1842, 68 (New Holland). *Trachurus trachurus* in part, of various writers, and apparently the most abundant type in the Mediterranean. We are unable to disentangle its synonymy entirely from that of the preceding into which it may perhaps be found to intergrade. We have collected numerous specimens of this type at Genoa and at Venice. A specimen collected by Mr. Xantus at Cape San Lucas is in the National Museum.

‡ *Seriola picturata* Bowdich, Excurs. Madeira, 1825, 123 (Madeira), *Trachurus cuvieri* Lowe, Trans. Zool. Soc. Lond. ii, 183, 1837 (Madeira). *Caranx symmetricus* Ayres, Proc. Cal. Ac. Nat. Sci. i, 1855, 62 (California). *Caranx amia* Risso, Ichth. Nice, 1810, 174 (not *Scomber amia* L.). *Caranx trachurus* "deuxième subdivision," C. & V. iii, 17 (specimens from various localities in the Mediterranean and from Valparaiso). *Trachurus fallax* Capello, Cat. Peix. Portugal, 1867, 318. *Trachurus rissoi* Giglioli, Catalogo degli Anfibi e Pesci Italiani, 1880, 27. Specimens examined by us from Monterey, Santa Barbara, and San Pedro, California, and Cape San Lucas.

These large examples were light brown above, silvery below, the pectoral creamy with a diffuse black blotch below; anal lobe and under side of tail deep yellow. Opercular spot jet black, sharply defined.

Head 3⅘; depth 3⅘; D. VI-I, 20; A. II-I, 16.

A portion of the true synonymy of this species has been detached to form a mythical "*Carangus chrysos*," by recent American authors. There is no doubt in our mind that the species called *carangus* Auct., *esculentus* Grd., and *defensor* Holbr. are identical with each other and with the original *Scomber hippos* of Linnæus. The original *Scomber chrysos* of Mitchill was probably the young of the same species.

Dr. Günther has identified the *Scomber hippos* of Linnæus with *Caranx fallax* C. & V. This must be erroneous, as *Caranx fallax* is rare at Charleston, whence Linnæus received his specimens, while the present species is very common. The two chief distinctive characters given by Linnæus "operculis postice macula nigra," and "dentium unica series, anterioribus duobus majoribus" apply, as Dr. Gill has shown, to the present species and not to the *fallax*.

55. Trachynotus carolinus (L.) Gill.—*Pompano.* P.

Generally abundant in summer; the most valuable food-fish of the Gulf coast. It reaches the weight of 10 or 12 pounds.

56. Trachynotus glaucus C. & V.—*Gaff-top-sail Pompano.* P.

Not rare; reaches a weight of two pounds; a food-fish of mediocre quality.

57. Oligoplites occidentalis (L.) Gill.—*Yellow-tail.* P.

Rather common in summer; not valued as food.

58. Seriola stearnsi Goode & Bean.—*Amber-fish.* P.

Not uncommon on the "Snapper Banks" about Pensacola; reaching a weight of about 10 pounds. One specimen was obtained and others were seen.

This species much resembles the "yellow-tail" of the Californian coast, *Seriola dorsalis* (Gill), which we have identified, with considerable doubt, with *Seriola lalandi* C. & V., a species originally described from Brazil.

S. stearnsi is, however, readily distinguished from the "yellow-tail" by its larger mouth, the maxillary reaching to the middle of the eye, about 2⅙ in head (in *S. dorsalis* barely to front of pupil, 2⅔ in head). *Seriola stearnsi* may be known from all the other Atlantic species, except *S. zonata*, by the greater number of rays in the soft dorsal. *S. zonata* has the occipital region carinated, while in *S. stearnsi*, as in *S. dorsalis*, this region is broadly rounded. *Seriola dubia* Poey seems to resemble *S. stearnsi*, and may be identical with it. In any event the name "*dubia*" could not be retained, as there is an earlier *Seriola dubia* Lowe. The description of *Seriola gigas* Poey does not indicate any character by which it may be separated from the true *Seriola lalandi*. The latter species has 2d D. I, 32 or 33; the Californian *dorsalis* I, 35.

The life coloration of *Seriola stearnsi* is light bluish above, whitish below; a very distinct stripe of brassy-yellow from snout through eye toward the tail. Caudal dusky, not yellow. Second dorsal and anal dusky; tip of dorsal pale. Pectoral dusky; ventral creamy, its inner edge somewhat dusky. Inside of mouth pale. D. V–I, 37; A. II–I, 21. Caudal keel unusually strong.

59. Seriola lalandi C. & V. P.

Seriola gigas Poey, Mem. Cuba.

A specimen weighing about 25 pounds, referred by us to this species, was seen in the New Orleans market. It was taken at Pensacola. This species appears to reach a larger size than *S. stearnsi*, and to have fewer rays in the dorsal.

60. Seriola falcata Cuv. & Val.—*Rock salmon.* P.

? *Seriola rivoliana* C. & V. ix, 207 (Mediterranean ?).
? *Seriola bosci* C. & V. ix, 209 (Charleston).
Seriola falcata C. & V. ix, 210 (Gulf of Mexico).
? *Seriola bonariensis* C. & V. ix, 211 (Buenos Ayres).
? *Seriola ligulata* Poey, Mem. ii, 231 (Cuba).
?? *Seriola coronata* Poey, Mem. ii, 232 (Cuba).
? *Seriola bonariensis* Günther, ii, 464.
Seriola falcata Günther, ii, 464 (Jamaica).
Seriola bonariensis Goode & Bean, Proc. U. S. Nat. Mus. ii, 129 (Pensacola).
Seriola rivoliana Lütken, Spolia Atlantica, 1880, 603 (considers *rivoliana, bosci, falcata,* and *bonariensis* as identical).

Not uncommon on the Snapper Bank at Pensacola, reaching a weight of 10 to 12 pounds. The synonymy of this species is badly confused on account of the imperfections in the earlier descriptions. If it be true, as supposed by Lütken, that all the *Seriolæ* with falcate dorsal constitute a single pelagic species, this species will stand as *S. rivoliana*. The only early description which applies well to our specimens is that of *Seriola falcata*. It is possible that the species with the black temporal band (which, according to Mr. Stearns, occurs in Southern Florida) may be different from *S. falcata*, in which case most or all the other synonyms referred to above might belong to it.

The life-coloration of *Seriola falcata* is as follows: Grayish above, paler but hardly silvery below. Fins blackish, the pectorals pale, the caudal not at all yellow. Eye white; lining of opercle pale; a very obscure olivaceous band from eye to front of dorsal, scarcely visible in fresh specimens. Preorbital and preopercle shaded with olive.

Head $3\frac{4}{5}$ ($4\frac{3}{5}$ in total); depth $3\frac{2}{3}$ (4). D. VII–I, 29; A. II–I, 21. Cœca 30.

Body rather deep and compressed. Head somewhat longer than deep, not conical. Snout $2\frac{3}{4}$ in head, maxillary reaching front of pupil, $3\frac{1}{2}$ in head, its tip broad, eye large, $5\frac{1}{4}$ in head, $1\frac{3}{4}$ in snout. Occiput somewhat carinated. Interorbital space wide, convex. Caudal keel little developed.

Dorsal high, somewhat falcate; its anterior lobe $1\frac{2}{3}$ in head, $2\frac{1}{3}$ in the base of the fin. Pectoral 2 in head; ventrals $1\frac{2}{3}$; anal lobe $1\frac{3}{4}$; anal spines small.

NOMEIDÆ.

61. Nomeus gronovii (Gmel.) Gthr. P.

One specimen obtained from the stomach of a Red Snapper at Pensacola.

POMATOMIDÆ.

62. Pomatomus saltatrix (L.) Gill.—*Blue-fish.* P.

Rather common about Pensacola, and valued as a food-fish. Rare or unknown at Galveston.

CENTROPOMIDÆ.

63. Centropomus undecimalis (Bloch.) C. & V.—*Robalo.* G.

A food-fish of large size and delicate flesh, much valued along the Mexican coast. It is occasionally taken about Galveston in summer. It becomes more abundant southward along the Texas coast, and is one of the staple food-fishes about Brazos Santiago. An individual, weighing 15 pounds, in the Galveston market, taken at Indianola, showed the following characters :

Dull pale olivaceous; lateral line black; caudal dull yellowish; lower fins pale. Maxillary $2\frac{2}{3}$ in head, extending to opposite posterior margin of pupil. Second dorsal spine reaching front of second dorsal, as long as from snout to edge of preopercle. All the dorsal spines strong. Second anal spine $2\frac{1}{3}$ in head. Lat. l. 70.

SERRANIDÆ.

64. Epinephelus morio (C. & V.) Gill.—*Red grouper.* P.

Common on the "Snapper Banks" about Pensacola, reaching a weight of about 30 pounds; rather less valued as a food-fish than the related species.

Color brownish-olive, everywhere flushed with light red, the lower parts nearly salmon-color; whole body marked with obscure round pale spots, these obsolete on the fins, and most distinct in the young. Dorsal, anal, and caudal edged with blackish; pectorals plain red. Inside of mouth deep scarlet.

65. Epinephelus drummond-hayi Goode & Bean.—*Spotted hind.* P.

Rather common on the banks about Pensacola, reaching a weight of 30 pounds; a beautifully colored species, probably the handsomest of the genus.

Dark brown, densely covered with small pearly-white spots; those below smaller and nearly round, all of them arranged somewhat in irregular series.

Fins all covered with similar spots, those of the paired fins chiefly on the inner suface. Lower side of head flushed, immaculate. Caudal more densely spotted than body, the terminal spots of a fine lavender. Pectoral with a submarginal band of orange.

66. Trisotropis stomias Goode & Bean MSS.—*Black grouper.* P.

This species is about equally common with the Red Grouper at Pensacola, and reaches a weight of about 40 pounds.

Color dark gray, each scale finely vermiculate with darker but without distinct spots; some specimens with the body everywhere marbled with darker in the form of large roundish blotches; old examples more uniform; belly pale; fins all grayish, their tips or edges blackish; pectorals with no yellow or pale edging. Lips blackish, not tinged with yellow. Head 3 in length; depth 4. A. III, 11. Lat. 1. 140. Scales smooth, with numerous accessory scales.

67. Trisotropis falcatus Poey.—*Scamp.* P.

Not rare on the "Banks"; a smaller species than the others, not exceeding 20 pounds. It is one of the best food-fishes, more delicate than the other "Groupers." It is called "*Scamp*" from its way of flapping when touched after lying apparently dead on the deck.

68. Serranus fascicularis C. & V.—*Squirrel-fish.* P. (30×31.)

A single fine specimen obtained from the "Snapper Banks" at Pensacola. Three smaller specimens were taken from the stomachs of Red Snappers. The upper lobe of the caudal fin in this species is sometimes produced in a long filament.

69. Serranus trifurcus (L.) J. & G. P.

Several young specimens, from 2 to 6 inches long, apparently belonging to this rare species, were taken from the mouths and stomachs of Red Snappers at Pensacola.

Color light olivaceous, the sides with about six dusky bars, which are rather broader than the interspaces. They are distinct only posteriorly and near the lateral line. No white band before the anal. A very small jet-black spot close behind eye in the young, becoming obsolete with age; opercle with a dark diffuse blotch; chin and upper parts of head somewhat soiled with dark points; lower parts plain white; cheeks with yellowish markings. Dorsal and caudal vaguely barred or spotted; no black blotch on last spines of dorsal; other fins pale.

Head 2⅓ (3¼ in total); depth 3⅓ (4⅖). D. X, 11 or X, 12; A. III, 7. Scales 5-52-12.

Body slender, little compressed; head long and somewhat pointed; lower jaw a little the longer; maxillary reaching to posterior border of pupil, nearly half length of head; teeth small, the canines little developed, those on sides of lower jaw largest. Eye large, 4 in head. Preorbital and interorbital space very narrow. Preopercle with its edge evenly and sharply serrate. Interopercle sharply serrate. Gill-rakers slender, rather long. Scales on cheek in about 6 series.

Proc. Nat. Mus. 82——18 **Aug. 15, 1882.**

Dorsal fin somewhat emarginate, the fourth spine highest, about $2\frac{1}{2}$ in head; this spine and some of the others, occasionally filamentous; soft dorsal rather high, rather higher than fourth spine, the longest rays more than half head. Caudal with the upper ray filamentous, $2\frac{1}{4}$ in length of body; middle rays also produced, $1\frac{1}{2}$ in head. Second anal spine stronger but shorter than third, 5 in head. Ventrals about as long as pectorals, $1\frac{3}{4}$ in head, not reaching front of anal.

Soft dorsal and anal scaleless. Scales moderate, ctenoid. Jaws, preorbital and top of head naked.

These specimens differ somewhat in form and color from others in the National Museum from Charleston, S. C. We cannot, however, separate them specifically.

70. Serranus subligarius (Cope) J. & G. P. (30859.)

(*Centropristis subligarius* Cope, Proc. Am. Philos. Soc. Phila. 1870.

Two young specimens, the largest 3 inches long, were obtained from the mouth of Red Snappers at Pensacola. Professor Cope's type, the only specimen of this species hitherto known, was also obtained at Pensacola.

Olivaceous, tinged with reddish above, paler below but not silvery; each scale on the sides with a blackish margin, these forming rather faint, continuous, dusky streaks. Posterior part of sides with faint traces of about 5 irregular cross-shades of darker along the sides. A large blotch of cream-color in front of the vent, extending upwards as an irregular cross-bar to near the middle of the side, its posterior edge sharply defined, its anterior fading into the color of the belly; a black ring around tail behind dorsal and anal; a large, black blotch on front of soft dorsal, extending downward on the body, where it is less distinct than on the fin. Cheeks yellowish; opercles darker; lower parts of head brown, the preopercle (below), interopercle, lower jaw, and branchiostegals covered by a network of wavy bluish streaks. Spinous dorsal dark gray, mottled; soft dorsal similarly and more distinctly marked. Pectorals, anal, and caudal grayish, with sharply defined narrow blackish bars, somewhat undulating. Ventrals faintly barred, mostly black.

Head $2\frac{1}{2}$ (3); depth $2\frac{2}{3}$ ($3\frac{1}{3}$), D. X. 13; A. III, 7. Scales about 6–42–17.

Body rather deep, compressed, the back elevated, the anterior profile nearly straight. Head long and low, slender, acuminate; its depth at middle of eye but half its length in the smaller specimen, in the larger proportionately deeper. Mouth rather small; lower jaw scarcely projecting; maxillary reaching to posterior margin of pupil, its length $2\frac{1}{4}$ in head; teeth small, the canines little developed; those on sides of lower jaw largest, as usual in *Serranus*. Eye rather large, 4 in head. Preorbital and interorbital space very narrow. Edge of preopercle subequally and rather sharply serrate; none of the teeth directed forwards. Gill-rakers short, rather few. Scales on cheek small, in about 10 series.

Dorsal fin scarcely emarginate, the fourth spine not elevated, about $2\frac{1}{2}$ in head, a little lower than the soft rays. Caudal subtruncate, a little more than half head. Second anal spine longer and stronger than third, $2\frac{1}{3}$ in head. Ventrals $1\frac{3}{4}$ in head; pectorals $1\frac{2}{5}$; neither reaching front of anal. Dorsal and anal fins, jaws, preorbital, and front of head scaleless. This species is technically close to the preceding, but is remarkably different in form and appearance, resembling somewhat a *Hypoplectrus*.

SPARIDÆ.

71. Lutjanus blackfordi Goode & Bean.—*Red Snapper; Pargo Colorado.* P.

This fish is at present taken in far greater numbers than any other on our Gulf coast. At Pensacola it is the most important food-fish, and in the New Orleans market it is sold in greater quantities than all other species combined. It is taken with hook and line on the "Snapper Banks" usually from 5 to 30 miles off shore. It reaches a weight of about 35 pounds, according to Mr. Stearns, to whom we are indebted for most of the statements of weights contained in this paper. It is a rather coarse fish, but bears transportation well.

This fish feeds on various small fishes—serranoids, eels, &c.—the species of which are thus far very little known. The systematic preservation of small fishes " spewed up" by the Snappers when caught, or found in their stomachs, has been begun by Mr. Stearns. We may in the future expect large results from this source, which has already yielded many of the most interesting forms contained in the present collection.

72. Lutjanus caxis (Bloch) Poey.—*Black Snapper; Lawyer.* P. (30848.)

Rather common about Pensacola, not reaching a large size. It is not often taken in nets, and its name "Lawyer" is said to be given in allusion to its skill in avoiding capture.

In life, young specimens are dark green, paler below; each scale above with a black spot which becomes gradually bronze down the sides; these spots forming distinct stripes along the rows of scales. Spinous dorsal with a blackish basal band, then a pearly band, a broad blackish band at tip. Soft dorsal yellowish, spotted at base. Ventrals and anal dark purplish red, darkest and spotted at base. Pectoral translucent yellowish. Caudal yellowish, tipped with reddish. A very distinct bright-blue stripe across preorbital and suborbital.

73. Lutjanus stearnsi Goode & Bean.—*Mangrove Snapper.* P.

Not uncommon on the " Snapper Banks " at Pensacola; one specimen obtained.

Color (in spirits) dusky brownish above, the sides and below paler, more or less flushed with salmon red; sides and lower parts of head flushed with red, especially behind jaws. Bases of scales on sides of breast and belly crimson; centers of scales on sides whitish. Vertical fins dusky; pectorals and ventrals largely rosy.

Scales above lateral line forming oblique series which are not through-out parallel with the lateral line.   Teeth on vomer in an anchor-shaped patch, prolonged backward on the median line ; outer pair of canines of upper jaw very strong ; inner small ; canines of lower jaw not much de-veloped ; maxillary reaching front of eye, 2¾ in head ; preopercle little notched ; band of scales on each side of occiput single, composed of about two series ; 5 or 6 rows of scales on cheek ; posterior nostrils ovate, pointed behind ; gill-rakers stoutish, not very long ; pectoral short, pointed, ⅔ length of head ; second and third anal spines subequal, short-ish, the soft rays rather low.   Caudal lunate, the upper lobe slightly longest.

74. **Diabasis formosus** (L.) J. & G.—*Red-mouth grunt.* P.

  (*Hæmulon arcuatum* Holbr. Ichth. S. C. 124, pl. xvii, f. 2.)

A single large specimen obtained at Pensacola.

Body and fins dull gray ; the middle of each scale paler ; second dor-sal, caudal, and tips of ventrals of a dusky slate-color ; front of head with narrow stripes of steel-blue alternating with bronze, these stripes covering maxillary, preorbital, suborbital, whole naked part of snout above cheeks, and temporal region behind eye ; the bands are faint or obsolete on opercle ; a dark, vertical blotch on opercle, near angle of preopercle, mostly hidden by the latter ; mouth orange within, the color fading anteriorly.

75. **Diabasis aurolineatus** (C. & V.) J. & G. P. (30869.)

  Hæmulon aurolineatum C. & V. v, 237.
  Hæmulon aurolineatum Gthr. i, 316.
  Hæmulon caudimacula Poey, Syn. Pisc. Cub. 319 (not of C. & V.).

Color light olivaceous, grayish-silvery below ; a dark-bronze band, narrower than pupil, darkest in the younger specimen, from snout through eye straight to base of caudal ; above this, two or three dark streaks, the middle one most distinct, from eye to above gill-opening ; another beginning on top of snout on each side, passing above eye, and extending parallel with the first-mentioned stripe straight to last ray of dorsal, where it meets its fellow of the opposite side ; a dark streak from tip of snout along median line to front of dorsal ; a large, rounded black blotch at base of the caudal, some obscure dusky shading below soft dorsal and at base of pectoral ; fins all plain ; upper slightly dusky ; anal nearly white ; pectorals, caudal, and ventrals light yellow ; lining of opercle pale orange ; inside of mouth scarlet.   In the large speci-men the dark stripes are fainter, paler, and more yellowish ; several fainter bands occur between the broader ones, and faint oblique streaks of light bronze follow the rows of scales, those above lateral line oblique.

Head 3 (3¾) ; depth 2⅕ (3¾). D. XIII, 15 ; A. III, 8.

Scales 7-52-13. Length of largest specimen 5 inches.

Body rather elongate, somewhat compressed, the back a little elevated. Head not deep, the snout short, but not blunt, 3 to $3\frac{1}{3}$ in head; preorbital very narrow, little wider than pupil; maxillary reaching middle of pupil, 2 in head; eye large, $3\frac{1}{3}$ in head; scales of cheek small, in about 11 rows; gill-rakers short, not one-third as long as pupil; preopercle sharply serrate.

Scales of moderate size, those above lateral line in very oblique rows, those below in horizontal rows.

Dorsal spines rather high, the longest $1\frac{3}{4}$ to $2\frac{1}{4}$ in head, longer than the second anal spine; caudal well forked, the upper lobe the longer, $1\frac{1}{3}$ to $1\frac{1}{2}$ in head; second anal spine strong, longer and stronger than third, $2\frac{1}{2}$ to $2\frac{4}{5}$ in head, reaching, when depressed, to base of last ray; ventrals $1\frac{2}{3}$ to $1\frac{3}{4}$ in head; pectorals $1\frac{1}{4}$ to $1\frac{1}{2}$.

Two specimens, in good condition, the largest $5\frac{1}{2}$ inches long, were taken from the mouth of a Red Snapper at Pensacola. Our specimens agree in color with *Hæmulon fremebundum*, described by Goode & Bean, from Clearwater Harbor. The latter species is, however, quite different, being less elongate, with much smaller mouth and much larger scales, there being but 9 or 10 series between the lateral line and the vent.

76. Pomadasys fulvomaculatus (Mitch.) J. & G.—*Pig-fish.* P. G. (31034.)

> *Orthopristis duplex* Grd. U. S. Mex. Bound. Surv. 1859, 15.
> *Pristipoma fasciatum* C. & V. v, 285; Günther, i, 301.

A common shore fish of small size and good quality. It has little economic importance.

Color in life light blue above, shading gradually into silvery below; preorbital and snout of a clear sky-blue; a dash of blue on side of upper lip; each scale on body with a blue centre, the edge with a bronze spot; these forming on back and sides very distinct orange-brown stripes along the rows of scales; those above the lateral line extending obliquely upward and backward, those below nearly horizontal. Snout with bronze spots; one or two bronze cross-lines connecting front of orbits; two or three oblique lines on preorbital; besides numerous bronze spots larger than those on the body; preorbital also with dusky shades, one of which extends on upper lip. Cheeks and opercles with distinct bronze spots, larger than those on the body. Inside of mouth pale; inside of gill cavity tinged with golden.

Dorsal translucent, with about three bronze longitudinal shades, composed of spots, those on soft dorsal most distinctly spot-like; edge of the fin dusky. Caudal plain, yellowish at base, dusky toward the tip. Anal whitish, its edge dusky, its base shaded with bronze. Pectorals and ventrals yellowish, the latter darker at tip.

Fresh specimens, so far as we have noticed, show no trace of vertical bands. On examples preserved in alcohol, the yellowish and blue markings gradually disappear, and dark cross shades become apparent. A specimen 5 years in alcohol shows the following coloration: Silver-gray, with faint streaks along the rows of scales. A distinct narrow dusky band from front of spinous dorsal through base of pectoral; behind this 7 or 8 cloudy, obscure bands, alternately broad and narrow; a horizontal dusky shade behind eye; spinous dorsal with a faint median pale shade, soft with three rows of faint spots; other fins nearly plain. This specimen evidently corresponds to the *Pristipoma fasciatum* of C. & V. and Günther, and as evidently belongs to *P. fulvomaculatus; Orthopristis duplex* Grd. does not seem to be at all different. Head 3⅓; depth 3. D. XII, 16; A. II, 13 or 14. Scales 8-54-16.

77. Lagodon rhomboides (L.) Holbr.—*Chopa Spina.* P. G. (31052.)

Exceedingly common everywhere along the shore. A fish of small size, little valued as food, and seldom brought to the market.

78. Diplodus probatocephalus (Walb.) J. & G.—*Sheephead.* P. G. (31041.)

Generally common, but less important as a food-fish than farther north. Specimens seen mostly small. Reaches a weight of about 12 pounds.

79. Stenotomus caprinus Bean MSS.—*Goat's Head Porgee.* P.

Two specimens, the larger partly digested, the smaller in good condition, were taken from stomachs of Red Snappers at Pensacola.

Color nearly uniform pale olive, silvery below; sides with faint traces of dark cross-bands; fins pale, the posterior margin of caudal blackish. Anterior teeth small, in a close-set band, the outer a little enlarged, compressed, and lanceolate. Two series of molars in each jaw. A well-developed antrorse spine before dorsal. Anterior profile irregular, abruptly depréssed above eye, the snout rather pointed. Scaly part of cheek as deep as long. Pectoral a little longer than head, reaching soft rays of anal; dorsal spines slender, rather high, the first two short the third somewhat filamentous. Head 3¼; depth 2. D. XI, 12; A. III, 12. Scales 7-47-14.

This species is strongly marked. It is distinguished from *S. argyrops* by the deeper cheeks and preorbital region and the less elongate form, as well as by the structure of the spinous dorsal.

80. Sparus pagrus L.—*Porgee.* P. (30838.)

(*Pagrus vulgaris* C. & V.; *Pagrus argenteus,* Good & Bean, Proc. Ac. Nat. Sci. Phila. 1879, 133.)

Not uncommon at Pensacola; two specimens obtained.

Color golden-olive, the middle of each scale largely pinkish-red, giving a general reddish hue to the fish; sides and below silvery, flushed with red; many scales of back and sides each with a small round spot of deep purplish-blue, these forming distinct longitudinal streaks on the sides below lateral line, the series somewhat irregular, running along the margins of the scales; above the lateral line these spots are somewhat scattered, forming very irregular oblique series, running upward and backward; a few of these spots on nape and upper part of opercle;

a dark blotch on upper part of orbital rim; snout tinged with purplish, occiput with olive; edge of opercle dusky; vertical fins largely orange, their edges translucent; spinous dorsal somewhat dusky; ventrals pale, with a pinkish blotch at base; pectorals yellowish, especially at base, the axil somewhat dusky; no antrorse spine before dorsal.

Our specimens agree with various descriptions of European examples of this species, except in the coloration. In none of these descriptions is there any allusion to the blue spots which form so striking a feature of the coloration of the American fish.

APOGONIDÆ.

81. Apogon maculatus (Poey) J. & G. P. (30863.)

A single specimen, 3 inches long, in perfect condition, "spewed up" by a Red Snapper at Pensacola.

Color intense scarlet, nearly uniform; a tinge of crimson about pectorals and on sides of head. A round, black, ink-like spot, a little larger than pupil, under second dorsal; another, smaller, on upper part of tail, on each side, just before root of caudal; tip of caudal whitish; iris red.

Head $2\frac{3}{4}$; depth $2\frac{4}{5}$. D. VI-I, 9; A. II, 8. Scales about $2\frac{1}{2}$-26-7 (some of them lost, so that the number cannot be exactly ascertained).

Maxillary $1\frac{4}{5}$ in head, reaching beyond pupil; eye very large, 3 in head; preopercle distinctly serrulate. Pectoral $1\frac{2}{3}$ in head, somewhat shorter than caudal.

This species has not been hitherto noticed north of Cuba.

82. Apogon alutus Sp. nov. P. (30874.)

A single specimen, $2\frac{1}{2}$ inches long, "spewed up" by a Red Snapper at Pensacola.

Color rusty-red with silvery lustre; sides of head little reddish. Body and fins everywhere much soiled and freckled with dark points. First dorsal blackish, thickly punctate; second dorsal, anal and caudal yellow, smutty with dark points, the posterior half of the caudal more dusky. Ventrals smutty yellow; pectorals colorless.

Head $2\frac{3}{4}$ in length; depth $2\frac{3}{4}$. D. VI-I, 9; A. II, 8. Lat. l. 21.

Head much compressed, short and high, its height at occiput six-sevenths its length; snout short and blunt, less than interorbital width, about half diameter of orbit; mouth very oblique, the maxillary reaching beyond pupil, but not to posterior margin of orbit; length of maxillary $1\frac{3}{4}$ in head; teeth in narrow villiform bands in each jaw, those on vomer and palatines minute; eye of moderate size, $2\frac{1}{5}$ in head; orbital rim elevated above and behind; interorbital width $3\frac{1}{3}$ in head, with a low median longitudinal ridge; both ridges of preopercle entire; opercle without spine; gill-rakers slender, the longest rather more than half diameter of orbit; 8 or 9 rakers on anterior branch of outer arch.

First dorsal low, of six rather weak spines, its base two-fifths length of head, and equal to greatest height of fin; second dorsal high, the longest ray $1\frac{1}{2}$ in head. Anal similar to second dorsal; second anal spine half length of longest ray, which is contained $1\frac{3}{4}$ in head; caudal $1\frac{1}{3}$; ventrals not reaching vent $1\frac{2}{3}$, and pectorals $1\frac{3}{4}$, in length of head.

Allied to *A. puncticulatus* (Poey), but with much larger scales.

MULLIDÆ.

83. **Mullus barbatus** L. Subsp. **auratus**; subsp. nov. P. (30828.)

One specimen $6\frac{1}{2}$ inches long, from the stomach of a Red Snapper, at Pensacola.

Head $3\frac{2}{5}$; depth 4. D. VII–I, 8; A. II, 6.

Form essentially as in *M. barbatus*, the profile a little less steep, the interorbital space a trifle broader, the maxillary extending exactly to opposite front of eye, its length $2\frac{3}{4}$ in head. Interorbital width $3\frac{1}{2}$ in head; barbels $1\frac{1}{4}$; eye $3\frac{2}{3}$; oblique length of snout $2\frac{3}{5}$. Teeth in lower jaw small; on upper jaw obsolete; on vomer and palatines coarse and granular, forming large patches. Gill rakers slender, a little shorter than pupil.

Dorsal spines slender, compressed, the longest about $1\frac{3}{5}$ in head ($1\frac{1}{2}$ to $1\frac{1}{3}$ in *M. barbatus*); height of soft dorsal half head; caudal as long as head. Pectoral $1\frac{2}{5}$ in head. Ventrals $1\frac{1}{3}$. Scales mostly lost, so that the number in the lateral line cannot be counted.

Color scarlet, becoming crimson where the scales are removed; snout scarlet; side with two distinct longitudinal yellow stripes. Caudal scarlet, first dorsal with an orange band at base and a yellow band higher up; the rest of the fin pale; no black on dorsal fin. Second dorsal mottled scarlet and pale; anal and ventrals plain, pectoral reddish; iris violet, dusky above; sides of head with silvery lustre.

This is the first authentic record of the European surmullet in our waters. Our specimen seems to indicate a third subspecies of *M. barbatus*, differing from subsp. *surmuletus* in the lower fins, and in the replacement of the black band on the spinous dorsal by light yellow; from subsp. *barbatus* it differs in the lower fins, less blunt snout, and in the presence of two yellow lateral bands.

EPHIPPIDÆ.

84. **Chætodipterus faber** (Brouss.) J. & G.—*Half-moon; Angel-fish; Spade-fish.* P. G. (31044.)

Generally common.

SCIÆNIDÆ.

85. **Pogonias chromis** (L.) C. & V.—*Drum; Tamboro.* P. G.

Common, a coarse fish of inferior quality, reaching a large size.

Head $3\frac{1}{3}$; depth $2\frac{7}{8}$. D. X–I, 23; A. II, 6. Lat. l. 47 (pores).

86. **Sciæna punctata** (L.) J. & G.—*Mademoiselle; Silver-fish; Bastard Croaker; Yellow Tail.* P. G.

A very common shore-fish of small size and good quality. It rarely reaches the weight of more than half a pound.

The specimens from the Gulf coast differ from those taken further north in the almost entire absence of the dark punctulations which are so conspicuous in the latter. They seem to be otherwise identical.

Color in life silvery, slightly bluish above, the scales of the opercles and middle of sides with some dusky points. Spinous dorsal light yellowish, dusky at tip. Second dorsal and caudal uniform dull yellow. Anal bright yellow in front, the color fading behind. Ventrals slightly yellowish, their axils orange. Pectorals yellowish above; axil silvery. Inside of mouth pure white ; an orange area punctulate with black on inside of opercle. Upper fins all with some punctulations. Head $3\frac{1}{6}$, depth $3\frac{1}{6}$. D. XI-I, 21; A. II, 9. Scales 7–52–11.

87. Scæna ocellata (L.) Gthr.—*Red-fish; Poisson Rouge; Pez Colorado.* P. G. (30845 Pens.; 31914 Galv.)

The most important food-fish of the Texas coast, the amount taken exceeding that of all other species combined. A good food-fish when not too large. It reaches a weight of 35 to 40 pounds, the large specimens being known as Bull Red-fish.

The serratures on the opercle, which are conspicuous in ordinary specimens, wholly disappear with age, the edge of the bone being finally entire and wholly covered by the skin. This change takes place gradually, being complete at a length of about 30 inches.

Color of adults deep brassy yellow above, verging towards orange on the sides; belly white; head bronze, darker above; a band of deeper bronze backward from eye. Young without bronze shades, all of which intensify with age; scales in the young with darker shades forming undulating streaks ; these obliterated with age; fins all pale, tinged with reddish, the pectoral most red ; second dorsal and caudal somewhat dusky. Mouth white within, lining of opercle black. Caudal ocella varying much in size, sometimes wanting; sometimes two or three or even 8 to 10 or more in number. About 19 out of 20 individuals have the normal single ocella on each side. Iris yellowish.

88. Liostomus xanthurus Lac.—*Chopa Blanca; Spot; Flat Croaker; Post Croaker.* P. G. (30836.)

Very abundant along the coast. A good pan-fish, but not very important from its small size. The color is rather paler and more silvery than usual in northern specimens, the humeral spot and dark oblique lines less distinctly indicated. Dorsal and caudal light brownish, the tips darker; soft dorsal yellowish tinge; anal yellowish, somewhat dotted ; paired fins pale.

D. X-I, 30; A. II, 12. Scales 9–48–13. Head $3\frac{1}{3}$; depth 3.

There is no evidence of the existence of two species of *Liostomus*. *Liostomus obliquus* is the species when well preserved. *Liostomus xanthurus* C. & V. is a faded Museum specimen. *Liostomus xanthurus* Lac. was so named from a confusion of the coloration of the species with that of the "yellow-tail," *Sciæna punctata.*

89. Micropogon undulatus (L.) C. & V.—*Croaker; Ronco.* P. G. (30840.)

Very common; a food-fish of considerable importance, although reaching but a small size.

The three species properly referable to this genus, after the removal of *Genyonemus* Gill, are very closely related, and might not improperly be taken as geographical varieties of one species. They may be thus compared:

a. D. X-I, 28; outer teeth of upper jaw evidently enlarged; snout projecting beyond premaxillaries; scales between front of dorsal and lateral line, in a vertical series 9, in an oblique series 12; in an oblique series from vent upward and forward 18. Head 3; depth 3½ ....UNDULATUS.

aa. D. X-I, 24; outer teeth of upper jaw scarcely enlarged; snout little projecting; Lat. l. 48 (oblique series; 53 pores). Scales between front of dorsal and lateral line, vertically, 6 or 7; obliquely, 8; 16 in an oblique series from vent. Head, 3⅔; depth, 3⅗ ....ECTENES *

aaa. D. X-I, 20; outer teeth of upper jaw scarcely enlarged; snout somewhat projecting; Lat. l. 42 (49 pores). Scales above lateral line, vertically, 5 or 6; obliquely, 8 : 12 in an oblique series from vent. Head 3¼; depth 3⅔ ....ALTIPINNIS.†

90. Menticirrus nebulosus (Mitch.) Gill.—*Whiting.* P.

One large specimen obtained at Pensacola, where it is said to be not uncommon.

We have carefully compared this specimen with others from the coast of Massachusetts, and unable to detect any differences.

This species has been hitherto supposed to be confined to the North Atlantic coast, from Cape Cod to Cape Hatteras.

This species is very close to *M. alburnus*, but differs constantly, so far as we have seen, in the smaller teeth, higher first dorsal and sharper coloration, a dark lateral shade always extending into the lower lobe of the caudal fin.

91. Menticirrus alburnus‡ (L.) Gill.—*Whiting; Ground Mullet.* G. (30917, 31051.)

(*Umbrina phalœna* Girard, U. S. Mex. Bound. Surv. 1859, 13.)

Generally common; a market fish of good quality but of small size.

Color in life, smutty-gray above, with strong reddish and bronze reflections. Sides with obscure traces of oblique bars; a short vertical bar below spinous dorsal; a U-shaped bar from nape and end of spinous dorsal surrounding the bar first mentioned; three or four other bars extending downward and backward behind it; a smutty stripe along each side of belly. Upper fins light yellowish; spinous dorsal and lower lobe of caudal tipped with black. Pectoral reddish, covered with

---

* *Micropogon ectenes* Jor. and Gilb. Proc. U. S. Nat. Mus. 1818. Mazatlan (Gilb.).

† *Micropogon altipinnis*, Günther, Proc. Zool. Soc. Lond.; Chiapam (*Gthr.*); San José (*Gthr.*); Panama (Gthr. Gilb.).

‡ The rude figure of Catesby (*Alburnus americanus* Catesb. p. 12, t. 12) has usually been referred to this species. In the eleventh edition of the Systema Naturæ, p. 321, this figure is the type of a "*Cyprinus americanus.*" If this figure is considered identifiable (which it really is not), this species should be called *Menticirrus americanus*, the name of *Perca alburnus* dating from the twelfth edition.

dark dots, so as to appear almost wholly black. Ventrals and anal creamy orange, somewhat soiled with black. Inside of opercle black.

D. X-I, 24; A. I. 7. Scales, 6–54–11; gill rakers almost obsolete; scales on breast not very small; outer teeth of upper jaw much enlarged.

92. Menticirrus littoralis (Holbr.) Gill.—*Surf Whiting*. P. G. (30815,30835,30837, 31046,31048.)

A common surf species, as abundant as the preceding, but less often brought to market. This species is very different from *M. alburnus*, with which it has been confounded. Its relations are with the two Pacific coast species, *M. undulatus* Grd., and *M. elongatus* Gthr., from the latter of which it is difficult to distinguish it. The following is a detailed description:

Color in life gray above, with some bluish and with very strong bronze reflections; a darker bronze shade along sides on level of pectorals, extending to tail and along cheeks, the belly below this abruptly white. No trace of dark bars. Dorsals light brown; spinous dorsal black at tip, the base narrowly white. Caudal pale, its tips usually black; anal creamy, sometimes dusky at tip. Pectoral whitish, only its upper rays with dark punctulations, especially on the inner side, which is sometimes quite dark. Ventrals pale, punctulate towards their tips, which are white. Lining of gill cavity pale.

Head $3\frac{1}{6}$ in length ($3\frac{2}{6}$ in total); depth $4\frac{2}{6}$ ($5\frac{1}{2}$). D. X-I, 23 (not 27 as stated by Holbrook); A. I, 7. Scales 6–50–11; 54 tubes in lateral line.

Body elongate, the caudal peduncle very slender, its least depth $3\frac{3}{4}$ in head. Head long, rather broad; the snout long, bluntish, 3 in head, projecting moderately beyond the premaxillaries (for a distance of about one-fifth its length), which project beyond lower jaw. Mouth rather small, wholly inferior, the maxillary reaching little beyond front of eye, $3\frac{1}{3}$ in head. Teeth in broad bands, the outer series in upper jaw a very little enlarged (very much smaller than in *M. alburnus*).

Posterior nostril a lanceolate slit, as long as barbel, or about half diameter of eye. Eye small, 5 to 6 in head, about one-fourth narrower than preorbital or interorbital space. Gill rakers about one-third diameter of pupil; about 7 on lower part of arch.

Dorsal spines rather slender and high, the longest about two-thirds length of head. Soft dorsal moderate, its longest rays about equal to snout. Lower lobe of caudal broader than upper, $1\frac{3}{8}$ in head. Longest rays of anal a little longer than snout; pectorals $1\frac{2}{8}$ in head, reaching slightly beyond tips of ventrals, which are about two in head. Axillary scale one-fourth length of pectoral; scales on breast very small; about 25 in a longitudinal series to front of ventrals, and about 15 in a cross series connecting outer margins of ventrals; 10 scales in a vertical series from vent to lateral line; 15 to 18 in an oblique series forward. No air bladder. Pyloric cœca 9.

The species of this genus are all American. Those known to us may be compared as follows:

a. Mouth comparatively large, the maxillary extending to below the eye; gill-rakers tuberculate or minute.

  b. Outer teeth of upper jaw much enlarged, more than half length of posterior nostril; snout protruding well beyond premaxillaries; scales on breast large, regularly arranged.

    c. Soft dorsal rather short (rays less than I, 23); coloration plain.

      d. Spinous dorsal elevated, its longest spines reaching past front of soft dorsal; snout very prominent, its tip slightly turned upward, projecting beyond premaxillaries for a distance about equal to the large eye; maxillary shortest, 3⅓ in head; posterior nostril oblong; upper caudal lobe elongate; tip of spinous dorsal black; lower fins pale or somewhat dusky. D. X-I, 22. Pacific coast of tropical America..NASUS.*

      dd. Spinous dorsal not elevated, the longest spines not reaching soft dorsal; snout bluntish, projecting beyond premaxillaries for about half diameter of eye; maxillary long, 3 in head; posterior nostril nearly round; upper caudal lobe not produced; pectoral large; lower fins mostly black. D. X-I, 18. Pacific coast of tropical America................PANAMENSIS.†

    cc. Soft dorsal rather long (D. X-I, 24); spinous dorsal moderately elevated, its tip reaching front of second dorsal; snout short, rather sharp, projecting beyond premaxillaries for a distance equal to about half eye; maxillary moderate, 3 in head; posterior nostril broad-ovate; lower caudal lobe longest; pectoral rather large; coloration nearly plain, or with faint oblique dusky bars; pectoral and lining of opercle black. South Atlantic and Gulf coasts of United States............................. ALBURNUS.

  bb. Outer teeth of upper jaw moderately enlarged, less than half length of posterior nostril; snout moderately protruding; scales on breast large; spinous dorsal high, the longest spine filamentous (in the adult) reaching past front of second dorsal, usually higher than body below it; gill rakers reduced to minute tubercles. Body always with distinct oblique bands, the anterior **V**-shaped; a dark lateral band, distinct posteriorly, and extending on lower lobe of caudal; lower fins blackish; lining of opercle mostly pale. D. X-I, 26. Cape Cod to Gulf of Mexico .... NEBULOSUS.

aa. Mouth comparatively small, the maxillary less than one-third head, barely reaching eye; outer teeth of upper jaw scarcely enlarged; snout little projecting; scales on breast small, irregular; the back and sides sometimes with faint undulating streaks. D. about X-I, 24.

  e. Pectorals, ventrals, and anal black; snout bluntish, scarcely projecting; posterior nostril oblong; pectoral large, 1⅓ in head; depth, 4¼ in length; scales, 9-60-14. Southern California ....UNDULATUS.

  ee. Pectorals, ventrals, and anal pale; lining of gill cavity pale. Snout somewhat projecting; pectorals shortish, 1⅗ in head.

    f. Upper lobe of caudal longer than lower; scales about 9-60-13; 25 scales in an oblique series forward from vent to lateral line; axillary scale one-third length of pectoral; posterior nostril short, one-third diameter of orbit; snout very little projecting; gill-rakers very short, 4 or 5 on lower part of arch; depth, 4⅗ in length. Pacific coast of tropical America.............. ....ELONGATUS.‡

---

* *Umbrina nasus* Günther, Fish. Centr. Amer. 1869, 426. Mazatlan (Gilb.); Panama (Gthr.; Gilb.).

† *Umbrina panamensis* Steindachner, Ichth. Beitr. iv, 9, 1875. Mazatlan (Gilb.); Panama (Steind.; Gilb.).

‡ *Umbrina elongata* Gthr. Proc. Zool. Soc. Lond. 1864, 148. Mazatlan (Gilb.); Chiapam (Gthr.); Panama (Steind.; Gilb.).

ff. Upper lobe of caudal not longer than lower; scales about 8–50–11; 15 to 18 scales in an oblique series forward from vent to lateral line; axillary scale less than one-fourth pectoral; posterior nostril lanceolate, half as long as eye; snout distinctly projecting; gill-rakers larger than in other species, about 7 on lower part of arch; depth 4⅔ in length. Caudal usually tipped with black. South Atlantic and Gulf coast of United States .... LITTORALIS.

Of these species, *nebulosus* and *alburnus* are closely related, as are also *littoralis*, *undulatus*, and *elongatus*, which appear to be representatives of one form in three different faunal areas; *nasus* and *panamensis* are better distinguished.

93. Cynoscion maculatum (Mitch.) Gill.—*Speckled Trout; Spotted Trout.* P. G. (30832, 30911, 31047.)

(*Otolithus carolinensis* C. & V.; *Otolithus drummondi* Richardson and Girard.)

One of the most abundant and valuable of the food fishes of the Gulf coast. Among the shore-fishes it ranks next in importance to the "Redfish" and its flesh is finer in quality. It reaches a weight of about 10 pounds.

Color in life grayish, with very brilliant reflections of violet, green, etc., becoming silvery below; sides of head iridescent. Back above lateral line and behind middle of second dorsal covered with round black spots, somewhat irregular in size and position, most of them smaller than the pupil; a few below lateral line. First dorsal blackish at tip, with some dark spots. Second dorsal yellowish, edged with dusky and with 2 or 3 series of round dark spots. Caudal creamy, edged and broadly tipped with blackish, the base and median parts of the fin with small round dark spots. Anal and ventrals creamy, slightly soiled with blackish. Pectorals light yellowish, immaculate, the axil somewhat dusky. Inside of mouth light orange-yellow. Inside of opercle slightly dusky.

Head 3⅓; depth 5. D. X–I, 25; A. II. 10. Scales 9–78–14; 68 tubes in lateral line.

Northern specimens have the spotted area extending usually farther forward, but do not otherwise differ. The *Otolithus drummondi* of Richardson is the same species, with some slight errors in the description. The anal rays are quite constantly II, 10.

POMACENTRIDÆ.

94. Chromis insolatus (C. & V.) J. & G. P.

A single small specimen " spewed up" by a Red Snapper, at Pensacola.

Steel brown; a curved blue streak between eyes in front; many scales on upper and anterior parts of body each with a blue spot; fins all plain dusky.

D. XIII, 13; A. II, 12. Scales 2½–25–9.

95. Chromis enchrysurus sp. nov. P. (30871.)

Several specimens in fine condition, the largest 3¼ inches long, "spewed up" by Red Snappers, at Pensacola.

Allied to *Chromis insolatus* (C. & V.) and *Chromis flavicauda* (Gthr.). Head 3⅛ in length; depth 2. D. XIII, 12; A. II, 12 (D. XII, 11; A. II, 11, in one specimen). Scales 3-26-9.

Body regularly ovate-oblong, the anterior profile evenly convex. Mouth small, oblique, the jaws equal, the maxillary extending little past front of eye, 3⅛ in head. Snout short, 4½ in head. Eye large, 2½ in head. Preorbital entire; preopercle with distinct obtuse serratures or crenations. Teeth slender, conical, in a moderate band, those of the outer series considerably enlarged. Gill-rakers long, not as long as pupil.

Dorsal somewhat emarginate, the longest spine 1½ in head, the longest soft ray about the same; caudal lunate, the upper lobe slightly longer, about as long as head. Anal about as high as soft dorsal, its second spine 1⅔ in head. Ventrals filamentous at tip, longer than head. Pectorals about as long as head. Vertical fins largely covered with small scales.

Color, when fresh, sooty gray, rather dark, a narrow blue stripe from tip of snout obliquely upward and backward across upper part of eye to above front of lateral line, where it ends in blue dots; sides paler posteriorly and below; fins dusky, the distal half of anal, most of soft dorsal, and the whole of caudal and pectorals of a very intense light yellow, deepest on the caudal; ventrals dusky-bluish, slightly tinged with yellow. A small black spot in upper part of axil.

LABRIDÆ.

96. Platyglossus caudalis (Poey) Gthr. P. (30817.)
  ? *Julis caudalis* Poey. Mem. Cuba. ii, 213.
  ? ? *Julis pictus* Poey. Mem. Cuba ii, 214.

A single fine specimen 6 inches long, taken from the stomach of a Red Snapper, at Pensacola. A second specimen badly injured was also obtained.

Head 3⅛ (3¾); depth 4¼ (5). D. IX–II; A. III, 12. Scales 2-25-6.

Body very slender, compressed, the snout rather pointed, 3⅛ in head. Eye moderate, 5½ in head. Posterior canine large. Dorsal spines low, rather slender, but pungent, lower than the soft rays. Caudal fin convex, its two outermost rays somewhat produced. Pectoral 1⅔ in length of head. Scales on breast small. Head naked.

Color, when fresh, olivaceous above; a row of round sky-blue spots along each side of back; a broad band-like area of orange intermingled with violet spots along sides from lateral line about to level of eye, extending backward about to middle of body; the lower edge of the orange band serrate. Below the orange a band of pale violet, becoming posteriorly deep violet. Still lower on level of lower edge of pectoral a deep yellow band about as wide as a scale, growing narrower and fainter behind. Belly pearly. Head above olivaceous, marked with blue; preorbital and suborbital region scarlet, with three violet-blue stripes, these margined with cherry red. Cheeks below lowest violet stripe translucent yellowish. Opercles bright red, with about 3 oblique violet stripes, the upper forming an oblique blotch behind eye, in the middle of which is a round black ink-like spot; no dark opercular spot; chin pearly. Iris red.

Dorsal light orange, the soft part with three rows of violet spots; caudal orange, with four rows of spots, the orange arranged in one longitudinal, two marginal, and two convergent orange bands, which are connected by reticulations around blue spots. Anal with a basal orange spot on each membrane, then a blue spot, then a broad yellow band, then a narrow blue band, and a terminal band of orange. Ventrals light red. Pectorals pale violet, yellow at base; a bluish oblique band below them. Blue spots of head and posterior parts clear, sky-blue; elsewhere of a violet shade and less bright.

This specimen agrees in many respects with Poey's "*caudalis*." Poey, however, had a deeper fish (depth $4\frac{1}{2}$ in total length), and he makes no mention at all of the broad orange lateral shade so conspicuous in our specimens. It is possible that the latter difference is sexual. Poey's "*pictus*" has the orange band, but the body is too slender (depth $5\frac{1}{2}$ in total), and the coloration is otherwise not quite like that of our specimens.

97. Platyglossus florealis sp. nov. P. (30839.)

Two specimens $3\frac{2}{3}$ inches in length were taken with a seine in the Laguna Grande, near Pensacola. They were found in shallow water in masses of *Zostera*.

Head $3\frac{1}{3}$ ($3\frac{1}{3}$); depth 4 ($4\frac{2}{3}$). D. IX, 11; A. III, 12. Scales $1\frac{1}{2}$-26-8.

Body rather slender, moderately compressed; snout not very sharp, $3\frac{1}{2}$ in head. Eye moderate, 5 in head. Posterior canines small. Dorsal spines rather low, stiff and pungent, lower than soft rays. Caudal truncate, $1\frac{2}{3}$ in head. Pectoral $1\frac{1}{3}$ in head. Scales on breast small; head naked.

Coloration in life: ground color olive-brown; a rather dull olive-green stripe from above snout along sides of back to tail, midway between lateral line and dorsal; a brownish area along lateral line; below this a distinct dark brown band from gill-opening to middle of caudal on level of eye, and about as broad as eye, ending in a small dark spot at base of caudal. Below this another light brownish area bounded by a dark bronze stripe on level of pectoral, the belly abruptly pale. Each scale of side with a narrow crescent of deep greenish-blue towards its base. These spots are very distinct, especially anteriorly, giving the whole fish a bluish cast. Sides of head pale orange; a bright blue wavy streak along preorbital, suborbital, and opercle, turning abruptly downward on the subopercle. A faint blue streak behind eye. Opercle with a deep indigo-black spot bordered by bluish and yellow; tip of opercle yellow; the color bounded by a <-shaped blue line. Lower jaw with two cross stripes of coppery orange, the interspaces white, the tip reddish. A small jet-black spot at base of last ray of soft dorsal.

Dorsal fin light cherry-red, with a row of translucent spots at base; a narrow translucent median band, the tips translucent. Caudal translucent, tinged with red toward the base. Anal with a row of pearly spots, and a cherry-red band, then a narrow pearly band, then a light yellow band, then a light red band, the tips translucent. Pectorals yellowish; ventrals white. Iris scarlet.

This gaily-colored little fish seems to be well distinguished from all thus far known in the West Indies.

TRIGLIDÆ.

98. Prionotus tribulus C. & V.   G.   (30910, 30201, 31053.)

Common; numerous specimens obtained at Galveston.

Coloration in life: light olive-green, the head and body everywhere reticulated with dark olive-green, in definite patterns, the dark lines on the head conspicuous, arranged in a series of curves and concentric circles; the dark streaks on the body mostly undulating and ascending backward. A diffuse band along side of bright orange. Belly white. Two faint diffuse dark bands downward and forward from soft dorsal, the hindmost ascending on the fin; a fainter band on spinous dorsal.

Spinous dorsal reddish, clouded with darker. A large dark blotch, not ocellated, between fifth and sixth spines. Second dorsal translucent reddish, with darker spots. Anal similar, paler, the spots almost obsolete. Caudal reddish, with three darker bands. Ventrals plain light reddish. Pectorals light clear green on the front side, grayish behind; with about 5 somewhat irregular dark cross-bands, the three median broadest and forked or Y-shaped above. Upper edge of pectorals pale. Pectoral appendages reddish, barred with darker.

Head 2½; depth 5. D. IX–I, 12; A. 11. P. 13–3. Lat. l. 49 (tubes).

99. Prionotus scitulus sp. nov. P.

> *Prionotus punctatus* Jor. & Gilb. Proc. U. S. Nat. Mus. 1878, 373 (not of C. & V.).

A single specimen taken from the stomach of a Red Snapper at Pensacola.

Head 3⅓ (4⅙); depth 6⅓ (7). D. IX, I, 13; A. 12; L—; Lat. l. ca. 70 (pores), about 60 scales. Length 5¾ inches; none larger.

Body much slenderer than in any other species; head small, low, rather pointed. Snout rather long, a little shorter than rest of head, its width between angles of mouth about 2½ in head. Maxillary not reaching front of eye, 2⅔ in head. Sides of snout finely and evenly serrate;

no spinous teeth on preorbital; preopercular spine simple, long, and slender, without tooth at base. Spines on top and posterior part of head about as in *P. palmipes*, but rather sharper, the furrow connecting orbits posteriorly not much marked. Opercular spine small. Eyes large, separated by a narrow concave space, the supraocular ridge prominent, serrate in front. Bands of palatine teeth narrow. Gill-rakers long and slender, as in *Prionotus punctatus*.

Dorsal high, its longest spine 1⅓ in head. Pectoral scarcely more than ⅓ length of body, reaching to base of fifth or sixth dorsal ray.

Coloration in life, dark olive above; back and sides covered with numerous round spots of different sizes, and not arranged in series; these spots bronze color in life, becoming brownish after death; spinous dorsal dusky, with lighter streaks; a distinct black spot on upper half of spinous dorsal, between the fourth and fifth spine, this spot being ocellated below and behind; a second black blotch on upper half of first spine and membrane, also ocellated behind; second dorsal and caudal spotted and finely blotched with black; anal largely black, with a pinkish border; pectorals blackish; ventrals pale; branchiostegals pinkish.

This species, formerly erroneously identified by us with *Prionotus punctatus*, differs from the latter in its much slenderer form, in color, in the absence of spinous teeth on snout, and in the short pectorals.

The original types were obtained by us at Beaufort, North Carolina, in 1877. Another specimen (15148) is in the National Museum, collected in West Florida, by Kaiser and Martin.

URANOSCOPIDÆ.

100. Astroscopus anoplus (C. & V.) Brev.—*Dog-fish; Electric Dog-fish.* P. G.
(30851, 30899.)

This species is rather common about Galveston, and is not rare about Pensacola. Two young specimens were obtained at each place. The fishermen at Galveston ascribe to it electric powers in life—a trait already noticed by Dr. J. A. Henshall in the closely allied *Astroscopus y-græcum*.

Coloration of young specimens in life: dark olive above, becoming abruptly white beneath, the sides with a darker shade. Back and top of head, as far back as front of soft dorsal and as low as upper edge of pectoral, covered with small, round, light-green spots, none of them as large as pupil, those on top of head light brown. Posterior part of body speckled with blackish dots. First dorsal black except at base; second dorsal plain, with a dark blotch in front; anal and ventrals immaculate; caudal with three black longitudinal stripes, the interspaces pale. Pectoral black at base, its edge pale. Lower jaw and median line of lower side of head yellow; a large oblong black blotch on each side of median line of lower part of head. Lips dusky. D. IV-I, 13; A. 13. Scales scarcely appreciable, visible only posteriorly.

The naked area behind the eyes is much smaller in this species than in *A. y-græcum*, its form being concavo-convex, its length barely twice that of the snout; the bony Y-shaped plate on top of head is much shorter and broader in *A. anoplus*, concave on the median line, and forked about half its length. The posterior, undivided part of the Y is broader than long. The bony bridge across the occiput is but little shorter than the part of the head which precedes it. In *A. y-græcum* the Y is forked for less than half its length, the posterior part is more than twice as long as broad, and not concave on the median line; the naked area behind the eyes is trapezoidal, longer than broad, and about 4 times the length of the snout. The bony bridge across the occiput is not half the length of the part of the head before it.

The coloration of the *A. y-græcum* is somewhat different. The pale spots on the body are larger; some of them are as large as the pupil, and each surrounded by a narrow ring of black. They extend backward to the end of the soft dorsal, and also cover the lower jaw. The second dorsal is black (the base paler), with two oblique stripes of white; the anal is white with a broad black band; the caudal black with two white bands, the corners also white; the pectoral brownish, with a broad black shade and a narrow edging of white; the two black blotches on the lower parts of the head are present as in *A. anoplus*, but less distinct.

OPISTOGNATHIDÆ.

101. Opisthognathus lonchurus sp. nov. (29671). P. (30864.)

Head not very large, rounded, and blunt anteriorly in profile; snout extremely short, shorter than pupil; eye large, $3\frac{1}{3}$ in head; maxillary $1\frac{1}{3}$ in length of head (in specimen 5 inches long), rather narrow at tip, with a well-developed maxillary bone; lower jaw included; teeth in both jaws cardiform, forming bands, the outer series enlarged, slender; vomer with 5 rather large teeth, forming a semicircle; palatines toothless; gill-rakers slender, of moderate length. Longest anal rays $1\frac{1}{2}$ in head; ventrals long, $1\frac{1}{6}$ in head; pectoral somewhat mutilated, apparently little more than half head.

Dorsal spines very slender, the longest about as long as head, slightly higher than soft rays. Caudal long, the middle rays longest, a little shorter than head. Scales entirely destroyed by the digestive process; head naked.

Head $3\frac{2}{3}$ in length; depth $4\frac{2}{3}$. D. ca. 25; A. ca. 15.

Color: head light olive, unmarked; rim of upper lip narrowly black; top of head and back rather darker; body apparently nearly plain light olive; caudal plain, with traces of three dark bars; breast white; eye dark.

A single specimen, 5 inches long (No. 29671, U. S. Nat. Mus.), in poor condition (the skin of the body having been digested), taken from the stomach of a Red Snapper, at Pensacola. A second specimen, in the U. S. Nat. Mus. (30712), since forwarded by Mr. Stearns, has the head $3\frac{1}{6}$; depth $3\frac{3}{4}$; lat. l. 67.

BATRACHIDÆ.

102. Batrachus tau (L.) C. & V.—*Sarpo.* P. (30811.)

Very common in grassy lagoons about Pensacola. Our specimens belong to the scarcely tangible var. *β*. of Günther. The "white" spots on the body and fins are bright olive-yellow in life.

103. Porichthys plectrodon sp. nov. G. (30894.)

Allied to *Porichthys margaritatus* (Rich.), but with the palatine teeth very different.

Head $3\frac{3}{4}$ ($4\frac{1}{4}$); depth $5\frac{2}{3}$ (6). D. II, 37; A. 34.

Body rather elongate, tapering and compressed behind. Head depressed, two-thirds as broad as long and half wider than deep; lower jaw considerably projecting, maxillary reaching to well behind eye, its length $1\frac{3}{4}$ in head. Teeth in single series on jaws, vomer, and palatines; those of upper jaw very small, a few of the anterior and two or three of the lateral teeth somewhat enlarged, the latter strongly hooked forwards. Teeth in lower jaw strong, rather weaker than in *P. margaritatus;* those in the front of the jaw hooked strongly inwards; the lateral teeth, which are larger, hooked backwards and inwards; one or two strong canines on each side of vomer, these curved backwards and outwards. Teeth on palatines distant, few in number (usually 4 or 5); among these are one to three very strong canines (usually, but not always, much larger than canines on vomer), strongly curved forwards and inwards. In *P. margaritatus** and *P. porosissimus*, the palatine teeth are not especially enlarged, subequal and more numerous; the canines on the vomer being much larger than any of the other teeth.

Gill openings extending from upper edge of pectoral to just below lower edge. Pectoral without axillary foramen.

Height of soft dorsal about 3 in head. Length of caudal nearly 2. Height of anal $3\frac{1}{3}$. Length of pectorals $1\frac{3}{5}$; of ventrals $2\frac{3}{5}$.

Color in life light brown above, the top of head much darker and clouded with dark brown; a row of about ten bar-like dark blotches along middle of side, each larger than eye; those anteriorly deeper than long, the others longer than deep. Each of these blotches is usually more or less confluent with a saddle-like dark blotch across the back. A crescent-shaped pale translucent area below the eye; below this a larger blue-black area, irregularly crescent-shaped, covering the preorbital and suborbital region, bounded below and behind by a row of shining mucous pores. On it are about four large pores, and above and behind it, close behind and below eye, is a large shining pore bordered with black. Cheek steel-bluish. Sides of body silvery, becoming golden below. Lower part of head and belly bright golden. A dark stripe along base of dorsal. Soft dorsal with 2 or 3 rows of small round dark olive

---

*The identity of the Pacific species (*margaritatus* Rich.=*notatus* Grd.) with the Surinam "*porosissimus*" is not yet proven, and is not very probable.

spots, the upper row posteriorly becoming a dark edging to the fin. Caudal dull red, edged with dusky. Anal very pale, edged with blackish. Pectorals light orange, usually with some small dark spots above. Ventrals orange, slightly darker anteriorly.

Numerous series of pores on the body, those of the lateral line accompanied by shining golden bodies, as in other species of the genus. According to fishermen, these bodies are phosphorescent, shining at night; a statement which is probably true, although we have been unable to verify it. Pores on sides of back not shining. Most of the pores, as in other species, accompanied by numerous small cirri or cilia.

The arrangement of the lines of pores and shining bodies is not very different from that found in *P. margaritatus*. It may be thus described in detail.

A series of pores beginning at tip of snout, extending down around preorbital region, bounding the dark subocular blotch and joining almost at a right angle with a series of pores which extends downward from lower posterior corner of eye to angle of mouth. Another series diverges from the first in front of eye, passing close below eye, then upward above cheek, ending in a large pore behind preopercle. A curved series of pores extending backward along opercle, and another parallel with it along subopercle.

Two obscure series from front of eye along top of head, becoming wide apart at the vertex, converging at the nape, then slightly diverging, converging in front of spinous dorsal, then again diverging to pass around the fin, each at last becoming straight at front of soft dorsal, extending close to its base to its last ray, there being about two pores to each ray. Just below this series, at front of soft dorsal on each side, begins a second series, with the pores wider apart and somewhat irregular, ceasing near the middle of the soft dorsal fin.

The lateral line proper next begins above upper posterior angle of preopercle, whence a short branch passes directly upward. Opposite front of soft dorsal, the lateral line is interrupted for a distance a little more than diameter of eye. A short branch arises at this interruption and passes upward and backward at an angle from the end of the anterior part. Thence the lateral line passes straight to base of caudal.

The next series arises just behind axil of pectoral, then curves abruptly downward and backward, becoming straight opposite third ray of anal, thence proceeding to base of caudal, the pores small and close-set, anteriorly bead-like and shining, becoming dull toward the tail. Next comes a double series on each side of base of anal, the two series converging behind and finally coalescing.

Another series begins at the middle of the base of the pectoral in front, curves downward, around the base of the fin, and, proceeding directly backward, ceases opposite vent. A series begins midway between gill opening and ventral and, extending straight backward, ceases opposite base of pectoral. Another begins, on each side, on lower side of head,

directly below angle of mouth, the two diverging slightly between ventrals, then converging a little behind ventrals, then abruptly diverging,
joining the series last mentioned, on each side, just in front of base of
pectoral.

A cross-series of pores extends straight across belly, between vent.
and anal fin. At each end of this cross-series a series of pores turns
abruptly forward, the two meeting in an acute angle on the belly just
in front of a vertical from base of pectorals. Finally, three parallel
series on each side of lower parts of head meet in front, the two anterior in obtuse curves, the posterior in an acute angle. The anterior
series along the mandible ends at the corner of the mouth. The next
just behind the mandible ends just below the corner of the mouth. The
next passes along the branchiostegal region, ending at the gill opening.
Mandible with two large foramina. A series of dark-colored pores along
each side of tongue.

This species is not rare about Galveston, where many specimens, the
largest 8 inches long, were obtained with the seine, in water of moderate depth. It seems to be unknown to fishermen at Pensacola.

GOBIESOCIDÆ.

104. Gobiesox virgatulus sp. nov. P. (30861.)

Three specimens, the longest about 1¼ inches in length, taken among
ballast rocks in Pensacola Bay.

Head 2¾ (3⅔); width of head 3¼; depth 6 (7). D. 10; A. 8.

Body rather slender, the head low and rather broad, broadly rounded
anteriorly; eyes very small, about 4 in head, their diameter two-thirds
to three-fourths the broad, slightly convex interorbital space. Cheeks
prominent; opercle ending in a sharp spine. Cleft of mouth extending
to below front of orbit; lower jaw somewhat shorter than upper.

Teeth of upper jaw in a narrow band of about two series; four teeth
of the outer series a little larger than the rest, somewhat canine-like.
Middle teeth of lower jaw incisor-like and partly horizontal, their edges
entire or somewhat concave. Ventral disk considerably shorter than
head. Distance from root of caudal to front of dorsal 2⅖ in length.
Pectoral short, about 2⅖ in head.

Color in life olivaceous, with numerous paler spots; the whole body
covered with rather faint, wavy longitudinal stripes or lines of a light
orange-brown color, about as wide as the interspaces, much as in some
species of *Liparis;* skin everywhere with dark punctulations. Caudal
dusky, slightly barred with paler, its tip abruptly yellowish. Dorsal
and anal dusky, somewhat barred. Body sometimes with traces of
darker cross-shades.

This species may be identical with *Gobiesox nudus* of Günther, but the
name *nudus* cannot fairly be retained, as the original *Cyclopterus nudus*
L. was an East Indian species, very different from this.

GOBIIDÆ.

105. Lepidogobius gulosus (Girard) J. & G. P. (30844.)

Three specimens obtained in the "Laguna Grande" at Pensacola, the longest 2¾ inches in length.

Coloration in life light, grayish olive, with rather sharply-defined markings of darker brown; head with a pale bluish stripe from behind the angle of the mouth upward and forward parallel with the gape to below the front of eye, then turning abruptly backward across suborbital region to upper edge of gill opening; another pale streak from snout along lower part of eye; between this and the first streak a dusky area; below the first-mentioned streak a dusky region on cheeks; opercle with an oblique blackish bar; top of head with dark marblings surrounded by paler reticulations; back with a series of black cross-blotches, mostly separated on the median line; two narrow vertical dark bars behind pectoral; middle line of side posteriorly with longitudinally oblong black blotches; besides these numerous other blotches not regularly arranged. First dorsal with two or three oblique black bands; second dorsal pale, with about four series of black dots; caudal spotted with black, pectoral yellowish, ventral black, its center yellowish; anal pale; lower side of head pale; jaws dusky.

Head 3¼ (4½ in total); depth 5 (6). D. VI–15; A. 16; Lat. l. about 42.

Body elongate, moderately compressed. Head long and large, low, rather sharp in profile. Eyes large, placed high and close together, 4 in head. Snout short, 4½ in head. Mouth large, very oblique, the lower jaw strongly projecting, the maxillary extending to below middle of pupil, its length 2⅓ in head.

Teeth in few series, those of the outer row very long, slender, and curved, those of the lower jaw longest.

Gill membranes not continued forward above opercle. Scales small, cycloid, imbedded. Head, nape, and breast scaleless; scales of anterior part of body not well developed.

Dorsal spines slender, the tips of the longest somewhat filamentous, although short, the longest about half head. Soft rays a little higher than the spines. Soft dorsal and anal unusually long. Caudal pointed, about as long as head. Pectorals about 1¼ in head, their upper rays not "silk-like." Ventrals about as long as pectorals, their insertion directly below front of pectorals.

106. Gobius lyricus (Girard) J. & G. G. (30897.)

A single specimen about 3½ inches long, taken with a dip-net in a brackish pool at Galveston.

Subgenus *Euctenogobius* Gill.

Color in life dark olive, with 4 or 5 irregular confluent blackish cross-bands, besides dark blotches and irregular markings. Head marbled with darker, the jaws, opercles, and branchiostegals blackish. First dorsal mostly dusky translucent, somewhat barred. Second dorsal and anal plain dusky. Caudal dark blue, with two longitudinal stripes of bright red. Pectoral finely barred or reticulated with blackish and pale. Head and belly yellowish. Female specimens probably duller and paler.

Head 4⅛ (5⅔); depth 4⅔ (6). D. VI–11; A. I, 10. Lat. l. 27.

Body rather elongate, moderately compressed. Head rather short, the profile very obtuse, descending abruptly from before the front of the eye to the snout. Eyes small, placed high, about as long as snout, and about 4½ in head. Mouth nearly horizontal, much below level of eye; the maxillary extending to beyond pupil, 2⅔ in head; jaws subequal; teeth strong, in one series in each jaw; in the lower jaw about 4 shortish, canine-like teeth behind the other teeth; anterior teeth of lower jaw small; of upper jaw rather large.

Gill opening not continued forward above opercle.

First dorsal with two or three spines filamentous, the longest reaching past the middle of the second dorsal, which is of moderate height and similar to the anal; caudal long and pointed, one-fourth longer than the head. Pectoral as long as head, about reaching front of anal. Upper rays of pectorals not silk-like. Ventrals somewhat shorter than head, their insertion below front of pectorals.

Scales large, rough, those on nape, pectoral region, and belly reduced in size; head naked.

Gobius würdemanni Girard is possibly identical with this species, although the scales are said to be smaller, and the teeth much smaller than in *G. lyricus*. The original types of *G. lyricus*, as of *G. würdemanni*, came from Brazos Santiago, Tex. The types of the latter are now lost.

107. Gobius boleosoma sp. nov. P. (30860.)

Subgenus *Coryphopterus* Gill.

Color in life: Male deep olive green, mottled with darker; middle of side with 4 or 5 vague darker blotches. A jet-black spot above gill opening, on side of back. Head mottled, dusky below; usually a dark bar below eye. Dorsals tipped with bright yellowish, each crossed by numerous narrow, somewhat oblique, interrupted bars or series of spots, these being of a rich reddish brown color. Caudal barred with black, its upper edge tinged with orange. Anal nearly plain, with a slight orange tinge. Ventrals bluish-black, their edges whitish.

Female paler and duller in color, more mottled, the black spot above gill opening obsolete or nearly so; a dark spot at base of caudal. Upper fins barred, as in the male. Lower fins mostly pale, tinged with orange.

Head 4 (5 in total); depth 4½ (5¾). D. VI–12; A. I, 11. Lat. l. about 33.

Body slender, subfusiform, little compressed. Head moderate, not very blunt, the anterior profile somewhat evenly decurved, the snout not very short, scarcely shorter than the large eye. Mouth not very large, horizontal, the lower jaw included, the maxillary extending slightly beyond front of pupil; its length about 3 in head. Teeth small, slender, in narrow bands, those of the outer series longer than the others. Eyes placed high, about 4 in head; interorbital space not wider than pupil.

Scales moderate, ctenoid; those on nape and belly not much reduced in size.

Gill openings not continued forward above opercle.

First dorsal with the spines slender but rather firm, none of them filamentous, the longest about three-fifths head. Second dorsal and anal rather large. Caudal long, pointed, slightly longer than head. Pectorals large, slightly longer than head, none of the upper rays silk-like. Ventrals slightly shorter than head, inserted below axil of pectorals.

Many specimens of this species, the largest about 2 inches in length were obtained in the Laguna Grande at Pensacola. It lurks in sea wrack on muddy bottoms in very shallow water (6 to 12 inches). In form, size, coloration, and movements, this little fish bears a remarkable resemblance to the percoid, *Boleosoma olmstedi.*

108. Gobius soporator C. & V. P. (30822.)

(*Gobius catulus* Grd.; *Gobius mapo* Poey; *Gobius carolinensis* Gill.)

Exceedingly abundant about the wharves at Pensacola, lurking under stones in ballast heaps, etc. It reaches a length of about 5 inches.

Color in life very deep olive-green, the back and sides obscurely barred and much marbled with different shades of olive-green; cheeks with the dark markings forming reticulations around pale spots. Whole under part of head blackish in the males; yellowish in the females.

First dorsal with an oblique median shade of blackish, the base in front and the distal part light orange. Second dorsal dusky at base with some spots, its margin light orange. Caudal reddish, with dusky cross-lines or spots. Anal and ventral dusky, yellowish at base in the female. Pectoral olivaceous, yellowish at base, reddish at tip; two dark spots on base of pectoral.

Head 3¼ (4 in total); depth 4 (5). D. VI, 10; A. I. 9. Lat. l. 30 to 38; 12 rows of scales from first dorsal downward and backward to anal. Scales on nape extremely small. Scales on sides firm, ctenoid.

Form robust. Head rather blunt and heavy, the snout less abruptly decurved than in *G. lyricus.* Mouth moderate, the jaws equal, the maxillary reaching about to front of pupil, 2⅔ in head. Teeth in moderate bands, the outer series somewhat enlarged. Cheeks full, tumid. Eyes moderate, placed rather high, much broader than the interorbital space.

Dorsal spines slender, the first longer than the other, but not filamentous, 1¾ in head; caudal rounded, 1¼ in head; upper rays of pectorals silk-like, the fin somewhat longer than ventral, 1¼ in head.

109. Gobiosoma alepidotum (Blocb & Schn.) Grd. P. (30854.)

(*Gobiosoma molestum* Grd.)

Rather common about Pensacola. Numerous specimens taken with the seine in the Laguna Grande.

Color in life light olive, closely punctulate with darker under the lens; sides of body with broad dusky shades which alternate with narrow paler bars, which are sometimes chain-like. A longitudinal series of small linear dark spots along the middle of side of the body; a dark space above and in front of base of pectoral; sometimes a dark area below eye. Dorsals, anal, and ventrals blackish, usually without distinct markings, sometimes faintly barred with reddish; pectorals pale, dusky, and speckled at base.

Head $3\frac{2}{5}$; depth 4. D. VII, 13; A. 10.

We are unable to distinguish our specimens from *G. alepidotum* of the Atlantic coast.

110. Ioglossus calliurus Bean MSS. P.

Body very elongate, slender, much compressed, of equal depth throughout; head compressed, without osseous crest; mouth very oblique; the lower jaw strongly projecting; premaxillaries in front on the level with pupil; maxillary extending to opposite front of pupil, its length $2\frac{3}{4}$ in head; upper jaw with a narrow band of about two series of conical cardiform teeth; those of the outer row much larger than the others; behind these are two small conical curved canines; lower jaw with a single row of smaller teeth, behind which are about 4 canines directed somewhat backward; the posterior pair largest and strongly curved; no teeth on vomer or palatines. Tongue narrow, pointed. Eye large, nearly twice length of snout, $3\frac{1}{2}$ in head, its diameter considerably more than depth of cheek, about half more than interorbital width; opercles unarmed. Pseudobranchiæ present. Gill openings wide, extending forwards below, the membranes attached mesially to the very narrow isthmus, across which they do not form a fold. Gill-rakers long and slender.

Dorsal fins separated by a short interval, the first of very slender somewhat filamentous spines, the longest about as long as head; second dorsal little more than half as high as first, apparently nearly uniform, separated from the caudal by an interval nearly half length of head; caudal lanceolate, its middle rays filamentous, about half the length of rest of body; anal rather high, similar to soft dorsal. Ventrals I, 4, inserted very slightly in advance of base of pectorals, the two fins very close together, but apparently quite separate and without basal fold of skin; the fin little longer than head, the inner rays filamentous. Pectoral with broad base, about $1\frac{1}{4}$ in head. Anal papilla very short, midway between tip of snout and base of caudal.

Body with very small, non-imbricate, imbedded scales, these a little larger and imbricate on the tail; cheeks with imbedded cycloid scales. Scales very weakly ctenoid, most of them appearing cycloid. No lateral line.

Head 5 in length; depth 7 to 7½. D. VI–22; A. 1, 21.

Color: light olive, top of first dorsal dusky; middle of caudal dusky (blue), with paler (perhaps red) edgings.

Two specimens of this remarkable species, the largest 4½ inches long, taken from stomachs of the Red Snapper at Pensacola.

BLENNIIDÆ.

111. Chasmodes saburræ Sp. nov. P. (30824.)

Allied to *Chasmodes bosquianus*, but with the mouth smaller, the form less elongate.

Head 3½ to 3¾; depth 3¼ to 3¾. D. XII, 17; A. II, 18.

Body rather deep and compressed, less elongate than in *C. bosquianus;* the back somewhat arched. Head comparatively short, much shorter than in *C. bosquianus*, not one-fourth longer than deep; profile forming a nearly even curve from the base of the dorsal to the tip of the snout, which is not blunt, although less acute than in other species of the genus.

Mouth low, nearly horizontal, the maxillary reaching to near the posterior margin of the pupil, its length 2⅛ to 2¾ in head (2 or less in *C. bosquianus*), lower jaw included; teeth rather short, equal; toothless posterior part of lower jaw occupying scarcely more than half the length of its side; oblique length of snout 3½ in head. Eye large, 5 in head, half wider than the interorbital space. Lower edge of gill-opening opposite base of third ray of pectoral, the height of the slit 4¼ to 5 in head. Lateral line extending as far as tip of pectoral. A minute cirrus (sometimes obsolete), not so long as pupil, above each eye, and a similar one over each nostril.

Dorsal fin continuous, the spines slender, but little lower than the soft rays, the longest of the latter being 1½ in head. Last ray of dorsal joined to base of caudal; anal free from caudal. Caudal 1⅛ in head. First two rays of anal short, in the males thickened and fleshy at tip, the short anal papilla close in front of them. Pectorals a little shorter than head; ventrals 1¾ in head.

Females (in spirits) with about 8 irregular blackish cross-bars extending on the dorsal fin, everywhere freckled with pale spots; a bar below eye, and two or three across the under side of head; fins all sharply barred with blackish, in fine pattern; the cross-bars on pectorals and ventrals usually very distinct.

Male in life: deep olivaceous, with traces of darker bars, and marbled with light and dark; a series of round greenish spots along middle of sides posteriorly, besides other series which form narrow undulating greenish lines converging backwards; a dark stripe downward and one forward from eye; lower side of head mostly dusky.

Dorsal fin dusky or greenish, the spinous part with a dark shade or one or two dark blotches near the base, and with a median longitudinal band of orange; usually a dusky blotch above this band between first and second spines, the margin of the fin somewhat dusky. Some specimens with the outer part of both dorsals and the top of head dusted with black spots; others with these spots obsolete; soft dorsal and caudal light orange, barred with light greenish; anal dull orange, with an obscure blackish median band, the exserted tips of the rays abruptly whitish. Pectorals dusky olive, strongly tinged with orange. Ventrals blackish, orange at tip.

The life colors of the female were not observed.

This species is very abundant in Pensacola Bay, where 14 specimens, the largest about 4 inches in length, were obtained. Some of these were taken with a seine in masses of *Zostera* in the Laguna Grande; others were caught with a pin-hook from the wharves, where it abounds among the ballast rocks (*saburra*) on which the wharves are built.

112. Isesthes* ionthas sp. nov. P. (30856.)

Head 4 (4⅔); depth 3⅘ (4½). D. XII, 13, or XII, 14; A. II, 13, or II, 14.

Body rather deep, moderately compressed, the back little elevated. Head short, blunt, but less so than in *I. punctatus;* the profile prominent above the eye, thence descending abruptly but not vertically to the tip of the snout; oblique length of snout 4 in head.

Mouth small, low, its cleft largely anterior, the short maxillary scarcely reaching past the front of the eye, 4 in head. Eyes large, placed high. 5 in head, the interorbital space about half their diameter. Orbital cirrus low, scarcely larger than nasal cirrus, which is about equal to diameter of pupil. Teeth moderate, equal; no posterior canines. Gill-opening extending downward to a point a little below middle of base of pectoral, the height of the slit 3 in head. Lateral line not reaching tip of pectoral.

Dorsal fin continuous, the spines low and not very stiff, slenderer than in *I. punctatus*, the longest spines a little lower than the soft rays, which are about 1⅓ in head. Caudal free from anal, slightly connected with dorsal; a little shorter than head; pectoral about as long as head; ventrals shorter than head.

Color clear olive-green, with only traces of darker bars; body everywhere densely freckled with small round blackish spots, smaller than the pupil; on the sides and lower part of head these spots are reduced to close-set dots; two dark lines, separated by a pale area, downward from eye; a vertical curved blackish line behind eye, in front of which is a golden area. Vertical fins all plain olive-green, their edges dusky; tips of anal rays pale; paired fins dusky-olive; lower parts of head tinged with golden, sometimes with dusky cross-bars; cirri green.

Four specimens, the largest about 2½ inches long, were obtained with hook and line from the wharves at Pensacola.

The small size of the orbital cirrus and the freckled coloration readily distinguish this species from its congeners.

---

* *Isesthes* J. & G. Syn. Fish. N. A. 757: type *Blennius gentilis* Grd.

113. Isesthes scrutator sp. nov. P.G. (30850, Pensacola.)

Head 4 ($4\frac{3}{5}$); depth $3\frac{3}{4}$ ($4\frac{1}{2}$). D. XII, 14; A. II, 16.

Body rather deep, compressed, the back not elevated; head short, very blunt, almost as deep as long, the profile abruptly descending before eye, the snout about one-fourth length of head. Mouth very small, anterior, the maxillary extending to opposite front of eye. $3\frac{1}{3}$ in head; teeth subequal, without canines. Orbital cirri very long, reaching when depressed about to the front of dorsal, their length more than half head in Pensacola specimen, in the other somewhat shorter; a short branch near its middle. Nasal barbel minute. Eye large, much broader than the concave interorbital space, about $4\frac{1}{2}$ in head. Lower edge of gill-opening a little below middle of base of pectoral, the depth of the slit $2\frac{1}{2}$ in head.

Dorsal fin scarcely emarginate, the spines rather stiff, lower than the soft rays, the longest spine 2 in head. Caudal slightly connected at base with dorsal, $1\frac{1}{3}$ in head. Pectoral about as long as head, reaching past front of anal. Ventrals $1\frac{3}{5}$ in head.

Lateral line extending to base of 8th spine, not to tip of pectoral.

Color in life deep olive-green, almost immaculate, or with faint traces of darker vertical bars; a golden blotch behind eye, behind which is a dusky crescent; two dark bars downward from eye, separated by a yellowish area. Fins all dusky greenish, nearly or quite immaculate. Front of spinous dorsal blackish. Colors of female, if different, unknown.

One fine specimen, 3 inches in length, taken with hook and line from the wharf at Pensacola. Another, which had been a long time in alcohol, and is discolored and somewhat shrivelled, was presented by Dr. August Galny. It was taken in Galveston Bay.

114. Blennius stearnsi sp. nov. P. (29669.)

Head, $3\frac{4}{5}$ ($4\frac{2}{3}$ in total); depth, $4\frac{2}{3}$ ($5\frac{2}{3}$). D. XI, 18; A. II, 21.

Body much elongate, compressed, tapering regularly behind; anterior profile moderately decurved; snout short and blunt; mouth large, oblique, the jaws even; maxillary reaching slightly beyond middle of orbit, $2\frac{1}{4}$ times in head; teeth in the front of the jaw only, occupying on each side a space equal to half length of maxillary; teeth $\frac{2}{2}\frac{6}{4}$, the lateral one on each side much enlarged and canine-like, rather short but strongly curved; canine in upper jaw, equaling about half diameter of pupil; eye moderate, equaling snout, $4\frac{1}{3}$ in head; interorbital space very narrow, not as wide as pupil; upper posterior rim of orbit with a long, slender filament, forked at base, its length equaling distance from tip of snout to posterior rim of orbit; no filament at the nape; gill membranes somewhat united to the isthmus in front, but forming a broad fold across it posteriorly, the gill openings of the two sides therefore continuous below.

Dorsal rather high; no notch between spines and soft portions, the membrane of last ray not reaching base of caudal; spines of nearly uniform height, all very slender and flexible, the tips almost filamentous; highest spine half length of head; highest soft ray $1\frac{3}{5}$ in head; anal lower than dorsal, its longest ray very slightly less than one-half length of head; length of caudal peduncle more than half its height, about equaling diameter of orbit; caudal about equal to pectoral, $1\frac{1}{5}$ in head; ventrals long, the inner ray much the longest, $1\frac{1}{5}$ in head, not quite reaching vent.

Color, light greenish-olive, somewhat mottled; sides with irregular dark bars formed of spots, these extending on the fin; skin everywhere finely punctate; dorsal dark olive, the spinous part darker at tip; anal blackish, with paler edge; ventrals dusky; pectorals and caudal olive.

Three specimens, the largest 3 inches long (No. 29669, U. S. Nat. Mus.), taken from the stomach of a Red Snapper, at Pensacola.

OPHIDIIDÆ.

115. Ophidium graëllsi Poey. P. (30868.)

Very light olive, somewhat punctate above, slightly silvery below; fins without trace of dark edging (but being mutilated they may have been dark-edged in life).

Head $4\frac{2}{3}$ in length, depth about 7. Head small, the profile not very obtuse; snout $4\frac{2}{3}$ in head; eye $3\frac{1}{4}$, more than twice the narrow interorbital space; mouth oblique, the maxillary reaching to posterior border of pupil, 2 in head; lower jaw slightly included; teeth small, in narrow bands in the jaws, the outer series in upper jaw somewhat enlarged; vomerine and palatine teeth small, subequal; head naked; snout spineless; opercle without spine; no evident pseudobranchiæ; gill-rakers rather long and strong, 4 below angle of arch; occiput nearly midway between origin of dorsal and front of eye. Air-bladder long and slender, occupying nearly the whole length of abdominal cavity, tapering backward.

Two specimens, one of which is in good condition and about 4 inches long, were taken from the stomach of a Red Snapper, at Pensacola. The type of *graëllsi* differed from the specimens before us in having a shorter head (more than 5 times in the length), and a larger maxillary (reaching posterior border of eye). But as the typical specimen of *graëllsi* was 8 inches long, the difference is probably due to increased size.

116. Genypterus omostigma sp. nov. P. (29670.)

Body comparatively short, highest at occiput; thence tapering rapidly to tip of tail; upper profile of head very convex; snout blunt; mouth horizontal, the lower jaw included; maxillary not quite reaching posterior border of orbit; teeth in jaws uniform, strongly incurved, in rather broad bands; a single series of small teeth in vomer; those on palatines minute; maxillary $1\frac{7}{8}$ in head; eye large, 3 in head, much larger than snout, equalling twice interorbital width; opercle terminating in a strong, compressed spine, the length of which is about two thirds diameter of

pupil; gill-rakers very small, 4 below on anterior arch. Longest ventral filament half length of head; the shorter three-quarters length of
longer. Distance from origin of dorsal to tip of snout 3⅓ in total length;
distance from origin of anal to snout 2⅘ in total length. Scales minute,
imbedded. Pseudobranchiæ not evident. Air-bladder short, thick, with
a large posterior foramen.

Head 4⅓ in length; depth about 6.

Color light olive-green, silvery on belly, cheeks, and lower side of
head; sides above with a few irregular, large, scattered, dark blotches;
about 9 of these along base of dorsal fin; an intensely black, round
blotch on scapular region, rather larger than pupil; dorsal with black
blotches; anal largely black; upper half of eye black, lower half bright
silvery.

A single specimen, 3⅓ inches long (No. 29670 U. S. Nat. Mus.), taken
from the stomach of a Red Snapper, at Pensacola.

As here understood the genus *Genypterus* differs from *Ophidium* in
the presence of a spine on the opercle, a character apparently of more
importance than that drawn from the dentition of the palatines. In
the latter respect *G. omostigma* agrees more nearly with *Ophidium*.

PLEURONECTIDÆ.

117. Paralichthys dentatus (L.) J. & G.—*Flounder.* P.: G. (31025.)

A common market-fish at Galveston, New Orleans, and Pensacola.
Our specimens agree with others from Washington market and other
northern localities.

The width of the interorbital space increases with age. In specimens
16 inches long, it is wider than the eye, and equal to the length of the
snout, without the premaxillary. In young specimens it is proportionately much narrower.

D. 88; A. 71. Gill-rakers narrowly triangular, 3 to 4 times as high as
broad; the mouth large, the maxillary reaching past eye, a little more
than half head.

The genus *Pseudorhombus* Bleeker is in all respects identical with the
prior *Paralichthys* Grd. *Ancylopsetta, Uropsetta,* and *Chœnopsetta* Gill,
as well as *Hippoglossina* Steindachner and *Xystreurys* J. & G. are
inseparable from *Paralichthys*.

118. Paralichthys albigutta sp. nov. P. (30818.)

Pseudorhombus dentatus ("*albigutta*") Goode & Bean, Proc. U. S. Nat. Mus. 1879,
125. (Specimen No. 4887, U. S. Nat. Mus.)

Body elongate, irregularly elliptical, the snout protruding, owing
to angulation of profile above front of upper orbit; caudal peduncle
short and high, its length two-fifths the height, which equals two-fifths
length of head; head large, 3⅓ in length; mouth large; maxillary reaching beyond lower eye, half length of head; teeth long, slender, conical;
those in lower jaw distant, 7 in number on each side, regularly and
rapidly decreasing in size towards angle of mouth; in front of upper

jaw are 3 or 4 canine-like teeth on each side, similar to those in lower jaw, but rather smaller; the lateral teeth all equally minute; interorbital space narrow, scaled posteriorly, not flat, the ridge of upper orbit prominent posteriorly; interorbital width $2\frac{1}{2}$ to 3 in eye; lower eye slightly in advance of the upper, $5\frac{3}{4}$ in head; gill-rakers moderate, broad, with 3 or 4 coarse serratures on inner margin; 10 rakers below angle, the longest $2\frac{1}{2}$ in orbit.

Fins all low; dorsal beginning slightly in advance of upper eye, the first two rays a little turned to blind or left side, the anterior rays not elevated or exserted; dorsal highest at beginning of last fourth of fin, the longest ray $2\frac{2}{3}$ in head. Anal similar to dorsal; distance from its origin to snout $2\frac{1}{5}$ in length of body; the highest ray $2\frac{3}{4}$ in head. Caudal rounded, $1\frac{1}{8}$ in head; pectoral long and slender, half head; ventrals long, reaching beyond front of anal, slightly less than one-third head.

Scales rather small, becoming somewhat larger on caudal peduncle; lateral line with a short, high, somewhat oblique, arch in front, the anterior end of arch much above axis of body; width of arch about $3\frac{1}{3}$ in straight portion of lateral line; scales all smooth and imbedded; minute accessory scales very numerous.

Head $3\frac{1}{3}$ in length, depth $2\frac{2}{5}$. D. 76 to 79; A. 59 to 61. Lat. l. about 90 (pores); about 60 oblique series behind curve of lateral line.

Color (in specimen from Pensacola) dark greenish, mottled with darker, and with many very small pale spots; fins all colored like the body. A specimen from Beaufort, N. C., is nearly uniform dark brown.

The types of the present species (No. 30818 U. S. Nat. Mus.) are two specimens, 7 to 8 inches long, obtained in the Laguna Grande, at Pensacola. A third specimen is known from Beaufort, N. C., and a few small specimens from Pensacola, in addition to the one mentioned above. There is also a small specimen (4887), which has been a long time in the National Museum, where it has received from unknown hands, the manuscript name "*Chænopsetta albigutta.*" This specific name we here adopt as our own.

119. Paralichthys squamilentus sp. nov. P. (30862.)

Sinistral. Body very deep, closely compressed, the greatest height at about the middle of the length; caudal peduncle very short, its length one-third its height, which is $2\frac{2}{3}$ in head; profile evenly arched, angulated at front of upper eye, the snout thus projecting; head short and high, the greatest height at occiput equalling the length, which is contained $3\frac{2}{5}$ times in length of body; snout $4\frac{3}{4}$ in head. Mouth large, very oblique, the lower jaw included; mandible with a sharp compressed knob at symphysis, its length $1\frac{2}{3}$ in head; maxillary narrow, reaching beyond pupil, but not quite to posterior margin of lower eye, its length very slightly more than half head; teeth in lower jaw of moderate size, the longest rather less than diameter of pupil, the largest next the symphysis, thence decreasing rapidly towards corner of mouth; the teeth are distant, few in number, 8 on each side; upper jaw with two or three rather large teeth on each side in front, these smaller than those in lower jaw; lateral teeth minute; an inconspicuous blunt tubercle on snout, in front of upper eye; interorbital space a narrow scaleless bony ridge, slightly concave anteriorly; interorbital width scarcely more than half diameter of pupil; upper eye slightly in advance of lower, its diameter about one-fifth head; gill rakers $\frac{3}{9}$, comparatively slender, compressed, the inner edge with a few distinct strong teeth; the longest raker nearly half diameter of eye.

Dorsals low, beginning over front of upper eye, the anterior rays not produced nor filamentous, but with free tips; the highest rays are at beginning of posterior third of fin, their length $2\frac{1}{8}$ in head; length of first rays $4\frac{1}{2}$ in head.

Anal spine weak; the fin similar to dorsal, but higher, the highest ray $2\frac{1}{8}$ in head; ventrals reaching front of anal, about one-third head; pectoral of colored side $2\frac{1}{5}$, of right side $2\frac{1}{2}$, in head, caudal about $1\frac{2}{3}$ in head.

Scales on head and body very small, cycloid, closely adherent, without free posterior edge; lateral line with a very short, high curve anteriorly, the width of which is contained $4\frac{1}{4}$ times in length of straight posterior part; snout, jaws, and preopercle scaleless, head otherwise scaly.

Head $3\frac{2}{3}$ in length; depth 2. D. 78; A. 59. Lat. l. 123 (pores).

Color (in spirits): very light grayish, with traces of several irregularly arranged, faintly ocellated, darker spots; lips dusky; fins all mottled with colors of body.

Two specimens, each about 5 inches long (No. 30862, U. S. Nat. Mus), were collected at Pensacola.

120. Hemirhombus pætulus Bean MSS. P.

Body elliptic-ovate, strongly compressed, not very deep; the anterior profile regularly decurved until just above the snout, where it forms an angle, the rather short snout thus abruptly projecting; mouth rather large, considerably arched; maxillary extending to below middle of lower eye, $2\frac{2}{3}$ in head; teeth in lower jaw in a single series; upper jaw with two distinct rows, those of outer series in front, enlarged, 2 to 4 of them forming small canines. Eyes large, the lower slightly longer than snout, about 4 in head, its front in advance of the upper eye, especially in adults, where half of it is thus in advance; interorbital space broad, concave, in old specimens as broad or broader than least diameter of orbit; the concavity caused by the prominent ocular ridges which converge backwards, the lower turning upward at an angle to join the other. Gill-rakers short, flattish, and stout, the longest about one-fourth diameter of orbit; the rakers are similar on all the arches, growing gradually shorter on the posterior ones.

Dorsal beginning over angle of snout, its first rays slightly turned to blind side, the longest rays $2\frac{1}{4}$ in head. Caudal short, rounded, $1\frac{1}{2}$ in head. Anal without spine, a little lower than dorsal. Left ventral $3\frac{1}{2}$

in head. Pectoral of left side with two filamentous rays, its length from 1¼ to nearly 2 times that of head; pectoral of blind side short, about 2½ in head.

Scales small, thin, weakly ciliate, with many smaller scales intermingled; about 7 series of scales on cheeks; lateral line straight, slightly raised anteriorly.

Head 2⅔ in length (4⅔ in total); depth 2⅔ (2⅞). D. 81; A. 63. Lat. l. 53 (pores on blind side).

Light yellowish-brown, with irregular blackish blotches, these most distinct along middle of sides; fins all grayish, mottled and spotted with black, the pectoral of left side distinctly barred; blind side white, immaculate.

Several specimens, only one of which was perfect, were taken from stomachs of the Red Snapper at Pensacola. The individual here described is 7 inches long, some of the imperfect specimens being nearly a foot long. As usual in the genus *Hemirhombus*, the adults show longer pectoral, wider interorbital space, and the upper eye farther back.

121. Etropus crossotus J. & G. N. O. G. (30980.)

One specimen found in the New Orleans market, it having been taken in Lake Pontchartrain. Three others were obtained at Galveston. We have compared these carefully with the original types of the species from Mazatlan and with others from Panama, and are unable to detect any difference whatever. The wide range thus shown for this species is remarkable.

122. Achirus lineatus (L.) Cuv. Subsp. browni (Gthr.).—*Sole.* P. G. (30847,36009, 31036.)

' Common; numerous specimens from Pensacola and Galveston. The Gulf form of this species ("*browni*") seems to differ from Northern specimens only in coloration, the dark bands being broader and the blind side wholly unmarked.

Color in life light brown, with 7 or 8 narrow black bands edged with brownish; these bands rather irregular and about as broad as the eye; between these bands irregular dark cloudings; the head spotted with blackish, fins with dark spots, the membranes largely black, the rays pale. D. 54; A. 40.

123. Aphoristia plagiusa (L.) J. & G. P. (30855.)

Abundant about Pensacola. Numerous small specimens taken in the Laguna Grande. The West Indian *Aphoristia ornata* (Lac.) Kaup has not yet been distinguished from the present species.

TETRODONTIDÆ.

124. Lagocephalus lævigatus (L.) Gill. G.

One specimen obtained at Galveston.

Aug. 15, 1882.

125. Tetrodon turgidus Mitch. Subsp. nephelus, Goode & Bean MSS.—*Blower-fish.* P. G.

Very abundant both at Galveston and Pensacola.

126. Chilomycterus geometricus (Bl. & Schn.) Kaup.   G.

Common about Galveston.

BALISTIDÆ.

127. Alutera sp. incog.   P. G.   (30849.)

Rather rare; a young specimen seen at Galveston.   Two very small ones collected by Mr. Stearns at Pensacola.

D. I.–30; **A.** about 30; dorsal spine somewhat barbed; body elongate; lower jaw projecting; no pelvic spine.

128. Balistes capriscus L.—*Leather Jacket.*   P.

One specimen obtained at Pensacola, where it is not uncommon.

OSTRACIIDÆ.

129. Ostracium quadricorne L.   P. G.

Not numerous; one specimen obtained at Galveston and another at Pensacola.

ANTENNARIIDÆ.

130. Pterophrynoides histrio (L.) Gill.   G.

Not uncommon about Galveston, where three specimens were seen.

MALTHEIDÆ.

131. Malthe vespertilio (L.) Cuv.   G.

One specimen obtained at Galveston, presented by Dr. **A.** Galny. Snout 8 in length to base of caudal.

The following species had not been recorded as occurring on the Gulf coast of the United States previous to the time when the present collection was made.   Several of them were, however, already in the National Museum.   Those in italics are described as new in the present paper; those marked with an asterisk have been previously recorded from points on the Atlantic coast of the United States.

Isurus dekayi.*

Carcharias platyodon.*

Scoliodon terrænovæ.*

Sphyrna tiburo.*

Clupea pseudohispanica.

Synodus intermedius.

Fundulus ocellaris.

Ophichthys macrurus.

Ophichthys chrysops.

Myrophis lumbricus.

Conger caudicula.

Exocœtus hillianus.

Siphostoma floridæ.

Siphostoma zatropis.

Hippocampus zosteræ.

Hippocampus stylifer.

Scomber ? grex.*

Caranx trachurus.*

Nomeus gronovii.
Serranus trifurcus.*
Stenotomus caprinus.
Diabasis aurolineatus.
Apogon maculatus.
Apogon *alutus*.
Mullus barbatus *auratus*.
Menticirrus nebulosus.*
Chromis insolatus.
Chromis enchrysurus.
Platyglossus caudalis.
Platyglossus florealis.
Astroscopus anoplus.*
Opisthognathus lonchurus.
Porichthys plectrodon.

Gobiesox virgatulus.
Gobius boleosoma.
Ioglossus calliurus.
Chasmodes saburræ.
Isesthes ionthas.
Isesthes scrutator.
Blennius stearnsi.
Genypterus omostigma.
Ophidium graëllsi.
Paralichthys albigutta.
Paralichthys squamilentus.
Hemirhombus pætulus.
Etropus crossotus.
Aphoristia plagiusa.

INDIANA UNIVERSITY, *May* 15, 1882.

A REVIEW OF THE SYNGNATHINÆ OF THE UNITED STATES, WITH A DESCRIPTION OF ONE NEW SPECIES

By JOSEPH SWAIN.

The number of species of Pipe-fishes on our coasts has been uncertain, owing to the fact that the fishes have not been carefully studied in large collections from their various localities. The writer has endeavored to go over the group critically, to ascertain the number of species and to find the limit of variation in the characters of each species. Nearly all the specimens studied by me have been collected by Professors D. S. Jordan and C. H. Gilbert; some of them belong to the United States National Museum, the others to the museum of Indiana University.

The writer wishes to express his great obligations to Professor Jordan for the use of his collection and library, and for many valuable suggestions.

ANALYSIS OF SPECIES.

a. Top of head strongly carinated.
    b. Breast shields not covered by soft skin; lower jaw slightly included; opercle with a prominent ridge; snout short; D. 23, covering 1+4 rings; rings 18+30; belly concave; twelve irregular brown cross-bands on body (*Corythroichthys* Kaup) .... ZATROPIS, 1.
    bb. Breast shields covered by soft skin; lower jaw included; D. 41; rings 19+39. (*Dermatostethus* Gill)...................................... PUNCTIPINNE, 2.

aa. Top of head with a slight carination, or with none; opercle without prominent longitudinal ridge. (*Siphostoma.*)

   c. Dorsal fin covering 1+9 rings; snout usually long.

      d. Rings 20 to 21+45 to 49; D. 39 to 46; top of head without keel; large, reaching a length of 18 inches.......................................CALIFORNIENSE, 3.

      dd. Rings 18 to 19+39 to 42; D. 36–41; top of head slightly keeled.

                                           GRISEOLINEATUM, 4.

   cc. Dorsal fin covering 1+7 (sometimes 1+6) rings.

      e. Rings 15+38; D. 29 to 30; top of head distinctly keeled; snout short.

                                           AULISCUS, 5.

      ee. Rings 16 to 19 before vent.

         f. Dorsal fin low, not longer than head.

            g. Rings 18+31; D. 34; snout short: body comparatively stout; tail short.

                                            BAIRDIANUM. 6.

         gg. Rings 17 to 19+36 to 41; D. 30 to 32; snout moderate, or rather short; body slender....................................LEPTORHYNCHUM, 7.

         ggg. Rings 17 to 18+31 to 32; D. 27; snout rather long.....FLORIDÆ, 8.

        ff Dorsal fin very high, not shorter than head; rings 16+30 to 33; D. 28 to 32; belly in female with black carina; snout rather short; sides of body with narrow vertical silvery streaks in life; dorsal spotted..AFFINE, 9.

   ccc. Dorsal fin covering 3+5 rings; rings 20 to 21+36 to 38; D. 32 to 37; belly flat or slightly concave; snout moderate........................LOUISIANÆ, 10

   cccc. Dorsal fin covering 5+4 or 4+5 rings; rings 18 to 20+36 to 40; D. 36 to 40; snout moderate ....FUSCUM, 11.

1. Siphostoma zatropis J. & G.

   ?? Corythoichthys albirostris Kaup, Lophobr. p. 25 (Bahia, Mexico).

   Syngnathus albirostris Günther, Cat. Fish. Brit. Mus. VIII, 170, 1870 (Mexico).

   Siphostoma zatropis J. & G. Proc. U. S. Nat. Mus. 1882 (Pensacola).

Head 9 in total length; D. 23; rings 18+30.

Body robust. Snout short, 2⅔ in head; a strong median ridge above on snout, two ridges below with a median groove, and on each side of the groove is a horizontal ridge running to lower part of orbit. Occiput and nuchal plates very sharply carinated; opercle with two horizontal ridges. Belly somewhat concave, little keeled. Dorsal much shorter than head, covering 1+4 rings. Caudal well developed, 1¾ in base of dorsal. Tail longer than rest of body, 1¾ in total length.

Color in spirits light olivaceous, with about twelve irregular brown cross bands, each covering from two to three rings; snout light, with two or three narrow cross-bands below; rest of head dusky.

Habitat.—Atlantic coast of America, Pensacola; Mexico.

Described from the original type, a specimen, 5⅔ inches in length, obtained by Prof. D. S. Jordan from the "Snapper Banks," near Pensacola, Fla.

2. Siphostoma punctipinne (Gill) J. & G.

   Dermatostethus punctipinnis Gill, Proc. Acad. Nat. Sci. Phila. 1862, 283 (San Diego, Cal.).

   Siphostoma punctipinne J. & G. Proc. U. S. Nat. Mus. 1880, 353, (name only); J. & G. Proc. U. S. Nat. Mus. 1881, 69; J. & G. Synopsis Fishes North America, 385, 1882.

Head 8 in total length; D. 41; rings 19+39; length 12 inches.

Body comparatively robust. Snout moderate. Occiput with a raised keel; joint between the occiput and the first dorsal shield more perfect than usual, so that the head can be placed at an angle with the body. Greatest depth about equal to length of post-orbital part of head. Skin on breast and anterior ventral plates thin, showing the striations of the bones. Tail twice as long as trunk. Only the original types are known.

Habitat.—Pacific coast of the United States; San Diego, Cal.

3. Siphostoma californiense (Stórer) J. & G.

Syngnathus californiensis Storer, Proc. Bost. Soc Nat. Hist. ii, 73, 1845 (California); Storer, Synopsis Fishes of North America, 524, 1846 (California); Gill, Proc. Acad. Nat. Sci. Phila. 1862, 283 (California). Duméril Hist. Nat. Priss. ii, 1870, 566.

Siphostoma californiensis Jor. & Gilb. Proceed. U. S. Nat. Mus. 453, 1880; J. & G. Proceed. U. S. Nat. Mus. 1, 69, 1881 (common south of San Francisco); J. & G. Synopsis Fishes North America, 384, 1882 (Pacific coast)

Head $6\frac{1}{6}$ to $8\frac{1}{2}$ in total length; D. 39–46; rings 20–21+47–49.

Trunk robust. Snout very long, $1\frac{1}{2}$ to $1\frac{1}{4}$ in head, with median ridge above and below. Occiput and nuchal plates not carinated in adults. Dorsal shorter than head, covering 1+9 rings. Distance to dorsal $2\frac{3}{4}$ in length. Pectorals as long as high, equaling in length the diameter of eye. Caudal pouch of males covering 21 to 25 rings, its length 3 in total.

Color in life " olivaceous, varying to brownish red, yellowish below; head and body variously marbled and speckled with whitish, the marking posteriorly taking the form of short horizontal grayish streaks, especially distinct on the top of the head; anteriorly often forming narrow bars." (*Jordan.*) This is much larger than the other American species, reaching a length of $18\frac{1}{2}$ inches. Described here from specimens taken at Santa Barbara and Monterey.

TABLE.

	Rings	D rays	Snout in head.
1	20 + 49	43	$1\frac{1}{2}$
2	21 + 47	45	$1\frac{1}{4}$
3	20 + 47	43	$1\frac{1}{5}$
4	21 + 49	46	$1\frac{1}{3}$
5	21 + 49	43	$1\frac{1}{3}$
6	20 + 47	43	$1\frac{1}{3}$
7	20 + 45	39	$1\frac{1}{3}$
8	21 + 49	43	$1\frac{1}{4}$
9	21 + 48	46	$1\frac{1}{4}$

The length of the snout is of but little value for specific distinction. Professor Jordan found specimens of *S. californiense*, at Santa Barbara and Monterey, with the snout no longer than the rest of the head.

As is usual in this group the females differ from the males, in a more robust trunk, in a longer snout, and in a greater keel on belly. These differences are not very constant.

Habitat.—Pacific coast of the United States; common south of San Francisco.

4. Siphostoma griseolineatum (Ayres) J. & G.

Syngnathus griseolineatus Ayres, Proc. Cal. Acad. Nat. Sci. 14, 1854 (San Francisco Bay); Gill, Proc. Acad. Nat. Sci. Phila. 1862, 284 (San Francisco, Tomales Bay, Fort Umpqua); Gunther, Cat. Fish. viii, 160, 1870 (Vancouver's Island, California).

Siphostoma griseolineatum J. & G. Proc. U. S. Nat. Mus. 69, 1881 (San Francisco to Puget Sound); J. & G. Syn. Fish. South America, 384, 1882).

Syngnathus abboti Girard, U. S. Pac. R. R. Surv. Fish. 346, 1858 (San Francisco). Duméril l. c. 567.

Syngnathus californiensis Girard, U. S. Pac. R. R. Surv. Fish. 344, 1858 (Tomales Bay, San Francisco, Monterey). (Not of Storer.)

D. 36 to 41; rings 18 to 19+39 to 42.

S. griseolineatum is closely allied to *californiense*, but it differs in a somewhat shorter snout, in the number of D. rays, in the number of rings, in its size, in the snout being slightly more keeled, and in the dorsal covering 0–1+9 rings.

TABLE.

	Rings.	D rays	Snout in head	Length
				Inches.
1	18+42	41	1 5/7	6
2	18+39	37	1¾	6¾
3	19+42	37	1 4/9	6½
4	19+42	37	1⅔	5½
5	18+42	37	1⅜	11
6	18+42	36	1¾	10

Habitat.—Pacific coast of the United States; Puget Sound, Fort Umpqua, Tomales, San Francisco, Monterey.

5. Siphostoma auliscus sp. nov.

Siphostoma dimidiatum J. & G. Proc. U. S. Nat. Mus. 453, 1880 (Santa Barbara, San Diego) (*not Syngnathus dimidiatus Gill*); J. & G. Synopsis Fishes North America (coast of California, chiefly south of Point Concepcion); Rosa Smith, San Diego Free Press, Nov. 5, 1880.

Head 9–9¼ in total length; D. 29–30; rings 15+37–38.

Trunk rather slender. Snout 2 in head, median ridge above distinct, below comparatively broad and blunt. Occiput and nuchal plates sharply carinated; belly weakly keeled. Opercle slightly keeled, very convex, making the head slightly broader than deep. Dorsal little longer than head, covering 1+7 rings. Pectorals scarcely higher than long, slightly exceeding diameter of eye. Tail longer than rest of body, 1⅝ in total length. Caudal pouch covering 21 rings. Color in spirits somewhat lighter than *S. californiense*, scarcely mottled or marbled.

TABLE

	Rings	D rays	Snout in head	Head in body	Length
					Inches.
1	15+38	30	2	9¼	6
2	15+37	29	2	9	4½

Habitat.—Pacific coast of the United States; San Diego, Santa Barbara.

6. Siphostoma bairdianum (Duméril) Swain. (31253.)

> ? *Syngnathus bairdianus* Duméril, Hist. Nat. Poiss. II, 574, 1870 (coast of Mexico, near California).

Body unusually stout, with short head, short snout, and short tail, the general appearance being much like *S. californiense,* but all the parts contracted. Snout short, compressed, just as long as the rest of the head (♀), its upper edge with a sharp, low keel, which is higher than in *S. californiense.* Top of head without keel. Opercle striate, with trace of a keel at base.

Keels of body not very sharp, the interspaces between the angles scarcely concave. Lateral line not continuous with the upper edge of the tail. Belly with a slight median keel. Dorsal fin low.

Rings 18+31. Dorsal rays 34, the fin inserted on 1+7 rings. Ten rings on the tail behind the caudal pouch. Head $7\frac{2}{3}$ in length; base of dorsal equal to head. Distance from snout to dorsal $2\frac{1}{4}$ in length; caudal pouch $2\frac{2}{3}$ in length of body; tail behind caudal pouch 6 times.

Color blackish, with fine pale vermiculations; top of head and neck with wavy longitudinal streaks; caudal dusky; dorsal somewhat mottled; a dusky blotch before eye.

A single male example, 9 inches long, was obtained by Mr. Andrea Larco at Santa Barbara, and is now in the National Museum. The caudal pouch in this specimen is full of eggs.

This species seems most nearly allied to *S. californiense,* differing in the stouter form, much shorter tail and snout, and in the smaller number of rings and of dorsal rays.

It agrees fairly with M. Duméril's account of *Syngnathus bairdianus,* the only discrepancy of importance being the statement that the dorsal covers 3+6 rings. The numbers of rings (17+31) and of dorsal rays (30), as given by M. Duméril, agree very closely with the specimen from Santa Barbara.

Habitat.—Coast of Southern California; Santa Barbara; Lower California.

7. Siphostoma leptorhynchum (Girard) J. & G.

> *Syngnathus leptorhynchus* Girard, Proc. Acad. Nat. Sci. Phila. VII, 156, 1854; Girard, U. S. Pac. R. R. Surv. Fish. 345, 1858 (San Diego); Gill, Proc. Acad. Nat. Sci. Phila. 1862, 284 (San Diego). Duméril l. c. 567.
> *Siphostoma leptorhynchus* J. & G. Proc. U. S. Nat. Mus. 23 and 453, 1880 (San Diego); Rosa Smith. San Diego Free Press, Nov. 5, 1880 (San Diego); J. & G. Proc. U. S. Nat. Mus. 1–69, 1881 (Santa Barbara to San Diego). J. & G. Synopsis Fishes North America, 384.
> *Syngnathus brevirostris* Girard, Proc. Acad. Nat. Sci. Phila. VII, 156, 1854; Girard, U. S. Pac. R. R. Surv. Fish. 345, 1858 (San Diego).
> *Syngnathus arundinaceus* Girard, U. S. Pac. R. R. Surv. Fish. 346, 1858 (coast of California); Gill, Proc. Acad. Nat. Sci. Phila. 1862, 284; J. & G. Proc. U. S. Nat. Mus. 23, 1880 (name only). Duméril l. c. 567.
> *Syngnathus dimidiatus* Gill, Proc. Acad. Nat. Sci. Phila. 1862, 284 (San Diego); Günther, Cat. Fish Brit. Mus. 165, 1870. Duméril l. c. 566.

Head $6\frac{1}{4}$ to $8\frac{1}{4}$ in total length; D. 30 to 32; rings 17 to 19+36 to 41.

Snout $1\frac{3}{5}$ to 2 in head; median line of snout above carinate; occiput and nuchal plates weakly keeled in young, the keels apparently disappearing in adults. Angle of belly less acute than in *S. californiense;* the keel sometimes wanting. Dorsal fin shorter than head, covering 1+7 rings; caudal pouch covering about 19 rings. Otherwise essentially as in *S. californiense.*

TABLE

	Rings	D rays	Snout in head	Head in length.	Length
					Inches.
1	18+40	30	$1\frac{5}{6}$	$6\frac{1}{4}$	5
2	19+38	31	2	$7\frac{1}{2}$	$6\frac{1}{4}$
3	18+38	31	$1\frac{4}{5}$	$7\frac{3}{5}$	8
4	18+39	32	$1\frac{1}{2}$	$7\frac{1}{5}$	$5\frac{3}{4}$
5	17+38	20	$1\frac{1}{4}$	$7\frac{3}{5}$	$5\frac{1}{4}$
6	18+41	31	2	$8\frac{1}{4}$	$4\frac{3}{4}$
7	18+37	30	$1\frac{3}{4}$	$7\frac{1}{4}$	5
8	17+36	..	$1\frac{3}{4}$	$7\frac{1}{4}$	$3\frac{1}{4}$

Habitat.—Pacific coast of the United States, San Diego, Santa Barbara.

3. Siphostoma floridæ J. & G.

> *? Syngnathus louisianæ* Goode & Bean, Proc. U. S. Nat. Mus. 333, 1879 (San Marco Island). (Not. of Günther.)
>
> *Siphostoma floridæ* J. & G. Proc. U. S. Nat. Mus. 1882 (Pensacola; Beaufort).

Head 6 to $6\frac{1}{2}$ in total length; D. 27; rings 17 to 18+31 to 32.

Snout rather short, about $1\frac{3}{4}$ in head; median line well keeled above and below, the ridge on both sides of median ridges above and below not so conspicuous. Occiput and opercle little keeled. Dorsal shorter than head, covering 1+6 to 7 rings, its height 5 times in its base. Caudal fin $2\frac{1}{2}$ in base of dorsal. Pectoral slightly higher than length of its base. Tail longer than trunk, $1\frac{5}{8}$ in total length, caudal pouch covering about 18 rings.

"Color in life, dark green; tail with faint darker bars, broader than the interspaces. Sides of tail, especially mesially, with many rough and oblong pale spots. Snout mottled, especially on side. Lower part of opercle nearly plain. Dorsal translucent, yellowish at base. Caudal yellow, dusky at tip. Anal plain." (*Jordan.*)

Here described from specimens from Beaufort, N. C., and from Pensacola, Fla.

TABLE

	Rings	D rays	Snout in head	Head in length	Length
					Inches.
1	18+32	27	$1\frac{5}{6}$	$6\frac{1}{4}$	6
2	17+31	27	$1\frac{3}{4}$	6	6
3	17+32	27	$1\frac{3}{4}$	6	6
4	17+33	27	$1\frac{3}{4}$	$6\frac{1}{4}$	6
5	17+33	27	$1\frac{3}{4}$	$6\frac{3}{4}$	$6\frac{3}{4}$
6	17+32	28	$1\frac{3}{4}$	$6\frac{1}{4}$	$6\frac{1}{4}$

Habitat.—South Atlantic and Gulf coasts of the United States; Beaufort, N. C.; San Marco Island, Fla.; Pensacola, Fla.

9. Siphostoma affine (Günther) J. & G.

Syngnathus affinis Günther, Cat. Fishes Brit. Mus. VIII, 163, 1870 (Louisiana.)

Siphonostoma sp. Jordan, Proc. U. S. Nat. Mus. 22, 1880 (Saint John's River, Fla.).

Siphostoma affinis J. & G. Synopsis Fishes North America, 383, 1882 (Saint John's River, Fla.); J. & G. Proc. U. S. Nat. Mus. 1882 (Pensacola, Fla.).

Trunk robust, very deep; width of adult females 2 in depth. Snout short, 2 to $2\frac{1}{4}$ in head; median ridge well defined above and below; a less conspicuous ridge above on each side of median line, from end of snout to nostrils, thence running over interorbital and temples. Occiput, nuchal plates, and opercle keeled. Belly, in females, sharply carinated. Fins well developed. Height of dorsal, $3\frac{1}{2}$ in its length; base of dorsal slightly longer than head, covering $3+4\frac{1}{2}$–5 rings. Caudal, $2\frac{1}{2}$ in base of dorsal.

Color in life, "deep olive green, varying to brown or blackish, or slightly reddish, according to surroundings; females with a black keel on the belly, which is obsolete in the male. Dark color of back forming about 15 dark cross-bars, very faint and much wider than the interspaces. Sides of head mottled, especially on lower half of opercles. Snout dark above, abruptly paler below. Dorsal high, having the dark color of the body with dark oblique shades, the paler color appearing like faint spots; vertical striæ on body plates, shining silvery, very distinct and bright in life. Caudal and anal colored like the dorsal, the latter conspicuous." (*Jordan.*)

Described from specimens taken at Pensacola, Fla.

TABLE

	Rings	D rays	Snout in head	Head in length	Length.
					Inches.
1	16+31	31	$2\frac{1}{5}$	8	$4\frac{1}{2}$
2	16+30	30	$2\frac{1}{5}$	8	4
3	16+30	31	$2\frac{1}{2}$	$7\frac{3}{4}$	5
4	16+31	30	2	8	$3\frac{3}{4}$
5	16+32	29	$2\frac{1}{3}$	8	$3\frac{1}{2}$
6	16+32	30	$2\frac{1}{4}$	$8\frac{1}{2}$	$4\frac{1}{4}$
7	16+32	28	2	8	$4\frac{1}{4}$
8	16+33	30	$2\frac{1}{3}$	8	$4\frac{3}{4}$
9	16+32	32	$2\frac{1}{6}$	$8\frac{1}{2}$	5

Habitat.—South Atlantic and Gulf coasts of the United States; Saint John's River, Fla.; Pensacola, Fla.; Louisiana.

10. Siphostoma louisianæ (Günther) J. & G.

Syngnathus louisianæ Günther, Cat. Fish. viii, 160, 1870 (New Orleans).

Siphonostoma louisianæ Jordan, Proc. U. S. Nat. Mus. 1880, 22 (Beaufort, N. C.).

Siphostoma louisianæ J. & G. Synopsis Fishes North America, 383, 1882 (Beaufort, N. C.).

Syngnathus fuscus Duméril, Hist. Nat. Poiss. ii, 574 (Savannah).

Head 7 to $7\frac{2}{3}$ in total length; D. 32 to 37; rings 20 to 21+36 to 38.

Trunk broader below. Snout moderate, about $1\frac{3}{4}$ in head; median ridge above and below, a ridge on each side of median ridge above and below. Occiput, nuchal plates, and opercle somewhat keeled. Belly flat or slightly concave, with a median ridge. Dorsal fin well developed, shorter than head, covering 3+5 rings. Caudal longer than pectoral, $2\frac{1}{2}$ in base of dorsal. Tail longer than trunk, $1\frac{7}{9}$ in total length.

Color in spirits brownish, lighter on lower part of trunk and below; brown of the side extends in a band through eye to middle of snout. Here described from specimens from Beaufort, N. C.

TABLE

	Rings	D. rays	Snout in head	Head in length.	Length.
					Inches.
1	20+38	32	$1\frac{3}{4}$	$7\frac{1}{4}$	8
2	20+36	33	$1\frac{3}{4}$	7	8
3	21+38	33	$1\frac{1}{2}$	$7\frac{1}{2}$	5
4	21+38	..	$1\frac{1}{2}$	$7\frac{3}{4}$	7
5	20+38	35	$1\frac{2}{3}$	$7\frac{1}{2}$	7

Habitat.—Atlantic coast of the United States; Beaufort, N. C.; Savannah, Ga.; New Orleans, La.

11. Siphostoma fuscum (Storer) J. & G.

"*Syngnathus typhle* Mitch. Trans. Lit. & Phil. i, 475, 1815." (Not of L.)

Syngnathus fuscus Storer, Report Fish. Mass. 162, 1839; De Kay, New York Fauna, 321, 1842 (coast of Mass.).

Siphonostoma fuscum J. & G. Proc. U. S. Nat. Mus. 1880, 22 (Wood's Holl, Mass.).

Siphostoma fuscum J. & G. Synopsis Fishes North America, 383, 1882 (Atlantic coast, northward).

Syngnathus peckianus Storer, Report Fishes Mass. 163, 1839; De Kay, New York Fauna, 321, 1842 (coast of Mass.); Storer, Synopsis Fishes North America, 490, 1846 (Mass. Conn. New York); Gill, Can. Nat. Aug. 1865, 21 (Bay of Fundy); Günther, Cat. Fishes Brit. Mus. 157, 1870 (Atlantic coast U. S.); Uhler & Lugger, Report Fishes Maryland, 76, 1876 (St. Mary's River).

Syngnathus fasciatus De Kay, New York Fauna, 319, 1842, pl. 54, fig. 176 (New York).

Syngnathus viridescens De Kay, New York Fauna, 321, 1842, pl. 54, fig. 176 (New York); Duméril, Hist. Nat. Poiss. ii, 570 (Cape Cod).

Syngnathus dekayi Duméril, Hist. Nat. Poiss. ii, 569, 1870, (after *S. fasciatus* Dek.).

Syngnathus milbertianus Duméril, Hist. Nat. Poiss. ii, 570, New York.

Head $7\frac{1}{2}$ to 9 in total length; D. 36 to 40; rings 18 to 20 + 36 to 40.

Snout short, about 2 in head; median line above and below well keeled, the ridge on each side of median ridges rather conspicuous. Occiput, nuchal plates, and opercle carinate, belly somewhat convex, scarcely keeled. Dorsal longer than head, covering 4–5+5–4 rings, its height 5–6 in length of its base. Tail much longer than trunk, $1\frac{3}{4}$ in total length.

Color in spirits, olivaceous or brownish, lighter below, especially on belly, lower half of opercles, and snout; sides mottled and blotched much as in other species.

TABLE

Specimens from Wood's Holl, Mass.

	Rings	D rays	Snout in head	Head in length.	Rings covered by dorsal	Length.
						Inches.
1	18+36	36	2	8¾	4 +5	5
2	19+38	40	2	7¼	5 +5	4½
3	18+36	36	2	7¾	4 +5	4¼
4	19+40	39	2	9	4½+5	4½
5	19+38	38	2	8½	5 +4	7¼
6	20+3	35	1⅞	8¾	5 +4	7¼
7	19+37	(?)	2⅛	7½	5 +5	3¾

Catalogue of nominal species, with identifications.

Nominal species	Date	Identification
Synngathus fuscum Storer	1839	Siphostoma fuscum
Syngnathus peckianus Storer	1839	Siphostoma fuscum
Syngnathus fasciatus De Kay	1842	Siphostoma fuscum
Syngnathus viridescens De Kay	1842	Siphostoma fuscum
Syngnathus californiensis Storer	1845	Siphostoma californiense
Syngnathus griseolineatus Ayres	1854	Siphostoma griseolineatum
Syngnathus leptorhynchus Girard	1854	Siphostoma leptorhynchum.
Syngnathus brevirostris Girard	1854	Siphostoma leptorhynchum
Syngnathus abboti Girard	1858	Siphostoma griseolineatum
Syngnathus arundinaceus Girard	1858	Siphostoma leptorhynchum.
Dermatostethus punctipinnis Gill	1862	Siphostoma punctipinne.
Syngnathus dimidiatus Gill	1862	Siphostoma leptorhynchum.
Syngnathus affinis Gunther	1870	Siphostoma affine.
Syngnathus louisianæ Gunther	1870	Siphostoma louisianæ
Syngnathus dekayi Duméril	1870	Siphostoma fuscum.
Syngnathus milbertianus Duméril	1870	Siphostoma fuscum
Syngnathus bairdianus Duméril	1870	Siphostoma bairdianus
Siphostoma zatropis Jor. & Gilb.	1882	Siphostoma zatropis
Siphostoma floridæ J. & G.	1882	Siphostoma floridæ
Siphostoma auliscus Swains.	1882	Siphostoma auliscus

Habitat.—Atlantic coast of the United States, Cape Cod to Virginia; Wood's Holl, Massachusetts; Connecticut; New York; Saint Mary's River, Maryland.

INDIANA UNIVERSITY, *May* 16, 1882.

NOTICE OF RECENT ADDITIONS TO THE MARINE INVERTEBRATA OF THE NORTHEASTERN COAST OF AMERICA, WITH DESCRIPTIONS OF NEW GENERA AND SPECIES AND CRITICAL REMARKS ON OTHERS.

PART IV.—ADDITIONS TO THE DEEP-WATER MOLUSCA, TAKEN OFF MARTHA'S VINEYARD, IN 1880 AND 1881.

By A. E. VERRILL.

The following article contains the species of Mollusca that have been added to our deep-water fauna since the publication of my former article on the same subject in these Proceedings (vol. iii, p. 356). This is intended as a supplement to that article, and I have, therefore, introduced here a number of the species previously recorded, of which the names have been changed, or which, on more careful study, have proved to be distinct from the European species with which they were, at first, identified. The names of such species are printed in italic type to distinguish them from species now recorded for the first time, which are in black-faced type. I have not given any general summary, because it is expected that dredging will be again carried on in the same region by the United States Fish Commission during the present season.

CEPHALOPODA.

Full descriptions and figures of all our Cephalopods may be found in the Transactions of the Connecticut Academy, vol. v, pp. 177–446, 1880–'81, and in the Report of the U. S. Commission of Fish and Fisheries for 1879, pp. [1–244], pl. i–xlvi, 1882.

DECACERA.

Lestoteuthis Fabricii (Licht.) Verrill.
> *Gonatus Fabricii* Steenstrup, Verrill, Trans. Conn. Acad., v, p. 291.
> *Lestoteuthis Fabricii* Verrill, Trans. Conn. Acad., v, p. 390, pl. 45, figs. 1–2*d*,
> pl. 49, figs. 1–1*f*, pl. 55, figs. 1–1*d*, 1881.
> Verrill, Report on the Cephalopods of the Northeastern Coast of America,
> in Rep. U. S. Com. of Fish and Fisheries for 1879 [p. 206], pl. 15, figs. 1–1*c*,
> 2–2*d*, 3–3*f*, pl. 45, figs. 1–1*d*, 1882.

Station 953; 715 fathoms; one rather large and perfect male specimen. Station 1031; 255 fathoms; one young specimen.

Chiroteuthis lacertosa Verrill.
> *Chiroteuthis lacertosa* Verrill, Trans. Conn. Acad., v, p. 408, pl. 56, figs. 1–1*f*,
> 1881; Rep. on Cephalop. [p. 209], pl. 46, figs. 1–1*f*, 1882.

Off Delaware Bay, station 1048, in 435 fathoms, 1881,—Lieut. Z. L. Tanner.

Brachioteuthis Beanii Verrill.
> *Brachioteuthis Beanii* Verrill, Trans. Conn. Acad., v, p. 406, pl. 55, figs. 3–3*b*,
> pl. 56, figs. 2–2*a*, 1881; Rep. on Cephalop. [p. 214], pl. 45, figs. 3–3*b*, pl. 46,
> figs. 2–2*a*, 1882.

Stations 1031 and 1033, in 255 and 183 fathoms, 1881.

Histioteuthis Collinsii Verrill.
> *Histioteuthis Collinsii* Verrill, Amer. Journ. Sci., xvii, p. 241, 1879; Trans.
> Conn. Acad., v, p. 234, pl. 22, pl. 27, figs. 3–5, pl. 37, fig. 5, 1880; Rep.
> on Cephalop. [pp. 121, 216], pl. 23, pl. 24, figs. 3–6.

Station 895; 372 fathoms. Jaws only.

Desmoteuthis tenera Verrill.
> *Desmoteuthis tenera* Verrill, Trans. Conn. Acad., v, p. 412, pl. 55, figs. 2–2*d*,
> pl. 56, fig. 3, 1881; Rep. on Cephalop. [p. 216], pl. 45. figs. 2–2*d*, pl. 46, fig. 3.

Station 952; 388 fathoms. Two specimens.

Stoloteuthis leucoptera Verrill.—Butterfly Squid.

> *Sepiola leucoptera* Verrill, Amer. Journ. Sci., vol. xvi, p. 378, 1878, vol. xix,
>   p. 291, pl. 15, figs. 4 and 5, April, 1880; Trans. Conn. Acad., v, p. 347,
>   pl. 31, figs. 4 and 5, pl. 54, fig. 4, June, 1881.
> *Stoloteuthis leucoptera* Verrill, Trans. Conn. Acad., v, p. 418, Oct., 1881: Rep.
>   on Cephalop. [p. 165], pl. 36, figs. 1, 1*a*, 2, 1882.

Stations 947, 952, 998, 999, 1026 (3 juv.); 182–388 fathoms.

OCTOPODA Leach.

Alloposus mollis Verrill.

> *Alloposus mollis* Verrill. 1880; Trans. Conn. Acad., v, p. 366, pl. 50, figs. 1, 1*a*,
>   2, 2*a*, pl. 51, figs. 3, 4, 1881; Rep. on Cephalop. [p. 181], pl. 39, figs. 1,
>   1*a*, 2, 2*a*; pl. 42, fig. 7; pl. 44, fig. 1, 1882.

This has occurred in 197 to 715 fathoms.

Two remarkably large female specimens of this species were taken in 1881, each weighing about 20 pounds. These occurred at stations 937 and 994, in 506 and 368 fathoms. The length was 812^{mm} (32 inches) to the tips of the arms. It was taken by Captain Tanner off Chesapeake Bay and off Delaware Bay, in 300 and in 197 fathoms.

Octopus lentus Verrill.

> Trans. Conn. Acad., v, p. 375, pl. 35, figs. 1, 2, ♀, pl. 51, fig. 2 ♂, 1881; Rep.
>   on Cephalop. [p. 191], pl. 43, figs. 1, 2, ♀, pl. 44, fig. 2, ♂, 1882.

Off the Carolina coasts, 464 to 603 fathoms, Blake Exp.,—A. Agassiz, 1880.

Eledone verrucosa Verrill.

> *Eledone verrucosa* Verrill, Bull. Mus. Comp. Zool., viii, p. 105, pl. 5, 6, 1881;
>   Trans. Conn. Acad., v, p. 380, pl. 52, 53, 1881; Rep. on Cephalop. [p. 183],
>   pl. 44, figs. 3, 3*a*, 1882.

South of George's Bank, 810 fathoms; off Nantucket, 466 fathoms, Blake Exp.,—A. Agassiz, 1880.

GASTROPODA.

RHACHIGLOSSA.

Marginella carnea Storer (?).

> *Marginella carnea* Storer, Journ. Boston Soc. Nat. Hist., i, p. 465, pl. 9, figs.
>   3, 4, 1837.
> *Marginella roscida?* Verrill, Amer. Journ. Sci., xx, p. 391, Nov , 1880: Proc.
>   U. S. Nat. Mus., iii, p. 369, 1880.

Our shell has a somewhat higher and more acute spire than the one figured by Storer, and the callus does not reach its summit. There are four prominent folds on the columella, the two anterior ones very oblique. The color is not preserved.

A single dead specimen was taken off Martha's Vineyard, at station 865, in 65 fathoms, 1880. Another specimen, also dead, but more perfect, was taken, in 1881, at station 949, in 100 fathoms. Key West, Florida,—Storer.

Buccinum Sandersoni Verrill.

Trans. Conn. Acad., v, p. 490, pl. 58, fig. 9 (nucleus), June, 1882.

Shell elongated, brownish, translucent, rather thin and delicate, with a high spire; well impressed suture; strongly convex, obliquely ribbed and strongly, spirally sculptured whorls; a large, smooth, mammillary nucleus; a small aperture; and a short, nearly straight columella.

Whorls, in our largest example, seven, a little flattened below the suture, strongly convex in the middle; the penultimate whorl with about 13 broadly convex, curved ribs or undulations, strongly excurved at the middle of the whorl; on the body-whorl the ribs are less prominent and fade out below the middle; on the three upper whorls they are absent. The spiral sculpture, on the lower whorls, consists of prominent, narrow, rounded cinguli, unequal in size and separated by narrow grooves; usually there are three or four smaller and lower cinguli between two of the larger ones, and sometimes a narrow groove appears on the larger ridges, dividing them into two; on the anterior part of the body-whorl the cinguli become more uniform in size and more numerous. The whole surface is covered with fine distinct lines of growth, which decussate the cinguli and mostly cross the ribs somewhat obliquely.

The nucleus is rounded and remarkably large for the genus (2^{mm} in diameter), translucent glossy brown, nearly smooth for about one turn and a half; the apex is regular and not obliquely raised.

The aperture is unusually small and short, elliptical, a little contracted posteriorly; outer lip thin, well rounded, the edge receding in a broad curve below the suture; canal short and narrow; columella rather straight, thin, with the folds slightly developed, the anterior end thin, rounded, and projecting quite as far as the lip; the upper part of the columella-lip is not excavated, nor distinctly excurved. The operculum is small, pale yellow, rounded-elliptical, with the nucleus at about the middle of the length, and a little to one side of the center. Epidermis thin and smooth. Color of the shell, with epidermis, yellowish brown to dark reddish brown, sometimes with small whitish spots on the larger spiral ridges; columella whitish, inside of aperture pale orange-brown or light amber.

Our largest example (female) is 46^{mm} long; breadth, 21^{mm}; length of body whorl, 29.5^{mm}; length of aperture, 21.5^{mm}; its breadth (lip broken), 12^{mm}; length of operculum, 11.5^{mm}; its breadth, 9^{mm}. A male has very nearly the same proportions.

Off Martha's Vineyard, station 939, in 258 fathoms; station 1032, in 208 fathoms, 1881, two living examples, male and female.

This species resembles some of the varieties of *B. undatum*, but besides its more slender and elongated form and more delicate texture, it differs decidedly in the character of the spiral sculpture, the shortness and small size of the aperture, and in the operculum; but the most striking differences are in the nucleus and upper whorls, for the nucleus is more than twice as large as that of *B. undatum*, and different in character; while on the second and third whorls the spiral cinguli are fewer and very much more prominent and coarser. The character of the nucleus and upper whorl will also distinguish it from all the other species of our coast.

I have named this interesting shell in honor of Mr. Sanderson Smith, of the U. S. Fish Commission parties during these explorations.

Sipho pubescens Verrill.

> *Neptunea propinqua* Verrill, Amer. Journ. Sci., xvi, p. 210, 1878
>
> *Neptunea (Sipho) propinqua* Verrill, Amer. Journ. Sci., xx, p. 391, Nov., 1880; Verrill, Proc. U. S. Nat. Mus., iii, p. 370, 1880 (*non* Alder, Jeffreys, etc.).
>
> *Sipho pubescens* Verrill, Tr. Conn. Acad., v, p. 501, pl 43, fig. 6, pl. 57, fig. 25, June, 1882.

Shell rather stout, fusiform, regularly tapered, obtuse at the tip of the spire, with the suture deep and canaliculate. Whorls about seven, broadly rounded and somewhat flattened, narrowly but distinctly channeled at the suture.

Sculpture over the whole surface, regular and numerous shallow, spiral grooves, or sulci, separated by slightly raised, flat, or somewhat rounded cinguli, usually but not constantly wider than the sulci; on the penultimate whorl there are about 14 to 16 of the sulci; slight but distinct curved lines of growth cover the surface. Aperture narrow ovate-elliptical; outer lip broadly and regularly rounded, the edge receding in the middle in a broad, concave curve; at the base of the canal the lip is decidedly incurved. Canal moderately long, somewhat contracted, spirally curved to the left and strongly bent backward at the tip. Columella very much bent, with a strong sigmoid curvature; portion opposite the middle of the aperture greatly receding. Epidermis thin, but firm, yellowish green to olive-green; when fresh and uninjured covered with fine, short, capillary processes, forming spiral lines along the cinguli.

Color of the shell white; inside of aperture translucent bluish white.

The nucleus is moderately large (diameter 2.15^{mm}), smooth, mammillary; its first whorl is strongly turned up obliquely, and incurved.

The median tooth of the radula is broad, with three denticles, the middle one largest; the lateral teeth are large, with three sharp curved denticles, the outer one much the largest, the middle one smallest; occasionally the inner one bears a small secondary denticle on its outer edge.

Operculum long, ear-shaped, with the nucleus at the tip of the small end, which is but little incurved; inner edge strongly convex beyond the middle; outer edge broadly rounded. A female of the ordinary adult size and form is 65^{mm} long; breadth, 28^{mm}; length of canal and body-whorl, 46^{mm}; breadth of body-whorl, 25^{mm}; length of aperture, 35^{mm}; its breadth, 14^{mm}; breadth of opening of canal at base, 5^{mm}.

An average male is 56^{mm} long; breadth, 26^{mm}; length of body-whorl, 40^{mm}; its breath, 17^{mm}; length of aperture, 31^{mm}; its breadth, 12^{mm}.

This species was first dredged by us, in 1877, on the United States Fish Commission steamer Speedwell, off Cape Sable, Nova Scotia, in 88 to 91 fathoms, fine compact sand, where it occurred in considerable numbers, living; and off Halifax, 42 fathoms, dead.

Off Martha's Vineyard this species is very common in deep water. It occurred at 48 stations in that region in 1880 and 1881; living specimens were taken in 86 to 410 fathoms, but it is most abundant between 200 and 410 fathoms; at station 998, in 302 fathoms, 154 specimens were taken, 140 of them living. Dead shells, inhabited by *Eupaguri*, occurred in 64 to 85 fathoms, and also in 458 fathoms. It was taken by Lieut. Z. L. Tanner, on the Fish Hawk, in 1880, off Chesapeake Bay, in 56 to 300 fathoms; and off Delaware Bay, in 156 and 435 fathoms, in 1881.

This shell is closely allied to *S. propinquus* (Alder) of Europe, to which I formerly referred it, with doubt. Our species is, however, a larger, more robust, and more hairy shell, and its nuclear whorls are totally different, for according to the descriptions, *S. propinquus* always has a regularly spiral nucleus, with the first whorl minute and not turned up; this is, also, the case with an authentic specimen, in my possession, received from the Rev. A. M. Norman.

Sipho Stimpsoni, var. *liratulus* Verrill.

> *Neptunea (Sipho) arata* Verrill, Proc. Nat. Mus., III, p. 370, 1880.

Specimens intermediate between this variety and the ordinary, nearly smooth, shallow-water form have been obtained. The name, *aratus*, having been used in this group, I propose to name the strongly spirally sculptured variety, *liratulus*.

Sipho glyptus Verrill.

> *Tritonofusus latericeus* Verrill, Amer. Journ. Sci.: xx, p. 391, Nov., 1880: Verrill, Proc. U. S. Nat. Mus., III, p. 369, 1880 (*non* Möll., Mörch).
>
> *Sipho glyptus* Verrill, Trans. Conn. Acad., v, p. 505, pl. 57, fig. 22, pl. 58, figs. 1, 1*a*, June, 1882.

Shell long-fusiform, with a high, tapering, acute spire; with an impressed, oblique, undulated suture; with convex, transversely ribbed and spirally grooved whorls; and with a narrow, rather long, nearly straight canal.

Whorls seven to eight, evenly rounded, crossed by about 13 slightly curved, regular, rounded and prominent ribs, separated by rather wider, regularly concave interspaces; the ribs are lower and a little excurved just below the suture, and fade out before reaching the base of the canal; sometimes they are mostly obsolete on the body-whorl. The raised spiral cinguli are numerous, regular and close, crossing equally the ribs and interspaces; they are mostly alternately larger and smaller, and are separated by narrow impressed grooves; the cinguli are crossed by very fine, close and delicate raised lines of growth, giving them a minutely wavy appearance. Aperture narrow-elliptical; outer lip evenly convex, incurved at the base of the canal, which is narrow and elongated, and but slightly bent to the left and a very little bent back at the tip; columella slightly sigmoid.

The nucleus is small, consisting of two whorls; the first whorl is minute and turned obliquely upward and inward, with a smooth glossy surface, crossed by a few small transverse grooves; the next whorl is regular, smooth at first, then with fine spiral lines; the normal sculpture begins on the third whorl. Color of shell, grayish white. No obvious epidermis.

The largest specimen is 30^{mn} long; breadth, 10.5^{mm}; length of body-whorl, 19^{mm}; its breadth, 9^{mm}; length of aperture, 15^{mm}; its breadth, 4.5^{mm}.

This species was dredged off Martha's Vineyard, by the United States Fish Commission steamer, Fish Hawk, in 1880 and 1881 (stations 894, 938, 951, 1028, 1029, 1032), in 219 to 458 fathoms.

This shell has a sculpture much like that of *S. cœlatus* V., 1880, but it has a longer and more acute spire, a longer canal, narrower aperture, and a different nucleus. In general appearance it resembles *S. latericeus*, but it is a more delicately sculptured shell, with a different nucleus. It also somewhat resembles *S. pellucidus* (Hancock) in general appearance, but the latter has a much shorter and wider canal.

According to the nature of the nucleus this shell would belong to the subgenus, *Siphonorbis* Mörch.

Sipho parvus Verrill and Smith.

> *Sipho parvus* Verrill and Smith, in Verrill, Trans. Conn. Acad., v, p. 504, pl. 57, figs. 20, 20*b*, June, 1882.

Shell small, thin, delicate, translucent, subfusiform, with a rather slender, acute spire; a short, straight canal; and few raised, revolving cinguli.

Whorls six, convex, usually with three (rarely five or six) prominent rounded cinguli or carinæ, separated by much wider, broadly concave interspaces; the uppermost one is usually some distance below the suture, which is impressed; on the last whorl there are about seven to nine principal carinæ, occasionally with a smaller one interpolated, and becoming more crowded anteriorly; delicate and close, raised lines of growth cover the interspaces and cross the raised cinguli.

The nucleus is very small, smooth and glossy; the first turn is minute and regularly spiral, not upturned; three spiral cinguli appear on the second whorl. Aperture elliptical; outer lip thin, rounded, incurved at the base of the canal, which is narrow, but very short and straight; columella nearly straight in the middle. The epidermis is thin, lamellose, but not ciliated.

Color yellowish or grayish white. Operculum ovate, with the smaller or left end rounded and incurved, forming a small lobe, defined by a notch, and with the nucleus central to this small lobe.

The radula is very slender; the outlines of the median plates are indistinct; they bear three very small, but distinct and nearly equal, denticles; the lateral teeth have only two denticles.

Length, 11^{mm}; breadth, 5^{mm}; length of body-whorl, 7.10^{mm}; length of aperture, 5^{mm}; its breadth, 2.15^{mm}.

Off Martha's Vineyard, in 312 to 506 fathoms (stations 937, 947, 994, 997, 1029), 1881, fourteen specimens.

This delicate species is liable to be confounded with the young of *S. pygmæus*, but it differs decidedly in its dentition, operculum, nuclear whorls, short and straight canal, and in the character of its spiral cinguli. The upper whorls of *S. pygmæus* are much more angular, with coarser and more prominent carinæ or cinguli, which are separated by narrower incised grooves.[*]

This species, by its regular spiral nucleus, would be referable to the group *Siphonorbis*. It also approaches *Mohnia* Friele, by the characters of its dentition and operculum.

Trophon clathratus (Linné) Möller.

Off Chatham, Mass.; stations 972, 976, in 16 fathoms.

Astyris diaphana Verrill.

> *Astyris rosacea* Verrill, Proc. Nat. Mus., iii, p. 408 (*non* Gould).
> *Astyris diaphana* Verrill, Trans. Conn. Acad., v, p. 513, pl. 58, fig. 2, June, 1881.

Shell thin, delicate, translucent, white, nearly smooth, elongated, with a long, tapering, acute spire. Whorls eight, broadly and evenly rounded; suture somewhat impressed, but not deep, frequently narrowly channelled. Surface, except anteriorly and on the canal, destitute of spiral lines, unless microscopic striations, and of any indication of ribs, but covered with very close, almost microscopic lines of growth, which give the surface a dull appearance, when dry; on the canal and extending to the anterior part of the body-whorl are a number of distinct spiral lines, becoming faint opposite the middle of the aperture. The nucleus is larger than in *A. rosacea*, rounded, depressed, and spiral, but somewhat mammillary. The aperture is small, oblong-ovate; the outer lip is sharp at the edge, but in adult shells has a distinct thickening a little back from the margin; the inner surface is usually smooth, but in a few adult examples it has a row of four or five small, transversely oblong

---

[*] There are two varieties of *S. (Siphonorbis) pygmæus* on our coast, which are often well-marked. The larger, typical form, from north of Cape Cod, has well-rounded whorls, covered with strong cinguli and sulci, and with a strongly ciliated epidermis; canal long and much curved. The other variety, which abounds off Martha's Vineyard, etc., in from 20 to 300 fathoms, on muddy bottoms, has the whorls flattened and much smoother, the cinguli often obsolete, in part, except on the upper whorls, and the epidermis dark green or olive, and only slightly ciliated, or often nearly or quite smooth; and the canal is perhaps a little shorter and less curved. This may take the variety name, *S. pygmæus*, var. *planulus*. The nucleus and apical whorls agree well, however, in the two forms. The generic names, *Neptunella* and *Siphonella*, formerly used by me for this shell, are both preoccupied.

tubercles, back from the margin, and a larger conical one at the base of the canal.

Columella signoid, a little excavated in the middle, and with a distinct, raised, spiral fold at its inner edge anteriorly; canal short, open, very slightly curved. Epidermis thin, closely adherent, minutely lamellose along the lines of growth, pale greenish or yellowish white, sometimes with microscopic spiral striations.

Length of one of the largest specimens, 12^{mm}; breadth, 4^{mm}; length of body-whorl and canal, 7^{mm}; length of aperture, 5^{mm}; its breadth, 1.8^{mm}. Some specimens are stouter and shorter.

• Off Martha's Vineyard, in 65 to 487 fathoms, 1880 and 1881,—U. S. Fish Commission. Taken at many stations. Off Chesapeake Bay, 300 fathoms,—Lieut. Z. L. Tanner.

It occurred in considerable numbers at stations 870, 876, in 155 and 120 fathoms.

The true *A. rosacea* occurs in shallow water from off Cape Cod northward to Nova Scotia. It differs much from *A. Hölbolli*, of Greenland, and if the latter is not a distinct species, it is, at least, a very marked variety.

Astyris pura Verrill.
> *Astyris zonalis, pars* (white var.), Verrill, Proc. Nat. Mus., iii, p. 408, 1881
> > (*non* Linsley).
> *Astyris pura* Verrill, Trans. Conn. Acad., v, p. 515, June, 1882.

This shell, formerly supposed to be a white deep-water variety of *A. zonalis* (=*A. dissimilis* St.), proves to be distinct from the latter.

It is a stouter shell with a narrower, blunter spire, a larger nucleus, and a wider aperture. It has a more distinct canal, which is a little curved at the tip. The surface is nearly smooth, except a few faint spiral lines on the canal. Shell pure white or pinkish, translucent, usually with the apex distinctly pink or yellowish. It is very common off Martha's Vineyard, in 100 to 487 fathoms. *

TOXOGLOSSA.

Pleurotoma Dalli Verrill and Smith.
> Verrill, Trans. Conn. Acad., v, p. 451, pl. 57, figs. 1, 1a, April, 1882.

A slender, transversely ribbed species, remarkable for the deep notch, widest within, and the deeply concave subsutural band.

Whorls ten, somewhat angular and shouldered, crossed by strongly marked, somewhat oblique, angular ribs, which are most elevated at the shoulder, below the strongly marked, concave, subsutural band; they do not extend on this band, and mostly fade out below, before reaching the suture; on the body-whorl the ribs are less distinct and sometimes absent; when present they extend only a little below the suture. The whole surface is covered with fine, wavy, spiral lines;

---

* The true *A. zonalis* also occurred from near the shore to 120 fathoms. Those from the deeper localities were highly colored and banded like the shore specimens.

fine, but rather conspicuous, lines of growth cover the surface, and recede strongly on the subsutural band.

Aperture small, ovate, rather narrow. Outer lip with a prominent, convex edge, which has a deep notch, situated a short distance below the suture. The notch is usually constricted or even nearly closed up at the edge of the lip, but is broadly rounded at its inner end; this gives it a button-hole like appearance. In some specimens it is but little constricted. Canal short, broad, slightly everted.

Color, brown of various tints; often brown, with one or two spiral bands of yellowish brown, and with streaks of light brown; or the ribs may be pale yellowish brown; aperture brown within; columella whitish in front. Operculum, and animal, not observed.

Length of the largest specimen, 19.5^{mm}; greatest diameter, 6^{mm}; length of body-whorl and canal, 10^{mm}; of aperture, 6^{mm}; breadth of aperture, 2.5^{mm}.

Off Martha's Vineyard, stations 1035, 1036, 1038, 1039, in 94 to 146 fathoms, 1881. Off Delaware Bay, station 1046, 104 fathoms, dredged by Lieut. Z. L. Tanner, Oct. 10, 1881.

Pleurotoma comatotropis Dall.

> Pleurotoma (Mangilia) comatotropis Dall, Bulletin Mus. Comp. Zoöl., ix, p. 58, 1881.

Differs from all our other species in having strong spiral ribs and grooves on the lower whorls.

One dead specimen. Off Martha's Vineyard, station 949, in 100 fathoms. Off Cape San Antonio, 640 fathoms (Dall).

Daphnella limacina Dall.

> Pleurotoma (Bela) limacina Dall, Bull. Mus. Comp. Zoöl., ix, p. 55, 1881.
> Pleurotoma (Daphnella) limacina Verrill, Am. Journ. Sci., xxii, p. 300, 1881.
> Daphnella limacina Dall, op. cit., p. 102; Verrill, Trans. Conn. Acad., v, p. 452.

Station 994, 368 fathoms. Gulf of Mexico, 447–805 fathoms, Blake Exp.,—Dall.

Bela Gouldii Verrill.

> Trans. Conn. Acad., v, p. 465, pl. 57, figs. 6, 6a, April, 1882.

Off Chesapeake Bay, station 898, in 300 fathoms,—Lieut. Z. L. Tanner. Common from Cape Cod to Nova Scotia and the Gulf of St. Lawrence, in 12 to 60 fathoms.

Bela harpularia (Couth.) H. and A. Ad.

> Fusus harpularius Couthouy, Boston Jour. Nat. Hist., ii, p. 106, pl. 1, fig. 10, 1838.
> Gould, Invertebrata of Mass., ed i, p. 291, fig. 191, 1841.
> Bela harpularia H. and A. Adams, Genera of Recent Mollusca, vol. i, p. 92, 1858.
> Gould, Invertebrata of Mass., ed. ii, p. 352, fig. 191 (non G. O. Sars).
> Verrill, Report Invert. Anim. of Vineyard Sd., in 1st Rep. U. S. Fish Com., pp. 508, 636, pl. 21, fig. 108 (after Gould), 1874 (auth. cop., p. 342); Trans. Conn. Acad., v, pl. 43, fig. 14, pl. 57, fig. 9, 1882.

This species ranges from Long Island Sound to Nova Scotia, but is less common northward. It is the most common species south of Cape Cod, in moderate depths (18 to 30 fathoms), where it is usually unaccompanied by any other species, and occurs of large size and typical form. We took it off Gay Head, Martha's Vineyard, 18 to 29 fathoms, in 1871, 1880, 1881; off Block Island, 20 to 28 fathoms, 1874, 1880; eastern end of Long Island Sound, 1874; Massachusetts Bay, 8 to 29 fathoms, 1873, 1877, 1878, 1879; Cape Cod Bay, and off Cape Cod, 15 to 34 fathoms, 1879; Casco Bay, 1873; Eastport, Me., and Bay of Fundy, 10 to 50 fathoms, 1870, 1872; Halifax harbor, 20 fathoms, and off Halifax 120 miles, 190 fathoms, 1877. Messrs. Smith and Harger, on the "Bache," in 1872, took it at various localities on George's and Le Have Banks, in 25 to 60 fathoms. Off Martha's Vineyard, 104 miles, 368 fathoms, 1881.

Bela pleurotomaria (Couthouy) Adams.

> *Fusus pleurotomarius* Couthouy, Boston Jour. Nat. Hist., ii, p. 107, pl. 1, fig. 9, 1838.
>
> *Fusus rufus* Gould, Invert. of Mass., ed. i, p. 290, fig. 192 (*non* Montagu).
>
> *Defrancia Vahlii* (Beck) Möller, 1842 (t. Lovèn).
>
> *Mangelia pyramidalis* Stimpson, Shells of New England, p. 49, 1851 (? *non* Ström, sp.).
>
> *Bela pleurotomaria* H. and A. Adams, Genera Recent Mollusca, i, p. 92, 1858.
> Gould, Invert. of Mass., ed. ii, p. 355, fig. 625.
>
>> Verrill, Report Invert. Anim. of Vineyard Sd., in 1st Rep. U. S. Fish Com., p. 637, 1874 (auth. cop., p. 343); Trans. Conn. Acad., v, p. 478.

This species is found from off Martha's Vineyard to Labrador! It is not uncommon in Eastport harbor and the Bay of Fundy, where I dredged it in 1864, 1865, 1868, 1870, in 15 to 80 fathoms. By the U. S. Fish Com. it has been dredged in Halifax harbor in 20 to 25 fathoms, 1877; George's Bank, 45 fathoms, 1872; Gulf of Maine, at Cashe's Ledge, 30 to 40 fathoms, 1874; off Cape Ann, 38 to 40 fathoms, 1874; Casco Bay, 1873; Massachusetts Bay, 31 to 48 fathoms, 1877, 1879; off Cape Cod, 30 to 122 fathoms, 1879; off Chatham, Mass., 16 fathoms, 1881.

Off Martha's Vineyard, 255 fathoms, 1881. It appears to occur on the coast of Greenland.

Whether it can be identified accurately with any European species is doubtful. Many writers have considered it identical with *B. pyramidalis* (Ström). But the shell figured under that name by Prof. G. O. Sars appears to be quite different.

Bela cancellata (Mighels) Stimpson.

> *Fusus cancellatus* Mighels, Proc. Boston Soc. Nat. Hist., i, p. 50, 1841; Boston Jour. Nat. Hist., iv, p. 52, pl. 4, fig. 18, Jan., 1842.
>
> *Bela cancellata* Stimpson, Check List, 1862.
> Gould, Invert. Mass., ed. ii, p. 355, description (but not the figure, 924), (*non* G. O. Sars).
>
>> Verrill, Proc. U. S. Nat. Mus., iii, p. 364, 1881; Trans. Conn. Acad., v, p. 475, pl. 43, figs. 10, 11; pl. 57, fig. 13.

This shell extends from off Martha's Vineyard, in 126 and 312 fathoms (stations 877, 947), north to Nova Scotia and Labrador; and probably to Greenland and Northern Europe. It is one of the most common species in the cold waters of the Bay of Fundy, near Eastport, Me., and Grand Menan I., in 10 to 100 fathoms, where I have often dredged it, in 1861, 1863, 1864, 1865, 1868, 1870, 1872. We have also taken it, on the various U. S. Fish Com. expeditions, off Nova Scotia; in the Gulf of Maine; Casco Bay; Massachusetts Bay; off Cape Cod, etc., in 12 to 92 fathoms.

Bela decussata (Couth.) H. and A. Adams.

> *Pleurotoma decussata* Couthouy, Boston Jour. Nat. Hist., ii, p. 183, pl. 4, fig. 8, 1839 (*non* Lam., *nec* McGilv.).
> Gould, Rep. on Invert. of Mass., 1st ed., p. 280, fig. 185, 1841.
> *Mangelia decussata* Stimpson, Shells New Eng., p. 49, 1851.
> *Bela decussata* Gould, Rep. on Invert. of Mass., Binney's ed., p. 354, fig. 623, 1870.
> Verrill, Trans. Conn. Acad., v, p. 472, pl. 43, fig. 13.

This shell is not uncommon on the New England coast, in moderate depths, mostly in 25 to 75 fathoms. Its range is from off Martha's Vineyard (station 991), in 34 fathoms, northward to Labrador. In the Bay of Fundy, where it is not rare, I have taken it in 20 to 100 fathoms, in 1868, 1870, 1872.

Bela pygmœa Verrill.

> *Bela tenuicostata* (*pars*) Verrill, Proc Nat. Mus., iii, p. 365, 1880 (*non* Sars).
> *Bela pygmaea* Verrill, Trans. Conn. Acad., v, p. 460, pl. 57, fig. 8, May, 1882.

Shell very small, fusiform, or subovate, with four or five convex whorls, a very short spire, and a large body-whorl; sculpture very finely cancellated or reticulated. The whorls are usually rather evenly rounded, moderately convex, but often have a very slightly marked, rounded shoulder; suture somewhat impressed, rather oblique. The nucleus is relatively not small, with the apex not prominent, so that it appears to be obtuse, or rounded, smooth, glassy. The whole surface below the nucleus is covered by fine, raised, revolving cinguli, separated by slight grooves of about the same width, and by equally fine, slightly sinuous, transverse riblets, coincident with the lines of growth, and receding in a distinct curve on the subsutural band; the crossing of these two sets of lines produces a finely cancellated sculpture over the whole surface, but the transverse lines are usually more evident on the convexity of the whorls, while the spiral lines are more conspicuous anteriorly, and on the siphon. Aperture relatively large, oblong-elliptical, slightly obtusely angled posteriorly; sinus shallow, but distinct, evenly concave; outer lip elsewhere evenly convex. Canal short and broad, not constricted at base by any incurvature of the outer lip. Columella strongly concave or excavated, in the middle, sigmoid anteriorly. Color of shell, pale greenish white, covered by a thin epidermis of similar color.

One of the largest shells is 5.5^{mm} long; 2.75^{mm} broad; length of body-whorl, 4^{mm}; of aperture, 3^{mm}.

Only a few specimens have been taken off Martha's Vineyard, at stations 892 and 894, in 487 and 365 fathoms, 1880; station 947, in 312 fathoms, 1881.

This little species bears some resemblance to *B. decussata*, but can be readily distinguished by the much finer and more uniform sculpture.

Bela incisula Verrill.

? Pleurotoma Trevelyana, var. *Smithii* Jeffreys, Ann. and Mag. Nat. Hist., 1876, p. 332 (*non Smithii* Forbes).

Bela impressa? Verrill, Proc. U. S. Nat. Mus., iii, pp. 365, 1880 (*non* Mörch.).

Bela incisula Verrill, Trans. Conn. Acad., v, p. 461, pl. 43, fig. 12; pl. 57, fig. 14.

The shell is small, subfusiform to short ovate, with about five or six turreted, flattened whorls, which are angularly shouldered just below the suture. The subsutural band arises abruptly from the suture, nearly at right angles, and its surface is flat or slightly concave, marked by strongly recurved lines of growth, but mostly without spiral lines. The shoulder is often nearly right-angled. The whorls are decidedly flattened in the middle. There are, on the last whorl, about twenty rather broad, flattened or rounded ribs, which are nearly straight, a little prominent and usually slightly nodose at the shoulder, but they disappear a short distance below it. They are separated by well excavated, concave grooves, deepest close to the shoulder.

The most characteristic feature of the sculpture is that the surface is marked by rather fine, but regular and distinct, sharply incised, narrow, revolving grooves, which are rather distant, with flat intervals. Of these there are usually about three to five on the penultimate whorl, and about twenty to twenty-eight on the last, the greater number being below the middle, on the siphon, where they become coarser and closer, with narrower rounded intervals. One of the sulci, just below the shoulder, is usually more distinct, and cuts the ribs so as to give their upper ends a subnodulous appearance; below this there is usually a rather wide zone, without grooves; usually no revolving lines above the shoulder. The apex is usually eroded; when perfect it is acute. The nucleus has a very small and slightly prominent, smooth apex; its first turn is marked with fine spiral lines; the next whorl has, at first, about three stronger, spiral, raised cinguli, which soon begin to be crossed by thin transverse riblets.

Aperture about half the length of the shell, narrow ovate, or elliptical, angulated above. Canal short, nearly straight, a little narrowed at the base by an incurvature of the lip. The outer lip has a decided angle at the shoulder, below which the edge is well rounded, and projects strongly forward, in the middle; the sinus, above the shoulder, is rather deep, wide, and evenly rounded within. Columella strongly excavated in the middle, obliquely receding at the end.

The shell is commonly greenish white and covered by a thin, close, greenish epidermis; but some specimens are clear white, rarely pinkish.

Ordinary specimens are about 6.5^{mm} long; 3.5^{mm} broad; aperture,

3^{mm} long. One of the largest, having six whorls, is 8^{mm} long; 4.5^{mm} broad; body-whorl, 6^{mm} long; aperture, 4.5^{mm} long.

This is one of the most common and generally distributed species of *Bela* found on the New England coast. It inhabits both muddy and sandy bottom, and sometimes is found among gravel and rocks. It occurs from the region off Newport, R. I., northward to Labrador, and from very shallow water. in the Bay of Fundy and Casco Bay, to 500 fathoms, off Martha's Vineyard. It is very common from Massachusetts Bay to the Bay of Fundy and Halifax, N. S., in 10 to 50 fathoms.

Bela concinnula Verrill.

   Bela exarata (pars) Verrill, Proc. Nat. Mus., iii, p. 366, 1880.
   Bela concinnula Verrill, Trans. Conn. Acad., v, p. 468, pl. 43, fig. 15; pl. 57, fig. 11.

Shell rather small and delicate, long-ovate, regularly turreted, with about six whorls, which rise almost at right angles from the suture, and have an angular, or squarish, nodulous shoulder, usually distinctly carinated by a thin, raised, spiral keel, which forms small, but prominent nodules where it crosses the ribs; below the shoulder the whorls are abruptly flattened. The subsutural band is usually little convex, or nearly flat.

The ribs are numerous (often 20 to 25), regular, nearly straight below the shoulder, separated by concave intervals of equal or greater width; they extend entirely across the upper whorls; above the shoulder they are slightly excurved on the subsutural band. Whole surface covered with regular and rather strong, rounded, elevated, revolving cinguli, which cross the ribs and produce on them small, rounded nodes, and give a pretty regularly and strongly cancellated appearance to the whole surface. On the penultimate whorl there are four or five cinguli below the angle. Aperture rather short, narrow-ovate, angulated posteriorly; sinus broad and shallow. Canal narrow, a little produced, and slightly curved; columella decidedly sigmoid, its inner edge excurved at the end.

Color of the shell white, or pale greenish white, covered with a thin, pale green epidermis.

A rather large male is 11.5^{mm} long; breadth, 5.25^{mm}; length of body-whorl, 7^{mm}; its breadth, 5^{mm}; length of aperture, 5^{mm}; its breadth, 2^{mm}. An ordinary specimen measures, in length, 10^{mm}; breadth, 4.5^{mm}; length of aperture, 5.5^{mm}.

This species is common and widely distributed on this coast. It ranges from the region south of Martha's Vineyard, in deep water, to Labrador. By the U. S. Fish Com. it was dredged, off Newport, R. I., and Martha's Vineyard, in 252 to 487 fathoms (stations 880, 892, 947, 994, 1038), 1880 and 1881; Cape Cod Bay and off Cape Cod, 25 to 122 fathoms, 1879; Massachusetts Bay, 20 to 29 fathoms, 1877; Gulf of Maine, many stations, 25 to 88 fathoms, 1873, 1874, 1878; 150 fathoms, 1872; Casco Bay, 1873; George's Bank, 50 to 65 fathoms, 1872; south of George's Bank, 430 fathoms, 1872; Halifax Harbor, 16 to 21 fathoms, and off Halifax, 42 fathoms, 1877.

Bela tenuilirata Dall.

Dall, Am. Journ. Conch., vii, p. 98, 1871.

Bela simplex Verrill, Proc. U. S. Nat. Mus., iii, p. 367, 1880 (*non* Middendorff).

A single immature specimen, referred to this species by Mr. Dall, was taken in 1880.

The nucleus, consisting of nearly three apical whorls, is chestnut-brown; the surface is finely decussated by equal lines running obliquely in opposite directions.

The shell is pale flesh-color, covered with a thin, smooth, glossy, pale yellowish brown epidermis. Length, 9^{mm}; breadth, 5^{mm}; length of body-whorl, 7^{mm}; of aperture, 6^{mm}.

One dead, but fresh, specimen, from station 894, in 365 fathoms, off Martha's Vineyard. Alaska,—Dall.

The nucleus of this shell is not like that of a *Bela*. It more nearly resembles *Pleurotomella*, in several respects.

TÆNIOGLOSSA.

Dolium Bairdii Verrill and Smith.

Dolium Baridii Verrill and Smith, in Verrill, Amer. Jour. Sci., xxii, p. 299, Oct., 1881 (description).

The apical or nuclear whorls are regularly spiral, yellowish brown, smooth, showing only faint lines of growth, and consist of nearly four turns. The color and character of the surface change abruptly beyond the nucleus, the normal sculpture suddenly appearing. The largest specimen taken (δ) is 68^{mm} long; breadth, 56^{mm}; length of aperture, 53^{mm}.

Of Martha's Vineyard, station 945; 202 fathoms, one large living δ. Stations 1032, 1036, 1038, 1040, 94 fathoms; young specimens and fragments of several large specimens.

Of Delaware Bay, station 1046, 104 fathoms, one living (δ), 1881,— Lieut. Z. L. Tanner.

Amaurpsis Islandica (Gmelin) Mörch.

Amauropsis helicoides Gould, Binney's ed., p. 348, fig. 161.

Off Chatham, Mass.; stations 965, 969, in 15 to 18 fathoms.

Lamellaria pellucida, var. Gouldii Verrill.

This differs from the original *L. pellucida* in having the mantle thicker, with more or less numerous, low verrucæ on the dorsal surface; color pale yelow or yellowish white, more or less blotched or specked with flake-white and bright yellow, and often with brown blotches. The verge appears to be different in form, the lateral papilla being larger and longer, and not so near the end, the portion beyond it forming a spatulate or olovate lobe, rounded at the end, but this may be due partly to the state of contraction. The shell, in the specimens examined, is very thin, delicate, and transparent, as in *L. pellucida*, but differs in being somewhat shorter, broader, with the spire a little lower, the apex less elevated, and the suture less impressed. In alcohol, a specimen is 18^{mm} long; breadth, 12^{mm}; height, 10^{mm}.

Off Martha's Vineyard, stations 925, 938, 939, 946, 1029, in 224 to 458 fathoms.

Several specimens of both sexes occurred at some of these localities. Off Delaware Bay, station 1047, 1881,—Lieut. Z. L. Tanner. It is usually associated with the smooth form originally described, and intermediate states, as to the number and size of the dorsal verrucæ occur, some being strongly verrucose, others nearly smooth.

Capulus Hungaricus (Linné).

> *Capulus hungaricus* Jeffreys, Brit. Conch., iii, p. 269, pl. 6, fig. 5; v, pl. 59, figs. 6, 6a.
>
> G. O. Sars, Moll. Arct. Norvegiæ, p. 145, pl. v, figs. 2a, 2b (dentition).

Stations 922, 1029, in 69 and 458 fathoms, off Martha's Vineyard, 1881; two living specimens.

On the European side of the Atlantic, this species is found from Iceland to the Mediterranean.

Torellia fimbriata Verrill and Smith.

> *Torellia fimbriata* Verrill and Smith, in Verrill, Trans. Conn. Acad., v, p. 520, p. 57, figs. 27, 27a, June. 1882.

Shell thin, fragile, translucent, broader than high, with a short, depressed spire, the apex small and a little prominent, the last whorl large and ventricose, with spiral carinæ, bearing divergent epidermal hairs. Whorls five, very convex, rapidly enlarging; suture deep, slightly channeled; nuclear whorls smooth and glossy, regularly spiral, the first whorl minute. Sculpture, several raised, angular, spiral carinæ separated by unequal intevals, on which are finer spiral lines, and numerous evident, thin, raised flexuous lines of growth, which cross both the intervals and carinæ, rendering the latter finely nodulous. On the last whorl there are about ten carinæ, each of which usually supports a spiral row of long epidermal hairs; the uppermost of these is just below the suture, and its epidermal processes are long and appressed against the preceding whorl; the next is separated by a wider space, while those on the conyex part of the whorl are nearer together; the last defines the border of the umbilicus, which is deep, but not broad. Aperture large, roundish, the lip continuous in adult shells; in the umbilical region the lip is somewhat reflected, so as to partially conceal the umbilicus; within the lip the columella has a very obtuse lobe, projecting inward.

Epidermis thick, pale yellowish or greenish yellow, more or less lamellose along the lines of growth, and rising into long and large divergent hair-like processes along the spiral carinæ.

Shell yellowish white.

Length of the largest specimen (δ), 14.5^{mm}; breadth, 17^{mm}; length of body-whorl, 13^{mm}; length of aperture, 10^{mm}; breadth, 10.2^{mm}; length of hairs, 2–3^{mm}.

Variety, tiarella Verrill.

A variety occurred in company with the typical form, at station 1026, 182 fathoms, in which the subsutural carina is well developed and crowned by its row of long hairs, but the other carinæ are nearly obsolete, and only bear rows of short, inconspicuous hairs; the epidermis is elsewhere thick and lamellose, not hairy. The spire is a little more elevated.

Off Martha's Vineyard, stations 869, 878, 939, 1025, 1026, 1033, 1038, in 142 to 258 fathoms, 1880 and 1881,—U. S. Fish Commission. A small specimen was taken in 1873, at station 21 B, 52 to 90 fathoms, near Cashe's Ledge, off the coast of Maine, by the party on the Bache.

Fossarus elegans Verrill and Smith.
> Verrill, Trans. Conn. Acad., v, p. 522, pl. 57, fig. 23, June, 1882.

Shell small, ovate, with a short, acute, turreted spire, and five angulated and sharply carinated whorls, elegantly latticed between the carinæ. The whorls increase rapidly, the last being relatively large. On the last whorl there is a sharp angular carina at the shoulder, often with a smaller one just below it, a larger and more prominent one around the periphery, and three or four smaller ones on the anterior slope, besides a spiral fold around the umbilical region; on the larger specimens there are, sometimes, two or three strong, raised varices on the last whorl, and the edge of the lip is thickened. The intervals between the carinæ are concave. On the preceding whorls the two larger carinæ are visible, often with a small intermediate one. The nucleus is minute, regular, smooth, a little prominent. The rest of the shell is covered, between the carinæ, with numerous, close, thin, oblique, raised lamellæ, or lines of growth; those on the subsutural band are flexuous. Aperture nearly round; lip continuous; outer lip thickened, and with denticles externally, where the carinæ terminate. Umbilicus spiral, very narrow, sometimes closed. Color white.

Length, 5.3^{mm}; breadth, 4^{mm}; length of aperture, 2^{mm}.

Off Martha's Vineyard, station 949, 100 fathoms, 1881; eight specimens, none living.

Velutina lævigata (L.) Gould.

Off Martha's Vineyard, stations 940, 949; in 100 to 130 fathoms.

Cerithiella Whiteavesii Verrill.
> Trans. Conn. Acad., v, p. 522, pl. 42, fig. 7, July, 1882.
> *Lovenella Whiteavesii* Verrill, these Proc., p. 375, 1880.

Cingula areolata (Stimp.) Verrill.
> Amer. Journ. Sci., xvii, p. 311, 1879.

Off Martha's Vineyard, station 940, in 130 fathoms.

Litiopa bombyx Rang.

Station 1038, clinging to floating *Sargassum*

Scalaria (Opalia) Andrewsii Verrill.

Scalaria, undetermined sp., Verrill, Proc. Nat. Mus., iii, p. 376, 1880.

Scalaria (Opalia) Andrewsii Verrill, Trans. Conn. Acad., v, p. 526, pl. 57, fig. 35, July, 1882.

Shell small, slender, elongated, with well-rounded whorls and deep suture. Whorls seven, crossed by about thirteen regular ribs, which are moderately elevated and evenly rounded, and, on the lower whorls, a little thickened, most so in the middle; their interstices are crossed by several distinct spiral cinguli, which also render the ribs a little nodulous; on the penultimate whorl there are about five cinguli; on the last whorl a strong, round, spiral carina surrounds the base or umbilical region, starting from under the upper margin of the outer lip and enclosing a space, on which two or more faint spiral grooves can be detected. Aperture round; lip continuous; margin of outer lip thickened by a rib; inner lip with the edge reflected in the umbilical region; no umbilicus. Color white. Length, 5.5^{mm}; breadth, 2^{mm}; diameter of aperture, 1^{mm}.

Station 873, off Newport, R. I., 100 fathoms, 1880. One specimen.

Dedicated to Mr. E. A. Andrews, of the U. S. Fish Commission parties, in 1880 and 1881.

Scalaria (Cirsotrema) Leeana Verrill.

Trans. Conn. Acad., v, p. 523, pl. 57, fig. 34, July, 1882.

Shell small, slender, elongated, with well-rounded whorls and deep, oblique suture (apex truncated). Whorls crossed by numerous small, little-elevated, oblique ribs, and on each whorl one large, strong, oblique varix-like rib, those on the three lower whorls nearly in one line, the last forming the greatly thickened margin of the lip. Both the ribs and the wider intervals between them are crossed by very numerous and fine spiral striæ. Aperture small, round-ovate, surrounded by a much thickened, continuous margin close to the edge; this rim around the outer lip is crossed by oblique striæ; base with spiral striæ, but without a distinct carina; no umbilicus. Size about the same as the preceding species.

Off Martha's Vineyard, station 1038, 146 fathoms, 1881.

Named in honor of Prof. L. A. Lee, of Bowdoin College, and of the U. S. Fish Commission party in 1881.

Acirsa costulata (Mighels) Verrill.

Turritella costulata Mighels. Proc. Boston Soc. Nat. Hist., i, p. 50, 1841; Boston Journal Nat. Hist., vol. iv, p. 50, pl. 4, fig. 20, 1842.

Gould, Invert. Mass., ed. ii, p. 318, fig. 587.

Scalaria Eschrichtii Möller, Kröyer's Tidsskr., iv, p. 83, 1842.

Acirsa borealis (Mörch) Verrill, Amer. Journ. Sci., iii, pp. 210, 281, 1872.

Crab Ledge, off the southern part of Cape Cod, stations 965 and 984, in 15 and 32 fathoms. Previously known from the Bay of Fundy, and northward to Greenland.

Aclis tenuis Verrill.
>Trans. Conn. Acad., v, p. 528, pl. 58, fig. 19, July, 1882.
>*Eulimella ventricosa (pars)* Verrill, these Proc., iii, p. 380,1880 (*non* Forbes sp.)

Shell very slender, smooth, white, acute. Whorls nine, evenly rounded; surface with few, faint, microscopic, raised, spiral lines; suture impressed; aperture elliptical, a little effuse in front. Nucleus small, regularly spiral, not upturned. Length, 3.8ᵐᵐ; breadth, 1ᵐᵐ.

Station 873, in 100 fathoms, 1880.

RHIPHIDOGLOSSA.

Machæroplax obscura, var. bella (Verk.).
>*Machæroplax bella* Friele; Verrill, Proc. Nat. Mus., iii, p. 378, 1880.

Station 1032, off Martha's Vineyard, 208 fathoms.

Doubtless this is only a strongly sculptured variety of *M. obscura*.

Machæroplax cinerea (Couth.) Friele.
>*Margarita cinerea* Gould, Invert. Mass., ed. ii, p. 279, fig. 539.

This species, which had not occurred south of Cape Cod previously, was taken at station 981, in 41 fathoms, off Chatham, Cape Cod.

Cyclostrema Dalli Verrill.
>Trans. Conn. Acad., v, p. 532, pl. 57, fig. 39, July, 1882.
>*Cyclostrema trochoides* Verrill, these Proc., iii, p. 378, 1880 (*non* Jeffr., Sars).

This shell differs from *C. trochoides* in having the base covered around the umbilical region with six to eight very distinct, incised, spiral lines. The umbilicus is closed, or represented only by a slight and narrow pit. The surface of the shell has only a little luster, and is slightly roughened by very faint and close lines of growth.

Color, yellowish white. Height, 2ᵐᵐ; breadth, 2.25ᵐᵐ.

Station 892, in 487 fathoms.

Cyclostrema rugulosum (Jeffreys, MSS.) Sars.
>G. O. Sars, Möll. Reg. Arct. Norvegiæ, p. 129, pl. 21, figs. 1, *a*, *b*.

Station 894, in 365 fathoms, 1880.

Northern Norway, 80–200 fathoms,—Sars.

Fissurella Tanneri Verrill, sp. nov.

Shell large, ovate, rather thin, with regularly and finely decussated sculpture. Apex nearer the anterior (smaller) end, moderately elevated. Perforation not large, round-ovate, conformable with the outline of the shell, but more rounded. Whole surface covered with rather fine, raised, radiating lines, with interstices of similar width or narrower; these are decussated by numerous concentric raised lines, which rise into nodules, or, towards the margin, form small, arched lamellæ in crossing the radii. Shell, externally, pale yellowish gray, internally lustrous bluish white; edge finely crenulated. Length, 46ᵐᵐ; breadth, 31ᵐᵐ; height, 16ᵐᵐ; longest diameter of apical foramen, 4ᵐᵐ; its breadth, 3ᵐᵐ.

Off Delaware Bay, station 1046, in 104 fathoms,—Lieut. Z. L. Tanner, 1881; one living specimen.

Scissurella crispata Fleming.

A single specimen was found by Mr. Dall in the aperture of a *Margarita*, from off Martha's Vineyard, 238 to 365 fathoms. Gulf of St. Lawrence,—Dawson.

Cocculina Beanii Dall.
> This volume, p. 403.
> *Acmœa rubella?* Verrill, Proc. Nat. Mus., iii, p. 391, 1880 (*non* Fabr., Sars).

Cocculina Rathbuni Dall.
> This volume, p. 403.

Off Martha's Vineyard, 100 to 365 fathoms. Several living young specimens were taken at station 997, in 335 fathoms. Mr. Dall, in a recent letter, informs me that he has received the same species from Mr. Jeffreys, taken by the "Porcupine" expedition, off the European coast. West Indies, 399 to 502½ fathoms (t. Dall).

Off Martha's Vineyard, 506 fathoms. West Indies, 399 and 502½ fathoms (t. Dall).

Addisonia paradoxa Dall.
> This volume, p. 405.

Off Martha's Vineyard, 69 to 130 fathoms, 1881.

Mr. Dall has recently informed me that he has received from Mr. Jeffreys a shell belonging to this genus, and perhaps identical with this species, judging from the shell only. Mr. Jeffreys identifies the shell referred to with *Gadinia excentrica* Tiberi.

POLYPLACOPHORA.

Chætopleura apiculata (Say) Carpenter.
> *Chiton apiculatus* Say; Gould, Invert. Mass., ed. ii, p. 258, fig. 522.

Off Martha's Vineyard, station 938, in 310 fathoms. One young specimen. Common in shallow water. Possibly the apparent occurrence in deep water was due to the accidental lodgment of the specimen in the seive, from some previous dredging.

GYMNOGLOSSA.

Stilifer Stimpsonii Verrill, 1872.

A living specimen of this species occurred at station 1028, in 410 fathoms, 1881. In 1880 it was taken in considerable numbers at stations 814, 823, 824, in 13 to 27 fathoms, off Block Island. These were living on the upper surface of the common sea-urchin (*Strongylocentrotus Dröbachiensis*). New Jersey to Nova Scotia!

Stilifer curtus Verrill.
> Trans. Conn. Acad., v, p. 535, July, 1882.

Shell broader than high, with a very low spire, nearly concealed by the ventricose body-whorl, which nearly envelopes the preceding whorls;

nucleus minute, only a little prominent. Aperture large, nearly as long as the shell, lunate; surface smooth, white.

Station 1028, in 410 fathoms; one living example. Host not known·

Turbonilla Emertoni Verrill.

Verrill, Trans. Conn. Acad., v, p. 536, pl. 58, figs. 14, 14a.

Shell small, white, lustrous, elongated, with a very slender, acute spire. Whorls eleven, not very oblique, broadly rounded, a little flattened at the periphery; suture strongly impressed; surface very smooth and glossy, without any spiral lines, but with slight, rather indistinct and irregular longitudinal furrows, which are often absent. Apical whorl small, strongly upturned.

Aperture small; outer lip flattened, projecting a little anteriorly (more or less broken in all my specimens). Columella nearly straight, with no trace of a fold.

Length, 4.8^{mm}; breadth, 1.2^{mm}.

Off Martha's Vineyard, station 895, in 238 fathoms, 1880.

This shell resembles *T. nivea* Stimpson, which also occurs in the same region, but the latter is a longer and larger shell, with a decidedly smaller and more prominent upturned nucleus, and is strongly and regularly longitudinally ribbed.

Named in honor of Mr. J. H. Emerton, for several seasons zoological artist of the Fish Commission.

Turbonilla Bushiana Verrill.

Trans. Conn. Acad., v, p. 537, pl. 58, fig. 16.
Turbonilla formosa Verrill and Smith, in Verrill, Amer. Jour. Sci., xx, p. 398, 1880; Proc. Nat. Mus., iii, p. 380, 1880 (*non* Jeffreys, Ad.).

The name *formosa* having been previously used, I propose to name this species *Bushiana*, in honor of Miss K. J. Bush, an excellent assistant in the conchological work of the U. S. Fish Commission.

Eulimella Smithii Verrill.

Trans. Conn. Acad., v, p. 538, pl. 58, fig. 18.
Turbonilla Smithii Verrill, Proc. Nat. Mus., iii, p. 380, 1880.

This species seems to belong to *Eulimella* rather than to *Turbonilla*, if the two groups be kept apart.

Menestho striatula (Couthouy) Verrill.

Menestho albula Gould, Invert. Mass., ed. ii, p. 333, fig. 604 (*non* Fabr., sp.).

Crab Ledge, off south side of Cape Cod, 10 to 15 fathoms.

Menestho Bruneri Verrill.

Menestho Bruneri Verrill, Trans. Conn. Acad., v, p. 539, July, 1882.

Shell small, white, with an elongated, acute-conical spire, the apical whorl very small, upturned, and incurved. Whorls six, with a rounded shoulder close to the suture, the portion next the suture rising abruptly, nearly at a right angle; periphery flattened or very slightly rounded;

suture little oblique, impressed, or subcanaliculate. Aperture narrowly contracted posteriorly, narrow ovate anteriorly; outer lip little convex, slightly produced anteriorly; columella excurved, flattened, with no fold nor tooth. Sculpture delicate, incised, spiral grooves, separated by wider intervals, and covering the anterior two-thirds of the body-whorl, extending a little back of the aperture, but mostly absent on the preceding whorls. No umbilicus.

Length, 5^{mm}; breadth, 2.5^{mm}; length of body-whorl, 3.5 ^{mm}; of aperture, 2.5^{mm}; its breadth, 1^{mm}.

Off Newport, R. I., station 892, in 487 fathoms, 1880.

Named for Mr. H. L. Bruner, an assistant, during the season of 1881, in the conchological work of the Fish Commission.

TECTIBRANCHIATA.

Actæon nitidus Verrill.

Auriculina insculpta Verrill, these Proc., iii, p. 381, 1880 (*non* Mont., sp.)
Actæon nitidus Verrill, Trans. Conn. Acad., v, p. 540, pl. 58, fig. 21.

Shell small, white, translucent, glossy, elongated, apex obtuse. Nuclear whorl rather large, regular. Whorls six, flattened at the periphery, gradually increasing, slightly roundly shouldered. Sculpture delicate, wavy, incised spiral lines, more distant and distinct on the anterior part of the body-whorl, becoming finer, closer, and more wavy behind the middle, obsolete near the suture, except one fine subsutural groove; suture impressed or slightly canaliculate. Aperture narrow-ovate, much contracted posteriorly, a little produced anteriorly; columella spirally twisted, the inner edge forming a slightly raised fold.

Length, 8^{mm}; breadth, 3^{mm}; length of body-whorl, 5.5^{mm}; length of aperture, 3.5^{mm}; its breadth, 1.8^{mm}.

Stations 892 and 947, in 487 and 312 fathoms, 1880 and 1881, south of Martha's Vineyard.

Cylichna Gouldii (Couth.) Verrill.

Bulla Gouldii Couthouy, Bost. Jour. Nat. Hist., ii, p. 181, pl. 4, fig. 6, 1838.
Utriculus Gouldii Stimpson; Gould, Invert. Mass. (ed. ii), p. 217, fig. 508.
Cylichna Gouldii Verrill, Proc. U. S. Nat. Mus., iii, p. 383, 1880.

Crab Ledge, off Chatham, Cape Cod, station 973. Stellwagen's Bank, Massachusetts Bay, in 15 to 25 fathoms, 1879.

Cylichna? Dalli Verrill.

Trans. Conn. Acad , v, p. 542, July, 1882.

Shell white, somewhat thickened when full grown, translucent when younger, elongated, broadest about the middle, narrowed to both ends, most so posteriorly; apex with a distinct pit, showing volutions within; no umbilicus; whole surface covered with fine, regular, wavy spiral lines, visible with a lens. Outer lip with a free, sharp edge, rising slightly above the body-whorl posteriorly, and separated from it by a deep, narrow slit; it is very slightly convex and a little flaring along the middle, anteriorly rounded and sharp to its union with the inner margin. Aperture very narrow posteriorly, suddenly enlarging to an ovate form anteriorly, by the decided excurvature of the inner margin. Animal unknown.

Length of largest example, 10^{mm}; breadth, 5.25^{mm}.

Stations 997 and 999, in 335 and 266 fathoms.

Philine tincta Verrill.

Trans. Conn. Acad., v, p. 544, July, 1882.

Shell very thin, rather large, irregularly oblong, broad, widest in the middle, not polished, tinged with smoky brown; surface without distinct spiral lines, covered with very evident, close, raised, wavy lines of growth. Apex rounded, neither spiral nor depressed. Outer lip rising a little above the body-whorl, and separated from it by a simple wide sinus, flaring, convex, and slightly angulated in the middle, a little narrowed and well-rounded anteriorly; a spiral fold where the inner lip passes into the shell, in front of the prominent body-whorl.

Length, 10.75^{mm}; breadth, 8^{mm}; breadth of aperture, 7^{mm}.

Station 921, in 65 fathoms; two living specimens.

CHORISTIDÆ Verrill.

The peculiar structure of the animal of the following species, and of its radula, will not allow it to be placed in any established family. Therefore, I propose to make it the type of a new family, *Choristidæ.*

This family may be characterized by the heliciform shell, with the periostraca continuous between the whorls; lip continuous; columella without a fold; operculum horny, paucispiral. Animal with frontal tentacles united by a fold, and with simple posterior tentacles. Jaws well developed; pharynx large, retractile.

Radula with three rows of rachidian teeth, the central ones small; with broad, bilobed, inner lateral teeth; and two rows of small, hook-shaped outer lateral ones. Gill composed of numerous lamellæ, attached to the inner surface of the mantle on the left side and over the neck.

The position of this family is doubtful. Its head, tentacles, pharynx, &c., resemble those of many *Tectibranchs.* Its dentition is, apparently, unique.

Choristes elegans Carp., var. **tenera** V.

Verrill, Trans. Conn. Acad., v, p. 541, pl. 58, figs. 27, 27a.

Choristes elegans Carpenter, Canadian Nat., p. 392, pl. 7, fig. 13, 1872.

Shell thin, fragile, short, heliciform, with a low spire, and a very large, ventricose body-whorl. Whorls, in our largest examples, four to five, very convex, evenly rounded; apical whorl small, spiral, oblique; suture impressed; surface smooth (the epidermis is destroyed and the surface of the shell is eroded in all the living examples). The whorls are in contact and united, but the epidermis continues around the whorls between or in the sutures. Aperture large, forming more than a half-circle; outer side well rounded, nearly straight on the columella-margin; lip continuous all around, raised up and with the edge slightly everted, in the umbilical region, so as to partially conceal the umbilicus, which is rather large and deep, nearly circular. Operculum spiral, thin, horny, round-ovate, with the nucleus excentric and with two to three rapidly increasing whorls.

The animals of several alcoholic specimens were examined. Head large, short, thick, rounded or truncate, with two short, flat, obtuse anterior tentacles, wide apart, but connected together by a transverse fold; posterior tentacles short, thick, conical, smooth; no eyes visible. Pharynx short, thick, retractile; jaws crescent-shaped, strong, black. Verge situated just below the right posterior tentacle, small, papilliform, swollen at base; below this and farther back, a larger and thicker papilla, with basal swelling; on each side, between the mantle and foot, at about mid-length of the foot, a small mammiform papilla; two small, flat cirri behind and beneath the operculum. Foot broad, ovate, with two tentaculiform processes in front.

The largest specimens are badly broken; some of them were about 10^{mm} in length; greatest diameter of operculum, 6^{mm}; its breadth, 4.5^{mm}. A perfect, but small, specimen is 6^{mm} long; breadth, 6^{mm}; length of body-whorl, 5.2^{mm}; length of aperture, 4^{mm}; its breadth, 3.2^{mm}.

Station 1031, off Martha's Vineyard, in 255 fathoms, 1881. About a dozen specimens, all living, were taken from the interior of an old egg-case of a skate (*Raia*, sp.). Most of them were badly broken.

I have compared these specimens directly with original specimens of the fossil *Choristes elegans*, found in the post-pliocene of Canada by Principal J. W. Dawson, who very kindly sent me specimens, both adult and young.

Our specimens agree very closely with the smaller fossil ones in form and structure. The principal difference is in the much thinner and more fragile texture of the recent shells. This may be due to mere local conditions. Therefore, until more specimens of the recent shells are obtained, I prefer to consider it a thin and delicate variety of the ancient type.

Koonsia Verrill.

Trans. Conn. Acad., v, p. 545.

Allied to *Pleurobranchæa*, with which it agrees in the character of the head, tentacles, proboscis, and gill. It differs in having the back swollen and overhanging, both on the sides and posteriorly, with a distinct mantle-edge all around, and with a wide groove between it and the foot posteriorly, as well as laterally; the foot is narrower and prolonged posteriorly, with a specialized glandular groove near the end, beneath, and a conical papilla above, near the tip.

The external reproductive organs appear less complicated than in *Pleurobranchæa*.

The verge is armed with small hooks, but the spicule, present in the latter genus, is not protruded in any of our specimens of *Koonsia*, if present.

Koonsia obesa Verrill.
    Trans. Conn. Acad., v, p. 545, July, 1882.

Body large, stout, broad, with a large swollen back, smooth and white in the preserved specimens, and defined by the mantle-edge, which forms a rim along the lateral and posterior borders. Head large and broad, with two short, flat, posteriorly grooved, anterior tentacles, one at each corner; the anterior mantle-border runs between them, and supports a row of small papillæ. Posterior tentacles short, stout, flattened, ear-like, with the outer edges incurved, forming a large groove.

Foot broad and rounded anteriorly, with small auricles; long, tapered and acute posteriorly, extending some distance beyond the mantle; a conical papilla, near the tip, above; under side, near the end, with a narrow, elongated, depressed, glandular area, surrounded by a raised border; this is sometimes tinged with bright red, in alcohol; the rest of the foot is usually tinged with chocolate-brown.

Gill large, bipinnate, deep purple.

This species grows to a great size. One, from station 939, was over 5 inches (128ᵐᵐ) long; 4 inches (102ᵐᵐ) wide; and about 2 inches (50ᵐᵐ) high, even after preservation in alcohol.

Off Martha's Vineyard, stations 895, 939, 946, 1025, in 216 to 258 fathoms. Off Delaware Bay, station 1045, in 312 fathoms. At station 946, in 241 fathoms, seven young specimens were taken, some of them not over 1 inch long; these were associated with *Pleurobranchæa tarda*.

This genus is dedicated to Mr. B. F. Koons, of the U. S. Fish Commission, in 1880 and 1881.

NUDIBRANCHIATA.

Issa ramosa Verrill and Emerton.
    Verrill, Amer. Journ. Sci., xxii, p. 301, 1881; Trans. Conn. Acad., v, p. 547, pl. 58, figs. 36, 36 a.

Stations 940 and 949, in 130 and 100 fathoms.

Heterodoris robusta Verrill & Emerton.
    Heterodoris robusta Verrill and Emerton, Verrill, Trans. Conn. Acad., v, p. 549, pl. 58, figs. 35, 35a, 35b.

Off Martha's Vineyard, station 1029, in 458 fathoms.

Dendronotus arborescens Alder & Hancock.
    Verrill, Proc. Nat. Mus., iii, p. 385, 1880.

Station 1038, in 146 fathoms, 1881; several specimens.

Fiona nobilis Alder & Hancock.
    Verrill, Amer. Journ. Sci., xxii, p. 301, 1881.

Abundant at stations 935, 995, among *Anatifers*, adhering to pieces of floating timber.

Eolis papillosa (Linné).
Station 1032, in 208 fathoms, 1881.

Coryphella, sp. nov.
Station 1038, in 146 fathoms.

PTEROPODA.
Triptera columnella Rang.
Station 947, about 89 miles south of Martha's Vineyard, 1881.

LAMELLIBRANCHIATA.

Xylophaga dorsalis (Turton) Forbes & Han.
  Verrill, these Proc., ii, p. 197, 1879; Trans. Conn. Acad., v, p. 559, pl. 44, fig. 9, July, 1882.
Off Martha's Vineyard, stations 880, 998, in 252 and 302 fathoms. North of Cape Cod, in 20 to 110 fathoms.

Mya truncata Linné.
Off Martha's Vineyard, station 991, in 34 fathoms; one, dead.

Pholadomya arata Verrill & Smith.
  Verrill, Amer. Journ. Sci., xxii, p. 301, 1881; Trans. Conn. Acad.,v, p. 567, pl. 58, fig. 37.
Stations 871, 940, 949, 950, in 69 to 130 fathoms, 1880, 1881.

Mytilimeria flexuosa Verrill & Smith.
  Verrill, Amer. Journ. Sci., xvii, p. 302, 1881; Trans. Conn. Acad., v, p. 567, pl. 58, fig. 38.
Station 947, in 312 fathoms, 1881.

Neæra perrostrata (Dall).
  Neæra ornatissima (D'Orbigny), var. *perrostrata* Dall, Bulletin Mus. Comp. Zool., ix, p. 110, 1881.
This shell has been examined by Mr. Dall and identified with those from the "Blake" expedition.
Stations 871, 874, 876, 1880, in 85 to 120 fathoms. Gulf of Mexico, 339 fathoms,—Dall.

Neæra obesa Lovén.
  G. O. Sars, Moll. Reg. Arct. Norvegiæ, p. 87, pl. 6, figs. 4, *a-c*, 1878.
Off Martha's Vineyard, stations 869, 891 to 895, 898, in 192 to 500 fathoms; stations 938, 947, 994, 997, 998, 1028, in 302 to 410 fathoms, 1881. Bay of Fundy, 1872; Gulf of Maine, 52 to 92 fathoms, 1873, 1874; off Cape Cod, 106 fathoms, 1879.

Verticordia cælata Verrill.
  Trans. Conn. Acad., v, p. 566, July, 1882.
Station 949, in 100 fathoms, 1881.

Syndosmya lioica Dall.
  Bulletin Mus. Comp. Zool., ix, p. 133, 1881.
Station 871, in 115 fathoms, 1880, one broken specimen; station 949,

100 fathoms, three specimens. Gulf of Mexico, 30 to 805 fathoms, "Blake" exp. (t. Dall).

I have compared our shell with specimens sent to me by Mr. Dall.

Spisula ovalis (Gould).

Stations 941, 950, off Martha's Vineyard, in 69 to 76 fathoms, dead; also at stations 965, 975, 976, 978, 981 to 983, off the south side of Cape Cod, in 15 to 41 fathoms.

Cardium (Fulvia) peramabilis Dall.

Dall, Bulletin Mus. Comp. Zool., ix, p. 132, 1881.
Cardium, sp. Verrill, Proc. Nat. Mus., iii, p. 407, 1880.

Station 871, in 115 fathoms, 1880; one valve. Gulf of Mexico 50 to 119 fathoms, "Bache" and "Blake" exp. (t. Dall).

I have identified our shell by direct comparison with specimens sent to me by Mr. Dall.

Diplodonta turgida Verrill & Smith.

Verrill, Amer. Journ. Sci., xxii, p. 303, 1881; Trans. Conn. Acad., v, pl. 58, fig. 42.

Station 950, in 69 fathoms, 1881.

Cryptodon subovatus? (Jeffr.). V.

Axinus subovatus Jeffreys, Proc. Zool. Soc. London, for 1881, p. 704, pl. 61, fig. 8, 1882.

A single specimen, from station 891, in 500 fathoms, appears to be this species. It is very thin and delicate, and very inequilateral.

Montacuta ovata Jeff.

Jeffreys, Proc. Zool. Soc. London, for 1881, p. 698, pl 61, fig. 4, 1882.
Verrill, Trans. Conn. Acad., v, p. 571, July, 1882.

Off Martha's Vineyard, 100 to 153 fathoms, living. These shells are encrusted with a thick coat of iron oxide. Perhaps the encrusted shells, recorded by me in 1880 as *Tellimya ferruginosa*, was the same species. The specimens were too much eroded for accurate determination.

Solemya velum (Say), var. *borealis* (Totten).

Off Chesapeake Bay, station 898, in 300 fathoms; one living specimen.

Dead shells of *S. velum* were taken off Martha's Vineyard, station 871, in 115 fathoms. I regard *S. borealis* as the adult of *S. velum*.

Leda unca Gould.

Verrill, these Proc., iii, p. 401, 1880.

Mr. Dall has identified our shells with those taken in the Gulf of Mexico by the Blake exp., in 54 to 640 fathoms.

He refers them to *L. Jamaicensis* D'Orbigny. I am not satisfied that this identification is correct, for D'Orbigny's figure is not very like our shells, of which we have taken large numbers.

Additional localities, in 1881, were stations 921, 949, 951, 1038, in 65 to 219 fathoms.

Leda tenuisulcata (Couth.) Stimpson.

Station 973, in 17 fathoms, off south side of Cape Cod. Off Chesapeake Bay, station 898, in 300 fathoms.

Leda pernula (Müller).

Station 1025, in 216 fathoms. Off Halifax, 59 fathoms.

Nucula tenuis (Mont.) Turton.

Stations 895, 943, 997 to 999, in 153 to 335 fathoms.

Modiolaria nigra (Gray) Lovén.

Station 921, in 65 fathoms, 73 miles south of Martha's Vineyard; also at stations 985, 986, 991, 993, off Martha's Vineyard, in 26 to 39 fathoms. Off Chesapeake Bay, station 900, in 31 fathoms.

Modiolaria corrugata (Stimpson) Mörch.

Station 918, in 45 fathoms, 61 miles south of Martha's Vineyard.

Modiolaria polita Verrill and Smith.

> *Modiola polita* Verrill and Smith, in Verrill, Amer. Journ. Sci., xx, pp. 392, 400. Nov., 1880; Verrill, Proc. U. S. Nat. Mus., iii, p. 402, Jan., 1881; Trans. Conn. Acad., v, p. 578, July, 1882.
> Dall, Bulletin Mus. Comp. Zool., ix, p. 116, 1881.
> *Mytilus luteus* Jeffreys, French Expl. in Bay of Biscay, in Rept. Brit. Assoc., 1880 (no description); Ann. and Mag. Nat. Hist., Oct., 1880, p. 315 (no description).
> *Modiola lutea* Fischer, Jour. de Conchyl., iii, vol. xxii, p. 52, Jan., 1882.

Two living specimens were taken at station 895, in 238 fathoms. Gulf of Mexico, 339 fathoms, "Blake" Exp. (t. Dall). Mr. Dall has compared his specimens with our original types. Bay of of Biscay, 677 to 960ᵐ,—Jeffreys, Fischer.

Mr. Dall has suggested that this species belongs to *Modiolaria*, rather than to *Modiola*. In this opinion I am disposed to concur. It forms a large nest of byssus-fibers and mud. The largest examples show fine radiating lines.

Idas argenteus Jeff., var. ? **lamellosus** Verrill.

> Trans. Conn. Acad., v. p. 579, July, 1882.
> *Idas argenteus* Jeffreys, Annals and Mag. Nat. Hist., Nov., 1876, p. 428; Proc. Zool. Soc. London, 1879, p. 570, pl. 45, fig. 3.

This shell is thin, translucent, covered with a yellowish epidermis; umbos and hinge reddish brown; inner surface iridescent; sculpture, distinctly raised thin concentric lamellæ, which are not crowded; no radiating lines. Some of the specimens have several horny, sharp, stiff, beard-like processes projecting from the posterior and dorsal surfaces. One of the largest specimens is 5.5ᵐᵐ long; greatest height, 2 2ᵐᵐ.

Station 997, in 335 fathoms; several living specimens.

Pecten glyptus Verrill.

> Trans. Conn. Acad., v, p. 580, July, 1882 (description).
> *Pecten*, sp., near *opercularis* Verrill, Proc. Nat. Mus., iii, p. 403, 1881.

Amussium fenestratum (Forbes) Jeffreys.

Jeffreys, Proc. Zool. Soc. London, 1879, p. 561.

Verrill, Trans. Conn. Acad., v, p. 582, July, 1882 (description).

Pecten fenestratus Forbes, Rept. Brit. Assoc. for 1843, pp. 146, 192, 1844.

Verrill, Proc. Nat. Mus., iii, p. 403, Jan., 1881 (description).

Pecten inæquisculptus Tiberi (t. Jeffreys).

This elegant species has been dredged, living, at several stations off Martha's Vineyard, in 86 to 310 fathoms. It was most numerous at stations 949 and 1040, in 100 and in 93 fathoms.

It occurs on the European coasts, off Portugal and in the Mediterranean Sea; from 50 to 250 fathoms.

DESCRIPTIONS OF SOME NEW NORTH AMERICAN BIRDS.

By ROBERT RIDGWAY.

1. Catherpes mexicanus punctulatus, subsp. nov.

CH.—In coloration, somewhat intermediate between *C. mexicanus* (*typicus*) and *C. conspersus* (paler than the former, darker than the latter), but in dimensions agreeing best with the latter.

Adult: Above dull rusty brown, less reddish anteriorly, the whole top of head, nape, back, and scapulars distinctly speckled with white, each white dot immediately preceded by an equally distinct one of dusky; rump and outer surface of wings ferruginous, the former nearly immaculate, the latter rather coarsely barred with black; upper tail-coverts chestnut-rufous, each feather with a white terminal and black subterminal dot. Tail clear rusty rufous, crossed by about seven or eight narrow, irregular bars of black, these less than .05 of an inch broad on the middle feathers, and about .10 of an inch wide on the outer pair. Chin, throat, and jugulum silky white (more or less tinged with ochraceous), passing gradually on the breast into soft ochraceous, this changing to rich ferruginous on sides, abdomen, and remaining lower parts, the parts thus colored marked, more or less distinctly, with black dots or bars, and, in some specimens, white terminal specks. Bill dusky, the mandible paler; iris brown; legs and feet brownish black or dark brown. Wing 2.25–2.40 (2.32), tail 2.00–2.20 (2.12). culmen .75–.85 (.81), bill from nostril .52–.65 (.60), tarsus .68–.72 (.70), middle toe .50–.58 (.53). (Five specimens.)

Hab.—California, north to San Francisco and the Calaveras River.

The Californian specimens of this species appear to differ uniformly from examples obtained in the Interior, in the characters indicated above. They are all decidedly darker in coloration, approaching in this respect the typical *C. mexicanus* of Mexico, but they are much smaller than the latter race. Compared with a series of seven examples of *C. conspersus*, as to dimensions, five examples of *punctulatus* average the same in length of wing, .05 of an inch less in length of tail, the middle toe and tarsus each .02 longer, and the culmen .09 of an inch longer. It is quite likely, however, that a larger series of each would negative these apparent slight differences.

Types, 82715, ♂ ad., Forest Hill, Placer County, California, October 7, 1862, F. Gruber, and 79154, ♀ ad., Calaveras R., 30 miles east of Stockton, L. Belding.

2. Lophophanes inornatus griseus, subsp. nov.

CH.—Differing from *L. inornatus* (*typicus*) in rather larger size and decidedly grayer colors. Above uniform brownish gray; beneath pale grayish, lighter on the middle of the abdomen. Wing 2.80–3.00, tail 2.40–2.70, culmen .40–.48, tarsus .80–90.

Hab.—Middle Province of United States, from Nevada, Utah, and Colorado to New Mexico and Arizona.

All specimens of this species from the Middle Province region differ from Californian examples as noted above, the difference being absolutely constant in the considerable series examined. The distinctions between the two races may be stated more precisely as follows:

Var. INORNATUS. Above grayish olive-brown, beneath grayish white. Wing 2.68–2.90, tail 2.20–2.60, culmen .38–.40, tarsus .80–.88. *Hab.*—California and Western Oregon.

Var. GRISEUS. Above brownish gray, beneath paler grayish. Wing 2.80–3.00, tail 2.40–2.70, culmen .40–.48, tarsus .80–.90. *Hab.*—Middle Province of United States.

3. Geothlypis beldingi, sp. nov.

SP. CH.—*Adult ♂* (No. 87685, U. S. Nat. Mus., San José del Cabo, Lower California; L. Belding): Entire lower parts very rich yellow (much deeper than in *G. trichas*), paler, but not inclining to white, on the anal region, the sides and flanks tinged with brownish; whole forhead, lores, malar region, and auriculars deep black, this having exactly the same limits and extent as in *G. trichas*, *G. melanops*, and *G. rostrata*, but *bordered behind for its whole extent with bright yellow*, inclining to whitish only in a very limited space, immediately back of the auriculars. Entire upper parts uniform olive-green (richer and browner than in the allied species), anteriorly fading gradually into the yellow behind the black mask, the occiput and nape somewhat tinged or indistinctly clouded with umber-brown. Bill wholly deep black; feet brownish. Wing 2.60, tail 2.70, its graduation .50, culmen .55, bill from nostril .40, tarsus .95, middle toe .65.

Adult ♀ (No. 87686, same locality, &c.): Above olive-green, the pileum and sides of head more brownish; lores, suborbital region, etc., brownish olive, mixed somewhat with yellowish; malar region and entire lower parts bright yellow, more ochrey-whitish about the anal-region. Bill black above, brownish below; feet pale brownish. Wing 2.35, tail 2.40, its graduation .30; culmen .55, tarsus .95, middle toe .65.

The two specimens described above have been compared with examples of all the known Mexican *Geothlypeæ*, excepting *G. speciosa* Scl., of which there is probably no specimen in any American collection. The latter, so far as I am able to judge from descriptions, seems to differ in "ochre-yellow" instead of intensely rich gamboge, lower parts,[*] in the smaller size (wing 2.40, tail 2.30, tarsus .85), and apparently in the absence of a light band bordering the hinder margin of the black mask, which it is said also occupies the top of the head, while in the present bird the black extends backward only .30–.35 of an inch from the frontal antiæ. From *G. trichas*, *G. melanops*, and *G. rostrata*, the only other related species, the differences are so great as not to need specification.

Since the above was written three more specimens (skins) have been received from Mr. Belding. The two males agree minutely with the one described above, except that the yellow of the lower parts is scarcely so intense, though still much deeper than in the allied species. In both there is the same very slight whitening (for the space of about .20 of an inch) just behind the auriculars, the feathers bordering the black mask being elsewhere entirely light yellow. These specimens measure as follows:

Number.	Locality.	Date.	Wing.	Tail.	Gradution of tail.	Culmen.	Bill from nostril.	Tarsus.	Middle toe.
87532	San José, Lower California ..	April 24, 1882	2 55	2 60	.40	.52	.35	.95	.65
87533	San José, Lower California...	April 29, 1882	2.60	2.70	.50	.52	.35	.95	.65

The female (No. 87534, San José, April 29) differs from the type chiefly in having the auriculars and lores quite distinctly darker, forming a slight indication of the mask of the male. Wing 2.40, tail 2.50, its graduation .45; culmen .50, tarsus .90, middle toe .60.

This fine new species is one well worthy to bear the name of the enthusiastic naturalist who has had the good fortune to discover it—Mr. L. Belding, of Stockton, Cal., already well known to ornithologists through his valuable contributions to our knowledge of Californian birds[†]—to whom I take great pleasure in dedicating it.

4. Rallus beldingi, sp. nov.

CH.—.Most resembling *R. elegans*, but darker and richer colored throughout, the sides and flanks with the white bars much narrower, and marked also with very distinct blackish bars. Size, smaller.

Adult ♀ (No. 86419, Espiritu Santo Islands, Lower California, February 1, 1882; L. Belding): Pileum and upper half of nape dark sooty brown or sepia; ground-color of other upper parts deep olive-brown (much as in *R. virginianus*—decidedly darker than in *elegans*), broadly

---

[*] *Cf.* BAIRD, Review Am. B. i, p. 223, and SALVIN & GODMAN, Biol. Centr. Am. Aves. i, p. 152.

[†] *Cf.* these Proceedings, vol. i, pp. 388–449.

striped with brownish black, about as in *R. obsoletus;* wing-coverts dull chestnut-brown, tinged with olive, the exterior feathers more rusty; supra-loral stripe light cinnamon, the feathers white at base; lores, continuous with a broad stripe behind the eye, dull grayish brown; under eyelid whitish; malar region, cheeks, entire foreneck, jugulum, and breast rich cinnamon, much deeper than in any of the allied forms; chin white, throat mixed white and cinnamon, the latter on tips of the feathers; entire sides and flanks rather dark hair-brown (less olivaceous than upper parts), rather distinctly barred with blackish and very sharply barred with pure white, the bars of the latter color about .05–.07 of an inch in width; lining of wing dark brown, with very narrow white bars; anterior and middle portion of crissum marked much like the flanks, the lateral and terminal lower tail-coverts pure white. Basal two-thirds of the mandible, and posterior portion of maxillary tomium deep orange; rest of bill dark horn-brown, the end of the mandible paler; feet dark horn-brown. Wing 5.70, tail 2.50, culmen 2.15, depth of bill at base .50, in middle .30; tarsus 1.92, middle toe 1.80.

Compared with specimens of all the allied species and races of the genus, the present bird is instantly distinguishable by the characters pointed out above. In intensity of coloration it most nearly resembles *R. virginianus;* but, apart from its much larger size, presents the following differences of coloration: The side of the head below the eye is chiefly cinnamon, whereas this portion is in *R. virginianus* very distinctly ashy; the breast, etc., are both deeper and redder cinnamon; the ground-color of the sides and flanks much paler (uniform black in *virginianus*); the black stripes of the upper parts are both narrower and less sharply defined, while the wings are much less rusty.

Compared with the larger species (*R. longirostris*, with its races, *R. elegans* and *R. obsoletus*), it is difficult to say to which it is most nearly related. None of the forms of *R. longirostris*, however, need close comparison, the darkest colored race of that species (*saturatus*, from Louisiana) having broader black stripes and a very different (ash-gray) ground-color above; the breast, &c., a very much duller and lighter cinnamon, and the flank-bars broader and on a uniform ground-color. *R. obsoletus* agrees best in the coloration of the upper parts, which, however, in all specimens (including one from San Quentin Bay, on the western side of Lower California) have a lighter, and in some a decidedly grayer, ground-color; but the white flank-bars are much broader, with unicolored interspaces, the breast very conspicuously paler, and the size considerably greater. *R. elegans* has also the breast paler, the ground-color of the upper parts a lighter and much more yellowish olive, and the black stripes much more sharply defined. Upon the whole, I see no other way than to consider the specimen in question as representing a very distinct species or local race, which I take great pleasure in naming after its collector.

DESCRIPTION OF A NEW SPECIES OF URANIDEA (URANIDEA RHOTHEA) FROM SPOKANE RIVER, WASHINGTON TERRITORY.

BY ROSA SMITH.

Head 3 (3⅗); depth 4½ (5½); length (30737) 3 inches. D. VIII–16; A. 11; V. I, 4; Br. 6.

Subgenus COTTOPSIS Girard.

Body of the usual form in the genus, widest anteriorly, gradually tapering to the tail, the greatest width just behind head, 1⅓ in greatest depth of body. Head wide, depressed, its depth half its width. Mouth moderate, maxillary reaching the vertical of posterior margin of pupil. Eye moderate, 1⅕ in snout. Snout 3½ in head, more pointed and the head broader than in *U. gulosa* or *U. aspera*. Interorbital space rather narrow, 2 in eye, slightly concave.

Villiform teeth on jaws and vomer, about as in *U. aspera*, the palatine teeth forming a broader and much longer band than in the latter species.

Opercular spines nearly as in *U. aspera;* a sharp spine at the angle of preopercle directed upward and backward, below which are two small and very blunt ones. A single spine directed forward at the inferior angle of opercle.

Skin of the head smooth to the touch, but there are numerous very minute tubercles on the nose, interocular width, and vertex. An appearance of prickles is observed on the space between occiput and origin of dorsal, but no roughness can be felt. Conspicuous prickles extend from the scapula and origin of dorsal fin almost to base of caudal, and below the lateral line a distance equaling the interorbital space; these prickles more prominent than in *U. aspera*. The lower surface of head, the abdominal region, and base of anal are smooth and without any trace of warts or prickles.

Isthmus rather broad, the gill-membranes not forming a fold across it. First dorsal low, its margin convex; fourth and fifth spines highest, about 4 in head. First ray of soft dorsal equaling highest part of spinous, increasing to the third, which is 2⅓ in head, the outline nearly straight from third to fifteenth rays. Caudal subtruncate. Anal similar to soft dorsal, its last ray inserted opposite insertion of fourteenth dorsal ray, the free tips not extending quite as far as those of dorsal. Ventrals not reaching vent, 2 in head. Pectoral attaining third ray of soft dorsal and barely to beginning of anal.

Caudal peduncle nearly 4 in greatest depth. In young examples the opercular spines and dermal prickles are more conspicuous than in the adult.

Color, in spirits, olivaceous with blackish markings. Upper part of head dark gray, with a darker area on occiput. Two blackish spots at base of spinous dorsal, and two larger spots, or almost bands, at base of soft dorsal, extending below lateral line, and a black connecting band below lateral line, which extends along middle of peduncle, spreading out, fan-like, at base of caudal; the general hue of the prickly region is dark gray; the thorax, abdomen, and base of anal yellowish-white with fine blackish punctulations on these regions in the larger specimen, the smaller ones without dots on this area. Fins with small black spots which tend to form waving horizontal lines in their arrangement. Ventrals almost plain yellowish-white. Branchiostegal membrane punctate with black in adult. Lining of mouth plain whitish. Peritoneum white.

This species is known to me from four individuals collected from the Falls of the Spokane River, in Washington Territory. The largest and smallest of these have been presented to the National Museum by Mr. W. G. W. Harford, and their catalogue number is 30737.

The smaller number of fin-rays separates this species readily from *U. aspera* and *U. semiscabra* (D. X, 21, A. 17 in *aspera;* D. VII, 18, A. 14 in *semiscabra*), while the very prickly skin at once distinguishes it from all others.

SAN DIEGO, CAL., *June* 1, 1882.

ON THE EASTWARD DISTRIBUTION OF THE BLACK-TAILED DEER (*CARIACUS COLUMBIANUS*).

By CAPT. CHAS. BENDIRE, U. S. A.

[Extracted from a letter to Prof. S. F. Baird.]

I have for the past two years carefully examined a great number of hides of the so-called black-tailed deer found in this region, and have sent several lots of tails to Judge J. D. Caton, who is the best authority on the *Cervidæ* we have. The judge is perfectly right in saying, "The most extraordinary fact in connection with this deer is the extremely narrow limits of its range, and this must be still farther restricted. I am now satisfied that it reaches no farther than the eastern slopes of the Cascade Mountains instead of the foot-hills of the Rockies." I have examined skins from various portions of this country, a great many taken near the eastern border of the Cascades, about Prineville, Oregon, also the Warm Spring Indian Reservation, The Dalles, Camp Harney, Oregon, Yakima Valley, and the Spokane Fort region, and I have to see the first true black-tailed deer skin (the *Cervus columbianus*) yet which comes from any point east of the Cascade range. I can speak positively about this, as I have made very careful examinations, and have looked over several thousands of hides, brought together from various places, at the instigation of Judge Caton. If this deer occurs at all east of the Cascade range, it will be found about Fort

Klamath, which is located on the eastern slope of this range, and well up in the mountains. But I do not think that it will even be found there, and it is strange to account for it, but it is true all the same. There are two gentlemen stationed at Fort Townsend, Washington Territory, who can and will get you specimens, I think. They are Col. Alexander Chambers, Twenty-first Infantry, and Capt. Stephen P. Jocelyn, same regiment. Come to think of it, they are likely to be removed any day, as their regiment is to go to Wyoming. I will try and see if I can't find some one to do this, and will write to a taxidermist in Portland about it, whom I know.

FORT WALLA WALLA,
Washington Territory, April 22, 1882.

DESCRIPTION OF A NEW SPECIES OF BLENNY (ISESTHES GILBERTI) FROM SANTA BARBARA, CALIFORNIA.

By DAVID S. JORDAN.

Head 4 in length (4⅔ with caudal); depth 4 (4⅔). D. XII, 19; A. II, 21. Length of largest specimen 4½ inches.

Body comparatively robust, deep, and compressed. Head large, rounded, the anterior profile less blunt than in *I. gentilis* and less rounded, nearly straight from tip of snout to above eye, thence again nearly straight to front of dorsal. Length of snout about equal to diameter of eye, 4¼ in head. Mouth rather small, terminal, the maxillary reaching to opposite middle of eye, 2⅔ in head. Teeth subequal, with no trace of posterior canines. Superciliary tentacle large, multifid, much branched from near the base, the principal division 3⅔ in head.

Gill-openings larger than in *I. gentilis*, extending downward to the level of lower edge of pectoral, the length of the slit, 1¾ in head.

Lateral line developed beyond the straight part, its posterior portion curved downwards.

Dorsal fin continuous, with a slight but distinct depression between the spinous and soft parts, the spines somewhat curved, but stiff and strong, the longest spine about 2⅛ in head; longest soft rays 2 in head. Caudal fin free from dorsal and anal, 1½ in head. Ventrals 1⅓ in head. Pectorals about as long as head.

Males, as usual in this genus, with the anal spines partly detached, and provided with fleshy tips.

Coloration olivaceous, the body and fins everywhere profusely mottled and reticulated with darker. Obscure dark shades extending downward from eye across, or partly across, lower side of head. Head without distinct spots, or other sharply defined markings; no pale bars on side of head in either sex. Some yellowish markings on anterior part of dorsal.

Numerous specimens of this species were obtained by Mr. Charles H.

Gilbert and myself in rock pools, at Santa Barbara, Cal., in the winter of 1880. It was at first supposed by us to be identical with *Blennius gentilis* Girard, a species of which we obtained no adult specimens. Numerous specimens of *Isesthes gilberti* have been distributed under the name of *Hypleurochilus gentilis* (number 26917, U. S. Nat. Mus.). Four specimens from Santa Barbara (26916) are the types of the present description; all of them are males.

The following description of specimens of *Isesthes gentilis* may be compared with the foregoing.

Head $3\frac{2}{3}$ in length ($4\frac{1}{2}$ with caudal); depth 4 ($4\frac{4}{5}$). D. XIII, 17; A. II, 19. Length (26645) $3\frac{4}{5}$ inches.

Body rather robust, deep and compressed, the head large, very bluntly and evenly rounded in profile, more obtuse and more evenly curved than in *I. gilberti*, the snout shorter, about equal to eye, $4\frac{1}{2}$ in head. Mouth rather small, terminal, the maxillary reaching to opposite middle of eye, its length 3 in head. Teeth subequal, the hindmost on each side of upper jaw shorter than the others, and a little apart from them but not forming "a small canine" as stated by Girard.

Superciliary tentacle long and simple in all specimens examined, its length about 3 in head. (Tentacles much smaller in the female, according to Steindachner.)

Gill-opening extending downward not quite to lower edge of pectoral, its length (vertical) $2\frac{1}{6}$ in head. Lateral line with only the straight anterior portion developed, not curved downward posteriorly.

Dorsal fin continuous, with scarcely a trace of emargination between the spinous and soft parts. Dorsal spines comparatively low and flexible, much less strong than in *I. gilberti*, the longest spines 3 in head; longest soft rays $1\frac{2}{3}$. Caudal free from dorsal and anal, $1\frac{2}{5}$ in head. Ventrals $1\frac{2}{3}$ in head; pectorals $1\frac{1}{3}$.

Coloration, in spirits, brown, the whole body closely mottled and blotched with darker brown, so that the light ground color forms, especially anteriorly, light reticulations around darker spots. On the head the dark spots are small and close together, smallest anteriorly, the lower parts of the head being immaculate. Extending from the curve of the preopercle downward, across the interopercle and branchial region, is a sharply defined white bar (said to be golden-yellow in life), edged with black. Behind this and parallel with it across subopercle and isthmus is a similar bar. These bars (which, according to Steindachner, are characteristic of the male) are present in all specimens examined. A few pale spots or bars in front of these. Back with about 6 dusky cross-shades; below each of these is an oblong dark blotch, the anterior placed along the lateral line, all together forming an interrupted dark stripe. A similar dark stripe near the median line of the body, interrupted by some pale blotches. Fins all blotched and spotted with light and dark colors, but without distinct markings (a blue spot on front of dorsal in life, according to Steindachner). Ventrals and anal nearly plain blackish in males, the base of the anal with a pale streak. Two specimens from Cape San Lucas, supposed to be the female of this species, have the tentacles much shorter, not longer than pupil, lack the pale stripes on the head, and have a very distinct blackish blotch on front of spinous dorsal.

The following specimens of this species (all of them, except 2481, apparently males) are in the National Museum:

489. (Girard's type.). Monterey. Trowbridge.

7859. (3). San Diego. A. Cassidy.

26645 (2). "California" (probably San Diego). Mus. Comp. Zool.

30742. San Diego. W. Cooper.

2481. Cape San Lucas. J. Xantus.

UNITED STATES NATIONAL MUSEUM, *June* 21, 1882.

DESCRIPTION OF A NEW SPECIES OF CONODON (CONODON SERRIFER), FROM BOCA SOLEDAD, LOWER CALIFORNIA.

By DAVID S. JORDAN and CHARLES H. GILBERT.

Conodon serrifer, sp. nov.

Head $3\frac{2}{5}$ in length to base of caudal; depth $3\frac{2}{5}$. D. XI, I, 12; A. III, 7. Scales 6–53–15. Length of largest specimen 8 inches.

Body comparatively elongate, elliptical, little compressed, the dorsal and ventral outlines regularly and nearly equally curved, the back not much elevated and not specially compressed. Head rather short, broad, not very acute anteriorly, the profile nearly straight from snout to base of dorsal; snout short, about equal in length to the large eye, $3\frac{1}{5}$ in head. Interorbital area broad and quite flat, its width $4\frac{1}{5}$ in head. Mouth moderate, terminal, oblique, the lips moderately developed. Maxillary extending to opposite front of eye, $2\frac{3}{4}$ in head. Premaxillaries in front on level of middle of eye.

Teeth in moderate bands, those in the outer series enlarged, but much less so than in *C. nobilis*, the teeth slenderer than in the latter; two teeth in front of lower jaw somewhat canine-like. Preorbital narrow, its least width about two-fifths diameter of eye. Jaws equal in front.

Preopercle with its posterior margin somewhat concave, armed with strong teeth, which are directed backward and somewhat upward. Angle of preopercle with a strong spine directed backward, its length about half length of eye. Lower limb of preopercle with strong spinous teeth (as in the species of *Plectropoma*), directed forward and downward, becoming gradually smaller anteriorly. Nostrils small, roundish, the anterior largest. Gill-rakers rather slender, of moderate length.

Scales rather irregularly arranged, those above lateral line forming series parallel with the lateral line, which are somewhat broken opposite the angulation of the lateral line. Small scales on soft parts of dorsal and anal.

Dorsal fin low, divided almost to base, the spines rather strong. First and second spines short and slender, the second little more than one-third the height of the third; the fourth or longest $2\frac{1}{6}$ in head; soft dorsal low, its longest rays 3 in head. Caudal subtruncate, the upper rays longest, $1\frac{3}{4}$ in head. Anal rather low, the second spine 2 in head, much longer and stronger than the third, which is little lower than the soft rays. Pectoral pointed, $1\frac{1}{10}$ in head; ventrals $1\frac{2}{3}$.

Color dusky bluish above, silvery below. Sides of back with about seven short black bars, each much narrower than the interspaces, the last under last rays of dorsal, all terminating below at the lower edge of the dark hue of the back. Fins all pale.

The types of this species (17546; U. S. Nat. Mus.), three adult specimens in good condition, were obtained by Dr. Thomas H. Streets at Boca Soledad, on the Pacific coast of Lower California. They have been mentioned by Dr. Streets (Bull. U. S. Nat. Mus., vii, 50, 1877) under the name of *Conodon plumieri*. They are closely related to the latter species, but distinguishable as follows:

COMMON CHARACTERS.—Body rather elongate; preopercle with strong antrorse teeth on its lower limb and a spine at its angle; series of scales above lateral line parallel with it; outer series of teeth in both jaws enlarged; dorsal deeply notched; soft rays of vertical fins scaly; second anal spine enlarged. (CONODON, C. & V.)

a. Back distinctly elevated and compressed, the depth equal to length of head, $3\frac{1}{5}$ in body; teeth of outer series very strong and thick; second dorsal spine more than half length of third; second anal spine more than half length of head; preopercular spine small; dark bars on sides extending to level of lower edge of pectoral ....NOBILIS.

aa. Back not elevated, the depth equal to length of head, $3\frac{2}{3}$ in body; teeth of outer series moderately enlarged, slender; second dorsal spine small, less than one-third length of third; second anal spine about half length of head; preopercular spine very strong; dark bars on sides not extending to level of pectorals...SERRIFER.

It may be here observed that of the two specimens referred to *Pristipoma leuciscus* by Dr. Streets, one (17539) belongs to *Pomadasys axillaris* (Steind.), the other (30746) to *Pomadasys nitidus* (Steind). The variety of *Pomadasys leuciscus* from Mazatlan and Panama mentioned by us (Proc. U. S. Nat. Mus. 1881, 387, foot-note) has received from Dr. Steindachner (Neue & Seltene Fische, aus. K. K. Museum, Wien, &c., 1879. 30, 52. taf 9. f. 2) the name of *Pristipoma leuciscus* var. *elongatus*. As it is apparently a valid species, although very closely related to *P. leuciscus*, it may stand as *Pomadasys elongatus*. It is much more abundant than the typical *leuciscus*.

UNITED STATES NATIONAL MUSEUM, *June 26, 1882.*

CATALOGUE OF THE FISHES COLLECTED BY MR. JOHN XANTUS AT CAPE SAN LUCAS, WHICH ARE NOW IN THE UNITED STATES NATIONAL MUSEUM, WITH DESCRIPTIONS OF EIGHT NEW SPECIES.

By DAVID S. JORDAN and CHARLES H. GILBERT.

Mr. John Xantus, when stationed at Cape San Lucas as a tidal observer for the Coast Survey, brought together a very large collection of objects of natural history, among which was a most excellent series of the fishes of the coast. The collections were formed under the auspices and direction of the Smithsonian Institution. They were studied by Professor Gill, who published descriptions* of most of the species in Proceedings of the Academy of Natural Sciences of Philadelphia in 1862 and 1863. Later, during a period of confusion in the Museum, this collection was scattered and many of the specimens lost or destroyed, and the study of the undescribed portion was abandoned by Professor Gill. The writers have gone over the entire collection again, and give here a catalogue of what remains. Even after the extensive collections studied by Günther, Steindachner, and the writers, there still remain in the Xantus collection several species new to science.

It may be observed that the descriptions published by Professor Gill are, for the most part, taken from immature fishes. This accounts for many discrepancies between these descriptions and those taken from adults of the same species. Most of the specimens obtained by Xantus were taken from tide pools and rocks, and few or none bought in the markets.

1. Elops saurus L.

2521. Small specimens.

2. Clupea thrissina sp. nov.

6388, 2524, 6339. Several specimens in fair condition, the largest $7\frac{1}{4}$ inches in length.

Allied to Clupea (Harengula) clupeola.

Head 4 in length; depth $3\frac{1}{3}$. D. I, 15; A. I, 13 or I, 14. Scales about 40–10. Ventral scutes 16 + 13.

Body rather deep, but more elongate than usual in the group called Harengula, to which this species belongs; rather strongly compressed. Head large, deep, rather blunt anteriorly. Mouth not large, rather oblique, the lower jaw projecting; the upper jaw scarcely emarginate in

---

* Catalogue of the Fishes of Lower California in the Smithsonian Institution, collected by Mr. John Xantus. By Theodore Gill. Part I, in Proc. Ac. Nat. Sci. Phila. 1862, pp. 140–151; Part II, op. cit. pp. 242–246; Part III, op. cit. 249–262; Part IV, op. cit., 1863, pp. 80–92. A few species were also described in other papers of Professor Gill, both earlier and later than those here mentioned.

Sept. 5, 1882.

front, its tip on the level of the pupil. Lower jaw very deep, its depth half its length. Maxillary extending to somewhat past the vertical from the front of the pupil, its length 2 in head.

Both jaws with small teeth, which appear to be permanent; teeth present also on palatines, pterygoids, and tongue, the teeth on the pterygoids very conspicuous, forming a large patch.

Eye large, 3 in head. Cheeks much longer than deep, not as deep as eye, the anterior margin of the preopercle very oblique. Opercle short and deep, shorter than eye, its posterior margin nearly vertical. Cheeks and opercles marked with fine, but distinct, branching striæ.

Gill-rakers rather short, slender, and close-set, about 30 below the angle of the arch. Longest gill-raker about half diameter of eye.

Scales firm and adherent, their posterior margins less convex than usual, rough with small fine teeth. Scales before dorsal similar to the others, but much smaller. Belly sharply compressed, the scutes strong, especially behind ventrals.

Distance from snout to dorsal $2\frac{3}{5}$ in length. Dorsal fin about as high as long, its free margin concave, its last ray slightly longer than that which precedes it. Length of anterior rays of dorsal $1\frac{1}{2}$ in head. Caudal well forked, the lower lobe slightly the longer, about as long as head. Anal low. Ventrals 2 in head; pectorals $1\frac{1}{3}$.

Color bluish above, silvery below; fins all pale; a round black spot behind upper part of gill-opening.

3. Clupea, sp. incog.

2534. A single young herring in poor condition, not belonging to any species known to us, but not in condition for description.

4. Pristigaster? sp. incog.

15443. A young specimen in very bad condition, which we are unable to identify with any of the known species of this type.

Body elongate, with a very distinct silvery stripe. Lower jaw strongly projecting, its teeth very strong, much stronger than upper teeth. Ventral outline not very prominent, strongly serrate. Ventral fins now wanting, but perhaps destroyed. It may possibly be a species of *Chirocentrodon*.

5. Synodus scituliceps Jor. & Gilb.

A single young specimen in bad condition, apparently belonging to this species.

6. Characodon furcidens, sp. nov.

9571, 30971. Many specimens, in fair condition, except that the coloration has faded; the largest $3\frac{1}{4}$ inches in length.

Head 4 in length; depth $3\frac{2}{5}$. D. 15 to 17; A 13. Scales about 50-15.

Body of a form different from that of the species of *Cyprinodon;* comparatively elongate, not greatly compressed, the head rather low and broad, depressed; the profile rising evenly from the tip of the snout to the nape, the region thence to the dorsal gibbous, especially in the larger examples, the caudal peduncle comparatively long and slender, about as long as head.

Anterior teeth large, firmly fixed, all bicuspid or Y-shaped, in a single series; a band of minute villiform teeth behind them, at least in upper jaw. Mandible not extending back to front of eye. Eye rather large, $3\frac{1}{3}$ in head. Interorbital area wide, very nearly half head.

Scales rather small, those on top of head not much larger than the others; humeral scale not enlarged. Opercle connected by membrane to shoulder girdle, from upper base of pectoral upward, as in *Cyprinodon*. Insertion of dorsal very far back, midway between base of caudal and base of pectoral. First ray of dorsal very slender and articulate, not at all spine-like. Dorsal fin low, not so high as long, its base $1\frac{2}{3}$ in head. Anal inserted below seventh ray of dorsal. Pectorals $1\frac{2}{3}$ in head; ventrals 2. Caudal obliquely truncate, very slightly emarginate, the upper lobe about one fifth longer than the lower, $1\frac{1}{4}$ in head; upper lobe usually more or less sharply angular; lower lobe rounded.

Coloration in spirits: Males with the sides profusely mottled with darker, sometimes nearly plain. Vertical fins each with several brownish bars and blotches and each with a dusky subterminal bar. A narrow dark line along middle of each row of scales on the back. Females with several short dark bars on the posterior half of the body, the fins colored as in the male. Some small dark specks on caudal peduncle.

7. **Fundulus parvipinnis** Girard.

7242. Numerous examples, precisely like others from San Diego.

8. **Fundulus vinctus**, sp. nov.

30973. One specimen, somewhat faded, but in fair condition. Length $2\frac{1}{2}$ inches. Head $3\frac{3}{4}$ in length; depth $4\frac{1}{3}$. D. 12; A. 11. Scales about 31–10.

Body little elongate, compressed posteriorly. Head large, very broad, and somewhat depressed above. Mouth moderate. Teeth in narrow bands, the outer much enlarged. Eye $3\frac{1}{4}$ in head. Interorbital space 2.

Scales comparatively large. Dorsal inserted moderately in advance of anal, its front midway between base of caudal and occiput; the fin of moderate height. Pectoral $1\frac{2}{3}$ in head; caudal $1\frac{1}{5}$.

Coloration, in spirits, olivaceous, with about 23 narrow silvery bars with undulating edges, the bars narrower than the darker interspaces. Fins now all plain.

This species is apparently related to *F. heteroclitus* and other Atlantic species. It may be distinguished from most of its relatives by its comparatively large scales.

9. **Fundulus extensus**, sp. nov.

30972. Two specimens, faded and rather soft, the longest nearly 3 inches long.

Head $3\frac{3}{4}$; depth $5\frac{2}{3}$. D. 15; A. 13. Scales about 47–12.

Body unusually elongate, moderately compressed, the caudal peduncle long, much longer than head. Head slender, not very broad, the interorbital width 2⅖ in head. Eye large, 3¼ in head. Mouth rather large; the teeth in a moderate band, the outer considerably enlarged.

Dorsal fin rather long, of moderate height, its insertion well in front of that of anal, at a point midway between eye and base of caudal.

Pectoral small, 1¾ in head. Caudal 1⅖.

Coloration, in spirits, plain, somewhat translucent, with no markings anywhere, except traces of some very narrow dark bars on the sides. Fins now plain.

This species resembles somewhat the Eastern *Fundulus diaphanus*, but it is more elongate.

10. Hemirhamphus unifasciatus Ranzani.

6320. An adult example, in fair condition.

11. Gymnomuræna nectura, sp. nov.

15442. One specimen, 6¼ inches in length, in good condition.

Body moderately elongate, the snout heavy, compressed, abruptly truncate in profile. Anterior nostril on the front of the snout, in a short tube; posterior nostril directly above the eye, without tube.

Eye rather large, about half as long as snout, which is 2⅔ in cleft of mouth. Cleft of mouth straight, its length 2⅔ in head. Jaws about even in front, the lower having little motion, but capable of completely closing the mouth.

Teeth rather strong, sharp, straight, erect, mostly in two series, and nearly all depressible; those on the vomer a little larger than the others. Teeth in outer series in each jaw small, much smaller than those of the inner series. Gill opening small. Head 2⅖ in trunk; head and trunk a little shorter than tail.

End of tail with a moderate fin, larger than usual in this genus; the fin more developed on the upper side, where its length is equal to that of the head.

Color dark brown, with ill-defined bars, blotches and reticulations of darker brown, the head and breast more distinctly marked.

Compared with *Gymnomuræna tigrina*, this species has the fin better developed, the snout and mouth longer, the teeth larger, the color different, &c.

12. Muræna pinta Jordan & Gilbert.

2324. One half-grown individual, in good condition. Young specimens of this species have an inner row of smaller teeth in the upper jaw.

13. Apterichthys selachops, sp. nov.

4391. One specimen, in good condition, about 14 inches long.

Body moderately elongate, the tail sharp-pointed. No trace of fins anywhere. Head tapering anteriorly to the long, sharp snout, which ends in a short flexible tip. Snout projecting much beyond the mouth; the form and position of the mouth and snout and the position of the nostrils giving a physiognomy remarkably shark-like. Cleft of the mouth oblique, somewhat curved downwards and backwards posteriorly. Teeth all small, pointed, their tips directed backward; apparently in about one series in each jaw and a narrow band on the vomer. Lower jaw anteriorly pointed, incapable of much motion. Width of lower jaw between angles of mouth, $1\frac{2}{3}$ in its length. Length of snout from eye, $1\frac{2}{3}$ in length of cleft of mouth. Cleft of mouth 4 in head.

Anterior nostrils without tube, posterior each in a short tube; both pairs on the lower side of the snout. Eyes minute, but evident, somewhat behind the vertical from the front of the lower jaw.

Gill-openings ventral, close together in front, slightly divergent behind, the slits about as long as snout. Lateral line conspicuous.

Head 5 in head and trunk; head and trunk $1\frac{1}{2}$ in tail.

Color uniform plain brown; the head slightly paler and mottled.

The specimen is a female full of ova; the ovaries extend backward in the abdominal cavity far behind vent.

14. Ophichthys miurus, sp. nov.

2304. Three specimens, in good condition, the largest about a foot long.

Body moderately elongate. Head long and slender, anteriorly pointed. Lower jaw included; cleft of mouth $2\frac{1}{5}$ in head.

Teeth all slender and pointed, directed backwards, most of them not depressible, those of the upper jaw in two widely separated series, those of the inner series largest, slender and close-set. Vomer with a median series of about 4 slender teeth. Lower jaw with a single series of rather long, slender teeth, wide apart, larger than the teeth of the upper jaw, but smaller than those of the vomer.

Snout very short, nearly twice the length of eye, 4 times in cleft of mouth. Eyes small, placed high and well forward. Nostrils without tubes. Lateral line conspicuous. Gill-openings small, placed very low, separated by an interspace, less than the length of one slit, which is about as long as snout.

Pectoral fin very small, pointed, about as long as snout. Gill-opening midway between tip of snout and beginning of dorsal. Fins very low; tip of tail pointed. Tail unusually short. Head $5\frac{3}{4}$ in head and trunk. Tail $1\frac{1}{4}$ in rest of body, a little shorter than trunk without head.

Coloration light yellowish; a series of roundish dark brown blotches on each side of body, the two series alternating; a series of small half-blotches on the back, these also mostly alternating. Head covered with small spots; dark spots on sides of lower jaw; fins all pale.

This would be a species of "*Herpetoichthys*" in Dr. Kaup's arrangement.

15. Mugil brasiliensis Agassiz.

2510, 3003, 7616. Numerous small specimens, mostly in poor condition, most or all of them belonging to the present species.

7090. Two large specimens in good condition.

16. Sphyræna argentea Girard.

(*Sphyræna lucasana* Gill, Proc. Ac. Nat. Sci. Phila. 1863, 86.)

6353. (Types of *Sphyræna lucasana* Gill.) Numerous young specimens, in rather poor condition, none of them more than 6 inches long. They agree in all tangible respects with *Sphyræna argentea.* Lat. l. about 142.

17. Lepidopus caudatus (Euphr.) White.

10115. One specimen, 10 inches long, in poor condition.

18. Decapterus hypodus Gill.

(*Decapterus hypodus* Gill, Proc. Ac. Nat. Sci. Phila. 1862, 261.

4005. (Types of *Decapterus hypodus.*) Four specimens, in good condition, 6 to 8 inches in length. This species is extremely closely related to *Decapterus macarellus* (C. & V.) Gill, of the Atlantic coast, of which it may well be taken as a geographical representative or variety. The only differences which we are able to appreciate are the following:

Body rather less slender in *D. hypodus* (depth $5\frac{1}{2}$ instead of $5\frac{3}{4}$); teeth rather stronger (distinctly seen on lower jaw and tongue; scarcely to be felt anywhere in *D. macarellus*); caudal armature stronger, about 30 plates having distinct keels (not more than 25 in *D. macarellus*); lateral line becoming straight more or less behind middle of trunk (near middle of body in *D. macarellus*).

It is possible that a large series would show that the two forms are absolutely identical.

19. Trachurus picturatus (Bowdich) J. & G.

(*Trachurus symmetricus* Gill, Proc. Ac. Nat. Sci. Phila. 1862, 261.)

8086. Two specimens, in good condition, of the usual Californian type.

20. Trachurus declivis (Jenyns) J. & G.

6351=31014. A single immature specimen, about 4 inches in length, evidently different from *Caranx picturatus* (*symmetricus* Ayres) and apparently identical with Mediterranean specimens of the species we have called *Caranx declivis.* Plates 36+36, those on anterior part of lateral line little lower than the others. Curve of lateral line $1\frac{2}{3}$ in straight part.

21. Caranx crumenophthalmus.

(*Trachurops brachychirus* Gill, Proc. Ac. Nat. Sci. Phila. 1862, 261.)

4007. (Types of *Trachurops brachychirus.*) Two specimens, in fair condition, each 8 to 9 inches in length.

We are unable to detect any difference between this species and the ordinary *crumenophthalmus.* The pectoral is not in the least shorter than usual, about $3\frac{2}{5}$ in length to base of caudal. Head $3\frac{1}{2}$; depth $3\frac{2}{3}$.

22. Caranx caballus Günther.

7570. Five young specimens, about 6 inches long, in fair condition.

23. Caranx crinitus Akerly.

(*Blepharichthys crinitus* Gill, Proc. Ac. Nat. Sci. Phila. 1862, 262.)

31012. One specimen, young, in fair condition.

24. Trachynotus carolinus (L.) Gill.

(*Trachynotus pampanus* Gill, Proc. Ac. Nat. Sci. Phila. 1862, 262 : *Trachynotus carolinus* Gill, Proc. Ac. Nat. Sci. Phila. 1863, 84.)

5085. Seven specimens, the largest 6 inches long. These are not distinguishable from the young of the Atlantic Pompano.

25, Trachynotus fasciatus Gill.

(*Trachynotus fasciatus* Gill, Proc. Ac. Nat. Sci. Phila. 1863, 86=*Trachynotus glaucoides* Günther, Proc. Zool. Soc. Lond. 1864, 150.)

9647. (Not original type.) An adult example, in good condition.

26. Seriola dorsalis (Gill) J. & G.

(*Halatractus dorsalis* Gill, Proc. Ac. Nat. Sci. Phila. 1863, 84 = *Seriola lalandi* Jor. & Gilb. Proc. U. S. Nat. Mus. 1881, 46. Not of C. & V.)

2511. (Type of *Halatractus dorsalis*.) A very young example, in good condition, $3\frac{1}{4}$ inches in length.

The banded coloration of this specimen is usual in immature *Seriolœ.* The large number of dorsal rays distinguishes this species from *Seriola mazatlana* Steind. It is apparently the young of the Californian "Yellow Tail," which we have formerly identified with *Seriola lalandi* C. & V. Until specimens of the two forms can be actually compared, it is better to retain the Pacific species under a separate name as *Seriola dorsalis.*

Head $3\frac{1}{2}$; depth 4. Tail scarcely carinated; vertical fins little elevated anteriorly, not falcate. Head about one-fourth longer than deep, somewhat carinated at the occiput; (this carina probably disappearing with age). Maxillary $2\frac{1}{4}$ in head, reaching nearly to the middle of the pupil. D. VII–I, 37; A. II–I, 21.

27. Rhypticus xanti Gill.

(*Rhypticus xanti* Gill, Proc. Ac. Nat. Sci. Phila. 1862, 250.)

30740. (Type of *Rhypticus xanti*.) One specimen, 5 inches long, in good condition.

28. Rhypticus nigripinnis Gill.

(*Rhypticus nigripinnis* Gill, Proc. Ac. Nat. Sci. Phila. 1861, 53, Panama : *Rhypticus maculatus* Gill, Proc. Ac. Nat. Sci. Phila. 1862, 251, Cape San Lucas; not *Rhypticus maculatus* Holbr.: *Promicropterus decoratus* Gill, Proc. Ac. Nat. Sci. Phila. 1863, 164, Panama.)

3689. (Type of *Rhypticus maculatus*.) One young specimen, about $2\frac{1}{2}$ inches long, in bad condition.

This specimen is undoubtedly the young of the species called *nigri-pinnis* and *decoratus* by Professor Gill, a species very closely related to *Rhypticus maculatus* Holbr. of the Atlantic, but distinct from it.

The number of dorsal rays is II, 25, not III, 24, as given by Professor Gill. The first soft ray having been detached and broken, was taken for a third spine, but its articulated tip is still attached.

29. Epinephelus sellicauda Gill.

(*Epinephelus sellicauda* Gill, Proc. Ac. Nat. Sci. Phila. 1862, 250=*Epinephelus ordinatus* Cope, Trans. Am. Philos. Soc. 1870, 466.) ·

7247. (Type of *Epinephelus sellicauda*.) A single specimen, very young and somewhat shrivelled.

30. Brachyrhinus furcifer (C. & V.) Poey.

(*Brachyrhinus creolus* Gill, Proc. Ac. Nat. Sci. Phila. 1862, 249.)

3688. Nine inches long, in fair condition. We have compared this specimen with one from Cuba, and, with Professor Gill, are unable to point out any differences likely to be permanent. The Californian specimen is somewhat deeper, with deeper and blunter head, and the pale spots on the sides are smaller than in the other, otherwise the two seem to be identical.

31. Anthias multifasciatus (Gill) J. & G.

(*Pronotogrammus multifasciatus* Gill, Proc. Ac. Nat. Sci. Phila. 1863, 81.)

2762. (Type of *Pronotogrammus multifasciatus*.) A very young example, about two inches long, the fore part of the head injured. It has a blunt head, forked caudal, scaly maxillary, large scales, high lateral line, and other characters of *Anthias*, to which genus it should probably be referred.

32. Xenichthys xanti Gill.

(*Xenichthys xanti* Gill, Proc. Ac. Nat. Sci. Phila. 1863, 83=*Xenichthys xenops* Jordan and Gilbert, Bull. U. S. Fish Commission, 1882, 325.)

5086. (Types of *Xenichthys xanti*.) Many small specimens, 3 to 4 inches in length, in lair condition. These evidently belong to the same species as the adult examples lately described by us from Panama as *Xenichthys xenops*.

The dorsal rays are XI–I, 17, instead of XII, 14, as stated by Professor Gill. The scales of the lateral line are perhaps a little more conspicuous than the others, but the difference is of no importance.

33. Lutjanus novemfasciatus Gill.

(*Lutjanus novemfasciatus* Gill, Proc. Ac. Nat. Sci. Phila. 1863, 251 = ? *Mesoprion inermis* Peters, Berliner Monatsberichte, 1869, 705 = *Lutjanus prieto* Jordan & Gilbert, Proc. U. S. Nat. Mus. IV, 1881, 353.)

4010. (Types of *Lutjanus novemfasciatus*.) Two specimens, about five inches in length, in fair condition.

The very young specimens on which this species was based, evidently belong to the species which we have lately described as *Lutjanus prieto*, an identification which could not be made from the description published. The dark bands are a character of extreme youth.

Serranus calopteryx Jor. & Gilb. (Proc. U. S. Nat. Mus. iv, 1881, 350) seems to be identical with *Prionodes fasciatus* Jenyns (Voyage of the Beagle, Fishes, 1842, 46). The absence of the vomerine and palatine teeth in Jenyns' type is, as has been suggested by Dr. Günther, purely accidental, and without significance. The name *fasciatus* is preoccupied in the genus *Serranus*, by *Holocentrus fasciatus* Bloch. This species may therefore retain the name *Serranus calopteryx*.

34. Diabasis sexfasciatus (Gill) J. & G.

(*Hæmulon sexfasciatus* Gill, Proc. Ac. Nat. Sci. Phila. 1862, 254=*Hæmulon maculosum* Peters, Berliner Monatsber. 1869, 705.)

3000. (Types of *Hæmulon sexfasciatus*.) One specimen, 4 inches long. 6467. About twelve specimens of similar small size.

This species reaches a very large size, and the adult examples are quite different in form and coloration from the little fish which served as the original type. As in related species, the black spots on the scales are developed with age.

35. Diabasis scudderi (Gill) J. & G

(*Hæmulon scudderii* Gill, Proc. Ac. Nat. Sci. Phila. 1862, 253 = *Hæmulon brevirostrum* Günther, Trans. Zool. Soc. Lond. 1869, 418 = *Hæmulon undecimale* Steindachner, Ichth. Beiträge, iii, 11, 1875.)

3683. (Types of *Hæmulon scudderii*.) Three young specimens, in good condition. The coloration is quite different from that of the adult or half-grown of this species, and is extremely similar to that of the young of *Pomadasys bilineatus*.

Grayish, the scales with inconspicuous darker spots. A broad black band through snout and eye, ending in a black blotch at base of caudal. A second band from between nostrils on each side, above eye straight to soft dorsal and upper edge of caudal peduncle. Fins, especially anal, a little dusky. A dark blotch hidden by angle of opercle. All these specimens have 12 dorsal spines, but most of those obtained by Mr. Gilbert have 11, as in the type of *Hæmulon undecimale*.

36. Diabasis steindachneri Jordan & Gilbert.

19879. Eight specimens, nearly adult, in good condition. These appear to have been received after the publication of Professor Gill's papers.

37. Diabasis flaviguttatus (Gill) Jor. & Gilb.

(*Hæmulon flaviguttatus* Gill, Proc. Ac. Nat. Sci. Phila. 1862, 254=*Hæmulon margaritiferum* Günther, Proc. Zool. Soc. Lond. 1864, 147.)

3681. (Type of *Hæmulon flaviguttatus*.) An adult example, in good condition.

38. Diabasis maculicauda (Gill) Jor. & Gilb.

(*Orthostœchus maculicauda* Gill, Proc. Ac. Nat. Sci. Phila. 1862, 255=*Hœmulon mazat-lanum* Steindachner, Ichthyol. Notiz, viii, 12, taf. vi, 1869.)

6557. (Types of *Orthostœchus maculicauda*.) Several immature speci-mens.

39. Pomadasys inornatus (Gill) J. & G.

(*Microlepidotus inornatus* Gill, Proc. Ac. Nat. Sci. Phila, 1862, 256 = ? *Pristipoma brevi-pinne* Steindachner, Ichth. Notiz, viii, 1869, 10 = ? *Pristipoma notatum* Peters, Berlin. Monatsber. 1869, 706.)

3684. (Types of *Microlepidotus inornatus*.) Two adult specimens, in good condition, 8 inches long.

2999. One young example.

6558. Numerous immature examples, from 1 to 4 inches long, showing lengthwise stripes.

7313. Four specimens, partly grown.

All the specimens examined have 14 spines in the dorsal, and the membranes of the soft dorsal and anal seem to be without scales.

The young of this species is silvery, with three regular parallel black-ish stripes, the lower from eye to middle of base of caudal, the next from above eye to upper part of caudal peduncle, the third higher up, to middle of soft dorsal. The adults are nearly plain with traces of about 6 narrow, dusky, wavy streaks, which do not follow the rows of scales.

The specimen from Guaymas (No. 29386), referred to by us in a pre-vious paper (Proc. U. S. Nat. Mus. 1881, 274) as *Pomadasys inornatus*, belongs apparently to *Pomadasys cantharinus* (Jenyns) J. & G.

40. Pomadasys ? bilineatus (Cuv. & Val.) J. & G.

(*Genytremus interruptus* Gill, Proc. Ac. Nat. Sci. Phila. 1862, 256 (young).

30927. (Types of *Genytremus interruptus*.) Nine young specimens, 3 to 4 inches in length. These young specimens resemble to a remarkable degree the young of the Atlantic species, *P. bilineatus*, with which they were compared by Professor Gill. Compared with specimens of the latter species they differ only in the larger size of the scales, above the lateral line mesially. In *bilineatus* there are usually 6 scales in a ver-tical series between the spinous dorsal and the lateral line. In the types of *interruptus* we find 4, 5, or 6 scales in such a series. In *Pomadasys fürthi* we find 4. *Fürthi* differs from *bilineatus*, so far as we can see, only in a slightly different color, more arched back, and rather larger scales between the spinous dorsal and lateral line. We are unable at present to decide whether the types of *interruptus* are the young of *fürthi* or of *bilineatus*. If the former, which is not unlikely, the occur-rence of the latter species in the Pacific is yet to be verified, although not improbable. All the definite records of *bilineatus* on the west coast of tropical America refer to young specimens, with lateral stripes like the types of *bilineatus*.

The coloration of the types of *Interruptus* is as follows:

Dull grayish, somewhat bluish above; scales anteriorly with inconspicuous darker spots. A wavy, sharply-defined black band through snout and eye, to opposite last ray of dorsal. where it ends abruptly. Behind it, at base of caudal, is a large oval black blotch. A similar black stripe from above eye straight to middle of base of soft dorsal. Ventrals black, other fins more or less tinged with dusky, the pectorals and spinous dorsal palest. If these prove to be the young of *Pomadasys fürthi*, the name *interruptus* is to be substituted for *fürthi*. This question cannot be settled with the material now at hand.

41. Girella nigricans (Ayres) Gill.

(*Girella nigricans* = *Girella dorsimacula* Gill, Proc. Ac. Nat. Sci. Phila. 1862, 244.)

20320. (Type of *Girella dorsimacula*.) A partly grown specimen, showing the pale blotch on the back by the side of the dorsal fin, characteristic of the young of this species.

42. Pimelepterus analogus Gill.

(*Pimelepterus analogus* Gill, Proc. Ac. Nat. Sci. Phila. 1862, 245 = *Pimelepterus elegans* Peters, Berliner Monatsber. 1869, 707.)

3001. (Types of *Pimelepterus analogus*.) In poor condition.

43. Apogon retrosella (Gill) J. & G.

(*Amia retrosella* Gill, Proc. Ac. Nat. Sci. Phila. 1862, 251.)

2454. (Types of *Amia retrosella*.) Seven specimens, in fair condition, 1½ to 3½ inches in length.

2997. Four specimens, in poor condition.

4001, 4002, 4003. (Types of *Amia retrosella*.) Three half grown specimens, in fair condition.

4413. (Types of *Amia retrosella*.) Three specimens.

44. Upeneus dentatus Gill.

(*Upeneus dentatus* Gill, Proc. Ac. Nat. Sci. Phila. 1862, 256.)

3699. (Types of *Upeneus dentatus*.) Three young examples, about 4 inches in length, in good condition. This species has not been obtained by any other collector. It is well distinguished from the common *Upeneus grandisquamis* Gill. Compared with the young of *grandisquamis* of the same size, *dentatus* is more slender, less compressed, with smaller scales, very much larger eye, much weaker teeth, and the dorsal outline less arched.

45. Umbrina dorsalis Gill.

(*Umbrina dorsalis* Gill, Proc. Ac. Nat. Sci. Phila. 1862, 257.)

3696. (Types of *Umbrina dorsalis*.) Ten specimens, the largest 4 inches long.

46. Umbrina xanti Gill.

(*Umbrina xanti* Gill, Proc. Ac. Nat. Sci. Phila. 1862, 257 = *Umbrina analis* Günther, Trans. Zoöl. Soc. London, 1869, 426.)

7156. (Types of *Umbrina xanti*.) Three young examples, the largest nearly 4 inches long.

2996. Two small specimens.

Compared with the young of *Umbrina dorsalis*, the young of *U. xanti* differ in the following respects:

The body is more slender and elongate (depth $3\frac{3}{4}$; $3\frac{1}{6}$ in *dorsalis*), the head is more elongate, the anterior profile much less blunt and rounded, the eye much smaller (not much longer than snout), the pectoral shorter, (2 in head; $1\frac{2}{3}$ in *dorsalis*), the anal spine shorter. The oblique streaks along the rows of scales are narrower and more sharply defined in *xanti* than in *dorsalis*. The number of dorsal rays in *dorsalis* is constantly greater.

Adult examples of the two species obtained by Mr. Gilbert show the following differential characters:

a. Snout very blunt, not longer than eye, 4 in head; preopercle with its membrana-ceous edges crenulate; pectorals more than two-thirds length of head. D. X–I, 30 to 33; A. II, 7. Scales 9–53–12. Dark stripes along rows of scales very faint, broader than the pale interspaces. Depth 3 in length................ DORSALIS.

aa. Snout rather acute, longer than eye, $3\frac{1}{4}$ in head; preopercle with its bony edge serrate; pectorals less than two-thirds length of head. D. X–I, 26; A. II, 6. Scales 6–48–10. Dark stripes along rows of scales very distinct, narrower than the pale interspaces. Depth $3\frac{1}{4}$ in length.............................XANTI.

Neither species appears to be very common along the coast.

47. Myriopristis occidentalis Gill.

(*Myriopristis occidentalis* Gill, Proc. Ac. Nat. Sci. Phila. 1863, 87 : ? *Rhamphoberyx leucopus* Gill, Proc. Ac. Nat. Sci. Phila. 1863, 88.)

6348. (Types of *Myriopristis occidentalis.*) Very many young speci-mens, 2 to 3 inches in length.

6350. (Types of *Myriopristis occidentalis.*) Many young specimens.

6304. (Types of *Rhamphoberyx leucopus.*) Two specimens, each about 2 inches in length.

These specimens appear to belong to the same species. In all the specimens called *occidentalis* the sides are dull and dusky with dark punctulations. In the types of *leucopus* the sides have a silvery luster. There is no tangible difference in form, so far as we can judge from these small specimens.

48. Myriopristis pœcilopus (Gill) J. & G.

(*Rhamphoberyx pœcilopus* Gill, Proc. Ac. Nat. Sci. Phila. 1863, 87: *Rhamphoberyx leucopus* Gill, Proc. Ac. Nat. Sci. Phila. 1863, 88.)

6273. (Types of *Rhamphoberyx pœcilopus.*) Three specimens, each about 2 inches in length, in good condition.

In these specimens the spinous dorsal is all black and the ventrals tipped with black. *Pœcilopus* is probably a species distinct from *M. occidentalis*, although the resemblance is remarkably great, the differ-ences, except in color, being scarcely appreciable.

Compared with *occidentalis* of the same size, *pœcilopus* has the lower jaw a trifle shorter and the eye a little larger. In *pœcilopus* the sides have a bright silvery luster, without dark punctulations, as in the specimens called *leucopus.*

There is no warrant for the generic name *Rhamphoberyx.* It is strictly synonymous with *Myriopristis.*

49. Holocentrum suborbitale Gill.

(*Holocentrum suborbitale* Gill, Proc. Ac. Nat. Sci. Phila. 1863, 86.)

2319. (Types of *Holocentrum suborbitale.*)

7312. Numerous specimens.

50. Polynemus approximans Lay & Bennett.

(*Polynemus approximans* Gill, Proc. Ac. Nat. Sci. Phila. 1862, 258.)

6418. Numerous young examples.

51. Prionurus punctatus Gill.

(*Prionurus punctatus* Gill, Proc. Ac. Nat. Sci. Phila. 1863, 242.)

3679, 4422, 9306. (Types of *Prionurus punctatus.*) Many specimens in good condition, mostly young.

52. Pomacanthus strigatus (Gill) J. & G.

(*Holacanthus strigatus* Gill, Proc. Ac. Nat. Sci. Phila. 1862, 243.)

3668. (Type of *Holacanthus strigatus.*) One specimen, about 3 inches in length, in good condition.

53. Chætodon nigrirostris (Gill) J. & G.

(*Sarothrodus nigrirostris* Gill, Proc. Ac. Nat. Sci. Phila. 1862, 243.)

3669. (Types of *Sarothrodus nigrirostris.*) Two specimens partly grown, in fair condition, but badly shriveled.

54. Pomacentrus rectifrænum Gill.

(*Pomacentrus rectifrænum* Gill, Proc. Ac. Nat. Sci. Phila. 1862, 148; 1863, 214: *Pomacentrus analigutta* Gill MSS, in Günther, Cat. Fish. Brit. Mus. iv, 27.)

3670. (Types of *Pomacentrus rectifrænum.*) Three partly grown specimens, in good condition.

3674. (Types of *Pomacentrus analigutta.*) Several specimens, in good condition, 1½ to 3 inches in length.

There seems little reason to doubt that the above-noticed specimens all belong to the same species.

55. Pomacentrus flavilatus Gill.

(*Pomacentrus flavilatus* Gill, Proc. Ac. Nat. Sci. Phila. 1862, 148; 1863, 214: *Pomacentrus bairdii* Gill, Proc. Ac. Nat. Sci. Phila. 1862, 149: *Pomataprion bairdii* Gill, Proc. Ac. Nat. Sci. Phila. 1863, 217.)

3677. (Type of *Pomacentrus flavilatus.*) One half-grown specimen, in fine condition, with the characteristic coloration of the species.

3656. (Type of *Pomacentrus bairdii.*) One very immature specimen, less than an inch long.

We are able to distinguish this species from *P. rectifrænum* only by the difference in coloration. No intermediate conditions have yet been noticed by us. According to Mazatlan fishermen, it reaches a larger size than as yet observed by collectors, still retaining its characteristic coloration.

56. Pomacentrus quadrigutta Gill.

(*Hypsypops dorsalis* Gill, Proc. Ac. Nat. Sci. Phila. 1862, 147 (adult): *Pomacentrus quadrigutta* Gill, Proc. Ac. Nat. Sci. Phila. 1862, 149: *Pomataprion dorsalis* Gill, Proc. Ac. Nat. Sci. Phila. 1863, 216: not *Pomacentrus dorsalis* Gill. Proc. Acad. Nat. Sci. Phila. 1859, 29; a Chinese species.)

3657. (Type of *Pomacentrus quadrigutta*.) A very young example, less than one inch in length.

The type of *Hypsypops dorsalis* (4369) has now gone to decay.

57. Glyphidodon declivifrons (Gill) J. & G.

(*Euschistodus declivifrons* Gill, Proc. Ac. Nat. Sci. Phila. 1862, 145, 146; 1863, 219: *Euschistodus concolor* Gill, l. c. 1862, 145, foot-note = *Euschistodus analogus* Gill, l. c. 1863, 219, Aspinwall.)

9332. (Types of *Euschistodus declivifrons*.) About ten young examples, 2 to 4 inches in length.

30744. A large example, 5½ inches in length, in good condition. On this specimen the dark bands have all disappeared.

58. Glyphidodon saxatilis (L.) Lac.

(*Glyphidodon troschelii* Gill, Proc. Ac. Nat. Sci. Phila. 1862, 150; 1863, 220.)

8173, 8180. (Types of *Glyphidodon troschelii*.) Many young specimens.

59. Chromis atrilobata Gill.

(*Chromis atrilobata* Gill, Proc. Ac. Nat. Sci. Phila. 1862, 149; 1863, 220.)

3675. (Type of *Chromis atrilobata*.) A half-grown specimen, in bad condition.

No second specimen of this species has yet been obtained. It may be identical with the Brazilian *Chromis marginatus*, as suggested by Dr. Günther, but it is certainly premature to unite the two on the basis of our present knowledge. A few species of shore-fishes are certainly common to the faunæ of Brazil and Lower California, but the supposition is against identity in any individual case. Much injury has been done to our knowledge of geographical distribution by the random identification of specimens with closely related species belonging to some other fauna. Of 50 species of marine fishes given by Dr. Günther (Trans. Zool. Soc. London, 1869, 385–392) as common to both sides of the Isthmus of Panama, at least 11 have been incorrectly identified and are not found on both coasts, the identity of 18 more is doubtful and must be verified, while but 21 of the list can be positively stated to be specifically identical. A large number not included in this list are also certainly identical, but in this case it is better to retain some doubtful species than to make many doubtful identifications.

We may notice that the green coloration of the type of *Chromis atrilobata* (Proc. Ac. Nat. Sci. Phila. 1863, 220) seems to have come from the copper tank in which it has been kept.

60. Harpe diplotænia Gill.

(*Harpe diplotænia* Gill, Proc. Ac. Nat. Sci. Phila. 1862, 140 (♀ ?): *Harpe pectoralis* Gill, Proc. Ac. Nat. Sci. Phila. 1862, 141 (♂).

4441. (Types of *Harpe diplotænia*.) One specimen, 9 inches long, in alcohol.

2986. Stuffed skin of adult; also one of the original types.

6430. (*Harpe pectoralis*; not type; record of locality and collector lost.) A specimen, about 10 inches long, in spirits.

2988, 8867. (Stuffed skins; types of *Harpe pectoralis*.)

These two forms have been well described by Professor Gill. We are unable to find any constant difference between them except in the color. It is not improbable that *pectoralis* is the male and *diplotænia* the female of the same species. The form called *pectoralis* is certainly the male.

61. Julis lucasanus Gill.

(*Julis lucasanus* Gill, Proc. Ac. Nat. Phila. 1862, 142.)

3676, 3677. (Types of *Julis lucasanus*.) Young and half-grown examples, in good condition.

4396. Two adult and one young example.

62. Xyrichthys mundiceps Gill.

(*Xirichthys mundiceps* Gill, Proc. Ac. Nat. Sci. Phila. 1862, 143.)

4370. (Types of *Xirichthys mundiceps*.) One half-grown and several small examples.

8082. (Types.) Very many young examples, in poor condition.

30929. Three adult males and one female (not types).

The large specimens last mentioned were received after the publication of Professor Gill's papers. The female example is plain light brownish like the original types. The males are darker, with a narrow vertical blue or violet line at the base of each scale, these most distinct and broadest on caudal peduncle. A conspicuous jet-black spot, rather larger than the eye, at base of caudal, just below lateral line. Three concentric blue curved lines on flap of opercle. Three narrow blue lines downward and forward from eye across cheek. Lower jaw and lower side of head with blue stripes and lines, the one connecting angles of the mouth below broader than the others. Fins pale; now plain.

In the male the body is deeper than in the female, and the anterior profile is steeper. The largest of the original types is a male, and still shows traces of the dark caudal spot.

63. Novacula mundicorpus (Gill) Günther.

(*Iniistius mundicorpus* Gill, Proc. Ac. Nat. Sci. Phila. 1862, 145.)

7388. One adult example, probably a male, 7 inches in length, evidently not the original type.

Color olivaceous, whitish below; three broad bars of dark olive on the back and sides, these bars nearly as wide as the interspaces.

Most of the scales of the back and sides with a vertical light bluish stripe, not so distinct as in *X. mundiceps*. In the middle of the first dark band, just above the lateral line, are one or two scales of a differ- ent color, the posterior half of each being jet black, the base light blue, the colors abruptly defined. Dorsal with narrow dark stripes running obliquely downward and backward. Anal pale, with a conspicuous light horizontal stripe near the tips of the rays; a narrower similar stripe near the middle of the fin. Some bluish clouds on opercle. Some vertical pale blue stripes below eye. Anterior dorsal dusky. A faint dusky streak below eye; tip of caudal a little dusky.

64. Caulclatilus princeps (Jenyns) Gill.

(*Caulolatilus affinis* Gill, Proc. Ac. Nat. Sci. Phila. 1865, 67.)

5789. (Type of *Caulolatilus affinis*.) One very young example, about 3 inches long, badly shrivelled. So far as we can see the number of fin rays in this specimen is not less than usual in the species to which it belongs.

65. Gobius soporator C. & V.

2466. One specimen.

66. Gobius banana Cuv. & Val.

2464. Several young examples.
2474. Adults.
2772. Adults.
30931. Three adult specimens.

67. Dormitator maculatus (Bloch) J. & G.

2491, 7350. Many examples.

68. Philypnus lateralis Gill.

(*Philypnus lateralis* Gill, Proc. Ac. Nat. Sci. Phila. 1860, 123.)

2435 to 2442. Types of *Philypnus lateralis*.
2492, 6283. Many specimens.

69. Porichthys margaritatus (Rich.) J. & G.

3004. Young examples.

70. Clinus xanti (Gill) Gthr.

(*Labrosomus xanti* Gill, Proc. Ac. Nat. Sci. Phila. 1860, 107.)

2334, 7050, 7314. Many specimens, of various sizes, some of them types of *Labrosomus xanti*.

This species is extremely close to the *Clinus nuchipinnis*, differing in the specimens examined, in the arrangement of the teeth on the vomer. In *xanti* there are three large bluntish teeth forming a triangle; in *nuchipinnis*, one large tooth and about six smaller ones forming a V- shaped figure. In *nuchipinnis* there is always a distinct black blotch on the opercle, which is faint or obsolete in *xanti*. In form, structure of fins, numbers of scales, &c., we are unable to find any differences.

71. Tripterygium carminale Jor. & Gilb.

2487. Two examples.

72. Salarias atlanticus C. & V.

2745, 7324, 7333, 7794. Many specimens, of various sizes.

73. Isesthes gentilis (Grd.) J. & G.

2481. Two examples, the largest 2½ inches long, answering entirely to the description of the female of this species given by Dr. Steindachner (Ichth. Beitr. v, 150). A male specimen of this species is in Mr. Lockington's collection, from La Paz.

74. Myxodagnus opercularis Gill.

(*Myxodagnus opercularis* Gill, Proc. Ac. Nat. Sci. Phila. 1861, 263.)

2531, 2532, 2533. (Types of *Myxodagnus opercularis*.) Three immature examples, faded.

75. Dactylagnus mundus Gill.

(*Dactylagnus mundus* Gill, Proc. Ac. Nat. Sci. Phila. 1862, 505.)

4915. (Type of *Dactylagnus mundus*.) One specimen, nearly 6 inches long.

76. Sebastopsis xyris, sp. nov.

30979. Six small specimens, somewhat discolored, the largest about 3 inches in length.

Head 2½; depth 3⅓. D. XIII, 10; A. III, 5. Lat. l. 24 (pores).

Body oblong, somewhat compressed, the back a little elevated. Head large, very strongly armed. Mouth rather large, oblique, the jaws subequal in front, the maxillary extending to beyond pupil, its length 1⅚ in head. No palatine teeth. Jaws naked. Preorbital narrow, its edge lobate, not spinous. Eye large, about 3¼ in head.

Cranial ridges very short, sharp, and high, their spines more or less hook-like and compressed. Interorbital space narrow, very deeply concave, with two curved longitudinal ridges, each armed with a small spine. Nasal spines sharp. Preocular, supraocular, postocular, tympanic, occipital, nuchal, and coronal spines present. Occipital ridge very short, spine-like. Coronal spines separating the naked frontal region from the scaly part of the head. A sharp temporal spine on each side; behind it two strong spines on the suprascapula; a spine on the shoulder-girdle. Opercle with two spines. Preopercle with about five spines, the largest with a smaller spine at its base in front, the two lowermost spines almost obsolete. Suborbital stay forming a sharp elevated ridge, with a sharp spine near its front, under the eye, and another near its junction with the preopercle. Gill-rakers very short, rather stout.

Dorsal fin rather deeply notched, the spines strong, the longest $2\frac{2}{3}$ in head. Longest soft ray about half length of head. Caudal truncate, $1\frac{2}{3}$ in head. Second anal spine $1\frac{2}{3}$ in head, very strong, much longer than third or than the soft rays. Pectoral $1\frac{1}{4}$ in head, the base rather broad, a little procurrent, the tip pointed. Ventral $1\frac{2}{3}$ in head, its insertion under anterior margin of base of pectoral.

Scales unusually large, ctenoid; 25 pores in lateral line, the number of rows of scales somewhat more.

Coloration faded, apparently light red or perhaps brown in life, with traces of darker shades. Caudal with bands and blotches of dark brown; traces of similar bands on anal and dorsal; in some specimens a large dark blotch on last dorsal spines. Pectoral faintly barred, with two dusky blotches near the base.

77. Dinematichthys ventralis (Gill) J. & G.

(*Brosmophycis ventralis* Gill, Proc. Ac. Nat. Sci. Phila. 1863, 253.)

2479, 2482, 2483. (Types of *Brosmophycis ventralis*.) Three specimens, the largest about 3 inches long, in fair condition.

78. Paralichthys adspersus (Steind.) J. & G.

7036. One specimen, about 8 inches long.

79. Tetrodon testudineus L.

12692. Young specimen. We are unable to distinguish the Pacific Coast form (*annulatus* Jenyns=*heraldi* Gthr.) from the West Indian *testudineus*.

80. Psilonotus punctatissimus (Günther) J. & G.

(= *Tetrodon oxyrhynchus* Lockington, Proc. Ac. Nat. Sci. Phila. 1881, 116.)

9899. Many specimens, the largest about 3 inches long.

81. Balistes mitis Bennett.

2990. Dried skin.

7318. Three adult specimens in spirits.

82. Antennarius strigatus Gill.

(*Antennarius strigatus* Gill, Proc. Ac. Nat. Sci. Phila. 1863, 92.=*Antennarius tenuifilis* Günther, Trans. Zool. Soc. Lond. 1869, 440.)

6267. (Types of *Antennarius strigatus*.) Two specimens, in fine condition.

83. Antennarius sanguineus Gill.

(*Antennarius sanguineus* Gill, Proc. Ac. Nat. Sci. Phila. 1863, 91. =*Antennarius leopardinus* Günther, Proc. Zool. Soc. Lond. 1864, 151.)

6393. (Types of *Antennarius sanguineus*.) Two fine specimens, one adult, the other nearly so.

18604. One half-grown example, in good condition.

The types of the following species described by Professor Gill appear to be lost or destroyed. Of all of these except *Doryrhamphus californiensis*, the Museum now possesses one or more examples in good condition, most of them being from the collection of Mr. Gilbert:

Dactyloscopus pectoralis	= *Dactyloscopus pectoralis* Gill.
Iniistius mundicorpus	= *Novacula mundicorpus* (Gill) J. & G.
Hypsypops dorsalis	= *Pomacentrus quadrigutta* Gill.
Diapterus californiensis	= *Gerres californiensis* (Gill) J. & G.
Diapterus gracilis	= *Gerres gracilis* (Gill) J. & G.
Hoplopagrus güntheri	= *Hoplopagrus güntheri* Gill.
Nematistius pectoralis	= *Nematistius pectoralis* Gill.
Cirrhitus betaurus	= *Cirrhitus rivulatus* Val.
Argyriosus brevoorti	= *Selene vomer* (L.) Lütk.
Trachynotus rhodopus	= *Trachynotus rhodopus* Gill. (*T. kennedyi* Steind.)
Trachynotus nasutus	= *Trachynotus rhodopus* Gill.
Doryrhamphus californiensis	= *Doryrhamphus californiensis* Gill.
Hippocampus gracilis Gill	= *Hippocampus ingens* Grd.

UNITED STATES NATIONAL MUSEUM, *June* 28, 1882.

LIST OF FISHES COLLECTED BY JOHN XANTUS AT COLIMA, MEXICO.

By DAVID S. JORDAN and CHARLES H. GILBERT.

About twenty years ago a considerable collection of fishes was made by Mr. John Xantus at Colima, on the west coast of Mexico, for the Smithsonian Institution. Much of this collection arrived at Washington in bad condition, and the greater part of it has gone to decay. In the present paper is given a catalogue of the specimens still remaining.

1. Ginglymostoma cirratum (Gmel.) Müller & Henle.

7332. Two young examples, each 10 inches long. This species has not hitherto been recorded from the Pacific coast of Mexico. A young specimen was seen by Mr. Gilbert at Mazatlan.

2. Arius guatemalensis Günther.

8144. Four specimens.

3. Characodon furcidens Jor. & Gilb.

5093. Very many examples in fair condition, the largest 3 inches long.

4. Muræna pinta Jor. & Gilb.

7328. One specimen, 8 inches long.

5. Rhypticus xanti Gill.

7740. One fine specimen, 8 inches long.

6. Epinephelus sellicauda Gill.

9583, 9587, 9589, 9601.

7. Diabasis sexfasciatus (Gill) J. & G.

30997. One half-grown specimen.

8. Diabasis steindachneri Jor. & Gilb.

9586, 9588, 9600, 19632.

9. Diabasis maculicauda (Gill) J. & G.

Three specimens, in very bad condition.

10. Pomadasys virginicus Subsp. **tæniatus** (Gill) J. & G.

31013. One specimen.

We have compared Pacific coast representatives of this species (*Anisotremus tæniatus* Gill) with specimens from the Bahamas. The former appear to have very slightly smaller scales (11–56–18 against 9–56–16), but we can find no other structural difference, and this may not be constant. The Atlantic form has the vertical bands much darker, almost black, instead of brown. The blue lateral stripes are wider and fainter, as broad as a scale and more than two-thirds the width of the interspaces; they are very faintly edged with darker. The additional smaller blue stripes between the broader stripes are more numerous than in the Pacific form. In the latter the blue stripes are much less wide than a scale and barely one-third the olive stripes. The coloration in Pacific coast specimens is very uniform, and the name *tæniatus* may be retained for the subspecies which they represent.

11. Pomacanthus strigatus (Gill) J. & G.

31008. A fine large specimen, 8 inches in length; pale bar downward from dorsal very distinct; dorsal and anal with a narrow edging of bright blue posteriorly. Blue stripes on head wholly obsolete.

12. Pomacentrus rectifrænum Gill.

Young specimens, in very poor condition.

13. Philypnus lateralis Gill.

8057. One example, 9 inches long.

14. Dormitator maculatus (Bloch) J. & G.

Specimens in bad condition.

15. Culius æquidens Jor. & Gilb.

5089. In bad condition.

16. Fierasfer arenicola Jor. & Gilb.

7531. Two specimens, the largest $4\frac{2}{3}$ inches long.

These specimens agree well with the typical example, but the mouth is larger, the maxillary extending much beyond orbit, its length nearly two-thirds that of head.

UNITED STATES NATIONAL MUSEUM, *June* 30, 1882.

LIST OF FISHES COLLECTED AT PANAMA BY CAPTAIN JOHN M. DOW, NOW IN THE UNITED STATES NATIONAL MUSEUM.

By DAVID S. JORDAN and CHARLES H. GILBERT.

About twenty years ago (1861–1865) several collections of fishes were forwarded to the Smithsonian Institution by Capt. J. M. Dow, from Panama and other points on the west coast of Central America. One of these collections has been studied by Professor Gill.* The others have hitherto remained unnoticed and many of the specimens have been allowed to decay. The present paper gives an account of what remains at present.

1. Mustelus dorsalis Gill.

(*Mustelus dorsalis* Gill, Proc. U. S. Nat. Mus. 1864.)

8068. (Types of *Mustelus dorsalis*.) Four half-grown specimens.

2. Anableps dowi Gill.

(*Anableps dowi* Gill, Proc. Ac. Nat. Sci. Phila. 1861, 4.)

8005. Five specimens, the largest nearly 11 inches long, from La Union, San Salvador.

3. Hemirhamphus poeyi Günther.

30953. Two fine adult specimens.

This species is very close to *H. unifasciatus*, if really distinct. It differs chiefly in the shortness of the lower jaw.

Four specimens of *Exocœtus*, representing three species, are also in the collection, but it is questionable whether any of them really came from Panama. We are informed by Captain Dow that the specimen which became the type of *Exocœtus albidactylus* Gill (Proc. Ac. Nat. Sci. Phila. 1863, 167) was taken off the northern coast of Brazil, and not at Panama.

4. Agonostoma nasutum Günther.

30966. One specimen.

5. Joturus stipes sp. nov.

31010. One large specimen, found in the same bottle as 30957 (*Pomadasys humilis*).

19915. Two still larger examples, in good condition, about 15 inches in length, from "Central America"; the exact locality and the collector unknown.

Head $4\frac{2}{5}$ in length; depth 4; D. IV—1, 9. A III, 9. Scales 45—13.

---

* Descriptive Enumeration of a Collection of Fishes from the Western Coast of Central America. Presented to the Smithsonian Institution by Capt. John M. Dow. By Theodore Gill. Proc. Ac. Nat. Sci. Phila. 1863, 162–174.

Body robust, a little compressed behind. Head heavy, little compressed, gibbous above and anteriorly. Snout thick, broad, protruding, blunt and tumid at tip, considerably overhanging the small inferior mouth, and entirely below the level of the eye. Length of snout 2⅔ in head. Maxillary reaching nearly to posterior margin of eye, 2½ in head, hidden entirely beneath the preorbital. Mouth broad, but without much lateral cleft. Lower jaw included. Upper lip thick, slipping beneath the snout. Lower lip very thick, its anterior edge forming a soft sharp-edged fold; outline of the lip very obtuse. Teeth rather strong, coarse, bluntly conical, forming a large ovate patch on each side of lower jaw, the two patches not confluent. A similar but smaller patch on the vomer. No teeth on the palatines. Upper jaw with a band of similar but rather smaller teeth.

Nostrils roundish, close together, in front of the small round eye, which is nearer angle of mouth than level of top of head. Interorbital space very broad, transversely convex. Eye 6 in head, 3 in interorbital width. No adipose eyelid. Neither lip with cirri or papillæ.

Scales of head each with many smaller ones at base; accessory scales on body largely developed. All the fins, including spinous dorsal, covered with small scales. Gill membranes largely united, free from the isthmus.

Dorsal spines compressed and curved, becoming rapidly shorter from the first, which is about two-thirds length of head. Second dorsal and anal with their free margins concave, the anal somewhat falcate, its longest ray 1⅙ in head. Caudal forked, as long as head. Pectoral as long as head, reaching middle of first dorsal.

Color dull olivaceous, without distinct markings, paler below.

6. Mugil brasiliensis Agassiz.

15121, 15122, 15128. Several young specimens.

7. Murænesox coniceps Jor. & Gilb.

30981. One large specimen, in poor condition.

8 Echeneis naucrates L.

30984. One half-grown specimen, in fair condition. Disk with 22 laminæ.

9. Scomber grex Mitchill.

30998. Two half-grown specimens, in poor condition. The air-bladder is present.

10. Oligoplites altus (Gthr.) J. & G.

30969. A young specimen, in good condition.

11. Oligoplites occidentalis (L.) Gill.

(*Oligoplites inornatus* Gill, Proc. Ac. Nat. Sci. Phila. 1863, 166.)

30959. (Type of *Oligoplites inornatus*.) One adult specimen, in good condition.

12. Trachynotus ovatus (L.) Lac.

30970. One partly grown specimen, in good condition.

15123. Three very young specimens.

Compared with Atlantic specimens of somewhat larger size, No. 30970 is somewhat deeper (depth $1\frac{2}{3}$ in length, instead of $1\frac{1}{2}$), and the dorsal and anal fins are much less elevated in front (anterior lobe of dorsal $4\frac{1}{3}$ in length; $2\frac{1}{10}$ in *T. ovatus* from Cuba).

13. Caranx dorsalis (Gill) Gthr.

(*Carangoides dorsalis* Gill, Proc. Ac. Nat. Sci. Phila. 1863, 166.)

4957. (Types of *Carangoides dorsalis*.) Two specimens, in good condition.

14. Caranx speciosus Lac.

(*Caranx panamensis* Gill, Proc. Ac. Nat. Sci. Phila. 1863, 166.)

30960. (Type of *Caranx panamensis*.) One adult specimen, in good condition.

15. Caranx fallax C. & V.

(*Carangus marginatus* Gill, Proc. Ac. Nat. Sci. Phil. 1863, 166.)

30958. (Type of *Carangus marginatus*.) One adult example, in good condition.

There can be no doubt that Dr. Günther's identification of *Scomber hippos* L. with this species is erroneous.

16. Caranx atrimanus J. & G.

30745. One specimen, $5\frac{1}{3}$ inches long, in good condition.

17. Rhypticus nigripinnis Gill.

(*Promicropterus decoratus* Gill, Proc. Ac. Nat. Sci. Phil. 1863, 164.)

30961. (Type of *Promicropterus decoratus*.) One specimen, 8 inches long, in good condition.

18. Alphestes multiguttatus (Gthr.) J. & G.

30988. One specimen, in fair condition, but somewhat faded.

30954. A young specimen, in good condition.

This species is closely allied to the West Indian *Alphestes afer* Bloch, (*Plectropoma chloropterum* C. & V.), but is readily distinguished by the more pointed snout and the totally different coloration.

19. Epinephelus analogus Gill.

(*Epinephelus analogus* Gill, Proc. Ac. Nat. Sci. Phila. 1863, 163.)

4944. (Type of *Epinephelus analogus*.) A half-grown specimen, in good condition.

30993. One fine young specimen.

20. Pomadasys humilis (Kner & Steindachner) J. & G.

30957. A fine adult specimen, and one young specimen.

The resemblance of this species to *Pomadasys crocro* (C. & V.) is very close.

21. Kuhlia xenura Jor. & Gilb.

(*Xenichthys xenurus* Jordan & Gilbert, Proc. U. S. Nat. Mus. 1881, 454.)

4356. (Types of *Xenichthys xenurus*.) Two specimens, in good condition. This species should be referred to the genus *Kuhlia* Gill (=*Moronopsis* Gill), rather than to *Xenichthys*. It has no enlarged scale in the ventral axil, and it has the naked snout, jaws, and fins, the compressed body, and high dorsal spines of the species of *Kuhlia*.

An examination of the Museum records shows that these specimens now bear a number originally given to one of the types of "*Euschistodus concolor*," from San Salvador. As the connection of these specimens with the Dow collection rests on the same records, we consider it doubtful whether they really came from San Salvador.* *Kuhlia xenura* appears to be a valid species distinct from *K. tæniura*, but it should be suppressed from the list of species inhabiting the Pacific coast of Central America, until its occurrence there is verified by some collector.

22. Centropomus unionensis Bocourt.

30991. One fine specimen, in good condition.

23. Apogon dovii Günther.

30990. Two specimens, in bad condition.

24. Polynemus approximans Lay & Bennett.

15129. One specimen, in good condition.

25. Sciæna oscitans Jor. & Gilb.

30967. Three fine specimens, two of them adult.

26. Sciæna armata (Gill) J. & G.

(*Bairdiella armata* Gill, Proc. Ac. Nat. Sci. Phila. 1863, 164 = *Corvina acutirostris* Steindachner Ichth. Beitr. III, 28, 1875.)

(Type of *Bairdiella armata*.) One specimen, in good condition.

27. Sciæna ophioscion (Gthr.) J. & G.

(*Ophioscion typicus* Gill, Proc. Ac. Nat. Sci. Phila. 1863, 165.)

22861. (Type of *Ophioscion typicus*.) One adult specimen, in good condition.

28. Pomacanthus zonipectus (Gill.) Günther.

(*Pomacanthodes zonipectus* Gill, Proc. Ac. Nat. Sci. Phila. 1862, 244 (adult) = *Pomacanthus crescentalis* Jor. & Gilb. Proc. U. S. Nat. Mus. 1881, 358, young.)

5922. (Type of *Pomacanthodes zonipectus*.) A large specimen, in good condition, from San Salvador.

29979. A young specimen (from Nicaragua), showing the coloration of the "*crescentalis*" stage, which is wholly different from that of the adult. The changes in coloration appear to be analogous to those of *Pomacanthus arcuatus*.

---

* There is some reason for thinking that these specimens belonged to Dr. Stimpson's collection, and came from the east coast of Asia.

29. Acanthurus tractus Poey.

30992. A young specimen, in good condition.

30. Holocentrum suborbitale Gill.

2765· Four specimens, in good condition.

31. Gerres dowi (Gill) Gthr.

(*Diapterus dowii* Gill, Proc. Ac. Nat. Sci. Phila. 1863, 162.)

30985. (Types of *Diapterus dowi*.) Three half-grown specimens, in good condition.

Two large specimens of *Gerres lineatus* (30982), from a fresh-water lake near Acapulco, Mexico, are also in the collection.

32. Glyphidodon declivifrons (Gill) Gthr.

(*Euschistodus declivifrons* and *concolor*, Gill, Proc. Ac. Nat. Sci. Phila. 1862, 145 : *Euschistodus analogus*, Gill, l. c., 1863, 219.)

30986. (Formerly 4356.) (Type of *Euschistodus concolor*.)
2757. One specimen.

33. Glyphidodon saxatilis (L.) Lac.

4360. Young specimens from San Salvador.

34. Pomacentrus quadrigutta Gill.

4365. One small specimen from San Salvador, having the coloration ascribed to *P. quadrigutta*.

35. Pomacentrus rectifrænum Gill.

30962. Small specimens from San Salvador.

36. Scorpæna plumieri Bloch.

One specimen, in bad condition, apparently belonging to this species.

37. Dormitator maculatus (Bloch) J. & G.

(*Dormitator microphthalmus* Gill, Proc. Ac. Nat. Sci. Phila. 1863, 170.)

4953. (Type of *Dormitator microphthalmus*.) A very large specimen, nearly a foot in length.

38. Philypnus lateralis Gill.

30994. Several specimens.

39. Gobius soporator C. & V.

2761. Many small specimens.

40. Clinus macrocephalus Günther.

30956. Two specimens, in bad condition.

41. Diodon liturosus Shaw.

(Shaw, General Zoöl. v. pt. 2, 437, 1804, after *Diodon tacheté* Lac. = *Diodon maculatus* Gthr.)

9876. One young specimen, in good condition.

The types of the following species described by Professor Gill, from the present collection, appear to be lost:

Centropomus armatus   = *Centropomus armatus* Gill.
Amblyscion argenteus   = *Larimus argenteus* (Gill) J. & G.
Exocœtus dowii   = *Exocœtus dowi* Gill.
Upeneus grandisquamis   = *Upeneus grandisquamis* Gill.
Trichidion opercularis   = *Polynemus opercularis* (Gill) Gthr.
Mugil guentherii   = *Mugil albula* L. (*Mexicanus* Steind).
Leptarius dowii   = *Arius dowi* (Gill) Gthr.
Sciades troschelii   = *Arius* Sp.
Aelurichthys panamensis = *Aelurichthys panamensis* Gill.
Atractosteus tropicus   = *Lepidosteus tropicus* (Gill) Gthr.
Urotrygon mundus   = *Urolophus mundus* (Gill) Gthr.

All these species are now represented in the National Museum, with the exception of *Sciades troschelii* and *Urotrygon mundus*, which remain unidentified.

UNITED STATES NATIONAL MUSEUM, *July* 4, 1882.

LIST OF A COLLECTION OF FISHES MADE BY MR. L. BELDING NEAR CAPE SAN LUCAS, LOWER CALIFORNIA.

By DAVID S. JORDAN and CHARLES H. GILBERT.

1. Muræna dovii Günther.

(*Muræna pintita* Jor. & Gilb.)

30486. A young specimen from Espiritu Santo Island, agreeing fairly with the original description of *Muræna pintita*, but the tail slightly shorter than the rest of the body. There are a few small yellowish spots on the posterior part of the head, similar to those on the body, which are not very numerous.

We have examined two very large eels (19893) collected by Captain Herendeen at the Galapagos Islands. They seem to be referable to *Muræna dovii*, agreeing as well with Günther's description as they do with each other, and there seems to be little room for doubt that our "*Muræna pintita*" is the young of the same species. There is considerable variation in the size and form of the small pale spots.

2. Leptocephalus* conger (L.) J. & G.

30930. A small specimen, 6¼ inches long, from near Cape San Lucas, does not show any variation from Mediterranean examples of this spe-

---

* The generic names *Leptocephalus* Gmelin (Syst. Nat. 1, 1150, 1788; based on *Leptocephalus morrisi*, a larval *Conger*) and *Echelus* Rafinesque (Caratteri di Alcuni Nuovi Generi, etc., 1810, 64; *E. macropterus* Raf.) have priority over *Conger* Cuvier. As *Leptocephalus* is the first generic name applied to this group, it should in our opinion be retained, in preference to *Echelus*, notwithstanding its common use for larval forms generally.

cies. No other specimen of this genus has been brought from the Pacific coast of tropical America.

3. Mugil albula L.

30932. Four small specimens, each about 5 inches long, from Cape San Lucas.

4. Mugil brasiliensis Agassiz.

30933. Three half-grown and numerous young specimens were collected in San José River, near Cape San Lucas.

5. Agonostoma nasutum Günther.

30934. Five specimens, the largest about 7 inches long, were collected at San José, where they are known as *trucha*, or trout. These do not differ essentially from the specimens described by Dr. Günther; the maxillary usually extends slightly beyond front of orbit; head $4\frac{1}{4}$ to $4\frac{1}{2}$ in length (to base of caudal); eye $4\frac{1}{3}$ in head; maxillary not longer than interorbital width, contained $2\frac{3}{4}$ to $3\frac{1}{4}$ times in head; a band of pterygoid teeth often but not always developed; dorsal spines very strong, not flexible, the origin of the fin nearer snout than tail; caudal well forked, the middle rays $1\frac{1}{2}$ in outer.

6. Remora squalipeta (Dald.) J. & G.

(*Echeneis remora* L.)

30941. A single specimen, 6 inches long, from San José.

7. Centropomus robalito Jor. & Gilb.

30940. Two small specimens, $3\frac{1}{2}$ inches long, were obtained at San José.

8. Gobius banana Cuv. & Val.

30935. Color light olivaceous, back and sides blotched and shaded with dark brown; a series of irregular roundish blotches along middle of sides; narrow black streaks radiating from eye, two of these running downwards and forwards to mouth, and one backwards to upper preopercular angle, with a similar parallel streak below it; a black streak running across upper margin of opercle, and extending on base of upper pectoral rays; dark markings on back, sometimes forming more or less distinct cross-bars; belly white; ventrals and anal immaculate; other fins all more or less distinctly barred with wavy black lines.

Head $3\frac{2}{3}$ in length; depth $5\frac{2}{3}$. D. VI-11; A. I, 10; scales 61–21.

Body subfusiform, long and low, scarcely or but little compressed. Head long and low, slender, much narrowed anteriorly, its greatest breadth but little more than its greatest depth, and $1\frac{1}{2}$ in its length; cheeks scarcely tumid; snout long, low, $2\frac{2}{3}$ in head, the profile very little curved. Upper jaw very protractile; lips thick; mouth low, narrow, subterminal, very variable in size, the maxillary from $2\frac{1}{6}$ to $2\frac{4}{5}$ in head, sometimes not reaching eye, sometimes to below middle of orbit; lower jaw included; scaly region of nape, beginning very close behind eye. Teeth in rather narrow bands, those in outer row in both jaws considerably enlarged, rather robust. Eyes very small, placed high, their range mostly vertical; eye 6 to 7 in head, somewhat greater than the narrow, flat, interorbital area. Isthmus moderate, its width $3\frac{1}{3}$ in head; gill-openings extending forwards but very little above opercle.

Head naked; scales on nuchal and antedorsal regions much reduced in size; nuchal patch of scales beginning close behind eyes; scales on body all regularly imbricated, roughly ctenoid, those on caudal peduncle largest.

Dorsal spines low, rather slender, the tips slender and slightly exserted, the longest spine not quite half head; soft dorsal moderate, the longest rays $2\frac{1}{4}$ in head; caudal slightly rounded behind, $1\frac{1}{4}$ in head; ventrals $1\frac{2}{3}$ in head, the basal membrane broad, moderately developed; vent midway between base of caudal and front of eye.

Two large specimens, each about 6 inches long, and five smaller ones were taken in fresh water near San José. Some (probably all) of the large-mouthed specimens are males, the others females.

9. Gobius sagittula (Günther) J. & G.

30936. Seven specimens were obtained from San José, the largest 4 inches long. The teeth in the upper jaw are not in a single series, as described by Dr. Günther, but form a narrow band, the outer series being much enlarged and separated from the band by a narrow interspace.

10. Philypnus lateralis Gill.

30937. Two specimens, the largest $4\frac{1}{2}$ inches long, collected at San José. This species differs very little from the Atlantic *P. dormitator*, the fins, formulæ, and general proportions being the same. The adult *lateralis* loses the dark bands along sides, but retains the black spot on base of upper pectoral rays; the depth of adult *lateralis* ($4\frac{1}{3}$ in length) is much greater than in *dormitator*, and the scales on cheeks and top of head are larger.

11. Dormitator maculatus (Bloch) J. & G.

30939. Very numerous specimens of this species, the largest 7 inches long, were procured at San José.

12. Culius æquidens Jor. & Gilb.

30943. Two specimens, one an adult $1\frac{1}{2}$ feet long, were taken in fresh water near San José. The adult has the mouth larger (maxillary reaching well beyond orbit) and eye smaller (contained nearly four times in interorbital space) than in the type specimens of this species.

13. Aphoristia atricauda Jord. & Gilb.

30942. A single small specimen, $1\frac{1}{2}$ inches long, has numerous small roundish light spots on the colored side, and the black of the tail ocellated with white.

14. Tetrodon testudineus L.

(*Tetrodon annularis* Jenyns; *Tetrodon heraldi* Gthr.)

30944. A single small specimen.

UNITED STATES NATIONAL MUSEUM, *June* 28, 1882.

LIST OF FISHES COLLECTED AT PANAMA, BY REV. MR. ROWELL, NOW PRESERVED IN THE UNITED STATES NATIONAL MUSEUM.

By DAVID S. JORDAN and CHARLES H. GILBERT.

At some time about the year 1860, a collection of fishes was sent from Panáma to the Smithsonian Institution, by Rev. Mr. Rowell. The following is an enumeration of the specimens belonging to this collection, now preserved in the United States National Museum:

1. Ælurichthys pinnimaculatus Steind.

31004. One specimen, 20 inches long.

2. Arius elatturus Jor. & Gilb.

30995. One specimen.

3. Arius insculptus J. & G.

30977. Two specimens, in fair condition.

4. Hemirhamphus poeyi Günther.

31019. One specimen, answering well to Günther's description of this species, which has not been hitherto noticed on the Pacific coast of Central America.

5. Muræna pinta J. & G.

7328. One specimen.

6. Rhypticus nigripinnis Gill.

(*Rhypticus nigripinnis* Gill, Proc. Ac. Nat. Sci. Phila. 1861, 53.)

3700. The original type of the species, $3\frac{1}{2}$ inches long, in bad condition, evidently identical with the adult specimen later described as *Promicropterus decoratus*, and with the young example called *Rhypticus maculatus*.

7. Centropomus armatus Gill.

One specimen, 7 inches long.

8. Pomadasys branicki (Steind.) J. & G.

7499. One specimen, 3 inches long.

9. Diabasis flaviguttatus (Gill.) J. & G.

31005. Two specimens, $8\frac{1}{2}$ inches in length.

10. Apogon dovii Günther.

6268. Two specimens, in bad condition.

11. Micropogon altipinnis Günther.

7010. A young specimen, in bad condition.

12. Gerres peruvianus C. & V.

5717. One specimen.

Two species allied to the present one occur in the West Indies, and all three have been called *Gerres rhombeus* by authors. One of these, evidently the *Gerres rhombeus* C. & V., has but two anal spines; the other, *Gerres rhombeus*, or *Mojarra rhombea* Poey (= *Gerres olisthostoma* Goode & Bean Mss.), has the ovate groove for the reception of the premaxillary processes completely covered with scales. In *Gerres peruvianus*, as in most species of *Gerres*, this region is entirely naked. There are also minor differences in the length of the fins. We have never seen a specimen with two anal spines on the Pacific coast of tropical America, but the two-spined species (*rhombeus*) is common at Aspinwall.

13. Citharichthys spilopterus Günther.

30996. Three specimens, in poor condition.

UNITED STATES NATIONAL MUSEUM, *June* 30, 1882.

ON A COLLECTION OF BIRDS FROM THE HACIENDA "LA PALMA," GULF OF NICOYA, COSTA RICA.

By C. C. NUTTING.

[WITH CRITICAL NOTES BY R. RIDGWAY.*]

Costa Rica, the southernmost of the Central American States, lies between the eighth and eleventh degrees north latitude, quite a considerable portion being actually south of Panama, owing to the peculiar curve of the continent between Costa Rica and South America proper.

Like all the Central American States, Costa Rica is characterized by comparatively low coast regions, with a rugged interior composed of mountains which reach an altitude of nearly 11,000 feet, as is the case with the volcanoes of "Irazu" and "Turrialba," and elevated valleys sometimes of considerable extent, as the valleys of San José and Cartago.

These physical characteristics render the region a most fertile one for the naturalist, who finds in this favored field vegetable and animal life varying with the altitude of his collecting ground, and embracing both tropical and temperate forms.

On the 13th of February, 1882, I landed in Punta Arenas, the only important point on the Pacific coast of Costa Rica. Although my instructions were to direct my efforts principally to the region of the

---

*The editor of this paper is responsible for the determination of the species, the nomenclature adopted, and all critical notes.—R. R.

Gulf of Nicoya, I found it necessary to go to San José, the capital, to present certain letters of introduction and confer with Señor Don José Zeledon, of that city, as to the best disposition of the short time at my disposal. This gentleman strongly advised me to spend some time collecting in the interior, more especially in the region of the volcano "Irazu," and I accordingly decided to spend a fortnight there; also, a few days in San José, after which I returned to the coast and spent a month in collecting in the region of the Gulf of Nicoya. These three fields of operation, embracing as they did the three distinctive avi-faunæ of low, middle, and high altitudes, seemed to me to be most likely to afford a representative collection of Costa Rican birds.

The collections from the interior having not yet been received, it became necessary to defer lists of the species therein included, but which, it is hoped, may be presented within a reasonable time.

The Gulf of Nicoya extends from northwest to southeast, and is 60 or 80 miles long, dotted with numerous conical islands (the largest being San Lucas, a convict island), and encircled by low hills closely covered with tropical vegetation.

La Palma, the hacienda of Don Ramon Espinach, was my home during my stay in that region, and it is to the courtesy of its kind proprietor that I owe whatever success has attended my visit there. Nothing could be more generous than his conduct toward me, an utter stranger, and it is with the greatest pleasure that I embrace this opportunity to express my sincere thanks, not only for a pleasant home for more than a month, but also for much practical assistance in the way of furnishing horses and men and all other facilities to aid my explorations and increase my collections. La Palma is situated about 10 miles northwest of Colorado, a little hamlet on the northern coast of the gulf.

The region is an exceedingly low one, and in the rainy season becomes a vast swamp, unhealthy and infested with numerous insects. My visit was at the end of the dry season, at which time the earth was exceedingly dry and hard, and checkered with deep cracks caused by the intense heat of the tropical sun.

Notwithstanding the fierce heat, the forests were green and the flowers were blooming luxuriantly, while birds and other animals were extremely abundant. The vegetation is, of course, entirely tropical in its nature. Among the fruit trees the palm, mango, plantain, banana, orange, and "marañon" are worthy of mention. This latter fruit I do not remember to have seen elsewhere. The fruit resembles a red pepper with a bean-shaped seed hanging from its lower end. The taste is slightly acid and very pleasant.

The rubber, red-wood, and mahogany trees are also abundant, although a market for them has not been opened in that region. The forests are composed of other strictly tropical trees, bound together and interlaced with a network of vines of every description and covered with orchids and parasitic cacti. The ground beneath is freer from obstruction in the way of undergrowth than might be expected, although numerous species of cacti and other thorn-bearing plants are sometimes exceedingly annoying to the collector.

The mammalian fauna is rich and varied. Three species of monkeys were noticed. The " Howling Monkey " (*Mycetes palliatus*) is most prominent to the ear, if not to the eye. Its cry is the most diabolical, in the estimation of the writer, of all sounds issuing from animate beings. The " Red Monkey " (*Sapajou melanochir*) is quite numero us, and is the largest in size of Costa Rican Quadrumana. One little domestic scene in connection with this monkey impressed the writer so forcibly that he cannot refrain from describing it. While hunting along a lagoon one day, I suddenly came under a tree in which a troop of these monkeys were disporting themselves. A female, with her "baby" clinging to her back, happened to be nearest me at the end of an overhanging branch. Upon seeing the strange-looking animal below, with true maternal solicitude for her offspring, she hastened to bear it out of danger. As she started for the main trunk of the tree, a male started from the trunk to go out and have a closer look at the intruder. They met about the middle of the branch, when she commenced to chatter and look down at me as if to implore his protection, upon which he put his arms around her and *embraced her*. After standing in that position for several seconds, they parted, each proceeding on its way. After such a scene of almost human affection it is needless to say that the writer could not find the heart to shoot one of the monkeys.

The most abundant by far is the White-faced Monkey (*Cebus hypoleucus*), which is black with the exception of the shoulders and sides of the face, which are covered with rather long white hair, thus giving the appearance of little bald-headed black men. They were often quite annoying from their habit of throwing sticks, nuts, etc., at the traveler passing below them. They soon discovered the place where I took my morning bath, and were so annoying in this particular that I appreciated as never before the pathetic story of the " Boys and the Frogs," and had to shoot one of them in pure self-defense. But I felt like a murderer for it.

The Felidæ are well represented in this region. The Jaguar (*Felis onca*) is quite common, but apparently of a smaller race than in South America. It is not considered dangerous by the natives. *Felis concolor*, the " Leon" of that country, is rather rare and much feared by the inhabitants. Several other animals of this family were seen, especially one entirely black (probably a melanism of the Jaguarundi) which I do not remember with sufficient distinctness to venture to identify.

A beautiful little species of Deer (*Cervus mexicanus*) is abundant, not at all timid, and easily approached. Its flesh is, of course, excellent food.

The Peccary (*Dicotyles torquatus*) is abundant, though usually seen in small droves of not more than eight or ten. I never heard of their attacking man, as they are said to do. The natives sever the scent-pouch from the animal as soon as possible after death. Otherwise it is extremely disagreeable both to the taste and smell.

The "Watousa" (*Dasyprocta cristata*) is also quite common, though very shy and mostly confined to the thick forests. Its flesh is, in the opinion of the writer, the most delicious meat he ever had the pleasure of eating.

The Tapir (*Elasmognathus bairdi*) is somewhat rare, and seldom seen, probably on account of its nocturnal habits.

The Coatimundi (*Nasua narica*) is abundant, and though eaten by the Indians is not considered eatable by the Spaniards. It somewhat resembles the Raccoon (*Procyon lotor*) but is diurnal, as a rule, and is frequently found in quite extensive troops of twenty or more.

Smaller mammals are numerous, but not having secured specimens, the writer will not venture to identify them.

Alligators are extremely abundant and constitute a source of constant annoyance, and sometimes of danger, to the collector while hunting along the rivers and lagoons of that region.

The avi-fauna, although strictly tropical, is not so varied as on the eastern coast.

Perhaps the most characteristic birds of the region are the Parrots (*Psittaci*). They are so numerous as to constitute a real source of annoyance to the collector. They are always noisy and apparently always quarreling. Their harsh, discordant cries make such a din that the faint twittering of the smaller birds is entirely drowned, and many rarities are doubtless unobserved by the naturalist who vainly attempts to trace their modest song among the clatter of their gaudy neighbors.

The *Falconidæ* are exceedingly numerous and easy to approach. It is by no means unlikely that novelties in this family will yet be reported from the Gulf of Nicoya.

The prevalence of the "zygodactyle" foot is a very marked feature of the birds of this region.

In concluding my remarks upon this region it may be well to mention some of the difficulties to be met by the naturalist, together with a few practical hints as to how they are best surmounted.

The climate is much more bearable than might be supposed. The heat is never so intense as that which we frequently experience in the United States. Indeed I never found it so oppressive as it is here in Washington as I write. The nights are always comfortably cool and one always finds use for his blanket before morning.

By far the most favorable time for collecting is during the dry season (October to May). The seasons (the wet and dry) are very distinctly defined, so that the collector may know what kind of weather to expect.

Ants are very troublesome to the collector. They attack the bills of his specimens and frequently ruin a rare bird in a very few minutes. But there is a sure remedy for this pest in the oil of bitter almonds, which, if rubbed on the bills of specimens, the ants will religiously leave the birds alone. This method was suggested to me by Don José Zeledon, of San José, and proved to be all that could be desired.

But there is another and far more terrible pest which attacks not the specimens, but the naturalist himself. I refer to a little insect known in those regions as the "Garrapata," which is a very minute species of Tick. During the dry season it is impossible to avoid being actually covered with these diminutive tormentors. They are so numerous that it is impossible to avoid them, and their bite is so aggravating that a man is actually panic-stricken when he finds himself literally alive with them. These insects constituted the most serious difficulty I met in Costa Rica. My body became entirely covered with their bites. The itching caused was frightful, and, although I could control myself during the day, I would wake up at night and find myself literally tearing my flesh in frantic though unconscious efforts to relieve the itching. Working daily with arsenic, this poison unavoidably found its way into the system through the bites of the "Garrapatas," and I was thus severely poisoned and my person covered with festering sores, making it dangerous to pursue my work further at that time. Any preventive for this evil would be a boon to tropical explorers. I regret to say that I know of none that is unobjectionable as well as effective. Moistening the lower part of the pantaloons and sleeves of the coat with a decoction of tobacco juice is a partial success, and so is anointing the entire body with kerosene. The best way to rid one's self of these pests after they have established themselves on the person is to follow the example of the natives, who first procure a piece of the black wax or "ĝera," which is abundant, and, after removing all their clothing, proceed to strike themselves with the wax. This they do systematically until every portion of the body has been struck. The wax, by adhesion, removes every "Garrapata" that it strikes. This simple and effective method of getting rid of these insects is the universal practice throughout the country.

The slow and inadequate means of transportation, especially in the less-settled portions of the country, is apt to cause long and annoying delays. When possible it is advisable for the collector to keep with him as many of his effects as he expects to need for a fortnight at least, including a large supply of *patience*.

1. **Merula grayi** (Bonap.).

Common. Habits and note similar to the common Robin of the United States, *M. migratorius*. Rather solitary and silent during the time I collected in that region, but this is probably due to the fact that it was the breeding season. Iris brown.

Three specimens obtained near La Palma, as follows:

No. 172. ♂ ad. April 4.

No. 242. ♀ ad. April 24.

No. 296. ♀ ad. April 30.

2. Polioptila bilineata (Bonap.).

[NOTE BY R. R.—The black-capped *Polioptilæ* of Central and South America are involved at present in so much confusion that the following remarks, based upon specimens in the collection of the United States National Museum, may not be out of place. The latest information which we have upon the subject is that contained in SALVIN and GOD-MAN'S *Biologia Centrali Americana* (Aves, vol. i, pp. 50–55), and this should be carefully consulted in the present connection. The authors of the work quoted recognize in Central America three species of the genus, besides *P. cærulea*, as follows:

(1.) P. NIGRICEPS Baird. *Hab.*—Southwestern Mexico (Mazatlan, Tepic, and States of Oaxaca and Tehuantepec); also, Colombia and Venezuela, but not recorded from any part of Central America proper, except San Salvador (La Union).

(2.) P. BILINEATA (Bonap.). *Hab.*—Guatemala to Colombia and western Ecuador.

(3.) P. ALBILORIS Scl. & Salv. *Hab.*—Southwestern Mexico (Sta. Efigenia and Tehuantepec City), Guatemala, and Nicaragua.

In their treatment of these three species, Messrs. SALVIN and GOD-MAN make some very interesting generalizations, based upon certain anomalies in their geographical distribution, but which appear to be somewhat negatived by the evidence afforded by additional specimens. Disclaiming, however, any intention of criticising the hypotheses offered by the authors of the great work in question, the following remarks are presented as perhaps throwing some additional light upon this more or less complicated subject.

Of the true *P. nigriceps* we possess specimens only from southwestern Mexico (Mazatlan and Tepic to Tehuantepec and Oaxaca). The seven examples before me may each be very readily distinguished from all black-lored specimens of the genus from more southern localities in the collection by the indistinct gray edgings to the tertials, all of the more southern black-capped forms, with the single exception of *P. bilineata*. having the tertials broadly and very distinctly edged with pure white. *P. bilineata*, however, may, in every plumage, be easily distinguished by the white lores and superciliaries.

The only other black-lored form of which the Museum possesses adult males is *P. leucogastra*. Of this, there are two adult males and two young males from Bahia, and a female from Venezuela. The young male has the crown plumbeous (darker than the back), the feathers darker in the center, and with a distinct postocular patch of glossy black. The Museum also possesses an adult male from Bogota, one from the Pacific coast of Central America (No. 30555, Capt. J. M. Dow), and another from Grenada, Nicaragua (No. 32556), which I cannot distinguish in any way from the Brazilian birds, or true *P. leucogastra*. The broad and conspicuous white edgings to the tertials at once separate them from *P. nigriceps*.

Of *P. albilora* there are in the collection two adult males and two females, as follows: No. 34101, ♂, Realejo (Pacific coast of Nicaragua), February, and 30554, ♀, same locality, July 16; No. 59584, ♂, Tehuantepec City, November 11, and 57470, ♀, Sta. Efigenia, Tehuantepec, December 25. Each of these specimens may be at once distinguished from any examples of *P. bilineata* by the broad and distinct white edging to the tertials; and if not a distinct species, must be a connecting link (possibly a hybrid) between *P. leucogastra* and *P. bilineata*, having the conspicuously white-edged tertials of the former and the white lores of the latter. That it is probably distinct from both these forms is suggested by the fact that its habitat is mostly to the northward of the district inhabited by *P. bilineata* and *P. leucogastra* together, although to the southward (*i. e.*, in Guatemala and Nicaragua) the three are found in the same localities.

P. bilineata is represented by a considerable series, embracing specimens from Venezuela, Panama, Veragua, Costa Rica, Nicaragua (Greytown), and Guatemala. The species is well defined, an adult male each from the first and last localities mentioned above being undistinguishable, except that one is in somewhat worn, the other in fresh and soft, plumage. An adult female from Venezuela is likewise undistinguishable from northern specimens.

Granting that *P. buffoni* has always the lateral tail-feather white except at the extreme concealed base, it may thus be distinguished from *P. leucogastra*, but of this species or race I have been able to examine but a single specimen, an adult female from Demerara (No. 55161, U. S. Nat. Mus.). This seems very distinct from the female of all the forms discussed above. The upper parts are a decidedly paler and bluer gray, almost exactly as in *P. cærulea;* the lateral pair of rectrices have the exposed portion entirely white, only the extreme concealed base being black; the greater wing-coverts are much paler gray than the back, and pass into grayish white at the tip; this white and also that on the outer webs of the tertials shows in very abrupt and striking contrast to the deep black of the primaries, primary coverts, and alulæ. The lores of this specimen are light grayish, and there is a distinct supraocular spot of white.

Upon the whole, it appears, from the material examined, that the following species, or at least well-marked races, of black-capped *Polioptilæ*, may be recognized as belonging to Central and South America:

A: Pileum and lores wholly black in fully adult males.

    a. Lateral tail-feather wholly white for exposed portion.

        1. P. BUFFONI. Tertials broadly edged with pure white. (Cayenne.)

    b. Lateral tail-feather black at base, this usually showing considerably beyond the coverts.

        2. P. LEUCOGASTRA. Tertials broadly edged with pure white. (Bahia, Bogota, Venezuela, Colombia, Nicaragua.)

3. P. NIGRICEPS. Tertials narrowly edged with dull gray. (Mazatlan, Tepic, Tapana, Tehuantepec, Quiotepec, Oaxaca.)

B. Pileum black, but lores white, in fully adult males.

4. P. ALBILORIS. Lores and eyelids white, but this scarcely passing beyond the eye; tertials broadly edged with pure white. (Realejo, Nicaragua; Sta. Efigenia, Tehuantepec.)

5. P. BILINEATA. Lores, eyelids, and superciliary stripe white; tertials narrowly edged with gray. (Venezuela, Panama, Veragua, Costa Rica, Nicaragua, Guatemala.)

ADDITIONAL NOTE.—Since the above was written, Mr. GEO. N. LAWRENCE has kindly forwarded for inspection his entire series of black-capped *Polioptilæ* from Middle and South America, embracing the following specimens: (1) *P. buffoni*: 1 ♀ from Guiana, agreeing with that described above. (2) *P. leucogastra*: 1 ♂ ad. from Bahia, 1 do. from Bogota, and 1 ♂ juv. from Venezuela, the latter being the type of *P. plumbeiceps* Lawr. (3) *P. nigriceps*: 1 ♂ ad. from Sta. Efigenia, Tehuantepec. (4) *P. albiloris*: 1 ♂ ad. from Sta. Efigenia, Tehuantepec, 1 ♀ ad. from Tapana, Tehuantepec, and 1 do. from Guatemala. (5) *P. bilineata*: 2 ♂ ad. and 1 ♀ ad., Panama, including the types of *P. superciliaris* Lawr., and 1 ♂ ad., said to be from Guatemala (but this on authority of a dealer only). This series so fully bears out the indications afforded by the the National Museum specimens that more extended remarks are unnecessary.—R. R.]

Habits similar to our *Mniotiltidæ*. Seems to prefer the open glades in the forests rather than the denser parts. Quite common near La Palma, although only one specimen was secured.

No. 248. ♀ ad. April 25, 1882.

3. Campylorhynchus capistratus (Less.).

[NOTE.—Five Costa Rican specimens of this species differ appreciably from two others from Guatemala and Honduras in much more distinctly streaked rump (even the feathers of the back being appreciably spotted with black beneath the surface), in having the light wing-bars much paler, in larger bill, and in some other characters. Without more specimens, however, from both regions, showing the differences observed to be constant, I hesitate to separate them as races. A single specimen of *C. rufinucha* Lafr. (which some authorities refer to *C. capistratus*) from Mirador (No. 30869, C. SARTORIUS), differs from all the above-mentioned specimens in having the whole back very conspicuously streaked, the abdomen buff instead of white, the crissum barred with black, and the flanks, sides, and breast minutely but sparsely dotted with the same. It seems to be quite distinct, but, *cf.* SALVIN & GODMAN, Biologia Americana Centrali, Aves, i. pp. 64, 65.—R. R.]

This handsome Wren is perhaps the most common and familiar bird of the Gulf region. Its song is very voluble and melodious. Less fond of low, dense shrubbery than most of its kind, it often nests at a considerable distance from the ground. It is fearless, almost impudent, in its manner, and somewhat inclined to play the bully, in a small way, and seems to take particular delight in tormenting the "Zopilotilla" (*Crotophaga sulcirostris*) when it approaches too closely the home of the former. This Wren seems to be particularly fond of solitary trees along the edge of the forest, where he can always be seen hunting his food much in the same manner as do the Titmice of the north. Their number is so great that the woods continually resound with their lively song, and the naturalist has no trouble in making their acquaintance and securing a full series of skins. Iris brown.

Three specimens secured.

No. 140.　♀ ad.　March 20.
No. 164.　♂ ad.　April 3.
No. 270.　♂ ad.　April 27.

4. Thryophilus rufalbus (Lafr).

On several occasions, while hunting in the dense forests near La Palma, I have been suddenly arrested by the enchanting song of this bird. Breaking suddenly upon the ear from the cool depths of the woods, it seemed to me to be the most exquisite melody I had ever heard. This song consists of three notes, the first low and sweet, the second about four notes above the first, and most exquisitely trilled and prolonged, the third high and clear. Sometimes this Wren varies the order of its song, sounding the high note first and the low one last. It also varies the pitch of each note about a semitone, thus producing a remarkably sweet minor strain.

The bird seems to be rather shy and retiring in disposition, and is usually seen in the deep shades and secluded nooks of the forest. Its song is usually stopped at the approach of a stranger, and the bird flits silently away and remains quiet until the danger is passed, thus making it a rather difficult species to secure.

One specimen obtained.

No. 190.　♂ ad.　April 15.

5. Thryothorus rutilus hyperythrus (Salv. & Godm).[*]

Only one specimen seen. This one was shot in a dense thicket along a stream which runs near La Palma.

No. 297.　♀ ad.　April 30.

6. Basileuterus semicervinus leucopygius (Scl. & Salv.).[†]

Common. Found always (so far as my experience goes) along the rocky bed of the stream mentioned under the last species. It is quite a sprightly little bird, and seems to have habits somewhat similar to

---

[*] *Thryothorus hyperythrus* SALV. & GODM. Biol. Centr. Am. Aves, i, p. 91; RIDGW. Proc. U. S. Nat. Mus., vol 4, p. 334 (Carrillos, Alajuela).—(R. R.)

[†] *Basileuterus leucopygius* SCL. & SALV. Nom. Neotr. 1873, p. 156; SALV. & GODM. Biol. Centr. Am. Aves, i, p. 172.—(R. R.)

· those of *Cinclidæ* or Dippers, at least so far as its habitat and manner of flitting along the rocks of water-courses is concerned. Iris brown.

Two specimens.

No. 191. ♂ ad. April 15.
No. 290. ♀ ad. April 29.

7. Vireosylvia flavoviridis Cass.

Apparently not very common. Iris red.

One specimen secured.

No. 250. ♀ ad. April 25.

8. Hylophilus decurtatus (Bonap.).

Common. Found in thick forest. Iris brown.

Two specimens.

No. 212. ♀ . April 17.
No. 273. ♀ . April 28.

9. Progne leucogastra Baird.

Abundant. Shot in early morning out of the top of a very high tree in an open field. Iris brown.

One specimen.

No. 104. ♂ ad. April 16.

10. Tanagra cana diaconus (Less.).

This Tanager seems to be pretty abundantly distributed throughout Costa Rica. It is found in small flocks, and its beautiful blue plumage renders it quite conspicuous. One of the fiercest and most stubbornly · prolonged bird-fights I ever saw was between two of this species. Indeed it is quite noticeable for its quarrelsome disposition.

One specimen from this locality.

No. 239. ♂ ad. April 24.

11. Ramphocelus passerinii Bonap.

Many specimens seen between San José and Punta Arenas, but having no ammunition with me it was impossible to secure specimens.

12. Embernagra superciliosa Salvin.

Rather common. Found in open woods. Iris brown.

Two specimens.

No. 195. ♂ . (Breeding.) April 16.
No. 291. ♂ . April 29.

13. Spiza americana (Gm.).

Only one large flock seen. They had settled upon a small tree near a cactus hedge, where they were literally gorging themselves upon a small black and yellow worm. They all seemed to have fared sumptuously, as the specimens killed were the fattest small birds I ever saw.

Eleven were killed at one shot, but, owing to the difficulty of making presentable skins of such unusually fat birds, I only saved five.

No. 306. ♀ ad. May 1.
No. 307. ♂ ad. May 1.
No. 308. ♀ ad. May 1.
No. 309. ♀ ad. May 1.
No. 310. ♂ ad. May 1.

14. Volatinia jacarina (Linn.).

These pretty little black sparrows were very abundant in small flocks, and seemed to prefer the cactus hedges along the cart-roads. They spend a great deal of their time upon the ground, and lead pretty much the same sort of a life as the little ground doves (*Chamœpelia rufipennis*) so abundant in that region.

Four specimens.

No. 170. ♂ ad. April 4.
No. 211. — juv. April 17.
No. 266. ♂ ad. April 27.
No. 317. ♂ ad. May 3.

15. Guiraca cyanoides concreta (Du Bus).

Only one specimen seen. That was shot when it was taking a drink from a running stream. Iris brown.

No. 316. ♀ ad. May 3.

16. Molothrus æneus (Wagl.).

Only one specimen secured. It was found associating with *Crotophaga sulcirostris* in an open field.

No. 210. ♀ juv. April 17.

17. Agelæus phœniceus (Linn.).

[The single specimen obtained by Mr. Nutting is an adult male, and agrees exactly with examples from Yucatan and other parts of Mexico. The middle wing-coverts are a rich brown-ochre tint, as in examples from the western United States, and the size is quite as large as in more northern skins.—R. R.]

Common at a large lagoon about 10 miles from La Palma, where it probably breeds. The Spaniards call it by a name which signifies " an officer," on account of its red shoulder patches.

No. 229. ♂ ad. April 20.

18. Icterus pectoralis espinachi Nutting (MS.).

[NOTE.—Three specimens of this species from the western coast of Costa Rica differ from more northern examples (one each from San Salvador, Guatemala, and Tehuantepec) in decidedly smaller size, the wing measuring only 3.70–4.05, and the tail 3.85–4.05, instead of 4.30–4.55 and 4.20–4.65, respectively. I am unable, however, to appreciate any tangible differences in coloration. Should the difference in size prove constant, the Costa Rican bird might form a local race, for which the name given above would be exceedingly appropriate.—R. R.]

A specimen of this beautiful Oriole was kept in a cage at La Palma, and as it hung near the place where I daily prepared my bird-skins, I had an excellent opportunity to observe its notes, and, in part, its habits.

He was the most accomplished vocalist I ever heard, had perfect command of every note in the scale, and apparently took great delight in his accomplishment.

Thinking him to be a promising subject, I undertook to instruct him in the art of whistling, but he scorned my services and went off into trills and harmonies of his own composition which put to shame the sample I had given in the shape of classic "Yankee Doodle." I did succeed in teaching him to run the scale perfectly, an exercise in which he reached great perfection, and gained the admiration of his hearers. He would whistle by the hour, not in a monotonous repetition of the same strain, but constantly varying his music from loud and lively to soft and sweet, reminding me of a flute-player running over bits of harmony from memory. He was an expert fly-catcher, though all sorts of food seemed to suit his taste. His greatest delight seemed to be in sitting on my finger and being "teetered" up and down. I afterward secured two specimens, which I shot in a wild state. Iris yellow.

No. 258. ♂ ad. April 17.
No. 259. ♀ ad. April 17.

19. Ocyalus wagleri (Gray & Mitch.).

A remarkable colony of these curious birds was observed on the road from San José to Punta Arenas. A very large dead tree standing in the road had been taken possession of, and from every limb their purse-like nests were suspended. There must have been over two hundred of these curious structures, and their occupants were swarming around them making a great clatter. A remarkable fact was noticed upon this occasion and greatly excited the curiosity both of the Spanish gentlemen of the party and myself. These birds had the novel habit of getting inside their nests and shaking them violently so as to produce a loud rattling noise. This we saw them do repeatedly, but could arrive at no satisfactory conclusion as to the object of so strange a performance.

20. Calocitta formosa (Swains.).

This fine Jay is not common, so far as my experience goes. Like all the rest of its kind it seems to like to make itself conspicuous, and is usually seen in the top of a tree calling loudly in a harsh voice. Its recurved crest is a prominent characteristic. It is said by the natives to talk like a parrot (*"habla como loro"*), but I never had an opportunity to assure myself of its accomplishments in this respect. Three specimens shot, but only one was in a fit condition to preserve.

No. 258. ♂ ad. April 26.

21. Megarhynchus pitangua (Linn.).

Common. A noisy and active bird, apparently not restricted to any particular altitude, as it was secured both near San José and on the coast. Iris brown.

No. 276. ♀ ad. April 28.

22. Pitangus derbianus Scl.

Abundant. Habits similar to the preceding species, but with even a greater range of altitude. Iris brown.

No. 165. ♂ ad. April 3.

23. Myiodynastes nobilis Scl.

Rather common. So far as observed this is rather a silent bird for its family, and it seems to attend to its own business more strictly than many of its relatives. Found usually at the edge of the woods, where it is actively employed in capturing the numerous insects of the region. Iris brown.

Two specimens.

No. 172. ♀ ad. April 4.
No. 251. ♂ ad. April 23.

24. Tyrannus melancholicus satrapa (Licht.).

[NOTE.—The specimen obtained by Mr. Nutting is peculiar in the very obtuse primaries, of which the outer ones are very slightly sinuated at the tip; the tail is very nearly truncated, but the two middle pairs of rectrices are wanting. The bird is apparently in molting condition, which may account for some of its peculiarities.—R. R.]

Not common. Single specimen shot near a stream. Iris brown.

No. 187. ♀ ad. April 11.

25. Myiarchus nuttingi Ridgway, sp. nov.

[SP. CH.—Similar in general coloration to *M. mexicanus* and *M. cinerascens*, but differing from both in the pattern of the tail-feathers, the inner webs of all the rectrices (except the intermediæ) being either wholly rufous or else with a very narrow stripe of dusky next to the shaft of the outer feather. *Adult:* Above brownish gray (exactly as in *M. cinerascens*), occasionally tinged with olive, the pileum much browner and with darker shaft-streaks; wings and middle pair of rectrices dusky brownish, the latter uniform; last row of lesser coverts, middle, and greater coverts, distinctly tipped with light brownish gray; tertials edged exteriorly with grayish white (tinged with sulphur-yellow in fresh plumage), the primaries edged with light rufous toward the base. Outer webs of rectrices dusky brownish (like both webs of the intermediæ), the outer pair with the exterior edge much paler (nearly white in some specimens); inner webs of all the rectrices excepting the middle pair clear rufous, including the extreme tip, and usually extending quite to the shaft, though in some examples separated from the shaft by a very narrow streak 'of dusky. Chin, throat, and jugulum very pale ash-gray (exactly the same shade as in *M. mexicanus* and *M. cinerascens*), the remaining lower parts sulphur-yellow (same as in *mexicanus* but deeper than in *cinerascens*). Bill black, the mandible sometimes brownish, paler at base; iris brown; legs and feet deep black. Wing 3.45–3.80, tail 3.20–3.80, culmen .60–.80, gonys .50–.60, width of bill at base .35–.40, tarsus .78–.90, middle toe .45–.52. (Six specimens.)

Hab.—Southwestern Mexico (Tehuantepec) to Costa Rica (Pacific side).

The above diagnosis is drawn up from six specimens of a *Myiarchus*, which cannot be referred to either *M. mexicanus* or *M. cinerascens*, though evidently very closely related to both of them. All the specimens hitherto seen are from the Pacific coast of Central America (Costa Rica to Tehauntepec, southwestern Mexico), a region where either *M. mexicanus* or *M. cinerascens*, or both, also occurs. It doubtless, however, represents the resident form specially characteristic of the district named, the other two occuring there as stragglers from other districts. In all respects, except the pattern of the tail-feathers, this form agrees to the minutest degree with the two species named above, except that in *M. cinerascens* the abdomen is a slightly paler sulphur-yellow. The most conspicuous specific character distinguishing *M. cinerascens* consists, however, in the terminal dusky space on the inner webs of the rectrices, of which there is no trace in *M. nuttingi. M. mexicanus* (in all its forms) has, on the other hand, a broad and very distinct stripe of dusky next the shaft on the inner webs of the rectrices, while in *M. nuttingi* there is never more than a mere indication of this stripe. Thus it may be seen that the present form, whether species or race, cannot be referred to either of the species named, and that it must, therefore, be considered quite as distinct from them both as they are from one another.—R. R.]

Rather common, more especially in open woods. Iris brown.

Two specimens.

No. 243. ♂ ad. April 24.
No. 256. ♀ ad. April 26.

26. Rhynchocyclus cinereiceps Scl.

Apparently not common. Two specimens secured in thick woods, near the water. Iris *white*. Nest secured.*

No. 227. ♂ ad. April 24.
No. 279. ♀ ad. April 28.

---

* The nest of this bird is a most remarkable structure, well worthy of description. It is a pendulous inverted pouch, suspended from a single twig, composed almost entirely of slender black filaments resembling horse-hairs (probably a vegetable fiber related to, if not identical with, the "Spanish Moss," or *Tillandsia* of the Southern United States), and so loosely built as to be easily seen through when held up to the light. The entrance is at the extreme lower end, the nest proper being a sort of pocket on one side, about 2 inches above the entrance. The total length of the entire structure is 10 inches, the greatest width 4 inches, the lower "neck," or wall of the entrance, being about 2¼ inches in diameter.—R. R.

27. Muscivora mexicana Scl.

This exquisitely ornamented Flycatcher is abundant in the vicinity of La Palma, especially along the water courses. Indeed, I never saw it away from the water. It builds its nest on a branch overhanging a stream, seems to be quite contented to remain in the immediate vicinity of its home, and is quiet and modest in manner.

Never having seen this bird before, my surprise and admiration were unbounded when I held one in my hand for the first time, and saw its wonderfully brilliant fan-shaped crest. The bird was only wounded, and the crest was fully spread, while the head was slowly moved from side to side, which gave it the appearance of a bright flower nodding in the wind. While admiring this new wonder, I heard a twitter of distress immediately above me and, looking up, was delighted to see the female perched on a twig not more than ten feet above me, with her crest erected and spread, and making the same waving motion of the head. Is it not possible that this bird is provided with its remarkable crest for the purpose of attracting its insect prey, and that the slow and regular waving motion is calculated to still further deceive by a simulation of a flower nodding in the breeze?

It is a singular fact that while this bird is quite common in that region, the natives had never discovered its peculiar ornamentation before I showed it to them.

Seven specimens secured and five preserved. Iris light brown.

No. 234. ♂ ad. April 29.
No. 287. ♀ ad. April 29.
No. 295. ♀ ad. April 30.
No. 298. ♂ ad. April 30.
No. 300. ♂ ad. April 30.

28. Myiobius atricaudus (Lawr.).*

Common. Prefers dense undergrowth, and is rather shy and noiseless. One specimen.

No. 285. ♀ . April 29.

29. Chiroxiphia linearis Bp.

Common. One of the most exquisite little birds of Costa Rica. It seems to prefer the dense thickets and underbrush. Its note closely resembles the discordant "meow" of the Cat-Bird, although it occasionally gives utterance to a clear, melodious whistle. Native name "Gallinita" or "Little Cock." Iris brown.

Four specimens secured.

No. 265. ♂ ad. April 27.
No. 304. ♂ ad. May 1.
No. 305. ♂ juv. May 1.
         ♂ juv. (Label missing.)

---

* The example obtained by Mr. Nutting agrees minutely with two from Panama city, which seem to me to differ much more from either *M. barbatus* or *M. sulphureipygius* (of both which the National Museum possesses numerous specimens) than these do from one another.—R. R.

30. Tityra personata Jard. & Selby.

Rare. At least I saw but one, and the natives appeared to be unacquainted with it. Shot in a large tree standing in an open field. Iris brown. Bill and orbital region carmine.

No. 202. ♂ ad. April 16.

31. Tityra albitorques fraseri (Kaup).

Common. Usually found in rather open country associating in small flocks of six or eight. Noisy and quarrelsome.

Two specimens secured.

No. 267. ♀ ad. April 27.
No. 268. ♂ ad. April 27.

32. Hadrostomus homochrous Scl. (?).

[NOTE.—The single specimen, an adult female, obtained by Mr. Nutting is almost certainly not referable to *H. aglaiæ*. It agrees much more closely in coloration with specimens of *H. atricapillus* from Ceara and Bahia, having, like them, the pileum slate-colored, the other upper parts a clear light rufous, and the lower parts ochraceous-white medially. In fact, I do not see how it can be distinguished by color alone. Geographical considerations, however, preclude the probability of its being *H. atricapillus;* and since *H. homochrous*, which is known to occur from Ecuador to Panama, may very likely extend its range still further along the coast to the Nicoya district, I with some doubt refer the specimen in question to the latter species, which is not represented in the collection of the National Museum.—R. R.]

Rare. But one seen, and that was shot near a large fresh-water lagoon.

No. 213. ♀ ad. April 17.

33. Picolaptes compressus (Cab.).

Common. A silent bird, as a rule. The nests, like those of Woodpeckers, are usually placed in a hollow tree. They usually hunt in pairs in the thick forests.

Two specimens.

No. 138. ♂ ad. March 20.
No. 249. ♂ ad. April 25.

34. Thamnophilus doliatus affinis (Caban.).

Habits similar to our Wrens. A quiet and industrious bird, usually seen in an active search for ants and other small insects. They seem to prefer the dense woods, but are occasionally seen in isolated trees. Iris white.

Five specimens.

No. 171. ♂ ad. April 9.
No. 189. ♂ ad. April 15.
No. 203. ♂ ad. April 16.
No. 247. ♂ ad. April 25.
No. 338. ♂ ad. April 24.

35. Myrmeciza immaculata Scl. & Salv. (?)

[NOTE.—The female from La Palma is referred doubtfully to this species. It differs markedly from three other Costa Rican specimens, from the Atlantic coast, in having the jugulum and breast bright chestnut instead of dull chestnut-brown, but I am unable to detect any other differences.—R. R.]

Not common. Only one specimen seen, and that was secured near a running stream.

No. 286. ♀ ad. April 29.

36. Amazilia fuscicaudata (Fraser).

Abundant. The period during which I collected at "La Palma" being the latter part of the dry season, most of the birds had gathered in the vicinity of the water courses. The Humming-birds seemed to be especially affected by the drought, but knowing that the *Trochilidæ* had been especially well worked up, I preferred to devote my time to groups more likely to yield novelties.

One specimen.

No. 303. ♀ ad. April 27.

37. Nyctidromus albicollis (Gmel.).

Exceedingly abundant in the vicinity of La Palma, where five or six may be heard at the same time. The Spaniards give it a name signifying "bird of the night."

Frequently in passing through the thick brush I have flushed this bird. It would flit silently ahead a short distance, and then apparently alight on the ground; but upon reaching the spot I would find that, like the "Irishman's flea," it was not there. Upon closer observation I found that the bird did not really alight when it appeared to, but would suddenly descend to the ground, over which it would hover for an instant as if in the act of alighting, and then glide silently on close to the ground for some little distance, and finally settle down in the dead leaves near a tree-trunk or bush.

One specimen.

No. 201. ♀ ad. April 16.

38. Campephilus guatemalensis Hartl.

Common. This handsome Woodpecker was not seen during the early part of my stay at La Palma, but it suddenly became quite common about the 27th of April, and from time to time until my departure. One of the commonest sounds of the forest was its quick, loud tap. It usually taps but twice in rapid succession, hunts in pairs, and seems to prefer the thick forests to the more open woods.

Five specimens secured.

No. 271. ♀ ad. April 27.
No. 272. ♂ ad. April 27.
No. 278. ♀ ad. April 28.
No. 282. ♂ ad. April 28.
No. 292. ♂ ad. April 29.

39. Centurus aurifrons hoffmanni (Caban.).

The common Woodpecker of Costa Rica. Found everywhere except on the more elevated mountains. Iris white.

Two specimens.

No. 156. ♀ ad. March 31.
No. 197. ♂ juv. April 16.

40. Momotus lessoni Less.

Not so common on the coast as in the interior. In the former locality it seems to prefer the thick woods, while in the latter it is often seen in the more open fields. Generally a silent bird, but not shy.

One specimen.

No. 320. ♂ ad. May 5.

41. Eumomota superciliaris (Sw.).

This exquisitely-colored Motmot is common throughout the coast region, where it bears the rather insulting name of "Bobo" (stupid). The natives account for this name by saying that the bird hasn't sufficient sense to fly away at the approach of the hunter. In truth it seems to be quite fearless, and seldom disturbs itself on account of human proximity. Although a very silent bird (I never heard its voice), its peculiar spatulate tail-feathers are apt to attract attention. It seems to be solitary in its habits, and not very industrious, as it is most often seen sitting on a limb not far from the ground apparently engaged in deep meditation, from which it is not aroused by the presence of the collector. Iris brown.

Three specimens.

No. 199. ♂ ad. April 16.
No. 207. ♂ ad. April 17.
No. 288. ♂ ad. April 29.

42. Ceryle torquata (Linn.).

Common. Habits almost precisely the same as *C. alcyon*, but not so noisy, as a rule.

One specimen.

No. 137. ♀ ad. March 30.

43. Ceryle americana cabanisi (Tschudi).

Abundant. The collector is sure to meet with them while following along the streams of that region. They are quite fearless and are not at all disturbed by the presence of man, but pursue their fishing after a short but emphatic expostulation at his approach. Iris brown.

Four specimens.

No. 168. ♂ ad. April 3.
No. 213. ♂ ad. April 17.
No. 240. ♀ juv. April 13.
No. 241. ♀ ad. April 13.

44. Ceryle superciliosa (Linn.)

Rare. This beautiful, diminutive Kingfisher is the smallest American species of its family, but is not a whit less spirited and courageous than the largest, of which it is almost an exact epitome except in coloration. Two seen and one secured.

No. 314. ♀ ad. May 2.

45. Trogon massena Gould.

Common. The largest Trogon of the coast region. I have never seen this species associating in flocks as the others do. On the contrary, it seemed to be rather a silent bird, preferring the deep recesses of the tropical forests. Its note is a kind of clucking noise hard to describe. Native name, "Bula." In common with all the smaller species of this genus it seems to be rather a stupid bird, hardly ever taking alarm at the approach of man.

Four specimens.

No. 179. ♂ ad. April 7.
No. 180. ♀ ad. April 7.
No. 196. ♀ ad. April 16.
No. 233. ♂ ad. April 22.

46. Trogon melanocephalus Gould.

Very abundant. Often seen in flocks of a dozen or more. Commonly seen in the dry open woods away from the water. It has a sort of a chattering note, low and soft. They are not startled at the report of a gun, and an entire flock may be shot out of the same tree. Iris brown orbital region sky-blue.

Nine specimens secured.

No. 185. ♀ ad. April 14.
No. 228. ♀ ad. April 20.
No. 231. ♀ ad. April 21.
No. 244. ♀ ad. April 25.
No. 254. ♂ ad. April 26.
No. 255. ♂ ad. April 26.
No. 262. ♀ ad. April 26.
No. 274. ♂ ad. April 28.
No. 313. ♂ ad. May 2.

47. Trogon caligatus Gould.

This elegant little bird, although not so common as the last, is frequently seen in this region. It is the only Trogon that I ever heard give utterance to a clear, distinct whistle. There is probably no bird more difficult to skin than this one, both on account of the looseness of the plumage and the extreme delicacy of the skin, especially about the head.

Five specimens.

No. 181. ♀ ad. April 9.
No. 188. ♀ ad. April 15.

No. 234.   ♂ ad.   April 24.
No. 236.   ♀ ad.   April 24.
No. 301.   ♂ ad.   May 1.

48. Galbula melanogenia Scl.

Only one specimen of this beautiful bird secured. It was shot in the thick forest while flitting through the undergrowth. Iris brown.

No. 246.   ♂ ad.   April 25.

49. Bucco dysoni Scl.

One specimen secured in open forest.

No. 260.   ♀ ad.   April 26.

50. Crotophaga sulcirostris Swains.

One of the most abundant and familiar birds in Costa Rica. Found everywhere and in great numbers. Habits remarkably similar to those of our common Cowbird (*Molothrus ater*). They are usually in flocks in the open fields. Native name "*Zopilotilla*" or "*Little Buzzard.*" They are said to destroy immense numbers of "Garrapatas" or ticks.

Two specimens from La Palma.

No. 269.   ♀ ad.   April 27.
No. 280.   ♂ ad.   April 28.

51. Piaya cayana mehleri (Bp.).

These graceful birds are also common throughout Costa Rica. Like the other true Cuckoos, it is a silent and solitary bird for the most part, although when disturbed it utters a loud, harsh note at regular intervals as it looks down upon the intruder and flirts its beautiful tail with angry jerks. Iris red.

Two specimens.

No. 169.   ♀ ad.   April 3.
No. 186.   ♀ ad.   April 14.

52. Coccyzus sericulus (Lath.).

Rare in the region of La Palma. Only one seen and shot out of a high tree. Iris brown.

No. 281.   ♂ ad.   April 28.

53. Pteroglossus torquatus (Gm.).

This is the only species of Toucan that I saw on the Pacific coast, although another species was described to me. It seems to prefer the open forest. Its uncouth bill would convey the idea that it is a clumsy bird, but on the contrary it is rather graceful and handles its immense beak with ease. The bill is very light, being cellular in its internal structure. I know from experience that it is capable of giving quite a severe bite, a fact to which a scar on my finger still testifies. I never heard its note although I observed several. Iris yellow, bill yellow, red, and black.

One specimen.

No. 157.   ♂ ad.   March 31.

Sept. 12, 1882.

54. Ara macao (Linn.).

Abundant. The size, gaudy colors, and loud voice of this bird make it, perhaps, the most noticeable one of the region. It feeds almost exclusively upon fruits and nuts, is strictly monogamous and, although matrimonial jars are of daily and hourly occurrence, is very affectionate. The bill is so enormous and strong and the bite so dangerous that the collector is sometimes at a loss as to the best manner of killing this bird when wounded. I solved the difficulty very quickly and satisfactorily by breaking its neck with a vigorous blow with the back of a "machete," the long heavy knife universally carried by the natives and absolutely indispensable to the collector. This is a sure and effective means of killing the Macaw. These birds generally sit in pairs close to each other, and both can usually be killed with one shot. Iris very pale yellow. Bare parts of the head pinkish white.

Five specimens.

No. 149.	♂ ad.	March 31.
No. 150.	♀ ad.	March 31.
No. 177.	♂ ad.	April 1.
No. 205.	♂ ad.	April 17.
No. 206.	♀ ad.	April 17.

55. Brotogerys tovi (Gm.).

Exceedingly abundant. The common Parakeet of the region. It is found in flocks varying from half a dozen to one hundred or more. It is a remarkably tough little bird and hard to kill. When struck by the shot it does not fly like most birds, but grasps the limb tightly with its strong feet and hangs on until quite dead. It seems to prefer solitary trees standing in open fields, although it is found in almost all situations except on the ground. Iris brown. Cere white.

Eight specimens.

No. 141.	♂ ad.	March 30.
No. 200.	♀ ad.	April 16.
No. 235.	♀ ad.	April 24.
No. 252.	♀ ad.	April 25.
No. 257.	♀ ad.	April 26.
No. 277.	♂ ad.	April 28.
No. 318.	♂ ad.	May 3.
No. 319.	♀ ad.	May 3.

56. Chrysotis auripalliata (Schleg.).

Abundant. Its harsh cry is always heard throughout that region from sunrise to sunset. As night comes on they begin to gather into some particular tree, coming always in pairs and making a great deal of noise in settling for the night. It seems difficult for them to suit themselves as to their quarters for the night, and they try a number of situations, discussing each with many querulous expressions of discontent. Finally they settle down, each crowding close to its mate. I have often seen them in moonlight nights remain quietly asleep for several hours, and then as if by a common impulse leave the tree with a loud whirring of wings, but otherwise in utter silence, to seek another tree. I was unable to discover the cause of these strange maneuvers. Iris orange.

Five specimens.

No. 151.　♀ ad.　March 31.
No. 167.　♂ ad.　April 3.
No. 174.　♀ ad.　April 4.
No. 245.　♂ ad.　April 25.
No. 275.　♀ ad.　April 28.

57. Pulsatrix torquata (Daud.).

Not common. Only one specimen seen and shot out of a high tree in the thick forest.

No. 144.　♂ ad.　March 30.

58. Tinnunculus sparverius (Linn.).

Apparently not common in the coast region. One specimen.

No. 145.　♂ ad.　March 31.

59. Regerhinus uncinatus (Temm.).

Commonly heard, especially in the evening. Note "*oóah!*" "*oóah!*" None secured.

60. Rupornis ruficauda (Scl. & Salv.).

Abundant. The commonest Hawk of the region. Seems to prefer rather open woods, although often seen in the thickest forests. Iris yellow. Cere orange-yellow.

Five specimens secured.

No. 142.　— juv.　March 30.
No. 148.　♀ ad.　March 31.
No. 166.　♂ ad.　April 3.
No. 175.　♀ ad.　April 4.
No. 243.　♂ juv.　April 26.

61. Asturina plagiata (Schleg.).

This bird is apparently not common in that region. While out hunting one day I shot at one of these Hawks, but it flew away apparently unhurt. Three days afterward, while in a different direction, I found this same bird dead in a hollow tree, where it had evidently just died of starvation on account of a single shot in the last joint of the wing. Iris brown. Cere and feet yellow.

No. 183.　— juv.　April 13.

62. Parabuteo* unicinctus harrisi (Aud.).

Abundant. Associates with the Carrion Crow, and eats offal. Notwithstanding this it is an inveterate poultry thief. The specimen obtained was shot with a revolver while carrying off one of Don Ramon's chickens. Iris brown. Cere and feet yellow.

No. 184.   ♀ ad.   April 13.

63. Buteo borealis costaricensis Ridgw.

One specimen shot, but so badly injured that it was not saved.

64. Urubitinga zonura (Shaw).

Common. Usually found in the vicinity of the water-courses. Iris brown. Cere and legs yellow.

Three specimens.

No. 217.   ♀ ad.   April 18.
No. 283.   ♀ ad.   April 29.
No. 312.   — ad.   May 2.

65. Urubitinga anthracina (Nitzsch.).

Common. Feeds largely upon reptiles. Iris nearly white.

Two specimens.

No. 143.   ♂ ad.   March 30.
No. 294.   ♀ ad.   April 13.

66. Spizætus ornatus (Daud.).

Not common.   Only one specimen secured.

No. 178.   ♂ juv.   April 4.

67. Busarellus nigricollis (Lath).

[The young specimen obtained by Mr. NUTTING is in plumage so different from that described by me in Bull. U. S. Geol. and Geog. Survey Terr. (vol. ii, No. 2, p. 143), that a detailed description seems desirable.

Young (No. 87446, La Palma, Costa Rica, April 11, 1882, C. C. NUTTING): Head and neck creamy buff, deeper posteriorly and becoming nearly white on frontlet, lores, and chin, each feather marked with a distinct lanceolate mesial streak of dusky, except on the whitish parts named above, where the feathers have merely narrow, dusky shaft-streaks. Lower parts, rump, and upper tail-coverts rusty ochraceous; lower part of throat crossed by a somewhat crescentic patch or bar of dull black, and breast crossed by a similar but broader band of chestnut-rufous, each feather having a central dusky, pointed spot; feathers of jugulum and lower part of breast marked with distinct mesial streaks

---

* The name *Antenor*, which was proposed by me in 1873 for this genus, is, as I have recently discovered, preoccupied in Conchology (MONTFORT, 1808); another name being therefore necessary, I have selected the one given above in preference to a new one, on account of its being already on record, in Hist. N. Am. Birds, vol. iii, 1874, p. 250, where, by an oversight in correcting proof-sheets, "*Parabuteo*" is allowed to stand instead of *Antenor.*—R. R.

of black; abdomen and flanks irregularly variegated with rusty and dusky; crissum nearly immaculate ochraceous. Rump and upper tail-coverts ochraceous, marked with arrow-heads and connected bars of dusky; basal half of tail rufous, crossed by several narrow bands of black, these narrower on the inner webs, which are ochraceous instead of rufous; terminal half of tail dusky black, the tip (narrowly) ochraceous. Back, scapulars, and wings rich chestnut-rufous, each feather dusky centrally; the tertials and inner secondaries crossed by narrow bars of dusky; primaries and outer secondaries nearly uniform black. Bill entirely black; "cere black; iris brown; legs and feet very pale flesh-color"; claws black. Wing 14.50, tail 8.50, culmen 1.10, tarsus 3.40, middle toe 1.80.—R. R.]

This bird I found to be abundant in the vicinity of the "Zapotal," a large fresh-water lagoon. It is exceedingly fearless, so far as man is concerned, although this may be due to the fact that it has not yet learned to fear him.

Two specimens.

No. 156. — juv. April 1.

No. 157. ♂ ad. April 1.

68. Gyparchus papa (Linn.).

Rather rare. Local name "Rey de Zopilotes," or King of the Vultures. One specimen, found dead. It was in such a condition that it would have been unsafe to attempt to skin it. The following notes were taken: Shoulders, lower neck, back, and below yellowish white. Tail, rump, and remiges black. Bare parts red. Iris white.

69. Cathartes aura (Linn.).

Common. None secured.

70. Catharista atrata (Bartr.).

The most efficient scavenger of tropical regions. These vultures are probably the most useful birds in existence. Indeed, they are absolutely indispensable in hot regions, where, in many instances, pestilence is doubtless averted by their valuable presence.

71. Tachypetes aquila (Linn.).

Abundant on the shores of the gulf.

72. Pelecanus fuscus Linn.

Abundant along the entire coast.

73. Sula leucogastra (Bodd.).

"Booby Gannet." Seems to be common all along the Pacific coast of Central America.

74. Plotus anhinga Linn.

Abundant, especially in the neighborhood of the "Zapotal," the lagoon where most of my water birds were secured. This bird has the smallest brain of any bird of its size that I ever dissected. It is expert at fishing, and may be seen sitting for hours at a time on a limb projecting over the water where it is watching for its prey. In habits it resembles the Kingfishers.

One specimen.

No. 223.   ♀ juv.   April 18.

75. Herodias egretta (Gm.).

Exceedingly abundant at the lagoon.   Iris yellow.

Three specimens.

No. 215.   ♀ ad.   April 8.
No. 226.   ♂ ad.   April 20.
No. 227.   ♀ ad.   April 20.

76. Florida cærulea (Linn.).

Not common.   Only one seen.

No. 222.   ♂ juv. tr.   April 19.

77. Butorides virescens. (Linn.).

Abundant wherever there is water.

No. 136.   ♀ ad.   March 30.
No. 225.   ♂ ad.   April 19.

78. Nycticorax griseus nævius (Bodd.).

Abundant.   Found at the "Zapatol."   Iris red.

Two specimens.

No. 216.   ♂ ad.   April 18.
No. 224.   ♂ ad.   April 19.

79. Tigrisoma cabanisi Heine.

Exceedingly abundant. The curious note of this Bittern is well calculated to startle the inexperienced collector in these regions. It is something between a bark and a growl, and sounds like the angry warning note of some fierce animal. At the lagoon I suppose a person could kill a wagon-load of these birds in a single day. Iris brown. Bare place in neck bright yellow.

Three specimens secured.

No. 176.   ♂ ad.   April 18.
          — ad.   (Label lost.)
          — juv.   (Label lost.)

80. Cancroma cochlearia Linn.

Common. This curious bird seems to have habits similar to the Herons. Its note is a harsh croak. They generally associate in small flocks. Iris brown. Sac under bill, and legs, flesh color.

Four specimens.

No. 192.   ♀ ad.   April 15.
No. 193.   ♂ ad.   April 15.
No. 194.   ♀ ad.   April 15.
No. 198.   — juv.   April 16.

81. Mycteria americana Linn.

[Juv. (No. 87485, La Palma, Costa Rica, April 21, 1882; C. C. Nutting): Pileum and occiput clothed with dusky black hair-like feathers, these longest on the occiput, where they form somewhat of a bushy crest; feathered portion of lower neck light brownish gray; rump, upper tail-coverts, and tail white; rest of upper part soft brownish gray, irregularly mixed with pure white feathers (of the adult livery?), these most numerous among the lesser wing-coverts and anterior scapulars; primaries white, tinged with gray at ends. Lower parts entirely white. Bill, all the naked portion of head and neck (except lower portion of the latter), legs, and feet black; "collar round lower neck bright scarlet; iris brown." Wing 24.50, tail 9.50, culmen 9.75, tarsus 11.25, middle toe 4.50.—R. R.]

Common. The natives have a name for this Stork which is extremely well chosen. It is "*Galan sin ventura*," or, literally, "Shabby Genteel." The fitness of this name can be appreciated only by one who has seen him in his native lagoon. The contrast between the gay red collar, stately bearing, and dignified movements and the general shabbiness of his dirty white coat and scaly legs is extremely ridiculous, and causes a realization of the appropriateness of its name.

The chief occupation of this bird is fishing, of course, although frogs and reptiles are by no means slighted.

One specimen.

— juv. (Label lost.)

82. Tantalus loculator Linn.

Abundant. The habits of this bird are so well known as to require no comment. Iris brown.

One specimen.

No. 155. — juv. April 1.

83. Eudocimus albus (Linn.).

Common. This Ibis is commonly seen in flocks, and seems less shy than the other water birds of the region. Iris blue; bill red; legs pale.

Two specimens.

No. 159. ♀ juv. tr. April 21.
No. 232. ♂ ad. April 21.

84. Ajaja rosea Reich.

This beautiful bird is quite common at the "Zapotal." It seems to prefer the small muddy branches of the lagoon to the main body, and delights in dabbling in the muddy water with its curious spoon-shaped bill, which it manages as the ducks do theirs. Iris red. Bill pinkish.

Two specimens.

No. 221. ♂ juv. April 19.
No. —, — ad. (Label lost.)

85. Dendrocycna autumnalis (Linn.).

Abundant. Found wild at the "Zapotal" and domesticated at La Palma. Sexes alike. Its note is loud, shrill, and discordant. Iris brown. Bill reddish. Legs flesh color.

Two specimens.

No. 154. — ad. April 1. (Wild.)
No. 311. ♀ ad. May 1. (Domestic.)

86. Cairina moschata (Linn.).

This magnificent Duck is common both at La Palma and the lagoon. I never saw more than four or five in a flock together. They seem to live a somewhat secluded life, and when not feeding on the water are usually seen perched in trees much after the manner of our Wood-Duck. It is the shyest and most difficult of any Costa Rican bird I have seen. Iris brown. Legs black. Excrescences on bill red and black.

One specimen.

No. 160. ♂ ad. April 1.

87. Melopelia leucoptera (Linn.).

Common in the dry season, but disappears in the wet season. Associates with *Engyptila verreauxi*. The song of this Dove is remarkably varied and melodious. Frequently seen near the houses and in rather open woods. Iris yellow. Feet red. Orbital region sky-blue.

One specimen.

No. 293. ♂ ad. April 29.

88. Engyptila verreauxi (Bonap.).

Abundant. The common Dove of the region. Iris yellow. Feet red,

No. 259. ♀ ad. April 26.
No. 315. ♀ ad. May 3.

89. Chamæpelia passerina (Linn.).

Common. Associates with *C. rufipennis*. Lives mostly on the ground. especially along the roads and cattle-paths.

One specimen.

No. 264. ♂ ad. April 27.

90. Chamæpelia talpacoti rufipennis (Bp.).

Very abundant. This beautiful little Dove is very similar in its habits to our common *Zenaidura carolinensis*, but is found in larger flocks. Iris red.

Two specimens.

No. 263. ♂ ad. April 27.
No. ——. ——. (Label lost.)

91. Crax globicera Linn.

This fine species was seen, but not secured. From what I could learn from the natives it is not very abundant, but well known on account of the excellence of its flesh.

92. Penelope cristata (Linn.).

Common. Found generally in the thick forest, perching in high trees. Local name "Pavo." As a game bird it seems to be a substitute for our Wild Turkey, and is much sought after for its finely-flavored flesh. Iris orange-yellow. Bare place on neck; front and back scutella on legs red. Bill black.

One specimen.

No. 182. ♂ ad. April 11.

93. Aramus pictus (Bartr.).

[NOTE.—There seems to be no essential difference between the La Palma specimen and some Floridan examples. It is rather darker-colored, however, than most northern specimens, though occasionally the latter approach it very closely in richness of coloration.—R. R.]

Abundant at the "Zapotal," where its harsh and rather mournful cry is often heard. Prefers marshy country to open water. Flesh very good eating.

No. 214. — ad. April 18.

94. Parra gymnostoma Wagl.

This remarkable bird is very abundant at the lagoon, where it may always be seen running over the lily-pads in search of its food. The alligators are its worst enemies, and are always on the watch for a chance to steal upon it unawares. The Jacana, on the other hand, is always on the lookout for its dreaded foe, and never alights without first hovering directly over the lily-pads and closely scrutinizing the water for alligators. The curious spurs on the wings of this bird are used as a weapon, and fierce fights are of frequent occurrence. Iris brown. Frontlet and spurs bright yellow.

Four specimens.

No. 161. ♂ ad. April 1.
No. 162. ♂ ad. April 1.
No. 219. ♂ juv. April 18.
No. 230. ♀ ad. April 20.

95. Larus (species undetermined).

Many Gulls were seen, but none secured.

96. Crypturus′sallæi Bonap.

Rather rare. Found in the thick forests, where they live on the ground and are quiet and secluded in their habits. Native name "Gallinos de las montañas," or "Wood-hens." Iris brown.

One specimen.

No. 153. ♀ ad. April 1.

97. Crypturus pileatus (Bodd.).

Common. Habits the same as the last. Iris brown. Legs greenish. Two specimens.

No. 163. ♀ ad. April 3.
No. 289. ♀ ad. April 28.

DESCRIPTIONS OF TWO NEW SPECIES OF FISHES (SEBASTICH. THYS UMBROSUS AND CITHARICHTHYS STIGMÆUS) COLLECTED AT SANTA BARBARA, CALIFORNIA, BY ANDREA LARCO.

By DAVID S. JORDAN and CHARLES H. GILBERT.

1. Sebastichthys umbrosus, sp. nov. (31140, 31141.)

Head $2\frac{3}{5}$ to $2\frac{3}{4}$ in length, without caudal; depth $2\frac{1}{5}$ to $2\frac{3}{6}$. D. XIII, 12; A. III, 6. Scales 40 (tubes in lateral line), the number of cross series about 50 (counted below lateral line).

Body moderately robust, little compressed, not specially elongate. Mouth moderate, oblique, the maxillary extending backward about to posterior margin of pupil, its length almost half head. Jaws about equal, the lower with a strong symphyseal knob, fitting into a broad notch in the upper. Premaxillary in front on level of lower margin of eye. Preorbital narrow, not more than half width of maxillary, armed with two retrorse spines. Eye large, 4 in head, somewhat longer than snout. Nasal spines strong.

Cranial ridges well developed, sharp, but not high, in form intermediate between those of *pinniger* and *constellatus*; as strong as in *constellatus*, but lower. Preocular, supraocular, postocular, tympanic, and occipital spines present. Preocular spine very conspicuous; supraocular ridge low, its spine smaller than postocular or tympanic. Occipital ridge about as long as supraocular, about two-thirds eye. Interorbital space much broader than in *constellatus*, its width two-thirds eye; it is concave, with two rather strong ridges diverging backward; between these posteriorly are two smaller ridges. Suprascapula with two spines. Space between occipital ridges slightly concave. Preopercular spines strong, all of them acute, the second longest and rather slender. Opercular spines well developed. Gill rakers rather long and slender, the longest $2\frac{1}{5}$ in eye; about 24 of them on lower limb of arch.

Dorsal fin deeply notched; neither the spines nor the soft rays very high. Longest dorsal spine $2\frac{1}{5}$ to $2\frac{1}{2}$ in head; longest soft ray scarcely shorter. Soft dorsal longer than high. Caudal very slightly emarginate, the middle rays $2\frac{1}{4}$ in head. Pectoral $3\frac{2}{3}$ in body, not reaching vent. Ventrals about half head.

Scales rough, the accessory scales numerous; small scales along bases of fins. Both jaws with some small smoothish scales, those on mandible mostly towards its base.

Ground color light orange, quite faint or obsolete on parts of the body. Upper parts overlaid with a dusky hue, formed largely of dark points so numerous as to give a dusty appearance. The dark color on the sides forms irregular vermiculations, the center of each scale being pale orange, the edge dusky. Some areas along the back, between the pale blotches, are quite blackish. Jaws and inside of mouth light orange, more or less soiled. Two or three dark shades from eye across cheek. A dusky shade along maxillary. Opercle dusky, its flap with a spot of pale pink or orange. Each side of back above the lateral line with 5 or 6 roundish pale blotches, of a light pink color, more or less tinged with orange. One of these just below base of fourth dorsal spine; two under base of eighth dorsal spine, the uppermost faint, the lower large, near the lateral line, and somewhat further back than the upper one. A large blotch under the last dorsal spine; a large one under last rays of soft dorsal, with sometimes a smaller one in front of it. These spots are rather less sharply defined and more yellowish than in *constellatus, rosaceus,* &c. They correspond in position nearly to those found in the latter species. Fins all pale orange, more or less shaded with blackish. Peritoneum black. In one specimen the orange shade is less intense than in the other.

Two specimens (31140, 31141), 10 and 11 inches in length, were taken by Andrea Larco at Santa Rosa Island, near Santa Barbara, and were forwarded by him to the National Museum.

The species is well distinguished from all its numerous congeners on our Pacific coast. It probably most nearly approaches *S. constellatus,* among the species thus far known.

2. Citharichthys stigmæus, sp. nov. (31099.)

Body moderately deep, the two profiles regularly and equally arched; the snout short, gibbous, projecting a little beyond the outline; caudal peduncle very short, not high, its length (from end of last vertebra to vertical from last anal ray) about two-fifths its height, which is three-sevenths length of head; caudal fin appearing sessile. Mouth moderate, very oblique, the maxillary reaching slightly beyond front of pupil, $2\frac{3}{4}$ in head; teeth in a single series, subequal in the two jaws, rather long, very slender and numerous, decreasing towards angle of mouth; about 40 teeth in the upper jaw, and 30 in the lower, on blind side. Eyes large, close together, separated by a narrow, sharp, scaleless ridge; the upper eye largest, slightly behind the lower, with considerable vertical range; diameter of upper eye, $3\frac{1}{3}$ in head. Snout and lower jaw scaleless; end of maxillary and rest of head scaled. Gill-rakers moderate, not strong, about 9 on anterior limb.

Dorsal fin beginning on the vertical from front of upper eye, the first three rays being somewhat turned to blind side; the fin low, highest at beginning of its posterior third, the longest ray nearly half length of head. Anal spine present, very small. Caudal rounded, about equaling length of head. Pectoral of colored side $1\frac{2}{5}$ in head, of blind side $2\frac{1}{5}$.

Scales moderate, those forming the lateral line persistent, the others deciduous; those on colored side with ciliated margins, on blind side smooth; lateral line without anterior curve; the scales are crowded and smaller anteriorly.

Head $3\frac{3}{4}$ in length, without caudal; depth $2\frac{1}{5}$. D. 87; A. 68; L. lat. 54 (pores).

Color in spirits uniform olivaceous, the scales dark-edged; lips and some of membrane bones of head margined with blackish. Fins dusky; each 7th (to 10th) ray of vertical fins with a very small but conspicuous black spot on its middle.

A single specimen (31099, U. S. Nat. Mus.) was collected at Santa Barbara, California, by Mr. A. Larco.

In the collection of which these specimens formed a part, are the following species not hitherto known from farther south than Monterey: *Oxylebius pictus, Ophidium taylori, Anarrhichthys ocellatus,* as also a single specimen of *Siphostoma bairdianum.*

UNITED STATES NATIONAL MUSEUM, *July* 11, 1882.

DESCRIPTIONS OF TWENTY-FIVE NEW SPECIES OF FISH FROM THE SOUTHERN UNITED STATES, AND THREE NEW GENERA, LETHARCUS, IOGLOSSUS, AND CHRIODORUS.

By G. BROWN GOODE and TARLETON H. BEAN.

The following budget of descriptions is presented as the result of a partial examination of the large collections of fishes from the Southern Atlantic States in the United States National Museum, some of which have been on hand for twenty years or more.

1. Tetrodon nephelus, n. sp.

The types are numbered 31427, 31428, and 26570. The first two were taken at Indian River, Florida, by Mr. R. E. Earll; the last at Pensacola, Florida, by Mr. Silas Stearns. The specimens range from $7\frac{1}{2}$ to 9 inches in length.

This species is the southern representative of *Tetrodon turgidus,* from which it differs in several particulars, as mentioned below.

The spines of the upper parts are much larger, farther apart, distinctly stellate with conspicuous roots; they extend backward not quite to front of dorsal above and to the vent below, the whole region behind these points being entirely smooth. There are less than 40 spines on the median line of the back between the eye and front of dorsal.

The dorsal is larger than in *T. turgidus,* its base one-fifth to one-sixth as long as the head, its largest ray three-sevenths as long as head.

The anal, also, is larger than in *turgidus,* its longest ray nearly one third as long as head.

The humeral process is somewhat longer than in *turgidus,* its length from axil of pectoral being a little more than half that of head.

Head contained $2\frac{5}{8}$ times, depth 4 times in length to caudal base.

D. 8; A. 6 (D. 7; A. 5 in *T. turgidus*).

Color somewhat variable, but distinguished from that of *turgidus* by the presence of paler blotches on the back and sides, around which the ground color often forms distinct reticulations. Dark bars on the sides placed as in *T. turgidus*, but much less distinct, the one in the axil of pectoral much smaller and less conspicuous than in *T. turgidus*.

2. Baiostoma brachialis, n. g. and n. s. Bean.

(*Achirus brachialis* Bean, in Goode & Bean, Proc. U. S. Nat. Mus., vol. v, p. 235. No description.)

Two specimens of this species have been sent to the Museum by Mr. Silas Stearns: No. 26605, 1.95 inches in length, from Appalachicola Bay; the other, No. 30463, 1.45 inches in length, from South Florida. Both were collected in 1879. In form and in most other characters *Baiostoma* is like *Achirus*, but it has a well-developed dextral pectoral fin. To this genus should, probably, be referred the *Monochir reticulata* of Poey, which has, however, a trace of a sinistral pectoral and a greater number of dorsal and anal rays than *B. brachialis*. The genus will be fully described in a later paper.

DESCRIPTION.—The body is ovate in form, being much more angular in its posterior than in its anterior outline. Its height is contained about $1\frac{2}{3}$ times in its standard length, and is nearly twice the length of the head, and about three times the greatest height of the dorsal and anal fins. Its least height, at the base of the tail, is slightly greater than one-third its height at the ventrals and one-fourth of its greatest height. Its greatest thickness is equal to about twice the diameter of the orbit and one-third the length of the head.

The scales of the nape, chin, and breast are larger than those of the body, and are armed with many more spinules. The scales upon the blind side are less strongly ctenoid than those upon the eyed side. The number of scales in the longitudinal series is about 60, about 53 of which are tube-bearing. The lateral line is almost straight upon the eyed side, very slightly arched in its anterior portion upon the opposite side; always conspicuous. The scales extend upon the dorsal, anal, and ventral fins almost two-thirds the length of the rays, but barely cover the base of the caudal rays.

The pectoral is scaleless. The head is short; its length is contained three times in the standard body length, and four times in the total. The length of the snout is slightly greater than that of the eye, and is contained four times in the length of the head. The mouth is small, oblique, the dorsal outline of the head projecting far beyond the upper jaw in a sickle-like expansion, which almost meets the tip of the lower jaw when the mouth is open; the contour of the head is consequently very peculiar. The width of the interorbital space is equal to half that of the eye. The teeth are inconspicuous, and upon the eyed side apparently absent; present, however, on the blind side, but very small in both upper and lower jaw.

The dorsal fin begins in advance of the tips of the jaws and contains 47–48 rays, the greatest height in its posterior portion equalling, as has been stated, one-third the length of the body.

The distance of the insertion of the anal from the tip of the snout equals the length of the head. The anal contains 35–37 rays, which correspond in general appearance to those of the dorsal.

The length of the caudal is one-third the standard length of the body, and one-fourth the total length. It is ovate-lanceolate in form, slightly pointed.

The distance of the ventral from the snout equals one third the height of the body. The right ventral is composed of five rays, and is connected with the anal by a low membrane. The left ventral is slightly smaller. Both are situated upon the ventral keel and are very close together, so that when expanded they are in contact throughout the entire surface, looking like one fin.

There is no trace whatever of a pectoral upon the sinistral, or blind side. The pectoral on the dextral side consists of five rays and its length equals one-third or two-fifths that of the head.

Color.—Grayish or brownish on eyed side with five or six faint dark vertical lines, and with scattered white spots, the largest nearly equal in size to the eye. The blind side is whitish.

Radial formula.—D. 47–48; A. 35–37; V. 5; P. 5; L. lat. 60; tube-bearing scales 53.

3. Hemirhombus pætulus, n. sp.

A single specimen, No. 30180, was obtained at Pensacola Florida, in 1882, by Mr. Silas Stearns.

DESCRIPTION.—Body sinistral; general form that of an ellipse, the caudal extremity being considerably produced. Its height is contained $2\frac{1}{4}$ times in its length, $2\frac{2}{3}$ times in its length to the end of the caudal fin, and is 4 times the height of the tail at its lowest portion. Its height at the ventrals is 3 times as great as at the tail. Its greatest width is equal to the diameter of the upper orbit.

The scales on the cheek of the blind side are arranged in thirteen series; those upon the nape and interorbital space of the eyed side are smaller than upon the body. The scales on the body are large, thin, deciduous, and cycloid. There are fifty-seven scales in the lateral line, fifty-four of which are tube-bearing. Lateral line straight, and over the axis of the body, save in its anterior fourth, within which it slightly ascends with a very gentle upward curve, to the upper angle of the gill-opening. Above the lateral line are thirteen scales; below, twenty-two. The vertical fins are scaly two-thirds of the distance to their tips.

The head is short, its length being contained $3\frac{2}{3}$ times in the standard body length, $4\frac{1}{2}$ times in the total length, and $1\frac{3}{4}$ times in the greatest height of body. The snout is short; its length, slightly less than the diameter of the lower eye, is contained 5 times in the length of the head. Mouth rather large, the upper edge somewhat curved, its cleft very oblique, the maxillary extending to below the middle of the lower eye. The lower jaw extends to the vertical from the anterior margin of the upper eye. Length of the upper jaw equals one-third the distance from the snout to the insertion of the anal. The lower jaw equals the distance from the tip of the snout to the posterior margin of the lower eye and is contained $2\frac{1}{3}$ times in the length of the head. Teeth moderate, equally developed on both sides, in two rows in the upper jaw, those of the outer row upon the blind side of the upper jaw and the anterior portion of the eyed-side considerably larger than those in the inner row. The teeth of the lower jaw uniserial, almost as large as in the outer row of the upper jaw.

The eyes are large, prominent, and far apart. Their longitudinal diameter equals the length of the snout, and is contained five times in the length of the head. Their vertical diameter is about three-fourths as great as their longitudinal diameter. The lower eye is far in advance of the upper, the vertical from the anterior margin of the upper orbit cutting the lower orbit at a point about two-thirds the distance from its anterior to its posterior margin. The upper eye is close to the dorsal profile, separated from it by a distance equaling about one-half its longitudinal diameter. The interorbital space is flattish and uneven, its width being contained four times in the length of the head. A prominent ridge extends from the upper posterior margin of the lower eye to the lower posterior margin of the upper eye, thence widening and curving downward to the upper angle of the branchial aperture. The margin of the pre-operculum is also somewhat elevated. The length of the operculum is very slightly greater than the width of the interorbital space.

There are eleven short and thick gill-rakers on the anterior arch, the longest equal in length to one-third the diameter of the eye.

The dorsal fin begins on the blind side of the body in advance of the anterior margin of the lower eye; its anterior rays are almost free, the longest rays behind its middle, its greatest height equal to the length of the upper jaw.

The anal is inserted under the anterior angle of the pectoral axilla. Its anterior rays are less free than are those of the dorsal, about two-thirds of their length being extruded from their membrane. Its outline similar to that of the dorsal, but greatest height somewhat less, being one-third the length of the head.

The greatest length of the caudal equals the length of the head without the snout, and one-fifth of the body length. Its middle rays are somewhat longer than the outer rays, giving to the posterior margin the outline of an obtuse angle.

The pectoral is inserted at the tip of the opercular flap; its second and third rays much produced in a filamentous extension. Its greatest length slightly exceeds $1\frac{1}{2}$ times that of the head. The pectoral on the blind side has no prolonged rays; its greatest length equalling that of the upper jaw.

The ventral on the eyed side is inserted on the ridge of the abdomen slightly behind its mate, which is a little removed from the medial line.

Distance between insertion of the ventral and the snout equals one-fourth the length of the body. The length of the ventral equals one-third that of the head. Vent, close to the origin of anal, and slightly removed from the medial point of the body on the blind side; behind it a small papilla, one-fourth as long as the eye.

Color: Eyed side, grayish brown; blind side, somewhat clouded with darker shade.

Radial formula.—D. 87; A. 67; C. 8 + 7; P. 11 sinistral and 9 dextral; V. 5; Scales 13—57—22.

4. Blennius asterias, n. sp.

The types of this species are the following: 2620, two specimens from Garden Key, Florida, collected by G. Würdemann; 2625, one specimen collected at the same place by Dr. Whitehurst; 6596, three specimens collected at Tortugas, by Dr. J. B. Holder. They vary from about $2\frac{1}{2}$ inches to 4 inches in length.

Length of head contained 4 times, depth 4 times in total without caudal. D. XI, 16; A. 19.

Body moderately elongate, compressed; the head very blunt and deep, almost as deep as long, its anterior profile straight or slightly concave, and nearly vertical. Mouth moderate, the maxillary reaching to past front of eye, its length contained 3 times in that of head. The lower jaw with two short, stoutish posterior canines; upper jaw without canines. Teeth about $\frac{32}{28}$. Preorbital deep, its depth equal to diameter of eye and contained $4\frac{1}{4}$ times in length of head. Interorbital space flat, narrow, two-thirds width of eye. Supraocular cirri small, fringed, their length about equal to that of pupil. Nape with a longitudinal dermal crest reaching to front of dorsal, provided with a series of about 20 filaments, the longest about as long as the eye. Gill-membranes forming a broad fold across the isthmus as in all species of *Blennius.*

Dorsal nearly continuous, the last spine a little lower than the first soft ray, not very high, beginning on the nape in front of the vertical of the preopercle; the spines all slender and flexible, the longest three-eighths as long as the head, the longest soft ray four-sevenths as long as the head. The caudal free from dorsal and anal, four-fifths as long as head. Anal moderate, four-ninths length of head. Pectoral somewhat shorter than head; ventral a little more than half length of head.

The lateral line forming the usual arch above pectoral, and continued backward on the median line to base of caudal, becoming indistinct posteriorly.

Color faded, apparently olivaceous, with about six dark cross-bars, which extend on the dorsal fin. Anal and posterior half of body with numerous round, whitish, stellate spots. probably bluish in life. Bluish streaks from eye across the cheeks. Anal edged with dusky; the other fins vaguely marked.

5. Blennius favosus, n. sp.

Of this new species there are two specimens, number 2629, collected at Garden Key, Florida, by Gustavus Würdemann; they are $3\frac{2}{5}$ inches and 3 inches long, respectively.

Length of head contained $3\frac{2}{3}$, depth $4\frac{3}{4}$ times in total to caudal base. D. XII, 18; **A.** II, 20.

Body comparatively elongate and compressed; anterior profile moderately decurved; head nearly one-half longer than deep; snout very short and blunt; mouth large, horizontal; jaws even; the maxillary reaches to posterior margin of orbit, its length contained $2\frac{1}{2}$ times in that of head. Each jaw with a long, curved, posterior canine, the canines of lower jaw largest. Preorbital two-thirds diameter of eye, which is contained $3\frac{3}{4}$ times in length of head, and equals more than twice interorbital width. An extremely long and slender supraocular cirrus, trifid to the base, the longest branch nearly as long as the head. No nuchal cirri. Gill-membranes forming a rather narrow fold across the isthmus.

Dorsal low, continuous; the spines very slender and flexible, the longest half as long as the head; the longest soft ray three-quarters as long as head; the last ray slightly joined to base of caudal. Caudal three-quarters as long as head. Anal rather high. Pectoral four-fifths as long as head; only the straight part of lateral line developed.

Color faded, brownish, finely reticultated, a series of obscure bluish blotches along the sides; front and sides of head marked with very distinct blue, reticulating lines surrounding honey-comb-like hexagonal interspaces; top of head with many small blue spots; dorsal with black dots and streaks; a black spot bordered with whitish between the first and second dorsal spines. Anal with oblique blue streaks; the fin margined with dusky; tips of the rays whitish. Base of pectorals with blue reticulations. The whole body was probably reticulated with blue in life.

6. Opisthognathus scaphiurus, n. sp.

The type of this species is a finely-preserved specimen, No. 5936, collected many years ago at Garden Key, Florida, by Dr. Whiteburst· Its length is 5 inches.

Body moderately elongate, somewhat compressed, its greatest depth contained 5 times in length to caudal base. Head rounded, blunt anteriorly in profile; snout very short, about as long as pupil; eye large, its length contained 4 times in that of head; maxillary reaching slightly past edge of preopercle, but not to end of head, its length contained $3\frac{3}{4}$ times in total to caudal base; ending in a flexible flap; lower jaw slightly included.

Teeth rather strong, wide set, forming two distinct series in front of each jaw, those of the inner series directed backward, especially in the upper jaw; the lateral teeth of lower jaw largest; a single vomerine tooth.

Anterior nostril with a short flap. Gill-rakers rather long and slender, the longest not quite half length of eye, nearly 20 below angle.

Sept. 18, 1882.

Head naked; scales of body very small, about 100 in a longitudinal series. Lateral line ceases near middle of trunk.

Dorsal fin low, continuous, the soft rays but little higher than the spines, which are slender and flexible, the longest contained 3⅓ times in length of head. Caudal short, rounded, its length 5⅔ times in total to its own base. Anal similar to soft dorsal. Pectoral half as long as head and a little longer than ventral, which does not quite reach the vent. Vent midway between front of eye and base of caudal.

Color grayish olive, much variegated with whitish and dark olive; about 6 irregular dusky bands on the body, which extend up on the dorsal fins; the bands are widest near the middle; the whitish markings on the body form roundish spots and are surrounded by reticulations of grayish olive. Head marbled, its posterior part as well as the sides of back and the pectoral base, with small blackish dots. Membrane lining the inside of the maxillary with two curved inky black bands on a white ground. Angle of mouth with a black spot. Lining of opercle inside black. Fins all variegated like the body. Pectorals pale, with small olive spots. Obscure blackish spots on the 6th and 7th dorsal spines; soft parts of vertical fins with a narrow dusky margin.

D. XI, 16; A. 18; V. I, 5; P. 17.

7. Gobius stigmaturus, n. sp.

Head contained 4, depth 6 times in total to caudal base. D. VI, 12; A. 12; V. I, 5; scales about 30 in lateral line.

Body rather elongate, little compressed. Head moderate, not very blunt, the anterior profile somewhat evenly decurved, the snout not very short, little shorter than the eye. Mouth rather large, nearly horizontal, jaws even, the upper jaw extending nearly to below middle of eye, its length 2⅔ times in that of head. Teeth in upper jaw in narrow bands, the outer series much enlarged, some of the anterior teeth canine-like. Teeth of lower jaw apparently in a single, somewhat irregular series, slightly smaller than those in the outer series of upper jaw. Eyes placed high, about 3¼ in head; interorbital space very narrow—a mere ridge. Scales large, ctenoid; those on the nape much smaller. Gill-openings not continued forward above opercle. Dorsal spines very slender, none of them filamentous, the longest three-fourths as long as the head, soft dorsal low, its longest ray two-thirds as long as head. Anal similar to soft dorsal. Caudal as long as the head. Pectorals slightly longer than the head. Ventrals about as long as head. Upper rays of pectoral not silk-like.

Color light olive, the sides marbled with whitish, the back with dark punctulations. A dark spot on opercle and one below eye, sides with about 5 dusky blotches along the median line, the last one forming a distinct round black spot at caudal base. Vertical fins with wavy, blackish bars; paired fins plain.

IOGLOSSUS, n. g., Gobiid, Bean.

Ioglossus Bean, in Jordan & Gilbert, Proc. U. S. Nat. Mus. V, 297.

DIAGNOSIS.—A genus closely allied to *Oxymetopon* Bleeker; but differing from it in the absence of a keel on the head and in the smoothness of nearly all of the scales. The body is moderately elongate and compressed, covered with small scales, which are all cycloid except a few at the caudal base; anteriorly the scales are not imbricated, posteriorly they are somewhat larger and regularly imbricated, mostly cycloid, a few in the tail weakly ctenoid; no lateral line; cheeks with imbedded cycloid scales. Head naked; mouth oblique, the lower jaw projecting. Teeth of the upper jaw in two rows, conical, slightly recurved, those in the outer row the largest. The two central teeth in the inner row enlarged, canine-like and much recurved. Teeth in lower jaw uniserial, with a pair of large canines on each side. Tongue free, slender, and elongate, sub-terete. Vomerine and palatine teeth absent. Eyes moderate. Gill-openings wide, the membranes attached mesially to the narrow isthmus, across which they do not form a fold. Gill-rakers long and slender.

Dorsal fins closely approximate; the first with six slender thread-like spines, the second with numerous rays, separated from the caudal by a considerable interval. Caudal very elongate, lanceolate, its middle rays filamentous. Anal similar to second dorsal. Ventrals inserted under the base of pectorals, closely approximate, very slightly connected by a basal membrane, inner rays filamentous; pseudo-branchiæ present; branchiostegals, four.

Etymology: *ἰός*, barb; *γῶσσα*, tongue.

8. Ioglossus calliurus n. s. Bean.

> *Ioglossus calliurus* Bean, in Goode & Bean, Proc. U. S. Nat. Mus. V, 236. Name only; also in Jor. & Gilb., op. cit., 297.

The museum has received from Mr. Silas Stearns, Florida, three specimens of a species of *Ioglossus* (No. 30198, one specimen; and No. 30797, two specimens) taken by him at Pensacola. Professor Jordan obtained specimens of the same species at Pensacola from the stomach of the red-snapper, *Lutjanus Blackfordii*.

DESCRIPTION.—The height of the body is contained 5½ to 6 times in its length to the origin of the middle caudal rays, and 8 to 9 times in the extreme length. Its greatest width equals half its height and is also about equal to the distance from the posterior ray of the second dorsal to the origin of the upper caudal rays. The least height of the tail is about equal to that of the head at the eye.

The greatest length of the head is contained 4½ times in the standard body length. The width of the interorbital area is equal to the diameter of the eye and considerably greater than the length of the snout, which is contained 3 times in the postorbital length of the head. The length of the postorbital region, including the opercular flap, is con tained 8 times in the standard body length. The upper jaw extends to the vertical through the anterior limb of the pupil, and its length equals the distance from the tip of the snout to the posterior margin of the orbit. The lower jaw equals the upper in length, slightly projecting beyond it, however, and, on account of the thickness of the chin, giving a heavy bulldog appearance to the head. The diameter of the eye is contained 4 times in the length of the head, the eye being inserted close to the upper profile. Nostrils minute, close to the upper anterior margin of the orbit, double; the two apertures placed side by side in a lateral line rather than longitudinally, as is usual. A pair of large pores near the upper posterior margin of the orbit, and a series of three or four similar pores along the posterior limb of the preoperculum; others scattered here and there over the head. Operculum membranous; gill-openings very wide, the upper angle of the branchial aperture located close to the upper angle of the pectoral base. About 10 teeth in the lower jaw in advance of the double canines. Teeth behind the canines minute; 14 or more on each side. Teeth in the upper jaw much more uniform than in the lower; at least 30 in the inner row. Slight granulations, or asperities, upon the vomer; palatines toothless.

The distance of the first dorsal from the snout is contained $3\frac{1}{2}$ to 4 times in the standard body length, the base of the first dorsal being equal to the height of the body; the distance between the fifth and sixth dorsal rays being double the distance between the other rays. The dorsal contains six slender filamentous rays, the greatest length of the longest being nearly or quite equal to the length of the head. The point of insertion of this fin is somewhat variable in its location, sometimes directly over that of the ventral and sometimes a little behind. The interspace between the first and second dorsal fins is equal to the width of the base of the pectoral. The second dorsal fin contains twenty-two or twenty-three slender, filamentous rays, those in the anterior portion being slightly longer than the others, and equal in length to the greatest height of the body.

Insertion of anal midway between the tip of snout and the base of caudal fin. The anal papilla large and located close to the vent. The anal fin contains 20 to 22 rays, about as long as those of the second dorsal, but much stouter, and apparently used in burrowing.

The caudal is lanceolate and extremely elongate. The length of the middle rays slightly greater than half that of the body, or one-third of the total length of the fish.

The pectoral is inserted directly over the ventrals. Its length is about equal to the height of the body, its base broad, vertically placed, and equal in width to the length of the operculum. The ventrals are composed of a spine and four filamentous rays; their length greater than that of the head, the tips reaching almost or quite to the vent; they are distinct, though slightly united by a basal membrane; their insertion is closer to the tip of the snout than to the vent.

Radial formula.—B. IV ; D. VI, 22–23; C. 9+11; P. 20; V. I, 5.

Color.—In alcoholic specimens, pale yellowish; in fresh condition, according to Jordan, light olive. Top of first dorsal dusky, middle of caudal dusky (blue) with paler (perhaps red) edgings.

9. Scorpæna Stearnsii, n. sp.

Body robust, little compressed, tapering posteriorly. Mouth moderate, oblique, the jaws equal when closed, the lower jaw with a small symphysial knob. The maxillary reaches to below posterior margin of orbit, and is half as long as head.

Height of body contained $2\frac{5}{6}$ times in length to caudal base; length of head, $2\frac{3}{5}$ times.

The preorbital has two strong diverging spines ; the suborbital without deep pit, its stay low, armed with two small spines. Nasal spines inconspicuous. Interorbital space deeply concave, with two longitudinal ridges, its width equalling three-fifths of the long diameter of eye. Eye 4 in length of head. The cranial ridges are rather low, moderately sharp, the following pairs of spines present : Preocular, supraocular, postocular, coronal, occipital, nuchal, besides three on the temporal region arranged in a right line behind the eye. Occipital region deep, a little broader than long.

Preopercular spines five, the two lower blunt and short, the upper much the longest, half as long as the eye, a small spine at its base. Opercular spines moderate. Scapular spines small. Supraocular flap very small, its length less than one-third that of eye. Preorbital, preopercle, cheeks, and nostrils with small dermal flaps. Opercular flap scaly ; a few rudimentary scales on cheeks and front of opercle. Breast with small scales. Gill-rakers short and thick, not twice as long as broad.

Scales large, smooth, their edges with a thin membrane, the radiating striæ conspicuous, but the concentric striæ inconspicuous. Scales of the belly smaller. A series of dermal flaps along the lateral line, and at the dorsal base.

Dorsal spines slender, the longest contained $2\frac{1}{4}$ times in length of head; the longest soft ray half as long as head. Caudal subtruncate, its angles rounded, its length four-fifths that of head.

Anal spines small, the second and third equal, contained $2\frac{1}{5}$ times in length of head. Soft anal rays high, the longest half as long as head.

Ventrals contained $1\frac{3}{4}$ times in length of head, the last rays joined to the belly by a broad membrane which extends nearly to their tips.

Pectorals reach to soft rays of anal, the longest ray slightly shorter than head. The base of the fin is a little procurrent, its length one-third that of head, the lowermost rays rapidly shortened.

D. XI, I, 9; A. III, 5; P. 20; V. I, 5. 32 series of scales in lateral line (31 tubes).

Color dusky olivaceous, whitish below. Head with some dark blotches, its lower and posterior parts with a few round black spots about as large as the nostril. The jaws dusky, marbled with whitish; sides of back with diffuse blackish blotches. Entire body sparsely covered with round dusky spots smaller than the pupil; these spots are most numerous and distinct in the axillary region, which is otherwise whitish. Skin of shoulder-girdle above marbled with black. Spinous dorsal with a broad, median, dusky band; tips of its membranes dusky, its base whitish, with black spots. Soft dorsal and anal irregularly marbled with blackish. Caudal with a broad median, and a terminal band of blackish. Pectorals blackish above, with dark spots; lower edge whitish; three obscure, broad, dusky cross-bands. Ventrals dusky towards the tips.

The type of this species is numbered 30,169; it is $6\frac{9}{10}$ inches long, and was obtained at Pensacola, Florida, by Mr. Silas Stearns, to whom the species is dedicated in appreciation of his services in adding to our knowledge of the fishes of the Gulf of Mexico.

A smaller *Scorpœna* $4\frac{7}{10}$ inches long (No. 30,185), from the same locality, agrees with the type of *S. stearnsii* in all respects, except that the preorbital and supraorbital flaps are very much longer, the latter reaching the front of dorsal, its length half that of head. The preorbital flap is as long as the pupil. The margins of both these flaps are without fringes. In the type of *S. Stearnsii* the supraorbital flap is nearly as broad as long, not so long as the pupil, and is distinctly trilobate; the preorbital is minute. Without additional material it is impossible to decide whether these differences are sexual or of specific value.

10 Scorpæna calcarata, n. sp.

The type of this species is numbered 23566; it is $2\frac{1}{4}$ inches long, and was taken in Clear Water Harbor, Florida, by Dr. J. W. Velie. The specimen is in poor condition.

Body moderately robust, the greatest depth slightly less than a third of length to caudal base, the lower jaw slightly projecting, with a small symphysial knob. The maxillary reaches to past the pupil; its length equal to half that of head.

The preorbital has three diverging spines; the suborbital without pit, the bony stay moderate, armed with two small spines. Nasal spines small. Interorbital space narrow, with two longitudinal ridges, its width two-fifths length of eye. The cranial ridges are rather low, with sharp spines, the following pairs present: preocular, supraocular, postocular, coronal, occipital, nuchal, besides three or four on the temporal region. Occipital cavity almost obsolete, represented by a slight depression.

Preopercular spines five, the lowermost stout, directed downward and forward, the uppermost rather long—more than half as long as the eye. Opercular and scapular spines moderate. Eye large, nearly one-third as long as the head. Supraocular flaps minute; a few other small flaps on the head.

Cheeks with rather large imbricated scales; opercle with some scales anteriorly and on its flap; breast scaly; scales of body large, not ctenoid, with few dermal flaps or none.

Pores of lateral line very conspicuous. Gill-rakers short and small. Dorsal spines rather slender, the longest contained $2\frac{2}{3}$ times in length of head; the longest soft ray $2\frac{1}{2}$ times in length of head.

Anal spines small, the second and third subequal, one-third as long as head. Soft anal rays moderate, the longest half as long as the head.

The ventrals reach past vent, their length contained $1\frac{2}{3}$ times in that of head, the last rays largely united to the belly by a membrane.

Pectoral long, contained $1\frac{1}{4}$ times in length of head, its base oblique, contained $2\frac{1}{3}$ times in length of head, the rays all simple.

D. XI, I, 9; A. III, 5; P. 19; V. I, 5.

Scales in about 28 series, the number being uncertain because many of them are rubbed off. There are about 25 tubes in the lateral line.

Color mostly obliterated, dusky grayish marbled with blackish; a black suborbital bar; a black bar at caudal base; axil of pectoral whitish with dusky specks, a black spot at its upper edge; ventrals mostly black.

11. Gerres olisthostoma, n. sp.

Mr. R. E. Earll, when engaged in the fishery-census investigation upon the coast of Florida, obtained at Indian River six specimens (No. 25118), of a new species of *Gerres*. They are known as the "Irish pompano" and "hog-fish." This species is one of the largest of the genus, and in general form resembles *Gerres gula* and *G. homonymus*, having short thick body, very protractile snout, elevated dorsal and elongate ventral fins. This species is reported to be rather common in the Indian River region; it is evidently the same as No. 12561, referred by Poey to *Gerres rhombeus*. The true *rhombeus* has, also, been sent to the Museum by Professor Poey.

DESCRIPTION.—A *Gerres*, with short, thick body, the greatest height of which, at the ventrals, is contained twice in its length and $2\frac{3}{4}$ times in the distance from its snout to the tip of the upper caudal lobe. Its least height at the base of the tail, being one-quarter of its greatest height. The greatest width of the body is equal to the greatest height of the tail.

The scales are large, somewhat loosely set; 39 in the lateral line; above it 7; below it 11. Jaws entirely naked, as well as the ordinary patch over the groove for the reception of the protractile snout. The greatest length of the head is contained $3\frac{1}{5}$ times in that of the body; the greatest width of the head is equal to half its length, and is half as wide again as the interorbital area. The length of snout is equal to the diameter of the eye; the length of the operculum, including the flap, equals one-fifth of the greatest height of the body. The length of the groove for the reception of the premaxillaries equals the length of the maxillary. The upper jaw when protruded extends beyond the tip of the maxillary a distance equal to the least height of the tail. Teeth brush-like, in bands; the band of the upper jaw more developed than that of the lower one, the length of the band equaling half that of the eye. The maxillary extends to the perpendicular through the anterior margin of the pupil; the mandible, to the vertical through the posterior margin of the eye. The length of the mandible equals half that of the head. The preoperculum is denticulated on its lower border and at the angle, the denticulations at the angle being slightly the largest. The gill-rakers are short, 12 in number on the anterior arch below the angle, the longest one-fifth as long as the eye. The eye equals the snout in length, and is contained $3\frac{2}{3}$ to 4 times in the length of the head. The naked space above the premaxillary groove in the majority of the types is prolonged backwards to an acute point, but in two of the types the acute point is replaced by a scaly space.

The distance of the spinous dorsal from the snout equals about $1\frac{1}{2}$ times the length of the head. It is inserted nearly over the middle of the pectoral base. The position of this fin varies very slightly in different individuals. The subsequent spines to the seventh are much stouter and longer than any of the others, and are so graduated in length that, when the fin is erect, the outline of the anterior portion presents nearly the figure of an isosceles triangle. The last two spines are nearly equal in length. The length of the first ray of the soft dorsal is nearly double that of the last dorsal spine; the last ray being about as long as the first. The spines and rays all protrude from one-half to one-third of their length beyond the membrane, giving to the fin a ragged appearance. The basal sheath of the dorsal fin is thick and prominent.

The insertion of the anal is equidistant between the tip of the snout and the tip of the upper caudal lobe, the fin being inserted under the perpendicular from the fourth dorsal ray. The first anal spine is short and stout, being half as long as the diameter of the eye; the second very stout, not quite so long as the third, being equal in length to the distance from the center of the eye to the end of the operculum. The first ray of the anal is also elongate, giving to this fin, when expanded, the appearance of an isosceles triangle, with base somewhat slenderer than that described in the first dorsal. The soft anal in its posterior part is lower than the soft dorsal.

The caudal fin is deeply forked, the longest ray of the upper lobe being equal to about one-third of the body length, and five times as long as the inner rays.

The pectoral is inserted at the tip of the opercular flap, and has its upper rays elongate, equalling the head in length.

The ventral is inserted beneath the axil of the pectoral at a distance from the snout equal to two-fifths of the length of the body. Its spine is as long as the fourth spine of the dorsal, but double as stout, the first ray being prolonged.

It seems desirable to state that the proportions in this species differ very considerably with individuals.

Radial formula.—D. IX, 11; A. III, 8; V. I, 5; C. 9, 8; P. 16. Scales: L. lat. 39; L. trans. $\frac{7}{11}$.

12. Calamus arctifrons, n. sp.

A species belonging to *Calamus* of Swainson. The type numbered 30163 is a specimen $9\frac{1}{2}$ inches long, collected at Pensacola, Florida, by Mr. Silas Stearns.

Body oblong ovate, more elongate than is usual in species of this group, deepest at origin of dorsal, the greatest depth being contained nearly $2\frac{1}{3}$ times in the total length to caudal base. Anterior profile evenly curved, unusually convex. A blunt protuberance before eye. Mouth comparatively large, the maxillary barely reaching the vertical from front of orbit. Length of upper jaw contained $2\frac{1}{6}$ times in head. The anterior teeth of both jaws are conical, rather strong and canine-like, 6 to 8 in each jaw. Behind these are bands of cardiform teeth. Molars rather large, in two rows anteriorly and three posteriorly in upper jaw; two rows in the lower; the molars of the inner series of both jaws much larger than the others. Lower jaw slightly included. Behind the upper lip on each side is an enlarged oblong pore, two-thirds as long as the posterior nostril; above it is a fleshy flap. The posterior nostrils slit-like and much larger than the circular anterior ones. Preorbital very deep, its depth, from eye to angle of mouth, contained $3\frac{1}{4}$ times in length of head. Eye rather small, placed very high, its diameter contained $3\frac{3}{4}$ times in length of head, and equaling four-fifths of the convex interorbital space. Cheeks with 5 to 6 series of scales. Four rows of scales on the opercle. Opercle very short, its length equaling two-thirds diameter of eye, and less than one-third of its height. Gill-rakers very short, thickish, few. Least depth of tail two-thirds length of caudal peduncle and contained $3\frac{1}{2}$ times in head.

The length of the head is contained $3\frac{1}{2}$ times in total to caudal base. Interorbital space contained $3\frac{1}{2}$ times in length of head. The oblique distance from snout to origin of spinous dorsal is slightly less than half the total length to caudal base. Dorsal spines slender, the longest (fourth) contained $3\frac{2}{5}$ times in length of head and about equal to longest ray of second dorsal.

Distance from front of anal to base of caudal contained $3\frac{1}{5}$ times in total. Anal spines small, graduated, the third one-fourth as long as head. The second spine somewhat stronger, but shorter, than the third.

Caudal deeply forked, its middle rays two-fifths as long as the outer. The upper lobe is nearly as long as the head.

Pectoral narrow, reaching slightly past vent, as long as head.

Ventrals inserted slightly behind pectoral origin, five-eighths as long as the head. A partly-concealed procumbent spine before the dorsal. First spine less than two-fifths as long as second, which is not much shorter than the third.

D. XII, 12; A. III, 11; P. 16; V. I, 5; Scales 5–46–14. Tubes in lateral line 46.

Color light olive, with bright reflections, paler below. Back and sides with 7 or 8 obscure dusky cross-bands, narrower than the interspaces; these, doubtless, disappearing with age. Head without distinct markings. Fins plain olivaceous; the ventrals and posterior edge of caudal slightly dusky, with faint traces of cross-bands on the lobes.

13. Stenotomus caprinus n. s., Bean.

> Stenotomus caprinus Bean, in Goode & Bean, Proc. U. S. Nat. Mus., Vol. v, p. 238, name only; also in Jor. & Gilb., op. cit., 278.

Two specimens, No. 30795, of a new species of *Stenotomus* were obtained, from the stomachs of red snapper at Pensacola, Florida, by Mr. Silas Stearns. Two were also similarly obtained by Professor Jordan at Pensacola. It is distinguished from *S. versicolor* by the presence of two short spines in advance of the elongate spines of the first dorsal, by the great elongation of the anterior dorsal spines, and by the greater depth of the cheeks and preorbital region.

DESCRIPTION.—Body irregular oblong-ovate. Its height is contained twice in its length. Its height at the tail is contained $4\frac{1}{3}$ times in its greatest height, and a little more than three times in the length of the head.

Scales in lateral line 45 to 47; above it, seven; below, fourteen. Anterior profile protuberant over the eyes; mouth moderate, maxillary arching almost to the vertical from the anterior margin of the orbit. Length of the upper jaw contained $2\frac{2}{3}$ times in that of head. There are ten narrow compressed incisors in the front of the upper jaw, and the same number in front of the lower jaw. Two rows of small molars in each jaw, the inner series very slightly larger than the outer.

Eyes circular, their diameter contained $3\frac{1}{4}$ times in the length of the head.

Distance between insertion of dorsal and snout contained $2\frac{1}{4}$ times in length of body. In front of the elongate dorsal rays are two upright and slightly curved spines, the height of which equals the diameter of the pupil, and a well-developed spine of about the same length projecting forward horizontally. The dorsal spines, from the third to the seventh, inclusive, are much elongated, filamentous, the length of the first being equal to the length of the pectoral fin, and contained $2\frac{1}{2}$ times in the body length. The base of the dorsal is equal to half the distance from tip of snout to the end of the middle caudal rays.

The anal is inserted in the perpendicular from the origin of the soft dorsal, almost equidistant between the tip of the snout and the tip of the upper caudal lobe. The three anal spines are stout, the second and third being the longest and of equal length, slightly shorter than the anal rays. The length of the base of anal equals the length of the sixth dorsal spine, and also the length of the ventral.

Caudal fin forked. The outer rays of the lower lobe twice as long as the middle rays, those of the upper lobe slightly less than those of the lower lobe.

Pectoral inserted in the vertical from the middle of the space between the third and fourth dorsal spines, its length being equal to the height of the body at the insertion of the anal; its longest ray reaches from the perpendicular to the fourth ray of the soft anal.

Ventral inserted in the perpendicular from the origin of the fifth dorsal spine; length almost equal to that of the sixth dorsal spine.

Color.—Silvery gray, slightly olivaceous above.   Professor Jordan states that in fresh specimens there are faint traces of dark cross-bands, and that the posterior margin is probably blackish.

Radial formula.—D. XI, 12; A. III, 10; C. 17; P. 15; V. I, 5. Scales: L. lat. 45 to 47; L. trans. $\frac{7}{14}$.

14. Trisotropis stomias n. s., Goode & Bean.

The species provisionally referred to by us, in the Proceedings of the National Museum, Vol. II, p. 143, as *T. brunneus* Poey, and which by previous writers was catalogued under the name *T. acutirostris,* having proved distinct from both of these species, we now propose to describe as new under the name *T. stomias.* The Museum has received five specimens, a tabulated list of which is here given.

	Number.	Locality.	Collector.
I	15462	New York market, Florida (?)	E. G. Blackford.
II	16902	Florida (?)	J. H. Richard.
III	21336	Pensacola, Fla	Silas Stearns (1878).
IV	26561	Key West, Fla	Silas Stearns.
V	26587	Pensacola, Fla	Silas Stearns.

This species is the black grouper of Pensacola, a fish of some commercial importance.   Specimens were also obtained at Pensacola in 1882 by Jordan, who states that it is almost as abundant as the red grouper, *Epinephelus morio,* and reaches a weight of 40 pounds.

DESCRIPTION.—A *Trisotropis* with body moderately compressed.   Its greatest height slightly more than one-fourth its length without caudal, and equal to or slightly less that 3 times as great as the least height of the tail; length of the head three-eighths length of the body and $3\frac{1}{2}$ times length of the snout; the lower jaw projects beyond the upper a distance equaling one-half the diameter of the eye.   The maxillary extends to the vertical from the center of the eye, and the mandible almost to the vertical from its posterior margin.   The distance of the eye from the upper profile of the head is about equal to half of the vertical diameter of the orbit. The horizontal diameter of the eye is contained $1\frac{2}{3}$ times in the length of the snout, almost 3 times in the postorbital portion of the head, and exactly twice in the length of the operculum to the tip of its flap. Lower jaw without canines.   The teeth in two rows, those in the inner row being double the length of those in the outer row and much less numerous. The teeth in the upper jaw very irregular in size, and hardly specialized, excepting in two patches at each side of the sympyhsis. Two moderate sized canines in advance of these patches. Vomerine teeth numerous and feeble. Palatine teeth very weak and with inconspicuous bands upon the crest of the bone. Preoperculum with minute denticulations, somewhat stronger at the angle. The length of the intermaxillary is considerably more than half that of the lower jaw.

Distance of insertion of dorsal from snout equals the greatest length of head including the opercular flap, the dorsal origin being very slightly in advance of the insertion of the ventral, which is located under the base of the third dorsal spine. Length of third dorsal spine is equal to that of the intermaxillary.

Distance of anal from snout about equal to twice the length of the head, the length of its base being slightly greater than the greatest length of the pectoral. The second anal spine is the stoutest, and is twice as long as the first, while the third, which is slender, is $2\frac{1}{2}$ times as long as the first.

Distance of pectoral from snout is one-third the standard body length.

Distance of ventral from snout is equal to twice the postorbital length of the head. The length of the ventral is slightly more than one-eighth of the standard body length (one-sixth or more in smaller specimens).

The length of the middle caudal rays is equal to the distance from the posterior margin of the orbit to the tip of the largest opercular spine. The upper and lower lobes of the caudal produce an incurving, giving to the space between the lobes a semicircular outline.

Scales in lateral line 130. Above lateral line 27–28; below, 60–61.

Radial formula.—D. XI, 16–17; A. III, 10–11; C. +17+; P. I, 16; V. I, 5.

Full measurements of three specimens will be found in Proceedings of the National Museum, Vol. II, p. 144.

15. Hypoplectrus gemma, n. sp.

A single specimen, No. 3422, of a new species of *Plectropoma*, has for many years been preserved in the museum. Name of collector unknown. In general appearance this species resembles *Hypoplectrus nigricans* of Poey, a specimen of which from the same locality was found in the same bottle and recorded under catalogue number 3423. In shape it is also similar to *Hypoplectrus puella*, but its coloration appears to have been much more uniform. The crescent-shaped caudal is a diagnostic mark, by which it can be distinguished from all other species now accessible to us.

DESCRIPTION.—Greatest height of the body is contained 3 times in its total length, and $2\frac{1}{3}$ times in its standard length. Greatest width equals length of second dorsal spine; least height of the tail is contained 3 times in the length of the head. The scales are small, weakly ctenoid, there being about 76 in the lateral line, 9 above it, and 29 below. The lateral line follows very closely the contour of the dorsal profile throughout its entire extent. Greatest length of the head is contained 3 times in the distance from the tip of the snout to the end of the middle caudal rays. Greatest width of head is about equal to the width of the body. Length of snout is contained 3 times in the length of the head. Length of the operculum to the end of the flap equals the length of the snout. The upper jaw extends to the vertical from the anterior margin of the orbit; its length is equal to half that of the head; the lower jaw is about the same length. The armature and squamation of the opercular bones are normal, as is likewise the dentition. The diameter of the eye equals one-fourth length of the head.

The distance of the dorsal fin from the snout is very slightly less than the greatest height of the body, the length of the dorsal base equalling the distance between its origin and the base of the posterior ray of the anal fin. Its fourth spine is the longest, its length equaling that of the base of the anal.

· The anal fin is inserted below the origin of the second dorsal ray, the base of its ultimate ray being beneath that of the ninth dorsal ray. Its third spine is very slightly longer than the second; their diameters are equal. The anal is higher than the dorsal, its greatest height being equal to the distance between the base of the ventrals and the origin of the anal fin.

The caudal is crescent-shaped, the external rays being much prolonged, especially those of the upper lobe, which are twice as long as the middle caudal rays.

Distance of pectorals from snout equals the height of the body at the ventrals, their length being equal to that of the superior caudal lobe. When extended horizontally these fins reach to the vertical from the insertion of the first anal ray.

Distance of the ventrals from the snout equals half the standard body-length. They extend to the insertion of the anal, and are equal in length to the rays of the lower caudal lobe.

Radial formula.—B. VII; D. X, 15; A. III, 7; C. 9 + 8; P. 14; V. I, 5. L. lat. 70; L. trans. 9–32.

Color.—In alcohol dull purple; in life probably deep purple, with cloudings of lighter color. Fins in alcohol colorless; in life probably pearly. The external rays of the caudal corresponding in hue with the deeper portions of the body color.

16. Menidia dentex, n. sp. .

The types of the present description, No. 18051, were taken at the mouth of the Saint John's River, Florida, by Prof. S. F. Baird. There are ten individuals in the lot, varying in length from $2\frac{3}{4}$ to $4\frac{1}{2}$ inches. The three which are made the special types of this description measure $3\frac{1}{2}$, $4\frac{2}{5}$, and $4\frac{1}{2}$ inches, respectively.

The species may be at once distinguished from *M. peninsulæ* and *M. vagrans* by the smaller number of dorsal spines and the larger number of anal rays, as well as by the stronger teeth; the teeth are much stronger than those of *Menidia notata*, the body is deeper, and there are fewer scales in a longitudinal series. From *M. boscii* of J. & G. it differs in its smaller eye, position of first dorsal, stronger teeth, and fewer scales in the lateral line.

The head, in shape and squamation, agrees with that of other species; its length is very slightly less than the greatest depth of the body, and is contained from $4\frac{1}{3}$ to $4\frac{1}{2}$ times in total length to caudal base (end of silvery band). The eye is a little shorter than snout, equals width of interorbital space, and is two-sevenths as long as the head. The snout is almost one-third as long as the head. Mouth rather large, the strongly curved and freely protractile intermaxillary being as long as the snout. Teeth in narrow bands, the outer series in both jaws much enlarged, a pair of canine-like teeth in the inner series at the symphysis of upper jaw. Lower jaw much longer than eye, as long as second dorsal base, contained $2\frac{2}{3}$ times in length of head. Jaws equal.

The greatest depth of the body equals the distance from origin of first dorsal to end of second dorsal, or nearly so. The origin of spinous dorsal is midway between tip of snout and end of middle caudal rays, immediately above anal origin. Longest dorsal spine one-third as long as head. Longest ray of soft dorsal equals length of head without postorbital part; it also equals two-thirds of length of pectoral. The ventral is inserted under the ninth scale in a longitudinal series; its length is about one-half that of the head; it does not quite reach the vent. The length of anal base equals twice that of head without postorbital part. The length of middle caudal rays equals one-eighth of total length to end of silvery band; the external rays are about $1\frac{1}{2}$ times as long as the middle rays. Only the caudal fin is scaly, and that for half its length.

Light olivaceous; minute brown punctulations on the jaws, top of head, and around the posterior margins of the scales of the back.

D. IV, I, 8; A. I, 22; V. I, 5; P. I, 12; Scales 8 to 9—39 to 40. Width of silvery band about two-thirds that of a scale in the series through which it runs, half length of snout, one-sixth length of head.

17. Tylosurus gladius, n. sp., Bean.

> *Tylosurus gladius* Bean, in Goode & Bean, Proc. U. S. N. M., v, 239. Name only.

The type of the species (No. 30151) was taken at Pensacola, Florida, by Mr. Silas Stearns. It is about 29 inches long.

Body robust, little compressed, its greatest breadth a little more than two-thirds greatest depth; caudal peduncle slightly depressed, a little broader than deep, with a slight dermal keel.

Head broad, broader above than below. Interorbital space nearly two-thirds length of post orbital part of head, with a broad, shallow, naked, median groove, which is wider behind and forks at the nape.

Supraorbital bones with radiating striæ. Distance between nostrils a little more than one-sixth length of snout.

Jaws comparatively short, strong, tapering; very stiff; lower jaw wider and longer than upper. Both jaws with broad bands of small teeth on the sides; within these is a series of very large knife-shaped teeth. The length of the longest teeth is a little more than 3 times the breadth.

Posterior teeth in both jaws directed backward; anterior teeth erect. Number of large teeth about $\frac{25+25}{23+23}$; length of large teeth about one-fifth diameter of eye; no vomerine teeth.

Upper jaw from eye about $1\frac{3}{4}$ times as long as the rest of head. Eye large, 7 in snout, $2\frac{3}{5}$ in post orbital part of head, and $1\frac{4}{5}$ in interorbital width. Maxillary entirely covered by the preorbital. Cheeks densely scaled; opercles scaled only along the anterior margin. Scales minute, especially on the back; somewhat larger below.

Dorsal fin rather high in front, becoming low posteriorly, the height of its anterior lobe equaling post orbital part of head; its longest ray is two-fifths of the length of the base of the fin. Caudal lunate, its lower lobe nearly one-half longer than the upper; middle rays about as long as eye. Anal falcate, low posteriorly, its anterior lobe equal to anterior dorsal lobe.

Ventrals inserted midway between base of caudal and middle of eye, their length a little less than that of pectoral, and equal to postorbital part of head.

• Upper ray of pectorals broad, sharp-edged; length of pectorals $3\frac{2}{5}$ in head, and slightly greater than postorbital part of head.

Head $3\frac{1}{3}$ in length to base of caudal; depth nearly 4 in head, about 13 in total to caudal base.

D. I, 22; A. I, 20; V. 6; P. 14.

Color dark green above, silvery below; dorsal and pectoral blackish; ventrals somewhat dusky; anal yellowish, the lobe slightly soiled; caudal dusky olivaceous. No suborbital bar and no scapular spot. A slight dusky shade on upper posterior part of cheeks, and a yellowish bar on anterior edge of opercle. Caudal keel black.

This species is closely allied to *T. fodiator* Jor. & Gilb., described from Mazatlan, differing from it chiefly in its longer jaws and greater number of fin-rays.

CHRIODORUS, new genus, SCOMBRESOCIDÆ.

Body moderately elongate and compressed, covered with large smooth scales. Lateral line extending along the lower side of the belly. Jaws short, equal, not produced. Teeth large, incisor-like, tricuspidate, close set, in two distinct series in each jaw, those of the inner series somewhat smaller than the outer. No teeth on vomer or palate. Premaxillary not protractile, slightly movable. Maxillary anchylosed with the intermaxillary. Gill-rakers rather long and slender, not very numerous. No pseudobranchiæ. Branchiostegals rather numerous. Gill-membranes separate and free from the isthmus.

Pectorals rather short, placed high. Ventrals small, median. Dorsal and anal far back, opposite and similar to each other, the anterior rays elevated. Caudal deeply forked, the lower lobe somewhat the longer.

18. Chriodorus atherinoides, n. sp.

The greatest breadth of the body is about three-fifths of its depth, which is contained $6\frac{3}{4}$ times in the length to caudal base.

The length of the caudal peduncle is one-half greater than its least depth and is a little less than half length of head.

Head rather long, contained $4\frac{3}{5}$ times in length to caudal base. The interorbital space is broad with a wide median ridge, on each side of which is a groove; the width of the space is about equal to eye. Eye large, very nearly median, equal to snout, contained $3\frac{1}{6}$ times in head. The area formed by the premaxillaries is fully 3 times as long as broad. Maxillary entirely concealed by the preorbital when the mouth is closed. Edge of premaxillary slightly concave and curved. The upper jaw extends to anterior nostril, its length contained 4 times in length of head. The lower jaw $2\frac{1}{4}$ times in head, its tip broadly rounded, without a symphysial projection. There are about 28 teeth in the outer series in each jaw.

The distance from snout to dorsal fin equals $3\frac{3}{4}$ times length of head; the dorsal base is a little greater than anal base and equals the distance from snout to posterior margin of preopercle. Dorsal elevated in front, but not falcate, its longest ray equals longest anal ray, which equals half length of head; the last dorsal and anal rays are very short. Anal entirely similar to dorsal, its insertion opposite front of dorsal. The upper lobe of caudal is nine-tenths as long as the head; the lower lobe is slightly longer than head; the length of middle rays is contained $2\frac{1}{4}$ times in head.

Ventrals midway between snout and caudal base, their length contained $2\frac{1}{6}$ times in length of head. Pectorals two-thirds as long as head, the upper ray broadened.

Vertical fins with small deciduous scales. Scales large, thin, deciduous. Top of head scaly.

B. 12; D. I, 14; A. I, 15; V. 6; P. 12. Scales 7—47—3.

Color very pale olivaceous, silvery below and on the sides of head; fine punctulations on the back, following the rows of scales; snout punctulate; a narrow, distinct, silvery lateral band, its width under dorsal origin nearly half length of eye, becoming much narrower anteriorly and on the caudal peduncle.

The single type specimen is $8\frac{1}{2}$ inches long, and is numbered 26593; it was collected by Mr. Silas Stearns at Key West, Florida.

19. Cyprinodon mydrus, n. sp.

Two specimens, No. 30479, were collected by Silas Stearns, at Pensacola, Florida. This species is most closely related to *C. gibbosus* Baird & Girard, from which it is distinguished by its much larger eye (the diameter of which is considerably greater than the length of the operculum, to which in *C. gibbosus* it is equal); by the greater number of dorsal rays, of which there are 13; by the smaller number of its anal rays, of which there are 29; by the smaller number of its scales, which in the lateral line is 24; by the smaller number of scales in the transverse line, 9 in number; by the longer tail and the greater size of the ventral fin.

The color of this species is silvery, the back being olivaceous, and the sides marked with seven or eight indistinct vertical bands. The scales are large, and their outlines are strongly marked, giving to the fish the appearance of a piece of hammered metal work; hence the specific name, which is derived from μύδρος, "a lump of metal."

DESCRIPTION.—Body short; similar in shape to females of *C. gibbosus*. Mouth small, terminal. Premaxillaries very protractile. Humeral scale scarcely as large as the contiguous scales; one-third as long as the head, and equal to the width of interorbital space.

Origin of the dorsal midway between the tip of the snout and the root of the caudal.

Origin of the dorsal and ventrals equidistant from the tip of snout. Dorsal fin, when depressed, not extending to the caudal. The longest dorsal ray equal to length of head without snout. Length of head is contained 3 times in the standard body length. Height of body is contained $2\frac{3}{4}$ times in the same length.

20. Zygonectes craticula, n. sp.

The types of this species, No. 31439, were obtained in a small branch of Elbow Creek, a tributary of Indian River, in East Florida, by Dr. J. A. Henshall, of Cynthiana, Kentucky, and a second lot, No. 28506, were obtained in July and August, 1880, at Nashville, Georgia, by Mr. W. J. Taylor. The relations of this beautiful species are with that described by Agassiz under the name *Z. dispar*.

Body stout, moderately compressed, especially posteriorly; head moderately broad and flattened above. Interorbital space flat, its width less than half the length of the head and $1\frac{1}{3}$ times the diameter of the eye. The distance between the eyes above is greater than below. Snout rather obtuse, its length equal to that of the eye. The total length of the head is contained $3\frac{1}{2}$ times in the standard length and $4\frac{1}{2}$ times in the total length of the body. Teeth in narrow bands, the outer series in both jaws enlarged and somewhat recurved. Scales moderate; 36 in lateral line, 10 in transverse series.

Fins small; dorsal smaller than anal, and inserted over the 20th scale in the longitudinal series, and slightly behind the anal insertion; somewhat nearer to the end of the tail than to the tip of the snout.

Sept. 29, 1882.

The ventrals and pectorals moderate, the length of the latter almost equaling the height of the body at the ventrals.

Color.—Brilliant, the ground color being yellowish-white with six deep black longitudinal stripes equidistant from each other; broader anteriorly; about as wide as the interspaces, and almost coalescing at the base of the caudal fin, an indistinct stripe of same width on each side of the dorsal line, about midway between the uppermost of the black lines and the center of the body. Cheeks brilliant white, a deep black blotch under and confluent with the eye. Dorsal and anal fins with indistinct blackish longitudinal lines. Sides of certain individuals, apparently males, with eleven or twelve distinct vertical bars, about equal in width to the longitudinal bars, the arrangement of these intersecting stripes suggesting the idea of a gridiron, whence the name "*craticula.*"

Length of the largest individuals about 2½ inches.

Radial formula.—D. 7; A. 9-10. Scales 36—10.

21. Stolephorus perthecatus, n. sp.

A single specimen 3.6 inches long, No. 30483, was collected by Mr. Silas Stearns, at Pensacola, Florida.

DESCRIPTION.—Body not carinated or serrated, somewhat compressed. Height of body contained 5 times in its length without caudal, and about 6 times in its total length. Length of head contained 3⅔ times in standard length. Diameter of the eye greater than length of snout and contained 3⅓ times in length of head. Width of eye equal to that of interorbital space. Snout conical, slightly compressed. Teeth minute in both jaws. Maxillary with acute tip, extending back almost to the gill-opening; toothed to the posterior angle of the straight inferior edge. Gill-rakers rather numerous, the longest two-thirds as long as the diameter of the eye.

Origin of the dorsal fin midway between root of caudal fin and the center of the pupil, and also between the tip of the snout and the end of the middle caudal rays.

Anal fin inserted vertically below penultimate ray of dorsal fin.

Pectorals considerably longer than ventrals and more than half as long as head, their tips falling short of reaching the origin of the ventrals by a distance almost equal to the diameter of the eye.

Ventrals half as long as lower jaw, inserted far in advance of the dorsal, their tips reaching to the perpendicular of the origin of the dorsal.

Axillary sheaths exceedingly large; in the case of ventrals and pectorals almost equaling the length of the fins.

Silvery stripe narrow, one-fourth height of body at the ventrals, not more than half as wide as the eye. Scales in lateral line about 38.

Radial formula.—D. ii, 11; A. i, 16.

22. Conger caudicula, n. sp., Bean.

Conger caudicula Bean, in Goode & Bean, Proc. U. S. N. M., v, 240, name only; also, in Jor. & Gilb., op. cit., 262.

Eye equal to snout, 4½ in head. Lips moderately developed. The posterior nostril is small, on a line with the lower edge of the pupil; the anterior nostril tubular near the intermaxillary symphysis. Upper jaw longer than the lower; the cleft of the mouth extends to the hind margin of the pupil. The patch of intermaxillary teeth subrectangular, scarcely a third as long as the eye; vomerine teeth in a patch one-third as long as the eye, tapering behind. The outer series of maxillary teeth contains 38 close-set, slightly truncate teeth, continued backwards after a slight interruption by six conical teeth which rake forward. In the mandible are about 36 close-set slightly truncate teeth, extended forward by a patch of conical teeth. The teeth in the main rows of both jaws are biserial. The length of the head is contained a little more than 1½ times in that of the trunk and 6 times in total length; the length of the tail exceeds that of the rest of the animal by the length of the head without the snout. The dorsal fin commences over the anterior part of the pectoral. The pectoral is a little more than one-third as long as the head. The width of the gill-opening equals one-half of the length of the postorbital part of the head. The color in the present partially digested condition of the fish is mainly light olive. The vertical fins are well developed; their coloration cannot be made out. The distance between the eyes is less than one-half of their diameter. The lower jaw extends beyond the hind margin of the eye and equals the distance from the end of the snout to the hind margin of the eye; it also equals the greatest depth of the head; the greatest width of the head equals one-half its length without the snout.

The type of the species, number 30709, was sent from Pensacola, Florida, by Mr. Silas Stearns. Its length is 13 inches.

23. Muræna retifera, n. sp.

Body moderately stout, somewhat compressed, its greatest depth equaling two-thirds length of head.

Teeth of upper jaw in two series, the outer series composed of a few short fixed teeth; the inner series of about 10 long compressed teeth, a few of which are depressible, all more or less directed backward, those of the middle of the jaw somewhat larger than those in front; vomer with one or two depressible canines. Teeth of lower jaw similar to those of upper; the large teeth rather shorter and broader, all of them entire; the lateral teeth as large as the anterior ones. The mouth does not close completely. Jaws subequal.

The tubes of anterior and posterior nostril about equal, slightly shorter than eye. Cleft of the mouth contained 2¾ times in length of head, the eye over middle of cleft. Eye 1⅔ in snout. Snout contained 6 times in length of head.

Head contained $2\frac{3}{4}$ times in length of trunk. Tail very little longer than rest of body. Height of body contained 6 times in distance from tip of snout to vent.

Dorsal moderately high, beginning at the middle of the length of the head. Gill-opening small, about as broad as the eye.

The color light brown regularly reticulated with blackish, the reticulations inclosing hexagonal or roundish spots of the ground color irregular in size, larger than the eye. The entire fish is thickly covered with small whitish spots, smaller than the pupil; these spots are smallest and most numerous anteriorly, nearly obsolete on the belly, and present on the inside of the mouth. Gill-opening surrounded by a small blackish blotch, whose diameter is less than twice that of the eye. Angle of the mouth black. Dorsal with about 5 longitudinal blackish lines, which become obsolete posteriorly. Anal with a single narrow blackish stripe extending along its whole length.

The single typical specimen is $20\frac{1}{4}$ inches long. It was collected at Charleston, S. C., by Mr. C. C. Leslie. The museum number is 31393.

24. Sphagebranchus teres, n. sp.

Body terete, moderately elongate, its greatest depth slightly more than two-fifths length of head.

Snout short, moderately pointed, projecting somewhat beyond lower jaw, its length contained $6\frac{1}{2}$ times in length of head, and contains the very small eye $2\frac{1}{2}$ times. Tubes of anterior nostril rather short—shorter than eye; posterior nostrils labial, not tubular. Cleft of mouth 4 in head; the front of eye behind middle of cleft. Teeth small, subequal, in moderate bands on jaws and vomer. Lower jaw rather short and weak. Tongue not free in front. Length of head contained $8\frac{2}{3}$ times in that of trunk. Trunk and tail equal in length.

The distance from the tip of snout to beginning of dorsal is contained $2\frac{2}{3}$ times in length of head. The dorsal is of moderate height, its longest ray slightly less than length of snout. Free end of tail acute, short. Anal well developed, lower than the dorsal. Pectorals minute, pointed at the upper edge of gill-opening, usually shorter than eye.

Gill-openings vertical, the length of one slit slightly more than breadth of isthmus, about equal to length of snout.

Lateral line distinct, the pores well separated, extending forward in a curve above the opercular region. Head with no conspicuous pores.

Color uniform, clear brown, paler below, whitish on the head. Front of head somewhat mottled. Fins all pale, without dark margins.

The museum possesses three specimens, number 31457, ranging in length from $18\frac{1}{2}$ inches to $21\frac{1}{4}$ inches; they were collected in West Florida many years ago, by Kaiser and Martin.

LETHARCHUS, new genus, OPHISURIDÆ.

This genus agrees with *Sphagebranchus* in most respects; it lacks, however, an anal fin; the anterior nostrils are not tubular, and the gill-openings are almost horizontal; the dorsal begins on the head; the tongue is largely free in front.

25. Letharcus velifer, n. sp.

Body rather robust, somewhat compressed; its greatest depth a little more than two-fifths the length of head. Head large, abruptly tapering anteriorly; snout very slender and pointed, projecting considerably beyond the lower jaw; its length contained nearly 10 times in that of the head, and equaling a little more than twice the diameter of the very small eye. Nostrils with nasal tubes rudimentary, posterior nostril labial, anterior under the tip of snout. Cleft of mouth, from tip of snout, 4 in head. Eye nearer tip of snout than angle of mouth. Lower jaw short and weak. Tongue short, free in front. Teeth small, pointed, subequal, in narrow bands on jaws and vomer.

Length of head contained $6\frac{1}{2}$ times in that of trunk; head and trunk equal $1\frac{4}{7}$ times length of tail.

The dorsal is unusually high, its height at the nape equaling distance from tip of lower jaw to angle of mouth; it begins at the end of the first third of the head. Free end of tail rather sharp. The anal fin is wanting or represented by a minute fold near the end of the tail. No trace of pectorals.

Gill-openings large, subinferior, oblique, convergent anteriorly; their length more than 3 times the breadth of the isthmus and equal to that of the lower jaw.

Lateral line very distinct, extending forward in a broad curve over the opercular region to below beginning of dorsal. Four conspicuous pores on each side of lower jaw, three behind each eye, three at the nape in front of dorsal, one on top of the head, and four on each side of upper part of snout, besides a few smaller ones about the lips.

Color dark brown, slightly mottled with darker, not paler below. Head paler than body. Dorsal fin pale below, with a broad blackish margin.

There are four individuals of this new fish, which were collected in West Florida by Kaiser and Martin; the types are numbered 31458; they vary in length from 15 inches to 18 inches.

UNITED STATES NATIONAL MUSEUM,
Washington, August 3, 1882.

DESCRIPTION OF A NEW SPECIES OF GOBY (GOBIOSOMA IOS FROM VANCOUVER'S ISLAND.

By DAVID S. JORDAN and CHARLES H. GILBERT.

Gobiosoma ios, sp. nov. (No. 29672.)

Head $4\frac{1}{5}$ in length to base of caudal; depth $6\frac{2}{3}$. D. VI–15; A. about 12.

Body comparatively long and slender, moderately compressed, the back not elevated. Head long and low, rather pointed anteriorly; the profile not at all convex; the premaxillaries projecting well beyond the front of the snout. Mouth very large, oblique, the jaws subequal, or the lower slightly projecting; maxillary extending far beyond the eye to nearly opposite the middle of the cheek, its length being a little more than half head. Teeth in moderate bands, slender, the outer series moderately enlarged. Eyes large, placed close together, as long as snout, about 4½ in head.

Body entirely scaleless. Fins all somewhat mutilated, so that the numbers of fin-rays are not readily ascertained, especially in the anal. Dorsal spines very slender and flexible; base of soft dorsal forming about two-fifths length of body; the fin well separated from the spinous dorsal. Caudal rather short, its tip apparently convex. Anal fin long. Pectorals and ventrals mutilated, apparently of moderate length.

Color light olivaceous; back, sides, and upper fins speckled with dark olive; caudal with 3 or 4 dark olive cross-bars; head with some dark markings; lower fins pale.

The type is a female specimen 2 inches in length, full of nearly ripe ova. It was obtained from the stomach of a specimen of *Hexagrammus asper*, captured by the writers in Saanich Arm, on the eastern shore of Vancouver's Island, in June, 1880. The specimen has been somewhat injured by the process of digestion, but all the distinctive characters can be readily made out. Its slender body and large mouth distinguish it at once from most species of the genus.

INDIANA UNIVERSITY, *August 10.*

DESCRIPTIONS OF NEW SPECIES OF REPTILES AND AMPHIBIANS IN THE UNITED STATES NATIONAL MUSEUM.

By H. C. YARROW, M. D. (Univ. Penn.),
Honorary Curator, Department of Reptiles.

Ophibolus getulus niger, subsp. nov.

In a valuable and interesting collection of reptiles, from Wheatland, Indiana, made by Mr. Robert Ridgway in 1881 were three specimens of *Ophibolus* which differ so materially from the ordinary *Ophibolus getulus* that it seems necessary to assign them a position as a subspecies, and the above name is therefore proposed.

DESCRIPTION.—Color entirely black with the exception of the under part of the head; upper and lower labials marked like the typical *O. getulus.* Head plates entirely black, not spotted, and in none of the specimens examined are light central spots on the head-scales to be seen. Verticals, occipitals, and superciliaries more elongated and narrower than in the normal type; frontals and prefrontals about the same. Abdominal scutellæ plumbeous white spotted, not yellow. A peculiarity of the type specimens from which this description is prepared is that the third, fourth, and fifth postabdominal scutellæ are entire, not divided; but this last trait has been noticed in other species of the genus. In the specimen described, on the second and third row of scales are a few sparsely scattered white spots resembling those of *O. sayi*, but there is no approach to regularity, nor is there any indication of a pattern. No. 12149, ad., Wheatland, Ind.; 7 upper labials on both sides, 9 lower on one, 10 on the other; 12 rows of scales, 1 anteorbital, 2 postorbitals; length 4 feet 6 inches.

A younger specimen 3 feet 4 inches long is similar in appearance to the older one, but there seems to be a tendency to a greater display of the white spots on the sides. Color of the back lustrous black; belly dull black, with milk-white maculations. Isolated and minutely punctulated spots on the back show a decided approach to a pattern of coloration as in *O. getulus*. In two specimens of *O. getulus*, Nos. 9109 and 8797, from Marietta, Ga., and Augusta, Ga., the cross-markings of white spots are almost obsolete, like those of the young specimens of *O. getulus niger;* but on the sides the white blotches are large and strongly defined, which is not the case in *O. getulus niger*. The heads, too, are yellow spotted. Mr. R. Ridgway and Mr. L. Turner inform me that the subspecies described is quite common in both Indiana and Illinois, and that it has doubtless replaced the normal *O. getulus*, which in those States has not been collected by either of these gentlemen.

Ophibolus getulus eiseni, subsp. nov.

A number of specimens of *Ophibolus getulus boyli* have been found in a collection of reptiles made by Mr. Gustav Eisen, at Fresno, Cal., and among them are three specimens differing so materially from the type that it is proposed to name them provisionally for their discoverer.

DESCRIPTION.—Plates of the head similar to those of *O. getulus boyli :* 1 anteorbital, 3 postorbitals, 7 upper labials, 10 lower labials ; 23 rows of scales on body ; difference, so far as the head is concerned, being an increase in number of postorbitals and lower labials.

Color markedly different from the type of *O. getulus boyli*. The first white annulus, three scales wide, commences 11 scales posteriorly to the occipitals; there is then an interval of 5 scales to second white annulus, another interval of 5 scales to the third white annulus, which instead of passing down towards the abdomen, expands at the base, and joins an oval ring 8 scales wide transversely to the body, and 8 scales long posteriorly. Eight scales behind this ring a white line 1½ scales wide commences, which extends the length of body to opposite the 207th abdominal scale, or about the 30th from the anus. This line is absolutely continuous, but breaks off into annuli at the place mentioned. There are then a triangular white blotch, 3 annuli, another blotch, and the dorsal line commences again and ends at tip of tail. At a distance of 4 scales below this dorsal line, on both sides, are indications of lines, and near the borders of the mentioned scales are broken and obsolete whitish blotches.

Color of head and upper part of body pitchy lustrous black; middle, lower third, and tail blackish brown. This subspecies placed side by side with *O. getulus boyli* presents a very different facies. The head and neck resemble *O. boyli* in markings and coloration, but the appearance of the body and tail is entirely different from any North American serpents with which we are familiar. It is hoped other specimens may come to hand and establish the validity of the subspecies.

Number.	Collector.	Date of collection.	Locality.
11787	G. Eisen	1879	Fresno, Cal.
11788	G. Eisen (type)	1879	do.
11744	G. Eisen		do.

Ophibolus getulus multicinctus, subsp. nov.

This name is proposed for a subspecies collected by Mr. Gustav Eisen, at Fresno, Cal.

DESCRIPTION.—Smaller in size than *O. getulus boyli*, to which the coloration gives it a similarity of appearance. Head smaller and more elongated than *O. getulus boyli;* neck compressed, scales in twenty-three rows, smooth and lanceolate. Rostral wider than broad, postfrontals very large; two nasals, nostril between; one loral; anteorbitals one; postorbitals two; vertical elongated without angles on sides, resembling an inverted cone; occipitals longer than broad, having each a small scale at the angular basal end. Upper labials 7, lower labials 9. Eye above notch between third and fourth upper labials. Postabdominal scutella entire, caudal all divided. There are 49 black bands from occipitals to end of tail, the 42d opposite the anus.

General color of body dirty white, the borders of the sides being brown. Upper part of head as far posteriorly as last third of occipitals, pitchy black. Behind this commences a white band extending and expanding on each side of the head, taking in the last two upper labials, passing completely around. Posterior to this are a black band six scales wide, and a reddish-brown band 4 scales wide. From the head, posteriorly, the black bands increase in width, being $5\frac{1}{2}$ scales wide on the middle of the body; the white bands are here narrower embracing 3 scales only. In many of the black bands there is a tendency at the bases to split up into reddish-white blotches, and one or two of them notably near the head are almost entirely divided by the running upwards of the blotches.

Number.	Collector.	Date of collection.	Locality.
11753	G. Eisen	1878	Fresno, Cal.

The National Museum has been very fortunate lately in receiving several collections containing a number of species which have long been desired. Among these collections is one made by Mr. L. Belding, near La Paz and Cape Saint Lucas, Lower California, and this contains not only many desiderata, but several new species, now to be described.

Bufo beldingi, sp. nov.

DESCRIPTION.—Head broader than long, muzzle acuminate and projecting. Canthus rostralis indefinite. Superciliary ridges small. Vertical gutter broad and small. Eyes very small, almost concealed by heavy overhanging lids which are densely tuberculated. Tympanum very small, one-half the size of parotid, which is subcircular and tuberculated. Skin smooth except on flanks. Toes a little more than two-thirds webbed, shovel very small and light colored. Two carpal tubercles, external large, both oval; one small rounded tubercle. Color in old specimens, bluish gray, darker on sides, with orange-colored tubercles. Legs banded with same color as on sides. Belly yellowish white, with bluish spots near insertion of both arms. In young specimens the color of the back is yellowish gray, the sides being darker, the tubercles being bright orange. Resembles somewhat *Bufo microscaphus*, in general outline, but the coloration is very different. Named in honor or Mr. L. Belding, the collector of the specimen.

Number.	Collector.	Locality.	Date of collection.	Number of specimens.
12660	L. Belding	La Paz, Cal	1882	6 (type).
1267	......do	......do		4

Crotaphytus copeii, sp. nov.

DESCRIPTION.—Head broader and longer than *C. wislizeni*. Superciliary ridges well developed. Anterior border of auditory aperture with one, two, or three larger scales than the surrounding ones. Scales anterior to orbits, and posterior to nostrils, on upper surface of head, larger than elsewhere. Scales on gular fold larger than those anteriorly or posteriorly. Upper and lower labials fifteen each to angle near base of jaw. Infraorbital chain consists of four plates, the second very large. Femoral pores large and distinct. First phalanx of hind leg extended reaches angle of jaw. Color dark gray, maculated with dark brown circular spots, each having a lighter center. Anterior to the lower extremities the spots become rhomboid in shape, and on the tail are oval. The head is densely and minutely punctulated with black spots. Belly white. This species is to be compared with *C. wislizeni*, from which it differs in certain particulars, the coloration being entirely different from any of the known species of *Crotaphytus*.

Number.	Collector.	Date of collection.	Locality.
12663	L. Belding	1882	La Paz, Cal. (type).

Uta elegans, sp. nov.

DESCRIPTION.—Dorsal scales smaller than ventral, carinated. Supraorbitals five, with one or two very much smaller ones anteriorly. The rows of large submental scales terminate in two or three sharply-pointed ones at anterior order of auditory aperture. Femoral pores fifteen. A sharp ridge of three scales runs from anterior margin of orbit to nostril. Gular fold bordered with large scales. Color greenish blue, light spotted. Seven oblong transverse black blotches from nape of neck to thighs. Under part of head indigo blue, with bright yellow markings near the jaws. Between auditory aperture and posterior to the axilla, an irregular series of cadmium yellow spots; posterior to this an indigo-blue blotch bordered with yellow above. Abdomen light indigo blue; tail unspotted, but with indication of dark bands on upper surface. This species in life must be very brilliant in coloration, resembling somewhat *Uta schotti.*

Number.	Collector.	Locality.	Date of collection.	Number of specimens.
12666	L. Belding.......	La Paz, Cal...........................	1882	11 (type).
12668	.....do	......do	....	4

Sceloporus rufidorsum, sp. nov.

This beautiful and characteristic species was discovered at San Quentin Bay, California, by Mr. L. Belding, who has forwarded other specimens of it from La Paz and Cerros Island.

DESCRIPTION.—Scales of dorsal region strongly carinated, as large as those of *S. clarki zoste romus*, in twelve rows between insertion of upper extremites. Abdominal scales smaller than labials. Cephalic shields not carinated, but slightly tuberculated. Prefrontal broader than long. Superoculars in three series, not in immediate contact with the superciliary series. Abdominal scales finely denticulate. Scale of base of tail larger than upon any other part of the body. Femoral pores fifteen. Color above, on dorsal ridge, light reddish brown, which in some specimens gradually fades towards the lateral region, and which in others is confined to three scales in width. Posterior to the upper border of the auditory orifice a light yellow line is seen, which extends seven scales backwards, and turns downwards at a right angle, continuing until the shoulder is reached; anterior to this, and in the angular space thus formed, is a patch of deep indigo blue. Sides of body and abdomen same color, many of the scales being spotted with malachite green. This color terminates abruptly at the line of femoral pores. Tail bluish brown above, bluish white beneath. This species is to be compared with *S. clarki zosteromus*, from which it differs principally in coloration.

Number.	Collector.	Locality.	Date of collection.	Number of specimens.
11981	L. Belding..................	San Quentin Bay, Cal..................	1882	Type.
12667	......do	La Paz, Cal......	....	5
11971	......do	Cerros Island, Cal	....	2

Phrynosoma douglassi pygmæa, subsp. nov.

In 1878 Mr. H. W. Henshaw forwarded to the National Museum, from the vicinity of Des Chutes River, Oregon, a number of horned lizards, which, though adults, are smaller than any known species of *Phrynosoma*. In 1881 Capt. Chas. Bendire, U. S. A., forwarded from Fort Walla Walla, Wash, Ter., the same species. A number of specimens have been found in the National Museum collection of reptiles from Fort Steilacoom. While resembling *P. douglassi* in many particulars, still there are many dissimilar characters, and the name is proposed as given above. Head more elongated and less flat above than in *P. douglassi*, superciliary ridges more strongly marked. Occipital and temporal spines, considering size, more acute and longer.

Body almost circular when viewed from above, not so long as in *P. douglassi;* limbs small in proportion to size, hind limbs extended, almost reaching axilla. Inframaxillary series of scales eight in number, not nine as in *P. douglassi*, separated from lower labials by two rows of subcircular scales, in each of which a well-developed pore may be seen. Femoral pores very minute.

Color above dark gray, with a double series of six black blotches, posteriorly margined with light gray. Chin and upper portion of breast minutely punctulated with black. The largest specimen, number 10918, from Fort Walla Walla, is from tip of tail to end of nose $3\frac{1}{10}$ inches in length, $1\frac{1}{2}$ inches in width across belly.

Number.	Locality.	Collector.	Date of collection.	Number of specimens.
10918	Fort Walla Walla, Wash............	Capt Chas. Bendire, U. S. A.	1878	5
11473	Des Chutes River, Oreg.............	H. W. Henshaw............	1878	2
11945	Oregon................................	......do	....	3
9199	Fort Steilacoom......................	......do	....	3

UNITED STATES NATIONAL MUSEUM,
Washington, August 14, 1882.

CONTRIBUTION TO THE MIOCENE FLORA OF ALASKA.

By L. LESQUEREUX.

The Miocene flora of Alaska is partly known by a memoir of Heer, published in the second volume of his Arctic Flora. The memoir was prepared from specimens collected by M. Furuhjelm, of Helsingfors, Finland, partly in the island of Kuiu, in the vicinity of Sitka, partly at Cook's Inlet, near the peninsula of Aliaska. The plants described by Heer, representing 56 species, are of marked interest by their intimate relation with those of Atane, in Greenland, on one side, and with those of Carbon, in Wyoming and of the Bad Lands in Nevada, on the other. They compose a small group which supplies an intermediate point of comparison for considering the march of the vegetation during the Miocene period from the polar circle to the middle of the North American continent, or from the thirty-fifth or fortieth to the eightieth degree of latitude. The remarkable affinity of the Miocene types, in their distribution from Spitzbergen and Greenland to the middle of Europe, had already been manifested by the celebrated works of Heer. But the Alaska flora has for this continent the great advantage of exposing, in the Miocene period, the predominance of vegetable types which have continued to our time and are still present in the vegetation of this continent.

To what was known until now of the Alaska flora a valuable addition has been procured by the collections made for the Smithsonian Institution by Dr. W. H. Dall, of the Coast Survey. A large number of finely preserved specimens of fossil plants were procured from Alaska and its vicinity—Coal Harbor, Unga Island, Shumagins (south side of Aliaska); Chugachik Bay, Cook's Inlet; and Chignik Bay, Aliaska Peninsula. In this valuable collection, which was intrusted to me for examination, I have found a number of species, already described by Heer, from Alaska, a few others described already from the Miocene of Greenland or of Europe, but not yet known from Alaska, and some new species. These last are described below with the enumeration of those described already, but not yet known in the flora of Alaska.

DESCRIPTION AND ENUMERATION OF SPECIES.

CRYPTOGAMEÆ.

EQUISETACEÆ.

Equisetum globulosum, sp. nov.

Rhizoma slender, thinly lineate, flexuous or rigid, distantly articulate, bearing simple opposite globular tubercles, more or less wrinkled by compression.

The branches from 1 to 6^{mm} in diameter, irregularly striate, straight, or flexuous, distantly articulate, bear at the articulations, simple opposite, globular appendages somewhat like those of *Physagenia Parlatorii* Heer (Fl. tert. Helv. 1, p. 109, pl. XLII, figs. 2–17), but globular and generally simple, very rarely appendiculate in two. These remains are much decomposed by maceration, and fragmentary, none of them continuous, and all without trace of sheath.

FILICES.

Osmunda Torelli Heer, Mioc. Fl. of Sakhalin, p. 19, pl. 1, f. 4, 4b.
Pecopteris Torelli Heer, Fl. Arct., 1, p. 88, pl. 1, f. 15.
Hemitelites Torelli Heer, *ibid.*, II, p. 462, pl. xl, figs. 1–5 a; lv. f. 2.

This species is represented by a very large number of specimens, mostly separate leaflets embedded in bowlders of carbonate of iron.

Most of the leaflets are simple, not lobate, oblong or ovate-lanceolate, entire or merely crenulate on the borders by the impressions of the veins. These leaflets are rarely preserved entire; the borders are often lacerated; they vary from 3½ to 6ᶜᵐ long and 1 to 2½ᶜᵐ broad. They evidently represent leaflets of an *Osmunda*.

Hab.—Coal Harbor, Unga Island.

CONIFERÆ.

Thuites (Chamæcyparis) Alaskensis, sp. nov.

Branchlets alternate, flattened, oblique; leaves imbricate on four ranks, the facial squamiform compressed, broadly rhomboidal quadrate, slightly narrowed to the base, inflated on the borders and in the middle toward the apex; the lateral flattened by compression, exposing half their face, and thus triangular, exactly filling the space between the base and the top of the facial leaves, all thick.

I find no distinct relation for this plant except with *Thuites Meriani* Heer. Fl. Arct., III, p. 73, pl, XVI, figs. 17, 18, a cretaceous species differing by the facial leaves ovate, narrower towards the apex.

Hab.—Same as the preceding.

MYRICACEÆ.

Comptonia cuspidata, sp. nov.

Leaves long, linear or very gradually tapering upwards to a terminal narrowly elliptical lobe, pointed or apiculate by the excurrent medial nerve, pinnately lobed; lobes coriaceous, convex, subalternate, free at base, irregularly trapezoidal or obliquely oblong, inclined upwards and sharply acute or cuspidate; primary nerves two, or three in the largest lobes, oblique, the upper curving in ascending to the acumen and branching outside, the lower parallel and curving along the borders, anastomosing with branches of the superior ones, generally separated by simple secondary short nerves.

Comparable to *Comptonia acutiloba* Brgt. and other European Tertiary species of this group, but distinct from all by the large cuspidate lobes turned upwards, &c.

Hab.—Same as the preceding.

Comptonia præmissa, sp. nov.

Leaves long, linear in their whole length, 5 to 10ᶜᵐ long, 12 to 15ᵐᵐ broad; deeply equally pinnate-lobate; lobes very obtuse or half round cut to the middle and slightly decurring in their point of connection, the terminal very obtuse; nervation obsolete, substance somewhat thick but not coriaceous.

The species has its greatest affinity to the living *Comptonia asplenifolia* Ait. It also appears related to *C. rotundata* Wat., as described by Schimper, Pal. veget., II, p. 555, a species known to me only by its description.

Hab.—Chignik Bay, Aliaska.

BETULACEÆ.

Betula Alaskana, sp. nov.

Leaves small, round in outline, rounded or truncate at base, deeply obtusely dentate all around, except at the base, turned back or recurved on a short petiole; medial nerve distinct, the lateral obsolete; catkins short. cylindrical, oblong or slightly inflated, in the middle erect.

Except that no glands are perceivable upon the stems, this species agrees in all its characters with *Betula glandulosa*, Michx. of Oregon. I consider it as identical.

Hab.—Chignik Bay, Aliaska.

Alnus corylifolia, sp. nov.

Leaves large, broadly ovate, rounded or cordate at base, acuminate or narrowly oblong-ovate, doubly dentate on the borders, primary teeth large, distant more or less sharply denticulate on the back, secondary nerves oblique, parallel, the lower pairs more open, all generally simple, except a few thin tertiary branches near the borders, passing to the points of the teeth; surface smooth; fibrilles rarely distinct; petiole comparatively long.

Resembles *Corylus M, Quarryi* Heer, differing by the smooth surface, the nervilles obsolete, the nerves not branching, the long petiole, &c.

Hab.—Chugachik Bay, Cook's Inlet, Alaska.

CUPULIFERÆ.

Carpinus grandis, Ung.

In numerous specimens.

Hab.—Same as the preceding. Described also from Greenland by Heer.

Fagus Deucalionis Ung.

The collection has a single specimen of this species. Heer has described it from Greenland.

Hab.—With the preceding.

Quercus Dallii, sp. nov.

Leaves subcoriaceous, oblong-lanceolate, acuminate, rounded or sub-cordate at base, 6 to 12^{cm} long, 4 to 8^{cm} broad, deeply equally undulate or obtusely dentate; lower lateral nerves nearly in right angle, branching, the others oblique, generally simple, all craspedodrome.

The secondary nerves are more or less distant according to the size of the leaves, being generally 14 pairs.

The relation of this species is to both *Q. Grœnlandica* and *Q. Olafseni* Heer, two species from Greenland, from which this one especially differs by the rounded or subcordate base and the lower nerves nearly in right angle. Except that the leaves are much larger, they may also be compared to *Paullinia germanica* Ung. (Sillog. plant., III, p. 52, Pl. XVI, fig. 8), and are possibly referable to this genus, mostly represented now in tropical America.

Hab.—Cook's Inlet, Alaska.

SALICINEÆ.

Salix Raeana Heer., Fl. Arct., I, p. 102, Pl. IV., figs. 11–13; XLVII, fig. 11.

Species described by Heer from specimens of Greenland.
Hab.—Cook's Inlet.

Populus Richardsoni Heer., U. S. Geol. Rep., VII, p. 177.

Species abundantly represented in the Miocene flora of Greenland and Spitzberg.
Hab.—Chignik Bay.

Populus arctica Heer., U. S. Geol. Rep., VII, p. 178.

Has the same distribution as the preceding, and is still more common in the Miocene of Greenland and North America.
Hab.—With the preceding.

ULMACEÆ.

Ulmus sorbifolia Ung., Schossnitz, Fl., p. 30, Pl. XIV, fig. 10.

Leaf oblong, with borders parallel in the middle; taper pointed or acuminate; secondary nerves numerous, close, parallel, half open (angle of divergence 60°), generally forking near the doubly dentate-crenate borders; primary teeth blunt, turned upwards.

The base of the leaf is destroyed. The preserved part is $4\frac{1}{2}^{cm}$ long, 2^{cm} broad, with 18 pairs of deeply marked secondary veins.

The species, which is not mentioned in Schimper's Veget. Paleont., is closely allied to *U. plurinervia* Ung., which has been found in Alaska.
Hab.—Chugachik Bay, Cook's Inlet.

NYSSACEÆ.

Nyssa arctica? Heer., Fl. Arct., II, p. 477, Pl. XLII, fig. 12 c; L. figs. 5, 6, 7.

The fruit which I refer to this species is of the same size and form as fig. 6, *l. c.*, but less distinctly striate lengthwise; the cross-wrinkles slightly marked by Heer, in fig. 6 b. (enlarged), being as prominent as the longitudinal striæ. The fruit somewhat deteriorated by maceration most probably represents the same species abundantly found in Greenland.

Hab.—Unga Island, Shumagin group, Alaska.

DIOSPYRINEÆ.

Diospyros anceps Heer., Fl. Tert. Helv. III, p. 12, Pl. CII, figs. 15–18; V, Sybir. Fl., p. 42, Pl. XI, fig. 7.

The leaves agree by all the characters with Heer's species especially similar to figs. 16, 17 of Fl. Helv. *l. c.*, the smaller leaf being of the same size as fig. 16. The other specimen, which is fragmentary, is much like fig. 7 of the Siberian Fl. The leaves are broader than in *D. Alaskana;* the lateral nerves more distant, &c.

Hab.—Cook's Inlet.

ERICINEÆ.

Vaccinium reticulatum, Al. Br., Heer., Fl. tert. Helv., III, p. 10, Pl. CI, fig. 30.

Leaves petiolate, oval, very entire, obtuse at the apex, narrowed at the base in rounding to a short alate petiole; lateral nerves open, few, interspersed with tertiary shorter ones; surface deeply reticulate.

The leaves from their size, shape, and nervation correspond with those described by Heer, *l. c.*, the only difference being that one of the leaves I had for examination, the largest, has the short petiole alate. In fig. 30 of Heer, the petiole seems also bordered in the upper part by the decurrent base of the leaf, but the appearance is less distinct. Moreover, there are other leaves in the same collection of Mr. Dall which are smaller and with naked petiole. The difference is not therefore of specific value.

Hab.—Cook's Inlet.

CORNEÆ.

Cornus orbifera Heer., U. S. Geol. Rep., VII, p. 243.

The specimen referable to this species has the lateral nerves curving inward along the borders, anastomosing with the upper ones by nervilles in right angles, as in Heer, Fl. tert. Helv., pl. CV, fig. 16. Heer has also described the species from Spitzbergen specimens.

Hab.—Cook's Inlet.

MAGNOLIACEÆ.

Magnolia Nordenskiöldi Heer., Beiträge zur foss. Fl. Spitzb. (Fl. Aret. IV), p. 82, Pl. XXI, fig. 3; XXX, fig. 1.

Leaves large, thickish, oval, obtuse, entire, emarginate, or shortly auriculate at base; secondary nerves distant, curved in traversing the blade, forking near the borders.

From the numerous well preserved specimens of this beautiful species, I have been able to complete the diagnosis of Heer, made from too fragmentary leaves. The leaves are longer than those of *M. ovalis,* Lesqx., to which Heer compares this species, and also sub-auriculate at base or emarginate; the surface is rugose, crossed in right angles to the

EXPLANATION OF PLATES VI-X.

PLATE VI.

FIGS. 1, 2. *Equisetum globulosum*, sp. nov., p. 444.
FIGS. 3, 4, 5, 6. *Osmunda Torelli*, Heer., p. 444.
FIGS. 7, 8, 9. *Thuites (Chamæcyparis) Alaskensis*, sp. nov., p. 445.
FIGS. 10, 11, 12. *Comptonia cuspidata*, sp. nov., p. 445.
FIG. 13. *Comptonia præmissa*, sp. nov., p. 445.
FIG. 14. *Betula Alaskana*, sp. nov., p. 446.

PLATE VII.

FIGS. 1, 2, 3, 4. *Alnus corylifolia*, sp. nov., p. 446.
FIGS. 5, 6. *Carpinus grandis*, Ung., p. 446.

PLATE VIII.

FIG. 1. *Fagus Deucalionis*, Ung., p. 446.
FIGS. 2, 3, 4, 5. *Quercus Dallii*, sp. nov., p. 446.
FIG. 6. *Salix Raeana*, Heer., p. 447.

PLATE IX.

FIG. 1. *Populus Richardsoni*, Heer., p. 447.
FIG. 2. *Populus arctica*, Heer., p. 447.
FIG. 3. *Ulmus sorbifolia*, Ung., p. 447.
FIG. 4. *Elæodendron Helveticum*, Heer., p. 449.

PLATE X.

FIGS. 1, 2. *Diospyros anceps*, Heer., p. 448.
FIGS. 3, 4, 5. *Vaccinium reticulatum*, Al. Br., p. 448.
FIG. 6. *Cornus orbifera*, Heer., p. 448.
FIGS. 7, 8, 9. *Magnolia Nordenskiöldi*, Heer., p. 448.

[Proc. Nat. Mus., 1882.]

PLATE VI.

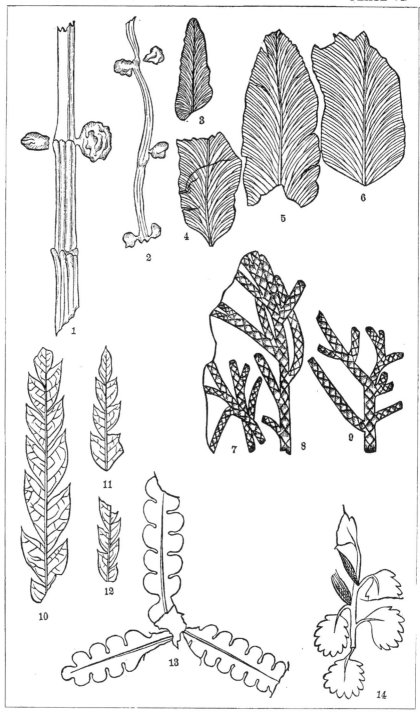

PLATE VII.

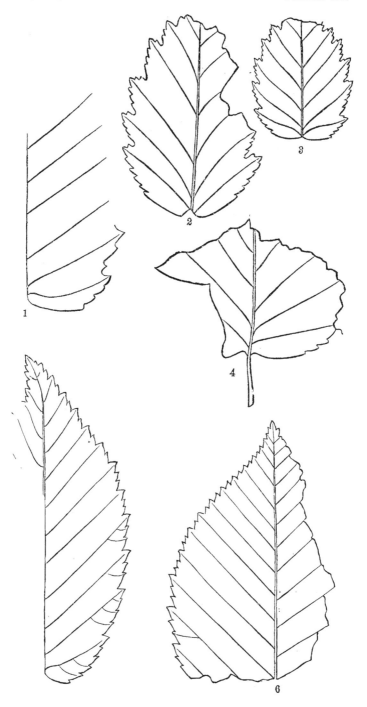

PLATE VIII.

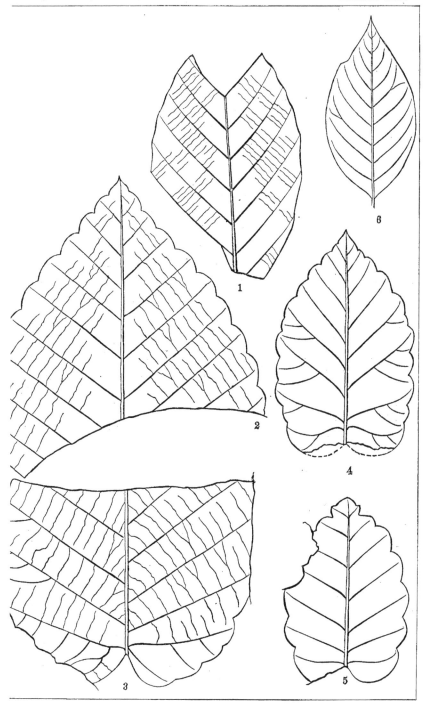

1

2

3

4

PLATE X.

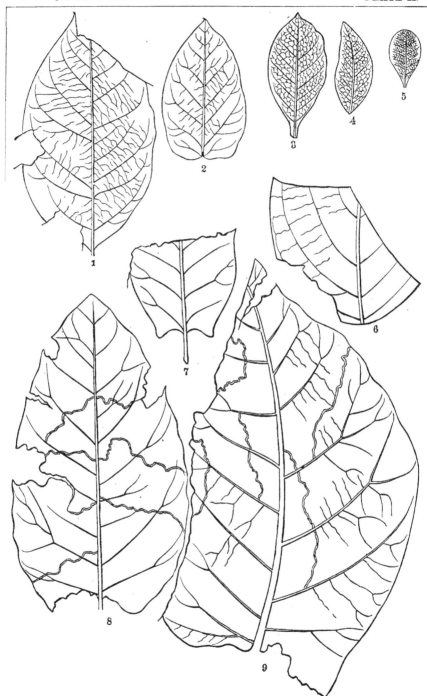

veins by simple or forked nervilles. The two lower pairs of veins are closer than those above. In a leaf of medium size, the two lower pairs of nerves are 8^{mm} distant, while those of the middle are nearly 2^{cm}· The angle of divèrgence in joining the midrib is open, but the nerves are much curved upwards in traversing the blade.

Hab.—Chignik Bay, Aliaska Peninsula, Alaska.

ELÆODENDREÆ.

Elæodendron Helveticum Heer., Fl. Tert. Helv., III, p. 71, Pl. CXXII, fig. 5.

Leaves coriaceous, oval, equally narrowed upwards to a blunt apex and downwards to a short petiole; secondary veins (seven), unequally distant, parallel, except the lowest, which are a little more oblique and ascending higher parallel to the borders; all camptodrome, arched at a distance from the margins, forming a double series of festoons by anastomising branches; surface rugose; borders undulate.

The leaves, according to Heer, are obtusely dentate on the borders, but part of the margin, near the base of the leaf described above, is destroyed, and Heer's fig. 5 loc. cit. shows from the middle upwards exactly the same undulations as the Alaska specimen. The only difference remarked on the leaf of Alaska is that it is more distinctly narrowed to the petiole. The specimen bears numerous fragments of *Taxodium distichum.*

Hab.—Coal Harbor, Unga Islandd, Shumagin group, south side of Aliaska.

JUGLANDINEÆ.

Juglans Woodiana Heer. Pflanz. v. Vancouver, p. 9, Pl. II, figs. 4–7. Two fragmentary specimens.

Hab.—Chignik Bay.

REMARKS ON THE SYSTEMATIC ARRANGEMENT OF THE AMERICAN TURDIDÆ.

By LEONHARD STEJNEGER.

The group here under consideration, the so-called "family" *Turdidæ,*[*] has given much trouble to those authors who have tried to arrange the genera naturally, and to define their limits distinctly. I do not intend to give here an analysis of their different essays, but as the last, viz, Mr. Seebohm's in the fifth volume of the "Catalogue of Birds in the British Museum," is very radical and opposed to commonly accepted

---

[*] I am not at all convinced that the groups of the *Passeres,* generally called families, are really equivalent to the family groups of the other orders of birds or other vertebrates; but as I am, for the present, unable to take up this question, I have contented myself with the generally adopted nomenclature.

Feb. 13, 1883.

views, I cannot pass it by in silence, inasmuch as the present study may be regarded as a reaction, provoked by the arrangement proposed in the above mentioned work.

It may then be proper to state first, that the definition of the group *Turdidæ* (=Seebohm's *Turdinæ*), given by Mr. Seebohm, seems to be a very proper one, and I think he has therein expressed the only chief character which really indicates the relationship of the birds to be included in this family. The peculiar spotted first plumage of the *Turdidæ* is a very striking feature, and its coincidence with booted tarsi very remarkable. A careful comparison with forms, which, without showing those characters, have at different times been referred to the *Turdidæ*, will convince us that the limits traced by Mr. Seebohm are the only reliable ones, and that the family thus defined is a very natural group, and, indeed, one of the best among the *Passeres*. It is only to be regretted that Mr. Seebohm did not include a few other forms which have the same peculiarities. I may especially allude to the *Myadestinæ*, the position of which will be discussed in full below. His concluding remarks on page 2 seem, however, to indicate that he himself has been aware of this fault.

It is not difficult to foresee that his definition of the family will be heartily accepted by ornithologists, but it is, on the other hand, probable that his peculiar generic arrangement will meet common opposition.

Mr. Seebohm states (p. viii) that he has "been obliged to fall back upon color or pattern of color as the only character which indicates near relationship."

To see how he has carried this out, let us first take his genus *Geocichla*, of which he says (p. 148), that it "on the whole must be considered one of the best defined of the family *Turdidæ*." One needs only to compare his plates X and XI in order to be convinced that he does not mean the general coloration of the bird, as the two plates represent birds, the general coloration of which is, at least, as different as that of a Robin and a Rock-Thrush, which he refers to different genera. The diagnosis of the genus shows, also, that special importance is attached only to the pattern of the under surface of the wing, these birds "having the outside web of all the secondaries and of many of the primaries white, occasionally tinted with buff, but abruptly defined from the brown of the rest of the quills," and the "axillaries parti-colored, the basal half being white and the terminal half black, slate-gray, or brown. Most of the wing-coverts are similarly parti-colored, but the relative position of the colors is reversed, the white portion being on the terminal half." But these characters do not hold good in all species, as Mr. Seebohm himself indicates. There are several exceptions, or, as he calls them, "aberrant species," which have the "axillaries and under wing-coverts uniform in color," and there are several species which he refers to other genera, but which possess the above characters at least as well determined as his "aberrant" *Geocichlæ.*

Thus, besides the cases mentioned by Mr. Seebohm himself, "*Turdus*" *pallasii* has the light color on the inner web "very abruptly defined;" his *Turdus albiventer* likewise has "the pale portions of the inner webs of the quills greatly developed and very abruptly defined," and so further on. But he gives no characters by which these aberrant species (aberrant of both genera) may be distinguished, and he also gives no reason why he places these aberrant forms in different genera. It would be interesting to know why *Turdus albiventer* is not as good a *Geocichla* as *G. sinensis,* when the pattern of coloration is the only character which has generic value; or, in other words, why he does not place *G. sinensis* in another genus. May there not, perhaps, be other characters of more importance and generic value, and which indicate a nearer relationship than the coloration? But one ought not to suspect that, as Mr. Seebohm, in another place, retains a species in the genus *Catharus,* for the reason that its " general style of coloration" so closely resembles the other species of that genus, although it "is a typical *Erithacus* so far as what are called structural characters are concerned." Here, again, "the general style of coloration" is the only generic character of value! In the one genus it thus apparently has no value at all, while in the other it is the only important one! It is curious to see that Mr. Seebohm, when neither pattern nor general style of coloration is sufficient to separate two genera, hastily takes refuge in a structural character; for example, p. 362, and p. 334, and especially the " Key," p. 146, *d'''*, *e'''*, *f'''*, and *g'''*. Curiously enough, he separates two genera, in either of which several species are simply inseparable as to general style and pattern of coloration from certain ones of the other genus, and yet such similarly colored species, he says, are typical members of the other genus, so far as structural characters are concerned! How, then, will Mr. Seebohm tell *Catharus gracilirostris* or *occidentalis* from "*Erithacus*" *luscinia* and *philomela?* In coloration *C. occidentalis* and *E. luscinia* agree so closely that it would be very difficult to separate them even specifically, if we had no structural characters, and he expressly makes the statement that *C. gracilirostris* is, as to structure, a typical *Erithacus.* We will attempt, by his " Key," to unravel this intricate question. In this the distinctive marks of each genus are given as follows:

a'''. General color of under parts slate gray, shading only into brown
    or white. Legs never black.. 5. CATHARUS.
b'''. Throat generally brilliant in color and frequently in violent con-
    trast to the cheeks; if not, legs pale .... 6. ERITHACUS.

Unfortunately, the "key" is of no use; both the species of *Catharus* and *Erithacus* mentioned above have *not* a brilliant colored throat; and further, the legs are in both pale, and never black. If no structural differences are to be found, the separation of the two genera would, in spite of Mr. Seebohm's statement, be hopeless; but, luckily, they may be distinguished by very recognizable and distinct characters; the different construction of the wing, in particular, rendering their separation easy.

A further examination of the birds included in the genus *Geocichla* shows that in several species the sexes are alike, while in others they are very differently colored; but it seems that Mr. Seebohm attaches no importance to this as a character of generic value. We cannot but indorse this view, being much surprised, however, to find that he makes this difference in coloration between the two sexes the chief, not to say the only, distinctive mark of the genera *Turdus* and *Merula;* in some instances carried out to the utmost, while on the other hand several species are included in *Merula* which have the sexes colored alike, and other species showing not unimportant differences between male and female are placed within the genus *Turdus.* In the one case the character is the only valuable one; in the other, again, it has no value at all!

Having adopted the singular theory that structural characters did not indicate natural relationship, while pattern of coloration was sufficient for the purpose, the author has given us a right to expect as the result of his investigations a more natural arrangement than any preceding it. Unfortunately, however, it must be said that he has not succeeded therein, for his own theory is so often and so violently ignored that most of his genera are quite void of definite limits.

It is hardly likely that anyone, be he ever so deeply enamored with the coloration theory, will consider it as according with natural affinities to arrange *T. nævius, wardii, pinicola*, and *sibiricus* together in one, and *T. maranonicus, dryas*, and *pilaris* in another subgeneric group, when, at the same time, such birds as *T. pilaris* and *torquatus* were separated generically. And as the natural relationship in these cases has been violated, so also have they in many others.

It being thus evident that the new mode of defining the genera does not lead to a more natural system than the rejected structural characters, it is to be doubly deplored that the generic groups resulting from its application are so indefinite and their limits so unstable, that Mr. Seebohm (p. 14) needs to appeal to "the instinct of the ornithological student," when he has not been "able to define the character of each genus." This instinct may in most cases be sufficient to "tell a Chat from a Redstart," but certainly it will be of no use when he shall separate a *Turdus* Seeb. from a *Merula* of the same author. The example of Mr. Seebohm himself proves that this instinct is often misleading.

The coloration and the pattern of coloration may, in many cases, be of very great value as indicating the relationship, but used as a distinctive mark for defining genera in the manner of Mr. Seebohm, who often only takes in consideration the colors of the male, it seems to me to have no scientific value at all.

It is an objection against the theory of coloration that in many genera of birds some species, in their colors, only represent the immature or young state of another species. And as the young and the old birds are frequently very unlike in their coloration, the species thus consequently also look very unlike in their various stages. They may, however, be very closely allied, and often more so, than very similarly colored species. This objection applies also to the matter here under consideration. The first species of *Turdus*, which Mr. Seebohm gives, is *maranonicus* Tacz., from South America. As I have had no opportunity to examine a specimen of this bird, I must content myself with the figure (*Proc. Zool. Soc. Lond.*, 1880, pl. xx). At first sight I was inclined to indorse the view of Mr. Seebohm, and was much perplexed to find a *Turdus* in South America. But, examining the structural characters given in the description, I soon became convinced that the species must belong to *Merula*. I had not to wait a long time before I obtained, to my satisfaction, an interesting proof that this opinion was correct. The same day Mr. Robert Ridgway called my attention to the pl. lxxv. of Sclater and Salvin's "Exotic Ornithology," and pl. xxix. in *Proc. Zool. Soc. Lond.*, 1867, representing the young *Turdus phæopygus* Cab. A comparison with the young bird on Taczanowski's plate shows that these birds are very closely allied and never should be placed in different, even subgeneric, groups. *Merula maranonica* (Tacz.) is evidently an immature bird,* which, perhaps, may later take a plumage more resembling that of the adult *phæopygus*, but I should not be surprised at all if future investigations would prove that it retains the immature-looking plumage also in the adult state. Its place near *M. phæopygus* must, however, be the same in both cases.

There is another work having a very important bearing on the subject to which it is necessary to refer in any dissertation on the arrangement of the American Thrushes, namely, Prof. S. F. Baird's "Review of American Birds." Written sixteen to eighteen years ago it is still the best treatment of the subject extant, and the views expressed therein vindicate their place above more recent essays. And I am glad to say that if I have succeeded in the following arrangement it is due to the most valuable hints which the work above mentioned contains.

As to the limits of the family, I have already remarked that I chiefly agree with Mr. Seebohm. It will, therefore, be perceived that I do not admit the *Miminæ*, which Professor Baird in 1864 placed as a subfamily with the *Turdidæ*. It seems to me that their proper place is near the Wrens, among which they also had been included by him in his work on the Birds of North America (1858). In fact the Mockingbirds are so closely allied to the *Troglodytidæ* that I am inclined to believe that the most natural arrangement would be to include them as a subfamily along with the *Troglodytinæ* within the same family.

---

* I have it on Mr. Lawrence's authority, that Mr. Sclater has determined *T. maranonicus* to be the young of *T. nigriceps*, Jelski, Dr. Stejneger's prediction being thus fully verified.—R. R.

The genus *Cichlherminia* Bp. has especially been regarded as an intermediate link between Thrushes and Mocking-birds. In 1854 Bonaparte divided the genus and made *C. fuscata* the type of the genus *Cichlalopia*, which name as untenable has been changed by Mr. Sclater into *Margarops*. Unfortunately, however, this separation later has been given up,* because the restricted genus *Cichlherminia* (type *herminieri* Lafr.) unquestionably belongs to the true Thrushes, group *Meruleæ*, while on the other hand the genus *Margarops* (including *fuscata*, *densirostris*, and *montana*) as undoubtedly belongs to the *Miminæ*. (See figs. on pages 457 and 476.) By separating and placing these genera in this manner, the limits between *Turdidæ* and *Miminæ* become very trenchant, and the arrangement of the families more natural.

Later investigations have confirmed the doubts expressed by Professor Baird (Rev. p. 410) about the validity of the family *Saxicolidæ*. Dr. Coues in his "Birds of the Colorado Valley" (1878), p. 76, still retains the term, but at the same time he very frankly confesses: "Recognition of the family *Saxicolidæ* is purely a conventional matter, in which most ornithologists tacitly agree to follow each other upon no better ground than that of precedent." Mr. Seebohm (*l. c.*) includes the genera *Saxicola* and *Sialia* among the *Turdinæ*. In fact, the *Saxicolidæ* are so closely allied to the Thrushes that they only can claim recognition as a special group within the same subfamily. Moreover, I have distinguished as a separate group the *Sialieæ*, which have formerly been united with the *Saxicoleæ*, but which certainly differ more from the Chats than from the Thrushes. The fact that I have found it necessary to unite with the Bluebirds a species which hitherto has been regarded as a *Turdus*,† shows where their true relations are to be found; the shape of their legs, tail, bill, their habits, and coloration prohibit their position within the same group as the Chats, showing the necessity of establishing for them a separate group, coequal to the groups occupied by the Thrushes and the Chats. Besides, the group *Saxicoleæ*, which only embraces one American species, viz, *Saxicola œnanthe*, by removing the Bluebirds becomes more natural and homogeneous, including, as I now believe, *Saxicola*, *Pratincola*, *Ruticilla*, etc. I cannot agree with the authors of the Catalogue of Birds in the British Museum, who keep the genus *Pratincola* within the family *Muscicapidæ* (vol. iv, p. 178), although it, on the other hand, may be regarded as a well-defined genus in contradistinction to the statement of Mr. R. Collett. (Chr. Vid. Selsk. Forh., 1881, No. 10, p. 3.)

It will be seen that the following arrangement differs from that of most systematists in separating *Turdus merula* and its allies as a group, *Meruleæ*, distinct from and coequal to the *Turdeæ* and *Luscinieæ*, and in

---

* See Sclater and Salv., Nomencl. Nr. Neotrop, p. 2, and Sharpe, Cat. Birds Brit. Mus., VI, p. 326.

† By Mr. Seebohm, however, included together with other heterogeneous elements among the genus *Geocichla*, subgeneric group *Hesperocichla* (*op. cit.*, p. 151).

including with the latter group the genus *Catharus*, which usually has been placed among the Thrushes.

At first sight it would seem that the *Merulæ* and *Turdeæ* are too closely related to constitute separate groups, the more so as there are few authors who distinguish the species of the two groups even generically. But the trouble of the prior attempts has been that the limit between the two groups has been so traced that each division has contained species really belonging to the other group. Thus, the *Turdus torquatus* has almost unanimously* been regarded as a true *Merula*, closely allied to the type of this genus, only because its color is black. A careful examination shows, however, that the Ring-Ouzel, so far from being an ally of *Merula nigra*, is a near relative of *Turdus viscivorus*, the type of the restricted genus *Turdus*. It is, then, a matter of course that it has been impossible to separate satisfactorily the two groups even generically or subgenerically. But if all heterogeneous elements be removed and put in their proper places, the differences between *Turdeæ* and *Merulæ* become very striking. In fact, the *Merula nigra* is at least as remote from the true Thrushes as is *Erithacus rubecula*, and the adoption of the group *Lusciniæ* (by most ornithologists admitted as family or subfamily on the same reasons as the *Saxicolidæ*) therefore necessitates the establishment of a co-ordinate group embracing the genus most nearly allied to *Merula*.

As has already been remarked, the genus *Catharus* will usually be found placed very near the true Thrushes, especially to the smaller North American species of the genus *Hylocichla*, and Mr. Seebohm goes even so far as to include *Catharus dryas* within the same subgeneric groups, embracing *Hylocichla musica, mustelina, Turdus viscivorus* and *pilaris*, chiefly, or rather only, on account of the dark spots on the under surface. I have found it, however, quite impossible to remove them from the *Lusciniæ* (genus *Erithacus*, Seebohm), with which they agree in the very important character of the structure of bill, wing, and legs, and also in the colors of the plumage. Notwithstanding the *Cathari* point towards the true *Turdeæ*, while many of the old world *Lusciniæ* show a similar tendency towards the *Merulæ*, so that the proper place of the group *Lusciniæ* will be between those two, thus fairly illustrating the gap between *Turdeæ* and *Merulæ*.

In 1866 Professor Baird (*op. cit.* p. 417) established the subfamily *Myiadestinæ* in the following words: "I am decidedly of opinion that, notwithstanding a close resemblance in general appearance, *Myiadestes*

---

* The only noteworthy exception is Prof. J. Cabanis, who, in his "Journ. für Ornith." 1860, p. 161, foot note, says: "*Turdus torquatus* should not be placed with *Merula*, but must, with respect to the shape of bill and wing, remain with *Turdus*."

I find no better place for correcting a very curious mistake in Gray's Handbook of Birds, i, p. 253, in which the subgenus *b* of genus *Turdus* has received the name "*Psophocichla, Heraug.* 1860." The memoir of Cabanis, quoted above, has the heading, "Eine neue Drossel-Gattung, *Psophocichla.* Vom Herausgeber"=a new genus of Thrushes, *Psophocichla*. By the editor, and hence the error.

and *Cichlopsis* should be removed from their usual association with *Ptilogonys*, among *Ampelidæ*, to, or at least very near, the *Turdidæ*, and ·form a subfamily with *Platycichla*. The latter genus is so closely related to *Cichlopsis* as almost to be the same; *Platycichla* forming the link with *Turdinæ* through *Planesticus*, while such species as *Myiadestes unicolor* show the affinities of *Cichlopsis* to *Myiadestes*." But so far as I can detect, Dr. Elliott Coues is the only author who, in his "Birds of the Colorado Valley" (1878), has adopted the view of Professor Baird, including the subfamily *Myiadestinæ* within the family *Turdidæ*. I have been much surprised to find those birds excluded by Mr. Seebohm, who has so nicely pointed out the value of the spotted plumage of the young *Turdidæ*, and of the coincidence of this character with smooth tarsi, and on the other hand to find them treated by Mr. Sharpe under the *Timeliidæ*.* The essential character of this latter group is their ·short and concave wing. But it is evident that the wing of the *Myadestinæ* does not in any respect differ from the structure of the wing of the *Turdinæ*, being rather longer than the average of the latter group, and as flat and straight. The relationship between the *Merulæ* of the true Thrushes, and the *Platycichleæ* of the "Flycatching Thrushes" (Coues) is so close, indeed, that several species, which really belong to the latter group, are usually found—also in the new "Catalogue of the Birds in the British Museum"—included in one of the genera composing the former division.

The earlier placing of these birds within the *Ampelidæ* is only due to their "resemblance in general appearance," and the differences have already been pointed out so exhaustively by Professor Baird, that it is unnecessary to repeat them here. The group will not, however, be ·naturally limited or clearly defined without removing the species *Myadestes leucotis* (Tschudi), which is widely different, from the *Myadestinæ*, being a true member of the *Ptilogonatidæ*. As its characters do not agree with those of any other genus, it will be necessary to make it the type of a new genus.†

It will be seen that in the following arrangement I have attached much importance to the form of the wing. It is certainly true, that in the *Passeres*, the more pointed wings very, often indicate migratorial

---

* As to the latter, it is proper to state that he himself is not content with the place thus attributed to the *Myadestinæ*. Here are his own words (*tom. cit.* p. 368): "The present position of the birds contained in this subfamily is not satisfactory to my mind. * * * Mr. Seebohm has not admitted them into his volume of the 'Catalogue.' I have, therefore, placed them near the Mocking-Thrushes, which they resemble in their power of song."

† ENTOMODESTES, *n. g.*

(ʾΕντομα=insects, ἐδεστής=an eater)
Type—*Entomodestes leucotis* (Tschudi)—

∴ Head without crest. Outer primaries broad, not attenuated nor pointed at end; 1st about half the 2d. Tail graduated, the feathers acute and acuminate at tips, the

habits, while the more rounded wings are oftener found in stationary birds. This fact, however, does not in any way diminish the value of the structural difference as a distinguishing mark, the purpose of which is to indicate the limits of the different genera; nor is it without importance in indicating the affinities of the different forms. In so far as it is connected with the migratory habits of certain species, it probably signifies the simultaneous immigration of those birds into the region to which they now belong, and indicates thus a geographical separation which, during the course of time, cannot have been without influence on their development.

That the more or less rounded or pointed form of the wing has not such an essential importance in regard to the migratorial phenomenon is evident from the general consideration, that not all migratory birds have long and pointed wings. It is also to be remarked, that in general the same species is migratory in some localities, while in other places it is stationary. Finally, we have in the group of birds here under consideration ample opportunity for showing instances which point to quite the opposite direction. So. for example, has *Ridgwayia pinicola*—which certainly is not a migratory bird, and the geographical range of which is remarkably restricted very pointed wings, with the 3d and 4th quills longest, and very short secondaries. We have also the genus *Sialia*, with its unusually lengthened and pointed wings.

On the other hand, the length of the secondaries and of the primary coverts seems to be of very great importance. Nor is their length in any way directly dependent upon the migratory or stationary habits of the birds, though it may certainly be admitted that longer secondaries

outer tapering from about its middle. Bill somewhat lengthened, rather weak, broad at base; nostrils large, rounded, much exposed; frontal feathers not reaching by far to the posterior margin. Tarsus scutellate anteriorly, as long as middle toe and claw.

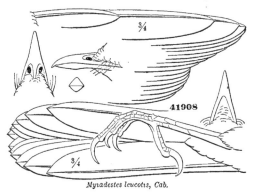

Myiadestes leucotis, Cab.

Professor Baird states that the tarsus is "without distinct scutellar divisions anteriorly except below," but a close examination shows that there is a well, marked division straight above the upper division of the outer side. The accompanying figure, No. 41,908, is also less correct in another respect, showing the nostrils too near the feathers of the forehead.

and shorter primary coverts usually are connected with rounded wings, and *vice versa*. We may also find many instances which prove that such a law for the construction of the wing does not exist.

Considering the great importance of the number of the primaries of the *Oscines*, and the deep-seated affinities expressed by the characteristic position of the middle wing-coverts, it is difficult to escape the impression that the construction of the wing is of especial importance in determining the relationships of the highest organized birds. As this difference in the construction usually consists in different development of one part in relation to others, it generally becomes a character rather easily expressed in words and represented by figures; thus being of great advantage to those who try to fix the limits of the different groups by means of structural characters.

It will be unnecessary to point out the impossibility of expressing all the manifold and intricate relationships of the genera by arranging them in a straight line. It is, consequently, a matter of course that the succession of the genera in the following synopsis expresses only to a certain degree their mutual relations. It may also be considered that the omission of the Palæogean forms makes the series incomplete. At first it was my intention to give a diagram showing the affinities, but, in view of the imperfectness of such an attempt, without including all old-world genera in addition to the American ones, I have thought it better to put it off to a later time.

On the other hand, the characters of the sections and genera given below are intended to embrace all forms belonging to them, and not only those occurring in America. If I have not always succeeded herein it is partly due to the relatively scarce material, which represents only a fraction of the extra-American birds.

As to the nomenclature and the manner of quotation, I only refer to my remarks in my paper, in Proceedings of the U. S. Nat. Mus., vol. 5, 1882, p. 29. It will be seen that examples strengthening the views there expressed are to be easily found in the present memoir. As a very striking one, I refer to the foot-note given under *Myadestes*, showing the character and the uselessness of philological "emendations" in ornithological nomenclature.

Before concluding these remarks, I take the opportunity of thanking my friend Robert Ridgway for his most valuable assistance, without which I should never have attempted the following essay.

I have also to acknowledge my indebtedness to the authorities of the Smithsonian Institution and United States National Museum for the opportunity of examining the collections upon which this paper is largely based.

WASHINGTON, D. C., *March* 20, 1882.

SYNOPSIS OF THE SUBFAMILIES AND GROUPS OF THE AMERICAN TURDIDÆ.

a^1 Gonys more than one-third the comissure; chin-angle not anterior to the line of the nostrils, or else the commissure very arched. Bill stouter, more lenthened, narrower at base and more compressed; width at base usually less than distance from nostril to tips; commissure very seldom more than twice the same distance ...A. TURDINÆ.

b^1 Wings not shorter than five times the tarsus. Tarsus very short, never longer than middle toe with claw, or commissure. Second primary often longer than the fifth; sometimes longer than the fourth. Wings covering more than two-thirds of the tail............ 1. *Sialieæ*.

b^2 Wings not more than four and three-fourths times the length of the tarsus. Tarsus moderate or long, never shorter than middle toe and claw, or commissure. Second primary seldom longer than fifth, never longer than fourth. Wings not covering more than two-thirds of the tail.

c^1 Culmen generally decidedly concave just before the nostrils, or, if straight, the commissure is also straight. Tail usually short, square, or emarginated .... 2. *Saxicoleæ*.

c^2 Culmen generally arched from the base; if straight at the base, the commissure very arched, or more or less abruptly bent downwards behind the nostrils.

d^1 Second primary more than four times longer than the first; usually longer than the sixth and equal to the fifth. Distance from the tip of the longest primary to that of the longest secondary generally longer, and not shorter, than the distance from the latter point to the tip of the longest of the greater wing-coverts.
3. *Turdeæ*.

d^2 Second primary not longer than four times the first, or else the tail three times the tarsus; usually shorter than the sixth. Distance from the tip of the longest primary to that of the longest secondary generally shorter, and not longer, than the distance from the latter point to the tip of the longest of the greater wing-coverts.

e^1 Tarsus more than twice the length of the exposed culmen.
4. *Luscinieæ*.

e^2 Tarsus not more than twice the length of the exposed culmen.
5. *Meruleæ*.

a^2 Gonys only one-third the commissure or less; chin-angle always anterior to the line of the nostrils; commissure rather straight; bill shorter, more depressed; mouth deeply cleft; width at base greater than twice the distance from nostrils to tip; commissure more than twice the same distance ....B. MYADESTINÆ.

b^1 Tail feathers never four times as long as the commissure.......6. *Platycichleæ*.

c^2 Tail feathers four times as long as the commissure or longer.....7. *Myadesteæ*.

Group SIALIÆ.

Synopsis of the genera.

a^1 Gonys very short, being shorter than two-fifths of the commissure, so that the chin-angle is considerably produced before the line of the nostrils. Tail double rounded ....1. *Ridgwayia.*

a^2 Gonys moderate, being longer than two-fifths of the commissure, so that the chin-angle does not reach before the line of the nostrils. Tail slightly forked...2. *Sialia.*

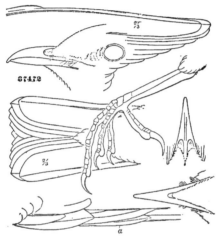

Ridgwayia pinicola.

RIDGWAYIA[*] Stejneger.

Type *Ridgwayia pinicola* (SCLAT).

Body of moderate size, with light spots on the fore parts. Wings proportionately long, and pointed, with long primaries and shorter secondaries; the first primary is placed in front of, but almost on the outside of the second, the inner web of it lying concealed between this and the primary coverts. Bill much arched, lower jaw decidedly concave; commissure with a distinct notch, and much curved, so that the whole mandible, with exception of the base, falls inside of the straight line between its tip and the angle of the mouth; lower jaw very weak; gonys very short, being shorter than two-fifths of the commissure, so that the chin-angle is considerably produced before the line of the nostrils. Bristles along gape proportionately few. Tarsi stout and exceedingly short, being shorter than the commissure, and shorter than the middle toe with claw, only making one-fifth of the length of the wings. Middle toe

[*]I have great pleasure in naming this remarkable genus in honor of Mr. Robert Ridgway, to whom the science is so highly indebted for his many eminent productions within all branches of American ornithology.

somewhat lengthened, the terminal joint especially so. Tail double rounded, the two outermost feathers being much shorter than the others.

REMARKS.—This genus embraces only one known species, the range of which is very restricted, being found only in the high table-land of southern Mexico.

The bird forming the type of the present genus has not been removed from the genus *Turdus* by any author except Mr. Seebohm. The place given to it by him within the genus *Geocichla*, "subgeneric group '*Hesperocichla*,'" is, however, by no means an improvement. *Geocichla* and the nearest allied forms are exclusively Old World and Australian birds, which have no true relatives within the Neogean part of the world, the *Hesperocichla nævia* being as badly placed among the Ground-Thrushes as the bird here under consideration. The main reason of Mr. Seebohm for placing the *R. pinicola* among these birds seems to have been the pattern of its wing, although he candidly admits that the pattern of the axillaries is not typical. Any one who will take the trouble of comparing the bird here under consideration with a young *Sialia*, will soon convince himself that the two genera should not be removed far from each other, even if he embraces the doctrine that the coloration is the only character of importance in regard to relationship. A close comparison of the structural features of both genera corroborates this view. The long and pointed wings, the short tail, and the exceedingly short tarsi, make the *Ridgwayia* widely distant from the *Turdeæ* and *Merulea*, closely resembling in these respects the *Sialia*. Besides, it will be remarked, that the geographical distribution of the two genera agrees very well, thus making the *Sialieæ* a nicely circumscribed group in this respect also.

From *Sialia* the *Ridgwayia* is easily distinguished by the more lengthened bill, the short gonys, and the double-rounded tail. Indeed it is one of the best defined genera of the whole family.

SIALIA Swains.

=. 1827.—*Sialia* Swains. Zool. Journ. III (p. 173). (Type *Motacilla sialis* L.) (*nec* Selby, 1831).

=. 1839.—*Sialis* Lafresn. Rev. Zool. 1839, p. 162. (Same type) (*nec* Latr., 1803).

Smaller size; predominant color blue and chestnut, in the adults unspotted. Wings very long and pointed, with long primaries and short secondaries; first primary normally placed, with tendency, however, to the same position as in *Ridgwayia*, very short, not one-fourth the second. Bill short, stout, compressed at the tip; commissure with a distinct notch, and more or less curved; gonys of ordinary length, so that the chin-angle is not produced before the line of the nostrils. Nasal fossæ filled with bristly feathers, only the openings of the nostrils being exposed; bristles along gape more or less developed. Tarsi stout and very short, being about of the same length as the commissure and the middle toe with claw, only making one-fifth of the length of the wings.

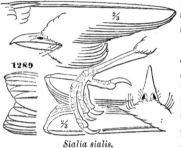

Toes stout, the middle one not unusually lengthened. Tail moderate; slightly forked.

REMARKS.—I have not been able to examine an example of *Grandala cœlicolor* Hodgs., which Mr. Seebohm includes within this genus. I have, however, very strong doubts as to the correctness of this arrangement, which seems mainly based on the blue color of the plumage. If the figure given by Wolf in Gray and Mitchell's "Genera of Birds" (I, pl. 50, fig. 3) is correct, the *Grandala* is a widely different genus, characterized, in contradistinction to *Sialia*, by the rictal bristles being obsolete, by the lengthened bill, and the exposed openings of the nostrils, the fore part of the membrane of which is not covered by feathers, also by the scutellated and lengthened tarsi, which are considerably longer than the middle toe. Besides, the toes are stated to be slender, and the tail to be strongly emarginated. It appears to me that *Grandala* is badly placed even within the same group as *Sialia*. I cannot think it will be impossible to find a more proper place near one of the Old World forms, although I shall not make any attempt without having examined the bird itself.

Sialia sialis.

Group SAXICOLEÆ.*

SAXICOLA Bechst.

<1803.—*Saxicola* Bechst. Orn. Taschb. p. 216 (*nec* Forster, 1817).
1816.—*Vitiflora* Leach. Cat. Mam. Birds Brit. Mus. p. —.
1817.—*Œnanthe* Vieill. Analyse, p. 43.
1822.—*Rupicola* Naumann. Nat. Vög. Deutschl. ii, p. iv (*nec* Briss).
1823.—*Ænanthe* Vieill. Faune Franç. p. 31.

Saxicola œanthe.

---

* Dr. Stejneger was not given time to prepare his remarks on this group. It embraces but one American genus, however (*Saxicola* Bechst.), the synonymy of which is given above. Other genera which he would refer to this group are the "Palæogean" *Pratincola* and *Ruticilla*, but whether he would include others, I do not know.—R. R.

Group TURDEÆ.

Synopsis of the American genera.

a^1. Fore part of the nasal fossæ bare, and nostrils never concealed with bristles.
   b^1. Wing never longer than three and a half times the length of the tarsus ....*Hylocichla.*
   b^2. Wing never shorter than four times the length of the tarsus..........*Turdus.*
a^2. Whole of the nasal fossæ feathered, and the nostrils nearly concealed by stiff bristles ....*Hesperocichla.*

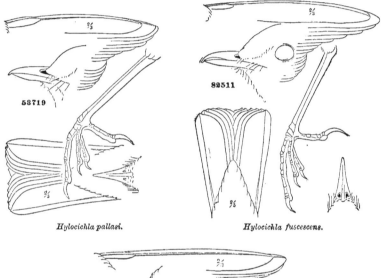

Hylocichla pallasi.   *Hylocichla fuscescens.*

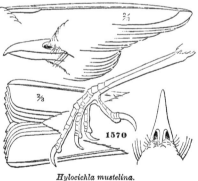

Hylocichla mustelina.

HYLOCICHLA Baird.

<1758.—*Turdus* Lin. Syst. Nat. x ed. i, p. 168.
×1860.—*Iliacus* Des Murs. Traité Ool. Ornith. p. 292. (Type *iliacus.*)
>1864.—*Hylocichla* Baird. Rev. American Birds, p. 12. (Type *mustelinus.*)

Small, spotted Thrushes, with long and pointed wings, the third and fourth primaries being the longest; with short first primary, arched culmen, moderate gonys, this being about half as long as the commissure, which has a distinct subterminal notch. The bill is short, broad at base, and much depressed. The fore part of the nasal fossæ naked, and the nostrils never covered by bristles. Tarsus long and slender, never shorter than two-sevenths of the wing, and always much longer than the commissure; outstretched legs reaching nearly to the tip of the tail.

REMARKS.—This group of smaller Thrushes is, I think, entitled to generic rank. Originally intended to embrace the North American species, it has later been shown that the *Turdus musicus* of the Old World is a true member of the group. Mr. G. R. Gray (Handb. of Birds, i, p. 254), unfortunately, however, at the same time included in it the *Turdus iliacus*, which only comes near to the *H. musica* in size and general appearance, thus embroiling the limits and discrediting the validity of the genus.

Not having seen any specimens, I am unable to decide whether we will have to enlist a *Hylocichla aurita* Verreaux or not. Verreaux's bird has been thought to be the eastern representative of the common European Song Thrush, and if such be really the case it is very likely that its proper place is within this genus.

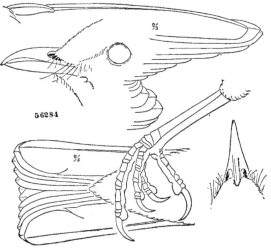

Turdus viscivorus.

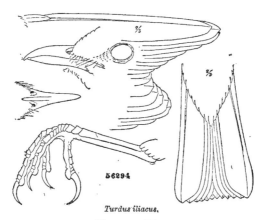

Turdus iliacus.

TURDUS Lin.

<1758.—*Turdus* Lin. Syst. Nat., x ed., i, p. 168.

×1816.—*Merula* Leach. Cat. Mamm. Birds, Brit. Mus., p. 20.

>1829.—*Copsichus* Kaup. Entwg. Eur. Thierw., p. 157 (nec *Copsychus* Wagl., 1847). (Type *torquatus.*)

>1829.—*Ixocossyphus* Kaup. Op. cit., p. 145. (Type *viscivorus.*)

>1829.—*Arceuthornis* Kaup. Op. cit., p. 93. (Type *pilaris.*)

>1829.—*Cichloides* Kaup. Op. cit., p. 153. (Type *atrogularis.*)

>1850.—*Thoracocincla* Reichb. Av. Syst. Nat., pl. liii. (Type *torquatus.*)

>1854.—*Cichloselys* Bonap. Nat. Coll. Delattre, p. 29.

>1856.—*Cychloselys* Bonap. Catal. Parzud., p. 5.

>1856.—*Planesticus* Bonap. ut supra (nec 1854).

×1860.—*Iliacus* Des Murs. Tr. Ool. Ornith., p. 292.

×1869.—*Hylocichla* G. R. Gray. Handb. of Birds, i, p. 253.

Larger, spotted Thrushes, with wings almost as in the foregoing genus. The feathering of the nasal region and the form of the bill are also the same, with the exception that the latter is stouter and higher. Tarsus stout and of moderate length, never being longer than two-eighths of the wing, but longer, however, than the commissure; out-stretched legs fall far short of the tip of the tail.

REMARKS.—The genus *Turdus* thus restricted forms a natural and rather well defined group, embracing, besides a few additional species from Eastern Asia, the following members of the west Palæarctic ornis: *T. viscivorus, pallidus, torquatus, pilaris, obscurus, iliacus, atrogularis, fuscatus, naumanni,* and *ruficollis.*

This genus, which is a strictly Palæarctic one, is entitled to admission into a synopsis of the American genera only on account of the accidental occurrence of *Turdus iliacus* in Greenland.

HESPEROCICHLA Baird.

=1858.—*Ixoreus* Baird. Birds of North Amer. p. 219 (nec Bp. 1854).

=1864.—*Hesperocichla* Baird. Rev. Amer. Birds, p. 12. (Type *nœvia.*)

Body stout, only very little spotted. Wing much as in the foregoing genera, the second primary, however, being considerably shorter than

Feb. 13, 1883.

the fifth. Bill more subulate, narrow at the base, with considerably curved commissure, and inflated tomia ; gonys long, being longer than half the commissure, which only very exceptionally has a subterminal

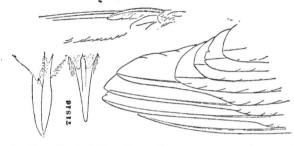

notch. The covering of the nasal fossæ is completely filled by feathers, and the openings of the nostrils concealed by a considerable number (about 7) of stiff bristles; besides, the bristles along the gape are much more developed than in other Thrushes. Tarsus stout, of moderate length, shorter than two-eighths of the wing, but still longer than the commissure ; outstretched legs fall far short of the tip of the tail.

REMARKS.—*Hesperocichla* is as well defined a genus as any within the family, and needs not to be degraded to the lower rank of a sub-

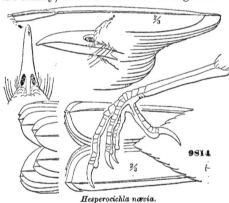

Hesperocichla nævia.

genus. It certainly only embraces one species, but I see no reason for the increasing displeasure at genera containing few species, as little as I take satisfaction in the not uncommon practice of subdividing a genus only on account of the great number of its species.

The main marks distinguishing this genus from the other members of the group *Turdeæ* are given above, these characters, indeed, as also the style of the coloration, being almost unique in the whole family. It is thought, however, that the relationship is rather with the true Thrushes than with any other genus. Their remoteness from the other forms is also expressed in the restricted geographical distribution of the present bird, which only inhabits the Pacific region of North America.

[Fig. 9814 gives an entirely erroneous view of the nostrils, which is corrected in the accompanying figure, in which the sinuation of the outer webs of the primaries and the form of the spurious primary are shown also.—R. R.]

Dr. E. Coues (Birds of the Colorado Valley, I, p. 15), remarks that the young is "like the adult female," and that "no speckled stage, like that of the very young Robin has been observed, though August specimens have been examined." But it is only apparently, what this statement seems to indicate, that the young is not speckled at all, thus differing from all the other members, and wanting the most essential character. I have now before me a specimen (U. S. Nat. Mus., No. 45897, Sitka, August, 1866) which differs considerably from the adult female. The under surface is of a much duller color, without white on the belly and under tail-coverts. All the feathers of the chin, throat, and upper breast, with well-marked, blackish edges, giving these parts a scaly appearance. In the adult female the feathers forming the collar are almost uniformly dark, the edges being lighter, if any, while the feathers of the above-mentioned parts of the young bird are gray and downy at their basal half, then ochraceous yellow, and, finally, narrowly edged with blackish. The feathers of the upper parts in the young have no light centers as usually among the Thrushes, except on the sides of the neck and on the head, where the middle of the feathers are more or less conspicuously marked with a lighter spot. Finally, we have a very striking difference between the adult and the young, showing the common Thrush-like feature of the plumage of the latter, the smaller wing-coverts having wedge-shaped, rusty spots towards the tip and dark edgings, while in the adult bird they are absolutely uniform in color. It will thus be seen that the speckled stage is not altogether wanting in this genus, although it may be admitted that it is not so conspicuous as in the young Robin. This fact seems to me to strengthen my view, that the present bird, notwithstanding a certain resemblance of the predominant colors and their tone, is widely remote from *Merula migratoria*, in the neighborhood of which it has been placed by many authors.

Group LUSCINIEÆ.

Synopsis of the American genera.

a^1 Feathers of the upper head elongated, forming a more or less distinct crest. Outer web of the outermost tail-feather not widened towards the tip, the shaft and the outline of the web being parallel. Toes more or less stout............... *Catharus*.

a^2 Head without crest; outer web of outermost tail-feather widening towards the tip. Toes very slender... *Cyanecula*.

CATHARUS Bonap.

= 1850.—*Catharus* Bonap., Consp. Av., I, p. 278. (Type *immaculatus*.)

> 1854.—*Malacocichla* Gould, Troc. Zool. Soc., Lond. 1854, p. 285. (Type *dryas*.)

> 1856.—*Malacocychla* Bonap., Compt. Rend., lxiii, p. 998.

Wing short, rounded, and concave, with long secondaries; first primary between four-eighths, and four-sevenths the second, which is always shorter than the seventh, the fourth and fifth being the longest. Culmen arched, seldom straight at the base; commissure arched, with a distinct subterminal notch; bristles more or less developed. Tarsi long, more or less stout, a little more than twice the length of the exposed culmen, and one and a half to one and three-fourths the length of the commissure, making about half the length of the tail. Toes more or less stout, the claws very arched and stout. Tail slightly rounded, the outer web of the outermost quill not widened towards the tip, the shaft and the outline of the web being parallel. Plumage soft and full, the feathers of the upper head being elongated, forming a more or less distinct crest.

REMARKS.—I have not been able to find any important difference between the species included within the genus *Malacocichla* Gould, and the typical *Cathari*. The difference is chiefly and alone to be found in the color, the former group having the throat and upper breast spotted, somewhat like the smaller species of *Hylocichla*, with which they, in fact, have been put together by Mr. Seebohm. They differ, however, widely from these in most respects, being structurally quite identical with the other species composing the genus here in queston.

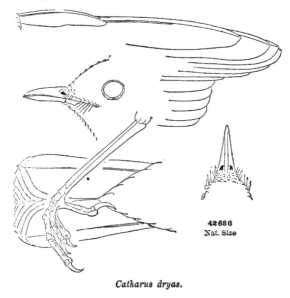

42686
Nat. Size

Catharus dryas.

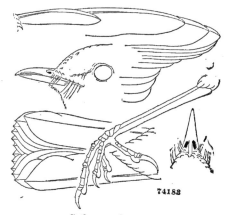

Catharus melpomene.

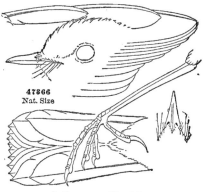

Catharus gracilirostris.

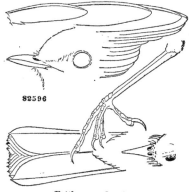

Erithacus rubecula.

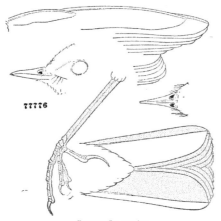

Cyanecula suecica.

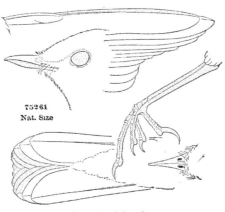

Luscinia philomela.

CYANECULA Brm.

<1758.—*Motacilla* Lin., Syst. Nat. x ed. I, p. 184.
<1760.—*Ficedula* Briss., Orn. III, p. 369.
<1769.—*Sylvia* Scop., Ann. I, Hist. Nat. p.
<1799-1800.—*Erithacus* Cuv. Leç. d'Anat. Comp. I, Tabl. ii.
<1822.—*Humicola* Naumann, Naturg. Vög. Deutschl. II, p. iii.
<1826.—*Dandalus* Boie, Isis, 1826, p. 972.
=1828.—*Cyanecula* Brm., Isis 1826 (p. 1280). (Type *suecica.*)
=1833.—*Tandicilla* Blyth, Renn. Field Nat. I (p. 291).
=1857.—*Cyanegula* Boie, Journ. Orn. 1857, p. 166.

Wing moderate, rather pointed, with proportionately short primaries; first primary less than one-third the second, which is about of the length of the sixth, and always shorter than the fifth and longer than the seventh, the third being the longest. Bill slender, with the culmen straight and the commissure arched at the base, and with the subterminal notch obsolete; bristles few and weak. Tarsi long and slender, two and two-third times the length of the exposed culmen, and one and four-fifths times the length of the commissure, making about two-fifths of the length of the tail. Toes long and very slender, the claws being unusually straight, small, and slender. Tail nearly even, the outer web of the outermost quill widening towards the tip. Plumage compact; head without crest.

REMARKS.—This genus is included here in account of the supposed occurrence of *Cyanecula suecica* (Lin.) in Alaska.

The characters, as given above, are sufficient to distinguish these birds from both *Erithacus*,* *Luscinia*,† and *Calliope*. Notwithstanding an external resemblance to *Phœnicurus*, it certainly belongs to this group and not to the *Saxicoleæ*.

Group MERULEÆ.

Synopsis of the American genera.

a^1 Tail not graduated; the tail-feathers considerably shorter than the wing.
   b^1 Tail-feathers more than two and a half times the length of the tarsus.
      c^1 Third, fourth, and fifth primaries largest, or else the tail square....*Merula.*
      c^2 Fourth, fifth, and sixth primaries largest, and the tail much rounded
                                         Semimerula.
   b^2 Tail-feathers only twice the length of the tarsus................ *Cichlherminia.*
a^2 Tail graduated; the largest tail-feathers about of the length of the wing
                                              Mimocichla.

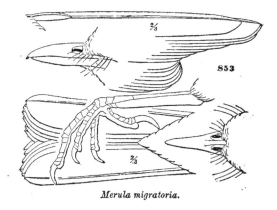

Merula migratoria.

Erithacus Cuv., Leç. d'Anat. Comp. I (1799–1800), tabl. ii. A true synonym of this is *Helminthophaga* Bechst. Orn. Taschb. (1803), p. 177 cf. pp. 507 and 548 (*nec* Cab. quæ *Helminthophila* Ridg.). It was an error when I informed Mr. Ridgway that Bechstein included the Nightingale within this subgenus. Cf. Bull. Nutt. Orn. Club, 1882, p. 53.

† *Luscinia* Forster, Syn. Cat. Brit. Birds (1817), p. 14, is prior to *Dandalus* of Boie.

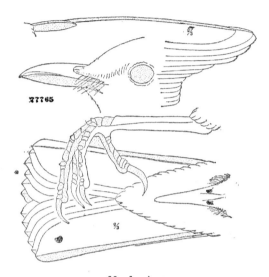

Merula nigra.

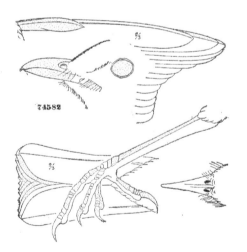

Merula jamaicensis.

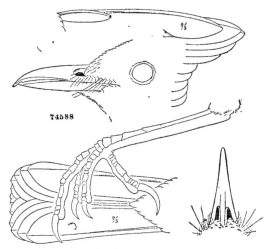

Merula (♀) aurantia.

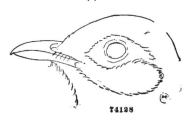

Merula gymnophthalma.

MERULA Leach.

<1758.—*Turdus* Lin., Syst. Nat. x ed. I, p. 168.

<1816.—*Merula* Leach, Cat. Mamm. Birds Br. Mus. (p. 20) *nec* Koch, 1816, quæ *Pastor*.

>1850.—*Hodoiporus* Reichb., Syst. Av. pl. LIII. (Type *jamaicensis*.)

>1854.—*Planesticus* Bonap., Coll. Delattre, p. 27 (*nec* 1856).

>1855.—*Cichlopsis* G. R. Gray, Cat. Gen. Birds, p. 43 (*nec* Cab. 1850). (Type *aurantius* Gm.)

+1859.—*Semimerula* Sclater, Proc. Zool. Soc. Lond. 1859, p. 332.

+1881.—*Merulissima* Seebohm, Cat. Birds Br. Mus. V, p. 232.

Size large or moderate; color more or less uniform, often black or blackish in both sexes; when streaked, only the throat is marked with dark streaks. Wing rounded, the third, fourth, and fifth primaries being longest, the third not commonly longer than the fifth; second primary not longer than four times the first; secondaries long, the distance from the tip of the longest primary to that of the longest secondary never being longer than the distance from the latter point to the tip of the longest of the greater wing-coverts. Bill stout; culmen arched from the base; commissure with a distinct subterminal notch, and not longer than two and a half times the moderate gonys; chin-angle not reaching before the line of the nostrils. Bristles along gape moderate in strength and number. Tarsus stout and moderate in length, never longer than twice the exposed culmen. Tail square or only slightly rounded; the tail-feathers more than two and a half times the length of the tarsus, but shorter than three and a half times the same length and considerably shorter than the wing.

REMARKS.—At the first sight this genus will appear somewhat heterogeneous, including such different looking birds, as *Merula nigra, migratoria* and *jamaicensis.* These differences are, however, only superficial and due to the color, but it will not be difficult to arrange the numerous species of this genus, which has representatives all the world over, but the chief range of which seems to be the tropical regions, in one series, showing nicely the transitions from the deepest black to the lightest rusty, and from the quite uniform to the most varied colored bird. As to the *M. aurantia* (Gmel.), from Jamaica, I have expressed my doubts under *Semimerula,* to which remarks I here refer.

SEMIMERULA Sclat.

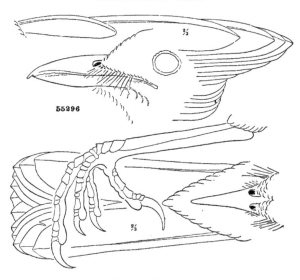

Semimerula gigas.

<1859.—*Semimerula* Sclater. Proc. Zool. Soc, Lond. 1859, p. 332.
<1881.—*Merulissima* Seebohm, Cat. Birds Br. Mus. V, p. 232.

Size large; color uniform blackish or dull brownish. Wing very rounded, the fourth, fifth, and sixth primaries being longest, the third never longer than the fifth; second primary never longer than two and a half times the first, never longer than the eighth; secondaries as in *Merula.* Bill very large and stout, being otherwise much like that of

Merula. Bristles along gape strong and numerous. Tarsus very stout and somewhat lengthened, never longer, however, than two and a half times the exposed culmen. Tail considerably rounded, the tail-feathers more than two and a half times the length of the tarsus, but shorter than three and a half times the same length, and decidedly shorter than the wing.

REMARKS.—As to which species should be included within this genus, authors have had different opinions. So has especially Professor Baird included within it the *Turdus aurantius* Gmel., although it seems that he is aware of the incongruity of this species and those which were considered typical by the founder of the genus, Mr. Sclater (see Rev. Amer. Birds. I, p. 4), and I think these birds are too heterogeneous to become members of the same genus. *T. aurantius* will be very difficult to separate from the genus *Merula.* The bird looks rather peculiar, and will probably require a separate genus for itself, although I have not succeeded in finding characters sufficient to separate it from the latter group, with which, for the present, I have been obliged to keep it.

Of the species which I have been able to examine, only the following belong to the genus *Semimerula*, restricted and defined as above: *Semimerula gigas, Semimerula xanthosceles,* and *Semimerula atrosericea.*

This genus does not occur anywhere else than in South America.

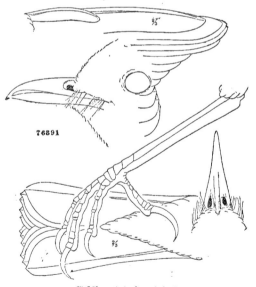

Cichlherminia herminieri.

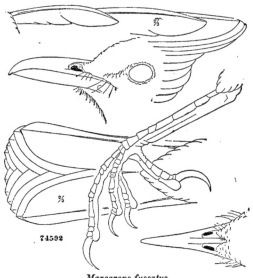

Margarops fuscatus.

CICHLHERMINIA Bonap.

<1854.—*Cichlherminia* Bonap., Coll. Delattre, p. 26.
=1859·—*Cichlerminia* Sclat., Proc. Zool. Soc. Lond., 1859, p. 335.

Size large. Plumage spotted and squamated underneath. Wing rounded, third, fourth, and fifth primaries being equal and longest; second primary about two and a half times the first; secondaries long. Bill very large and stout; culmen arched from the base; commissure with very distinct subterminal notch, only a little more than twice the length of the long gonys. Bristles along gape numerous, and very long and strong; on the *apex malaris* a tuft of numerous stiff bristles. The booted tarsus stout and lengthened, not being, however, more than two times the exposed culmen. Tail very slightly rounded and short, so that the outstretched legs are reaching nearly to the tip of tail; tail-feathers scarcely more than twice the length of the tarsus, and very much shorter than the wing. Below and behind the eye a large naked space.

REMARKS.—As has already been stated, the genus *Cichlherminia*, restricted as above, belongs to the *Turdidæ*, whereas the other species, generally admitted to it, form a well-defined genus, *Margarops* Sclat., and belong to quite a different family, being characteristic members of the *Miminæ*. All those specimens which I have had opportunity of examining have booted tarsi, *Merula*-like bill, and a very characteristic Thrush-like plumage, while in *Margarops* these parts are characteristically Mimine. (See fig. of *M. fuscatus* given above.) Unfortunately, however, I have not been able to procure a young specimen, and consequently I cannot tell whether its markings agree with those of the young of the other *Turdidæ*, although I have but little doubt that this will be found to be the case. The relationship to the *Miminæ* seems to be a very remote one, and even the external spotted appearance, which appears to have been the chief reason for uniting it with those birds, shows only a slight and very superficial resemblance to the genus *Margarops*.

This genus is peculiar to the lesser Antilles.

MIMOCICHLA Sclat.

<1850.—*Galeoscoptes* Cab., Mus. Hein., I, p. 82.
=1859.—*Mimocichla* Sclat., Proc. Zool. Soc. Lond., 1859, p. 336.
=1865.—*Mimokitta* Bryant, Proc. Bost. Soc. IX, p. 371.
=1866.—*Mimocitta* Newton, Ibis, 1866, p. 121.

Size, moderate; prevalent color, bluish gray; the outer tail-feathers having a large white patch at the tip. Wing somewhat rounded, the third, fourth, fifth, and sixth primaries being longest; the third never longer than the sixth and considerably shorter than the fifth; second primary shorter than the seventh, and never longer than two and three-fourth times the first; secondaries rather long. Bill large and rather slender; the commissure with a more or less distinct notch, only very little larger than two times the gonys; chin-angle not protruding before the nostrils. Rictal bristles inconspicuous. Tarsus somewhat lengthened, but less than twice the exposed culmen. Tail graduated and long, the outstretched legs falling far short of its tip; the largest tail-feathers about five times the tarsus and about as long as the wing. Below and behind the eye a naked space.

REMARKS.—The few species composing this genus, which is confined to the West Indian Islands, form a well circumscribed group. It shows some relationship towards the *Miminæ*, but as neither its position among the *Turdidæ* nor its validity as a distinct genus has been disputed, it needs no further remarks at this place.

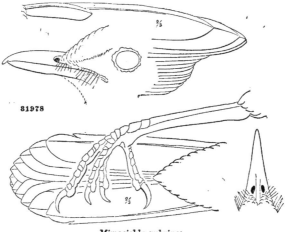

Mimocichla rubripes.

Group PLATYCICHLEÆ.

Synopsis of the genera.

a^1 Outermost tail-feathers longer than the inner ones; second primary shorter than the seventh... *Cossyphopsis.*

a^2 Outermost tail-feathers shorter than the inner ones; second primary longer than the seventh.

$\quad b^1$ First primary two-sixths to two-sevenths the second; tail slightly rounded.
$\qquad\qquad\qquad\qquad\qquad\qquad\qquad\qquad\qquad\qquad\qquad\qquad$ *Platycichla.*

$\quad b^2$ First primary about two-fifths the second; tail emarginated and rounded.
$\qquad\qquad\qquad\qquad\qquad\qquad\qquad\qquad\qquad\qquad\qquad\qquad$ *Turdampelis.*

REMARKS.—This group shows a near relationship towards the *Merulæ*, with which some of the species of the two first genera always have been treated. The characteristic shortness of the gonys, however, and the statement of Professor Baird of the very close relationship between the genus *Platycichla* and *Turdampelis* (*Cichlopsis*), which I myself have never seen, and between the latter and *Myadestes,*[*] led me to the conclusion that their proper place will be here within the *Myadestinæ,* forming an intermediate link between the true Thrushes and the more aberrant looking *Myadestes.*

COSSYPHOPSIS[†] Stejneger.

Type *Cossyphopsis reevei* (Lawr).

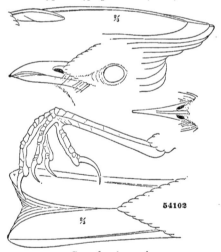

Cossyphopsis reevei.

Size moderate; color uniform; throat marked with black streaks. Wing rounded, the third, fourth, and fifth primaries being longest, the

---

[*]See Rev. Amer. Birds, I, p. 417: "The latter genus [*Platycichla*] is so closely related to *Cichlopsis* as almost to be the same," and *op. cit.,* p. 434: "The relationship of this genus [*Cichlopsis*] to *Myiadestes* is very close. * * * In fact, the only tangible differences are in the stouter bill, rather more united toes, more compact plumage, and absence of wing-pattern." In these respects the *Cichlopsis* agrees with the *Cossyphopsis* and *Platycichla,* thus forming, as it seems, a very natural group.

[†] Κόσσυφος=*merula,* ὄψις=*facies.*

third not longer than the fifth; second primary three and one-fourth times the first, and shorter than the seventh; secondaries very long, the distance from the tip of the longest of the greater wing-coverts to the tip of the longest secondary twice the distance from the latter point to the tip of the longest primary; bill Thrush-like, attenuated at the end; culmen arched from the base; commissure rather straight, with a distinct subterminal notch, three times the length of the short gonys; chin-angle reaching considerably before the line of the nostrils, the openings of which are large and oval, the overhanging membrane being rather narrow; bristles along gape weak and short; tarsus moderate, longer than middle toe and claw, and considerably longer than one-fourth the longest tail-feather, but shorter than twice the exposed culmen; tail fan-shaped, emarginated, the feathers gradually becoming longer from the middle pair outwards, the outer pair decidedly the longest; the outer web of the outermost tail-feathers broadens conspicuously toward the somewhat acuminate tip; longest tail-feathers less than four times the commissure.

REMARKS.—Of this genus only the type species is as yet known, but this bird is so peculiar as to show its difference from all other Thrushes at once. The shortness of the gonys, and several other features, point towards its position among the *Platycichleæ*, but the tail, with its emarginate shape, is, so far as I am aware, unique among those birds which can claim any relationship with it.

PLATYCICHLA Baird.

< 1854.—"*Myiocichla* Schiff," Bonap. Coll. Delattre, p. 30. (Type *Cichlopsis leucogenys*, Cab.)

= 1864.—*Platycichla* Baird, Rev. Am. Birds, I, p. 32. (Type *P. brevipes*.)

Size moderate; color uniform; wing rounded, the third, fourth, and fifth primaries being longest, the third about equal to the fifth; second

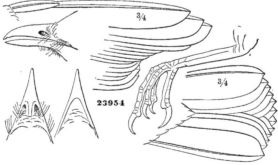

Platycichla "*brevipes*."

primary not longer than three and a half times the first, and longer than the seventh; secondaries moderate, the distance from the tip of the longest of the greater wing-coverts to the tip of the longest secondary only a little longer than the distance from the latter point to the tip of the longest primary; bill much like that of the foregoing genus, the commissure being only a little more arched, and the gonys still shorter, lower mandible much weaker and narrower; rictal bristles stronger and much longer; tarsus short, rather shorter than middle toe and claw, less than one-fourth the longest tail-feather, and much shorter than twice the exposed culmen; tail rounded, the feathers grad-

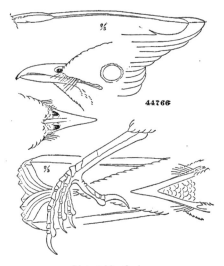

Platycichla flavipes.

ually becoming shorter from the middle pair outwards, which is the longest one; the tips of the tail-feathers very acuminated, the outer web not conspicuously broader towards the tip; longest tail-feathers never four times the commissure.

REMARKS.—This genus and its type species have had a somewhat peculiar fate. Although twice mentioned in one of the most admired and admirable works of modern ornithology (Rev. Am. Birds, I, pp. 32 and 436), it has been passed by in silence by all authors, and almost forgotten, until lately Mr. Sharpe (Cat. Birds, Brit. Mus., VI., p. 379) has reprinted the original definition and description. Even in Sclater and Salvin's *Nomenclator Avium Neotropicalium* this bird is omitted, and nobody has been able to obtain a second specimen besides the type.

When examining the specimens of "*Turdus*" *flavipes* and *T. carbonarius* I felt soon convinced that they did not belong to the true *Merulea*, but that their proper place would be somewhere in the neighborhood of *Myadestes*, and had just decided to make them types of a new genus, the name of which I had already composed, as I was struck by the agreement of their peculiar characters with those of *Platycichla*. I consequently very eagerly compared specimens of the two species mentioned above with the type specimen of Professor Baird's P. brevipes, and found them, to my great surprise, agree so well that I am convinced that the latter species is only the female of flavipes, or of a very nearly allied species. I have no female of flavipes at hand for comparison, but the structural features are so nearly the same, and the descriptions of the females of this species agree so well with the color of brevipes, that I have no doubt that my identification is right. The only difference which I can detect is the somewhat shorter tarsus of brevipes, but this is perhaps only an individual variation, although it possibly may turn out to be a different species.

We would then have the following species within this genus:

1. *Platycichla brevipes* (Baird).
2. *Platycichla flavipes* (Vieill.).
3. *Platycichla carbonaria* (Licht).

As to the generic name, it may be remarked that flavipes has been formerly united with its near relative, *Cichlopsis leucogenys*, Cab. within the genus *Myiocichla* "Schiff.", as the type of which it has usually been regarded. But it will seem from the following remark of Bonaparte, the first author by whom the genus *Myiocichla* was published, that the leucogenys is the true type. He says (Notes Coll. Delattre, p. 30), "*Turdus flavipes*, Vieill. (*carbonarius* Ill., *ardesiacus* Cuv. nec Auct.!) est pour Schiff une Myiocichla; mais y est-il bien placé si le type de ce genre est, *comme nous le croyons*, sa *Myiocichla ochrata*, du Bresil (*Turdus brunneus!* Freyreiss, nec Anglorum et Bodd.)". In this case the name Professor Baird has given it will stand.

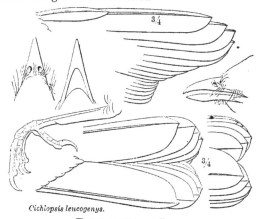

Cichlopsis leucogenys.

TURDAMPELIS Less.

= 1884.—*Turdampelis* Less., Echo du Monde Sav., 1844, p. 156. (Type *T. lanioides*.)

= 1850.—*Cichlopsis* Cab., Mus. Hein., i, p. 54. (Type *C. leucogenys*.)

< 1854.—"*Myiocichla* Schiff" Bonap. Coll. Delattre, p. 30. (Same type.)

Size moderate; color uniform. "Wing rather longer than tail," "fourth and fifth primaries longest;" "third, between fifth and sixth;" "first

Feb. 28, 1883.

quill about two-fifths the second;" "second intermediate between sixth and seventh;" "wings quite pointed." "Bill rather stout and somewhat Thrush-like;" "the lower mandible is rather deeper and stouter" than in *Platycichla*, "the upper less attenuated, viewed from above;" "gonys about two-fifths the lower edge of lower mandible." "Frontal and rictal bristles well developed. Feet short; tarsus about equal to middle toe." "Tail emarginated and still more rounded." (Baird, Rev. Amer. Birds, i, pp. 433–435.)

REMARKS.—As I have not had the opportunity of examining any specimen, I have nothing to add to Professor Baird's description (*l. c.*), of which I have given extracts above, showing the essential characters in the same manner as are given the marks of the other genera here defined and described.

Although it may be admitted that there is some doubt as to the identification of the species of Lesson, belonging to his genus *Turdampelis*, I think that this name is the same as *Cichlopsis* of Cabanis.

The genus is only known to embrace two species, one of which is but lately described, viz: *Turdampelis leucogenys* (Cab.), and *Turdampelis gularis* (Salvin & Godman), Ibis, 1882, p. 76.

Group MYADESTEÆ.

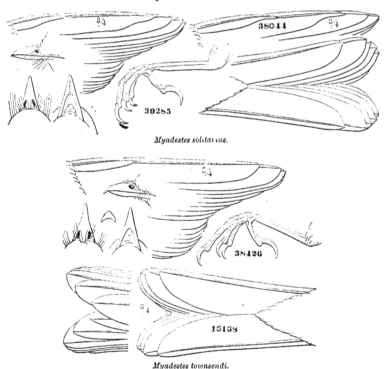

Myadestes solitarius.

Myadestes townsendi.

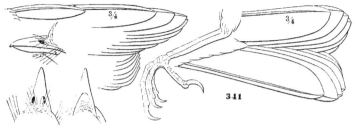

Myadestes elisabethæ.

MYADESTES.* Swains.

= 1838.—*Myadestes* Swains., Nat. Libr., xiii, p. 132.
= 1838.—*Myidestes* Swains., op. cit., p. 134.
= 1842.—*Myiadestes* Agass., Nomencl. Zool. Aves, p. 49.
= 1846.—*Myiesthes* Agass., Index Univers., p. 241.
= 1854.—*Myiadectes* Bonap., Not. Coll. Délattre, p. 27.

Size rather small; color unspotted and unstreaked. Wing rounded, the third, fourth, fifth, and usually, also, the sixth primaries longest; second primary never longer than three and a half—usually only two and a half—times the first, and usually shorter than the seventh; secondaries rather long. Bill weak, short, and broad, much depressed; commissure nearly straight, with distinct subterminal notch, and three times as long as the short gonys; chin angle reaching considerably before the line of the nostrils, which are oval, with overhanging membrane. Bristles rather well developed. Tarsus rather short, seldom exceeding in length the middle toe and claw, usually a little shorter, but about twice the exposed culmen, and about one-fourth or less the length of the longest tail-feathers. Tail rounded, or double rounded, the outermost pair of tail-feathers always considerably shorter than the longest; feathers rather narrow, tapering gently from base to tip, the shafts of the outermost converging towards the tip; longest tail-feathers never shorter than four times the commissure. Feathers of occiput full and somewhat lengthened.

REMARKS.—The relationship of this genus has already been pointed out. I will here only remark that I am inclined to believe that the "Flycatching Thrushes," besides their close affinities to *Turdampelis* and the *Platycichleæ*, on the other hand are somewhat related to the *Sialeæ*. That the group, besides, show some relationship towards certain African forms is not so very surprising, when we remember several other affinities of South American and West Indian birds with forms from Africa. A very striking instance is in this respect the close relationship between members of the genus *Merula*—especially those composing the division *Planesticus*—inhabiting the two continents.

---

* The recently adopted spelling is *Myiadectes* (see Sharpe, Cat. Birds Brit. Mus., vi, p. 368, where Salvin and Godman are erroneously given as the original authors). This is intended to be an "improvement" or "correction" of Swainson's original name, being, however, quite unnecessary, as the derivation of *Myadestes* is from Μόα (att. for μυία), a fly, and 'Εδεστής (Hdt. 3, 99)=an eater, devourer.

BY THEODORE GILL.

The genus *Centropomus* of the tropical American seas and rivers has generally been referred to the family Percidæ. As long ago as 1865, however, I was struck by the remarkable differences in its osteology from any other fishes known to me, and communicated the results of my examination to Professor Poey, who agreed with me that the type was entitled to family distinction. Both Professor Poey and myself have, therefore, isolated the form in question as a peculiar family. That family has, however, not yet been characterized, and the object of this communication is to indicate some of the most peculiar features which distinguish the form from those with which it has been usually associated. The want of an accessible large collection of skeletons precludes a detailed comparison with many types, but most of the American genera of Percidæ (typical), Labracidæ, Serranidæ, and Sparidæ have been examined as to their skulls at least. It is possible that the genus *Lates* and even *Niphon* may be more nearly related, but no skeletons of those fishes are available. It is to be hoped that the present notice may attract attention to their relations.

CENTROPOMIDÆ.

Synonyms as families.

= Centropomidæ, *Gill*, MSS., 1865.
= Centropomatidi, *Poey*, Repertorio Fisico-Natural de Cuba, v. 2, p. 280, 1868 (not defined).
= Centropomidæ, *Gill*, Arrangement Families of Fishes, p. 11, 1872.
Percoides and Percidægen., authors generally.

As will be seen, Professor Poey was the first ichthyologist to publish a name for the family.

Typical Acanthopterygians with the postorbital portion of the skull longer than the oculo-rostral; the parietals behind the constriction continuous with the epiotics and transverse laminæ arising from the suproccipital crest, the three together forming a well differentiated posterior oblong pentagonal or hastiform area; the re-entering parietal sinus, with its anterior margin, produced *forwards nearest the opisthotics*; the exoccipitals well developed and contiguous above the foramen magnum; the vertebræ in typical number (10 + 14) and longish; the anterior two partly co-ossified and the first with selliform apophyses extending backwards and embracing the second vertebra; the vertebræ mostly with foveæ or pits for the ribs and only with developed parapophyses for the posterior (6–10) pairs of ribs; the second neural spines suberect, and with laminiform extensions which embrace the first; the neurapophyses and neural spines of the other vertebræ depressed at their bases, continuous with the zygapophyses in front, and slightly curved upwards at their tips; the hæmal spines resembling the neural.

PLATE XI.

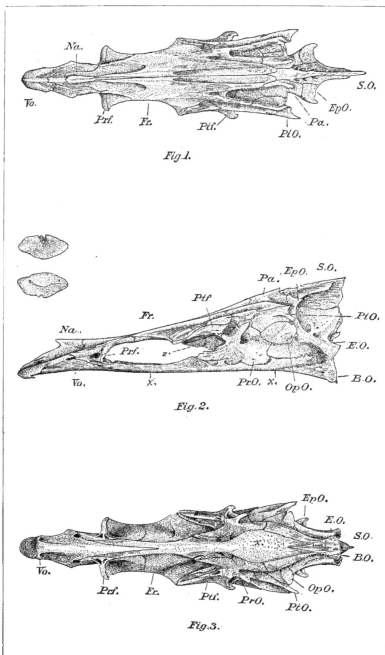

Fig.1.

Fig.2.

Fig.3.

CENTROPOMUS.

(Plate VI.)

< Centropomus, *Lacépède*, Hist. Nat. des Poissens, t. 4, p. 248, 1802.
< Centropome *Duméril*, Zool. Anal., pp. 133 (Centropoma), 333, 1806.
< Les Centropomes, *Cuvier*, Regne Animal, t. 2, p. 294, 1817.
= Les Centropomes, *Cuv. & Val.*, Hist. Nat. des Poissons, t. 2, p. 102, 1828.
= Oxylabrax, *Bleeker*, Arch. Néerland. Sc. Ex. et Nat., t. 11, p. 264, 1876.
Sciæna sp., *Bloch.*
Platycephalus sp., *Bloch-Schneider.*
Perca sp., *Lacépède.*
Sphyraena sp., *Lacépède.*
Not Centropomus, *Bleeker (op. cit.*, p. 265), 1876. (= Stizostethion Raf. = Lucioperca, *Cuv.*)

NOMENCLATURE OF THE XIPHIIDS.

BY THEODORE GILL.

The synonymy of the family Xiphiidæ and its subfamilies has been partially given in Professsor Goode's excellent article on "The Taxonomic relations and geographical distribution of the members of the Sword-fish family" (Proc. U. S. Nat. Mus., v. 4, pp. 415–433, 1882), and may be supplemented by the following exhibit. As the characters of the several groups have been already well given by Professor Goode, it is unnecessary to repeat them here. It may be stated, however, that skeletal differences confirm those used for the diagnoses, and the vertebræ especially are even characteristic for the distinction of two families.

The old family Xiphiidæ has been differentiated by Dr. Günther, as a "division" from the Scombridæ and the latter contradistinguished in a "division" of Acanthopterygians containing many very heterogeneous forms. Nevertheless, a careful study of the type renders it evident that the family is closely related to the Scombridæ, and the genus *Acanthocybium*, a representative of that family, manifests an incipiency of the characteristics of the Xiphiidæ in the structure of the gills as well as the projection of the snout, the development of the dorsal fin, and, to a less extent, other features. As Lütken and Goode have expressly contended, and as I indicated in 1873, by the sequence in the "Catalogue of the Fishes of the Eastern Coast of the United States" (pp. 9, 3), the Tetrapturinæ or Histiophorinæ are the most generalized forms of the family and deviate least from the Scombridæ while the Xiphiinæ are highly specialized, and by the inferior position of the pectorals and investment of the fins mimic the sharks, the largest of which they almost rival in size.

XIPHIIDÆ AUCT.

Synonyms as families.

× Pantoptères, *Duméril*, Zoöl. **Anal.**, p. 114, 1806.
× Atractosomes, *Duméril*, Zoöl. Aual., p. 124, 1806.
> Istioforidi, *Raf.*, Indice d'Ittiol. Sic., p. 30,* 1810.
> Zifidi, *Raf*, Indice d'Ittiol, Sic., p. 39,† 1810.
× Lophionota, *Raf.*, Analyse de la Nature, 11ᵉ fam., 1815.
× Pantopteria, *Raf.*, Analyse de la Nature, 23ᵉ fam., 1815.
= Xiphirhynques (Xiphirhynchi), *Latreille*, Fam. Nat. du Regue An., p. 131, 1825.
< Xiphoides, *Risso*, Hist. Nat. de l'Europe merid., t. 3, p. —, 1826.
= Xiphiidæ, *Bonaparte*, Nuovi Annali delle Sci. Nat., t. 2, p. —, 1838; t. 4, p.—, 1840.
= Xiphioides, *Agassiz*, Recherches sur les Poissons Fossiles, v. 5, p. 89, 1843.
= Xiphioidei, *Agassiz*, "Pisc. Ad.", 1843.
= Xiphioidæ, *Agassiz*, Nom. Zoöl. Index Universalis, 12° ed., p. 1123, 1848.
= Xiphioidei, *Bleeker*, Enum. sp. Piscium Arch. Ind., p. xxii, 62, 1859.
= Xiphioidæ, *Gill*, Cat. Fishes E. Coast N. A., p. 38, 1861.
= Xiphiadidæ, *Cope*, Trans. Am. Phil. Soc., n. s., v. 14, p. 459 (Oct. 7, 1870), 1871.
= Xiphiæ, *Fitzinger*, Sitzungsber. k. Akad. der Wissensch. (Wien), b. 67, 1. Abth., p. 33, 1873.
= Xiphiidi, *Poey*, Anal. Soc. Esp. Hist. Nat., t. 4 (Enum Pisc. Cub., p. 7, 70,) 1875.

HISTIOPHORIDÆ.

Synonyms as subfamilies.

= Istiophoria, *Raf*, Analyse de la Nature, p. —, 1815.‡
= Tetrapturinæ, *Gill*, Rep. U. S. Fish Comm., v. 1, p. 787, 1873.
= Tetrapturini, *Poey*, Anal. Soc. Esp. Hist. Nat., t. 4 (Enum. Pisc. Cub., p. 7), 1875.
= Histiophorinæ, *Lütken*, Videnskab. Meddel. Naturhist. Forening Kjobenhavn, 1875, p. 18, 1875.
= Tetrapturinæ, *Goode*, Proc. U. S. Nat. Mus., v. 4, pp. 416, 417, 1882.

XIPHIIDÆ.

Synonyms as subfamilies.

< Xyphidia, *Raf*, Analyse de la Nature, p.—, 1815. §
< Xiphiadini, *Bon.*, Giorn. Arcad. di Scienze, v. 52, p. — (Saggio Distrib. Metod. Animal Vertebr. a sangue freddo, p. 34, 1832).
< Xiphianæ, *Swainson*, Nat. Hist. and Class. Fishes, etc., v. 2, p. 175, 1839.
< Xiphinæ, *Swains.*, Nat. Hist. and Class. Fishes, etc., v. 2, p. 239, 1839.
= Xiphiini, *Poey*, Anal. Soc. Esp. Hist. Nat., t. 4) Enum. Pisc. Cub., p. 7), 1875.
= Xiphinæ, *Gill*, Canad. Nat., (2,) v. 2, p. 250, 1867.
= Xiphinæ, *Lütken*, Videnskab. Meddel. Naturhist. Forening Kjobenhavn, 1875, p. 18.
= Xiphiinæ, *Goode*, Proc. U. S. Nat. Mus., v. 4, pp. 416, 427. 1882.

---

* XXI. Ordine. Gli Istioforidi "Corpo, e mascelle allungate, Ale toracine con 1 raggi riunite senza membrana intermedia."—RAF.

† XLVII. Ordine. I Zifidi. "Corpo conico, ò lanceolato, nessun' ala di meno, muso colla mascella superiore multo prolungata, o spinosa."—RAF.

‡ 1re subfam. of 11ᵉ fam (Lophionota).

§ 2e sub-fam. of 23ᵉ fam (Pantopteria).—The subfamily is a heterogeneous group, containing *Anarhichas*, *Comephorus*, "*Opictus, R.*," *Xiphias*, and *Macrognathus*.

ON THE FAMILY AND SUBFAMILIES OF CARANGIDÆ.

BY THEODORE GILL.

The family of Carangidæ, as limited by me in the "Arrangement of the families of Fishes" (also as in the Proc. Acad. Nat. Sc. Phila., for 1862, p. 430, after the exclusion of *Pomatomus*) is an exceedingly natural one, notwithstanding the differences in external form. I have examined skulls of representatives of all the groups hereinafter named, and their common characters are so numerous, while their severally peculiar ones are so insignificant that the so-called subfamilies are scarcely entitled to that rank. The most characteristic skull is manifested in *Trachynotus;* in that form, the orbito-rostral portion is shorter in comparison, the post-frontal bones larger and more projecting, the inner lateral crests more produced forwards than in any others, and the ethmoid is abbreviated and markedly and abruptly declined. Analagous characters in many families, however, are of inferior systematic value. It is to be also remarked that the Caranginæ and Seriolinæ are especially nearly allied, so far as their crania are concerned, and there is even less superficial difference between the skull of *Seriola* and most Caranginæ—e. g. *Carangus*—than between it and the related genus *Elagatis*.

Greatly as the elongated *Trachurus* and the high *Selene* differ, even they essentially agree as to the structure of the skull, that of *Selene* differing from the Carangine chiefly in being compressed, with its crest elevated and extended backwards and its rostral portion attenuated and produced forward. Its ethmoid especially is characteristic in being much compressed and carinated above instead of flattened and double-headed. If, therefore, the subfamilies already indicated are retained in the present communication, it is rather in order to epitomize the history expressed in their nomenclature than because I insist on or persist in their retention. The hæmal canal is perhaps more characteristic.

CARANGIDÆ.

Synonyms as families.

× Centronotides, *Risso*, Hist. Nat. de l'Europe Mérid., t. 3, pp. 110, 426, 1826.

> Carangoidei, *Bleeker*, Enum. Sp. Piscium Archipel. Indico, p. xxiii, 1859.

× Lichioidei, *Bleeker*, Enum. Sp. Piscium Archipel. Indico, p. xxii, 1859.

× Serioloidei, *Bleeker*, Enum. Sp. Piscium Archipel. Indico, p. xxiii, 1859.

× Carangidæ, *Günther*, Cat. Fishes Brit. Mus., v. 2, p. 417, 1860.

× Carangidæ, *Günther*, Archiv für Naturg., 28. Jahrg., B. 1, p. 59, 1862.

< Carangoidæ, *Gill*, Proc. Acad. Nat. Sci. Phila., [v. 14,] p. 430, 1862.

< Carangidæ, *Cope*, Proc. Am. Assoc. Adv. Sci., v. 20, p. 342, 1872.

= Carangidæ, *Gill*, Arrangement Families Fishes, p. 8, 1872.

= Carangidæ, *Poey*, Anal. Soc. Esp. Hist. Nat., t. 4 (Enum. Pisc. Cub., p. 7), 1875.

> Caranges, *Fitzinger*. Sitzungsber. K. Akad. der Wissensch. (Wien), B. 67, 1 Abth., p. 33, 1873.

> Chorinemi, *Fitzinger*, Sitznugsber. K. Akad. der Wissensch. (Wien), B. 67, 1. Abth.,
  p. 33, 1873.
× Naucratæ, *Fitzinger*, Sitzungsber. K. Akad. der Wissensch. (Wien), B. 67, 1. Abth.,
  p. 33, 1873.
Zeidæ gen., *Swainson*.

Dr. Günther (*op. cit.*, p. 417) has claimed special merit for his family
of Carangidæ, remarking "that several authors have also distinguished
a family *Carangidæ*, but if they defined it at all they have applied characters very different from those given above [his remarks], and have
not paid attention to the structure of the skeleton." I am not aware that
any author except Bleeker had previously distinguished a family Carangidæ; the name "Carangidæ," suggested by Agassiz, being merely
an orthographical substitute for subfamily names of the Caranginæ.
As is too often the case with that author, Dr. Günther has withheld all
definite information and means of verification of his statement. It
may be added, too, in this connection, that Dr. Günther had evidently
also "not paid attention to the structure of the skeleton" further than
as to the number of the vertebræ, for had he done so he would have
avoided the remarkable combination of genera he has assembled as
constituents of his "Carangidæ."

The family may be briefly diagnosed as follows:
Scombroidea* with the vertebræ in typical (10 +14), or nearly typical, number, the skull not expanded backwards and outwards, but with
the internal as well as external lateral crests continued backwards to
the exoccipital condyles, and the frontal bones coalesced; the body
moderately elongated and more or less compressed; a short spinous dorsal more or less developed, and a long soft dorsal and anal fins, the latter preceded by a more or less detached and distinct finlet of two spines
(sometimes atrophied).

The more detailed characteristics are as follows:
Body oblong, compressed, generally subfusiform (sometimes fusiform,
sometimes elevated), highest below the first dorsal fin, and with a slen-
der caudal peduncle. Anus antero-median.
Scales small, generally cycloid, and regularly imbricated.
Lateral line continuous to and ending at the base of the caudal fin.
Head compressed, oblong or short, and with the crown generally de-
curved or arched. Eyes moderate and submedian or anterior.
Suborbital bones small and not articulated with the preoperculum.
Opercular bones normally developed; suboperculum forming most of
the posterior border and the angle.
Nostrils double, in front of each eye.
Mouth moderate, with the cleft lateral and little oblique, generally
partly extending under the eyes.
Upper jaw not protractile, formed above by the premaxillary bones,

---

'The Scombridæ and Coryphænidæ exhibit the peculiarities of the vertebræ (as compared with spariform and perciform fishes, *e. g.*) manifest in the Carangidæ.

whose posterior or ascending processes are short, and on the sides by the supramaxillary bones, which are expanded towards the ends.

Teeth acute, variable in position, and sometimes entirely obsolete or lost in old age.

Branchial apertures very large and ample. Branchiostegal membrane deeply emarginated, sustained generally by 7 rays on each side (rarely by 5, 6, 8, 9, or 10.)

Spinous dorsal fin short, generally fully developed, but sometimes represented by free spines, which may be very small or even obsolete.

Soft dorsal fin commencing near the middle of the length, and little less than half as long as the trunk.

Anal fin opposite to, and generally nearly equal to, the soft dorsal, with two (rarely obsolete) spines in front, detached from the fin.

Caudal fin forked, and with its lobes slender and pointed.

Pectoral fins inserted at the normal moderate height above the breast on the scapular arch ; they are generally pointed.

Ventral fins thoracic and usually normally developed, each having a spine and five branched rays, which are regularly graduated. (In the *Paropsinæ* they are obsolete.)

The vertebræ are in normal (10 + 14) number, with few deviations (*e. g., Naucrates*, with 10 + 16); they are much contracted at the middle (like an hour-glass), and most (the costiferous and last caudal excepted) have anterior as well as posterior zygapophyses above and below, and the anterior pair of one vertebra are frequently interposed (or so tend) between the posterior pair of the preceding; the neurapophyses and hæmapophyses spring from near the middle or contracted portion of the vertebræ, and are moderately curved backward ; the costiferous vertebræ have pits behind or above the parapophyses for the reception of the ribs ; the parapophyses are obsolete on the anterior vertebræ, and only moderately developed backwards.

The skull is oblong, inclining to triangular, seen from above ; the brain-case is not expanded backwards or outwards, but provided with extensions from the lateral external and internal crests towards the exoccipital condyles ; the internal crests are continued forwards in a nearly or quite parallel direction ; the frontal bones are co-ossified ; the vomer projects forwards and downwards ; the post-frontals are more or less excavated or impressed on their inferior surface.

SERIOLINÆ.

Synonymy.

> Centronotini, *Bonaparte,* Giorn. Arcad. di Scienze, t. 52 (Saggio Distrib. Metod. Animali Vertebr. a Sangue Freddo, p. 34), 1832.

< Centronotinæ, *Swainson,* Nat. Hist. and Class. Fishes, etc., v. 2, pp. 176, 243, 1839.

> Centronotini, *Bonaparte,* Nuovi Annali delle Sci. Nat., t. 2, p. 133, 1838 ; t. 4, p. 275, 1840.

< Seriolinæ, *Gill,* Cat. Fishes E. coast N. A., p. 36, 1861 (n. d.).

> Centronotinæ, *Gill*, Cat. Fishes E. coast N. A., p. 36, 1861 (n. d).
= Centronotinæ, *Gill*, Proc. Acad. Nat. Sci. Phila., [v. 14,] p. 431, 1862.
< Seriolinæ, *Poey*, Anal. Soc. Esp. Hist. Nat., t. 4 (Enum. Pisc. Cub., p. 7), 1875.

The chief genera are the following:

SERIOLA.

Synonymy.

< Seriola, *Cuvier*, Règne Animal, 2e éd., t. 2 p. —, 1829. (Not *Seriola* Cass.)
< Seriola, *Cuv. & Val.*, Hist. Nat. des Poissons, v. 9, p. 200, 1833.
< Seriola, *Günther*, Cat. Fishes in Brit. Mus., v. 2, p. 462, 1860.
= Halatractus, *Gill*, Proc. Acad. Nat. Sci. Phila. [v. 14], p. 442, 1862.
Scomber sp., *Mitchill, etc.*

In returning to the name *Seriola* and abandoning *Halatractus*, I defer to the majority of naturalists, who consider that the same name may be used without interference in zoology and botany.

NAUCRATES.

Synonymy.

= Centronotus, *Lacépède*, Hist. Nat. des Poissons, t. 3, p. 311, 1802. (Not *Centronotus* Bl., Schu., 1801.)
= Naucrates, *Cuvier*, Règne Animal, 2e éd., t. 2, p. —, 1829.
= Naucrates, *Cuv. & Val.*, Hist. Nat. des Poissons, t: 8, p. 312. 1831 (adult).
> Nauclerus, *Cuv. & Val.*, Hist. Nat. des Poissons, t. 9, p. 247, 1833 (very young).
= Naucrates, *Günther*, Cat. Fishes in Brit. Mus., v. 2, p. 374, 1860 (adult).
> Nauclerus, *Günther*, Cat. Fishes in Brit. Mus., v. 2, p. 469, 1860 (very young).
= Naucrates, *Gill*, Proc. Acad. Nat. Sci. Phila. [v. 14,] pp. 262, 440, 1862.
Gasterosteus sp., *Linn., Daldorf, etc.*
Scomber sp., *Bloch, Mitchill, etc.*
Thynnus sp., *Gronow.*
Seriola sp., *Cuv. & Val.*, *Günther* (moderately young).

Even the partial synonymy of the pilot-fish is remarkable, viz:

NAUCRATES DUCTOR.

" *Pilot-fish.*"

1st c.—Pompilus, *Oridius*, Halieutica, l. 5.
1st c.—Pompilus, *Plinius*, Historiæ Mundil. ix, c. 61; xxxii, c. 11.
2d c.—Πόμπιλος, *Oppianus* ἀλιευτῖκων βιβλία, i.
3d c.—Πόμπιλος, *Athenæus* Deipnosophisticarum, vii.
3d c.—Πόμπιλος, *Aelianus*, Περὶ ζώων ἰδιότητος, ii. c. 15; xv, c. 2.
1558—Pompilus, *Gesner*, Historiæ Animalium l. iv.
1613—Pompilus, *Aldrovandus* De Piscibus l. iii, c. 19.
1667—Pilote, *Dutertre*, Hist. Gen. des Antilles, 2e éd., t. 2, p. 233.
1686—Pompilus, *Willoughby*, De Hist. Piscium. lib. p. 215, app. pl. 8, f. 2.
1713—Pompilus, *Ray*, Synopsis Methodica Piscium, p. 101.
1714—Pompilus, *Feuillée*, Journal d'Observations de Physique, etc.
1738—Coryphæna No. 3, *Artedi*, Genera Piscium, p. 16.
1754—Gasterosteus spinis dorsalibus quatuor, *Linnæus*, Museum Adolph. Friederici, p. 88.

1755—Scomber ductor, *Osbeck*, Acta Stockholmense, p. 71 ?, (fide Linnæi).
1757—Scomber ductor, *Hasselquist*, Iter Palistinense, p. 336.
1758—Gasterosteus ductor, *Linnæus*. Systema Naturæ, ed. x, t. 1, p. 295, 1758; (ed. xii, t. 1, p. 489, 1766).
1763—Scomber sp., *Gronow*, Zoophylacium No. 309.
1768—Scomber, *Læfling*.
1768—Gasterosteus ductor, *Brunnich*, Ichthyologia Massiliensis, p. 67.
1770—Scomber sp., *Koelreuter*, Novi. Commentar. Petrop., t. 9, p. 464, tab. 10, f. 4, ? 5.
1771—Scomber ductor, *Osbeck*, Voyage to China.
1782—*Duhamel du Monceau*, Traité Gén. des Pesches, t. 2, sect. 4, pl. 4, f. 4, pl. 9, f. 3.
1792—Gasterosteus ductor, *Walbaum*, Artedi Genera Piscium, p. —.
1793—Scomber ductor, *Bloch*, Ausländische Fische, p. —, taf. 338.
1800?—Gasterosteus antecessor, *Daldorf*, Skrivt. Nat. Selskab. Kjobenhavn, t. 2, p. 166.
18 —Gasterosteus antecessor, *Geoffroy St. Hilaire*, Annales Mus. d'Hist. Nat., t. 9, p. 469.
18 —Pilote, *Bosc*, Dict. d'Hist. Nat. de Deterville.
1801—Scomber ductor, *Bloch*, Systema Ichthyologiæ, Schneider ed., p. 32.
1801—Scomber Koelreuteri, *Bloch*, Systema Ichthyologiæ, Schneider ed., p. 570.
1802—Centronotus conductor, *Lacépède*, Hist. Nat. des Poissons, v. 3, p. 311.
1803—Scomber ductor, *Shaw*, Gen. Zoology, v. 4, p. 586.
1810—Naucrates fanfarus, *Rafinesque*, Caratteri de Alcuni Nuovi Generi e Nuove Specie di Animali e Piante della Sicilia, p. 45.
1810—Naucrates conductor, *Rafinesque*, Caratteri de Alcuni Nuovi Generi e Nuove Specie di Animali e Piante della Sicilia, p. 44.
1810—Centronotus conductor, *Risso*, Ichthyologie de Nice, p. 428.
1814—Scomber ductor, *Mitchill*, Trans. Lit. and Phil. Soc. New York, v. 1, p. 424.
1825—Centronotus conductor, *Couch*, Trans. Linn. Soc., v. 14, p. 82.
1827—Centronotus conductor, *Risso*, Hist. Nat. Europe Mérid., t. 3, p. 193.
1829—Naucrates indicus, *Lesson*, Voyage sur la Coquille, Zoologie, p. 157, pl. 14.
1831—Naucrates ductor, *Cuv. & Val.*, Hist. Nat. des Poissons, v. 8, p. 312.
1831—Naucrates noveboracensis, *Cuv. & Val.*, Hist. Nat. des Poissons, v. 8, p. 325.
1831—Naucrates indicus, *Cuv & Val.*, Hist. Nat. des Poissons, v. 8, p. 326.
1831—Naucrates Keolreuteri, *Cuv. & Val.*, Hist. Nat. des Poissons, v. 8, p. 327.
1833—Seriola Dussumieri, *Cuv. & Val.*, Hist. Nat. des Poissons, v. 9, p. 217.
1833—Seriola succincta, *Cuv. & Val.*, Hist. Nat. des Poissons, v. 9, p. 218.
1833—Nauclerus compressus, *Cuv. & Val.*, Hist. Nat. des Poissons, v. 9, p. 249
1833—Nauclerus abbreviatus, *Cuv. & Val.*, Hist. Nat. des Poissons, v. 9, p. 251.
1833—Nauclerus brachycentrus, *Cuv. & Val.*, Hist. Nat. des Poissons, v. 9, p. 253.
1833—Nauclerus triacanthus, *Cuv. & Val.*, Hist. Nat. des Poissons, v. 9, p. 253.
1833—Nauclerus annularis, *Cuv. & Val.*, Hist. Nat. des Poissons, v. 9, p. 254.
1833—Nauclerus leucurus, *Cuv. & Val.*, Hist. Nat. des Poissons, v. 9, p. 255.
1834—Naucrates ductor, *Cuvier*, Animal Kingdom, Griffith ed., v. 10, p. 189, pl. 47, f. 1.
1835—Centronotus ductor, *Jenyns*, Syst. Cat. Brit. Vertebr. Animals, p. 365.
1839—Naucrates ductor, *Swainson*, Nat. Hist. and Class. Fishes, v. 2, p. 412.
1839—Naucrates cyanophrys, *Swainson*, Nat. Hist. and Class. Fishes, v. 2, p. 412.
1839—Naucrates serratus, *Swainson*, Nat. Hist. and Class. Fishes, v. 2, p. 413.
1840—Gasterosteus ductor, *Bennett*, Narrative of a Whaling Voyage, v. 2, p. 274.
1840—Nauclerus abbreviatus, *Lowe*, Proc. Zool. Soc. London, v. 8, p. 36; *reprinted in* Trans. Zool. Soc. London, v. 3, p. 3.
1841—Naucrates ductor, *Yarrell*, Brit. Fishes, 2d ed., v. 1, p. 170; (3d ed., v. —, p.—)
1842—Naucrates noveboracensis, *DeKay*, Nat. Hist. of New York, Fishes, p. 112.
1842—Naucrates ductor, *DeKay*, Nat. Hist. of New York, Fishes, p. 113.
1846—Naucrates indicus, *Richardson*, Rep. 15th Meeting Brit. Assoc. Adv. Sci., p. 269.
1846—Naucrates fanfarus, *Bonaparte*, Cat. Metod. Pesci Europei, p. 72.

1846—Naucrates ductor, *Bonaparte*, Cat. Metod. Pesci Europei, p. 72.
1846—Naucrates ductor, *Storer*, Mem. Am. Acad. Arts and Sci. (2), v. 2, p. 349; Syn. Fishes N. Am., p. 97.
1846—Naucrates noveboracensis, *Storer*, Mem. Am. Acad. Arts and Sci. (2), v. 2, p. 349; Syn. Fishes N. Am., p. 97.
1846—Naucrates indicus, *Cuvier*, Règne Animal, éd. de luxe, t. 2, p. —, pl. 54, f. 1.
1850—Naucrates ductor, *Guichenot*, Exploration Scient. de l'Algérie, Poissons, p. 60.
1854—Thynnus pompilus, *Gronow*, Systema Ichthyologicum, publ. Gray, p. 123.
1860—Naucrates ductor, *Gunther*, Cat. Fishes in Brit. Mus., v. 2, p. 374.
1860—Seriola Dussumieri, *Gunther*, Cat. Fishes in Brit. Mus., v. 2, p. 468.
1860—Seriola succincta, *Günther*, Cat. Fishes in Brit. Mus., v. 2, p. 462.
1860—Nauclerus compressus, *Gunther*, Cat. Fishes in Brit. Mus., v. 2, p. 469.
1860—Nauclerus abbreviatus, *Günther*, Cat. Fishes in Brit. Mus., v. 2, p. 469.
1860—Nauclerus brachycentrus, *Gunther*, Cat. Fishes in Brit. Mus., v. 2, p. 470.
1860—Nauclerus triacanthus, *Günther*, Cat. Fishes in Brit. Mus., v. 2, p. 470.
1860—Nauclerus annularis, *Gunther*, Cat. Fishes in Brit. Mus., v. 2, p. 470.
1860—Nauclerus leucurus, *Gunther*, Cat. Fishes in Brit. Mus., v. 2, p. 470.
1862—Naucrates ductor, *Gill*, Proc. Acad. Nat. Sci. Phila. [v. 14], pp. 262, 440. (*Naucrates* recognized as old and *Nauclerus* as young of same fish.)
1868—Naucrates ductor, *Poey*, Repertorio Fisico-Natural de la Isla de Cuba, t. 2, p. 374.

Habitat.—High seas. •

It will be thus seen that twelve nominal species were based on specimens of this one by Cuvier and Valenciennes, and nine by Dr. Günther, who referred some to the family Carangidæ because they were supposed to have 24 (10 + 14) vertebræ, and one to the family Scombridæ, because the skeleton in the B. M. had 26 (10 + 14) vertebræ, I demonstrated in 1862 that all such forms belonged to one species, and the truth of this has been generally recognized since.

SELENINÆ.

Synonymy.

>Selenidi, *Rafinesque*, Indice d' Ittiolog. Siciliana, p. 15, 1810.
<Vomerini, *Bonaparte*, Nuovi Annali delle Sci. Nat., t. 2, p. 133, 1838; t. 4, p. 276, 1840.
<Vomerini, *Bonaparte*, Giorn. Arcad. di Scienze, v. 52 (Saggio Distrib. Method. Animali Vertebr. a Sangue Freddo, p. 34), 1832.
=Vomeriinæ, *Gill*, Proc. Acad. Nat. Sci. Phila., [v. 14,] pp. 431, 436, 1862.
≞Vomerini, *Poey*, Anal. Soc. Esp. Hist. Nat., t. 4 (Enum. Pisc. Cub., p. 7), 1875.

CARANGINÆ.

Synonymy.

× Caranxia, *Rafinesque*, Analyse de la Nature, p. —, 1815.*
< Carancini, *Bonaparte*, Giorn. Arcad. di Scienze, t. 52 (Saggio Distrib. Method. Animali, Vertebr. a Sangue Freddo, p. 34), 1832.*

---

* Corrected to "Carangidæ" (not Carangoidæ) by Agassiz (Nom. Zool. Index Un., p. 188, 1848), but without intending to adopt the group as a family.

< Carangini, *Bonaparte*, Nuovi Annali delle Sci. Nat., t. 2, p. 133, 1838; t. 4, p. 275, 1840.*

< Carangina, *Günther*, Cat. Fishes in Brit. Mus., v. 2, pp. 417, 419, 1860.

= Caranginæ, *Gill*, Proc. Acad. Nat. Sci. Phila. [v. 14], p. 431, 1862.

< Carangini, *Poey*, Anal. Soc. Est. Hist. Nat., t. 4 (Enum. Pisc. Cub., p. 7), 1875.

< Centronotinæ gen., *Swainson*.

The synonomy of the genus Trachurus is as follows:

TRACHURUS.

Synonymy.

= Trachurus, *Rafinesque*, Caratteri di Alcuni Nouv. Genere e Nuov. Specie di Animali e Piante della Sicilia, etc., p. 41, 1815.

= Caranx (Trachurus), *Cuv. & Val.*, Hist. Nat. des Poissons, t. 9, p. 6, 1833. (Section.)

< Selar, *Bleeker*, Natuurkundig Tijdschrift voor Nederlandsch Indie, v. 1, pp. 343, 352, 1850.

< Trachurus, *Girard*, Expl. and Surv. for R. R. Route to Pac. Oc., v. 10, Fishes, p. 107, 1858.

= Trachurus, *Günther*, Cat. Fishes in Brit. Mus., v. 2, p. 419, 1860.

Scomber sp., *Linn.*

Caranx sp., *Lac. et al.*

Caranxomorus sp., *Lac.*

Seriola sp., *Bowditch.*

CHLOROSCOMBRINÆ.

Synonymy.

= Chloroscombrinæ, *Gill*, Proc. Acad. Nat. Sc. Phila. [v. 14,] p. 431, 1865.

= Chloroscombrini, *Poey*, Anal. Soc. Esp. Hist. Nat., t. 4 (Enum. Pisc. Cub., p. 7), 1875.

TRACHYNOTINÆ.

Synonomy.

= Trachynotinæ, *Gill*, Proc. Acad. Nat. Sc. Phila., [v. 14,] p. 431, 1862.

< Trachynotini, *Poey*, Anal. Soc. Esp. Hist. Nat., t. 4 (Enum. Pisc., Cub., p. 7), 1875.

CATALOGUE OF A COLLECTION OF BIRDS MADE IN THE INTE-RIOR OF COSTA RICA BY MR. C. C. NUTTING.

BY ROBERT RIDGWAY.

On page 383 of the present volume, reference is made to a collection of birds from the interior of Costa Rica, but which at the time of writing had not been received. This collection has lately come to hand, and a list of the species is presented herewith. The better to aid our knowledge of the geographical distribution of Central American birds, the specimens collected at the two principal points of San José and the Volcan de Irazú, are given in separate lists. The prominent characteristics of these two localities having been given in the paper above referred to, we will proceed at once with the enumeration.

All notes on habits, color of eyes, etc., are by Mr. Nutting.

I.—*Species collected on the Volcan de Irazú.*

1. Catharus frantzii (Cab.).

One specimen secured. Iris brown; legs very pale.

No. 78. ad. March 11, 1882.

2. Merula grayi (Bp.).

Apparently not so common at this altitude as lower down.

No. 61. ♀ ad. March 7.

3. Merula plebeia (Cab.).

Common at a high altitude.

No. 22. February 28.

No. 47. ♀ ad. March 4.

4. Merula nigrescens (Cab.).

The single specimen secured was shot on the summit of the Volcano Irazú.

No. —. ♀ ad. February 24.

5. Thryophilus modestus (Cab.).

One specimen shot in the thick forest.

No. 63. ♀ ad. March 8.

6. Henicorhina leucophrys (Tsch.).

This pretty little wren seems to prefer the cool shade of the dense woods to more open country, and is a voluble songster, although most of its time seems to be passed in silence.

One specimen. Iris, reddish brown.

No. 82. ♂ ad. March 11.

7. Parula gutturalis (Cab.).

Abundant, rather high on the mountain.

Two specimens.

No. 4. February 23.

No. 5. ♀ ad. February 23.

8. Dendrœca virens (Gm.).

Common. Two specimens.

No. 19. (Sex ?) February 27.

No. 52. ♀ ad. March 6.

9. Myiodioctes pusillus (Wils.).

This sprightly and familiar warbler is one of the most common representatives of its family in Costa Rica, especially in the more elevated portions of the country. Five specimens.

No. 17. (Sex?) February 27.
No. 18. ♂ —. February 27.
No. 28. (Sex ?) February 28.
No. 29. ♂ ad. February 28.
No. 50. ♂ ad. March 4.

10. Setophaga torquata (Baird).

Abundant in thick forest, at a high elevation.

No. 6. February 23.
No. (?) March 1.

11. Pyranga bidentata (Sw.).

Only one specimen seen, and that shot from a hedge-row in the open country.

No. 23. ♂ juv. February 28.

12. Buarremon brunneinucha (Lafr.).

Common. Habits very like our *Zonotrichia albicollis*, at least so far as a marked preference for brush heaps and tangled thickets of under-growth is concerned. Iris brown.

No. 44. March 3.
No. 72. ♂ ad. March 9.

13. Buarremon chrysopogon (Bp.).

Common. Habits like the preceding. Iris reddish brown.

No. 64. ♀ ad. March 8.
No. 77. ♂ ad. March 10.

14. Pheucticus tibialis (Lawr.).

Rather common. A shy and silent bird, found in thick growths of tall reeds.

No. 51. ♀. March 6.

15. Phonipara pusilla (Sw.). .

Not common. Found generally in open country.

No. 60. ♀. March 7.

16. Passerina cyanea (L.).

Rare in Costa Rica. Sr. Zeledon informs me that my specimen was the first he had seen, although he had heard of its occurrence in the region.

No. 26. ♂. ad. April 28.

17. Junco vulcani (Bouc.).

A special trip to the top of the volcano was made for the purpose of securing specimens of this rare bird, which has been reported from no other locality. There is a belt of sandy soil studded with clumps of thick bushes surrounding the volcano near its summit, and in this belt *Junco vulcani* is abundant. In fact, it seems to be more abundant than any other bird in that exact locality. It is gregarious in its habits, like the rest of the genus, but seems to be rather more timid than the others.

Iris yellow. Legs pale.

Five specimens were secured February 23.

18. "Zonotrichia" pileata (Bodd.).

Very abundant, particularly along the hedge-rows that border the lanes.

No. 24. ♂ ad. February 28.
No. 27. ♀ juv. February 28.
No. 48. ♀ ad. March 4.

19. Psilorhinus mexicanus (Rüpp).

Abundant. The common Jay of the region. Very noisy and impudent. Found generally in open country. Iris brown.

No. 21. ♀ ad. February 27.
No. 30. Ad. February 28.
No. 31. ♀ ad. February 28.
No. 48. ♂ ad. March 4.

20. Elainea frantzii (Lawr.).

Very abundant along the hedge-rows. Six specimens. Iris brown.

No. 20. February 27.
No. 25. ♀ ad. February 28.
No. 57. ♂ ad. March 6.
No. 79. ♀ ad. March 11.
No. 80. March 11.

21. Tyrannus melancholicus satrapa (Licht.).

Abundant in open country.

No. 9. (Sex ?). February 24.
No. 56. ♂ ad. March 6.
No. 67. March 8.
No. 68. March 8.

22. Milvulus tyrannus (Linn.).

Common. At times these elegant Flycatchers associate in flocks, generally preferring the open fields.

No. 59. ♀ . March 7.

23. Chasmorhynchus tricarunculatus (Verr.).

Rather common in a restricted range of elevation on the volcano. The note of this bird seems to me to be anything but musical, being a curious compound of a croak, whistle, and creak, at somewhat lengthy intervals. I was unable to ascertain whether the curious wattle-like appendages were erectile or not.

No. 35. ♂ ad. March 1.

24. Picolaptes affinis (Lafr.).

Common, especially in thick forests.

No. 70.   ♂ ad.   March 9.
No. 83.   March 11.

25. Melanerpes formicivorus (Sw.).

No. 38.   ♂ ad.   March 1.
No. 39.   — ad.   March 1.
No. 40.   ♂ ad.   March 2.
No. 41.   ♂ ad.   March 2.
No. 42.   ♂ ad.   March 2.
No. 81.   ♂ ad.   March 11.

26. Selasphorus flammula Salv.

Rather common on Irazú at about the same altitude in which *Junco vulcani* is found.

Two specimens.

No. —.   ♂ .   February 23.
No. —.   ♀ .   February 23.

27. Pharomacrus mocinno costaricensis (Cab.).

NOTE.—In commenting upon Dr. Cabanis's proposed separation of the Costa Rican "Quezal" from that of Guatemala, Mr. Salvin points out (Proc. Zool. Soc. Lond., 1870, pp. 202, 203) the apparent unstability of the characters adduced. So far as my own experience goes, however, it is usually, if not always, quite easy to distinguish between birds from the two countries at first glance. I have just measured 19 adult males of the Costa Rican form, and find that in none of them do the longest upper tail-coverts exceed 30 inches in length from their insertion, the average being only 25½ inches, the minimum 19 inches. In *none* of them are there more than two of these feathers greatly elongated. The Guatemalan specimens which I have examined are unfortunately fewer in number,* but they could all be very readily distinguished not only by the very much longer and broader, but also more compact-webbed covert-plumes, while the shade of green was also appreciably more golden. I cannot at present give measurements of the Guatemalan bird, but am quite satisfied that the differences alluded to will be found reasonably constant.*

---

* I have handled altogether probably nearly 100 males of the Costa Rican bird.

* Since the above was written I have had an opportunity of measuring three specimens of the Guatemalan bird with the following result: Longest tail-coverts, 34–35.50 (average, 35.00); tail proper, 8–8.75 (average, 8.45); wing, 8.20–9 (average, 8.63).

Feb. 28, 1883.

Following are the extreme and average measurements of the series of adult males of the Costa Rican "Quezal" which I have just examined:

	Minimum.	Average.	Maximum.
Longest plumes, from point of insertion	19.00	25.50	29.75
Length of tail proper	7.50	7.74	8.50
Length of wing (11 specimens).................................	8.00	8.14	8.50

Common on Irazu at an altitude of about 8,000 feet. Note resembles that of a parrot. A shy and, for the most part, silent bird, much sought after by native hunters for its brilliant plumage.

Twelve specimens secured during the month of March, 1882.

28. Crotophaga sulcirostris Sw.

Here, as elsewhere in Costa Rica, this is among the most familiar of all birds.

No. 55.   March 6.

No. 62.   March 8.

No. 73.   ♂ ad.   March 10.

No. 74.   ♀ ad.   March 10.

No. 75.   March 10.

No. 76.   ♂ ad.   March 10.

28. Piaya cayana mehleri (Bp.).

Common. Generally silent, but it occasionally utters a loud, clear cry. Iris red.

No. 34.   ♂ ad.   March 1.

No. 45.   March 3.

29. Tinnunculus sparverius (Linn.).

Probably the most abundant hawk of the region.

No. 54. March 6.

30. Columba albilineata Gray.

Apparently not common, as only one specimen was seen. That was shot in a group of trees in a pasture near "Cot."

No. 43.   ♂ ad.   March 3.

31. Engyptila verreauxi (Bp.).

Rather common along the roads. Iris yellow. Legs red.

No. 16.   February 27.

No. 71.   ♀ juv.   March 9.

32. Geotrygon costaricensis Lawr.

Not very common. Found only in the densest parts of the forest, on the mountain side. Habits terrestrial. Iris and legs red.

No. 32.   ♀ ad.   March 1.

No. 33.   ♂ juv.   March 1.

II.—*Species collected in the vicinity of San José.*

1. Merula grayi (Bp.).

Abundant.

No. 87.   ♀ ad.   March 14, 1882.

No. 93.   ♀ ad.   March 15.

2. Thryophilus modestus (Cab.).

Common.   A fine songster.   I once heard a pair of these wrens singing together in a remarkable manner.   The male would utter two or three notes, and the female would take up the strain and finish it in perfect time.   This I heard repeated on several occasions.

No. 111.   ♀ ad.   March 19.

3. Dendrœca æstiva (Gm.).

Common.

No. 92.   ♀ ad.   March 15.

4. Basileuterus mesochrysus Scl.

Common in open country.

No. 90.   ♀ ad.   March 14.

No. 116.   (? ?)   March 20.

5. Hirundo erythrogastra Bodd.

Abundant.

No. 98.   March 15.

6. Tanagra cana diaconus (Less.).

A very abundant and familiar bird.   Often seen in the trees which surrround the plaza in San José, where it seems to make itself as much at home as the English Sparrow does in our public parks.

No. 88.   March 14.

No. 89.   March 14.

No. 112.   ♂ ad.   March 19.

No. 119.   ♂ ad.   March 20.

7. Ramphocelus passerinii Rp.

Apparently not very common.

No. 120.   (sex ?).   March 20.

8. Phonipara pusilla Sw.

Rare in this vicinity.   Only one seen and that was shot in a thicket bordering a stream.

No. 99.   ♂ ad.   March 15.

9. Pyrgisoma cabanisi Scl. & Salv.

Common a little lower down than San José.

No. 130.   ♂ ad.   March 25.

10. "Zonotrichia" pileata (Bodd.).
Abundant. Legs quite pale.
No. 97. ♂ juv. March 15, 1882.

11. Icterus galbula (Linn.).
Abundant around San José during our winter.
No. 86. ♂ ad. March 12.
No. 95. ♀ ad. March 15.

12. Elainea pagana (Licht.).
Common along the hedge rows.
No. 94. ♀ ad. March 15.
No. 96. March 15.
No. 118. ♂ ad. March 20.

13. Myiozetetes texensis (Giraud).
Common.
No. 109. March 19.
No. 110. ♀ ad. March 19.
No. 113. ♀ ad. March 19.

14. Pitangus derbianus Kaup.
Abundant.
No. 104. ♀ ad. March 19.

15. Megarhynchus pitangua (Linn.).
Sr. Don José Zeledon, who has collected for many years around San José, informs me that this is the only specimen which has been reported from the vicinity of San José. They usually are found at a considerably less elevation, where they are common. This specimen was found associating with the preceding species.
No. 108. ♂ ad. March 19.

16. Chiroxiphia linearis Bp.
This specimen was given to me, and I cannot vouch for its being secured near San José.
No. 127. March 25.

17. Tityra personata Jard. & Selby.
Common.
No. 131. March 25.

18. Petasophora cyanotis (Bourc.).
Common.
One specimen. Label list.

19. Oreopyra calolæma Salvin.
Bought in San José.
No. 134. ♂ ad. March, 1882.

20. Campylopterus hemileucurus (Licht.).
Common. Bought in San José.
No. 126. ♂ ad. March 25.
No. 128. ♂ ad. March 25.
No. 129. ♂ ad. March 25.

21. Chlorostilbon caniveti salvini (Cab. & Heine).
Only one specimen seen, though they are said to be abundant.
No. 117. ♂ ad. March 20.

22. Nyctidromus albicollis (Gm.).
Common.
No. 3. ♀ ad. February 21.

23. Centurus aurifrons hoffmanni (Cab.).
Abundant. The common Wood-Pecker of the region. Iris yellowish-brown.
No. 91. March 15.
No. 105. March 19.

24. Momotus lessoni Less.
This specimen was presented to me by Sr. Zeledon, who says they are common in the region, though I shot none myself.
No. 84. ♂ ad. March 13.

25. Ceryle americana cabanisi (Tsch.).
Abundant, especially in the lower parts of the country.
No. 125. ♂ ad. March 25.

26. Pharomacrus mocinno costaricensis (Cab.).
Brought to me at San José by native hunters. These gorgeous birds are only found in the elevated mountains in the interior, where they have a restricted and perfectly defined range of elevation.
No. 86. ♂ ad. March 14.
No. 102. ♂ ad. March 18.
No. 103. ♂ ad. March 18.
No. 132. ♂ ad. March 25. Presented by Dr. Van Patten.
No. 133. ♂ ad. March 25. Presented by Dr. Van Patten.

27. Conurus finschi Salvin.
The single specimen obtained is a female, perhaps immature. The plumage is entirely green, but with a few small red feathers on the fore-head and a very faint tinge of red on the under wing-coverts; under surface of remiges and rectrices, yellowish olive, appearing more yellow in certain lights; wing, 6.30; tail, about 5.00 (allowing for worn-off portion of the tip.).
No. 1. ♀ juv. February 19. Presented by Dr. Van Patten.

28. Glaucidium phalænoides (Daud.).
Rather rare; only one specimen seen; iris yellow, legs and cere greenish-yellow; secured in open country. ·
No. 107. ♂ ad. March 19.

29. Tinnunculus sparverius (L.).
Exceedingly abundant.
No. 2.   February 19.
No. 106.   ♀ ad.   March 19.

30. Chamæpelia passerina (L.).
Common; iris orange.
No. 85.   ♀ ad.   March 14.
No. 115.   March 20.

31. Engyptila verreauxi (Bp.).
Common; iris yellow; legs red.
No. 114.   ♀ ad.   March 19.

32. Geotrygon costaricensis Lawr.
Presented by Dr. Van Patten, of San José.
No. 135.   March 25.

33. Butorides virescens (L.).
One specimen.   Said to be common.
No. 100.   Guv.   March, 15.

In closing this list, justice requires an acknowledgment of the efficient aid of Sr. Don José Zeledon, who left nothing undone in the way of cheerful and painstaking assistance and genuine hospitality. Indeed, whatever of success has attended my trip to Costa Rica is due largely to his thoughtful generosity.

C. N.

BRIEF DESCRIPTIONS OF FOSSIL PLANTS, CHIEFLY TERTIARY, FROM WESTERN NORTH AMERICA.

BY J. S. NEWBERRY.

The following brief characterizations of fossil plants from the West are supplementary to the descriptions issued in the "Notes on Our Later Extinct Floras", published in the Annals of the Lyceum of Natural History of New York, 1868. Fuller descriptions, with figures of all the species enumerated in both series, with others yet to be added, will soon appear in a volume which is to form one of the Reports of the United States Geological Survey. Most of the fossil plants here enumerated were collected by Dr. F. V. Hayden, but a large number have also been obtained by Prof. Thos. Condon, State geologist of Oregon, by Prof. J. J. Stevenson and his assistant, Mr. I. C. Russell, and by others whose names are indicated in connection with their contributions.

Most of the originals of these descriptions will be placed in the National Museum and the annotated catalogue now issued finds an appropriate place in the Proceedings of the Museum.

J. S. NEWBERRY.

COLUMBIA COLLEGE, NEW YORK,
August 15, 1882.

1. EQUISETUM OREGONENSE, n. sp.

Stem robust, 3 centimeters wide; longitudinal flutings numerous, about 24 in a half-circumference; joints 5 centimeters distant; teeth triangular, short.

Formation and locality.—Miocene? Tertiary beds, Currant Creek, Oregon. Collected by Prof. Thos. Condon.

2. LASTREA (GONIOPTERIS) KNIGHTIANA, n. sp.

Frond large, tripinnate; pinnae linear, 2 centimeters wide, 14 to 16 centimeters long; pinnules diverging at a large angle, united for two-thirds of their length, upper third free, pointed and curved upward; venation clear and exact, midrib reaching the extremity of the pinnule; the lateral nerves about ten on either side, parallel, curved upward.

Formation and locality.—Tertiary strata, Currant Creek, Oregon, where it occurs matted together in masses. Collected by Prof. Thomas Condon.

3. ACROSTICHUM HESPERIUM, n. sp.

Frond large, pinnate; pinnae linear, 1½ to 2 inches wide, 6 to 12 inches long, rounded at remote extremity, those in lower part of frond rounded or wedge-shaped at base, those above united by the entire base to the rachis and with each other; rachis of frond and midrib of pinnae strong, smooth, somewhat sinuous; nervation reticulated, lateral nerves numerous, diverging from the midrib at an acute angle, anastomosing to form elongated six-angled areoles; fructification unknown.

Formation and locality.—Eocene Tertiary, Green River, Wyoming. Collected by Dr. C. A. White.

4. PTERIS ELEGANS, n. sp.

Pinnae linear, 25 millimeters wide; nervation remarkably strong and uniform; lateral nerves springing from the midrib at an angle of 45°, simple, strong, parallel from midrib to margin.

Formation and locality.—Tertiary strata, Currant Creek, Oregon. Collected by Prof. Thomas Condon.

5. PTERIS RUSSELLII, n. sp.

Frond large, pinnate; pinnae crowded, linear in outline, narrow, long-pointed above, attached to rachis by entire base; decurrent; length 16 to 20 centimeters; width 10 millimeters; margins undulate, irregularly-toothed; nervation fine, but distinct; branches leaving midrib at an angle of about 45°, all twice or three times forked.

Formation and locality.—Laramie Group, Vermejo Cañon, N. Mex. Collected by Mr. I. C. Russell.

6. PECOPTERIS (PHEGOPTERIS) SEPULTA, n. sp.

Frond small, delicate, pinnate; lower pinnae straight; broadly linear in outline, rounded above, attached to rachis, by the whole breadth of base; margins strongly lobed by the confluent pinnules, 1 centimeter wide by 5 centimeters long; upper pinnules crowded, conical in outline, gently curved upward, with waved or lobate margins; pinnules united by one-third of their length, oblong, obtuse; basal ones on lower side round, on the upper side flabellate, both attached by all their lower margin to the rachis of the frond; nervation strong and wavy, consisting of one many-branched nerve stem in each pinnule, each branch once or twice forked; fructification unknown.

Formation and locality.—Eocene Tertiary strata, Green River, Wyoming. Collected by Dr. C. A. White.

7. SEQUOIA SPINOSA, n. sp.

Branches slender, foliage open, rigid; leaves narrow, acute (acicular), arched upward, appressed or spreading; spirally divergent; staminate flowers in slender terminal aments 2 inches long, two lines wide, anthers few, under peltate connective scales; cones ovate or subcylindrical, composed of rhomboidal or square peltate scales.

Formation and locality.—Cook's Inlet, Alaska. Collected by Captain Howard, U. S. N.

8. SABAL POWELLII, n. sp.

Leaves of medium size, 4 or 5 feet in diameter, petiole smooth, unarmed, terminating above in a rounded or angular area, from which the folds diverge; beneath concavely narrowing to form a spike 3 to 4 inches in length; rays about fifty, radiating from the end of the petiole, perhaps sixty in the entire leaf, compressed to acute wedges where they issue from the petiole, strongly angled and attaining a maximum width of about 1 inch; nerves fine, about twelve stronger ones on each side of the keel, with finer intermediate ones too obscure for enumeration.

Formation and locality.—Eocene strata, Green River Station, Wyoming.

9. MANNICARIA HAYDENI, n. sp.

Frond large; leaves primately plicated, folds 1½ centimeters in width above, slightly narrowed below; flat or gently arched, smooth, springing from the midrib at an angle of 25° above, 30° below (in the specimens figured); folds attached to the midrib obliquely by the entire width and to each other by their entire length (?); the nervation fine, uniform (?), parallel.

Formation and locality.—Eocene strata, Green River Station, Wyoming. Collected by Dr. F. V. Hayden.

10. QUERCUS GRACILIS, n. sp.

Leaves narrow, lanceolate, long-pointed, acute, wedge-shaped at the base; margins set with remote, low, acute teeth; nervation regular and fine; nerve branches 15 to 20 on each side, curved gently upward, and terminating in the marginal teeth.

Formation and locality.—Laramie group, Point of Rocks, Wyoming.

11. QUERCUS CONSIMILIS, n. sp.

Leaves petioled, lanceolate, acuminate, wedge-shaped or rounded at base, where they are often unequal; margins usually dentate, occasionally only undulate, sometimes entire below, denticulate above; teeth acute, often spinous, sometimes short and closely appressed; nervation fine and regular; lateral nerves slender, parallel, generally arched upward, below; where margin is entire, camptodrome, above, craspedodrome, the branches terminate in the marginal teeth; tertiary nervation consisting of minute branches connecting the lateral nerves either directly or anastomosing, with fine quadrangular net-work filling the intervals. Fruit ovoid; when mature 2 centimeters in length by 15 millimeters in breadth; cupule scaly, covering nearly half of the glans.

Formation and locality.—Miocene ? strata, Bridge Creek, Oregon. Collected by Prof. Thomas Condon.

12. QUERCUS SIMPLEX, n. sp.

Leaves lanceolate, long-pointed, narrowed, and slightly rounded at the base; margins entire; nervation fine and regular.

Formation and locality.—Miocene ? strata, Bridge Creek, Oregon. Collected by Prof. Thomas Condon.

13. QUERCUS CASTANOPSIS, n. sp.

Leaves oblong-elliptical, rounded at the base; nervation regular; midrib straight, branches parallel, simple, terminating in the principal teeth of the margin; margin doubly dentate, the larger teeth receiving the extremities of the nerve branches, and each carrying a minor denticle; upper surface smooth; texture of the leaf coriaceous.

Formation and locality.—Argillaceous limestone, Yellowstone River. Collected by S. M. Rothhammer.

14. QUERCUS PAUCIDENTATA, n. sp.

Leaves oblanceolate, 6 inches in length by 1½ in breadth, narrowed to the base, sometimes unsymmetrical, long-pointed, and acute at the summit; margins entire below, coarsely toothed above; nervation strong and regular, about ten branches on each side of the midrib, which curve upward, festooned below, terminating in the teeth above.

Formation and locality.—Miocene ? Tertiary, Bridge Creek, Oregon. Collected by Prof. Thomas Condon.

15. QUERCUS LAURIFOLIA, n. sp.

Leaves petioled, lanceolate, 6 inches in length by 1½ inches in width, equally narrowed to the point and petiole; margins entire, or faintly toothed, or undulate; nervation regular; midrib strong, straight, lateral branches, about ten pairs, arching gently upward, terminating in the margins.

Formation and locality.—Burned shales, over lignite beds, Fort Berthold, Dakota. Collected by S. M. Rothhammer, on the expedition of General Alfred Sully, U. S. A.

16. QUERCUS DUBIA, n. sp.

Leaf ovoid in outline, unsymmetrical; margins strongly and remotely toothed; teeth subacute or obtuse; nervation delicate; midrib flexuous; lateral branches, about six on a side, somewhat waved, branched, and interlocking, and terminating in the marginal denticles; surface smooth, consistence probably somewhat coriaceous.

Formation and locality.—Tertiary strata, Tongue River, Wyoming. Collected by Dr. Hayden.

17. QUERCUS SULLYI, n. sp.

Leaves ovate, pointed, wedge-shaped, or rounded at the base; margins set remotely or closely, with acute, spiny-pointed teeth; nervation strong, somewhat flexuous; lower pair of lateral nerves giving off numerous branches, middle and upper pairs simple below, forked at the summit.

Formation and locality.—Burned shales over lignite beds, Fort Berthold, Dakota. Collected by S. M. Rothhammer, on the expedition of General Alfred Sully, U. S. A.

18. QUERCUS CASTANOIDES, n. sp.

Leaf linear-lanceolate, acute, 6 inches long by 1 inch broad; margins remotely and somewhat irregularly set with coarse, in some cases spinous, teeth; nervation strong; midrib straight, sharply defined; lateral branches unequally spaced, simple, forked near the extremity, terminating in the marginal denticles.

Formation and locality.—Eocene Tertiary, Green River, Wyoming. Collected by Dr. C. A. White.

19. POPULUS POLYMORPHA, n. sp.

Leaves petioled, ovate, rounded or slightly wedge-shaped at the base, acute or blunt-pointed at the summit; margin coarsely and irregularly crenate, dentate, or crenate-dentate; nervation strongly marked, pinnate; in the more elongated forms, about eight branches on each side of the midrib given off at an acute angle; in the broader forms the lower nerves issue at nearly a right angle; the upper ones at an angle larger than in the preceding form.

Formation and locality.—Tertiary strata, Bridge Creek, Oregon. Professor Condon.

20. POPULUS ROTUNDIFOLIA, n. sp.

Leaves of small size, rarely more than an inch in diameter, approximately circular in outline, either quite round or transversely or longitudinally elliptical; slightly wedge-shaped at the base, and decurrent on the long petiole; basal margin entire; upper half of leaf coarsely crenate, dentate, and usually short pointed at the summit; nervation flabellate, consisting of a median and two principal lateral nerves, which give off numerous branches.

Formation and locality.—Tertiary strata, Yellowstone River, Wyoming. Collected by Dr. Hayden.

21. JUGLANS DENTATA, n. sp.

Leaves large and relatively broad, 7 inches long by 2½ inches wide; short petioled ; rounded, narrowed or unsymmetrical at base, marked with remote, appressed, somewhat coarse, teeth ; nervation distinct and regular; midrib straight, strong; lateral nerves about 12 pairs on each side, arched upward, much curved toward the extremities, deflected along the margin, finally terminating below in the marginal teeth, above, camptodrome ; tertiary nervation forming a complicated and irregular but sharply defined net-work.

Formation and locality.—Eocene strata, Green River Station, Wyoing. Collected by Dr. C. A. White.

22. JUGLANS OCCIDENTALIS, n. sp.

Leaves somewhat variable in form and size, from 3 to 8 inches in length and 1 to 2 inches in width, but generally 6 inches long by 1½ inches wide, broad-lanceolate in outline, widest in the middle, summit acute, base rounded, often unsymmetrical ; margins entire ; nervation delicate; midrib straight; lateral nerves, about twenty on each side, gently curved upward, the lower ones branched and anastomosing near their extremities, the upper simple and terminating in the margins ; tertiary nervation very delicate, or obscure from being buried in the parenchyma of the leaf, forming an open and irregular network. Fruit small, elongated. somewhat prismatic ; divisions of the envelope lenticular in outline, narrow, thin.

Formation and locality.—Eocene Tertiary, Green River, Wyoming. Collected by Dr. C. A. White.

23. CRATEGUS FLAVESCENS, n. sp.

Leaves small, about 1 inch in length and breath ; lobed ; lobes rounded and bearing a few teeth or crenulations ; the summit of the leaf trilobed, with two lateral lobes below on either side.

Formation and locality.—Miocene? Tertiary, Bridge Creek, Oregon. Collected by Prof. Thomas Condon.

24. ULMUS SPECIOSA, n. sp.

Leaves 4 to 6 inches in length by 2 inches in width ; petioled, long-ovoid, or elliptical in outline, pointed at summit ; margins coarsely and doubly serrate; nervation strong, regular, 15 to 20 parallel branches on either side of the midrib. Fruit large, 27 centimeters in diameter, sub-circular, emarginate.

Formation and locality.—Tertiary strata, Bridge Creek, Oregon. Collected by Prof. Thomas Condon.

25. ULMUS GRANDIFOLIA, n. sp.

Leaves large, 16 centimeters long by 8 centimeters wide, ovate, often unsymmetrical; nervation strong, regular; midrib straight; lateral nerves, about thirteen on each side, strong and simple, except at summit, where they give off numerous branches; margins sometimes entire at base, but oftener simply serrate-dentate throughout.

Formation and locality.—Tertiary strata, Tongue River, Wyoming. Dr. Hayden.

26. PLANERA VARIABILIS, n. sp.

Leaves lanceolate to broad ovate; usually unsymmetrical, petioled; summit acute, sometimes long-pointed; base rounded or wedge-shaped; margins coarsely crenulate-dentate, or serrate, with remote, appressed teeth; midrib straight, strong; lateral nerves delicate, frequently alternating stronger and finer, gently arched upward, terminating in the teeth of the border; the finer intermediate ones sometimes fading out before reaching the margin.

Formation and locality.—Eocene Tertiary, Green River Station, Wyoming. Collected by Dr. C. A. White.

27. PLANERA NERVOSA, n. sp.

Leaves ovate or lanceolate, pointed, wedge-shaped, or rounded at the base, petioled; margins set with coarse, appressed teeth; nervation strong, crowded, regular; lateral nerves simple, parallel, terminating in the teeth of the margins.

Formation and locality.—Eocene Tertiary strata, Green River, Wyoming. Collected by Dr. C. A. White.

28. PLANERA CRENATA, n. sp.

Leaves oblong, ovate; short petioled, 5 centimeters long by 25 millimeters wide; base rounded; summit blunt pointed; margins coarsely crenate; nervation simple, delicate, six simple branches on each side of the midrib terminating in the crenations of the margin.

Formation and locality.—Tertiary strata, Tongue River, Wyoming. Collected by Dr. Hayden.

29. BETULA ANGUSTIFOLIA, n. sp.

Leaves petioled, oblong-lanceolate, 3 inches long by 1 inch wide; wedge-shaped or slightly rounded at the base, acuminate at summit; margins finely serrate below, coarsely and doubly serrate above; nerves slender, about eight branches on each side of the midrib.

Formation and locality.—Miocene ? Tertiary strata, Bridge Creek, Oregon. Collected by Prof. Thomas Condon.

30. BETULA HETERODONTA, n. sp.

Leaf 2 to 4 inches in length, long petioled, ovate, acuminate, rounded at the base; margins coarsely and irregularly serrate, the principal denticles receiving the terminations of the nerve branches; the sinuses between these sometimes plain, sometimes set with a few small teeth; nervation delicate, about 8 branches given off from each side of the midrib.

Formation and locality.—Miocene? Tertiary strata, Bridge Creek, Oregon. Collected by Prof. Thomas Condon.

31. ALNUS ALASKANA, n. sp.

Leaf large, oblong-ovoid, acuminate, rounded, or slightly heart-shaped at base; nervation crowded, 16 to 18 branches on each side of the midrib; margins set with very numerous, small, uniform, acute teeth.

Formation and locality.—Tertiary strata, Kootzanoo Archipelago, latitude 57° 35', longitude 134° 19', Alaska Territory. Collected by U. S. steamer Saginaw, February 18, 1869.

32. ALNUS GRANDIFLORA, n. sp.

Leaves 4 to 5 inches in length by 3 inches in width; ovate; rounded or wedge-shaped at the base; blunt pointed at the summit; margins coarsely dentate; nervation strong, crowded; 12 or more parallel branches on either side of the midrib, the intervals between these crossed by numerous parallel, mostly straight nervules, dividing the surface into oblong, quadrangular areoles.

Formation and locality.—Tertiary strata, Cook's Inlet, Alaska. Collected by Captain Howard, U. S. N.

33. PLATANUS ASPERA, n. sp.

Leaves attaining a diameter of 1 foot or more; petioled; rounded at the base more or less; three-lobed, sometimes nearly ovoid; nervation strong, about 9 branches on each side of the midrib; margins deeply, and often compoundly, toothed.

Formation and locality.—Miocene? Tertiary, Bridge Creek, Oregon. Collected by Prof. Thomas Condon.

34. FRAXINUS INTEGRIFOLIA, n. sp.

Leaves short-petioled or sessile; lanceolate; broadest near the base, which is abruptly narrowed and wedge-shaped; summit narrowed, extremity rounded; margins entire; nervation reticulate, camptodrome; lateral branches connected in elegant festoons near the margins; intervals filled with net-work of roundish, polygonal meshes.

Formation and locality.—Tertiary strata, Bridge Creek, Oregon. Collected by Prof. Thomas Condon.

35. PRUNUS VARIABILIS, n. sp.

Leaves short-petioled, very variable in form; lanceolate or broadly lance-ovate, 2 to 3 inches long by 1 to 2 inches wide; acuminate at the summit, wedge-shaped at base; margins thickly set with minute, acute, appressed teeth.

Formation and locality.—Tertiary strata, Cook's Inlet, Alaska. Collected by Captain Howard, U. S. N.

36. ILEX MICROPHYLLA, n. sp.

Leaves small, short-petioled, ovate, slightly decurrent on the petiole, abruptly pointed above, often unsymmetrical; margins set with 3 to 5 spiny teeth on each side; nervation distinct, but open, about 4 pairs of branches springing from each side of the midrib, arching upward, terminating in the teeth of the margin; tertiary nervation consisting of a coarse, irregular reticulation.

Formation and locality.—Tertiary strata, near Fort Union, Dakota. (Dr. Hayden.)

37. CELTIS RUGOSA, n. sp.

Leaf long-ovoid to lanceolate, rounded and slightly heart-shaped at the base, long-pointed at summit, 7 to 12 centimeters long by 3 to 5 centimeters wide; margins set with coarse, obtuse teeth, undulate or rarely entire; nervation strong, flexuous; midrib undulate; lateral branches about six on each side, branching and interlocking near the margins; tertiary nervation transverse, parallel, strong.

Formation and locality.—Tertiary strata, Tongue River, Wyoming. Collected by Dr. Hayden.

38. CELTIS PARVIFOLIA, n. sp.

Leaves small; oblong-ovate in outline; rounded and unsymmetrical at the base, pointed at the summit; margins, except at the base, coarsely dentate; nervation sparse; two principal branches on each side of the midrib, one pair springing from the base and throwing off branchlets, another strong pair issuing from the midrib at the middle of the leaf, other delicate branches given off near the summit.

Formation and locality.—Tertiary strata, Tongue River, Wyoming. (Dr. Hayden.)

39. CERCIS BOREALIS, n. sp.

Leaves small, orbicular, or roundish ovate; blunt pointed, cordate at the base; margins entire; nervation delicate; midrib flexuous, about three lateral branches on each side, the basal pair throwing off several branchlets on the lower side and reaching to or above the middle of the leaf.

Formation and locality.—Tertiary beds, valley of the Yellowstone River, Wyoming. Associated with *Platanus Raynoldsii*, N, *Rhamnus parvifolius*, N, and *Aristolochia crassifolia*, N. (Dr. Hayden.)

40. FRAXINUS AFFINIS, n. sp.

Leaves petioled, lanceolate, long pointed, attenuate at base; margins coarsely and irregularly toothed at and above the middle.

Formation and locality.—Miocene (?) Tertiary strata, Bridge Creek, Oregon. Collected by Prof. Thomas Condon.

41. RHAMNUS PARVIFOLIUS, n. sp.

Leaves short-petioled, 2 to 3 inches long, elliptical or obovate, rounded at the summit, narrowed to the petiole below; margins dentate, except at base; teeth coarse, acute, appressed near the summit; nervation uniform, rather open, six to seven branches on each side of the midrib.

Formation and locality —Tertiary strata, associated with *Platanus Raynoldsii*, &c., valley of Yellowstone River, Wyoming. Collected by Dr. Hayden.

42. LAURUS ACUMINATA, n. sp.

Leaves about 40 millimeters in length by 16 millimeters wide; long-ovate or ovate-lanceolate in outline, rounded at the base, long-pointed, acuminate at summit; nervation camptodrome; midrib straight, strong, about five pairs of lateral nerves, strongly arched upward, forming festoons near the margin; the lower pair opposite strongest, and reaching the middle of the leaf; secondary nervation open, forming irregular, chiefly quadrangular spaces, filled with minute uniform areoles.

Formation and locality.—Yellowstone Valley, Wyoming. Collected by Dr. Hayden.

43. VIBURNUM GRANDIDENTATUM, n. sp.

Leaves ovate (?) long-pointed, very coarsely dentate, with triangular teeth; nervation fine, lateral branches terminating in the marginal teeth; the lower pair reaching above the middle of the leaf and throwing off branchlets, which enter the marginal denticles.

Formation and locality.—Tertiary strata, Tongue River, Wyoming. Collected by Dr. Hayden.

44. VIBURNUM CUNEATUM, n. sp.

Leaves petioled, long-obovate, 10 centimeters or more in length by 4 centimeters in width; margins entire below the middle, above, set with coarse subacute or acute teeth; nervation strong, simple; midrib straight, giving off at an acute angle 7 or 8 simple, strong nerve branches on either side, which terminate in the teeth of the margin.

Formation and locality.—Tertiary beds, Tongue River, Wyoming. Collected by Dr. Hayden.

45. VIBURNUM PAUCIDENTATUM, n. sp.

Leaves petioled; 4 inches long by 1½ inches wide; ovate-lanceolate, pointed; narrowed and slightly rounded at base; margins below the middle entire, above bearing three large obtuse teeth; nervation strong, simple; midrib straight, about 4 strong, simple branches on either side of the midrib, issuing at an acute angle, the lowest terminating in a rounded tooth in the middle of the leaf, the others in the three large teeth above.

Formation and locality.—Tertiary rocks, valley of Tongue River, Wyoming. Collected by Dr. Hayden.

46. FICUS ALASKANA, n. sp.

Leaves large, reaching 8 to 10 inches in length and breadth; trilobed, generally unsymmetrical; lobes pointed, usually obtuse; margins entire or locally undulate; nervation strong, conspicuously reticulate; principal nerves, three, giving off branches, which divide near the margins, sometimes connecting in festoons, sometimes craspedodrome; tertiary nervation forming a coarse net-work of usually oblong meshes filled with a fine polygonal reticulation; upper surface of the leaf smooth and polished, lower roughened by the reticulation of the nerves.

Formation and locality.—Tertiary strata, Cook's Inlet and Admiralty Inlet, Alaska. Collected by Captain Howard, U. S. N.

47. FICUS MEMBRANACEA, n. sp.

Leaves sessile, 4 to 6 inches in length, by $2\frac{1}{2}$ to $3\frac{1}{2}$ in width; ovate, abruptly and usually blunt pointed, narrowed to the base, generally unsymmetrical, margin entire, nervation delicate, open, camptodrome; 10 or more branches given off on either side of the midrib, curving upward, and forming a festoon near the margin.

Formation and locality.—Tertiary strata, Cook's Inlet, Alaska. Collected by Captain Howard.

48. FICUS CONDONI, n. sp.

Leaves large, sometimes nearly 2 feet in length, three to five lobed, slightly decurrent, and the petiole sometimes stipulate; margins entire, or gently undulate; nervation very strongly marked and closely reticulate, roughening the surface, camptodrome, but nerve branches sometimes terminating in the margins of the middle lobe.

Formation and locality.—Tertiary beds, Bridge Creek, Oregon. Collected by Prof. Thos. Condon.

49. FICUS (PROTOFICUS) NERVOSA, n. sp.

Leaves large, 8 to 10 inches in length by 5 inches wide, oval in outline, pointed at the summit, rounded at the base; nervation crowded, remarkably exact and regular; midrib strong and straight, 12 or more branches on either side, nearly equidistant, simple, strongly arched upward, forming a festoon along the margin; tertiary nervation consisting of numerous nearly simple and straight cross-bars, connecting the secondary branches at right angles, and short nervules running off from the midrib at right-angles; margins entire.

Formation and locality.—Light grey sandstone, Laramie Group, Evanston, Utah.

50. PROTOFICUS INEQUALIS, n. sp.

Leaves 4 to 5 inches long, by 3 inches wide; oval, pointed at the summit, narrowed and rounded at the unsymmetrical base; margins entire or in part undulate; nervation strongly defined but open; about 7 branches on each side of the midrib, the lower two or three giving off branches below, the upper simple, arched upward, terminating in the margin, the intervals between the branches spanned by numerous, generally simple, tertiary nerves.

Formation and locality.—Tertiary strata, Tongue River, Wyoming. Collected by Dr. Hayden.

51. VITIS ROTUNDIFOLIA, n. sp.

Leaf broadly rounded or sub-triangular in outline, cordate at the base, and with an acute point at the summit, and at the extremity of each of the angles; intermediate portions of the margin coarsely and bluntly toothed; strongly three-nerved; tertiary nervation distinct and flexuous.

Formation and locality.—Tertiary strata, Admiralty Inlet, Alaska. Captain Howard, U. S. N.

52. MAGNOLIA ROTUNDIFOLIA, n. sp.

Leaves petioled, large (8 inches in length by 6 inches in width), round-ovate in outline, rounded or blunt-pointed above and slightly wedge-shaped below; margins entire; nervation open and delicate; 4 to 6 lateral branches given off from the midrib at remote and irregular distances, curving gently upward, and forming festoons near the margin.

Formation and locality.—Laramie group; Fisher's Peak, New Mexico. Collected by Dr. Hayden.

53. MAGNOLIA ANGUSTIFOLIA, n. sp.

Magnolia attenuata. Web. Lesq. Tert. Flor., p. 250. Pl. XLV. Fig. 6.

Leaves petioled, 1 foot or more in length, by 2 to 3 inches wide in the middle; lanceolate, pointed above, gradually narrowed to the base; margins entire; nervation sparse; midrib straight, lateral nerves few, thin, gently arched, camptodrome.

Formation and locality.—Laramie group, Fisher's Peak, N. Mex.

54. ZIZYPHUS LONGIFOLIA, n. sp.

Leaves four to seven inches long by six to twelve lines wide; lanceolate, long-pointed, wedge-shaped at base and long-petioled; margins waved, or more or less distinctly toothed; midrib well defined from base to summit; basal pair of lateral nerves approaching closely to the margin near the middle of the leaf, then curving gently inward, and anastomosing with the higher lateral nerves, of which there are three or more set alternately and curving upward, forming a festoon near the margin; tertiary nerves very finely reticulated.

Formation and locality.—Eocene Tertiary, Green River, Wyoming. Collected by Dr. C. A. White.

55. ARALIA MACROPHYLLA, n. sp.

Leaves large, long-petioled, palmately five parted from the middle upward, divisions conical in outline, sometimes entire, often remotely

March 21, 1883.

occasionally coarsely toothed; nervation strong and regular; the midribs of the divisions strong and straight, those from the second lateral lobes springing from near the bases of the first lateral lobes; secondary nerves numerous, distinct, curved gently upward; where the margins are entire, partially camptodrome, where dentate, terminating in the teeth; tertiary nerves anastomosing to form quadrangular and very numerous areoles.

Formation and locality.—Eocene Tertiary, Green River, Wyoming. Collected by Dr. C. A. White.

56. BRASENIA ANTIQUA, n. sp.

Stems long, flexuous, cylindrical (now flattened), smooth, many times branched toward summit, bearing pedunculate spheroidal capitula consisting of numerous club-shaped pods.

Formation and locality.—Eocene Tertiary, Green River, Wyoming.

57. CABOMBA GRACILIS, n. sp.

Stem slender, smooth; submerged leaves, set at intervals of half an inch to an inch apart on the stem, opposite dichotomously and frequently branched, segments narrowly linear, or filiform, flattened, smooth, truncated, scarcely distinguishable from the stems and leaves of *C. Caroliniana.*

Formation and locality.—Tertiary strata, Fort Union, Dakota. Collected by Dr. Hayden.

58. CABOMBA GRANDIS, n. sp.

Stems smooth, originally cylindrical, now flattened; leaves opposite, many times dichotomously forked, spreading 4 to 6 inches long; segments flat, 2 to 3 millimeters wide, smooth, truncated or slightly rounded at the extremities. Resembles *C. gracilis* in all respects, but very much larger.

Formation and locality.—Tertiary strata, Fort Union, Dakota. Collected by Dr. Hayden.

59. BERBERIS SIMPLEX, n. sp.

Leaves pinnate with three or more pairs of leaflets: leaflets ovoid, rounded or emarginate at base, acute, with two to four large spiny teeth on each side.

Formation and locality.—Tertiary strata, Bridge Creek, Oregon. Collected by Prof. Thos. Condon.

60. CARPOLITHUS SPINOSUS, n. sp.

Fruit enclosed in an exocarp composed of three elliptical or lentiform segments, furrowed along the middle line of the dorsum and bristling with erect, acute spines 6 to 8 millimeters long; peduncle cylindrical, strong, 1 inch or more in length.

Formation and locality.—Upper Cretaceous? North branch of Purgatory River, New Mexico. Collected by Mr. I. C. Russell.

NOTE ON THE LEPTOCARDIANS.

BY THEODORE GILL.

To complete the series of notes on the nomenclature, etc., of the inferior vertebrates, I add the synonyms of the class Leptocardii and its subordinate terms.

THE CLASS LEPTOCARDIANS.

Synonyms as class names.

=Myelozoa (*Is. Geoffroy St. Hilaire*), *Bonaparte*, Comptes Rendus hebd. seances Acad. Sci., t. 43, p. 1022, 1856.
= Acrania *, *Häckel*, Generelle Morphologie der Organismen, B. 2, p. cxix, 1866.
= Leptocardia, *O. Schmidt*, Handbuch der Vergl. Anat., 6. aufl., p. 259, 1872.
= Leptocardia, *Cope*, Proc. Acad. Nat. Sci. Phila., [v. 20], p. 256?, 1868.
= Leptocardii, *Gill*, Arrangement Fam. Fishes, pp. ix, 25, 1872.

Synonyms as subclass names.

= Leptocardii, *Müller*, Abhandl. K. Akad. Wiss. zu Berlin, 1844, p. —, 1846.
= Entomocrania, *Huxley*, Proc. Zool. Soc. London for 1876, p. 58, 1876.

THE ORDER AMPHIOXI.

Synonyms as ordinal names.

=Amphioxi, *Bonaparte*, Catalogo Metodico dei Pesci Europei, pp. 9*, 92*, 1846.
= Cirrostomi, *Owen*, Anatomy of Vertebrates, v. 1, p. 9, 1866.

BRANCHIOSTOMIDÆ.

Synonyms as family names.

=Amphioxidæ, *Gray*, Synopsis of the Brit. Mus., p. 150, 1842.
= Amphioxini, *Müller*, Abhandl. K. Akad. Wiss. zu Berlin, 1844, p. 198, 1846.
= Branchiostomidæ, *Bonaparte*, Catalogo Metodico dei Pesci Europei, pp. 9*, 92*, 1846.
=Cirrostomi, *Owen*, Anatomy of Vertebrates, pt. 1, p. 9, 1846.
= Amphioxidæ, *Gray*, List Specimens Fish in Brit. Mus., pt. 1, p. 149, 1851.
= Amphioxoidei, *Bleeker*, Enum. Sp. Piscium Archipel. Indico, p. xxxiii, 1859.
= Branchostomoidæ, *Gill*, Cat. Fishes E. Coast N. Amer., p. 63, 1860.
= Cirrhostomi, *Günther*, Cat. Fishes Brit. Mus., v. 8, p. 513, 1870.

A representative of this family (apparently *Branchiostoma lanceolata*) has been found on the coast of the United States, on the Atlantic side, as high north as the Chesapeake Bay, and on the Pacific Coast a species also occurs (Branchiostoma, *Cooper*, Nat. Wealth Cal. by Cronise, p. 498,

---

* The "class" Acrania is coequal with the "*subphylum*" Leptocardia of Häckel ("Erstes subphylum der Wirbelthiere: LEPTOCARDIA, *Röhrenherzen* [=] Einzige classe der Leptocardier: ACRANIA *Schädellose*").

1868), three specimens having been dredged at "San Diego in ten fathoms' water; they here were yellowish, translucent, with a brown streak near the back."—(*Cooper MSS.*) It may be that this form belongs to the genus *Epigonichthys*.

BRANCHIOSTOMA.

Synonymy.

= Branchiostoma, *Costa*, Cenni Zoologici Napol., p. 49, 1834.
= Amphioxus, *Yarrell*, Hist. Brit. Fishes, [1st ed.,] p. 468, 1836.
= Branchiostoma, *Günther*, Cat. Fishes Brit. Mus., v. 8, p. 513, 1870.
Limax sp. *Pallas.*

EPIGONICHTHYS.

Spnonymy.

= Epigonichthys, *Peters*, Monatsberichte K. Akad. Wissensch. Berlin, 1876, pp. 322-327.

NOTE ON THE MYZONTS OR MARSIPOBRANCHIATES.

BY THEODORE GILL.

Inasmuch as it has been stated by Dr. Günther, in his recent "Guide to the Study of Fishes" (p. 1), that "according to the views generally adopted at present, all those vertebrate animals are referred to the class of fishes" which are below the Amphibians, the following note is added in connection with the succeding papers.

The heterogeneity of the combination formerly regarded as the class of fishes is now so evident to any one who has familiarized himself with the anatomy of the vertebrates that it is unnecessary to detail the points of difference. Suffice it to state that the differences between the Lepto-cardians, Marsipobranchiates, and typical Fishes at least are far greater than those between any adjoining classes of terrestrial vertebrates. To still confound them in a single class is therefore a taxonomic falsehood, without any justification from either a scientific or "practical" stand-point. The degree of divergence of the branchiferous vertebrates has been aptly recognized by Häckel, Gegenbäur, Lankester, etc., in their classifications, by the differentiation of *Amphioxus* from all the other verte-brates and of the Marsipobranchiates from all those thereafter remain-ing. *Amphioxus* has even been excluded from the true vertebrates by Semper (1874), Hoppe-Seyler (1877), and Balfour* (1880). In the United States there is not a single active ichthyologist who does not admit at least three classes of branchiferous vertebrates—the Leptocard-ians, Marsipobranchiates, and Fishes. The remark of Dr. Günther, therefore, finds no illustration in the United States, and the exceptions are conspicuous and brilliant even in England.

---

* There is only a nominal difference between the views of Lankester and Balfour, the former enlarging the term Vertebrata to include the Tunicates, and the latter limiting it to exclude *Amphioxus*.

The progress toward the recognition of the class of Myzonts or Marsipobranchiates is indicated in the following synonymy:

THE CLASS OF MYZONTS, OR MARSIPOBRANCHIATES.

Synonyms as class names.

< Pisces, *Artedi*, Genera Piscium, 1738.

× Pisces, *Linnæus*, Systema Naturæ, ed. x, t. 1, p. 239, 1758; ed. xii, t. 1, p. 419, 1766.*

< Pisces, *Gmelin*, Linn. Systema Naturæ, t. 1, p. 1126, 1788.

< Ichthyoderes (Ichthyodera), *Geoffroy St. Hilaire, Latreille*, Familles Nat. du Règne Animal, p. 107, 1825.

< Pisces, *Costa*, Cenni Zoologici Napol., p. 49, 1834. (Includes the genus *Branchiostoma* < class *Leptocardii*—in the class.)

< Myzontes, *Agassiz*, Contrib. to Nat. Hist. of U. S., v. 1, p. 187, 1857.

= Marsipobranchia, *Häckel*, Generelle Morphologie der Organismen, B 2, p. cxx, 1866.

= Dermopteri, *Cope*, Proc. Acad. Nat. Sci. Phila., [v. 20,] p. 256, 1868.

= Cyclostoma, *Gegenbaur*, Grundriss der vergleich. Anat., p. 577, 1870.

= Cyclostomata, *O. Schmidt*, Handbuch der vergleich. Anat., 6. Aufl., p. 259, 1872.

= Dermopteri, *Cope*, Proc. Am. Assoc. Adv. Sci. 1871, v. 20, p. 320, 1872.

= Marsipobranchiates, *Gill*, Arrangement Fam. Fishes, ix, 25, 1872.

= Marsipobranchiates, *Jordan*, Man. Vertebrates Northern U. S., p. 199, 1876.

= Myzonts (Marsipobranchii), *Wilder*, Proc. Am. Ass. Adv. Sci., v. 24, § B, p. 185, 1876. †

> Hyperotreta, † *Lankester*, Quart. Journ. Micr. Sc., v. 17, p. —, 1877.

> Hyperoartia, † *Lankester*, Quart. Journ. Micr. Sc., v. 17, p. —, 1877.

Synonyms as subclass names.

= Marsipobranchii, *Bonaparte*, Nuovi Annali delle Sci. Nat., t. 2, p. —, 1838; t, 4, p. 277, 1840.

= Marsipobranchii, *Müller*, Abhandl. K. Akad. Wiss. zu Berlin, 1844, p. —, 1846.

< Dermopteri (*Owen*), *Gill*, Cat. Fishes E. Coast N. A., p. 24, 1861 (Not Dermoptères, *Dum.*, q. = Salmonidæ.)

< Holocrania, *Huxley*, Proc. Royal Soc. London, v. —, 1874; Ann. and Mag. Nat. Hist., (4,) v. 15, p. 225, 1875.

NOTE ON THE BDELLOSTOMIDÆ AND MYXINIDÆ.

BY THEODORE GILL.

In 1872, in my "Arrangement of the Families of Fishes," I have recognized two families in the order Hyperotreta, and recently have communicated to Professors Jordan and Gilbert a new generic name for the many-gilled species. I herewith give diagnoses of the family and genera. The distinctions in each case, from the nearest allies, are so evident that the groups do not really need justification. Nevertheless, as others have thought differently, it is not superfluous to add that characters analogous to such as have been used to differentiate the family and genera would be regarded as of great systematic value in the

---

* The genus *Myxine* was referred to the class Vermes.

† These "classes" are regarded as constituents of a "grade" ("Cyclostoma") contrasting with another (Gnathostoma) including all the vertebrates with jaws.

classes of selachians and true fishes, and the only comprehensible reason why they should not be so regarded in the present case is because the species are few in number. Inasmuch, however, as the function of taxonomy is to express morphological truths, and not the quantities under which a given type is manifested, such reasons appear to be very insufficient.

The synonymy of the order Hyperotreta is appended.

HYPEROTRETA.

Synonyms as orders.

< Cyclostomi, *Bonaparte,* Giorn. Accad. di Scienze, v. 52 (Saggio Distrib. Metod. Animali Vertebr. a Sangue Freddo, p. 41), 1832.

< Helminthoidei, *Bonaparte,* Nuovi Annali delle Sci. Nat., t. 2, p. 133, 1838; t. 4, p. 277, 1840.

= Hyperotreta, *Müller,* Abhandl, K. Akad. Wiss. zu Berlin, 1844, p. —, 1846.

= Hyperotreti, *Bonaparte,* Cat. Metod. dei Pesci Europei, pp. 9*, 92,* 1846.

< Cyclostomata, *Fitzinger,* Sitzungber. K. Akad. der Wissensch. (Wien), B. 67, 1, Abth., p. 57, 1873.

BDELLOSTOMIDÆ.

Synonymy.

= Bdellostomidæ, *Gill,* Arrangement Families Fishes, p. 25, 1872.

Petromyzontidæ gen. *Bonaparte,* etc.

Myxinidæ gen., *Günther, Putnam,* etc.

Hyperotreta with the branchial apertures separate (6–11) and lateral, debouching directly on the sides.

POLISTOTREMA.

Synonymy.

= Polytrema, *Girard,* Report U. S. Naval Expd. to Chili, v. 2, p. 251, 1854, (provisional name, not of Ferussac, 1822, nor Risso, 1826.)*

= Polistotrema, *Gill* with *Jordan & Gilbert,* Proc. U. S. Nat. Mus., v. 4, p. 30, 1881.

Gastrobranchus sp., *Lacépède.*

Heptatrema sp., *Cuvier.*

Bdellostoma sp., *Müller, Girard, Günther, Putnam.*

Bdellostomids with an increased number of branchiæ (about 10 or 11 on each side) and the base of the tongue between the seventh or eighth pair of gills.

---

* "It is to be regretted that Duméril's appellation of *Heptatrema,* by referring to a point of organic structure subjected to variations, could not be retained to designate these fishes generically. If that name be restricted to the species provided with seven respiratory apertures, then each species would constitute a genus by itself; that with six of these apertures ought accordingly be called *Hexatrema;* then *Heterotrema* when six are observed on one side and seven on the other; *Heptatrema* when seven; and finally *Polytrema* for the species described farther on.

"Considering, however, the structure of the mouth, both internally and externally, we would not hesitate in uniting them all under the well appropriate name of *Bdellostoma,* suggested by Professor Müller."—*Girard, op. cit.*

Two species have been indicated, but whether they are really distinct remains to be established. Both are represented in the U. S. Nat. Mus., but the Chilian form is in very poor condition.

POLISTOTREMA DOMBEYII.

Synonymy.

1798—Gastrobranchus Dombey, *Lacépède*, Hist. Nat. des Poissons, t. 1, p. 531, pl. 23.
1829—Heptatremes Dombeyii, *Cuvier*, Règne Animal, 2. ed., t. 2, p. 405.
1834—Bdellostoma Dombeyii, *Müller*, Abhandl. Akad. Wiss. zu Berlin, p. 80.
1851—Bdellostoma Dombeyii, *Gray*, Cat. Specimens Fishes in Brit. Mus., pt. 1, p. 149.
1854—Bdellostoma polytrema, *Girard*, Proc. Acad. Nat. Sci. Phila., v. 7, p. 199.
1854—Bdellostoma polytrema, *Girard*, Report U. S. Naval Exped. to Chili, v. 2, p. 252, pl. 33.
1870—Bdellostoma polytrema, *Günther*, Cat. Fishes. Brit. Mus., v. 8, p. 512.
1874—Bdellostoma polytrema, *Putnam*, Proc. Bost. Soc. Nat. Hist., v. 16, p. 166 (156).
1881—Polistrotrema Dombeyi, *Gill with Jordan and Gilbert*, Proc. U. S. Nat. Mus., v. 3, p. 458, 1881.

POLISTOTREMA STOUTII.

Synonymy.

1878—Bdellostoma Stoutii, *Lockington*, Am. Nat., v. 12, p. 793.
1881—Polistotrema Dombeyi pt., *Gill with Jordan and Gilbert*, Proc. U. S. Nat. Mus., v. 3, p. 458; v. 4, pp. 18, 29, 1881.

Habitat.—California.

" Eleven gill openings on each side; ten teeth on the anterior and nine in the posterior series. 15½ inches long. Eel River, Humboldt County.

" It is rather singular that this fish, which is abundant in Eel River, and is sold for food, and also occurs in this harbor, should hitherto have escaped notice. I believe it to be the only species of its genus hitherto found on the Pacific coast of North America; and it differs from *Bdellostoma polytrema*, a species which occurs along the coast of Chili, both in the number of the gill openings and that of the teeth, *B. polytrema* having fourteen of the former and twelve of the latter in each series."— *Lockington.*"

A specimen was received, 1866, at the Smithsonian Institution from Dr. Canfield, and on comparison with *B. polytrema* could not be satisfactorily diagnosed.

HEPTATREMA.

Synonymy.

= Heptatrema, *Duméril fide* authors; * (possibly in his "Dissertation sur les Poissons Cyclostomes," which I have not seen.)
= Les Heptatrèmes, *Duméril*, Cuvier Règne Animal, 2é ed., t. 2, p. 405. 1829.
= Bdellostoma, *Müller*, Abhandl. K. Akad. Wiss. zu Berlin, 1834, p. 79, 1836; 1838, p. 173, 1839; 1841, p. 111, 1844.
< Bdellostoma, *Girard*, Report U. S. Naval Exped. to Chili (provisional name), v. 2, p. 251, 1854.

---

* This genus was not proposed in the "Zool. Anal." as Gray and Girard indicated, nor in the 1st ed. of Cuvier's Règne Animal.

> Hexatrema, *Girard*, Report U. S. Naval Exped. to Chili (provisional name), v. 2, p. 251, 1854.
> Heterotrema, *Girard*, Report U. S. Naval Exped. to Chili (provisional name), v. 2, p. 251, 1854.
> Heptatrema, *Girard*, Report U. S. Naval Exped. to Chili (provisional name), v. 2, p. 251, 1854.

Bdellostomids, with typically 7 (sometimes 6) pairs of branchial apertures and the base of the tongue between the anterior pair of gills.

MYXINIDÆ.

Synonymy.

= Missinidi, *Rafinesque*, Indice d'Ittiologia Siciliana. p. 49 (order), 1810.
< Cyclostomia, *Rafinesque*, Analyse de la Nature, p. 94, 1815.
= Suceurs, *Curier*, Règne Animal. t. 2, p. 116, 1817.
= Diporobranchia, *Latreille*, Fam. Nat. du Règne Animal, p. 112, 1825.
< Myxinidæ, *Gray*, List Spec. Fish Brit. Mus., part 1, Chondropteygii, p. 145, 1851.
< Myxinidæ, *Bonaparte*, Cat. Metod. dei Pesci Europei, pp. 9*, 92*, 1846.
< Myxinoidei, *Bleeker*, Enum. sp. Piscium archipel. Indico, p. xxxiii, 1859.
< Myxinidæ, *Gunther*, Cat. Fishes Brit. Mus., v. 8. p. 510, 1870.
= Myxinidæ, *Gill*, Arrangement Fam. Fishes, p. 25, 1872.
< Gastrobranchi, *Fitzinger*, Sitzungsber. K. Akad. der Wissensch. (Wien), B. 67, 1. Abth., p. 58, 1873.

Hyperotreta with the branchial sacs (6 pairs) opening by ducts confluent behind into one which opens below on each side of the median line.

MYXININÆ.

Synonyms as subfamily.

= Myxinia, *Rafinesque*, Analyse de la Nature, p 94, 1815.
= Gastrobranchini, *Bonaparte*, Giorn. Accad. di Scienze, v. 52 (Saggio. Distrib. Metod. Animal. Vertebr. a Sangue Freddo, p. 41), 1832.
= Gastrobranchini, *Bonaparte*, Nuovi Annali delle Sc. Nat., t. 2, p. 133, 1838; t. 4, p. 277, 1840.
= Myxinnii, *Bonaparte*, Cat. Metod. dei Pesci Europei, p. 92*, 1846.

MYXINE.

Synonymy.

Myxine, *Linnæus*, Syst. Nat., t. 1, p. —,* 1758.
Myxine, *Retzius*, K. Vet. Acad. Nya Handl. Stockholm, v. 11, pp. 110–114, 1790.
Gastrobranchus, *Bloch*, Allg. Naturgeschichte der Fisches, t. 12, p. 66, 1795.
Gastrobranchus, *Bloch*, Systema Ichthyologiæ, ed. *Schneider*, p. 534, 1801.
Myxine, *Muller*, Abhandl. K. Akad. Wissensch. Berlin, 1834–1843 (passim).

For the most recent and important information respecting the species of this genus, the "Notes on the genus Myxine, by F. W. Putnam" (Proc. Boston Soc. Nat. Hist., v. 16, pp. 127–135, 1874), should be consulted.

---

* Myxine is referred by Linnæus to the class Vermes.

NOTE ON THE PETROMYZONTIDS.

BY THEODORE GILL.

The purpose of the present note is to make known the characteristics of a genus (*Entosphenus*) proposed by me long ago, and, in connection with it, diagnoses of the genera of the family Petromyzontidæ are given, as well as the synonymy of the several groups of the order HYPERO-ARTIA.

THE ORDER HYPEROARTIA.

Synonyms as orders.

< Cyclostomi, *Bonaparte*, Giorn. Accad. di Scienze, t. 52 (Saggio. Distrib. Metod. Animali Vertebr. a Sangue Freddo, p. 41), 1832.

< Helminthoidei, *Bonaparte*, Nuovi Annali delle Sci. Nat., t. 2, p. 133, 1838; t. 4, p. 277, 1840.

= Hyperoartia, *Müller*, Abhandl. K. Akad, Wiss. zu Berlin, 1844, p. —. 1846.

< Hyperoartii, *Bonaparte*, Cat. Metod. dei Pesci Europei, pp. 9*, 91,* 1846.

< Cyclostomata, *Fitzinger*, Sitzungsber. K. Akad. der Wissensch. (Wien), B. 67, 1. Abth., p. 57, 1873.

The order 10, "Helminthoidei" is identical with the "Sectio 6, Cyclostomi" and the "Subclassis 4, Marsipobranchii" of Bonaparte.

PETROMYZONTIDÆ.

Synonymy.

= Lampredini, *Rafinesque*, Indice d'Ittiolog. Siciliana, p. —, 1810.

< Cyclostomia, *Rafinesque*, Analyse de la Nat., p. 94, 1815.

< Suceurs, Cyclostomes, *Cuvier*, Règne Animal [1re éd.], t. 2, p. 116, 1817; 2e éd., t. 2, p. 49, 1829.

= Anloedibranchia, *Latreille*, Fam. Nat. du Règne Animal, p. 111, 1825.

= Petromyzides, *Risso*, Hist. Nat. de l'Europe Merid., t. 3, p. 99, 1826.

< Petromyzonidæ, *Bonaparte*, Giorn. Accad. di Scienze, v. 52 (Saggio Distrib. Metod. Animali Vertebr. a Sangue Freddo, p. 41), 1832.

< Petromyzonidæ, *Swainson*, Nat. Hist. and Class. Fishes, etc., v. 2, pp. 196, 337, 1839.

= Petromyzonidæ, *Bonaparte*, Nuovi Annali delle Sci. Nat., t. 2. p. 133, 1838; t. 4, p. 277, 1848.

< Petromyzidæ, *Gray*, Syn. Brit. Mus., pp. 148, 150 (fide Gray), 1842.

= Petromyzontidæ, *Girard*, Expl. and Surv. for R. R. Route to Pacific Oc., v. 10, Fishes, p. 376, 1858.

= Petromyzontoidei, *Bleeker*, Enum. Sp. Piscium Archipel. Indico, p. xxxiii, 1859.

= Petromyzontoidæ, *Gill*, Cat. Fishes E. Coast N. America, p. 62, 1861.

= Petromymyzonini, *Siebold*, Süsswasserfische von Mitteleuropa, p. 366, 1863.

= Petromyzontidæ, *Günther*, Cat. Fishes Brit. Mus., v. 8, p. 499, 1870.

= Petromyzontes, *Fitzinger*, Sitzungsber. K. Akad. der Wissensch. (Wien), B. 67, 1. Abth., p. 58, 1873.

PETROMYZONTINÆ.

Synonymy.

< Lampredia, *Rafinesque*, Analyse de la Nature, p. 94, 1815.

< Petromyzonini, *Bonaparte*, Giorn. Accad. di Scienze, v. 52 (Saggio Distrib. Metod. Animali Vertebr. a Sangue Freddo, p. 41), 1832.

< Petromyzonini, *Bonaparte*, Nuovi Annali delle Sci. Nat., p. 133, 1838; t. 4, p. 277, 1840.

< Petromyzontiformes, *Bleeker*, Enum. Sp. Piscium Archipel. Indico, p. xxxiii, 1859.

Young.

> Ammocoetina, *Gray*, Proc. Zool. Soc. London, 1851, p. 235, 240; List Fish. B. M.; pp. 137, 145, 1851.

> Ammocoetini, *Bonaparte*, Cat. Metod. dei Pesci. Europei, p. 92,* 1846.

> Ammocoetiformes, *Bleeker*, Enum. Sp. Piscium Archipel. Indico, p. xxxiii, 1859.

Petromyzontids with the suproral lamina median and undivided.

Analysis.

I. Lingual teeth of anterior row coalesced into one; suproral lamina bicuspid or tricuspid.

  1. Discal and peripheral teeth in obliquely decurved continuous rows; suproral lamina contracted.

    1a. Suproral lamina bicuspid; anterior lingual tooth with a deep re-entering median groove terminating in a point .... *Petromyzon.*

    1b. Suproral lamina tricuspid; anterior lingual tooth with a slight median groove .... *Ichthyomyzon.*

  2. Discal and peripheral teeth differentiated; former vertically uniserial and bi- or tri- cuspid; latter uniserial and minute; suproral lamina extended laterally.

    2a. Suproral lamina bicuspid; anterior lingual tooth with a crescentiform dentated edge and the median denticle enlarged .... *Ammocœtes.*

    2b. Suproral lamina tricuspid; anterior lingual tooth wedge-shaped, and with an almost straight, finely serrate edge................................ *Entosphenus.*

II. Lingual teeth of anterior row two, distinct and conic; suproral lamina quadricuspid .... *Geotria.*

III. Lingual teeth three, standing on the same base, pointed and curved; the median smallest.. *Exomegas.*

PETROMYZON.

Synonymy.

= Petromyzon, *Linnæus*, Syst. Nat., ed. 10.

< Petromyzon (*Duméril*), *Cuvier*, Règne Animal, t. 2, p. 404, 1817.

< Petromyzon, *Gray*, Proc. Zool. Soc. London, pt. 19, p. 235, 236; List Specimens Fish Brit. Mus., pt. 1, p. 137, pl. 1, f. 1 (mouth), 1851.

= Lampetra, *Malm*, Forhandl. Skand. Naturf. 8 möde, p. 580, 1860.

< Petromyzon, *Gunther*, Cat. Fishes Brit. Mus., v. 8, p. 500, 1870.

Petromyzontinæ with the suproral lamina contracted and with two converging teeth ; introral lamina multicuspid; disco-peripheral teeth numerous, and in arched series (of 4 to 6 each) declining downwards;

and lingual teeth three, pectinate, the anterior deeply impressed and curved back at middle, and the posterior correspondingly curved backwards at inner anterior angles.

ICHTHYOMYZON.

Synonymy.

< Ichthyomyzon, *Girard*, Expl. and Surveys for R. R. Route to Pacific Oc., v. 10, Fishes, p. 381, 1858.
< Ichthyomyzon sp., *Günther*, Cat. Fishes, Brit. Mus., v. 8, p. 506, 1870.
Petromyzon sp., *auct.*

Petromyzontinæ with the suproral lamina rather contracted and tricuspid; infroral lamina multicuspid; disco-peripheral teeth numerous, and in arched series (of 4–7 each) declining downwards; and of lingual teeth, anterior little impressed and incurved at middle, and posterior (shed or wanting).

Type Ichthyomyzon castaneus, *Girard.*

AMMOCŒTES.

Synonymy.

> Lampreda, *Rafinesque*, Analyse de la Nature, p. 94 (not described), 1815.
> Pricus, *Rafinesque*, Analyse de la Nature, p. 94 (not described), 1815.
> Ammocœtes *A. Duméril*, in *Cuvier* Régne Animal [1re ed.], t. 2, p. 119, 1817 (based on larval form).
= Lampetra, *Gray*, Proc. Zool. Soc. London, pt. 19, pp. 235, 237, 1851; List Specimens Fish Brit. Mus., pt. 1, pp. 137, 140, pl. 1, f. 2 (month), 1851.
? Scolecosoma, *Girard*, Expl. and Surveys for R. R. Route to Pacific Oc., v. 10. Fishes, p. 379? 1858 (based on larual form).
= Petromyzon, *Malm*, Forhandl. Skand. Naturf. 8 möde, p. 580, 1860.
Ichthyomyzon sp., *Girard.*

Petromyzontinæ, with the suproral lamina transversely extended, and with two cusps, one at each end; infroral lamina multicuspid; discal teeth uniserial, broad and bi- or tri-cuspid; peripheral teeth uniserial; and very small lingual teeth.

ENTOSPHENUS.

Synonymy.

= Entosphenus, *Gill*, Proc. Acad. Nat. Sci. Phila., [v. 14,] p. 331, 1862.
< Ichthyomyzon, *Günther*, Cat. Fishes Brit. Mus., v. 8, p. 506, 1870.

Petromyzontinæ with the suproral lamina transversely extended and tricuspid; infroral multicuspid; discal teeth uniserial, broad, and bi- or tricuspid; peripheral teeth uniserial and minute; lingual teeth two, an anterior wedge-shaped, with an almost straight, finely serrate edge, and a posterior horse-shoe-shaped, and with a double finely serrate keel on the sides.

A partial synonymy of the type is as follows:

ENTOSPHENUS TRIDENTATUS.

Synonymy.

1836—Petromyzon tridentatus, *Gairdner, MSS.*
1836—Petromyzon tridentatus, *Richardson,* Fauna Boreali-Americana, v. 3, p. 293, 1836.
1842—Petromyzon tridentatus, *DeKay,* Nat. Hist. New York, v. 5, p. 381. (Extralimital.)
1846—Petromyzon tridentatus, *Storer,* Synopsis of the Fishes of North America (p. 266) in Mem. Am. Acad., v. 2, p. 518.
1851—Petromyzon tridentatus, *Gray,* Proc. Zool. Soc. London, pt. 19, p. 240; List Specimens Fish in Brit. Mus., p. 144.
1858—Petromyzon lividus, *Girard,* Expl. and Surv. for R. R. Route to Pacific Oc., v. 10, Fishes, p. 379.
1862—Eutosphenus tridentatus, *Gill,* Proc. Acad. Nat. Sci. Phila., [v. 14,] p. 331.
1870—Ichthyomyzon tridentatus, *Günther,* Cat. Fishes Brit. Mus., v. 8, p. 506.

Habitat.—California.

·GEOTRIA.

Synonymy.

> Geotria, *Gray,* Proc. Zool. Soc. London, pt. 19, p. 239, pl. 4, f. 3, (mouth,) pl. 5, (fish,) 1851; List Specimens Fish in Brit. Mus., pt. 1, p. 142, pl. 1, f. 3 (mouth), pl. 2 (fish), 1851.
> Velasia, *Gray,* Proc. Zool. Soc. London, pt. 19, p. 239, pl. 4, f. 4 ; List Specimens Fish in Brit. Mus., pt. 1, p. 143, pl. 1, f. 4 (mouth), 1851.
= Velasia, *Günther,* Cat. Fishes Brit. Mus., v. 8, p. 508, 1870.

Petromyzontinæ with the suproral lamina transversely extended and arched, armed with "four sharp flat lobes," the outer of which are enlarged ; the infroral lamina crescent-like, sinuous, or denticulated on the edge ; the discal teeth numerous and in diverging series, and the lingual teeth elongated, conic, and two in number (Gray and Günther).

EXOMEGAS.

Synonymy.

= Exodomegas, *Gill MSS.*
Petromyzon sp. *Burmeister.*

Petromyzontidæ with the discal teeth in concentric series, the outer containing the largest teeth (about 24 on each side), lingual teeth three, large pointed and curved, the median smallest, all standing on the same base.

Type *Exomegas macrostomus = Petromyzon macrostomus Burmeister,* Anal. Mus. Buenos Aires, pt. 5, 1868, Acta Soc Palæont., p. xxxvi.

CARAGOLINÆ.

Synonymy.

Caragolinæ, *Gill, MSS.*

Petromyzontids with two lateral suproral laminæ.

CARAGOLA.

Synonymy.

< Caragola, *Gray*, Proc. Zool Soc. London, pt. 19, p. 239, pl. 4, f. 5; List Specimens Fish in Brit. Mus., pt. 1, p. 143, pl. 1, f. 5 (month,) 1851.
< Mordacia, *Gray*, Proc. Zool Soc. London, pt. 19, p. 239, pl. 4, f. 6; List Specimens Fish in Brit. Mus., pt. 1, p. 143, pl. 1, f. 6 (month), 1851.
= Mordacia, *Günther*, Cat. Fishes Brit. Mus., v. 8, p. 507, 1870.

Caragolinæ with the suproral laminæ entirely separated, triangular, and each with three conic teeth; the infroral lamina "crescent-shaped, with about nine acute conical cusps;" the discal teeth "in somewhat distant series, radiating from the center," those of a series more or less confluent, and the lingual teeth serrated and in two pairs. (Günther.)

DESCRIPTION OF A NEW WARBLER FROM THE ISLAND OF SANTA LUCIA, WEST INDIES.

BY ROBERT RIDGWAY.

DENDRŒCA ADELAIDÆ DELICATA, subsp. nov.

CH. –Differing from *D. adelaidæ*, from Porto Rico, in very much more intense yellow superciliaries and lower parts, the former much broader (occupying the whole of the forehead except a central line), more decided plumbeous of the upper parts, more distinct black mark on sides and fore part of the crown, larger size, and other particulars.

Adult (♂?): Middle of crown, occiput, auriculars, sides of neck, and upper parts in general, clear uniform plumbeous; crown bordered anteriorly and laterally by a broad ∧-shaped mark of deep black, the apex of which is continued in a narrow streak to the base of the culmen, dividing two very broad frontal patches of intense yellow, which extend backward, more narrowly, to just behind the eye; a large suborbital spot of clear yellow, separated from that of the forehead by a narrow dusky loral streak. Malar region, chin, throat, and entire lower parts, except anal region and crissum, very rich, pure gamboge yellow, the first separated from the yellow suborbital spot by a dusky rictal stripe, which gradually blends posteriorly into the plumbeous of the auriculars; anal region, crissum, and lining of the wing pure white. Wings dusky black, the feathers edged with plumbeous, and both rows of coverts very broadly tipped with pure white, forming two conspicuous bands; tail feathers black, edged with plumbeous, the lateral feather with about .80 of an inch of the end of the inner web white, the next two similarly marked, but the white areas gradually reduced in size, the fourth feather with merely a broad terminal edging and small subterminal spot of white. Bill brownish black; feet horn-brown. Wing, 2.30; tail, 2.30; culmen, .42; tarsus, .70; middle toe, .40. (Type No. 80909; Sta. Lucia, W. L; F. A. Ober.)

This new form may be compared with *D. adelaidæ* (the type of the latter being used for comparison), as follows:

1. D. ADELAIDÆ. Above plumbeous-gray, slightly tinged with pale olive on the back, the crown narrowly but distinctly streaked with black; middle of forehead plumbeous-gray, this bounded on each side by an interrupted black streak, continued back along sides of crown; a very small or barely appreciable yellow suborbital spot; sides of forehead (or more properly, a broad supraloral stripe) with lower parts clear lemon-yellow, the belly much paler posteriorly. Remiges and rectrices edged with olive-gray. Wing, 2.10; tail, 1.80; culmen, .39; tarsus, .70; middle toe, .40.

Hab.—Porto Rico.

2. D. DELICATA. Above clear plumbeous, without trace of olive tinge on back or of dusky streaks on crown; middle of forehead (narrowly) black, this forming a broad **V**-shaped mark bordering the crown anteriorly and laterally; rest of forehead intense yellow, almost orange; a very conspicuous suborbital spot of pure yellow; lower parts intense gamboge yellow, scarcely paler on lower part of abdomen. Remiges and rectrices edged with pure plumbeous-gray. Wing, 2.30; tail, 2.30; culmen, .42; tarsus, .80; middle toe, .40.

Hab.—Sta. Lucia.

DESCRIPTION OF A SUPPOSED NEW PLOVER FROM CHILI.

BY ROBERT RIDGWAY.

ÆGIALITES ALBIDIPECTES, sp. nov.

SP. CH.—*Adult* (No. 26997, U. S. Nat. Mus.; "Chili, S. Am.;" collector and donor unknown): Forehead (broadly) superciliary stripe (extending back to end of auriculars), cheeks (up to lower eyelid) and lower parts generally, pure white, the whole breast strongly tinged with light pinkish cinnamon, this growing gradually deeper cinnamon on the sides of the neck and across the nape. A distinct stripe from the rictus to the eye, across the lores, the whole crown, and auriculars, black. Occiput and upper parts in general, deep brownish gray, tinged, especially on the back, with light yellowish fulvous. Primaries dusky, with white shafts; greater wing-coverts distinctly tipped with white; inner secondaries chiefly white. Lateral upper tail-coverts white. Two outer tail-feathers wholly white, the others dusky.

Wing, 4.10; tail, 1.90; culmen, .60; tarsus, 1.05; middle toe, .55.

HAB.—"Chili."

This species resembles somewhat the *Æ. ruficapillus*, Temm., of Australia, but the latter has the whole crown and nape deep cinnamon-rufous, bounded anteriorly by a line of black, widening somewhat on the side of the breast, and the breast and jugulum snowy white, besides other minor differences.

There is nothing known as to the history of the type specimen, the Museum record giving simply the locality as above quoted.

I have for a long time hesitated to describe this bird as a new species, having an idea that it might perhaps prove an abnormal example of *Æ. collaris*, but that such is not the case I am now quite convinced, since I have had the opportunity to examine large series of the latter species, including specimens from Chili.

CATALOGUE OF A COLLECTION OF BIRDS MADE AT VARIOUS POINTS ALONG THE WESTERN COAST OF LOWER CALIFORNIA, NORTH OF CAPE ST. EUGENIO.

BY L. BELDING.

[Edited by R. Ridgway.]

[The most important result of Mr. Belding's explorations along the western coast of Lower California is the discovery of the fact that, as far south at least as Cerros Island and Sta. Rosalia Bay (or to latitude about 28° north) the bird-fauna presents no differences whatever from that of the southern coast of California, in the vicinity of San Diego. The coast between Sta. Rosalia Bay and Cape St. Lucas being entirely unexplored zoölogically, it is of course unknown where the San. Diego fauna merges into, or gives way to, that of the cape district. Since, however, we are accustomed to look to topographical indications as a probable solution of questions of this kind, it seems not unlikely that the promontory of Cape St. Eugenio, which, as a lateral offshoot from the main range extends quite to the sea-coast on the south side of the bay in question, may form the dividing line; but it is, of course, quite possible that the San Diego fauna may extend much farther toward the cape. This question, however and that of how far northward along the eastern side of the peninsula the peculiar fauna of the cape district extends, can only be decided by further investigation.—R. R.]

I. Coronados Islands, about 20 miles south and west of San Diego.

These islands are small and rocky, and situated about 10 miles off the coast of Lower California. The larger island is inhabited by several species of reptiles which are found also on the mainland, thus affording good evidence of former connection with the peninsula.

A few birds only were taken on and in the vicinity of these islands, May 16 and 17, the more important of which are the following:

1. **Hæmatopus palliatus** Temm.

[A single specimen, collected May 17, is the only example of this species in the National Museum from any locality on the Pacific coast of North America north of Mazatlan. Not mentioned in Mr. Belding's notes.—R. R.]

2. Hæmatopus niger Pall.
A few of these birds found here.

3. Phalacrocorax penicillatus (Brandt).
Very abundant here, at Cerros, and several intervening points.

2. San Quentin Bay, west coast of Lower California, latitude about 30° 23′.

I arrived at this place May 2, and remained until the 11th. Much of the bay is shallow, consequently at low-tide very extensive mud flats are exposed, making this a favorite resort for water birds, which upon our arrival were very abundant, but previous to May 11 most of them had taken their departure for their northern breeding grounds. So far as observed, the land birds of this locality are about the same as those found in the vicinity of San Diego.

1. Harporhynchus redivivus (Gamb.).
Rare; apparently the same as the San Diego bird. [No specimens.]

2 Thryomanes bewicki spilurus (Vig.).
Not common.

3. Passerculus anthinus (Bp.).
Very common in grassy meadows ; probably resident here and at San Diego.

4. Melospiza fasciata samuelis Baird.
But one individual noticed. This was found in *tules* by a pond of fresh water.

5. Pipilo fuscus crissalis (Vig.).
Specimens shot but not preserved appeared to be identical with others shot at San Diego in April and May.

6. Guiraca cærulea (Linn.).
Rare ; observed May 8, in an extensive willow thicket. [No specimens.]

7. Calypte costæ (Bourc.).
Common. [No specimens.]

8. Lophortyx californica (Shaw).
Moderately common.

9. Squatarola helvetica (Linn.).
Common as late as May 10.

10. Ægialites semipalmatus Bp..
Rare.

11. Ægialites alexandrinus nivosus (Cass.).
Rare, but mated, and probably breeding.

12. Pelidna alpina americana (Cass.).
Abundant May 2, but rare by the 10th of the month.

13. Ereunetes pusillus occidentalis (Lawr.).

The same remarks apply to this as to the last.

14. Limosa fœda (Linn.).

Abundant early in May; perhaps a few breed here, as single birds were several times flushed from the meadows, and their actions led me to believe they had nests in the vicinity.

15. Rallus obsoletus Ridgw..

Although only two of these birds were seen here, considering their habits this would not determine the question of rarity or abundance; for marshes, such as birds of this family delight in, are here numerous, extensive, and difficult of exploration. One of the birds was flushed from her nest, which was well concealed in and beneath rank marsh grass on the bank of a slough, the eggs were eight to ten in number (some of them being crushed by the foot of a companion), and contained large embryos.

16. Bernicla nigricans (Lawr.).

Several small flocks were apparently much at home in the bay until May 9 or 10, when they disappeared. One of the specimens shot was astonishingly fat, and had been feeding on eel-grass, their usual food at this locality. The species was also seen at San Diego in April.

17. Dytes nigricollis californicus (Heerm.).

Very common May 16.

3. *Santa Rosalia Bay, west coast of Lower California, latitude* 28° 28.*

Crossing from Cerros Island to this locality, after waiting a day for the surf to subside, a landing was effected April 28.

In a brisk walk of five hours five species of land birds were observed. At the end of this time a fresh, increasing breeze made a hurried departure from this barren, arid region necessary.

I have been informed by persons who appeared to be acquainted with the coast that there is no fresh water between Magdalena Bay and some point not far from Sacramento Reef, and on this coast a collector's greatest danger is that he may perish from thirst.

1. Polioptila californica Brewst.

Common.

[NOTE.—In the absence of specimens it would be quite hazardous to conjecture whether the above-named species or *P. plumbea*, Baird, was the one seen. The latter, only, occurs in the vicinity of Cape St. Lucas, but it may be that on the Pacific side of the peninsula it is replaced by *P. californica.*—R. R.]

2. Amphispiza bilineata (Cass.).

Moderately common; not seen at San Quentin Bay nor San Diego.

3. **Amphispiza belli** (Cass.).
Common.

4. **Otocorys alpestris chrysolæma** (Wagl.).
One small flock seen.

5. **Calypte costæ** (Bourc.).
Common.

6. **Ægialites alexandrinus nivosus** (Cass.).
Paired.

7. **Larus heermanni** (Cass.).
Very common.

4. *Cerros Island, west coast of Lower California (latitude just north of the parallel of 28°.)*

Arriving at this island April 14, twelve days were spent in exploring it.

The total length of Cerros Island is about 20 miles, its greatest width about 8 miles. The highest peak reaches an altitude of about 4,000 feet, while much of the land is more than 1,500 feet above sea level. Like the western side of the peninsula, it is mostly rocky or sandy, and sparsely covered with, or in places entirely destitute of, vegetation. On the western side, from the crest-line downward, between 1,500 and 2,000 feet altitude, there is a considerable forest of pines (*Pinus muricata*). This forest, from which much was expected, proved to be a very poor collecting ground, although a few beetles, spiders, and ants, not noticed elsewhere on the island, were procured here. Wild goats (descendants of domestic animals) were the only quadrupeds seen; but deer, no doubt, inhabit the island, since two pairs of discarded antlers were found. There were also indications of the presence of a rodent, probably a species of *Neotomys* (Cave Rat).

Sixteen species of land-birds were noted, five of them being represented, so far as my observation is concerned, by a single specimen each; some of them may have been stragglers from the main land, as they were seen during the period of migration.

A "horned toad" (*Phyrnosoma*), three or four species of lizards, a tree-frog (*Hyla*), a few insects, and a land snail (*Helix*), comprised, apparently, with the other creatures herein enumerated, all there is of animal life on the island.

The plants of Cerros Island are partly Californian and partly Lower Californian, some of the species of the southern part of the peninsula growing there. The California Holly (*Heteromeles arbutifolia*) and *Juniperus californicus* were found, as well as the above-mentioned pine, the former growing in the pine forest. Fishermen go from all points of the compass to get the water from the spring on Cerros Island, designated on the charts "Watering place." This water is not good, but will do in the absence of better. It trickles out of a rock a few feet above sea level, and is marked by a bunch or two of bulrushes. There are also other bunches of bulrushes on the east side of the island, but this, I believe, is the southernmost of them all.

1. **Thryomanes bewicki spilurus** (Vig.).
Common.

2. **Troglodytes aëdon parkmanni** (Aud.).
Rare.

3. **Carpodacus frontalis rhodocolpus** (Cab.).
Three examples seen.

4. **Astragalinus psaltria** (Say).
Rare.

5. **Zonotrichia leucophrys** (Forst.)
Only one specimen seen.

6. **Amphispiza bilineata** (Cass.).
Common.

7. **Zamelodia melanocephala** (Sw.).
Only one seen.

8. **Sturnella neglecta** (Aud.).
One or two observed only.

9. **Corvus corax carnivorus** (Bartr.).
Common.

10. **Tyrannus vociferans** (Sw.).
Three or four seen on the edge on the pine forest; altitude about 2,000 feet.

11. **Sayornis sayi** (Bp.).
Common in cliffs near the beach; paired.

12. **Empidonax difficilis** Baird.
Only one seen. A bird of this species came aboard the sloop one foggy morning during the trip, when ten miles from the peninsula. It was secured, seemingly uninjured, but died soon afterwards.

13. **Calypte annæ** (Less.).
But one example was seen. This was shot at an altitude of about 2,000 feet.

14. **Calypte costæ** (Bourc.).
Common. A nest found April 19, contained recently hatched young.

15. **Pandion haliaëtus carolinensis** (Gm.).
Very common, nesting on the cliffs.

16. **Cathartes aura** (Linn.).
Very common.

17. Heteroscelus incanus (Gm.)

Seen on several occasions.

18. Phalacrocorax dilophus cincinnatus (Brandt).

A colony of about one hundred of these cormorants were breeding on almost inaccessible cliffs which rose perpendicularly from the water. Seven nests, examined from above, contained either three or four eggs each. Thousands of this species were observed at Elida and St. Martin's Islands, and San Quentin Bay.

[NOTE.—The specimens obtained are in full breeding plumage, and, so far as the skins indicate, can only be distinguished from the eastern forms, *dilophus* proper and *floridanus*, by the entirely white superciliary tufts, these being wholly black or but slightly mixed with white in the above-named races.—R. R.]

19. Larus heermanni Cass.

Common.

20. Thalasseus regius Gamb.

Common in April, rare in May.

CATALOGUE OF A COLLECTION OF BIRDS MADE NEAR THE SOUTHERN EXTREMITY OF THE PENINSULA OF LOWER CALIFORNIA.

BY L. BELDING.

[Edited by R. RIDGWAY.]

This paper is merely intended to give what are supposed to be the most interesting results of a winter's work in Lower California. Sickness in May, occasioned by exposure to the hot sun, prevented a contemplated visit to one or more high mountains near the village of Miraflores, and also prevented the securing of specimens of some well-known birds at San José—consequently the list is not as complete as it might otherwise have been.

Some of the species found by Xantus do not appear in the list. Perhaps the most important of these is the rare Cape Robin (*Merula confinis*), which, with *Columba erythrina*, is probably a bird of the mountains.

Most of the Cape species are, as in 1859, abundant and tame.

It appears quite likely Mr. Xantus neglected to report some very common, well-known residents, since these do not appear in his list.*

------

* The only papers relating specially to the birds of the vicinity of Cape Saint Lucas are the following:

(1.) XANTUS, JOHN.—Descriptions of supposed new species of birds from Cape Saint Lucas, Lower California. <Proc. Philad. Acad. Sci., Nov., 1859, pp. 297–299. (New species: *Picus lucasanus*, p. 298; *Campylorhynchus affinis*, p. 298; *Harporhynchus cinereus*, p. 298; and *Brachyrhamphus hypoleucus*, p. 299.)

(2.) BAIRD, S. F.—Notes on a collection of birds made by Mr. John Xantus, at Cape Saint Lucas, Lower California, and now in the Museum of the Smithsonian Institu-

Others which are recorded from Cape Saint Lucas were probably obtained at other though not very distant localities.

The fauna and flora of La Paz, Cape Saint Lucas, and San José del Cabo, are quite identical, allowing for difference in surroundings and in variety of collecting grounds.

For the purpose of determining the resident species a later stay at San José was desirable, but I was compelled reluctantly to leave the field.

Collections were made at La Paz from December 15, 1881, to March 21, 1882; at San José from April 1, 1882, to May 17, of the same year.

tion. <Proc. Philad. Acad. Sci., Nov., 1859, pp. 299–306. (The new species described in this paper are, *Cardinalis igneus*, p. 305; *Pipilo albigula*, p. 305; *Chamœpelia passerina?* var. *pallescens*, p, 305; the name *pertinax* being proposed, on p. 303, for the Cape Saint Lucas *Myiarchus*, if distinct. Only 42 species are enumerated in this list.)

(3.) RIDGWAY, ROBERT.—On two recent additions to the North American bird-fauna, by L. Belding. <Proc. U. S. Nat. Mus., vol. 4, 1882, pp. 414, 415. (*Motacilla ocularis*, Swinhoe, La Paz, Lower California, January 9, 1882; *Dendrœca vieilloti bryanti*, Ridgway, same locality, common in July.)

(4.) RIDGWAY, ROBERT.—Descriptions of some new North American birds. <Proc. U. S. Nat. Mus., vol. 5, 1882, pp. 343–346. (The Cape Saint Lucas birds described in this paper are, *Geothlypis beldingi*, p. 344, San José del Cabo; and *Rallus beldingi*, p. 345, Espiritu Santo Islands.)

The first two papers cited above were based upon an early installment of Xantus's collections, no list of the species subsequently obtained by him having ever been published. I have, therefore, gone over the record books of the National Museum and made a careful enumeration of all the species collected by Xantus in Lower California. The total number of species amounts to 130, of which only 42 are included in Professor Baird's list. Of the species collected by Xantus 34 were not found by Mr. Belding, who, however, obtained or observed 39 species not represented in Xantus's collections. The latter are distinguished in the present catalogue by a * prefixed to the number, while a list of those obtained only by Xantus is given herewith.

1. HYLOCICHLA UNALASCÆ (Gm.). Cape Saint Lucas, January.
2. MERULA CONFINIS (Bd.). Todos Santos, "summer, 1860."
4. MYIODIOCTES PUSILLUS PILEOLATUS (Pall). Agua Escandida, November; Sierra San Gertrude, January.
5. PYRANGA LUDOVICIANA (Wils.). Cape Saint Lucas, September 27, 28; October 20; November 5–17.
6. PASSERCULUS SANDWICHENSIS ALAUDINUS (Bp.). Cape Saint Lucas, September 13.
7. ZONOTRICHIA GAMBELI INTERMEDIA (Ridgw.). San José del Cabo, November 15; San Nicolas, October.
8. ALUCO FLAMMEUS AMERICANUS (Aud.). San José del Cabo, December 6–15; January; Caduana, November 25.
9. ASIO ACCIPITRINUS (Pall). Mira flores, November 25.
10. SCOPS TRICHOPSIS (Wagl)? Cape Saint Lucas (no date).
11. HIEROFALCO MEXICANUS (Schleg.). Mira-flores, November 25; Cape Saint Lucas December 14; San José del Cabo, December, January.
12. ACCIPITER COOPERI (Bp.). Cape Saint Lucas, October 26–31; San Nicolas, October.
13. COLUMBA FASCIATA (Say). Cape Saint Lucas, November 25; Mira-flores, November 25.
14. OREORTYX PICTA PLUMIFERA (Gould). Cape Saint Lucas, April.
15. BOTAURUS LENTIGINOSUS (Montag.). Cape Saint Lucas, November 4; San José del Cabo, November 29, 30.

Several days were spent at Cape Saint Lucas, and other localities were visited, as shown by the specimens which were forwarded to the National Museum.

The people whom I met in Lower California were invariably courteous and kind, and to Señor Grinda, collector at La Paz, I am indebted for substantial favors.

The California and Mexico steamship line deserves credit for free transportation of specimens.

The species enumerated in the following catalogue were common to most or all of the localities where collections were made, those observed at only one locality being given afterwards in separate lists.

1. **Oreoscoptes montanus** (Town.).
Rare.

2. **Mimus polyglottus** (L.).
Abundant.

3. **Methriopterus cinereus** (Xant.).
Very common.

4. **Phainopepla nitens** (Sw.).
Rare on the coast, common in the interior.

5. **Polioptila cærulea** (L.).*
Common.

---

16. OXYECHUS VOCIFERUS (L.). Cape Saint Lucas, October 20–31; November 19–22.
17. GALLINAGO WILSONI (Temm.). San José, November 23.
18. ACTODROMAS MINUTILLA (Veill). Todos Santos (no date).
19. CALIDRIS ARENARIA (L.). *Fide* BAIRD.
20. TOTANUS MELANOLEUCUS (Gm.). San José, December; Cape Saint Lucas.
21. SYMPHEMIA SEMIPALMATA (Gm.). (Locality not stated; January.)
22. HIMANTOPUS MEXICANUS (Müll.). Sierra de Santiago, January; Cape Saint Lucas; San José del Cabo, February.
23. ANAS BOSCAS (L.). San José del Cabo, December.
24. CHAULELASMUS STREPERUS (L.). San José del Cabo, December, February.
25. DAFILA ACUTA (L.). San José del Cabo, January.
26. NETTION CAROLINENSIS (Gm.). San José del Cabo, January, February.
27. ERISMATURA RUBIDA (Wils.). San José del Cabo, December, February; Laguna de Santiago, January; Saint Lazaro Mountains, January.
28. LOPHODYTES CUCULLATUS (L.). San José del Cabo, February.
29. PELECANUS ERYTHRORHYNCHUS (Gm.). San José del Cabo, January, February; Cape Saint Lucas (no date).
30. PUFFINUS GRISEUS (Gm.). Cape Saint Lucas, August 18.
31. HALOCYPTENA MICROSOMA (Coues). San José del Cabo, May.
32. CYMOCHOREA MELÆNA (Bp.). Cape Saint Lucas.
33. BRACHYRHAMPHUS HYPOLEUCUS (Xantus). Cape Saint Lucas, San José.
34. BRACHYRHAMPHUS CRAVERI (Salvad). Cape Saint Lacas.

*The only specimen sent by Mr. Belding agrees with other western examples in being decidedly darker above and in having the white of the tail-feathers more restricted than eastern birds of this species. The lores are also darker, while there is little if

6. Polioptila plumbea Baird.

Very common.

7. Auriparus flaviceps (Sund.).

Common.

8. Campylorhynchus affinis Xant.

Very common.

9. Salpinctes obsoletus (Say).

Not rare.

10. Catherpes mexicanus conspersus Ridgw.

Not rare.

11. Troglodytes aëdon parkmanni (Aud.).

*12. Motacilla ocularis Swinh.

Accidental. A single specimen shot January 9, 1882, during a cold gale from the north. It was found on a drift of sea-weed on the beach. (See vol. 4 of these Proceedings, page 414.)

*13. Helminthophila celata (Say).

A single specimen (No. 86272 U. S. Nat. Mus.), collected in January, appears to be referable to this form.—R. R.

---

any trace of the light superciliary streak. Should these differences prove constant, I propose the name *P. cærulea obscura* for the western race.

An equal number of eastern and western adult males measure as follows:

EASTERN SPECIMENS.

Catal. No.	Locality.	Date.	Wing.	Tail.	Culmen.	Tarsus.	White on outer tail-feather.
82109	Wheatland, Ind	April 23.	2.20	2 20	.42	.70	1.20
82604	Mount Carmel, Ill	April...	2 15	2 15	.40	.70	1.25
55485	....Do.	do	2 15	2 15	.42	.70	1.20
82599	Laurel, Md	May 2..	2.10	2 20	.40	.68	1.25
82600	District of Columbia	April 22.	2 10	2 05	.40	.70	1.20
81993	Milton, Fla	April 4	2.05	2 15	.40	.70	1.10
Average			2.12	2.15	.41	.70	1 20

WESTERN SPECIMENS.

80466	Calaveras County, California.	April 5.	2 00	2.15	.38	.68	1.05
80821	......do.	May 28	2 00	2.10	.40	.70	1.10
73884	......do.	April 5.	2 05	2.20	.40	.67	1.15
52608	Fort Whipple, Ariz.	May 19.	2.00	2 15	.38	.70	.....
69395	Apache, Ariz	July 19	2.05	2.20	.40	.70	.....
87530	San José, Lower California..	April 17.	2.00	2.10	.40	.68	0.95
Average			2.02	2.15	.39	.69	1 06

R. R.

14. Helminthophila celata lutescens, Ridgw.

Common.

15. Dendrœca æstiva (Gm.).

Rare.

* 16. Dendrœca vieilloti bryanti, Ridgw.

(See vol. 4, these proceedings, page 444.—R. R.)

Common in the shrubbery around the Bay of La Paz; also seen at Pichalinque Bay and Espiritu Santo Island. It frequented almost exclusively the mangroves (*Rhizopora mangle*), and is probably resident.

17. Dendrœca auduboni (Towns.).

Common.

18. Siurus nævius notabilis Grinnell. ?

Rarely seen.

[I am now inclined to believe that the type specimen of *S. nœvius notabilis* was an unusually large one, especially as regards the bill, since I have not yet met with any other western example of the species which agrees with it in dimensions. The two examples obtained by Mr. Belding are decidedly smaller, but both are females. I give below their measurements, as well as those of the type of *S notabilis* and two other examples, which I refer provisionally to the same form. The type specimen of *notabilis* is so very different in proportions from any of the many eastern specimens of *S. nœvius* which I have examined, that I am not yet prepared to yield its claim to recognition as the representative of a geographical or local race of the species; but whether other western specimens belong to the same form or not, is a question which can only be determined by more abundant material.

	Wing.	Tail.	Culmen.	Bill from nostril.	Tarsus.	Middle toe.	Remarks.
a Type of *S. notabilis* (sex not determined)	3.25	2.50	....	.50	.83	.56	Beneath yellowish white.
b. 329, coll F Stephens, Tucson, Ariz ♂	3.20	2.25	.52	.40	.85	.60	Do.
c ——. coll G. H Ragsdale, Gainesville, Tex ♂	3.20	2.45	....	.42	.90	.55	Beneath white, faintly-tinged, and pale, buffy yellow.
d 86268, U. S Nat Mus , La Paz, Lower California. ♀	3.15	2.40	.52	.40	.88	.52	Beneath decidely sulphur yellow.
e 86269, U S Nat. Mus , La Paz, Lower California. ♀	3.05	2.30	.55	.40	.90	.55	Beneath white, scarcely tinged with yellowish.

* Mr Brewster's measurements of this specimen (*cf.* Bull. Nutt Orn. Club, July, 1882, p. 138) are so different that our methods of measurement must vary. The dimensions by Mr. Brewster are as follows Wing, 3 10, tail, 2.32; culmen, .64 (!), tarsus, .85.

19. Geothlypis macgillivrayi (Aud.).

Rare, on mountain cañons.

20. Geothlypis trichas (L.).

Common.

21. Icteria virens longicauda (Lawr.).

Rare.

22. Vireo pusillus. Coues.

Rare.

23. Lanius ludovicianus excubitorides (Sw.).

Common.

24. Ampelis cedrorum (Vieill.).

Very rare; like *Phainopepla nitens*, it feeds upon the berries of the mistletoe.

*25. Tachycineta bicolor (Vieill.).

Often seen in winter.

26. Tachycineta thalassina (Sw.).

Often seen in winter.

27. Carpodacus frontalis rhodocolpus (Cab.).

Abundant.

[The fine series collected by Mr. Belding shows the character of this well-marked race as given in Hist. N. Am. Birds (vol. i, pp. 460, 468) with wonderful uniformity. In five adult males the forehead, superciliary stripe, cheeks, throat, breast, upper part of abdomen and sides, and in some even the belly and flanks, also, are clear, soft, rose-red; the rump similar, but brighter—more of a carmine shade; the crissum in all strongly tinged or washed with rose-pink; the crown, occiput, nape, and whole back overlaid or very strongly tinged with deep wine-red.—R. R.]

28. Astragalinus psaltria (Say).

Common.

*28. Chrysomitris pinus (Wils.).

Only one observed; this in a flock of *A. psaltria*, with which, in California, the species frequently associates.

29. Passerculus rostratus (Cass.).

[The series of 10 specimens obtained by Mr. Belding I find puzzling in the extreme. The majority of them agree exactly with typical specimens from San Diego and other parts of Southern California, while others differ in darker colors, thus forming an approach to *P. guttatus*, the unique type of which came from San José del Cabo. In fact, one specimen (No. 86292, ♀, Feb. 25) agrees exactly with the latter in coloration, and I had unhesitatingly referred it to the same species or race; but I now find, after a very close examination of a large amount of material, that it either cannot be *guttatus* or else the latter is nothing but an abnormally small, slender-billed individual of *rostratus*, thus destroying the validity of *guttatus* altogether. Great as is the variation in size and form of the bill in *rostratus*, I am unable to find among 27 examples of the latter a single one having the bill nearly so slender as in the type of *guttatus*, as the following measurements of the latter compared with the minimum corresponding measurements of *rostratus* will show :]

	Wing.	Tail.	Tarsus.	Middle toe.	Culmen.	Bill from nostril.	Depth of bill at base.
Type of *P. guttatus*	2.55	1.95	.82	.62	.45	.32	.22
Minimum measurements of *P. rostratus*..................	2.55	1.90	.80	.60	.45	.36	.27
Maximum measurements of *P. rostratus*.................	2.95	2.30	.92	.68	.51	.40	.29

From the above figures it is evident that the difference between specimens of *P. rostratus* is greater than between smaller individuals of the same and the so-called *P. guttatus, except in the thickness of the bill.* In fact, all other measurements intergrade, as does also, unquestionably, the coloration. It therefore follows that if *P. guttatus* is allowed to rank as a species, or even subspecies, its claim to such rank rests solely upon the peculiarly slender bill.

Two specimens collected on the island of San Benito, Pacific coast of Lower California, by Dr. T. H. Streets, U. S. N. (Nos. 70636 and 70637, U. S. Nat Mus.), are labeled in the handwriting of Dr. S. " *Passerculus sanctorum* Coues (Type)," and are quite appreciably different in coloration, and also in the form of the bill, from ordinary *P. rostratus*, including an example (No. 70635) obtained by Dr. Streets at Todos Santos. The coloration of these two specimens being essentially identical with that of the type of *P. guttatus* I had referred them to that supposed species; but I now find that so far as measurements go, they belong rather to *P. rostratus.* Certain it is, that among Mr. Belding's La Paz specimens are some which I cannot *in any way* distinguish from these San Benito specimens, while others of the same series lead the way by the most gradual but complete transition to " typical " *rostratus*. In order to make the matter more plain, I have divided the collection before me into three series, as follows : (*a*) Specimens typical of *rostratus* as to coloration ; (*b*) types of " *P. sanctorum* Coues ;" (*c*) specimens agreeing with " *P. sanctorum*" in coloration ; (*d*) specimens agreeing with the type of " *P. guttatus*" in coloration, and (*e*) the type of " *P. guttatus*" itself.

a. Typical *P. rostratus.*

Catalogue number.	Sex and age.	Locality.	Date.	Wing.	Tail.	Culmen.	Bill from nostril.	Depth of bill at base.	Tarsus.	Middle toe.
83464	— ad.	San Pedro, Cal	November..	2.60	2.00	.50	.38	.28	.90	.60
31896	— ad	San Diego, Cal	....do	2.80	2.25	.50	.40	.28	.85	.60
6339	♂ ad	......do......	....	2.65	2.10	.50	.40	.28	.85	.60
6640	♂ ad	......do......	....	2.95	2.25	.51	.40	.29	.90	.65
59189	♂ ad.	Mouth Colorado River	October.....	2.90	2.30	.50	.40	.28	.95	.65
59191	♀ ad.	......do......	...do	2.75	2.15	.49	.38	.28	.85	.65
		Average	....	2.77	2.17	.51	.39	.28	.88	.62
16969	♂ ad.	Cape Saint Lucas	September..	2.70	2.20	.50	.40	.28	.85	.62
26605	♂ ad.	.... do...........	...do	2.60	2.00	.48	.38	.27	.88	.60
26611	♀ ad	......do...........	....	2.65	2.15	.48	.38	.28	.85	.65
26604	♀ ad	......do...........	November..	2.80	2.25	.50	.40	.29	.90	.65
16940	♀ ad	......do...........	September..	2.70	2.00	.50	.40	.28	.90	.62
26614	♀ ad.	......do...........	... do	2.60	......	.50	.40	.28	.82	.60
26608	♂ ad.	.... do...........	... do	2.70	2.00	.50	.38	.27	.85	.60
32508	— ad	......do...........	....	2.55	1.90	.50	.38	.28	.80	.60
		Average.............	....	2.66	2.07	.49	.39	.28	.86	.62
86294	— ad	La Paz, Lower California ..	January	2.95	2.20	.50	.40	.29	.92	.68
86297	♀ ad.	......do...	December ..	2.70	2.00	.48	.38	.28	.85	.65
86295	— ad..	......do...........	January	2.75	2.15	.50	.40	.28	.90	.65
86299	♀ ad	......do...........	February ..	2.60	1.90	.45	.36	.26	.85	.60
86301	♀ ad.	......do...........	January	2.55	1.95	.49	.40	.27	.85	.67
86293	— ad.	......do...........	....	2.70	2.10	.50	.40	.28	.85	.60
		Average.............	....	2.71	2.05	.49	.39	.28	.87	.64
		Mean average	....	2.71	2.10	.50	.39	.28	.87	.63

b. Types of *P. sanctorum* COUES.

70636	♂ ad.	San Benito Island, Lower California..............	....	2.75	2.25	.50	.40	.28	.85	.65
70637	— ad	......do	....	2.80	2.10	.50	.40	.27	.90	.62
		Average.............	....	2.77	2.17	.50	.40	.27	.87	.63

c. Specimens resembling "*P. sanctorum*" in coloration.

86300	♀ ad	La Paz, Lower California...	February ...	2.75	2.25	.50	.40	......	.80	.68
86298	♀ ad	......do...........	January	2.70	1.95	.50	.40	.28	.85	.62
86296	— ad.	......do...........	February ...	2.60	2.05	.50	.40	.28	.88	.68
		Average	....	2.68	2.08	.50	.40	.28	.84	.66

d. Specimens resembling *P. guttatus* in coloration.

86292	— ad	La Paz, Lower California...	February ...	2.70	2.10	.48	.38	.28	.90	.65
70635	— ad	Todos Santos Island, Lower California....	....	2.80	2.30	.52	.40	.29	.90	.65

e. Type of *P. guttatus.*

26615	♂ ad.	San José, Lower California..	December ..	2.55	1.95	.45	.32	.22	.82	.62

* 30. **Coturniculus passerinus perpallidus** Ridgw.

Rare, but seen at several localities.

31. **Chondestes grammica strigata** (Sw.).

Common.

32. **Zonotrichia leucophrys** (Forst.).

Common at and south of La Paz, on May 1, during and after very hot weather. This species, *Spizella breweri*, and *Pipilo chlorurus* were missed from San José, and not afterward seen. This is the only *Zonotrichia* yet detected at and south of La Paz*, and the only one found by me in central California in summer south of 39°. It breeds regularly in the mountain meadows of Calaveras and Alpine Counties†, at an altitude of 7,000 feet or more, where I have seen it unmated as late as July 9. I have not yet found it below 7,000 feet.

33. **Spizella pallida** (Sw.).

Common; found also at San José.

34. **Spizella breweri** Cass.

Abundant.

35. **Amphispiza bilineata** (Cass.).

Common.

36. **Pipilo chlorurus** (Towns.).

Common.

37. **Pipilo fuscus albigula** (Baird).

Not often seen at any locality.

[The specimens obtained by Mr. Belding are in fine winter plumage, and all exhibit distinctly the characteristic features of the race, as distinguished from *P. mesoleucus*, in much more distinctly ashy breast, decidedly more trenchant definition of the buffy throat-patch, smaller size, etc. As in *P. mesoleucus*, however, the throat is occasionally entirely unicolored, three of the seven examples showing no difference in intensity of the buff on different parts of the throat; therefore, the phrase "ochraceous of throat palest posteriorly, where it becomes nearly white," as given in the diagnosis of the race on p. 122, vol. ii, *Hist. N. Am. B.*, requires some modification, to this extent, that *when it does vary in intensity*, it is palest posteriorly, instead of the reverse, as in *P. mesoleucus.*—R. R.]

---

* In the Xantus collection, from "Cape Saint Lucas," are a few examples of *Z. intermedia*, but most of the specimens obtained by Xantus are *Z. leucophrys*.

† *Z. leucophrys* is undoubtedly the species found by me breeding abundantly at Mountain Meadows, summit of Donner Lake Pass, in the Sierra Nevada, in July, 1867. In my report on the ornithology of the fortieth parallel (p. 471), the birds of this locality were erroneously referred to *Z. intermedia*, for the reason that no specimens were obtained, while at that time *Z. leucophrys* was not supposed to occur west of the Rocky Mountains.—R. R.

38. Cardinalis virginianus igneus (Baird).

Common.

[A fine series, including nine males and three females of this excellent race.—R. R.]

39. Pyrrhuloxia sinuata Bp.

Rather rare, less so in the interior.

[The five adult males from Lower California differ from an equal number of Texan examples in having the red of an appreciably lighter, less rosy, tint, and that around the fore part of the head very much clearer. An example from Camp Grant, Ariz., agrees with the peninsular specimens in this respect. It is possible that a larger series will show these differences to be constant. In specimens obtained in April the bill is now, after the lapse of three months' time, still deep orange-colored, while in winter specimens the bill is horn-colored, the mandible paler, but scarcely inclining to yellowish.—R. R.]

40. Zamelodia melanocephala (Sw.).

Not common.

*41. Passerina amœna (Say.).

Not common.

42. Calamospiza bicolor (Towns.).

Abundant.

43. Icterus parisorum Bp.

Rare in winter.

44. Icterus cucullatus Sw.

Common.

45. Corvus corax carnivorus Bartr.

Common.

46. Aphelocoma californica (Vig.).

Common.

[Specimens from Lower California appear to be quite identical with those from the coast region of California proper. If any different, they are purer white beneath, the crissum having absolutely no tinge whatever of blue.—R. R.]

47. Tyrannus vociferans Sw.

Common.

48. Myiarchus cinerascens Lawr.

Common.

49. Sayornis sayi (Bp.).

Rare.

50. Sayornis nigricans (Sw.).

Rare.

51. Empidonax difficilis Baird.

Rare.

52. Empidonax obscurus (Sw.).

Very common in winter—more rare in summer.

53. Pyrocephalus rubineus mexicanus (Scl.).

One specimen only.

54. Calypte costæ (Bourc.).

Abundant in winter; not common at San José, Cape Saint Lucas, or Miraflores in April and May.

[Two nests of this species, collected at La Paz, by Mr. Belding, are quite different in size and shape. Both are ordinary looking structures, composed of dull gray lichens and small pieces of thin bark, held together with spiders' webs, the interior containing a few soft small feathers, in one nest, apparently of the summer yellow bird (*Dendœca œstiva*). The larger nest measures about 1½ inches in diameter by a little more than 1 inch in depth, the cavity being about .1 × .80; the smaller one measures about the same in diameter across the top, but is much narrower at the bottom, is less than 1 inch high, and has a shallower cavity with much thinner walls. Each contains a single egg, one of which measures .30 × .50, the other .32 × .50. The identification is positive, the parent bird accompanying each nest. One of these females has a very large spot or patch of metallic violet on the throat, while the other has instead only a few dusky specks.—R. R.]

55. Basilinna xantusi (Lawr.).

In winter, found only in mountain cañons. It was common at the western base of Cacachiles mountain in February, more so, in fact, than *C. costæ*. It was not observed at San José until some time after my arrival, though it occurred in cañons only two or three miles to the westward. About the last of April it was common in orchards at San José.

While incubating, this species is very confiding and courageous, sometimes remaining upon the nest until removed from it by the hand. A nest taken April 23 was placed underneath an awning or shade of boughs and weeds in front of a farmhouse. It was surrounded by downy heads of composite plants and could scarcely be distinguished from them, having, as usual, been made of raw cotton.

[The two nests of this species obtained by Mr. Belding are very neat structures. quite different in appearance from the nest of any other North American Hummer, though they differ much from one another. The finer of the two (No. 18563, San José, April 23) is a compactly felted mass composed chiefly of raw cotton, but this coated exteriorly with spiders' webs and light brown fine fibrous materials. It is securely fastened to two forks of a twig, and rests between them. The shape is very irregular, owing to the manner in which it is secured to the twigs, but on top the transverse diameter is about 1.50 inches, the cavity being about 1 inch across and about .60 of an inch deep. The two eggs measure respectively .32 x .50 and .34 x .49, being essentially identical in size and shape with those of *Calypte costæ*, from which it is apparently quite impossible to distinguish them. The other nest (No. 18564, Arroyo, north of Santiago Peak, May 9) is quite different both in shape and material. It is very regularly but shallowly cup-shaped, averaging a little over 1.50 inches in external diameter, but only about .80 of an inch in extreme height. The cavity is about 1 inch across by a little over .50 of an inch in depth. The material is chiefly raw cotton, but this much mixed, especially outwardly, with fine leaf-stems, seed-capsules, spiders' webs, etc., besides one or two small soft white feathers. Like the other nest, this one is supported between two twigs. The eggs measure respectively .34 x .49 and .32 x .50.—R. R.]

56. Chordeiles acutipennis texensis (Lawr.).

Rarely seen at La Paz, but abundant at San José after April 23. Common at San Diego in May, 1881.

57. Picus scalaris lucasanus (Xantus).

Very common.

58. Centurus uropygialis Baird.

Abundant.

59. Colaptes chrysoides Malh.

Very common.

60. Ceryle alcyon (L.).

Common.

61. Geococcyx californianus (Less.).

Common.

62. Bubo virginianus subarcticus (Hoy).

Rarely seen.

*63. Speotyto cunicularia hypogæa (Bp.).

Rare.

64. Tinnunculus sparverius (L.).

Common.

65. Polyborus cheriway (Jacq.).

Abundant.

66. Pandion haliaëtus carolinensis (Gm.).

Common.

67. Circus hudsonius (L.).

Common.

68. Accipiter fuscus (Gm.).

Rare.

69. Parabuteo unicinctus harrisi (Aud.).

Common. Frequently met with in May along the route from San José to Miraflores.

70. Buteo borealis calurus Cass.

Common.

[The single specimen collected by Mr. Belding cannot by any means be referred to the so-called var. *lucasanus*, the tail being marked not only by a very distinct subterminal narrow black band, but with more or less distinct narrow bars entirely to the base. The under plumage is very light colored, the usual abdominal belt of dusky markings being indicated only by very small hastate streaks; the tibiæ are creamy white, barred with light rufous, and the sides more distinctly barred with dark brown and rufous. It is somewhat doubtful whether the principal character assigned to "*lucasanus*" (the uniform rufous tail without subterminal black bar) will prove constant, even in birds from the cape.—R. R.]

*71. Buteo abbreviatus Caban.

Very rare.

*72. Cathartes aura (L.).

Abundant.

73. Zenaidura carolinensis (L.).

Very abundant in winter; rare at other localities in April and May.

74. Melopelia leucoptera (L.).

Abundant.

75. Chamæpelia passerina pallescens Baird.

Abundant.

76. Lophortyx californica (Shaw).

Common.

77. Herodias egretta (Gm.).

78. Dichromanassa rufa (Bodd.).

Common; *white plumage not seen.*

*79. Hydranassa tricolor ludoviciana (Wils.).

Less common than the preceding.

*80. Butorides virescens (L.).

81. Nycticorax griseus naevius (Bodd.).

*82. Nyctherodius violaceus (L.).

83. Tantalus loculator (L.).

*84· Eudocimus albus (L.).

85. Plegadis gua:auna (L.).

*86· Squatarola helvetica (L.).

*87. Ægialites alexandrinus nivosus Cass.

*88. Ochthodromus wilsonius (Ord.).

Very common.

*89· Ereunetes pusillus occidentalis (Lawr.).

*90· Limosa lapponica novæ-zealandiæ Gray.

91. Tringoides macularius (L.).

92. Numenius longirostris Wils.

93. Numenius hudsonicus Lath.

94. Phalaropus fulicarius (L.)

*95· Rallus beldingi Ridgw.

Rare. (See vol. 5, p. 345.)

96. Fulica americana Gm.

97. Tachypetes aquila (L.).

Abundant.

98. Pelecanus fuscus Linn.

Very abundant until nearly exterminated by disease in February. The stomachs of several examined were full of small worms. A great many died at Cape St. Lucas and San José. I was informed that this mortality occurs every winter. I copy the following from my notes: "February 24. ♂ P. fuscus. Back of neck dark brown,* bare skin around eye, brown; base and much of pouch deep red; specimen in breeding plumage, and condition."

*99· Phalacrocorax dilophus cincinnatus (Brandt).

100. Phalacrocorax penicillatus (Brandt).

*101· Phaëthon æthereus Linn.

Only three individuals seen; one of them several hours' sail from Mazatlan; one obtained at Esperitu, Santo Islands, February 1.

102. Larus occidentalis And.

*103. Larus delawarensis Ord.

104. Larus heermanui Cass.

*105. Larus philadelphiæ (Ord.).

106. Thalasseus regius Gamb.

---

* In the single specimen sent the back of the neck is a rich brownish black, quite different from the seal-brown or chestnut of all eastern specimens I have seen. Audubon describes the color of the naked orbits as pink, the naked skin about base of the bill as deep blue, and the pouch greenish black. Thus it would seem that the soft parts are very differently colored. Should this difference prove constant, the western bird would have to be separated as a race.—R. R.

March 21, 1883.

*107· Sterna forsteri Nutt.

*108· Dytes nigricollis californicus (Heerm.).

*109· Podilymbus podiceps (L.).

Additional species found at San José del Cabo from April 1 to May 17.

*1. Telmatodytes palustris paludicola Baird.

Rare.

2. Anthus ludovicianus (Gm.).

A flock remained until about May 3, or later.

*3. Geothlypis beldingi Ridgw.

(See vol v., p. 344.)

Common in the few suitable localities around San José, Miraflores, and cañons of the Miraflores and Santiago Peaks. At Agua Caliente a pair were noticed feeding their young just out of the nest May 7. The only note traced to this species was a loud *chip*. I listened long, when in the neighborhood of one or more of these birds, for the familiar song of the Maryland Yellow-throat (*G. trichas*), but failed to hear it. Their habits are quite like those of *G. trichas*, and the eggs are not materially different, if a nest found by my guide on the Miraflores and Todos San-tos trail May 6 belonged to this species, as I supposed it did, having seen a fine male near the spot from which it was taken.

4 Lanivireo solitarius cassini (Xantus)

Found breeding; common at Miraflores.

5. Guiraca cærulea (L.)

Only two specimens seen.

6. Passerina versicolor (Bp.).

Rare.

7. Molothrus ater obscurus (Gm.).

Common in the streets and on buildings, associated with *Scolecophagu cyanocephalus*.

8· Xanthocephalus icterocephalus (Bp.).

Rare; not seen in May.

*9. Scolecophagus cyanocephalus (Wagl.).

Common, breeding.

*10· Crotophaga sulcirostris Sw.

Only four individuals seen. A nest found April 29 contained eight eggs. It was fastened to upright reeds, and was composed of coarse weed stalks and mesquit twigs, lined with green leaves.*

---

* I was informed several years since, by Sr. Don José C. Zeledon, of San José, Costa Rica, that he has found nests of this species in Costa Rica, which were lined with green leaves of the lemon tree. It would be an interesting fact should this prove to be a regular habit of the species.—R. R.

The female, while incubating, was very wary, slipping quietly away from the nest and returning to it very stealthily, below the tops of the reeds.

The 1st of April I discovered four of these birds in a marsh, in which was a rank growth of *tule*, flags, and reeds. Having shot one of them, and the others were not molested, they remained in the marsh until May 15, or later. This marsh, the only one seen during the winter, harbored several species not elsewhere noticed, among them *Porzana carolina*. On one side of the marsh a lagoon or pond of fresh water, of 10 or 15 acres extent, was the resort of numerous gulls, ducks, and waders.

April 29 I noticed, for the first time during this visit to Lower California, *Progne subis*, *Petrochelidon lunifrons*, and *Cypselus saxatilis* circling over the lagoon. It is a question whether these birds came from the mountains of the peninsula or from the mainland.

May 1, while on the beach, a very large, compact flock of *Numenius hudsonicus* was, with the aid of a field-glass, discovered in the distance, rapidly approaching from the south. After sweeping in large circles over the lagoon, thus enabling me to shoot several of them, they alighted. They appeared to be weary as well as strange birds. The following morning, as I could not find the flock, my impression was it had resumed the journey northward.

That the birds of Lower California breed regularly in spring I have no doubt. The first nests observed were *Pandion haliaëtus carolinensis*, Espiritu Santo Island, February 1; *Auriparus flaviceps*, La Paz, February 27 (nest in this case unfinished); *Calypte costæ*, La Paz, March 2, (bird setting).

With the single exception of a juvenile or dwarfed *Lophortyx californica*, shot January 25, at Pichaliuque Bay, and which was apparently but six or eight weeks old, no young birds were seen until April 14 when a brood of *Polioptila plumbea* just out of the nest were observed.

May 17, the last day of my stay at San José, I saw the following species, besides well-known residents:

1. Progne subis (L.).
Common.

2. Molothrus ater obscurus (Gm.).
Rarely seen in May.

3. Scolecophagus cyanocephalus (Wagl.).
Rarely seen in May.

4. Cypselus saxatilis Woodh.
Rare; no specimen taken here.

5. Polyborus cheriway (Jacq.).
Abundant.

6. Pandion haliaëtus carolinensis (Gm.).
Common.

7. **Parabuteo unicinctus harrisi** (Aud.).
Common.

8. **Cathartes aura** (L.).
Abundant.

9. **Ardea herodias** L.
Rare.

10. **Herodias egretta** (Gm.).
Several seen.

11. **Garzetta candidissima** (Gm.).
Several seen.

12. **Dichromanassa rufa** (Bodd.).
Rare.

13. **Hydranassa tricolor ludoviciana** (Wils.).
Rare.

14. **Nyctherodias violaceus** (L.).
Very common.

15. **Tantalus loculator** L.
A pair seen in April and May.

16. **Plegadis guarauna** (L.).
A flock present in April and May.

17. **Tringoides macularius** (L.).
Rare.

18. **Mareca americana** (Gm.).
A flock of about a dozen.

19. **Spatula clypeata** (L.).
Mated.

20. **Querquedula discors** (L.).
Mated; common.

21. **Querquedula cyanoptera** (Vieill.).
Mated; rare.

22. **Tachypetes aquila** (L.).
Common.

23. **Pelecanus fuscus** L.
Common.

24. **Phalacrocorax pencillatus** (Brandt).
Abundant.
P. cincinnatus was very common at La Paz in the winter months; rare in March.

*25. Larus californicus Lawr.

Moderately common.

26. Larus occidentalis Aud.

Moderately common.

27. Thalasseus regius Gamb.

Common.

Most of these twenty-seven species were breeding.

The following additional species were found at the village of Miraflores, which lies two or three miles east of a peak of the same name. It is on a branch of the San José River and is about twenty-five miles north of the town of San José. The trail leading to it follows the gradually ascending sandy bed of the river. The altitude of the village is about 700 feet.

It was probably here that Xantus obtained his specimens marked " Miraflores " instead of getting them from the high and quite inaccessible, sharp, rocky peak of the same name, which has an altitude of more than 6,000 feet.

There is some very fertile bottom land here and numerous fine, large evergreen oaks grow on the uncultivated portion of it.

In these oaks were found, with other species—

*1 Virosylvia gilva swainsoni Bd.

Moderately common May 9.

*2· Dendrœca townsendi (Nutt.).

An individual, male, seen April 4.

*3. Pipilo maculatus megalonyx Baird.

Rare in April and May.

4. Melanerpes formicivorus angustifrons Baird.

Common, burrowing in oaks, whereas all? the other Woodpeckers of Lower California, including *Colaptes chrysoides*, as far as I have observed, burrow in the Giant Cactus (*Cereus giganteus*).

*5. Micrathene whitneyi (Coop.).

Common, if not abundant.

Whitney's Pigmy Owl utters monotonous calls or whistlings, faint, tremulous notes, and when perched within a few feet of an intruder expresses its anxiety by complaining cries.

As an attempt to describe the notes of three other obscurely known owls may not be out of place here, I transcribe the following from my journal:

" Big Trees, August 16, 1880. Bright moonlight.

" *Scops flammeolus* has a firm single note, which is often repeated after short intervals; shot specimen while calling."

Same locality, June 30, 1882:

"*Scops flammeolus* utters•frequently a single quite unvarying rounded note.

"Murphy's, October 2, 1880. This morning shot *Glaucidium gnoma,* which I heard calling, and at first supposed it was the Yellow-billed Cuckoo (*Coccyzus americanus*). The specimen shot was perched on the dead limb of a pine tree about 50 feet from the ground. Its calls varied but little in the fifteen or twenty times I heard them. They may be nearly represented thus: '*Coo-coo-coo-coo-coo-coo—cow—cow.*' The first six or seven guttural notes were equidistant, and uttered at the rate of about two in a second; then, after a pause of about two seconds, the longer notes followed. It was occasionally answered in similar notes by an unseen bird.

"Big Trees, July 13, 1881. *Strix occidentalis.*—Listened to its call about sunset; the bird in sight. Its call resembles the barking of a dog, the first three or four notes lasting about one second each; these succeeded by long, harsh, whining notes."

ON THE GENUS TANTALUS, LINN., AND ITS ALLIES.

BY ROBERT RIDGWAY.

The only species of *Tantalus* given by Linnæus in the tenth edition of "Systema Naturæ" is *T. loculator*, which may, therefore, be properly regarded as the type of the genus. In the twelfth edition *T. ibis* also appears, along with several true Ibisis of the genera *Eudocimus*, Wagl., and *Plegadis*, Kaup. So far as I am able to ascertain, the *T. ibis* and other Old World species related to it have never been separated generically from *T. loculator;* but a recent careful comparison* has convinced me that they all belong to quite a distinct genus from *T. loculator*. No generic name having, to my knowledge, been yet given specially to the Old World species, I propose for this group the term *Pseudotantalus*. The main differential characters of the two genera may be expressed as follows:

TANTALUS.—Adult with the whole head and upper half of neck naked, the skin hard and scurfy; crown covered by a quadrate, or somewhat shield-shaped, smooth horny plate, and skin of nape transversely wrinkled or corrugated. Nostrils subbasal; tertials longer than primaries, and with compact or normal webs. (Type, *T. loculator* Linn.).

PSEUDOTANTALUS.—Adult with only the fore part of the head naked, the hinder half and entire neck densely feathered; nostrils strictly basal; tertials shorter than primaries, and with their webs somewhat

---

* Although I have been able to actually examine only *T. ibis*, the excellent plates and descriptions of the remaining species which have been consulted leave no doubt that all the Old World Wood Ibises are strictly congeneric.

decomposed. Bill, legs, and tail very much longer, and basal outline of bill of different contour. (Type, *Tantalus ibis* Linn.)

The species belonging to *Pseudotantalus*, besides the type, are, so far as known, the following :

P. leucocephalus (Gm.). India.

P. longuimembris (Swinh.). Southern China.

P. lacteus (Temm.). Java and Sumatra.

SUPPLEMENTARY NOTE ON THE PEDICULATI.

BY THEODORE GILL.

In the proceedings of the United States National Museum for 1878 (v. 1, pp. 215–232), I have given the characteristics of the families, subfamilies, and genera of the Pediculate fishes. The present communication will supplement the article in question by detailing the synonyms of the families and subfamilies. The generic synonyms have been already indicated.

I.

Since the publication of the "Note on the Antennariidæ" (op cit., pp. 221–222), a "new genus" has been added to the family by Dr. A. Günther, which may be distinguished as follows :

TETRABRACHIUM.

= Tetrabrachium, *Günther*, Zool. Challenger, part 6, p. 44, 1880.

Antennariids with the body oblong conic from the head backwards; the skin naked; a compressed cuboidal head; small vertical mouth; dorsal spines (3) isolated and dwarfed, but exserted; second largest, "wide and fringed;" dorsal and anal fins low, long, and free behind; and pectorals with the upper portion (4 rays) detached from the lower.

Type, Tetrabrachium ocellatum, *Günther*, Zool. Challenger, part 6, p. 45, pl. 19, 1. c.

Ocean south of New Guinea (specimen obtained at a depth of 28 fathoms).

All that has been suggested as to this interesting form is what may be implied by the reference to the "Pedicalidæ," which, as appears by the "Systematic List" at the end of the volume (p. 78), is simply a misprint for Pediculati. Possibly the nearest ally of the genus is *Histiophryne*, but it appears to be quite an isolated form.

Dr. Lütken (Vidensk. Medd. fra den Naturhist. Foren. Kjobenhavn, 1879–'80, pp. 67–68) has objected *inter alia* to the generic differentiation of *Corynolophus*, but has not traversed or even met the reasons and arguments in favor thereof submitted by me (Proc. U. S. Nat. Mus., v. 1, p. 230). I therefore need only refer again to my original statement.

The differences alleged to exist between *Himantololophus* and *Corynolophus* are very marked, and if they really *do* exist, as stated, there can be no doubt that the two should be kept apart. I know of no reason, except the singularity and greatness of the differences specified, for doubting the correctness of Reinhardt's observations on *Himantolophus*, and prefer to assume their reliability rather than to discredit them, but at the same time admit the desirability of confirmation. The burden of proof meanwhile lies on those who would keep the forms together, and not on those who would separate them.

Finally, as to the genus *Lophius*, it seems to me that the *L. setigerus* should be generically distinguished from the *L. piscatorius*, notwithstanding the close external resemblance of the two. The two groups may be diagnosed as follows :

LOPHIUS.—Lophiids with vertebræ in considerable number, *i. e.*, 27—31,* and toothed vomer.

Type, *Lophius piscatorius.*

LOPHIOMUS.—Lophiids with vertebræ in diminished number, *i. e.*, about 19, and toothed vomer.

Type, *Lophiomus setigerus=Lophius setigerus,* Wahl.

It is surprising that the two have not been differentiated by Dr. Günther, inasmuch as he sometimes considers a difference of one or two vertebræ to be sufficient to distinguish *families.*

II.

PEDICULATI.

Synonyms as family names.

= Brachioptères, *Blainville,* Journ. de Physique, t. 83, p. —, 1816.

= Percoides à pectorales brachiformes, *Cuvier,* Régne Animal, t. 2, p. 308, 1817.

< Acanthoptérygiens à pectorales pédiculées, *Cuvier,* Régne Animal, 2 éd., t. 2 p. 249, 1829.

< Lophidæ, *Bonaparte,* Giorn. Accad. di Scienze, v. 52, (Saggio Distrib. Metod. Animali Vertebr. a Sangue Freddo, p. 111). 1832.

< Loñdi, *Bonaparte,* Fauna Italica, fol. 105, 1835.

< Lophiidæ, *Bonaparte,* Nuovi Annali delle Sci. Nat., t. 2, p. 130 ? 1838; t. 4, p. 185,? 1840.

= Lophidæ, *Girard,* Expl. and Surv. for R. R. Route to Pacific Oc., v. 10, p. 133, 1858.

= Pediculati, *Günther,* Cat. Fishes, Brit. Mus., v. 3, p. 178, 1861.

Synonym as subordinal name.

= Pediculati, *Gill,* Can. Nat., n. s., v. 2, p. 246, Aug., 1865.

Synonyms as ordinal names.

? Loñdi, *Rafinesque,* Indice d'Ittiolog. Siciliana, p. 42, 1810.

< Chismopnés, *Duméril,* Zool. Anal., p. 105, 1806.

< Plectognathes, *Swainson,* Nat. Hist. and Class. Fishes, etc., v. 2, pp. 193, 323, 1839.

* 28 is the number of vertebræ in two skeletons of *L. piscatorius* examined by myself.

= Antennarii. *Bleeker*, Enum. Sp. Piscium Archip. Ind., p. xvi. 1859; Atlas Ich. des Indes Néerland., t. 5. p. 1, 1865.
= Pediculati *Cope*, Proc. Am. Assoc. Adv. Sci.. v 20. p. 335, 1872.
= Pediculati, *Gill*, Arrangement Families Fishes, pp. xli, 2, 1872.
= Pediculati. *Fitzinger*, Sitzungsber. K. Akad. der Wissensch. (Wien). B. 67. 1. Abth., p. 48, 1873.

This order ("*ordo* 15") was also isolated by Bleeker as a distinct phalanx ("*phalanx* 1. *Herpetoichthyes seu Pediculati Cur.*") of the second subseries ("*Kanonikodermi*") of the second series ("*Isopleuri seu Homosomata*") of Pectinibranchiate fishes. The use of the term Pediculati in such sense has determined its retention as the ordinal name.

I. ANTENNARIIDÆ.

Synonyms as family names.

<Chironectidæ. *Swainson*. Nat. Hist. and Class. Fishes, etc.. v. 2, p. 195, 1839.
=Chironectidæ, *Swainson*. Nat. Hist. and Class. Fishes. etc., v. 2, p. 330, 1839.
=Chironecteoidei, *Bleeker*, Enum. Sp. Piscium Archip. Ind., p. xvi, 1859 : Atlas Ich. de Indes Néerland., t. 5. p. 4, 1865.
=Antennarioidæ, *Gill*, Proc. Acad. Nat. Sci Phila.. [v.15.] pp. 89, 90, 1863.
<Antennariidæ, *Cope*, Proc. Am. Assoc. Adv. Sci.. v. 20, p. 340, 1872.
=Chironectæ, *Fitzinger*, Sitzungsber. K Akad. der Wissensch. (Wien). B.67, 1.Abth., p. 48, 1873.
>Chaunacidæ. *Lutken*. Vidensk. Selsk. Skr.. (5.) v. 11, p. 325, 1878.
>Antennariidæ, *Lutken*, Vidensk. Selsk. Skr., (5,) v. 11. p. 325, 1878.
= Antennariidæ, *Gill*. Proc. U. S. Nat. Mus.. v. 1, pp. 215. 221, 1878.

ANTENNARIINÆ.

Synonyms as sub-family names.

=Antennarinæ. *Gill*, Cat Fishes. E. Coast N. A.. p. 47, 1861.
=Antennariinæ, *Gill*, Proc. Acad. Nat. Sci. Phila.. [v. 15,] p. 90, 1863.
<Chironecteiformes. *Bleeker*, Atlas Ich. des Indes Néerland.. t. 5, p. 5, 1865.

BRACHIONICHTHYINÆ.

Synonyms as subfamily names.

=Brachionichthyinæ, *Gill*. Proc. Acad. Nat. Sci. Phila.. [v. 15,] p. 90, 1863.
<Chironecteiformes. *Bleeker*. Atlas Ich. des Indes Néerland.. t. 5, p. 5, 1865.

CHAUNACINÆ.

Synonym as sub-family name.

= Chaunacinæ, *Gill*. Proc. Acad. Nat. Sc. Phil.. [v, 15,] p. 90, 1863.
= Chaunacinæ. *Gill*. Proc. U. S. Nat. Mus., v. 1. p. 222, 1878.

Synonym as family name.

= Chaunacidæ *Lutken*, Vidensk. Selsk. Skr.. (5,) Nat. og Math. Afd.. v. 11, p. 325, 1873

II. CERATIIDÆ.

Synonyms as family names.

= Ceratiidæ, *Gill*. Proc. Acad. Nat. Sc. Phil., [v. 15,] pp. 89, 90, 1863.
= Ceratiadæ. *Lutken*, Vidensk. Selsk. Skr., (5.) v. 11, p. 325, 1878.
= Ceratiidæ, *Gill*, Proc. U. S. Nat. Mus., v. 1. pp. 215. 216, 227, 1878.
Chironecteoidei subfam., *Bleeker*, 1865.

CERATIINÆ.

Synonyms.

= Ceratianæ, *Gill*, Cat. Fishes E. Coast N. A., p. 47, 1861.
= Ceratiaeformes, *Bleeker*, Atlas Ich. des Indes Néerland., t. 5, p. 6, 1865.
= Ceratiinæ, *Gill*, Proc. U. S. Nat. Mus., v. 1, pp. 217, 227, 1878.

ONEIRODINÆ.

Synonym.

= Oneirodinæ, *Gill*, Proc. U. S. Nat. Mus., v. 1, pp. 217, 227, 1878.

HIMANTOLOPHINÆ.

Synonyms.

= Himantolophinæ, *Gill*, Cat. Fishes E. Coast N. A., p. 47, 1861.
= Himantolophiformes, *Bleeker*, Atlas Ich. des Indes Néerland., t. 5, p. 6, 1865.
= Himantolophinæ, *Gill*, Proc. U. S. Nat. Mus., v. 1, pp. 218, 227, 1878.

ÆGÆONICHTHYINÆ.

Synonym.

= Ægæonichthyinæ, *Gill*, Proc. U. S. Nat. Mus., v. 1, p. 227, 1878.

MELANOCETINÆ.

Synonym.

= Melanocetinæ, *Gill*, Proc. U. S. Nat. Mus., v. 1, p. 227, 1878.

III. LOPHIIDÆ.

(See, also, under Pediculati, p. 552.)

Synonyms as family names.

< Lofidi, *Rafinesque*, Indice d'Ittiolog. Siciliana, p. 42, 1810.
< Branchismea, *Rafinesque*, Analyse de la Nature, p. —, 1815.
< Lophides, *Latreille*, Fam. Nat. du Régne Animal, p. 139, 1825.
< Baudroies, *Risso*, Hist. Nat. de l'Europe Merid., t. 3, p. 101 ? 1826.
= Lophidæ, *Swainson*, Nat. Hist. and Class. Fishes, etc., v. 2, p. 195, 1839.
< Lophidæ, *Swainson*, Nat. Hist. and Class. Fishes, etc., v. 2, p. 330, 1839.
< Lophoidei, *Bleeker*, Enum. Sp. Piscium Archip. Ind., p. xvi, 1859
= Lophioidei, *Bleeker*, Atlas Ich. des Indes Néerland., t. 5, p. 2, 1865.
= Lophioidæ, *Gill*, Proc. Acad. Nat. Sci. Phila., [v. 15,] pp. 89, 90, 1863.
< Lophiidæ, *Cope*, Proc. Am. Assoc. Adv. Sci., v. 20, p. 340, 1872.
= Lophii, *Fitzinger*, Sitzungsber. K. Akad. der Wissensch. (Wien), B. 67, 1. Abth., p. 48, 1873.
= Lophioidæ, *Lütken*, Vidensk. Selsk. Skr., (5,) Nat. og Math. Afd., v. 11, p. 325, 1878.
= Lophiidæ, *Gill*, Proc. U. S. Nat. Mus., v. 1, pp. 215, 219, 1878.

LOPHIINÆ.

Synonyms as sub-family names.

< Lophidia, *Rafinesque*, Analyse de la Nature, p. —, 1815.
< Lophini *Bonaparte*, Fauna Italica, fol. 105, 1835.
< Lophiinæ, *Bonaparte*, Nuovi Annali delle Sci. Nat., t. 2, p. 130 ? 1838; t. 4, p. —? 1840.

< Lophiinæ, *Bonaparte*, Catal. Metod. Pesci Europei, pp. 9, 89, 1846. †
= Lophiinæ, *Bleeker*, Atlas Ich. des Indes Néerland., t. 5, p. 5, 1865.

IV. MALTHEIDÆ.

Synonyms as family names.

< Chironectidæ, *Swainson*, Nat. Hist. and Class. Fishes, etc., v. 2, p. 195, 1839.
< Lophidæ, *Swainson*, Nat. Hist. and Class. Fishes, etc., v. 2, p. 330, 1839.
< Mathæoidei, *Bleeker*, Enum. Sp. Piscium Archip. Ind., p. xvi, 1859.
> Lophioidei, *Bleeker*, Enum. Sp. Piscium Archip. Ind. p. xvi, 1859.
< Mathæoidei, *Bleeker*, Atlas Ich. de Indes Néerland., t. 5, p. 3, 1865.
= Antennarioidæ, *Gill*, Proc. Acad. Nat. Sci. Phila., [v. 15,] pp. 89, 90, 1863.
< Lophiidæ, *Cope*, Proc. Am. Assoc. Adv. Sci., v. 20, p. 340, 1872.
> Halieutheæ, *Fitzinger*, Sitzungsber. K. Akad. der Wissensch. (Wien). ?. 67, 1. Abth.,
    p. 48, 1873.
< Malthæ, *Fitzinger*, Sitzungsber. K. Akad. der Wissensch. (Wien), B. 67, 1. Abth.,
    p. 48, 1873.
= Maltheidæ, *Lütken*, Vidensk. Selsk. Skr., (5,) v. 11, p. 325, 1878.
= Maltheidæ, *Gill*, Proc. U. S. Nat. Mus., v. 1, pp. 215, 219, 231, 1878.

MALTHEINÆ.

Synonyms.

= Maltheinæ, *Gill*, Cat. Fishes E. Coast N. A., p. 47, 1861.
= Maltheinæ, *Gill*, Proc. Acad. Nat. Sc., Phil., [v. 15,] p. 90, 1863.
= Maltheinæ, *Gill*, Proc. U. S. Nat. Mus., v. 1, pp. 220, 231, 1878.

HALIEUTÆINÆ.

Synonyms.

= Halieutæinæ, *Gill*, Proc. Acad. Nat. Sc., Phil., [v. 15,] p. 90, 1863.
= Halieutæinæ, *Gill*, Proc. U. S. Nat. Mus., v. 1, p. 231, 1878.

III.

In "Descriptions of Some New Species of Pediculati," published in 1863 (Proc. Acad. Nat. Sc. Phila., 1863, pp. 88–92), I have made known, in addition to the *Halieutichthys reticulatus* Poey, four species of *Antennarius*. These have not been re-described under the names then given, but two, originally found in Lower California, have been found elsewhere, and described by Dr. Günther under other names, as has just been recognized by Messrs. Jordan and Gilbert (Proc. U. S. Nat. Mus., v. 5, p. 370). Messrs. Goode and Bean have also recently recognized the two species described as inhabitants of the waters about the Florida Keys, in their "List of the Species of Fishes, recorded as occurring in the Gulf of Mexico" (Proc. U. S. Nat. Mus., v. 5, p. 235, 1882). The present status of the species is therefore as follows:

1. Antennarius sanguineus *Gill*, o. c., p. 91 = Antennarius leopardinus *Günther*, Proc. Zool. Soc. London, 1864, p. 151.
2. Antennarius annulatus *Gill*, o. c., p. 91; *Goode & Bean*, o. c., p. 235 (name only).

3. Antennarius pleurophthalmus *Gill*, o. c., p. 92; *Goode & Bean*, o. c., p. 235 (name only).

4. Antennarius strigatus *Gill*, o. c., p. 92 = *Antennarius tenuifilis Günther*, Trans. Zool. Soc. London, v. —, p. 440, 1869.

The *Halieutichthys reticulatus* Poey, it appears, was described as early as 1818, by Dr. Mitchill. The author was acquainted with Dr. Mitchill's paper, but did not think of connecting his description of the new "*Lophius*" with the *Halieutichthys*. The species referred to has the following history, and Mr. G. B. Goode first recognized the identity of the two. The history of the species may be epitomized as follows:

Halieutichthys aculeatus = Lophius aculeatus *Mitchill*, Am. Monthly Mag. and Crit. Rev., v. 2, p. 325, 18.8 = Halieutichthys reticulatus (*Poey*, MSS.) *Gill*, Proc. Acad. Nat. Sc. Phil., 1863, p. 91 = Halieutichthys aculeatus *Goode*, Proc. U. S. Nat. Mus., v. 2, p. 109, (with *Bean*) p. 333, 1879; (with *Bean*) v. 3, p. 467, 1881; v. 5, p. 235, 1882.

IV.

References to illustrations of osteology of the Pediculates.

ANTENNARIIDÆ.

BRACHIONICHTHYS HIRSUTUS.

Chironectes punctatus, *Cuvier*, Mém. Mus. Hist. Nat., t. 3, p. 434, pl. 18, f. 5, 1817.

PTEROPHRYNE HISTRIO.

Chironectes lævigatus, *Cuvier*, Mém. Mus. Hist. Nat., t. 3, p. 423, pl. 18, f. 4, 1817.

CERATIIDÆ.

CERATIAS HOLBOLLII.

Ceratias Holbolli, *Lütken*, Vidensk. Selsk. Skr., (5,) Nat. og Math. Afd., v. 11, p. 328 (f. 2 = vert. col.), 330 (f. 3 = interspinals), 331 (f. 4 = cran. behind, f. 5 = cr. lat.), 332 (f. 6 = cr. above), 334 (f. 7 = extracr. bones), 337 (f. 8 = sh. girdle). 1878.

LOPHIIDÆ.

LOPHIUS PISCATORIUS.

Lophius piscatorius, *Agass.*, Récherches Poiss. Foss., t. 5, (2,) p. 111, pl. M.

Lophius piscatorius, *Mettenheimer*, Disq. anat.-comp. membro pisc. pect. pl. 1, f. 4, (Sh. girdle and base P.), 1847.

Lophius piscatorius, *Hollard*, Ann. Sc. Nat. (5), Zool. et Pal., t. 1, pp. 241-256, passim, pl. 10, f. 1 (op. pieces), 1864.

MALTHEIDÆ.

MALTHE VESPERTILIO.

Lophius histrio! *Rosenthal*, Ichthyotom. Tafeln, pl. 19, f. 2, (Skel) 1822.

NOTE ON THE POMATOMIDÆ.

BY THEODORE GILL.

In 1862, in a "Synopsis of the Carangoids of the Eastern Coast of North America" (Proc. Acad. Nat. Sci. Phila , 1862, pp. 430–443), the family of Carongoids was limited, the chief subfamilies defined, and one established for the blue-fish and named "Pomatominæ," but the statement was made that "although the genus *Pomatomus* Lac. (*Temnodon* Cuv.) is here retained in the family, I am not certain that it truly belongs to it" (p. 430). Two years later I proposed and defined a peculiar family for the genus. In the catalogue of the fishes of the eastern coast of North America, in 1873, it was intended to have been placed, as were all the families 60–65, among the Acanthopteri "incertæ sedis," but through some inadvertence the word "incertæ sedis" was omitted. In order to determine the affinities of the doubtful form, I have re-examined the fish and its skeleton, and am now satisfied that the approximation of the type to the Carangidæ was correct, but still believe that it should be regarded as a peculiar family group. Pending a more detailed comparative study of the Scombroids, this may be briefly diagnosed as follows:

POMATOMIDÆ.

Synonym as family.

= Pomatomidæ, *Gill*, Can. Nat., n. s., v. 2, p. 246 (defined), 249, Aug., 1865.
= Pomatomidæ, *Gill*, Cat. Fishes E. Coast N. Am., p. 10 (name only), 1873.

Synonym as subfamily.

=Pomatominæ, *Gill*, Proc. Acad. Nat. Sci. Phil. [v. 14,] pp. 431 (defined), 443, 1862.

Scombroidea of Carangoid aspect, with the lateral line nearly parallel with the back (not angulated toward the middle) and elevated behind, and continuous on the base of the caudal; soft vertical fins densely scaly, and anal spines inseparable from the rayed portion.

NOTE ON THE AFFINITIES OF THE EPHIPPIIDS.

BY THEODORE GILL.

The Ephippiids, although presenting a superficial resemblance to the Chætodontids, otherwise exhibit such peculiarities as to have made me doubtful respecting their affinities, and to consider the family as incertæ sedis. The post-temporal bones were found to be bifurcated and thus failed to fulfill the requisites of Professor Cope's suborder Epilasmia wherein the Chætodontidæ were arranged. To satisfy myself as to their relations, I have examined their osteology, and am now convinced that, notwithstanding this deviation, they are most nearly connected with Chætodonts. They exhibit the following external and skeletal characteristics.

EPHIPPIIDÆ.

Synonymy.

= Ephippioids, *Gill*, Proc. Acad. Nat. Sci. Phila. [v. 14,] p. 238 (not defined), 1862.
= Ephippiidæ, *Gill*, Arrangement Families Fishes, p. 8 (named only), 1872.
= Ephippidi, *Poey*, Anal. de la Soc. Esp. de Hist. Nat., t. 4, p. 7 (named only), 1875.
Squamipennes gen., *Cuvier, etc.*
Chaetodontidæ gen., *Bonaparte, etc.*

EPHIPPIINÆ.

Synonymy.

< Ephippiformes, *Bleeker*, Enum. Sp. Piscium Archipel. Indico, p. xx, 1859.
< Chætodipteriformes, *Bleeker*, Archives Néerlandaises, t. 11, p. 300 (s. f. of Chaeto-
      dontoidei), 1876.
Chaetodontinæ gen., *Bonaparte, Günther, etc.*

Body much compressed and elevated, highest under the dorsal spines, and with the caudal peduncle short.

Scales of small or moderate size, either very finely ciliated or smooth, covering the whole body and head, and encroaching on uninterruptedly and more or less investing the vertical fins.

Lateral line continuous, parallel with the back and ending at the base of the caudal fin.

Head moderate, much compressed, short and high; eyes moderate, high and lateral, situated nearly midways between the snout and occiput.

Infraorbital chain with the bones decreasing backwards, and none articulated with the preperculum; preorbital moderately developed.

Opercular bones normal.

Nostrils double.

Mouth moderate, terminal, with the cleft lateral, scarcely extending to the vertical of the eyes. Upper jaw not or little protractile.

Teeth setiform, in a band in each jaw; none on the palate.

Branchial apertures lateral and separated from each by a wide, scaly isthmus.

Branchiostegal rays seven on each side.

Dorsal fin commencing some distance behind the nape, and thence extending nearly to the caudal; its spinous and soft portions are un-equally developed; spinous portion highest about the third spine, and emarginated behind; the soft long and elevated in front.

Anal fin similar to and opposite the soft dorsal, and armed in front with three spines.

Caudal fin expanded vertically and with its margin concave.

Pectoral fins normally situated on the scapular cincture, and with its lower rays branched.

Ventral fins thoracic, each with a spine and five rays, the first or second of which is longest.

The vertebræ number 24—10 abdominal and 14 caudal; their bodies are compressed and higher than long. The first two are specially modified: (1) The first has its central portion directed downwards, and its articular facets for the exoccipitals nearly vertical and directed forwards; its spine fits into the second neural spine. (2.) The second vertebra has a very short body, compressed antero-posteriorly, and its spine is erect, and with the basal portion expanded forwards. The other vertebræ are gradually modified.

The anterior zygapophyses are well developed, as are also the posterior of the caudal.vertebræ, and about the middle the posterior partly overlap the anterior of the succeeding; inferior zygapophyses are rudimentary; the neurapophyses and neural spines arise direct from the anterior margins of the vertebræ, and those of the middle of the column (e. g., 7 to 16) are erect, while the hindmost gradually decline backwards; the parapophyses of the third to ninth vertebræ arise near the inferior surface of the vertebral bodies, are well developed, spiniform, and are all directed downwards and outwards, and partly (7 to 10) with a hæmal canal; those of the tenth are expanded at their base externally, and their points converge and repose in the first hæmal spine; the first hæmal spine is grooved in front and somewhat expanded mesially. The sockets for the ribs are on the sides of the centra and at the external bases of the (third—eighth) parapophyses.

These characters have been formulated on comparison of specimens in alcohol and skeletons of *Chætodipterus faber* with those of Chætodontids, Serranids, Pristipomatids, &c. The resemblance to the Chætodonts (e. g., *Chætodon* or *Pomacanthus paru*) is much greater than to any other. *Chætodipterus* differs from most fishes, and resembles the typical Chætodontids in the specialization of the two foremost vertebræ, the great development of the parapophyses, and the inferior position of the sockets for the ribs. The. skull likewise resembles that of the Chætodontids in general characters, and especially in the oblique occipitosphenoid axis and the development of the exoccipital condyles. In fine, the Ephippiids are very closely related to the Chætodontids, but may be distinguished as follows:

Chætodontoidea with a wide scaly isthmus extending from the pectoral region to the chin and separating the branchial apertures; spinous partially differentiated from the soft portion of the dorsal; upper jaw scarcely protractile; ethmoid cariniform above (not sunk and concave) and vomer declivous (not projecting forwards or retuse), parapophyses spiniform and, posteriorly inclosing a hæmal canal, and post-temporal bones bifurcated.

Only two genera certainly belong to the family, *Ephippus* Cuv.

(not Blkr) and *Chætodipterus* Lac. (=*Parephippus* Gill). *Drepane*, according to Cape, is a Carangid, and *Scatophagus*, judging from the figure of its skeleton (Agassiz's *Poissons Fossiles*, t. 4, pl II. f 1), belongs to a peculiar family—the *Scatophagidæ*—the ribs of which are simple and received in sockets comparatively high on the centra, and, apparently,* the post-temporal is forked. In fact, *Scatophagus* appears to have no direct affinity with the Chætodontids.

ON THE RELATIONS OF THE FAMILY LOBOTIDÆ.

BY THEODORE GILL.

Among those families which are "incertæ sedis" has been that designated as Lobotidæ. Its type—*Lobotes surinamensis*—has been almost universally placed with the Pristipomids except by American authors. There was, however, nothing in its physiognomy or characteristics, except the unarmed palate, to justify such a reference, and recent examination shows that the skepticism as to the propriety of such association was amply warranted. On the whole it appears to be most nearly related to the Serranidæ of the families whose characters are to some extent known, and may be provisionally defined as follows:

LOBOTIDÆ.

Synonymy.

< [Lobotoidæ] *Gill*, Proc. Acad. Nat. Sc. Phila., [v. 14,] p. 238 (not named or defined †), 1862.

? Lobotidi *Poey*, Repertorio Fisico Nat. de Cuba, t. 2, p. 324 (not defined), 1868.

= Lobotidæ, *Gill*, Cat Fishes E. Coast N. Am , p. — (not defined), 1873.

Sciénoides gen , *Cuvier*, etc.

Pristipomidæ gen., *Günther*, etc.

Percoidei gen , *Bleeker*.

Percoidea with an oblong compressed body equally developed above and below, a short snout and anterior eyes, edentulous palate, dorsal and aval with the soft portions equal and opposite, the former preceded by a much larger spinous portion, the latter with three spines, vertebræ 24, 12 abdominal and 12 caudal ;‡ the fifth to eleventh with short but gradually lengthening parapophyses projecting sideways and behind downwards, and the twelfth with the parapophyses elongated, converging at their extremities, and fitting into a groove of the first hæmal spine, the costiferous pits excavated obliquely in the developed parapophyses, and gradually ascending forwards on the vertebræ, and finally

---

*The figure given by Professor Agassiz is ambiguous.

† "*Lobotes* Cuvier and *Datnioides*, Blkr., rather represent a family, perhaps, somewhat allied to the Naudoidæ." Gill, *op. cit.*

‡ Dr. Günther has attributed to the "*L. auctorum*" "Vert. 13 | 11 " (Cat. Fishes B. M., i, 338).

on the neurapophyses; the skull with its frontal portion broad, expanded forward and outward, and entering into the posterior borders of the orbits, which are advanced far forwards; the post-frontals elongated forwards and underlying the frontals; ethmoid short, decurved and expanded sideways.

The abbreviated orbital and ante-orbital regions and ensuing modifications contrast strongly with the corresponding parts in all the forms with which the genus *Lobotes* has been associated. With the exceptions noted, the vertebræ are essentially similar to those of the Serranidæ.

Lobotes is the only certainly known member of the family.

NOTE ON THE RELATIONSHIPS OF THE ECHENEIDIDS.

BY THEODORE GILL.

Among those forms that have been most shifted from place to place in the ichthyological systems is the genus *Echeneis* of Artedi and Linnæus.

By Artedi (1738) as well as by Linnæus, at first, it was placed in the order MALACOPTERYGII next to *Coryphæna*, the last a true acanthopterygian fish.

By Linnæus, in the later editions of the Systema Naturæ (1758, 1766), it was placed in the order THORACICI, but still kept by the side of *Coryphæna*.

By Cuvier (1817) it was referred to the order of "Malacoptérygiens subbrachiens" and the family "Discoboles" after *Lepadogaster* and *Cyclopterus* (R. A., t. 2, p. 227, 1817).

By Swainson (Nat. Hist. and Class. Fishes, etc., v. 2, 1839) the genus *Echeneis* was raised to family rank and the family (Echeneidæ) referred to the order "Acanthopteryges" and the tribe "Microleptes," in which it was supposed to constitute an "aberrant family" (p. 30), which "represented" the Acanthopterygian "tribe Blennides" (p. 32) and the "order Apodes" (p. 31).

It was preceded by the "typical" families (1) "Scomberidæ" and (2) "Zeidæ," and followed by the "aberrant" families (4) "Centriscidæ" and (5) "Coryphænidæ."

Subsequently all reference to the family as well to the genus was omitted (apparently through forgetfulness) by Swainson in the later and synoptical portion of the work. His eccentric classification is only noticed here because a similar or still more extreme view as to the affinity of the genus became long afterwards quite prevalent.

By Müller (1844) the genus was put in the order Acanthopteri and in the family Cyclopodi, but as the representative of a peculiar "group" ("3. Gruppe. Echeneiden").

March 23, 1883.

By Agassiz and Holbrook, and later by Günther* (1860), it was trans-ferred to the family Scombridæ, next to Elacate.

By Bleeker (1859) the genus was entitled with family rank (Echenoi-dei) and also ordinally distinguished (with the name "ordo 38. Disco-cephali") and interposed between "ordo 37. Fistulariæ," and "ordo 39. Cyclopteri."

By Cope (1870) it has been retained next to some Scombroid fishes (the Carangidæ), but as a distinct family, and placed in his order "Per-comorphi" and suborder "Distegi."

In later years the views of Müller, and subsequently of Swainson and Günther, have been generally adopted by European ichthyologists. In my "Arrangement of the families of fishes" the family Echeneididæ has been relegated to the category of Teleocephali "*incertæ sedis.*" A de-sire to reach some definite conclusion has induced me to examine its osteological as well as other characteristics, and has resulted in the fol-lowing conclusions :

The ventral fins being furnished with true spines, the fish is not a Malacopterygian, but an Acanthopterygian of Artedi, Cuvier, etc. The opposite reference to the Malacopterygians was due, in the first place, to the failure of Artedi and the older naturalists to appreciate the differ-ence between slender spines and "soft rays," and subsequently to the assumption, without attempt at verification, by Cuvier, of the correct-ness of his predecessors' statements.

The "basis cranii" is not double but simple, and there is no "tube." The type, therefore, is not at all related to the Scombridæ, Carangidæ, and other typical fishes, and consequently does not belong to the sub-order "Distegi" of Cope.

The contrary statement implied by Professor Cope is due, doubtless, to the preoccupation of his mind with the idea as to the affinity claimed to exist between Echeneis and the Scombridæ, and the consequent as-sumption that the former had a *basis cranii* like the latter. Inasmuch as the cranial cavity is partly closed, the true state of affairs can only be seen on opening or bisecting the skull, and this has probably been neglected. The group would really be referable to the suborder Scypho-branchii in Professor Cope's system, were it not for the form of the third pair of upper pharyngeal bones.

But what could have been the reason for referring the fish to the family Scombridæ (as contradistinguished from the Carangidæ) as a simple genus?

The family of "Scombéroïdes" was constituted by Cuvier for certain forms of known organization, among which were fishes evidently related to *Caranx*, but which had free dorsal spines. In the absence of knowl-edge of its structure, the genus *Elacate* was approximated to such be-cause it also had free dorsal spines. Dr. Günther conceived the idea

---

*On the History of Echeneis. By Dr. Albert Günther. <Ann. and Mag. Nat. Hist. (3), v. 5, pp. 386–402. 1860.

of disintegrating this family, because, *inter alias*, the typical Scombéroïdes (family Scombridæ) had more than twenty-four vertebræ and others (family Carangidæ) had just 24. The assumption of Cuvier as to the relationship of *Elacate* was repeated, but inasmuch as it has "more than 24 vertebræ" (it has 25 = 12 + 13) it was severed from the free-spined Carangidæ* and associated with the Scombridæ. *Elacate* has an elongated body, flattish head, and a colored longitudinal lateral band; *Echeneis* has also an elongated body, flattened head, and a longitudinal lateral band; therefore *Echeneis* was considered to be next allied to *Elacate* and to belong to the same family! The very numerous differences in structure between the two were entirely ignored, and the reference of *Echeneis* to the Scombridæ is simply due to assumption piled on assumption. The collocation need not, therefore, longer detain us.

The possession by *Echeneis* of the anterior oval cephalic disk in place of a spinous dorsal fin would alone necessitate the isolation of the genus as a peculiar family. But that difference is associated with almost innumerable other peculiarities of the skeleton and other parts, and in a logical system it must be removed far from the Scombridæ, and probably be endowed with subordinal distinction. In all essential respects it departs greatly from the type of structure manifested in the Scombroidea and rather approximates—but very distantly—the Gobioidea and Blennioidea. In those types we have *in some* a tendency to flattening of the head, or anterior development of the dorsal fin, a simple basis cranii, etc. Nevertheless there is no close affinity nor even any tendency to the extreme modification of the spinous dorsal exhibited by *Echeneis*. In view of all these facts *Echeneis*, with it subdivisions, may be regarded as constituting not only a family but a suborder, which is definable as follows:

Suborder DISCOCEPHALI.

Synonymy.

=Discocephali, *Bleeker*, Enum. sp. Piscium archipel. Ind., p. xxvi, (order; not defined), 1859.
=Echeneidoidea, *Gill*, Arrangement Fam. Fishes, p. 12, (super family; not defined), 1872.

Teleocephali with a suctorial transversely laminated oval disk on the

---

* "This family [Carangidæ] forms a very natural division, widely [*sic!*] differing from the Scombridæ in the structure of the vertebral column, which is composed of ten abdominal and fourteen caudal vertebræ. The *only exception* is found in the genera *Chorinemus* and *Temnodon*." (*Gthr.* Cat. Fishes B. M., v. 2, p. 417.) Besides the genera specially excepted, according to Dr. Günther's own figures, the following falsify his generalization, viz: *Caranx goreensis* (p. 457)—"Vert. 10 | 16"; *Psettus argenteus* (p. 488)—"Vert. 9 | 14"; *Platax arthriticus* (p. 491)—"Vert. 11 | 13"; *Zanclus cornutus* (p. 493)—"Vert. 9 | 13"; *Capros aper* (p. 496)—"Vert. 10 | 12–13"; *Equula fasciata* (p. 498)—"Vert. 10 | 13." There are a number of other exceptions, but their consideration is not called for in this place.

upper surface of the head, (homologous with a first dorsal fin*,) thoracic ventral fins with external spines, a simple basis cranii, intermaxillary bones flattened, with the ascending processes deflected sideways, and with the supramaxillary bones attenuated backwards, flattened, and appressed to the dorsal surface of the intermaxillaries; hypercoracoid (or scapula) perforated nearly in the center; and with four short actinosts (" carpals ").

Family ECHENEIDIDÆ.

Partial Synonymy.

< Eleutheropodes, *Duméril*, Zool. Anal., p. 123, 1806.
= Echeneidi, *Rafinesque*, Indice d'Ittiolog. Sicihana, p. 29, 1810.
< Cephoplia, *Rafinesque*, Analyse de la Nature, 13. fam., 1815.
< Encheliosomes, *Blainville*, Journal de Physique, t. 83, p. 255 ? (Includes *Echeneis*, *Cépoles*, and *Gymnètres*). 1816.
< Discoboles, *Cuvier*, Règne Animal, t. 2, p. 227, 1817.
< Discobola, *Latreille*, Fam. Nat. du Règne Animal, p. 127, 1825.
= Echeneides, *Risso*, Hist. Nat. de l'Europe Merid., t. 3, p. 269, 1826.
= Echeneididæ, *Bonaparte*, Giorn. Accad. di Scienze, v. 52. (Saggio Distrib. Metod. Animal. Vertebr. a Sangue freddo, p. 38,) 1831-'32.
= Echeneididæ, *Bonaparte*, Nuovi Annali delle Sc. Nat., t. 2, p. 133, 1838.
= Echeneidæ, *Swainson*, Nat. Hist. and Class. Fishes, etc., v. 2, pp. 31, 32, 42, 43, 44, 1839.
= Echeneisidæ, *Gray*, Syn. Brit. Mus., p. 143, 1842.
< Cyclopodi, *Muller*, Archiv für Naturgeschichte, Jahrg. 1843, v. 1, p. 297, 1843.
= Echeneididæ, *Gray*, *White*, List Spec. Brit. Animals Brit. Mus., Fish, p. 55, (placed between Callionymidæ and Lophiidæ.) 1851.
= Echeneididæ, *Richardson*, Encyclopædia Brit., v. 12, p. 272, (271,) 1856.
= Echeneoidæ, *Bleeker*, Enum. Sp. Piscium Archipel. Indico, p. xxvi, 1859.
= Echeneidæ, *Cope*. Proc. Am. Assoc. Adv. Science, v. 20, p. 342, 1872.
= Echeneididæ, *Gill*, Arrangement Fam. Fishes, p. 12, 1872.
= Echeneides, *Fitzinger*, Sitzungsber. k. Akad. der Wissensch. (Wien), B, 67, 1. Abth., p. 43, 1873.
Scombridæ gen., *Günther*, (Int. to Study of Fishes, p. 460,) 1880.

Sub-family ECHENEIDINÆ.

Synonymy.

= Echenidia *Rafinesque*, Analyse de la Nature, 1. s. f. of 13. fam., 1815.
= Echeneidini, *Bonaparte*, Nouvi Annali delle Sc. Nat., t. 2, p. 133, 1838; t. 4, p. 275, 1840.
= Echeneiden, *Müller*, Archiv für Naturgescicthte, Jahrg. 1843, p. 297, ("group" of Cyclopodi), 1843.
Scombrina gen., *Günther*.

External characters. (See plate VII, showing skull).

Body elongated, subcylindrical, diminishing backwards gradually from the head and into the slender caudal peduncle. Anus subcentral.

---

* Baudelot (E.) Étude sur le disque céphalique des Rémores (Echeneis) <Annales des Sciences Naturelles, (5e série, Zoologie et Paléontologie,) t. 7, pp. 153–160, pl. 5, 1867; (tr. pt.) Ann. and Mag. Nat. Hist., (4,) v. 19, pp. 375–376, 1867.

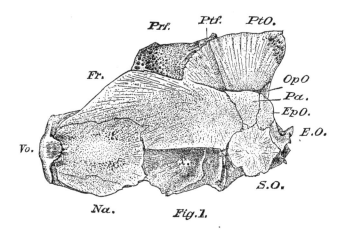

Fig. 1.

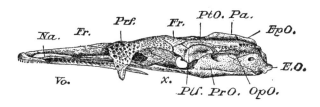

Fig. 2.

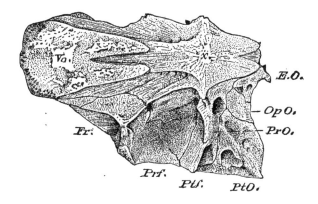

Fig. 3.

Scales, cycloid, very small, and not or scarcely imbricated.

Lateral line nearly straight and very faint.

Head above oblong and with a flattened straight upper surface fur-nished with an adhesive oblong or elongated laminated disk. The eyes are rather small, submedian, and overhung by the disk.

Suborbital bones forming a slender infraorbital chain; the first or preorbital triangular and thick.

Opercular apparatus normally developed and unarmed.

Nostrils double, close together.

Mouth terminal or, rather, superior, the lower jaw projecting, but with the cleft nearly horizontal and not extending laterally to the eyes.

Teeth present on the jaws and palate.

Branchial apertures ample and fissured forwards. Branchiostegal rays seven (or eight) on each side.

The adhesive disk on the upper surface of the head is a modified first dorsal fin and from the snout generally extends more or less posteriorly on the nape and back; it is oblong or elongated and of an oval or ellip-tical form, divided into equal halves by a longitudinal septum, and with more or less numerous transverse pectinated or spinigerous transverse laminæ in each division, the laminæ being slightly erectile and depres-sible.

Dorsal fin oblong or elongated, on the posterior half of the body (including head), ending some distance from the caudal.

Anal fin opposite and similar to the dorsal.

Caudal fin rather small, variable in outline but never deeply forked.

Pectoral fins moderate, inserted high on the sides.

Ventral fins thoracic; each with a spine and five branched rays.

The vertebral column has vertebræ in slightly increased number, the abdominal vertebræ being about twelve to fourteen and the caudal fif-teen or sixteen.

The stomach is cæcal and the pyloric cæca are present in moderate numbers. The air bladder is obsolete.

Who can consistently object to the proposition to segregate the Echeneididæ as a suborder of teleocephalous fishes?

Not those who consider that the development of three or four inar-ticulated rays (or even less) in the front of the dorsal fin is sufficient to ordinally differentiate a given form from another with only one or two such. Certainly the difference between the constituents of a disk and any rays or spines is much greater than the mere development or atro-phy of articulations.

Not those who consider that the manner of depression of spines, whether directly over the following, or to the right and left alternately, are of ordinal importance; for such differences again are manifestly of less morphological significance than the factors of a suctorial disk.

Nevertheless there are doubtless many who will passively resist the proposition because of a conservative spirit, and who will vaguely recur to the development of the disk as being a "teleological modification," and as if it were not an actual fact and a development correlated with radical modifications of all parts of the skeleton at least.

But whatever may be the closest relations of *Echeneis*, or the systematic value of its peculiarities, it is certain that it is not allied to *Elacate* any more than to others of the hosts of Scombroid, Percoid, and kindred fishes, and that it differs *in toto* from it, notwithstanding the claims that have been made otherwise.* It is true there is a striking resemblance, especially between the young—almost as great, for example, as that between the placental mouse and the marsupial antechinomys— but the likeness is entirely superficial, and the scientific ichthyologist should be no more misled in the case than would the scientific therologist by the likeness of the marsupial and placental mammals.

NOTE ON THE GENUS SPARUS.

BY THEODORE GILL.

Messrs. Jordan and Gilbert propose to restore the Linnæan name *Sparus* to *Sparus boops*, after the example of Swainson (Nat. Hist. and Class. Fishes, etc., v. 2, pp. 171, 221), instead of to the *Sparus aurata*, as I have done. This course is inadmissible, as those naturalists will doubtless recognize when they become conversant with the facts of the case.

Linnæus, after Artedi and the older authors, employed the name for Sparoid and other fishes of diverse kinds, and including *Sparus aurata*, *Sparus boops*, etc. Both Artedi and Linnæus placed the *S. aurata* at the head or as first of the genus.

Blóch and Lacépède variously restricted the genus, but still retained the forms just noted.

Cuvier, in 1817, subdivided the old genus into "tribes" and "genera," distinguishing for the *Sparus boops*, etc., the "second tribe," and the genus "Boops Cuv.," and for the *Sparus aurata* and related forms the "third tribe" and the restricted genus "Sparus Cuv." The "genus" was subdivided into subgenera, viz: "Les Sargues (Sargus. Cuv.)," "les Daurades" (without a latin equivalent), and "les Pagres (Pagrus. Cuv.)."

The name *Sparus* must, therefore, be retained for a section of the genus as restricted by Cuvier.

Risso, in 1827, supplied a Latin name "Aurata" for "les Daurades" of Cuvier.

Cuvier, in 1829, retained the genus *Sparus* with the same limits as in 1817, but with a slightly different subdivision of subgenera, viz: "Les Sargues (Sargus)," "les Daurades (Chrysophris N.)," "les Pagres" (without a Latin name), and "les Pagels (Pagellus Cuv.)."

---

*"This genus [*Echeneis*] is closely allied to the preceding [*Elacate*], from which it differs only by the transformation of the spinous dorsal fin into a sucking organ." (*Günther*, Int. to Study of Fishes. p. 460, 1880.)

Bonaparte, in 1832, revived the name *Sparus*—"Sparus *N*. (Aurata *Riss.*, Chrysophrys *C*.)," for the *Sparus aurata*.

For the *Sparus aurata* and its relations, the Linnæan name must therefore be retained and the subsequent applications of the name in no wise affect the legitimacy of this application.

Whether the groups designated as *Pagrus* and *Chrysophrys* are, or are not, generically distinct is not a matter for present consideration. It is certain, however, that the group as proposed by Cuvier, and adopted by later writers (*e. g.*, Günther), is artificial and heterogeneous, and doubtless the typical species of *Chrysophrys* and *Pagrus* are more nearly allied to each other than are such types to forms with which they have been associated. For the present, the genus *Sparus* may be retained as distinct from *Pagrus* and with the eliminations required.

SPARUS.

Synonymy.

<Sparus *Linnæus*, Syst. Nat., 10. ed., t. 1, p. 277, 1758; 12. ed., t. 1, p. 467, 1766 Gmelin, ed., t. 1, p. 1270.

<Sparus *Bloch*, Systema Ichthyologiæ, ed. Schneider, p. 269, 1801.

<Sparus *Lacépède*, Hist. Nat. des Poissons, t. 4, p. 26, 1803.

<Les Spares (Sparus Cuv.) *Cuvier*, Régne Animal, t. 2, p. 271 (genus), 1817.

<Les Daurades *Cuvier*, Régne Animal, t. 2, p. 272 (subgenus of *Sparus*), 1817.

<Aurata *Risso*, Hist. Nat. de l'Europe Merid., t. 3, p.—, 1827.

<Les Daurades (Chrysophrys,) *Cuvier*, Régne Animal, 2. ed., t. 2, p. 181, 1829.

<Sparus *Bonaparte*, Giorn. Acad. di Scienze, t. 52 (Saggio Distrib. Metod. Animali Vertebr. a sangue freddo, p. 33), 1832.

=Chrysophrys *Swainson*, Nat. Hist. and Class Fishes, etc., v. 2, pp. 171, 221, 1839.

ON THE PROPER NAME OF THE BLUE FISH.

BY THEODORE GILL.

The propriety of the substitution of the name *Pomatomus* in place of *Temnodon* for the blue-fish of the Americans has been questioned by (1) those who contend that a generally accepted name should not be disturbed, and by (2) those who would go to an extreme in the application of the law of priority. A brief history of the nomenclature of the genus seems therefore to be desirable.

The blue-fish had been referred to genera with which it has little affinity (to *Gasterosteus* by Linnæus and *Scomber* by Bloch) till the close of the last century.

In 1802 Lacépède described as a new species, and as the first of a new genus, a form which was evidently identical with the *Gasterosteus saltatrix* of Linnæus and the blue-fish of the United States, but which was obtained by Commerson in the "Océan Équatorial." With this species were associated eight others and the including genus was defined as follows:

> " *Cent sixième genre.*

> " LES CHEILODIPTÈRES.

"La lèvre supérieure extensible; point de dents incisives, ni molares; les opercules des branchies dénués de piquans et de dentelures; deux nageoires dorsales."

In 1828, the genus *Cheilodipterus* was amended by Cuvier and Valenciennes (Hist. Nat. des Poissons, t. 2, p.162), and restricted to the third species of Lacépède (le *C. rayé*) and related species. Inasmuch as (1) the Cuvierian genus had received no previous name, (2) the Lacépèdian name and diagnosis were as applicable to it as to any other of the species, and (3) it did not conflict with any other rights, there appears to be no sufficient reason for transferring the name to any other genus. Nevertheless, because the first species of the genus was the Blue-fish, Dr. Bleeker has proposed to revive the name *Cheilodipterus* instead of *Temnodon*, and given a new name (*Paramia*) for the genus *Cheilodipterus* Cuv. & Val. ex Lac. Common usage does not seem to justify such a procedure.

But in 1803 Lacépède described a supposed previously unknown form of fish, the Pomatome skib (*Pomatomus skib*), in the following terms:

[P. 435.]   *Cent Vingt-Quatrième Genre.*

LES POMATOMES.

L'opercule entaillé le haut de son bord postérieur, et couvert d'écailles semblables à celles du dos; le corps et la queue alongés; deux nageoires dorsales; la nageoire de l'anus très adipeuse.

ESPÉCE.	CARACTÈRES.
LE POMATOME SKIB. (*Pomatomus Skib.*)	Sept rayons aiguillonés à la première dorsale; trois entailles à chaque opercule; la mâchoire inferieure plus avancée que la supérieure; la caudale très fourchue.

[P. 436.]   LE POMATOME SKIB.[1]

Nous devons la connoissance de ce poisson à notre savant confrère M. Bosc, qui a bien voulu nous communiquer un dessin et une description de cette espèce, dont il a observé les formes et les habitudes, avec son habileté ordinaire, pendant le séjour qu'il a fait dans les États-Unis.

---

[1] "*Pomatomus skib.*

Skib jack, *dans la Carloine.*

Perca skibea pinnis dorsalibus distinctis, secundâ viginti-quatuor radiis, corpore argenteo, cauda bifurcâ."

Ce pomatome[2] habite dans les baies et vers les embouchures des rivières de la Caroline. On ne l'y trouve cependant qu'assez rarement. Il saute et s'élance fréquemment à une distance plus ou moins grande; et cette faculté ne doit pas surprendre dans un poisson dont la queue est conformée de manière à pouvoir être agitée avec rapidité. La chair du skib est très-agréable au goût.

Les mâchoires sont garnies chacune d'une rangée de dents aplaties, presque égales, et un peu séparées les unes des autres. La seconde dorsale est plus longue que la première, et d'une étendue à peu près égale à celle de la nageoire de l'anus. Celle-ci est si adipeuse [p. 437] qu'on peut à peine distinguer les rayons qui la composent.

L'animal est verdâtre dans sa partie supérieure, et argenté dans sa partie inférieure. L'iris est jaune ; et l'on voit une tache noire sur la base des pectorales, qui sont jaunâtres.*

As this description clearly applies to the ordinary bluefish, and, in fact, is well known to be based on that species, the name *Pomatomus* should have been used for it as the earliest given therefor. But Risso, in 1826, referred to Lacépède's genus, a deep-sea fish, which he considered to be congeneric with the *" P. skib."* Cuvier recognized that Risso's fish was generically distinct, but notwithstanding (1) revived the name *Pomatomus* from Risso for the latter fish, (2) suppressed Lacépèdes genus, and (3) proposed a new designation (*Temnodon*) in place of Lacépède's *Pomatomus*. Each step in this procedure was inadmissible. In 1862, I therefore restored the name *Pomatomus* to the bluefish in lieu of *Temnodon* and this revived name has been generally adopted since by American naturalists, as it undoubtedly will be by all others when they have learned that obedience to law (laws formulated by the British and American Associations for the Advancement of Science) is more conducive to stability of nomenclature than deference to the whim or prejudice of any " authority."

The synonymy of the genus is quite voluminous, as will be evident from the following exhibit :

POMATOMUS.

< Cheilodipterus *Lacépède*, Hist. Nat. des Poissons, t. 3, p. 542, 1802.

= Pomatomus *Lacépède*, Hist. Nat. des Poissons, t. 4, p. 436 (yg.), 1803.

= Gonenion *Rafinesque*, Caratteri alc. n. gen. e. n. sp. An. e. Piante Sicilia, p. 53 (pl. 10, f. 3 = yg.), 1810.

---

[2] " Ce nom générique désigne la forme de l' opercule: *poma*, en grec, signifie opercule, et *tome*, incision."

* 7 rayons à la membrane branchiale du pomatome skib.

24   à la seconde dorsale.
15   à chaque pectorale.
·6   à chaque thoracine.
26   à la nageoire de l'anus.
18   à celle de la queue.

=Temnodon *Cuvier*, Règne Animal, t. 2, p. 346, 1817.
=Sypterus *Eichwald*, Fauna Caspio-Caucasica, p. —? (fide Bonaparte), ? 1841.
=Chromis *Gronow*, Systema Ichthyologicum (1780), publ. by J. E. Gray, p.—, 1854.
=Pomatomus *Gill*, Proc. Acad. Nat. Sc. Phila., [v. 14,] p. 443, 1862.
=Cheilodipterus *Bleeker*, Nat. Verhandel. Holl. Maatschapij Wetenschappen (3), v. 2, no. 1, p. 74, 1874.
=Sparactodon *de Rochebrune*, Bull. Soc. Philomathique Paris (7), t. 4 ? pp. 159-169 (yg.), 1880 (identified with "Temnodon" by *Steindachner*, Denkschr. k. Akad. Wiss., Math.-Nat. Cl., v. 14, p. 51, 1882.

DOES THE PANTHER (FELIS CONCOLOR) GO INTO THE WATER TO KILL FISH?

BY LIVINGSTON STONE.

[Letter to Prof. S. F. Baird.]

My mind has been quite exercised lately on the question whether panthers go into the water to kill fish. They are so numerous and bold here this year, that they come to our very doors and kill pigs and fowls under our windows. We estimate that they have killed a hundred dollars' worth of hogs here this season, besides calves, colts, and full grown cattle and horses. As far as boldness is concerned, they are fully equal to jumping into our trout ponds and killing our trout. And if you think they are likely to do this, we will take special precautions against it. They easily jump over any obstacle not more than 15 feet high, so that our fences are no protection from them.

They frequently swim the river, which made me think that perhaps they might get into the trout ponds sometimes for a meal of fish.

UNITED STATES FISH COMMISSION,
Baird, Shasta County, California, September 21, 1882.

ON CERTAIN NEGLECTED GENERIC NAMES OF LA CÉPÈDE.

BY DAVID S. JORDAN AND CHARLES H. GILBERT.

In the Histoire Naturelle des Poissons (1799–1803) of La Cépède a considerable number of generic names are proposed, some of them founded on errors of various sorts, others properly defined. About one-fourth of these were adopted by Cuvier and Valenciennes, and have come into general use. A large number are simple synonyms. The remainder, for different reasons, were set aside by Cuvier and Valenciennes, and new names proposed in their places. As the laws of priority are constantly becoming more and more urgent, we find ourselves obliged to go behind Cuvier, and to adopt these earlier names.

The present paper contains a discussion of some of these names, the adoption of which would affect the nomenclature of American fishes.

1. HIATULA.

In Gmelin's edition of the Systema Naturæ, p. 1287, under the genus *Labrus*, the following description appears:

* * CAUDA INTEGRA.

Hiatula. 12. L. pinna anali nulla. Br. 5, D. $\frac{17}{15}$, P. 16, V$\frac{1}{5}$, **A**. 0. C. 21.

Habitat in Carolina, *fascus nigris, 6–7 pictus.* D. Garden. Labium retractile, intus rugosum; dentes *in mandibulis lamarii, in palato orbiculati;* branchiarum *operculum anterius margine punctatum;* pinna *dorsalis fere longitudinalis, radiis spinosis æqualibus, posterius nigra.*

With the exception of the two characters, absence of the anal fin, and presence of rounded teeth on the palate, which belong to no fish of this type, this description applies well to a young tautog, and to no other fish which Dr. Garden could have obtained at Charleston. The specimen most likely was one in which the anal fin had been bitten off, an accident to which fishes are not unfrequently subject. The rounded teeth on the palate must be either the posterior teeth of the premaxillaries, which are bluntish, or possibly the papillæ which cover the membrane before the vomer.

In the second volume of La Cépède's work (ii, 522, 1800), this species appears under the name of *Hiatula gardeniana,* as the type of a new genus, *Hiatula,* distinguished from *Labrus* by the absence of the anal fin.

As this character was merely the accident of a mutilated specimen, this genus is a virtual synonym of *Labrus,* and by many writers would be suppressed as such. The name *Hiatula,* however, stands on the same footing as that of *Micropterus,* which was likewise based by La Cépède on a mutilated fish. As *Micropterus* has now come into general use, we suggest that *Hiatula* be substituted for *Tautoga.*

2. GOBIOMORUS.

The genus *Gobiomorus* was proposed by La Cépède (Hist. Nat. Poiss. ii, 583, 1800) as a subdivision of the Linnæan genus Gobius, with the following definition:

"Les deux nageoires thoracines non réunies l'une à l'autre; deux nageoires dorsales: la tête petite; les yeux rapprochés; les opercules attachés dans une grande partie de leur contour."

In definition and in intention, this group corresponds to the genus *Eleotris* of Bloch and Schneider, as revised and restricted by Cuvier, for Bloch and Schneider seemed to have no clear idea of the group, and very few of the species referred by them to *Eleotris* are related to *Eleotris gyrinus.*

Four species are referred by La Cépède to *Gobiomorus,* viz, *G. gronovii* (=*Nomeus gronovii* (Gmelin) C. & V.) *G. taiboa* (=*Eleotris strigata* (Broussonet) C. & V.) *G. dormitor* Lac. (later called *Platycephalus dormi-*

tator by Bloch & Schneider = *Philypnus dormitator* (Lac.) C. & V.) and *G. kœlreuteri* (= *Periophthalmus kœlreuteri* (Gmelin) Bloch & Schneider).

Of these species, the first, *gronovii* has no relation to *Gobius*, and does not correspond to the definition of *Gobiomorus*, as the gill membranes are free from the isthmus. Its association with the Gobies is an error which originated with Gmelin. It may therefore be omitted from consideration. The remaining modern genera included in *Gobiomorus*, viz, *Eleotris* Bloch & Schneider, 1801 (Subgenus *Valenciennea* Bleeker, 1856), *Philypnus* Cuv. & Val., 1837, and *Periophthalmus* Bloch & Schneider, 1801, are all subsequent to *Gobiomorus*, and in place of one of them the latter name must be retained. It has not as yet been restricted by any author, so far as we know. It seems to us best to consider as the type of *Gobiomorus*, *G. dormitator* LaCépède, and therefore to use the name *Gobiomorus* instead of *Philypnus*. A serious practical objection to the consideration of *taiboa* (*strigatus*) as the type of *Gobiomorus* lies in the uncertainty whether this species is really congeneric with *Elcotris gyrinus*, (which species must, we think, as "*Eleotris pisonis*," be considered the type of *Eleotris*). In Bleeker's system, *strigatus* is made the type of a distinct genus (*Valenciennea* Bleeker) and placed at a distance from *Eleotris*, but no diagnostic features of importance have been made known by which it may be distinguished.

3. GOBIOMOROIDES.

The genus *Gobiomoroides* was proposed by La Cépède (Hist. Nat. Poiss., ii.. 592, 1800), with a definition identical with that of *Gobiomorus* except that "une seule nageoire dorsale" is substituted for "deux nageoires dorsales." Its type is *Gobiomoroides piso* La Cépède, a species which is considered by La Cépède identical with *Gobius pisonis* Gmelin, the "*Eleotris*" of Gronow.

Gobius pisonis Gmelin is identified by Cuvier & Valenciennes with *Eleotris gyrinus*, with considerable doubt, however, as the descriptions and figures of the former species are both incomplete and erroneous. The identity is probably too doubtful to warrant the use of the specific name *pisonis* for *gyrinus*. La Cépède's description of *G. piso*, is, however, not taken from Gmelin, but from a dried fish "given by Holland to France." This specimen has 45 rays in the dorsal which is continuous, 23 in the anal, and the lower jaw has a series of canines besides the cardiform band. Whatever this fish may be, it is not an *Eleotris*, and the name *Gobiomoroides* cannot be used for *Eleotris gyrinus*, even if it be shown that this species is identical with *Gobius pisonis* Gmelin.

4. KYPHOSUS.

The genera *Kyphosus* (La Cépède, Hist. Nat. Poiss., iii, 114, 1802), *Pimelepterus* (l. c. iv., 429, 1803): *Dorsuarius* (l. c. v., 482, 1803), and *Xyster* (l. c. v. 484, 1803), are identical, as has been shown by Cuvier and Valenciennes, vii, 254. The earliest of these names should be used, and *Kyphosus* should therefore supersede *Pimelepterus*. The word should however be spelled with an initial C, as *Cyphosus*.

5. MONODACTYLUS.

The genera, *Monodactylus* La Cépède (Hist. Nat. Poiss., iii, 131, 1802, *M. falciformis* Lac.), *Centropodus* La Cépède (iii, 303, *C. rhombeus* Lac.), and *Acanthopus* (iv, 558; *A. argenteus* (Gmelin) and *A. Boddærti* (Gmelin)) are all based on species of the genus afterwards called *Psettus* Cuv. & Val. This genus should therefore receive the name of *Monodactylus*.

6. SCOMBEROMORUS.

Scomberomorus (iii. 293; *S. plumierii* La Cép.) is based on a drawing by Plumier. The genus is distinguished from *Scomber* by the supposed continuity of the dorsal fins, a fallacious character. The species is identical with *Scomber regalis* Bloch, and the name *Scomberomorus*, if accepted, must supersede *Cybium* Cuv. & Val.

7. CEPHALACANTHUS.

It appears to be reasonably certain that the small fishes which have received the name of *Cephalacanthus* La C. (iii, 323, 1802; *C. spinarella* L.) are the young of, or, at least, not generically different from, the Flying Gurnards (*Dactylopterus* La C. iii, 325). The name *Cephalacanthus* has two pages priority, and should in strictness supersede *Dactylopterus*. The application of the law of priority to different parts of the same work is often as important for the avoidance of confusion as its application to different works. The law of primogeniture applies to twins.

8. DIPTERODON.

The genus *Dipterodon* La C. (Hist. Nat. Poiss., iv, 165, 1803) is based on six species, mostly unrelated, belonging to *Lutjanus*, *Apogon*, *Aspro*, and *Sciæna*. The first of this species, *D. plumieri*, is identical with *Lutjanus synagris*, and the name may be considered as a synonym of *Lutjanus*.

The sixth species mentioned, " *Dipterodon chrysourus*," is evidently identical with *Sciæna argyroleuca* (Mitch.), the second of the two species called "*Perca punctata* " by Linnæus in the Systema Naturæ. If the duplicated Linnæan name be restricted to the first species to which it was given (*Epinephelus punctatus*), the name *chrysura* must take the place of *argyroleuca*, and the species stand as *Sciæna* (*Bairdiella*) *chrysura*.

The name *Dipterodon* has been used by Cuv. & Val. for a genus unknown to La Cépède. This transfer of the name is not allowable, and the *Dipterodon* C. & V. should receive a different name, that of *Coracinus* Gronow (1854).

9. CHÆTODIPTERUS.

Chætodipterus (iv., 503 ; *Chætodon plumieri*, Gmelin.) is correctly distinguished from *Chætodon*, by the separation of the dorsal fins. Its type
is identical with *Zeus faber* Broussonet. The name *Chætodipterus* must
therefore supersede *Parephippus* Gill, as Bleeker has already shown.

10. POMADASYS.

Pomadasys (iv. 515) is based on *Sciæna argentea* Forskål, which is a
species of Cuvier's genus *Pristipoma*, according to Günther and Cuvier.

The generic description is not altogether correct, but is copied from
the specific description of Forskål. The name *Pomadasys* must therefore take the place of *Pristipoma*, a change already made by Cantor and
Bleeker.

11. CLUPANODON.

The genus *Clupanodon* was proposed by La Cépède (Hist. Nat. Poiss.,
v. 468, 1803) for those species of *Clupea* which had no teeth in the
jaws, and with the following definition :

" Plus de trois rayons à la membrane des branchies, le ventre carenè ;
la carène du ventre dentelée ou très-aigus ; la nageoire de l'anus separée
de celle de la queue ; une seule nageoire sur le dos ; point de dents
aux mâchoires."

Six species are referred by La Cépède to this genus, viz :

thrissa (L.). (*Opisthonema* Gill.)
nasica Lac. (*nasus* Bloch). (*Dorosoma* Raf.)
pilchardus L. (*Sardinia* Poey.)
sinensis L. (*Clupeonia* C. & V.)
africanus Bloch. (*Pellona*, C. & V.)
jussieu Lac. (*Clupeonia* C. &. V.)

One of these, *Pellona africana*, does not conform to the definition and
should be excluded. All the others (except *Dorosoma nasus*) are very
closely related, and are probably all representatives of sections of the
genus *Clupea* rather than of distinct genera. The name of *Clupañodon*
is prior to all of these and must take the place of one of them. So far
as we know, it has never been formally restricted. It seems to us best
to consider *C. jussieui* as the type of *Clupanodon*, and to substitute
Clupanodon for *Clupeonia*.

12. GYMNOMURÆNA.

The genus *Gymnomuræna* La Cépède (Hist. Nat. Poiss., v. 648, 1803),
was defined as follows :

" Point de nageoires pectorales ; une ouverture branchiale sur chaque
côté du poisson ; le corps et la queue presque cylindriques ; point de
nageoire du dos, ni de nageoire de l'anus ; ou ces deux nageoires si

basses et si enveloppées dans une peau épaisse, qu'on ne peut reconnoître leur présence que par la dissection."

Two species are mentioned, *Gymnomuræna doliata* La C. (=*Echidna zebra* (Shaw) Bleeker) and *Gymnomuræna marmorata* (=*Muræoblenna marmorata*), both of which agree fairly with the generic definition.

The first restriction of the genus *Gymnomuræna* is that of Kaup (Apodes, 1856, 103), in which *zebra* (*doliata*) is regarded as the type; and the group is recognized (probably correctly) as distinct from *Echidna* Forster (=*Pœcilophis*, Kaup).

Later Dr. Günther (Cat. Fish, Brit. Mus., viii, 133, 1870) has restricted the name *Gymnomuræna* to the second species of La Cépède (*marmorata*). This arrangement seems to us not allowable. The first proper restriction must hold, and the name *Gymnomuræna* henceforth go with *G. doliata*.

13. MURÆNOBLENNA.

The group called by Dr. Günther *Gymnomuræna* should stand as *Muræoblenna* La Cépède (Hist. Nat. Poiss., v. 652, 1803). This genus is based on a single species, *M. olivacea* La C., and is defined as follows:

" Point de nageoires pectorales ; point d'apparence d'autres nageoires ; le corps et la queue presque cylindriques ; la surface de l'animal repandant en très grande abandance, une humeur* laiteuse et gluante."

14. MACRORHAMPHOSUS.

The genus *Macrorhamphosus* La Cépède (v. 136) is based on *Silurus cornutus* Forskål=*Centriscus scolopax* L. In the tenth edition of the Systema Naturæ, Linnæus refers to his genus *Centriscus* but one species, *C. scutatus*. This species should, therefore, properly be taken as the type of *Centriscus* (=*Amphisile* Cuv.), while the name *Macrorhamphosus* should be used for *C. scolopax* and its relatives, the group usually called *Centriscus*.

The following is a summary of the changes in nomenclature suggested in the present paper:

HIATULA La Cépède for *Tautoga* Mitchill.
GOBIOMORUS La Cépède for *Philypnus* Cuv. & Val.
CYPHOSUS La Cépède for *Pimelepterus* La Cépède.
MONODACTYLUS La Cépède for *Psettus* Cuv. & Val.
SCOMBEROMORUS La Cépède for *Cybium* Cuv. & Val.
CEPHALACANTHUS La Cépède for *Dactylopterus* La Cépède.
SCIÆNA (BAIRDIELLA) CHRYSURA (La Cép.) Jor. & Gilb. for *Sciæna* (*Bairdiella*) *argyroleuca* (Mitchill), J. & G.
CHÆTODIPTERUS La Cépède for *Parephippus* Gill.
POMADASYS La Cépède for *Pristipoma* Cuv. & Val.

---

* Hence the name ; " Blenna, en grec, signifie *mucosité*." (La Cépède.)

CLUPANODON La Cépède for *Clupeonia* Cuv. & Val.

GYMNOMURÆNA La Cépède for *Murœna zebra* Günther and affines.

MURÆNOBLENNA La Cépède for *Gymnomurœna* Günther.

MACRORHAMPHOSUS La Cépède for *Centriscus* Auct.

CENTRISCUS L. for *Amphisile* Auct.

INDIANA UNIVERSITY, *October* 4, 1882.

<hr>

ON THE SYNONYMY OF THE GENUS BOTHUS RAFINESQUE.

BY DAVID S. JORDAN AND CHARLES H. GILBERT.

In the Caratteri di Alcuni Nuovi Generi, etc., 1810, 23, the genus *Bothus* is established by Rafinesque for flounders, which are allied to the European turbot. Three species are referred to this genus : *B. rumolo* Raf., *B. tappa* Raf., and *B. imperialis* Raf. The first of these is, according to Bonaparte (Cat. Metod. dei Pesci Europ., 1846, 49) identical with *Pleuronectes rhombus* L.; the third, with the Turbot *Pl. maximus* L., and the second has not yet, so far as we know, been identified. The relations of these fishes to the Linnæan *Pl. rhombus* seems to have understood by Rafinesque, who observes that he should have called the genus *Rhombus*, had not La Cépède removed the latter name to another genus. It will be, therefore, not unfair to take the first species mentioned by Rafinesque, and which is really identical with *Pleuronectes rhombus* L., as the type of his genus *Bothus*. A group substantially identical with this had been previously outlined by Klein under the name of *Rhombus*. This name was afterwards accepted by Cuvier for the Turbot and its relatives, and has now come into general use. If we adopt the pre-Linnæan and non binomial generic names proposed by Klein, as has been done by Bleeker, and formerly by Professor Gill, the name *Rhombus* must be used for this group. If we reject these pre-Linnæan names, as is now the custom of most writers, the *Rhombus* of Cuvier is antedated by *Rhombus* of La Cépède (=*Peprilus* Cuvier), and moreover, it is not the earliest name of the group in question.

In the Indice d' Ittiologia Siciliana, 1810, p. 53, a few months later than the "Caratteri," a genus "*Scophthalmus*" is thus defined: "Ale giugulari ed ale caudale sciolte, occhj alla sinistra."

Three species are referred to this genus (p. 14) : *Pleuronectes maximus* L., *Pleuronectes rhombus*, L., and a new species based on an erroneous and indeterminable figure of Rondelet, which receives the name of *Scophthalmus diurus*. Rafinesque's genus *Scophthalmus* is therefore equivalent to his own *Bothus*, the sole difference between them being, according to Bonaparte (l. c., p. 49), that *Bothus* was founded on actual specimens ("ex natur") and *Scophthalmus* on the descriptions of others ("ex auct").

Later, as already stated, both these fishes, with others, received the name of *Rhombus*, a name not tenable under the rules of nomenclature followed by us.

In 1839 the genus *Psetta* was proposed by Swainson (Nat. Hist. Classn. Fishes, etc., ii, 302) in the following words:

" *Psetta Aristotle*,* *Cuv.*—Body rhomboidal; dorsal fin commencing at the edge of the upper jaw, and extending, as well as the anal, almost to the caudal; eyes approximating, with a short, crest-like cirrus.

"P. MAXIMUS, *Bloch*, pl. 49."

This name *Psetta* is adopted by Bonaparte (Catalogo Metodico di Pesci Europei, 1846, 49) for the entire group called *Rhombus* by Cuvier, while the name *Bothus* is transferred to a different genus which had been previously called *Platophrys* by Swainson, and later *Rhomboidichthys* by Bleeker. The name *Scophthalmus* is likewise diverted from its original meaning, and is used for the genus previously named *Zeugopterus* by Gottsche.

In 1862 (Proc. Acad. Nat. Sci., Phila. 1862, 216) an American species (*Pleuronectes maculatus*, Mitchill) which, from any point of view, is strictly congeneric with *Pleuronectes rhombus* L., was recognized by Professor Gill as the type of a distinct genus (*Lophopsetta* Gill). In 1882 (Syn. Fish. N. Am., 815) the present writers have referred this species to the genus *Bothus*, recognizing as the type of *Bothus*, *Bothus rumolo* Raf., = *Pleuronectes rhombus*, L.

Whether the extremely rudimentary or obsolete condition of the scales of *Pleuronectes maximus* L., justifies its separation from *Bothus* as a distinct genus we are not yet prepared to say. At present we may regard it as the representative of a distinct subgenus, for which the name *Psetta* must apparently be retained. The three species noticed in the present paper may therefore stand as

1. *Bothus (Bothus) rhombus* (L.).
2. *Bothus (Bothus) maculatus* (Mitch.).
3. *Bothus (Psetta) maximus* (L.).

INDIANA UNIVERSITY, *October* 9, 1882.

DESCRIPTION OF A NEW SPECIES OF ARTEDIUS (ARTEDIUS FENESTRALIS) FROM PUGET SOUND.

BY DAVID S. JORDAN AND CHARLES H. GILBERT.

Artedius fenestralis sp. nov.

Closely allied to *Artedius notospilotus* Girard.

Head, 2⅖ in length to base of caudal; depth, 4⅓. D. IX–17. A. 12. Lat. l. 36.

Length (27206), 5 inches.

---

* "I see no reason for substituting *Rhombus* Cuv., for the more ancient and classic name of *Psetta* imposed by Aristotle upon this group."—*Swainson.*

General form of *A. notospilotus,* the body rather robust; the head large and broad. Lower jaw included. Maxillary extending to posterior part of eye, 2⅓ in head. Eyes rather large, 5 in head, about one-third broader than the concave interorbital space. Nasal spines strong, with a conspicuous cirrus behind them. Top of head less depressed and less concave than in *A. notospilotus ;* its lateral ridges smooth and covered by skin, without spine-like projections. No tubercular prominences behind eye. Preopercle ending in a short process, which has usually three spines at its tip, the two uppermost hooked upward. The three prominences below this spine are small, entire, covered with smooth skin. (In *A. notospilotus* these projections are much larger, and more or less coarsely serrate.) A few small dermal flaps on top and sides of head. Head with small stellate, non-imbricate scales, arranged much as in *A. notospilotus,* but extending lower on the sides of the head, covering the suborbital and postorbital regions, as far down as the suborbital stay. Scales on body cup-shaped, arranged, as in *A. notospilotus,* in a broad band along each side of the back; each band about 9 scales in breadth. This band extends much further back than in *A. notospilotus,* meeting its fellow across the back of the tail behind the dorsal fin. A small but distinct pore-like slit behind the fourth gill (wholly wanting in *A. notospilotus*).

Fins low, the dorsal much lower than in *A. notospilotus;* the longest dorsal spine about equal to snout; 3½ in head (in the female), probably higher in the males. Ventrals about reaching vent; pectorals past front of anal.

Color, in spirits, essentially as in *A. notospilotus,* but paler; olivaceous, the head mottled and barred with blackish; back with about 4 saddle-like black bars. Base of caudal blackish. Fins all, except the ventrals, which are pale (probably dusky in males), with cross-bars and series of spots. A black blotch bordered by orange between first and second dorsal spines, and another between 7th and 8th.

This species is evidently the northern representative of *Artedius notospilotus,* but has apparently become so thoroughly differentiated from the latter as to be worthy of a distinct specific name. In *A. notospilotus,* the head is more uneven, the body and head less completely scaled, the fins larger, the armature of the preopercle different, and especially there is no trace of slit behind the last gill.

Several specimens of this species were obtained by the writers in Commencement Bay, near New Tacoma, Washington Territory, in June, 1880. These are numbered 27206 and 27146, and some of them have been distributed by the National Museum as "*Artedius notospilotus.*" The latter species was found by us in abundance only at Santa Barbara. Girard's original types apparently included both species, but his description applies best to the southern form.

If we include in the genus *Artedius* all the species (*lateralis, fenestralis, notospilotus, quadriscriatus, pugettensis, megacephalus*) from the west coast of the United States, which have been referred to it, it becomes practically impossible to separate it from the genus *Icelus* of Kröyer. Some of the different members of *Artedius* are more like *Icelus hamatus* than they are like each other. At present we are unable to draw any satisfactory dividing line among these species, and elsewhere (Syn. Fish. N. A., 689) we have referred all of them to *Icelus*. In the same memoir we have mentioned the specimens which here become the types of *A. fenestralis* as "Northern specimens," representing "a marked variety" of *Icelus notospilotus*.

INDIANA UNIVERSITY, *October* 11, 1882.

DESCRIPTION OF A NEW SPECIES OF UROLOPHUS (UROLOPHUS ASTERIAS), FROM MAZATLAN AND PANAMA.

BY DAVID S. JORDAN AND CHARLES H. GILBERT.

UROLOPHUS ASTERIAS sp. nov.

Disk almost round, a little broader than long; its length just about equal to length of tail. Anterior margins of disk nearly straight, the tip acute, slightly exserted, much less prominent than in *U. aspidurus*, longer in the male specimen than in the females. Distance from eye to tip of snout, about one-fourth length of disk and a little more than twice interorbital width. Interorbital space somewhat concave. Eyes small, much smaller than the large spiracles, the diameter about half the interorbital width. Width of mouth $2\frac{1}{6}$ in its distance from tip of snout. Teeth conic and sharp in the males, blunter and somewhat pavement-like in the females. Nostrils directly in front of angles of mouth; nasal folds forming a broad continuous flap, the edges of which are slightly fringed.

Ventrals projecting a little beyond outline of disk. Caudal spine very long, somewhat longer than snout, its insertion considerably in front of middle of tail. Caudal fin moderate, the upper lobe deepest, inserted opposite tip of caudal spine, the lower lobe beginning farther forward, the depth of the tail with caudal fin, about half the interorbital space.

Skin above everywhere rather sparsely covered with small stellate prickles, these larger and more numerous toward the median line of the back and head; wanting on the ventral fin. Males and females about equally rough. Median line of back with a series of rather strong, sharp recurved spines, 18 to 32 in number, extending from the shoulders to the front of the caudal spine, these usually becoming much larger and sharper backward, but the largest much smaller than the spines in *U. aspidurus*.

Color, light brown, without distinct markings; tail, faintly edged with dusky; lower side white.

This species is not rare at Mazatlan, where it is known as *Raia*. It is also occasionally taken at Panama.

Three females and one male specimen, from 12 to 16 inches in length, were brought from Mazatlan, and one young male from Panama.

Specimens in United States National Museum.

28204.   Mazatlan, Gilbert.
29524.   "   "
29542.   "   "
29580.   "
29318.   Panama,   "

The species of *Urolophus* thus far known from the Pacific coast of tropical America may be distinguished by the following analysis:

a. Anterior margins of disk nearly straight; insertion of caudal spine in front of the middle of the tail; the spine longer than snout.

  b. Disk everywhere perfectly smooth; no spines or prickles; disk broader than long, considerably longer than tail; teeth obtuse in both sexes; length of snout less than one-fourth disk; brown above, with many yellowish spots..HALLERI *

  bb. Disk smooth, or with a few minute prickles; upper part of tail with a few (2 to 8) large buckler-like spines on the median line; disk slightly longer than broad, slightly shorter than tail; teeth in males acute; length of snout, about one-third disk; brown above, nearly plain ....ASPIDURUS †

  bbb. Disk covered with small stellate prickles; a series of small spines along median line from shoulder to caudal spine; disk a little broader than long, as long as tail; length of snout, about one-fourth disk; teeth in males acute; brown above, nearly plain...... ....ASTERIAS

aa. Anterior margins of disk convex; insertion of caudal spine in front of middle of tail; the spine not longer than snout; tail rather longer than body; teeth sharp in both sexes; skin beset with stellate tubercles ....MUNDUS‡

NOTES ON A COLLECTION OF FISHES FROM CHARLESTON, SOUTH CAROLINA, WITH DESCRIPTIONS OF THREE NEW SPECIES.

By DAVID S. JORDAN AND CHARLES H. GILBERT.

Four weeks during the months of July and August, 1882, were spent by Mr. Gilbert in collecting and studying the fishes of Charleston and vicinity in the interests of the United States National Museum. One hundred and twenty-three species of marine fishes were observed; of these twenty one had not been previously recorded from our South Atlantic coast, nineteen being additions from the West Indies and the Gulf of Mexico.

---

* *Urolophus halleri*, Cooper, Proc. Cal. Acad. Nat. Sci., 1863, III, 95. Point Concepcion to Panama (Santa Barbara, San Pedro, San Diego, Mazatlan, Panama.)

† *Urolophus aspidurus* Jor. & Gilb., Bull. U. S. Fish Com., 1881, 307. Panama.

‡ *Urotrygon mundus* Gill, Proc. Acad. Nat. Sci., Phila., 1863, 173. Panama. (Dow collection; the type now lost.)

Especial acknowledgments are due to Mr. Charles C. Leslie for aid of various kinds rendered Mr. Gilbert while in Charleston. It was only through his co-operation that the present collection was made possible. Dr. G. E. Manigault kindly gave free access to the collections in the museum of Charleston College, and also presented several interesting specimens.

1. Ginglymostoma cirratum (Gmel.) Müller and Henle.

A stuffed skin of this species, about 5 feet long, is in the Charleston Museum.

2. Mustelus canis (Mitch.) De Kay.

A single specimen seen; probably not common.

3. Scoliodon terrænovæ (Rich.) Gill.

Very abundant in the harbor.

4. Sphyrna tiburo (Linn.) Raf.—*Shovel-head Shark.*

Probably the most abundant shark in Charleston Harbor. It is skinned and eaten by the negroes.

5. Sphyrna zygæna (Linn.) Raf.

No specimens obtained. A large skin of this species is in the Charleston Museum.

6. Odontaspis littoralis (Mitch.) J. & G.

A stuffed skin is in the Charleston Museum. A large pair of jaws was also obtained from Mr. Leslie.

7. Hypoprion brevirostris Poey.

Body robust, its greatest height equaling the greatest breadth of the very depressed, flattened head; snout short, flat, broadly rounded anteriorly, the greatest height before mouth equaling distance from tip of snout to nostril; nostril midway between tip of snout and posterior edge of pupil; distance from snout to front of eye two-thirds the broad interorbital space; preoral portion of snout about one-half interorbital width; flap of anterior nostril very short, ending in an acute angle; width of mouth equaling distance from tip of snout to posterior margin of orbit, and slightly less than twice distance from tip of lower jaw to line connecting angles of mouth; angle of mouth with a short, deep, fold, half diameter of orbit, extending on upper lip only.

Teeth, $\frac{2\,2}{3\,2}$; those in upper jaw from a broadly triangular base, which is distinctly though minutely serrulate on outer side only; above the base the cusp is abruptly constricted, narrowly triangular, with entire edges, the point turned but little towards the side. Teeth in lower jaw much narrower and shorter than those in the upper, erect, with base and edges of cusp entire.

Eye small, its diameter about one-fifth interorbital width.

Gill openings very wide; width of first slit nearly equaling its distance from fourth gill slit.   Branchial area about as deep as long.

First dorsal inserted posteriorly, its origin nearly midway between anterior insertions of pectorals and ventrals; the fin but little higher than long, the upper margin lunate, the greatest height one-half length of head from snout to third gill slit.   Second dorsal similar to the first, the height but little less.   Distance between dorsals twice the base of the first, 2½ times base of second.

Anal smaller than second dorsal, the margin very deeply incised; its origin slightly posterior to that of second dorsal, the two fins terminating about on the same vertical.

Caudal with a deep groove-like pit at base above, and a shallow, inconspicuous one below.   Length of upper caudal lobe rather more than length of head from snout to last gill slit.

Pectorals short and very broad, their posterior margins crescentic; tips reaching nearly half way to middle of ventral base, scarcely to end of first third of dorsal base.   Outer edge of ventrals one-third length of caudal, one-half that of pectoral.

Color greenish olive, dusky above; fins all, except first dorsal, with black margins, which are very wide on anal and caudal; eyes very light grayish; tongue and inside of mouth generally, brilliant white.

A single specimen, about 2½ feet long, was taken in Charleston Harbor.   It was not recognized by the fishermen as a common shark.

The description given by Professor Poey is so short that we cannot consider the present identification of it as certain.

8. Pristis pectinatus Latham—*Saw-fish.*

A large skin of this species and several " saws " are in the Charleston Museum, having been taken on the coast of South Carolina.

9. Rhinobatus lentiginosus Garman.

Olive-brown above, everywhere, except on dorsal and caudal fins, and on sides of snout covered with small, round, bluish-white spots, about one-third diameter of pupil; these spots extend on rostral cartilage two-thirds distance to tip; lateral margins of snout, as well as rostral cartilage, dusky below; ventrals margined posteriorly with whitish, very distinctly white in the young; entire coloration distinct in young before birth.

Body narrow, the greatest width of disk one-half distance from snout to origin of first dorsal; snout very long and narrow, its length from front of eye equaling one-third its distance to vent; rostral ridges wholly united below for their entire length; above, the ridges are very narrow, uniting to form a spatulate tip, thence separated by a very narrow groove, which becomes wider on posterior fourth; sides of snout semi-translucent.

Eye equal to the concave interorbital space, which is contained 4⅔ times in snout. Greatest width of spiracle two-thirds eye; posterior margin of spiracle with two folds. Nostrils about one-sixth wider than the interspace; anterior valve with a narrow wing-like membrane reaching outer angle, the valve not reaching inner angle by nearly one-third width of nostril.

Mouth perfectly straight, the lower jaw with a very inconspicuous projection, fitting into a slight emargination of the upper; width of mouth 2⅓ times in distance to tip of snout; teeth not pointed, in about 75 vertical series in each jaw.

Distance from snout to end of pectoral 2⅔ in total length; distance to vent, 2⅓ in total.

Dorsals equal, the interval between them three-fourths length of snout (to eye); their base one-half their height, which equals length of snout and eye. Distance from first dorsal to root of ventrals, 1¼ in snout.

Caudal broad and short, the two lobes of nearly equal width, the upper pointed; posterior margin of fin obliquely truncate, without notch; upper lobe five-sixths length of snout.

Skin very minutely granular; a group of six large tubercles at tip of snout; a series of smaller tubercles on anterior rim of orbit, and a few on upper rim posteriorly; a series of similar small tubercles, compressed, and with backward-directed spine, running from head along median line of back to dorsal; those between dorsals obsolescent; a single tubercle on each shoulder.

Sides of tail with a very conspicuous wide fold, extending to lower lobe of caudal.

A single specimen, a female, about 2 feet long, with five well-developed young, was obtained (July 26) in Charleston Harbor. This species is well known to fishermen, but is said to be not abundant.

10. Torpedo occidentalis Storer.

Traditions of the electric fish being taken at Charleston are current among the fishermen. No specimens were seen, and the fish is doubtless rare.

11. Pteroplatea maclura, (Le Sueur) M. & H.

Abundant in the harbor, where numerous specimens were taken. None of these had any trace of the caudal spine, though the largest seen was 18 inches long. At what size, if at all, is the caudal spine developed?

12. Trygon sabina Le Sueur—Sting Ray.

Agreeing well with Garman's account of the species (in J. & G., Syn. Fish. N. A. 68), but with the snout somewhat produced and acute. Teeth about 30/30, those in sides of upper jaw enlarged. Width of mouth equaling that of interorbital space; nasal flap broadly concave behind. Length of disk greater than its width, contained 1⅖ times in length of tail. Caudal spine long, nearly equaling snout; a short, rather high, cutaneous fold, beginning immediately behind its tip, and extending for a distance rather less than length of spine; a much longer, rather higher fold on under side of tail, beginning slightly in advance of base of spine, and extending beyond end of upper fold.

Top of head between eyes rather sparsely covered with small stellate prickles (these almost wanting in one specimen), which do not extend backward on body; body naked with exception of the median dorsal series of very strong backward-hooked prickles, each arising from a long narrow base; a single prickle on each shoulder (sometimes wanting); upper surface of tail behind the fold with numerous minute backward-hooked prickles, arising from stellate bases; a few also on lower surface of tail towards tip.

Very abundant in the harbor.

13. Stoasodon narinari (Euphrasen) Cantor—*Clam-cracker.*

Not rare. A single large specimen seen.

14. Manta birostris (Walb.) Jor. & Gilb.—*Devil-fish.*

Two stuffed skins in the Charleston Museum. The " devil-fish " is said to be abundant off Port Royal, S. C., each year, about the last of August.

15. Lepidosteus osseus. (Linn.) Agassiz.—*Gar.*

Two specimens were taken in the salt water of the harbor.

16. Amia calva Linn.

A specimen in Charleston Museum from Black River, South Carolina.

17. Arius felis (Linn.) J. & G.—*Small-mouthed cat-fish.*

Exceedingly abundant in the harbor, but eaten only by the poorer classes. In this species the maxillary barbel frequently extends beyond base of pectoral spine, thus agreeing in all respects with " *A. equestris*" Bd. & Grd. In July many males were captured with mouths full of their young.

18. Ælurichthys marinus, (Mitch.) B. & G.—*Large-mouthed cat-fish.*

Very abundant, although much less so than the preceding.

19. Elops saurus Linn.—*Jack Mariddle.*

Common in the harbor, but not eaten, the flesh said to be tasteless.

20. Brevoortia tyrannus (Latrobe) Goode.—*Menhaden.*

The young are very abundant in the harbor during the summer months. A study of the material in our possession, comprising specimens from Beaufort, N. C., Charleston, S. C., Saint John's River, Florida, Pensacola, Fla., Mobile, Ala., and Galveston, Tex., convinces us that the Gulf menhaden (*B. patronus*, Goode) should be considered a scarcely tangible variety of *tyrannus*, rather than a distinct species. We are unable to appreciate any constant differences in proportions of head and fins, or in the serration of the scales. The length of the head in our specimens is about one-third length of body, sometimes a little more, sometimes less, and without reference to locality.

21. Dorosoma cepedianum (Le Sueur) Gill.—*Gizzard-shad.*

Comparison with specimens from White River, Indiana, and from Charleston, where the species is abundant, fails to show any difference between them. Examples from Galveston, however, as has been already noted (Proc. U. S. Nat. Mus., 1882, 248), differ conspicuously in appearance from the ordinary type because of much slenderer body, the depth being $2\frac{5}{6}$ in length (instead of $2\frac{1}{2}$); in the Galveston form the caudal peduncle is notably longer and slenderer, and the head slenderer. This Galveston form seems to us worthy of being distinguished as a subspecies, and may be called *Dorosoma cepedianum* subsp. *exile.*

22. Stolephorus mitchilli (C. & V.) J. & G.

(Jor. & Gilb. Proc. U. S. Nat. Mus., 1882, 248.)

Very common in Charleston Harbor, and agreeing perfectly with specimens from Wood's Holl, Galveston and Pensacola. Head, $3\frac{3}{4}$; depth, $3\frac{5}{8}$; D. 14; A. 27.

23. Stolephorus browni (Gmel.) J. & G.

Several specimens in Charleston Museum.

24. Synodus fœtens (Linn.) Gill.—*Providence Whiting.*

Common in the harbor and on the Black-fish banks. Cautiously handled by the fishermen because of its supposed poisonous properties.

25. Fundulus majalis (Walb.) Günther.

Several specimens in the Charleston Museum, collected on the South Carolina coast.

26. Fundulus similis (Girard) Jordan.

Many young specimens caught in tide-pools in the harbor.

27. Fundulus heteroclitus (Linn.) Günther.—*Mud fish.*

Many specimens from Charleston enable us to make a more detailed comparison with specimens from the Gulf, and to demonstrate the permanence of the characters separating the two forms. Of these the east-coast form (typical *heteroclitus*) has all the fins conspicuously larger, and the white spots on vertical fins, in the male, smaller and more numerous. Other details of form and coloration are the same in both, and it will probably be better to consider the Gulf form as a subspecies.

In adult male *heteroclitus* the longest dorsal ray is contained $1\frac{2}{3}$ times in head (in *grandis* 2); longest anal ray $1\frac{1}{3}$ (in *grandis* $1\frac{1}{6}$); caudal $3\frac{1}{2}$ in length (in *grandis* 4); ventrals reaching front of anal, 2 in head (in *grandis* $2\frac{2}{5}$ in head, barely reaching vent); base of dorsal 2 in head (in *grandis* $2\frac{2}{5}$). The young of both sexes, one inch long, are conspicuously barred with darker; in females the bars narrower than the interspaces, in males much wider than the interspaces and less numerous.

28. Zygonectes cingulatus (C. & V.) Jordan.

> *Fundulus zonatus* et *cingulatus* C. & V., xviii, 196, 197 (not *Esox zonatus* Mitch.)
> ? *Hydrargyra luciæ* Baird, Ninth Smithson. Rept., 1855, 344. (♂ ?)
> *Haplochilus chrysotus* Günther, vi, 317.

A single specimen from Black River, South Carolina, presented by Dr. G. E. Manigault, agrees in most respects with Günther's description of *H. chrysotus.* It differs in having all the vertical fins dotted with brown, the dots not forming distinct cross bands on the caudal, and in having the dorsal inserted rather more anteriorly (opposite the third anal ray instead of the fifth). The following is a detailed description of our specimen

Body short and robust; the caudal peduncle high and compressed, its least height $1\frac{3}{4}$ in head; head short, wide, and flat, the interorbital width one-half its length. Teeth in jaws in a narrow band, the outer series much enlarged, those in the lower jaw larger and more numerous than those of the upper. Snout very short and blunt, the two jaws nearly equal in closed mouth; length of snout nearly two-thirds diameter of orbit, which is contained $1\frac{1}{2}$ times in interorbital width, and $3\frac{1}{3}$ times in head.

Origin of dorsal midway between tip of caudal and posterior rim of orbit; its distance from base of caudal one-half distance from front of orbit; base of dorsal $2\frac{1}{2}$ in head, its height $1\frac{1}{6}$ in head; its origin is opposite the nineteenth scale of lateral line, and the third ray of the anal fin.

Base of anal fin rather less than half length of head, its greatest height somewhat more than half; dorsal and anal not nearly reaching caudal when depressed. Caudal mutilated, apparently broadly rounded.

Pectorals reaching ventrals, $1\frac{2}{3}$ in head. Ventrals short, not nearly reaching vent, one-half head.

Head, $3\frac{1}{2}$ in length; depth, $3\frac{4}{5}$; D. 9; A. 11; scales, 32–12. L. $1\frac{3}{4}$ inches.

Color in spirits: Light olive-brown, top of head and a narrow median streak in front of dorsal fin darker; middle of sides, especially behind, with rather indistinct pearl-colored dots; middle of sides of trunk and tail with about 14, not clearly defined, narrow half-bars; an elongate dark area above base of pectorals. Vertical fins with small, black specks, less numerous on caudal fin; other fins plain.

29. Gambusia patruelis holbrooki (Agassiz) J. & G.

A specimen from Black River, South Carolina, was presented by Dr. Manigault.

Head 3⅔ in length; eye 2⅞ in head, 1⅙ in interorbital width; D. 7; A. 9; lat. l. 30. The dark bar across cheeks is distinct, and the vertical fins are marked with blackish dots, which form two very distinct cross-bands on caudal fin. In *G. patruelis* from Galveston, these bars on the caudal are either indistinct or altogether wanting, and the dark bar on cheeks is often obsolete; in all specimens of *holbrooki* seen by us, these markings are conspicuous.

Three young males (less than 1 inch long), from Eutaw Springs, S. C., show dark spots on dorsal fin, and a dark shade across cheeks. The specimens agree in proportions and fin rays with adult females.

30. Ophichthys chrysops Poey—"*Sea-serpent.*"

A single specimen, 20 inches long, evidently of the same species as our mutilated specimen from Pensacola (Proc. U. S. Nat. Mus. 1882, 261), and answering well Poey's description of *Ophichthys chrysops*, shows the following characters:

Olive-brown above, thickly dusted with dark points; pectorals wholly dusky; dorsal and anal translucent, with blackish margins; body white below; mucous pores on head conspicuous, black; lower jaw with dusky cross-blotches; no dark lines on throat.

Head and trunk 1¾ in tail; head 2½ in trunk; eye equaling interorbital space 1¾ in snout, 9⅓ in head; a series of about eight mucous pores along side of each mandible; numerous pores on nape and top of head, three in a vertical series behind eye, about four along sides of upper jaw below eye; cleft of mouth 2¾ in head. Teeth conical, short and strong, not blunt, uniform in size, none of them enlarged; in two very distinct series on all the dentigerous bones. Anterior nostrils not elongate, the tube less than diameter of eye. Gill openings broadly crescent-shaped, separated by a distance 1½ their width, which is about one-third gape of mouth.

Dorsal beginning over last fourth of pectoral, the distance of its origin from snout equaling two-fifths distance from snout to vent; pectoral about equaling gape of mouth. Free portion of tail sharp, compressed, about two-thirds diameter of orbit.

The description of *Ophisurus gomesii* Castelnau is possibly based on a specimen of this species, and the specific name would, in that case, supersede *chrysops*. But the description is inadequate and might refer to one of several other species. We think it best to retain Poey's name.

31. Tylosurus marinus (Bl. & Schn.) J. & G.

Numerous specimens seen swimming about in the harbor, where it is doubtless abundant.

32. Hemirhamphus unifasciatus Ranzani.

The single example obtained agrees in all respects with specimens of *unifasciatus* from Beaufort, N. C., but has the anterior rays of dorsal and anal, and the upper and lower rays of caudal, jet black. In these respects it agrees with specimens obtained at Mazatlan, Mexico.

33. Exocœtus mesogaster Bloch.

(*Exocœtus hillianus* Gosse.)

Evidently not rare in the open sea off Charleston Harbor. Two specimens were brought in by a fishing smack, having flown on board during the night. A third specimen was presented by Mr. Leslie. D. 11–12; A. 12–13.

34. Siphostoma louisianæ (Günther) J. & G.—"*Gar-fish.*"

Abundant. Dorsal on 9 or 10 rings; snout exceedingly variable in length, sometimes half longer than rest of head, and longer than base of dorsal; often much shorter than this; occiput and belly more or less strongly carinate; rings 16+9+31 or 15+10+31. D. 33 to 35. An adult female has the dorsal whitish, with oblique dusky bands about as broad as the interspaces.

35. Hippocampus stylifer Jor. & Gilb.—*Sea-horse.*

(J. & G., Proc. U. S. National Museum 1882, 265.)

A single specimen, nearly 2 inches long, was obtained. The characteristic coloration is well shown at this age, the light gray cross-bands with dark brown borders being very distinct. The body is very slender, its greatest depth about two-thirds length of head; snout somewhat shorter than in the specimen from Pensacola, its length equaling post-orbital part of head. Dorsal inserted on four rings, a half only of the first and fourth rings being covered; dorsal rays, 16; plates of body, 12+35.

Still another specimen of this species, collected in the Gulf of Mexico by Prof. O. P. Hay, has 18 rays in the dorsal, and the body plates 12+34.

36. Mugil albula Linn.—*Mullet.*

Abundant.

36(b) Mugil brasiliensis Ag.—*Mullet.*

Rather more abundant than the preceding.

37. Querimana harengus (Gthr.) J. & G. (*Gen. nov.*)

(*Myxus harengus* Gthr., iii, 467.)

Several specimens, about two inches in length, were taken, agreeing entirely with others from Mazatlan, Panama, and Zorritas, Peru. The wide distribution of this little mullet is remarkable. It probably does not reach a greater length than 2 or 3 inches. It is not a true *Myxus*, as it has but two anal spines (instead of three), fixed teeth in the upper jaw only. We therefore consider it as the type of a distinct genus *Querimana* (from *Queriman*, a Portuguese or Spanish name of *Mugil liza*, in Surinam). The stomach is gizzard-like as in *Mugil*. Eyelid not adipose.

38. Menidia laciniata Swain.—*Silver-fish.*

Four young specimens were obtained, in all of which the anal rays are 1, 19, thus agreeing with specimens from Beaufort, N. C., and differing from typical *vagrans* from Galveston, which has the anal rays usually 1, 17 (1, 14 to 1, 17). These young specimens show the following coloration: Clear translucent, greenish above; back with two or more rather regular series of minute black dots, usually not more than one on each scale; snout and lower jaw dusky; lateral silvery streak rather wide, covering the third row of scales, not bounded above by a dark line, but the entire band dusted with dark points. A few minute dots on base of anal; caudal dusky.

39. Menidia bosci (C. & V.) Swain.—*Silver-fish.*

(*Menidia dentex* Goode & Bean, Proc. U. S. Nat. Mus., 1882, 429.)

Two young specimens, one having the anal rays 22, the other 23, are colored as follows: Greenish-yellow on back, very thickly covered with fine dots, as are also the snout and lower jaw; lateral streak very narrow, bordered above with a conspicuous greenish-black line; the stripe about as wide as pupil, covering the middle of the fourth series of scales. Caudal conspicuously light yellow; dorsal and pectoral fins less so; base of anal dusky.

40. Sphyræna picuda (Bloch) Poey.—*Barracuda.*

Rare off Charleston Harbor; said to be very infrequently seen. A single specimen, about 18 inches long, was taken on the bottom in 10 fathoms of water. It shows the following characters:

Color, dusky bluish above and on sides, silvery white below; about 20 dusky bars, much wider than the interspaces, descend from back not quite to lateral line; lower part of sides with a few black blotches, irregular in shape and position, usually little larger than pupil; top of head blackish; opercular membrane above black; soft dorsal, anal, ventrals and caudal black; the dorsal and anal with tips of first and last rays white; pectorals and spinous dorsal fin dusky, the axil black; ventrals margined with white posteriorly.

Head very large, the lower jaw especially strong and heavy, the snout rather bluntly conical; maxillary about half length of head, reaching front of pupil. Premaxillary series of teeth small, compressed, of uniform size, about 40 in number; vomer with two pairs of very large, compressed teeth, triangular in shape, their length more than half diameter of pupil; the anterior pair directed downwards, the posterior downwards and backwards, the two pairs separated by an interspace equal to their length; palatines with a close set series of about 8 teeth similar to those on the vomer, but rather smaller; a large compressed tooth at symphysis; those of the lateral series of lower jaw small anteriorly, increasing constantly backwards, where they equal those of the palatine series. Eye large, $2\frac{1}{2}$ in snout, 2 in postorbital part of head, nearly equaling interorbital area. Interorbital space concave, with a shallow median groove, divided by a ridge in front and behind; supraocular ridge bony, striate.

Distance from snout to front of dorsal $1\frac{1}{4}$ in distance from latter to root of caudal; second dorsal spine longest, one-half length of snout and eye; space separating dorsals $5\frac{1}{2}$ in length of body; second dorsal and anal opposite and equal, their margins concave, the longest ray about $2\frac{2}{5}$ in head. Caudal broad, moderately forked; the middle rays half the outer; the two lobes equal, concave. Pectoral reaching somewhat beyond front of dorsal, one-third head. Ventrals inserted slightly in advance of dorsal; their distance from snout two-fifths length of body; their length $3\frac{2}{5}$ in head. Scales large, uniform in size; head naked, except cheeks and opercles, which are covered with small embedded scales. ·

Head three in length; depth equaling snout, $2\frac{4}{5}$ in head. D. V—1, 9; A. II, 8. Scales 10–78–10 (the cross series counted from lateral line to front of dorsal, and anal fins respectively). About 12 series of scales on the cheeks.

41. Polynemus octofilis (Gill) J. & G.

In appearance much resembling *P. approximans*, the body comparatively little elongated, with short head and small mouth; snout heavy, projecting beyond mouth for a distance nearly equal to its own length, posterior margin of orbit midway between preopercular margin and anterior nostril; mouth small, the maxillary extending beyond orbit, for a distance equaling two-thirds diameter of orbit; maxillary $2\frac{1}{5}$ in head; snout three-fourths diameter of orbit; eye slightly less than interorbital space, $4\frac{1}{2}$ in head; preorbital two-fifths vertical diameter of orbit; longest-gill raker five-sixths diameter of eye; 18 on lower limb.

Interval between dorsals $1\frac{3}{4}$ in head; third dorsal spine highest, $1\frac{1}{4}$ in head, nearly reaching origin of second dorsal when depressed; second dorsal falcate, its highest ray $1\frac{3}{4}$ in head.

Anal not falcate, the tips of anterior rays not projecting beyond the gently concave outline of the fin; longest ray $1\frac{3}{4}$ in head; insertion of anal opposite second soft ray of dorsal; anal spines comparatively well developed, the third equaling diameter of orbit.

Lower caudal lobe $3\frac{1}{4}$ in body.

Ventrals inserted under fifth dorsal spine, their length nearly $\frac{1}{2}$ head.

Pectorals reaching vertical from tips of ventrals $1\frac{1}{5}$ in head. Filaments slender; 8 in number; the length of the upper one one-third distance from tip of snout to fork of caudal fin, reaching slightly beyond the vent; the lowermost filament two-thirds head.

Head $3\frac{3}{5}$ in length; depth $3\frac{1}{3}$; D. VIII—1, 12; A. III, 13. Lateral line forking at base of caudal; thence continued to margin of fin; 62 tubes from shoulder to fork; $5\frac{1}{2}$ series above lat. l., 10 below.

Color very light olivaceous, tinged with light yellow; scales on back, with wide dusky margins formed by dark punctulations; belly white; tip of snout with numerous coarse black points; a few of these on maxillary also; vertical fins yellowish and dusky, with black points; tip of anterior anal rays white; ventrals whitish; the outer rays dusky; pectorals almost uniform deep black, the color formed by closely approximated coarse black points; filaments translucent, slightly dusky.

It is probable that all species of Polynemus have three anal spines and not two, although this latter number has been assigned to various species by different authors. The first spine is very short, and usually largely enveloped in the scales. Our specimen differs from young specimens of *P. octonemus* Grd. (no adults being known) from the Gulf of Mexico in its shorter pectoral filaments, shorter ventral fins, and in the pectoral fins being black.

One specimen only was obtained at Charleston, where it is evidently very rare. It was wholly unknown to the fishermen.

42. Echeneis naucrates Linn.—*Pilot-fish.*

Of frequent occurrence. The specimen obtained has 22 laminæ, the length of disk being $4\frac{2}{5}$ in total, and the greatest width between pectorals one-half length of disk. A specimen from Pensacola has 22 laminæ in the disk, which is contained $4\frac{4}{5}$ times in total, and a third specimen, from Saint John's River, has the disk also with 22 laminæ, but the length only $4\frac{1}{4}$ in total.

43. Remora squalipeta (Dald.) J. & G.

According to Lütken (Contributions Ichthyographiques V, 5) *Echeneis squalipeta* Daldorf is based on the young of *Echeneis remora* Linné. In case, then, it is considered desirable to give generic rank to *Remora*, Daldorf's name will be the oldest available for the species.

Numerous specimens from the vicinity of Charleston are in the Charleston Museum.

44. Phthirichthys lineatus (Menzies) Gill.

Body with the general form and appearance of *Echeneis naucrates*, the head much more narrowed anteriorly, the tip of lower jaw thus forming a very narrow, linguiform projection, out of line with the rounded profile of sides of head. Mouth with wide gape, the maxillary about $\frac{2}{5}$ head (from tip of snout). Teeth comparatively large and few in number, somewhat recurved, not forming a close-set band; those laterally in upper jaw in about 2 distinct series, forming a narrow patch in front; no external series of compressed, close-set teeth as is found in *Remora*, and no distinct canines, though the outer series are larger than the inner; teeth in lower jaw similar to those in the upper, arranged in about three series laterally, and forming a narrow wide-set patch in front; teeth on vomer, palatines, and tongue similar to those in jaws, but much smaller; vomerine patch broad, concave, with two lateral backward processes; on each side of this is the short, narrow palatine band (wholly lacking in specimens examined of *Remora squalipeta*, and *Echeneis naucrates*) of about 3 irregular series. Eye $3\frac{2}{3}$ in head, half width of interorbital space. Disk wide, covering all of top of head, its width $1\frac{2}{3}$ in its length, which is one-fifth total length with caudal; lamellæ but 10 in number, very strongly pectinate.

Origin of dorsal midway between base of caudal and third cephalic plate; the shape of dorsal, anal, and caudal as in *Echeneis naucrates*, the median caudal rays being, in our young specimen, produced. Pectoral pointed, the rays all normal, about 18 in number; its tip not quite reaching tip of ventral, which is $\frac{4}{7}$ head.

D. X–30; A. 30. Head $5\frac{1}{4}$ in length; depth about $\frac{1}{2}$ length of head. Length 4 inches.

Color, slaty-black, a darker band along middle of sides, bounded above and below with a narrow white streak, the upper beginning on snout, the lower below eye, the two slightly converging backwards; under side of head lighter; anterior lobes of dorsal and anal, upper and lower caudal rays, and pectoral fins, broadly margined with white; ventrals and posterior dorsal and anal rays with narrow white margins.

The genus *Phthirichthys* is evidently most nearly related to *Echeneis*, from which it may be separated, as well by the peculiar dentition as by the reduced number of plates on the head.

A single small specimen, 4 inches in length, was taken at Charleston. This agrees well with descriptions given by Poey, of *Echeneis apicalis* and *Echeneis sphyrœnarum*, but has not the conspicuously enlarged teeth in sides of lower jaw, assigned to the latter.

45. Elacate canada (Linn.) Holbrook.—*Cobia.*

Not infrequently taken in the summer months. A single specimen was obtained.

46 Trichiurus lepturus Linn.— "*Sword-fish*"; *Silver-eel.*

Very abundant in Charleston Harbor, being brought in by every seine-boat.

47. Scomber colias Gmelin.

A single specimen of this species, captured at Charleston in the fall of 1880, was presented by Mr. Chas. C. Leslie.

The three species of *Scomber*, known to occur on our coasts, may be thus distinguished.

a. Air bladder none.
    1. S. SCOMBRUS Linn.
      Scomber scombrus Cuv. & Val. ix, 6.
      ? *Scomber vernalis* Mitch. Trans. Lit. and Philos. Soc. New York, 1815, 423.
      Scomber vernalis DeKay. N. Y. Fauna, Fishes, 101.

Sides silvery below, immaculate; top of head almost uniformly dark, the cranium without conspicuous transparent area.

Eye small, slightly less than interorbital space, 5 in head. Maxillary 2⅖ in head, the distance from tip of snout to angle of mouth 2⅖; preopercle very wide, the posterior margin strongly convex, little oblique, the angle very bluntly rounded; a single series of evident pores along lower margin of preopercle; subopercle moderate, the greatest width 1¾ in orbit; head 3⅜ in length (without caudal).

First dorsal normally with 12 spines.

Scales minute, not forming a corselet.

Specimens examined from the coasts of New England and Virginia, and from Venice and Genoa, Italy.

aa. Air bladder present, well developed (*Pneumatophorus*, subgen. nov.).
   b. Sides below silvery, immaculate.
    2. S. PNEUMATOPHORUS De la Roche.
        Scomber pneumatophorus. De la Roche, Ann. du Mus. d'Hist. Nat. xiii, 335·
           Cuv. & Val. ix, 36. Gervais et Boulart, Poissons de France, ii, 119. Giglioli, Elenco Sistematico dei Pesci di Italia, 24. Günther, ii, 359.
        Scomber grex Mitch., Trans. Lit. and Phil. Soc., N. Y. 1815, 422.
        Scomber grex Cuv. & Val., ix, 46.
        ? Scomber vernalis Cuv. & Val., ix, 48.
        Scomber diego Ayres, Proc. Cal'a Acad. Nat. Sci. 1856, 92.

Top of head with a very conspicuous transparent area, appearing whitish in alcoholic specimens. Eye somewhat larger, its diameter greater than interorbital space, 4⅖ in head. Maxillary 2⅘ in head, the distance from tip of snout to angle of mouth, 2⅘ in head; posterior margin of preopercle straight or even slightly concave, the angle much less blunt, and the inferior margin more nearly straight than in *scombrus;* many very minute pores along lower part of preopercle, not arranged in series; subopercle wider than in *scombrus*, the greatest width 1½ in diameter of orbit; opercle with a deeper emargination opposite base of pectoral. Head 3½ in length (without caudal).

First dorsal normally with 9 spines; pectoral 2⅘ in head.

Scales larger; those in lateral line more conspicuous; those around pectoral fin enlarged, forming a distinct corselet.

Specimens from Santa Barbara, Cal., from the coast of New England, and from Venice, Italy, show no appreciable differences.

  bb. Sides below with very numerous, roundish, or oblong, dusky-olive blotches.
    3. S. COLIAS Gmelin.
        ?? Lacerto or *Colias*, Cetti, Hist. Nat. Sard. iii, 190.
        ?? Gmelin, Syst. Nat. 1788, 1329.
        Risso, Ichthyologie de Nice, 1810, 171.
        ? Rafinesque, Indice d'Ittiologia Siciliana, 1810, 20.
        ? Walbaum, Art. Pisc. 1792, 209.
        ? Bloch & Schneider, Syst. Ichth. 1801, 22.
        Cuv. & Val. Hist. Nat. des Poiss. ix, 39.
        Storer, Synop. Fish. N. A. 342.
        DeKay, N. Y. Fauna Fishes, 104.
        Day, Fishes Great Britain, 91.
        Günther, ii, 361.

    83. April 25, 1883.

Gervais et Boulart, Poiss. de France 118.
'Giglioli, Elenco, &c. 24.
Moreau, Hist. Nat. Poiss. de la France, 412.
? *Scomber lacertus* Walb., Art. Pisc. 209, 1792 (= *Lacerto* Cetti).
Scomber dekayi Storer, Hist. Fish. Mass. 52.

Top of head with a conspicuous transparent area, whitish in spirits; eye very large, wider than interorbital area, 4 in head. Maxillary $2\frac{3}{4}$ in head; the distance from tip of snout to angle of mouth $2\frac{1}{7}$ in head; posterior margin of preopercle straight, and rather less oblique than in *pneumatophorus;* the lower margin longer and less rounded than in *scombrus;* subopercle very long and narrow; its greatest width rather less than one-half diameter of orbit; opercle with a deep emargination opposite base of pectorals. Head $3\frac{4}{5}$ in length (without caudal).

Dorsal fin normally with 9 spines, a 10th sometimes present.

Scales still larger than in *pneumatophorus;* those on sides in about 175 oblique series; lateral line very conspicuous; corselet conspicuous, composed of large scales.

Our specimen from Charleston, S. C., one from Pensacola, Fla., and several from Venice and Genoa, Italy, agree in all respects.

The *Scomber colias* of Gmelin was founded on the fish called by Cetti *Lacerto* or *Colias*, and it can probably never be known with certainty which of the three species found in the Mediterranean was thus designated. There can, however, be no doubt as to the species called *colias* by Risso. The name may therefore be retained for the present species, inasmuch as no other name had been given prior to this definition.

Steindachner considers *S. pneumatophorus* as the young of *S. colias*. We have specimens young and old of both. We are not yet fully convinced, however, that the two forms are really distinct species. .

48. Scomberomorus maculatus (Mitch.) J. & G.—*Spanish mackerel.*

Numerous specimens were observed in the market.

49. Scomberomorus (?) caballa (C. & V.) J. & G.—*King-fish.*

A large species of *Scomberomorus*, known as *King-fish*, and having inconspicuous dusky spots on sides, is, during the summer months, very abundant off shore, from Cape Hatteras southward. Coasting steamers catch them with trolling lines on every trip, the fish averaging from 3 to 5 feet long. A single specimen was seen, about $3\frac{1}{2}$ feet long, captured off Cape Lookout, but no description taken sufficient for the positive identification of the species. The fishermen at Charleston are well acquainted with the *King-fish*, though they seldom capture it.

50. Caranx chrysus (Mitch.) DeKay.—*Jack-Crevalle.*

Scomber crysos Mitchill, Trans. Lit. & Philos. Soc. N. Y. I, 424, 1815.
Caranx chrysos DeKay, N. Y. Fauna, Fish. 1842, 121.
Caranx pisquetus Cuv. & Val., IX, 98.
Caranx hippos Holbrook, Ichth. S. C., 1860, 90.
Paratractus pisquetus Gill, Proc. Acad. Nat. Sci. Phila., 1862, 432.

There can be little doubt that the species described by Mitchill as

"*Scomber crysos*," the "yellow mackerel," is the *Caranx pisquetus* C. & V., and not the *Caranx hippos* Linn. The only reason that can be urged for the identification of "*crysos*" with *hippos*, is the depth assigned to the former (3¼ in total length), this being greater than that usually found in *pisquetus* (3½ to 4 in total). It is to be noted, however, that Mitchill's specimen was only 6½ inches long, and the young of all the species of *Caranx* have the depth appreciably greater than do the adults. Furthermore, Mitchill's measurements, taken as they were in inches, would easily permit the slight inaccuracy necessary to account for this difference in depth. The figure given by Mitchill, if sufficiently accurate to be of value, would seem to be based on a young specimen of *hippos*. It differs, however, too widely from the accompanying description to allow us to consider it identical with the specimen used by Mitchill for the type of the species.

The following characters, given by Mitchill, leave little doubt as to the species he had in mind; "a neat, compact, handsome fish, about ¾ inch thick. He is plump, generally. Back forms a neat regular curve. Belly an opposite corresponding sweep. Head neither rostrated nor blunt." "A black spot frequently at the edge of the gill cover." "D. 8, 24, A. 20." These characteristics are exactly those of *pisquetus*, while *hippos*, on the contrary, is a high compressed fish, not at all plump, with the back forming a high uneven curve, and the belly not at all arched, but running in a straight oblique line from chin to front of anal; the head is also blunt, the rostral profile being sub-vertical, and the fin formula is 2d D. 21–22 : A. 16–17. In addition we have the fact, of little importance, perhaps, that the *pisquetus* is by far the commoner form northward, and is generally known as the "Yellow Mackerel."

Caranx chrysus is the only species of *Caranx* brought in much abundance to the market of Charleston, during the summer months. The name *Jack-Crevalle* is there applied to all species of Caranx without distinction.

51. Caranx hippos (Linn.) J. & G.

But few specimens seen.

52. Caranx setipinnis (Mitch.) J. & G.

Many specimens taken in the harbor are in the Charleston Museum. A single immature example was seen in the market. As has been noticed by Bleeker and Steindachner, this species has the armed lateral line of *Caranx*, from which genus we do not see how it can be separated.

53. Selene vomer (Linn.) Lütken.—*Hog-fish*.

The young form of this species, with filamentous dorsal and elongate ventrals, was very abundant in the harbor. No adults were seen.

54. Chloroscombrus chrysurus (Linn.) Gill.—*Bumper*.

Very abundant.

55. Trachynotus carolinus (Linn.) Gill.—*Crevalle.*

The most highly prized of the fishes of Charleston. Not brought into the market in great numbers; known universally as *Crevalle*, the name *Pompano* being seldom used.

56. Seriola carolinensis Holbrook.—*Jack-fish; Amber-jack.*

　　Seriola carolinensis and *zonata* Holbrook, Ichth. S. C. 72 and 75.
　　Seriola stearnsi Goode & Bean, Proc. U. S. Nat. Mus. 1879, 48.
　　(?) *Seriola dubia* Poey, Memorias de Cuba, II, 228.

Two young specimens, each about 1 foot long, were obtained in the market, and many very young (3 or 4 inches long) were seen swimming on the surface, on the fishing grounds outside the harbor. These latter had the lateral bands intensely black and very conspicuous. The following is the color shown by the two larger specimens, when fresh :

Back dusky bluish, becoming dull white on sides and dull silvery below; five rather faint, broad, dark, half-bars downward from back to axis of body, about as wide as the interspaces; a light yellow streak from eye back along axis of body to tail, most distinct where it crosses the vertical bars; an irregular yellowish area on lower half of sides anteriorly; an oblique dusky band from front of dorsal to eye, and one from eye forward to suborbital; a broad dusky streak above base of anal; soft dorsal and anal blackish olive, margined with white, the margin broad anteriorly; spinous dorsal blackish; caudal dusky olive; ventrals silvery white, within dusky yellowish-green; pectorals with olive tinge; a horizontal blackish streak on opercle.

This species is exceedingly close to *S. zonata* Mitch., the number of fin rays, the pattern of coloration, and the general proportions of head and body being the same. The northern form, *zonata*, has, however, the bands on the sides appearing jet black at all ages, while in *carolinensis* of the same size these are merely darker shades. *Zonata* has also the depth much greater, and the body more compressed; in specimens 1 foot long, the depth is contained $2\frac{5}{6}$ times in length (to base of caudal), while in *carolinensis* of the same size the depth is $3\frac{1}{4}$ in length.

A detailed comparison of *carolinensis* from Charleston with a specimen of *stearnsi* from Pensacola fails to show any differences. In young specimens the occiput is more or less sharply keeled, as in *zonata*, this disappearing with age, the occiput becoming very broadly and obtusely rounded.

57. Stromateus paru Linn.

Very common during the summer months.

Above, light bluish; below, silvery; everywhere with iridescent and brilliant silvery reflections; sides often with chocolate-brown blotches; head light olive, translucent, without silvery reflections above; snout and sides of head with much coarse, black speckling; anal yellowish-silvery, more or less dusky on the falcate rays, everywhere with iridescent reflections; the falcate rays margined anteriorly and above with black, the posterior rays densely punctulate with black, especially towards tips. Dorsal rays pinkish or purplish, with bright reflections, margined with blackish; the posterior rays thickly dusted with dark points; pectorals and caudal with slight yellowish tinge, and much black specking towards tip, the caudal margined very narrowly above and below with white. Iris silvery.

Head, 3 to $3\frac{1}{5}$ in length; depth, $1\frac{1}{5}$ to $1\frac{1}{4}$; pectoral, $2\frac{1}{4}$ to $2\frac{1}{2}$; eye, $2\frac{2}{3}$ to 3 in head; D. III—I, 44 to 47; A. III, 43 to 45.

There is nothing to indicate that the West Indian form (*Rhombus xanthurus* C. & V., IX, 405) constitutes a species distinct from the above, unless it be the small number of fin rays attributed to the latter (D. IV, 40; A. III, 39). This is probably due either to a miscount or to the great variability of the species in this respect. Cuvier and Valenciennes identify with "*xanthurus*" the figure of Sloane, on which Linnæus founded his *Stromateus paru*. The latter name must then supplant "*alepidotus*," and "*gardeni*," unless it be shown that the form from the West Indies is really distinct.

58. Stromateus triacanthus Peck.

A single specimen obtained; evidently not abundant.

59. Coryphæna hippurus Linn.

> *Lampugus punctulatus* Dekay, N. Y. Fauna, Fish. 134—not of C. & V.
> *Coryphæna globiceps* Dekay, N. Y. Fauna, Fish. 132.
> *Coryphæna sueurii* Cuv. & Val., IX, 302.
> *Coryphæna dorado* Cuv. & Val., IX, 303.
> *Coryphæna guttata* Poey, Mem. de Cuba, II, 245.
> *Lampugus punctatus* Poey, Mem. de Cuba, II, 419.
> *Coryphæna hippurus* Lütken, Spolia Atlantica, 1880, 45.

Two female specimens of the common dolphin of our Atlantic coast, each about two feet long, were caught with trolling lines off Cape Lookout, during a trip from Baltimore to Savannah. Later in the summer a larger, mutilated, specimen was examined; captured by a fishing-smack in the vicinity of Charleston. Still later, two young specimens were sent by Mr. Stearns, from Pensacola. This material has enabled us us to make a careful review of the history of our Dolphins, which has convinced us that all names hitherto applied to Dolphins from North America are synonyms of one species, the *Coryphæna hippurus* Linnæus. It is not improbable that the *Coryphæna immaculata* of Poey is the *C. equisetis* Linn., as it has the fin rays, the inconspicuous spots, and the short pectorals of that species, but the name *equisetis* should not be introduced into our faunal lists until a *bona fide* example of the species is taken on our coast. From our own experience in counting the fin rays of the dolphin, it seems evident that a synonym cannot be referred either to *hippurus* or to *equisetis* on the basis of the count alone, even though, as Lütken concludes, there probably are but two species, distinguished by different fin-formulæ. If the fins have become hardened or dried by exposure or by being immersed in too strong alcohol, it is impossible to obtain the correct count except by dissection.

C. hippurus is very abundant off our South Atlantic coast in summer, being caught south of Cape Hatteras by coast steamers on nearly every trip. North of the cape it is said to be rarely taken. The species reaches a length of 4 or 5 feet.

When first caught, the head, body, and tail, are greenish olive, or dark greenish olive-brown, lighter below; a series of about 15 round blue spots on back along each side of base of dorsal, these placed at nearly uniform distances apart, and about one-third size of pupil; sides below with numerous blue spots irregular in size, shape, and position, but none of them so large as those along back; lower lip largely blue; about three concentric blue lines around snout above. Dorsal purplish blue, with irregular areas of lighter and darker, and with some greenish reflections; in one specimen the dorsal and caudal are sparsely covered with blue spots similar to those on body. Caudal yellow; anal yellowish, with translucent border; pectorals translucent, with brownish axil; ventrals outwardly greenish olive, within of an indeterminate dark brownish, with olive cast. The play of color in the dying dolphin has been largely exaggerated, judging from our own observation. Such change as there is, seems to consist in the apparent rapid development of an external bright silvery pigment, with some blue and green reflections, this development being accompanied with partial restorations of the ground color, thus affording some real play of shades, which are, however, not brilliant. At death the fish is largely silvery, the intense deep lustrous blue of the spots remaining meanwhile unchanged; afterwards appear large irregular patches of the ground color, yellowish on sides, yellowish olive-brown on back.

D. 59 to 63; A. 29.

Head little elevated, its height at origin of dorsal $1\frac{1}{8}$ to $1\frac{1}{4}$ in its length; maxillary scarcely reaching middle of orbit, $2\frac{1}{8}$ in head; teeth recurved, in broad cardiform patches, those on vomer uniform, the patches on jaws and palatines with an external series of larger conical teeth. Eye $5\frac{2}{3}$ in head, $1\frac{7}{8}$ in snout. Head $4\frac{2}{3}$ in length to base of caudal; depth 5. Dorsal beginning slightly in advance of posterior margin of orbit; the longest ray about $\frac{5}{3}$ snout, slightly more than $\frac{1}{2}$ head; upper lobe of caudal $3\frac{1}{8}$ in body. Ventral inserted slightly posterior to base of upper pectoral ray, its length $1\frac{1}{4}$ in head, less than 6 in body; pectoral $1\frac{1}{2}$ in head, 7 in body.

60 Centrarchus macropterus (Lac.) Jor.

Many young specimens of this species, with the ocellated dorsal spot very conspicuous, are in the Charleston Museum, from Black River, South Carolina.

61. Enneacanthus simulans (Cope) McKay.

Several seen, taken in Black River, near Georgetown, S. C. D. IX, 11; A. III, 10. Depth, ½ length; head, 2⅗; longest dorsal spine, half head. Ventral spine reaching vent, the longest ray reaching base of last anal spine. Lateral line continuous, complete. Dark bars on body evident, about five in number (specimen 3 inches long); ear flap small, little wider than pupil.

˙ 62. Mesogonistius chætodon (Baird) Gill.

Many specimens seen from Black River, South Carolina. A comparison with specimens from New Jersey fails to show any differences. This seems to be as yet the southernmost record for the species. D. X, 11; A. III, 12.

63. Lepomis pallidus (Mitch.) Gill & Jor.

A single specimen seen, from fresh water near Charleston.

64. Perca americana Schranck.

Many specimens seen from the Santee River. Evidently not rare in the southern streams.

65. Pœcilichthys barratti (Holb.) J. & G.
  Pœcilichthys butlerianus Hay, in Jor. & Gilb. Syn. Fish. N. A., 519.

A specimen from Black River answers well the existing descriptions of *barratti*.

Head, 4 in length; depth, 5¼; eye, 3⅓ in head. Lateral line on 18 to 20 scales. Cheeks and opercles completely scaled. Maxillary reaching anterior margin of pupil. D. X — 12; A. II, 7. Scales 46.

Olivaceous very profusely tessellated with brownish on the sides; middle of sides with a series of about 10 blotches alternating with an equal number of square blotches on the back. A brown band below and one before eye. Vertical fins more or less barred with brown.

We can find nothing in the description of *butlerianus* to indicate that it is a distinct species from *barratti*.

66. Roccus saxatilis (Bloch & Schn.) J. & G.

This is the favorite game fish of the coastwise streams and inland lakes, but, according to fishermen, does not visit the salt water. Is it not possible that this difference in the habit of the fish in the North and in the South may have developed varietal or specific differences? No specimens were obtained at Charleston, so we are unable to make the comparison. The name *lineatus* ought not to be retained for this species, as *Sciæna lineata* Bloch, was apparently the European species.

67. Serranus formosus (Linn.) J. & G.—*Squirrel-fish.*
  Perca formosa Linn., Syst. Nat. Not *Hæmulon formosum* C. & V., 230.
  Serranus fascicularis Cuv. & Val., II, 245.

Very abundant, both in the harbor and on the fishing banks outside. D. X, 12; A. III, 7.

Perca formosa Linnæus, consists of the fin-formulæ and a description of the color of a "*Squirrel-fish*" received from Dr. Garden. The fin-rays ("D. X, 13; A. III, 7") are sufficient to show that Linnæus's specimen was not a *Diabasis* (D. XII, 17; A. III, 9), as has been generally supposed, while the color and the common name given leave no doubt as to the species in his possession. Catesby's *Perca marina capite striato* (= *Diabasis plumieri* Lac.) was wrongly identified by Linnæus with his *formosa*, apparently on the ground that it also had the head striped.

68. Serranus atrarius (Linn.) J. & G.—*Black-fish*.

The most abundant food-fish at Charleston, forming probably more than nine-tenths of all the fish caught on the banks by the smack-men. It is not considered a choice fish, and is bought mostly by the poorer people. It is caught on the bottom with hook and line, and is found abundantly at all seasons, though in much greater numbers in winter.

69. Serranus philadelphicus (Linn.) J. & G.—*Rock-fish*.

.   *Perca philadelphica* Linn., Syst. Nat., Ed. x & xii.
Perca trifurca Linn., Syst. Nat. Ed. xii.

Not rare in Charleston Harbor, though never abundantly taken. Five specimens were obtained during the summer. The fish is usually caught with hand-lines among the rocks. Better specimens than those noted by us (Proc. U. S. Nat. Mus. 1882, 273) must be examined from the Gulf of Mexico before the range of this species can be confidently extended to those waters, as the Gulf specimens are peculiar in several respects. Specimens from Charleston show the following traits:

Color in life, olivaceous above, whitish below ; seven broad brown bars from back obliquely forwards to level of middle of pectorals, these almost obsolete along lateral line; the color of the bars is not intense, and is formed by shadings along the base and margins of the scales; the anterior bar crosses the nape, and is very indistinct. Snout and upper part of head with numerous brownish red spots and lines, three or four of these parallel and running from eye to snout, the interspace usually light blue ; upper lip reddish brown; tip of lower jaw broadly purplish; a dark blotch on opercle anteriorly, and sometimes a small dark spot behind eye ; lining of opercle and throat lemon yellow ; a large jet black blotch behind pseudobranchiæ. Spinous dorsal translucent, with indistinct whitish and dusky longitudinal streaks; a large blackish blotch on membrane of last spines, immediately above fourth vertical bar of sides ; some dark spots on the spines form two irregular lengthwise series ; dorsal filaments bright scarlet; the fin usually with light bluish shading. Soft dorsal, with a series of bluish white spots near margin (one between each two rays); one or more incomplete series above and below this; the fin is margined with reddish brown, and has usually several series of reddish-brown spots, these most numerous posteriorly; some irregular olive-brown spots towards base ; a small black spot on base of membrane between 8th and 9th and one between 10th and 11th rays, the former frequently absent. Caudal translucent, with irregular cross-series of round brownish-red spots, the space between them often with bluish-white spots; the fin margined above with brownish red; lower lobe whitish, unspotted. Anal white, with a median sulphur-yellow streak, and a terminal dark bar; ventrals whitish, with dusky areas, often uniform blackish; pectorals translucent; peritoneum silvery.

Head 2⅔ to 2¼ in length; depth 3⅓ to 3⅔. D. X, 11; A. III, 7; P. 17; C. 18. Scales 5–55–15. Length 9½ inches.

Maxillary reaching posterior margin of pupil, 2¼ in head; mandibular band of teeth becoming a single series laterally; a few inner teeth in the front of each jaw enlarged; lower jaw with the inner series laterally, and the outer series anteriorly of enlarged conical teeth, the lateral teeth but little larger than those in front; outer series of upper jaw much enlarged, becoming smaller laterally, those in front larger than any in lower jaw; patch on vomer crescent-shaped; on palatines long and narrow. Head naked forwards from occiput, including suborbital ring, snout, preorbital, top of head, maxillary and lower jaw; scales on cheeks small, in 9 to 11 very regular oblique series; scales on opercles as large as those on body, in 8 or 9 oblique series, those on the flap again smaller; least interorbital width about four-sevenths diameter of eye, which is 4¾ in head; serræ on and below preopercular angle slightly enlarged and more distant than those above; subopercle and interopercle finely, evenly serrate. Gill-rakers one-half eye, three above angle, ten below.

First two dorsal spines short, the third and fourth nearly equal, the fourth one-half or nearly one-half head; the last spines are then much shortened, forming a notch much as in species of "*Paralabrax;*" the last spine 3⅗ in head, two-thirds the ray following; membrane deeply incised between the spines, the upper angles produced beyond the spines in long, narrow filaments, very variable in length, usually less than diameter of orbit; the spines themselves are acute, and not at all filamentous as figured by Holbrook (Ichth. S. C. pl. VII, fig. 1); the structure of the dorsal thus does not differ from that of *S. atrarius*, which has also a trifurcate tail; this latter character does not however seem sufficient to warrant the retention of the genus *Centropristis.*

Caudal with the upper and middle rays much produced and nearly equal, the lower lobe but little lengthened; median rays nearly as long as head (seven-eighths to eleven-twelfths), the lower rays about two-thirds head. A young specimen, 5 inches long, has caudal nearly evenly convex behind, with the upper rays only slightly projecting.

Anal spines short, graduated, the second the strongest, the third slightly longer, about one-fourth head; longest rays nearly one-half head.

Middle ventral rays longest, not nearly reaching vent, four-sevenths head; pectoral sub-truncate, reaching vent, 1⅔ in head.

Scales very strongly ctenoid, running well up on caudal fin, and in narrow series on membranes of soft portions of vertical fins; ventrals with series nearly half-way to tip.

The description of *Perca philadelphica* given in the 10th edition of Linnæus could not have been identified with this species had not Linnæus himself, in his 12th edition, revised his description, correcting his count of fin rays, and adding numerous details. The first description stands : " Dorsal fins connate, with 11 spines and 9 soft rays. D. XI–9; P. 16; V. 6; A. III, 5; C. 11. Habitat in America." In the 12th edition the number of dorsal and anal rays is changed to : " D. X–11; A. III, 7," while the other counts are left uncorrected. The coloration given is characteristic and leaves no doubt as to the species described : "A black spot on middle of dorsal fin; sides with black spots and bands; red below; scales and opercles ciliate; opercle mucronate posteriorly; first two dorsal rays (spines) shorter. Habitat in North America. *Chub.* Dr. Garden."

70· Pomadasys fulvomaculatus (Mitch.) J. & G.—*Sailor's choice.*

Taken daily during the summer, but not in large numbers. Considered an excellent food-fish.

71· Diabasis aurolineatus (C. & V.) J. & G.—*Red-mouthed Grunt.*

  ? *Perca marina gibbosa* Catesby.
  Hæmulon aurolineatum Cuv. & Val., v. 237.
  Hæmulon chrysopteron Cuv. & Val., v. 240—not *Perca chrysoptera* Linnæus.
  Hæmulon chrysopteron Holbrook, Ichth. S. C. 121.
  ? *Hæmulon quadrilineatum* Holbrook, S. C. 195—not of C. & V.
  ? *Perca striata* Linn. Syst. Nat.

The *Perca chrysoptera* of Linnæus is not identifiable with this species, probably not with any other. The description is based on a specimen which was received from Charleston through Dr. Garden, and which was identified by Linnæus with Catesby's figure of *Perca marina gibbosa*, this latter evidently some species of *Diabasis*. But as not a sentence in the description of *chrysoptera* agrees with Catesby's figure, we cannot admit the identification to have been correct, and denying this, there is nothing in the description of *chrysoptera* to indicate that it is any *Diabasis*, much less the species at hand.

Brownish-olive above, lighter on sides and below ; scales of back with central portions olivaceous, the bases and margins brownish olive; bright specimens show narrow yellow streaks on margins of scales on back, following the series upwards and backwards, these, however, seldom visible ; several longitudinal yellow streaks on sides; one midway between dorsal outline and lateral line, beginning on snout and running to last rays of soft dorsal; one on head just above eye, usually not continued on body; a third very distinct streak along median line of body, beginning on snout and running through eye to tail; several fainter streaks above and below the median one, following the series of scales, in bright specimens a streak on each series below median line; snout very dark brown; sides of head more or less silvery, with yellowish tinge; in bright specimens showing seven or eight yellow stripes, two of which are forward continuations of the two principal body stripes, the others smaller and not joining body stripes; head white below; a dusky bar at base of pectoral; mouth, within, bright brick-red, becoming yellowish red on lining of opercles; fins all plain dusky olive, somewhat darker towards tips; the lower fins more distinctly yellowish; a diffuse black blotch at base of caudal. The color is very variable, differing much with the surroundings and condition of the fish. Specimens are frequently seen of a plain silvery cast, the yellowish lines indistinct or wholly wanting, and the caudal blotch obsolete. It is without doubt from such a specimen that Holbrook drew his description of "*Hæmulon chrysopterus,*" while his "*Hæmulon quadrilineatum*" is quite evidently a somewhat careless description of a brightly-colored specimen of the same. The stripes vanish in spirits.

Head $2\frac{7}{8}$ in length; depth $2\frac{3}{4}$ to $3\frac{1}{2}$. D. XIII, 15; A. III, 9. Scales $\frac{7}{14}$; 55 pores or oblique series; 70 vertical series.

Body moderately elongate; snout $2\frac{3}{5}$ in head; maxillary reaching below middle of eye, $1\frac{5}{6}$ in head; teeth in a villiform patch anteriorly, with an outer enlarged series, which is continued singly on sides of jaws. Eye much more than half length of snout, less than greatest width of preorbital, $4\frac{1}{3}$ in head. Gill-rakers short and weak, $\frac{11}{14}$ in number.

Scales above lateral line in oblique, below in horizontal series. None of the scales conspicuously enlarged; those on middle of sides anteriorly somewhat wider and less closely imbricated. Head scaled forwards to front of eyes, the snout above and the upper jaws largely naked; some imbedded scales on preorbital and mandible. Soft parts of vertical fins wholly-enveloped in fine, thin scales. Spinous dorsal high, the fourth spine highest, 2 to $2\frac{1}{4}$ in head; the outline of the fin rather evenly rounded; last spine the shortest, about four-fifths longest soft ray, and two-fifths longest spine.

Upper lobe of caudal subfalcate, longer than the lower, $1\frac{2}{5}$ in head; the middle rays, $\frac{4}{7}$ the upper.

Second and third anal spines not very unequal in length, but the second evidently longer and much stronger, about equal to length of longest soft ray, and $\frac{1}{6}$ head.

Ventrals reaching to or slightly beyond vent, $1\frac{2}{5}$ in head; pectorals equaling distance from snout to preopercular margin.

This species is very abundant on the fishing banks outside the harbor, where it is taken in much greater quantity than any other species except the *Black-fish.*

72. Diabasis plumieri (Lac.) J. & G.—*Black Grunt.*

    Perca marina capite striato Catesby.

    Hæmulon formosum Cuv. & Val.; not *Perca formosa* Linn. = *Serranus fascicularis* C. & V.

    Hæmulon arcuatum C. & V.

Frequently taken on the fishing banks, though not abundant. Compared with a specimen from Aspinwall, the stripes on sides of snout are much narrower, and the color of body and fins is much darker; the preopercular denticulations are stronger, the snout longer and the eye smaller. It is possible that the southern form may represent a tangible variety, but our material is not sufficient to enable us to characterize it.

The Charleston specimens showed in life the following coloration: The basal half of each scale dark brown, the terminal half silvery, with bluish tinge; snout and lower jaw dark chocolate-brown, the end of the snout and the tip of lower jaw white; sides of head with brassy luster, and marked with about 18 very narrow, often wavy, blue lines, the widest on the snout being less than half width of interorbital space; a few of these lines are extended on the body for a very short distance (less than diameter of eye); two or three stripes run concentrically around snout above, joining anterior margins of orbits; mouth very bright scarlet. A dark brown bar across base of pectoral, continued half way down on axil; fins brownish olive; ventrals and anal blackish, the ventrals margined externally with white. Scales below pectorals with numerous very short and narrow, horizontal, black lines; scales on lower part of sides, and above lateral line, with dendritical clusters of dark lines diverging from the base. No blue streaks on lower part of sides.

73. Lobotes surinamensis (Bloch) Cuv.—*Black Perch; Sea Perch.*

Occasionally taken; a single specimen seen during the summer.

74. Calamus bajonado (Bloch & Schneider) Poey. *White-bone Porgy.*

A well-known food-fish at Charleston, averaging much larger than the common *Porgy*, specimens 18 inches long being of not infrequent occurrence.

Our specimens fail in many respects to answer the incomplete description given by Poey (Monogr. des Sparini, 176), notably in the number of canines in each jaw (8 instead of 6); but this is in all probability the species described by him.

Head $3\frac{1}{4}$ in length; depth $2\frac{1}{4}$; pectoral 3; snout two-thirds head; eye two-ninths; maxillary three-sevenths, and ventral five-eighths head. D. XII, 12; A. III, 10. Pores in lateral line 44.

The young (5 inches long) is olivaceous, with white longitudinal lines above and on sides, formed by series of spots, one on each scale; sides of body with many irregular narrow dusky blotches, with a tendency to form bars on lower half of sides; belly whitish; vertical fins and ventrals with irregular wavy bars of dusky and whitish; pectorals with a dusky bar at base. Adults have all the markings less evident, with usually no trace of vertical bars on sides; the dusky and whitish bands on fins persisting.

75. Stenotomus chrysops (Linn.) Bean.—*Porgy.*

An abundant food-fish, usually not reaching a length of more than 8 inches. The second ray of the dorsal is frequently filamentous. The young show a broad, dusky, vertical bar on middle of sides. Both the *argyrops* and *chrysops* of Linnæus are based on this species as is also *Chrysophrys aculeatus* C. & V.

76. Lagodon rhomboides (Linn.) Holbrook.—*Brim.*

Rather less abundant in Charleston Harbor than at other points along our Atlantic and Gulf coasts.

77. Diplodus probatocephalus (Walb.) J. & G.—*Sheepshead.*

A fine food-fish, not taken in great abundance.

78. Diplodus holbrooki (Bean) J. & G.—*Salt-water Brim.*

Taken abundantly with hook-and-line on the banks outside the harbor. None were seen in the harbor, although this species is very abundant around the wharves at Beaufort, N. C. On the banks it reaches a length of 12 inches.

Color in life : Body dark brassy-olive ; the large black blotch across caudal peduncle often not intensely black ; naked part of head dark olive-brown ; opercular membrane black ; a black blotch above and below at base of pectoral, that above continued around on upper half of axil of fin ; margins of all membrane-bones of head black, this often conspicuous on membrane of opercle only ; ventrals black, the rays with greenish tinge ; other fins uniform olive-brown.

Although by no means satisfied that this species is distinct from *D. caudimacula* (Poey), we think it preferable to retain the name given to specimens from our own waters until comparison can be made with a sufficient series from the West Indies.

79. Pogonias chromis (Linn.) C. & V.—*Drum.*

Esteemed as a food-fish, but not very abundant.

80. Sciæna lanceolata (Holbrook) J. & G.

Rather uncommon ; taken occasionally with hook-and-line on the margins of deeper channels in the harbor.

Color: grayish-olive above, silvery below ; fins all nearly uniform, dusky ; the ventrals margined with white ; much coarse, black specking along middle of sides, base of anal fin, and inner lining of opercle.

Head = depth, 3 to $3\frac{1}{4}$ in length. D. XI, I, 22–23 ; A, II, 7–8. Scales, $\frac{4}{12}$, 47–50 (pores).

Mouth large, maxillary reaching beyond middle of eye, sometimes to posterior margin of orbit, $2\frac{1}{5}$–$1\frac{1}{8}$ in head ; teeth in lower jaw uniform, in a very narrow band ; the upper jaw with the outer series enlarged ; eye medium, its long diameter oblique, $1\frac{3}{5}$ in interorbital width, 4 to $4\frac{2}{5}$ in head.

Preopercle evenly rounded, the serrations gradually increased in size towards the angle, which rarely shows three radiating spines larger than the others, the lowermost spine turned downward and backward. Pseudobranchiæ well developed. Gill-rakers rather long and slender, about 18 on lower limb.

Second dorsal spine two-thirds to three-fifths the third spine, which is $5\frac{2}{3}$ to 6 in length of body. Second anal spine $1\frac{4}{5}$ length of longest soft ray, $2\frac{2}{5}$ in head. Caudal, $1\frac{1}{4}$; pectoral, $1\frac{7}{10}$; ventrals, $1\frac{2}{5}$ in head.

This species differs in numerous respects from *S. trispinosa* (C. & V.) (?? = *Bodianus stellifer* Bl.) and from *S. microps* (Steind.), as can be seen from the comparative descriptions of the two latter by Steindachner (Ichth. Notiz. I, 6).

81. Sciæna chrysura (Lac.) J. & G.—*Yellow-tail.*

Very common.

82. Sciæna ocellata (Linn.) Günth.—*Red Bass.*

Of frequent occurrence.

83. Liostomus xanthurus Lac.—*Chub.*

Very abundant.

84. Micropogon undulatus (Linn.) C. & V.—*Croaker.*

Abundant.

85. Menticirrus alburnus (Linn.) Gill.—*Whiting.*

Very abundant, forming one of the most valuable food-fishes of Charleston.

86. Menticirrus littoralis (Holbr.) Gill.—*Surf Whiting.*

Abundant, but less so than the preceding, and not reaching so large a size.

87. Larimus fasciatus Holbrook.—*Bull-head.*

Not uncommon in the harbor; numerous specimens procured, the largest about 8 inches long.

In life the color is grayish-olive above, with some silvery; below, clear silvery-white; back with 7 to 9 rather inconspicuous darker bars downwards and backwards to below middle of sides, the bars about as wide as the interspaces; fins, dusky-olive; the anal fin and lower rays of caudal yellow; ventrals, orange-yellow, dusky towards tips; lower side of head very bright silvery; inside of mouth, and lining of gill-cavity, as well as cheeks and opercles, with some light yellow.

Head, $3\frac{1}{3}$ in length; depth, 3. D. X, 1, 25–26; A. II, 5–6. Scales: 41 oblique series, 54 vertical series, 5 above lateral line, 10 below.

Head rather larger and less compressed, with less oblique gape, than in *L. breviceps*, and the body much less compressed and elevated. Gape placed at an angle of about 25°. Teeth, uniserial, uniform, very small. Mandible less projecting than in *breviceps* and *argenteus*, the symphyseal knob little marked. Head above cavernous, spongy, as in *Sciæna lanceolata*. Preorbital narrow, but little widened below, its width rather less than diameter of pupil; maxillary about reaching posterior margin of pupil, 2 in head. Eye large, $3\frac{2}{3}$ to $3\frac{3}{4}$ in head, much longer than snout, equal to interorbital width. Preopercular margin nearly vertical, entire, inconspicuously striate. Gill-rakers $\frac{14}{22}$, very long, $1\frac{1}{3}$ in orbit.

First dorsal spine short, the spines thence increasing to the fourth, which is $2\frac{1}{10}$ times in head; longest soft ray equals length of snout and eye. Caudal double-truncate, $1\frac{1}{6}$ in head. Base of anal fin rather less than diameter of eye, the second spine strong, about one-third head, and two-thirds the longest anal ray. Pectorals as long as head, not quite reaching vertical from vent. Ventrals, $1\frac{1}{6}$ in head.

Series of scales run nearly to tips of all the vertical fins, except spinous dorsal; ventrals likewise scaly.

88. Cynoscion maculatum (Mitch.) Gill.—*Salmon Trout.*

An abundant food-fish, caught with seines in muddy channels in the harbor.

89. Cynoscion regale (Bl. Schn.) Gill.—*Trout; Shad-Trout.*

Less abundant than the preceding.

90. Cynoscion nothum Holbrook.—*Bastard Trout.*

Caught mostly outside the harbor, where it can be found at all seasons, though most abundant in summer. It is never so abundant as the other species of the genus, and though occasionally reaching a length of 3 feet, the specimens caught are mostly of small size; the largest seen by us was about 12 inches long. It is universally known to the fishermen as "bastard trout," the belief prevailing that being unspotted and still evidently a "trout," it must be a cross between *maculatum* or *regale*, and some silvery species, as the "whiting."

Color in life, grayish-silvery above, and on sides to lower level of pectorals; then abruptly silvery; upper parts thickly punctulate with darker; inconspicuous dark streaks follow the rows of scales above, formed by the darker centers of the scales. Snout and tip of lower jaw blackish; mouth white within. Anal and ventrals white, other fins dusky.

Head $3\frac{1}{6}$ to $3\frac{2}{3}$ in length; depth, 4. D. IX, I, 29; A. I, 9. Scales, 58 (oblique series or pores); 70 vertical series. L. 8 inches.

Body well compressed; lower jaw distinctly projecting; maxillary $2\frac{1}{6}$ in head, reaching slightly beyond pupil; teeth in lower jaw anteriorly small, in a very narrow band; laterally enlarged and in a single series; upper jaw with a very narrow band in front and on sides, the outer row enlarged; one or two long, slender canines in front. A deep pit on each side of vomer.

Eye large, longer than snout, or than interorbital width, 4 in head. Preopercle very broad, the angle much produced backwards, the thin membranaceous portion with conspicuous radiating striæ. Opercle terminating in two very long, slender spines, the membrane continued beyond them. Gill-rakers long and strong, $\frac{8}{8}$ in number, the longest one-half orbit.

Spinous dorsal not high, connected with the second by a low membrane; the third spine the highest, about $2\frac{3}{5}$ in head, the upper margin of the fin descending obliquely in a straight line; soft dorsal little lower than spinous. First three or four spines with series of scales behind them; a well-developed scaly sheath at base of soft dorsal and anal; both fins being thickly scaled to tip. Anal spine small, firmly imbedded in the scaly membrane. Pectorals and ventrals also scaled, the former $1\frac{2}{3}$ in head; the ventrals reaching about half-way to vent, $1\frac{4}{5}$ in head. Caudal with median rays produced.

91. Gerres gula C. & V.

The young found abundantly in small tide-pools in the harbor.

92. Platyglossus radiatus, (Linn.) J. & G.—Butter-fish.

> *Sparus radiatus* Linn. Syst. Nat. not *Chærojulis radiatus* Goode = *Julis cyanostigma* C. &. V.
> ? *Labrus bivittatus* Bloch, taf. 284, fig 1.
> *Labrus psittaculus* Lac. iii, 522.
> *Julis humeralis* Poey, Mem. Cub. ii, 212.
> *Chærojulis grandisquamis* Gill, Proc. Acad. Nat. Sci. Phila. 1863, 206.
> *Platyglossus florealis* Jor. & Gilb. Proc. U. S. Nat. Mus. 1882, 287 (young).

Common in the harbor.

It is undoubtedly to this species that we must refer the *Sparus radiatus* of Linnæus, received from the coast of Carolina through Dr. Garden. As Garden made most of his collections at Charleston, it is most highly probable that this, rather than *Pl. cyanostigma*, was the species sent by him, the latter never having been recorded north of Key West. In addition, we have points in Linnæus's description (" Green above, sides purple, head with blue lines, variegated with greenish-yellow. *Opercle with a purple and a yellow spot.*") which answer very well to our Carolina specimens, but could not well apply to *P. cyanostigma*.

Color in life: Pinkish olive above, whitish below, a narrow, vertical, vivid blue or green line across the middle of each scale, the line usually convex forwards. Adults in life with traces only of two broad dark longitudinal bands on sides, the upper running from opercular spot to base of tail, the lower from below base of pectoral, very narrow and obscure, vanishing on middle of body ; the young show this marking much more plainly; it is occasionally very conspicuous in adults, though usually appearing as indistinct darker shades on back and sides. An olive green streak nearly as wide as eye running upwards and backwards from orbit to sides of nape, thence along back and parallel with it; another green streak above this from eye to nape, where it meets its fellow. Head pinkish-bronze, overlaid with greenish-yellow on cheeks and opercles below eye; the latter area is bounded above and behind by a narrow blue line passing from snout in a wavy course below eye to near posterior margin of opercle, where it turns abruptly downwards and forwards; opercular flap greenish, ocellated with light blue; an intense dark blue spot at upper angle of opercle, surrounded above and below with some greenish bronze margined with a light blue line; the green streak backwards and upwards from eye tapers to a point anteriorly, and is margined by a <-shaped blue line; lower jaw with two blue cross-bands; subopercle with two or three blue spots or streaks; branchiostegal membrane blue mesially.

Dorsals narrowly margined with blue; below this a broad streak of orange red; then one of greenish-yellow, bordered below with blue (this median streak distinctly black in one specimen); then a streak of purplish, separated from the greenish-yellow area at base by an oblique line running downwards and backwards on the membrane between each two rays. Caudal with five concentric, more or less irregular, bars alternately of reddish and of greenish-yellow margined with blue, the bars strongly convex posteriorly; a terminal blue-black bar, much widest at corners of fin; anal pinkish, with a broad median greenish bar, margined above and below with blue; the fin with a narrow blue margin, and a blue spot at base of each ray; in a second specimen the anal is greenish-yellow, with the median band lighter, and the blue markings as described. Ventrals translucent, pinkish-brown towards tips, with a blue streak before each ray; pectorals light bluish or greenish, without decided markings, light at base.

Head = depth, $3\frac{1}{3}$ to $3\frac{1}{2}$ in length; 26 or 27 pores in lateral line. D. IX, 11; A. III, 12.

Maxillary $3\frac{1}{4}$ in head; teeth large, the posterior canines well developed. Eye 6 to $6\frac{2}{3}$ in head.

Last rays of dorsal highest, scarcely reaching base of caudal, $2\frac{1}{2}$ in head. Caudal (from true base of rays) $1\frac{2}{3}$ in head; pectorals $1\frac{3}{4}$; ventrals $1\frac{3}{5}$.

Seven specimens were procured, from 6 to 7 inches in length.

93. Xyrichthys lineatus (Linn.) J. & G.

> *Coryphæna lineata* Linn., Syst. Nat. (not *Xyrichthys lineatus* Cuv. & Val., xiv, 50).
>
> ? *Xyrichthys martinicensis* Cuv. & Val., xiv, 49.
>
> *Xyrichthys vermiculatus* Poey, Mem. ii, 215.

This species was not seen in life, but numerous specimens are in the museum at Charleston, having been taken in the harbor, where it is said to be not rare. We cannot doubt that this, and not *X. lineatus* C. & V., is the species described by Linnæus as *Coryphæna lineata*. The specimens described by Linnæus were sent from Charleston by Dr. Garden; the color given agrees well with our fish, while "*lineatus* C. & V." has not "the dorsal and anal fins painted with lines." *X. lineatus* C. & V. has apparently not been seen since the original description, and there is no probability that it reaches our coasts. No differences have ever been pointed out between our species and the Mediterranean *X. novacula*, but Labroids are not as a rule fishes of wide distribution, and it is wiser to retain our name until a comparison of the two forms has been made.

May 12, 1883.

Our alcoholic specimen shows about six narrow blue lines on the snout, these angulated below and continued across the interopercle; scales with very narrow vertical blue lines, with some interspersed blue dots; anal fin with very evident broad vertical blue streaks.

Head 4 in length; depth 3⅗. D. IX, 12; A. II, 13. Pores in lateral line 25.

Last rays of dorsal and anal equal, reaching beyond base of caudal, two-thirds head; caudal 1⅓ in head; pectoral 1⅓: ventral 1⅖.

94. Chætodipterus faber (Brouss.) J. & G.

Less abundant than at other points along our South Atlantic coasts.

95. Astroscopus y-græcum (Cuv. Val.) Gill.

> ? *Uranoscopus anoplos* C. & V. viii, 493.
> *Astroscopus anoplus* Jor. & Gilb. Proc. U. S. Nat. Mus. 1882, 289.
> *Astroscopus guttatus* Abbott, Proc. Acad. Nat. Sci. Phil. 1860, 473.

Frequently taken in the harbor. The comparison made by us (Proc. U. S. Nat. Mus. 1882, 289) between examples from the Gulf ("*anoplus*") and *y-græcum* was with specimens of very unequal size, those of the former being 3½ inches long, while those of the latter were nearly adult. Specimens now in our possession from Charleston, less than 5 inches long, and evidently the same as an adult of *y-græcum* from the same locality, enable us to make a more satisfactory comparison, and show that the characters supposed to distinguish our Gulf specimens are due to their immature condition only. Thus the bones cuirassing the top of the head become narrower with age; the Y-shaped process becomes much narrower, and has the fork proportionally shorter; the profuse black-specking on body behind, still visible in specimens 5 inches long, entirely disappears in adults; and the white spots on body become much larger in proportion to size of eye. We strongly doubt the existence of a second species of *Astroscopus* in our waters. In case such should be demonstrated, it would still be very probable that the *anoplos* of C. & V., based on a specimen two inches long, was the young of *y-græcum*, everywhere common on our southern coasts.

96. Culius amblyopsis Cope.

A single specimen, 4 inches long, was taken in the harbor.

Color in spirits; brown, lighter above and below; each scale on middle of sides with a dusky streak, these forming obscure lengthwise lines; back anteriorly with a few small, black spots; under parts, including sides of head, very thickly punctulate with black. Lips black; a dark streak from snout through eye to upper angle of preopercle; two dusky streaks from eye downwards and backwards across cheeks; a very conspicuous black blotch as large as eye in front of the upper pectoral rays. Pectorals and ventrals transparent, dusky; vertical fins all barred with light and dark in fine pattern.

Body slender, compressed, the head depressed, becoming very narrow anteriorly; a notable depression above orbits, the premaxillary processes protruding before it; lower jaw longest; maxillary reaching vertical behind pupil, $2\frac{2}{5}$ in head.

Teeth in the jaws in narrow, villiform bands, becoming a single series on sides of lower jaw; those of the outer and inner series in each jaw are somewhat enlarged, the largest being the single series in sides of lower jaw. Preopercular spine as usual in the genus.

Scales smooth above and below, ctenoid on sides.

Head $3\frac{1}{4}$ in length; depth $4\frac{1}{4}$. D. VI–9; A. 9. Lat. l. 48. Eye $6\frac{3}{4}$ in head; pectoral $1\frac{1}{5}$; ventral $1\frac{1}{2}$; highest dorsal ray 2; highest anal ray 2; caudal $1\frac{1}{4}$.

97. Gobius encæomus sp. nov. (29673.)

Three specimens, two males and one female, were obtained in tidepools in the harbor, the largest $1\frac{3}{4}$ inches in length. The type is numbered 29,673 on the register of the United States National Museum.

Colors in life: ♂ light olivaceous, mottled above with darker olive brown; a series of about 4 obscure oblong dark blotches along middle of sides; a dark spot at base of caudal; each side of nape with an intense blue-black spot larger than eye; an obscure dusky streak from eye forward to mouth; a small dusky spot sometimes present on upper portion of base of pectorals. Both dorsals translucent, with series of bright reddish-brown spots, as large as pupil; upper lobe of caudal light reddish, the lower lobe blue-black. Anal and ventrals dusky-bluish; pectorals slightly dusky, with a narrow, bright pink border behind.

♀ without bright markings; body light olive, with 5 oblong dark blotches on sides, the last on base of caudal; from each of the three middle blotches a V-shaped bar runs to the back (these visible also in males); back somewhat mottled with dusky; a black blotch on scapula; a small one on opercle; a dark bar from eye forward to mouth. Vertical fins with dusky streaks, these appearing on caudal in the form of crossbars. Ventrals light, with two lengthwise dark streaks; pectorals plain.

Head 4 in length; depth $5\frac{3}{4}$. D. VI–11; A. 12. Lat. l. about 37 (a few of the anterior scales gone, the count, therefore, not certain).

Body very elongate, much tapering backwards; head compressed, the cheeks high and vertical; snout very short, compressed, obtusely rounded vertically. Mouth nearly horizontal, low, large, the maxillary one-half head, nearly reaching vertical from posterior margin of orbit. Teeth in very narrow bands in both jaws, those of the outer series in the upper jaw much enlarged; eyes inserted high, the interorbital space very narrow, about as wide as pupil; diameter of orbit much greater than snout, nearly one-third head. Gill-opening $2\frac{1}{3}$ in head; the isthmus wide. Dorsals contiguous, the membrane of spinous dorsal reaching nearly to base of soft dorsal; dorsal spines high, of nearly uniform length, the last reaching well beyond origin of soft dorsal when depressed; the longest spine about half length of head. Soft dorsal and anal long and high, the posterior rays of both fins reaching at least to base of caudal when depressed. Caudal lanceolate, the middle rays produced, $2\frac{2}{3}$ in body. Ventrals reaching vent, somewhat longer than pectorals, which about equal length of head; ventral sheath well developed, its length two-sevenths that of fin.

Body wholly covered with large, strongly ctenoid scales, which are much reduced in size anteriorly; head, ante-dorsal region, and breast naked.

In the female specimen, the mouth is evidently smaller, and the caudal less elongate.

99. Gobius thalassinus sp. nov. (29574.)

Closely allied to *G. emblematicus* J. & G.

Head and body translucent, overlaid by brilliant green luster, which is formed by exceedingly minute close-set green points; the luster is intense towards the head, where it assumes a blue tint, and becomes hardly noticeable on caudal peduncle; three conspicuous translucent bars wider than the interspaces, crossing body immediately behind head; head with two brilliant narrow blue or green lines running obliquely across cheeks below eye; opercle with greenish luster; branchiostegal membrane white. Dorsals whitish, with two or three lengthwise series of large reddish-brown spots; spinous dorsal blackish at base. Upper caudal rays marked with red, the lower portion of caudal, and the most of the anal fin blackish, anal whitish at base, the anterior rays tipped with brilliant white. Ventrals light buff. Pectorals translucent. In spirits the body appears dusted with dark points; two light cross-bars towards head; lower part of caudal and anal black.

Head $3\frac{1}{2}$ in length; depth $4\frac{3}{4}$. D. VII–16; A. 15.

Body elongate, much compressed, highest in front of ventrals, thence tapering regularly to a very narrow, short, caudal peduncle; the body with a peculiar, translucent, fragile appearance, common also to *G. emblematicus*. Head compressed, much higher than wide; snout very short, acute, the preorbital not as wide as pupil; mouth terminal, very wide and oblique, the jaws equal; maxillary reaching vertical from middle of orbit, one-half length of head; teeth in a narrow band in each jaw, the outer series enlarged, canine-like (under a microscope the band of small teeth behind the outer series seems evident, but the size of our specimens does not enable us to verify it with certainty); eyes placed high, separated by a narrow ridge, the diameter about one-third length of head.

Dorsals very closely contiguous; spines very slender, the fifth slightly produced and filamentous, reaching (in our specimens) to base of third soft ray when depressed; caudal lanceolate, very long and pointed, the middle rays produced, $2\frac{2}{5}$ in body; pectorals as long as head; the upper rays not silk-like; ventrals with basal membrane well developed; the fin long, reaching to or slightly beyond front of anal, somewhat longer than head.

Body covered with rather small cycloid scales; head naked; the scales are very readily deciduous; as they have in our specimens mostly fallen off, the count cannot be given.

Two specimens, the largest 1½ inches long (No. 29674, U. S. Nat. Mus.), were taken in muddy tide-pools in Charleston Harbor. The species has thus much the habit of its congener, *G. emblematicus*, from Panama.

100. Gobionellus oceanicus (Pallas) J. & G.

> *Gobius lanceolatus* Bloch., Fische Deutsch. II, 12, pl. 38, 1784.
> *??Gobius lanceolatus* C. & V., XII, 114.
> *Gobionellus hastatus* Grd., U. S. Mex. Boun. Surv. 1859, 24.

A single specimen, 11 inches long, was taken in the harbor.

Color in spirits, reddish-olive; a distinct, round, blackish blotch below spinous dorsal, twice as large as orbit; an indistict dusky shade along middle of sides, terminating in a distinct dusky blotch on base of caudal; middle of sides with a series of <-shaped marks, formed by very narrow veiny lines widely diverging backwards; a similar narrow line from eye to maxillary, and one from eye backwards to upper angle of preopercle; evident traces of the emerald spot at base of tongue; two small dark spots on first dorsal spine; spinous dorsal dusky, with a light and a dusky streak at base; soft dorsal dusky, a light (? bluish in life) area behind each ray; anterior rays barred with light and dark; anal and ventrals whitish (probably blue in life), the ventrals without dark markings; pectorals dusky, the base lighter, and with some indistinct dusky bars; a dusky half-bar on upper part of axil.

Head 6 in length (8¼ in total); depth 8½. Eye 5 in head; ventrals= pectorals=head; D. VI—14; A. 1, 14.

Upper part of opercle with a broad patch of about 20 scales, arranged in 4 series; head otherwise naked. Scales on body very small, becoming much larger behind; arranged in 80 cross-series.

All the dorsal spines more or less filamentous. Caudal fin nearly one-third total length.

There is apparently another species very closely related to *oceanicus*, and occurring with it in the West Indies. This is represented in our collection by a specimen from Colon, U. S. C., and appears to be characterized by a longer head (5 in length, 7 in total), by the much larger scales (60 in lateral line), by the obsolescence of the patch of scales on opercles, and by different coloration. *Gobius lanceolatus* C. & V. and *Gobionellus lanceolatus*, Poey, Syn. Pisc. Cub., 393, seem to refer to this latter species.

101. Gobiosoma bosci (LaC.) J. & G.

Very abundant along the muddy shores of lagoons, hiding in oyster-shells and holes in the mud.

102. Scorpæna stearnsi Goode & Bean.

A single specimen obtained. This may be identical with *S. brasiliensis* C. & V. The description of the latter given by Kner (Novara Fische, 114) applies well to our specimen.

103. Prionotus palmipes (Mitch) Storer.

?? *Trigla carolina*, Linn., Mantissa.

Trigla palmipes, Mitch. Trans. Lit. and Phil. Soc. N. Y., I. 431.

Prionotus carolinus, C. & V., iv, 90.

Evidently not abundant in Southern waters, no specimens being obtained by us during the summer. Several examples are, however, preserved in the Charleston museum, from the coast of South Carolina. Linnæus' description of *Trigla carolina* applies almost equally well to any of our species. The fin formula given by him ("D. X—13; A. 12") is found commonly only in the present species and in *scitulus*, but this does not lend any high degree of probability to the identification. We must, therefore, make use of Mitchill's name *palmipes*, it having been given prior to the use of *carolinus*, definitely for this species, by Cuvier and Valenciennes.

104. Prionotus scitulus J. & G.—*Sea Robin.*

Not rare, several specimens having been obtained. Probably not reaching as large a size as other species of the genus, the largest examples seen being but 5½ inches long.

The coloration given by us (Proc. U. S. Nat. Mus., 1882, 288) was apparently drawn from a female specimen. The following is the life color of the male:

Light olive brown, with four saddle-like dark blotches on back, one downwards and forwards from middle of spinous dorsal to humeral spine; a second from front of soft dorsal; a third from end of dorsal downwards and forwards to below lateral line, thence continued forwards as a narrow horizontal streak; a fourth on caudal peduncle; sides everywhere with reddish-brown spots, as in the female. Opercle reddish-brown; branchiostegal membrane, and palatine region largely jetblack. Spinous dorsal olive-brown, with two irregular lengthwise translucent streaks and an intense well-defined black spot on membrane above, between fourth and fifth spines. Second dorsal olive-brown vermiculated with whitish translucent, and without round spots. Caudal reddish-brown, blackish towards tip, with a conspicuous white longitudinal streak on upper lobe. Anal blackish, with white base and margin. Pectoral dark brown, irregularly barred and blotched with greenish and light brown. Free rays of pectorals, and inner face of ventrals dusky, tinged with orange.

Head 2¾ to 3; depth 5⅓ to 6; D. X— 13; A. 12. Longest dorsal spine (in ♂) 1¾ in head; pectoral fin 2¼ to 2⅔ in body. Preopercular spine with an inconspicuous cusp above and one below its base; small specimens show also inconspicuous spinous teeth on preorbital.

105. Prionotus tribulus C. & V.—*Sea Robin*.

Abundant.

106. Prionotus sarritor sp. nov.—*Sea Robin* (29675.)

> *Prionotus evolans* J. & G. Proc. U. S. Nat. Mus., 1878, 374. (Not *Trigla evolans* L.)
>
> *Prionotus evolans* J. & G. Syn. Fish N. A., 735.

This form is in many respects intermediate between *P. strigatus* C. & V., and *P. tribulus* C. & V. The color is in most particulars like that of *tribulus*, but the white spots on back and sides are much less numerous, or wholly wanting, and the brown bar backward from humeral spine is present, as in *strigatus*, and the dorsal fin is not barred; the gill rakers are, as in *strigatus*, slender and fine, 18 to 20 developed on lower limb; the spines on the head are not strong as in *tribulus*, that above orbit behind not conspicuously raised above surface of head; in two specimens from Beaufort, N. C., the pectorals are much lengthened, reaching nearly to base of caudal, but this seems to be here, as in *tribulus*, a very variable feature, as specimens from Charleston have the pectorals but one-half length of body.

Head $2\frac{3}{5}$ in length; depth $4\frac{3}{4}$. D. X—12; A. 11.

Lat. l. 53 (pores). Soft dorsal high, the longest ray = longest spine, $2\frac{1}{2}$ in head; caudal $3\frac{1}{2}$ in length.

Color in life, olive-brown above, becoming light olive on sides, white below; back with three brown cross-bars, the first under spinous dorsal, the second under first third of second dorsal, the third under its end, all of these bars extending downwards and forwards to lateral line, the posterior forming a brown blotch on base of last dorsal rays; back and sides with numerous small white spots, irregular in shape and size; these often wanting; a lateral line running in a narrow brown streak; distinct broad reddish-brown streak from humeral spine backwards to opposite end of anal; traces of a narrow streak above this. Branchiostegal membrane yellowish above; a dark brown streak from angle of mouth to base of preopercular spine; opercle dusky brown without, deep reddish-brown within. Caudal with a light brown bar at base, then a broad translucent bar, the terminal two-thirds orange-yellow, narrowly margined behind with white.

Spinous dorsal dusky, with a diffuse black blotch between fourth and sixth rays above; soft dorsal translucent brownish, without streaks of any kind; anal wine-color, translucent at base and tip. Ventrals light reddish. Pectorals glaucous green within, the lower rays reddish, the upper white; the outer side dark greenish-brown, unbarred, with a very narrow blue margin behind.

The description given by Linnæus of "*Trigla evolans*," is too meager to permit identification, and the name should therefore not be used for any of our species. We are obliged also to reject the name *lineata* as applied to our northern species, the *Trigla lineata* of Mitchill being merely a mistaken identification of *Trigla lineata* Bloch, as described by

Shaw. The oldest name available for the northern form will therefore be *Prionotus strigatus* Cuv. & Val.

107. Cephalacanthus volitans (Linn.) J. & G.—"*Flying Fish.*"

But few specimens seen.

108. Batrachus tau Linn.

Very common.

109. Porichthys plectrodon J. & G.

Rare in Charleston Harbor; a single specimen obtained.

110. Chasmodes bosquianus (La Cépède) J. & G.

Common in muddy tide pools in the harbor.

111. Isesthes scrutator J. & G.

Two specimens obtained.

112. Isesthes punctatus (Wood) J. & G.

Blennius hentz Le Sueur, Jour. Acad. Nat. Sci. Phila., iv, 363.

A single specimen obtained, 4 inches in length.

Color in spirits: Olivaceous, back and sides of head and body everywhere covered with brown spots, very irregular in size and shape; on posterior part of body the spots are larger, and show a tendency to form vertical bars; cheeks dark; lower side of head with traces of three crossbars; spinous dorsal with an elliptical black spot on membrane of first three spines; soft dorsal and caudal obscurely barred; anal, ventrals, and lower rays of pectorals dusky; pectorals olivaceous, spotted with brown.

Head $3\frac{2}{5}$; depth 3. D. XII, 15; A. 18. Pectoral $1\frac{1}{4}$ in head; ventral $1\frac{3}{4}$; gill slit $2\frac{1}{4}$; eye $4\frac{1}{2}$; maxillary $2\frac{2}{3}$. Orbital tentacle very slender, once forked, 3 in head.

Tip of each dorsal spine with a filiform, articulated, ray-like appendage.

**113. Phycis earlli, Bean.—"*Tom-cod.*"

Two specimens seen in the Charleston Museum. Said by the fishermen to be not uncommon in the harbor during the winter.

Head 4 in length; depth $5\frac{1}{3}$. Eye $5\frac{1}{4}$ in head; maxillary 2. Gill-rakers 2+9. D. 10–59; A. 46.

114. Paralichthys ommatus Jor. & Gilb.—"*New York Flounder.*"

Abundant in Charleston Harbor, where many specimens were obtained.

The ground color is usually light olivaceous, rather than olive brown; the ocellated spots are frequently furnished with a bright white center; and the sides and vertical fins have often a few scattered white spots. A small, indistinct, dark spot on middle of each 8th or 10th ray of dorsal and anal.

Head $3\frac{3}{4}$–$3\frac{4}{5}$ in length; depth $1\frac{2}{3}$. D. 70 to 76; A. 57 to 59. Pores in lateral line, 83 to 90; vertical series of scales, 70. Gill-rakers very short, $2 + 6$. Fourth or fifth dorsal ray longest, nearly two-thirds length og head. Caudal $1\frac{1}{5}$ in head; ventral of colored side, $1\frac{2}{3}$.

115. Paralichthys ocellaris (DeKay) J. & G.—*Flounder.*

Platessa ocellaris De Kay, N. Y. Fauna, Fish, 1842, 300, pl. 47, fig. 152.
Platessa oblonga Storer, Hist. Fish. Mass. 1867, 395, pl. 31, fig. 2.
Paralichthys ophryas Jor. & Gilb. Syn. Fish. N. A. 822.

Abundant in the harbor, but much less so than the following species. It does not reach as large a size as *dentatus*, and is much less valuable as a food-fish. The largest specimen obtained is about 1 foot long. This species has by recent writers been confounded with *dentatus*, along with which it occurs on both northern and southern portions of our East coast. From *dentatus* it is readily distinguishable by the ocellated spots, the narrow, interorbital, and especially by the slender, more numerous gill-rakers.

Color in life: Light olive-brown; adults with very numerous small white spots on body and vertical fins; sometimes a series of larger white spots along bases of dorsal and anal fins; about 14 ocellated dark spots on sides, these sometimes little conspicuous, but always present; a series of 4 or 5 along base of dorsal, and 3 or 4 along base of anal, those of the two series opposite, and forming pairs; two pairs of smaller less distincts spots midway between these basal series and lateral line anteriorily, with a small one on lateral line in the center between them; a large distinct spot on lateral line behind middle of straight portion; fins without the round dark blotches characteristic of *dentatus*.

Head $3\frac{1}{2}$ to 4 in length; depth $2\frac{2}{5}$; eye 6 in head; maxillary 2; pectoral $2\frac{1}{5}$; ventral $3\frac{1}{2}$; caudal peduncle 4; caudal $1\frac{1}{4}$. D. 86 to 91. A. 65 to 71. Lat. l. 108 (tubes). Curve of lateral line $3\frac{3}{3}$ to $4\frac{1}{3}$ in straight portion.

Gill-rakers comparatively long and slender, $5 + 15$ to $6 + 18$ in number. Length 12 inches.

Teeth as in *dentatus*, very long, in a single series, those in lower jaw larger.

Scales smooth; the posterior margin of each scale of colored side beset with a row of minute accessory scales.

116. Paralichthys dentatus (Linn.) J. & G.—*Flounder.*

The only flounder of much value for food found at Charleston. It is much more numerous than other species, and reaches a larger size. Specimens were seen $2\frac{1}{2}$ feet long.

This species is readily distinguished by the nearly uniform dark olive-brown coloration, without a trace of ocellated spots; the fins are plain, with characteristic round, dusky blotches; the interorbital space is, in adults, wide and flat; the gill-rakers are comparatively short and strong, very constantly $2 + 9$ in number, rarely $3 + 10$.

There is nothing in Linnæus's description of *dentatus* to indicate that he had the present species rather than *ocellaris*. But as the original type of *dentatus* is still preserved by the Linnæan Society of London, judgment may be suspended until a re-examination of this has been made.

117. Citharichthys spilopterus Günther.

Very common in the harbor, where numerous specimens were obtained. Compared with a large series from Mazatlan, Mexico, our Atlantic form differs constantly in having the interorbital space consisting of a single sharp, knife-like ridge, while those from the Pacific have the interorbital space broader, three-fourths width of pupil, and composed of two ridges with a groove between them. The Charleston specimens have constantly the depth slightly less, and the coloration much lighter. In spite of the slightness of the differences noted it is probable that the Pacific form is worthy of separation as a subspecies.

Head $3\frac{2}{3}$–$3\frac{3}{4}$ in length; depth $2\frac{1}{3}$–$2\frac{1}{2}$. D. 76; A. 58. Lat. l. 45 (pores). Eye $5\frac{1}{2}$ in head; maxillary $2\frac{2}{3}$; pectoral $1\frac{2}{3}$. Gill-rakers 14 on lower limb.

117. Etropus crossotus Jor. & Gilb.

Abundant. Specimens observed differ from those from Panama in slightly greater depth, which is more than half length of body.

118. Achirus lineatus (Linn.) Cuv.
(*Solea brownii* Günther, iv, 477.)

Very abundant. Specimens with the left side plain whitish, and those having it covered with dusky spots, are equally common. Some were also observed with the eyeless side dusky, but not spotted. As no other differences could be appreciated, it is not probable that this difference in coloration is significant of specific distinctness.

119. Aphoristia plagiusa (Linn.) J. & G.—*Tongue-fish.*

Not rare.

120. Pterophrynoides histrio (Linn.) Gill.

Two specimens seen.

121. Balistes capriscus Gmelin.—*Old-wife.*

Common on the fishing banks where it is often caught with hook-and-line, and used for bait.

122. Monacanthus hispidus (Linn.) J. & G.

? ? *Monacanthus setifer* Bennett, Proc. Comm. Zool. Soc. 1830, 112.
Monacanthus broccus Mitch. Trans. Lit. and Phil. Soc. N. Y., 1, 467.

A single specimen obtained on the fishing bank, caught with hook-and-line in 10 fathoms of water. The caudal peduncle shows the characteristic lengthened setæ, and the first dorsal ray is produced and filiform. The species is evidently not abundantly found at Charleston.

The description given by Linnæus of *Balistes hispidus* is based on a specimen, evidently of the present species, received by him from South Carolina through Dr. Garden. The body was "hispid, roughened towards the tail with setæ." A reference is also made by Linnæus to the account given by Seba of *Monacanthus longirostris.* This, however, cannot invalidate a description made from a specimen in hand, and the name *hispidus* must be used for our species.

123. Ostracium quadricorne Linn.—*Cow-fish.*

Very common.

124. Lagocephalus lævigatus (Linn.) Gill.

Two specimens seen.

125. Tetrodon turgidus Mitch.

Common.

From specimens of *T. nephelus* in our collection from Pensacola and Galveston, this species differs conspicuously in its coloration, in having the snout entirely covered with spines, and in having all the spines short and immovable. *T. nephelus* has a broad space below eye, and the snout, with exception of a small median patch above, naked, and the spines of head and body are longer, slender, and erectile.

126. Chilomycterus geometricus (Bl. & Schn.) Gill.—*Pin-cushion.*

Very abundant.

Very young specimens have the body soft and flabby, with the spines admitting of considerable movement because of the looseness of the skin; the caudal peduncle is scarcely noticeable; the belly is often of purplish black, with pink spines.

ADDITIONS.

The following species, included in the present list, are here for the first time authentically recorded from our coast north of Key West, Fla.

1. Ginglymostoma cirratum.
2. Hypoprion brevirostris.
3. Rhinobatus lentiginosus.
4. Fundulus similis.
5. Ophichthys chrysops.
6. Exocœtus mesogaster.
7. Hippocampus stylifer.
8. Querimana harengus.
9. Sphyræna picuda.
10. Phthirichthys lineatus.
11. Calamus bajonado.
12. Xyrichthys lineatus.
13. Culius amblyopsis.

14. Gobius encæomus sp. nov.
15. Gobius thalassinus sp. nov.
16. Gobius oceanicus.
17. Scorpæna stearnsi.
18. Porichthys plectrodon.
19. Isesthes scrutator.
20. Etropus crossotus.

Additional facts are also made known with regard to the distribution of *Scomber colias, Coryphæna hippurus, Mesogonistius chætodon,* and *Pœcilichthys barratti.*

In a list given by us of the fishes of Beaufort Harbor, North Carolina (Proc. U. S. Nat. Mus. 1878, 365), the following errors of identification are made:

Siphonostoma fuscum=*Siphostoma louisianæ* and *floridæ.*

Pseudorhombus ocellaris=*Paralichthys dentatus, ocellaris* and *albigutta.*

Prionotus punctatus=*Prionotus scitulus.*

Prionotus evolans=*Prionotus sarritor.*

Carangus chrysus=*Caranx beani* type (probably young of *Caranx ruber.*)

Chirostoma menidium=*Menidia bosci* and *laciniata.*

Belone hians=the young, probably, of *Tylosurus caribbæus.*

Under the heading of Lophopsetta maculata, it should have been stated that the species was admitted to the list on the authority of Dr. Yarrow, but was not seen by us.

INDIANA UNIVERSITY, *November 6, 1882.*

LIST OF FISHES NOW IN THE MUSEUM OF YALE COLLEGE, COLLECTED BY PROF. FRANK H. BRADLEY, AT PANAMA, WITH DESCRIPTIONS OF THREE NEW SPECIES.

By DAVID S. JORDAN and CHARLES H. GILBERT.

About the year 1866 a considerable collection of fishes was made at Panama and in the neighboring Pearl Islands, by the late Prof. Frank H. Bradley. These specimens are now preserved in the museum of Yale College. By the courtesy of Prof. A. E. Verrill they have been placed in our hands for determination. We give here a list of the species contained in the collection, with remarks on some of the more interesting forms. Three species appear to be still undescribed, and a very large proportion of the others were unknown at the time the collection was made. A series of duplicates has been presented by Professor Verrill to the National Museum. Unless otherwise stated, all the species mentioned were obtained at Panama, by Professor Bradley.

1. **Ginglymostoma cirratum** (Gmel.) M. & H.

A single young example.

2. Mustelus sp.

? ? Mustelus mento Cope, Proc. Am. Philos. Soc. XVII, 1877, 47.

Three specimens, each about 10 inches long, in poor condition. The fins are much larger than in *M. lunulatus*, the space between dorsals being but twice base of first dorsal and 2½ times base of second. The color is also much darker, that of the fins nearly uniform dusky, with lighter edges. *M. lunulatus* was hitherto known only from Mazatlan, unless indeed *Mustelus mento* Cope, from Peru, should prove to be the same.

3. Urolophus halleri Cooper.

Two specimens. The species has now a recorded range from Panama to Point Conception, Cal. It is abundant only along the northern part of this range, from San Diego to Santa Barbara.

4. Syrrhina exasperata (Jor. & Gilb.) Garman.

Two adult female specimens, each over 20 inches long, are in the collection made by Mr. Bradley at Panama. The species was hitherto known only from Southern California, and was represented in collections by numerous immature males collected by ourselves at San Diego, and by a single adult male (type of *Trygonorhina alveata* Garman) in the Museum of Comparative Zoology at Cambridge.

The following points in regard to these female specimens are worthy of note : The general plan of coloration is the same as in males, including the large black blotch covering posterior angles of pectorals below ; the upper side of disk has, however, several round yellowish spots as large as pupil, each spot ocellated with blackish ; a very distinct spot on each side of shoulder; a second on pectoral fins near posterior angle; and a third midway between the latter and median line of back ; several other less conspicuous spots near middle of back anteriorly. The disposition of spines and prickles above is the same as in males ; but below, the entire surface of body and tail is covered with uniform fine shagreen, instead of being largely naked.

Disk somewhat broader than long, the length slightly greater than that of tail.

5. Arius brandti Steind.

6. Arius alatus Steind.

Two specimens, each about 16 inches long. Head 3¾ in length; maxillary barbel reaching nearly to tip of pectoral spine.

7. Arius kessleri Steind.

A single specimen shows the following characters: Head very coarsely granular, the occipital process narrowly triangular and sharply keeled, rounded posteriorly ; the antedorsal shield very narrow, about half diameter of orbit. Humeral process with few granulations. Maxillary barbel barely reaching base of pectoral spine. Vomerine patch of teeth much narrowed toward median line, and divided by a furrow. Fontanelle club-shaped.

8. Arius insculptus Jor. & Gilb.

Three specimens. Head with very fine and numerous granulations; occipital process very wide, truncate posteriorly, sometimes with fluted margin into which fit projections from the antedorsal shield; the latter is wide. Humeral process with very fine numerous granulations. Fontanelle tapering to a point posteriorly. Barbels much longer than in *kessleri*, the maxillary barbel reaching beyond first third of pectoral spine. Vomerine patch of teeth not divided on median line.

9. Arius planiceps Steind.

A male and a female of this species, each about 10 inches long, differ somewhat from those examined by Dr. Steindachner, and from each other. In the male the head is very long, $3\frac{1}{2}$ in body; in the female, 4 in body. The maxillary barbels in the male are short, not reaching base of pectoral spine, and the granulation of the cephalic plates is much less marked, the granules on occipital process scarcely larger or more thickly set than on rest of head. In both specimens the occipital process is broader at the base, and much more tapering posteriorly than is represented in the figure given by Dr. Steindachner. None of the specimens examined by us show any distinct trace of a median furrow through the vomerine patch of teeth.

10. Arius dasycephalus Gthr.

11. Ælurichthys panamensis Gill.

12. Ælurichthys pinnimaculatus Steind.

12. Albula vulpes (L.) Goode.

13. Elops saurus Linn.

14. Opisthonema libertate (Günth.) J. & G.

This species differs apparently from *thrissa* in the absence of dark spots on the scales of the back, in the longer and more numerous gillrakers, and in the longer head. In *libertate*, the head varies from $3\frac{2}{3}$ (in young) to $4\frac{1}{3}$ (in adults); in *thrissa*, from 4 to $4\frac{4}{5}$. *Libertate* is bluish or greenish above, silvery on sides and below, a yellowish-olive streak on level of orbit. A small indistinct black spot at upper angle of preopercle, and a larger more distinct one on scapula. Dorsal olive-yellow, with dusky margin; caudal dusky, the lobes tipped with jet black; upper rays of pectorals dusky. Tip of snout and lining membrane of opercle black.

A specimen of *O. thrissa* is also in the collection, reputed to have been taken by Professor Bradley at Panama. We prefer not to admit it to the list from the Pacific coast until its occurrence there is verified.

15. Stolephorus panamensis (Steind.) J. & G.

Two specimens, about 5 inches long, with anal rays respectively 33 and 37 in number.

16. Stolephorus miarchus Jor. & Gilb.

Many small slender anchovies collected by Professor Bradley in the

Pearl Islands belong to this species. They are of the same size and general appearance as the original types from Mazatlan. The anal rays are quite constantly 13, and the body is exceedingly slender, the depth being about $\frac{1}{7}$ the length.

17. Pœcilia elongata Günther.

18. Ophisurus xysturus Jor. & Gilb.

Three fine examples, the · longest 28 inches long, from Mazatlan, Acapulco, and Panama, respectively. These specimens vary from the original types from Mazatlan in the following respects: The vomerine patch of teeth is broader, with a well-marked constriction anteriorly, with teeth arranged in about three irregular series; the eye is contained twice in snout, which is $\frac{4}{5}$ interorbital space; length of pectoral less than width of gill-opening. The dark spots are arranged more regularly, those of the upper two series nearly equal in number. The specimen from Mazatlan has the spots of the upper two series corresponding, while in the other two specimens they alternate. Spots on dorsal fin distinct, not confluent. In the smallest specimen (from Panama) the head is contained but three times in the trunk.

19. Ophichthys zophochir Jor. & Gilb.

A fine specimen, about 2 feet long, collected by Mr. J. A. Sutter at Acapulco. The species was hitherto known only from Mazatlan Harbor.

20. Sidera panamensis (Steind.) J. & G.

> *Murœna panamensis* Steindachner, Ichth. Beitr. V, 19 ; not *Sidera panamensis* J. & G., Bull. U. S. Fish. Com., 1882, 106 = *Sidera castanea* J. & G., MSS.

Three specimens from Pearl Islands, the largest 10 inches long, answer perfectly to Steindachner's description of this species.

21. Sidera verrilli sp. nov.

A single specimen in the Yale College Museum, $17\frac{1}{2}$ inches long, collected by Professor Bradley at Panama, serves as the type of the following description :

Body comparatively slender, the tail about equal to the rest of the body. Head $3\frac{1}{4}$ in length of trunk. Cleft of mouth moderate, 3 in head. Mandible somewhat curved, and the teeth very long, so that the mouth does not admit of being completely closed.

Teeth everywhere uniserial, those on sides of mandible strong, compressed, hooked backwards, about 13 in number on each side, the teeth growing gradually smaller backwards, those next angle of mouth very small ; 4 or 5 anterior teeth on each side very large, subequal. Teeth of upper jaw in all respects similar to those in the lower, and in equal number. A short row of very small teeth on vomer posteriorly ; the anterior canines wanting in our specimen (perhaps lost); teeth all apparently entire.

Eye rather large, somewhat nearer angle of mouth than tip of snout, its diameter about half length of snout. Gill-opening small, scarcely wider than orbit. Tube of anterior nostril rather short, less than half eye. Posterior nostril above front of eye. Occipital region little prominent.

Dorsal fin rather high, commencing nearly midway between gill-opening and eye, its greatest height rather more than half greatest depth of body.

Color, in spirits, light chestnut brown, finely freckled, but without distinct spots of any kind. Dorsal with a conspicuous edge of blackish, the margin narrowly white. Anal edged with white. No black about eye or gill-opening

22. Muræna ? melanotis (Kaup.) Gthr.

A specimen, 22 inches long, has the teeth everywhere uniserial, otherwise agreeing with descriptions of *melanotis*. Body and fins dark brown, marbled with blackish, everywhere with small yellowish spots much more numerous anteriorly, those on the tail narrowly oblong. Angle of mouth, and a large roundish blotch around gilt-slit black, this blotch nearly four times as wide as orbit.

Eye over middle of gape, which is $2\frac{3}{8}$ in head. Head $2\frac{1}{2}$ in trunk. Tail slightly longer than rest of body.

23. Tylosurus pacificus (Steind.) J. & G.

24. Hemirhamphus ? brasiliensis (Linn.) C. & V.

Two adults, about 15 inches long, agree with specimens collected by Mr. Gilbert at Panama, and differ from Atlantic representatives of the species in their longer pectoral fins, and in the more anterior insertion of the ventrals. It is probable that the Pacific form is a distinct species or subspecies, but our material from the Atlantic is too limited to warrant the separation of the former.

The specimens before us have the pectoral nearly six-sevenths length of head (three-fourths in Atlantic specimens) and greater than depth of body; the distance from root of ventrals to base of caudal is slightly less than one-third distance to front of snout, and measured from base of ventrals forwards reaches a point nearer base than tip of pectorals. D. 14; A. 11 or 12. Scales 58. Head $4\frac{2}{5}$ in length; lower jaw 5 in total length (including caudal). Eye $4\frac{1}{5}$ in head, equaling interorbital space. The first 3 to 6 rays of dorsal and anal with series of scales, these fins otherwise naked.

25. Hemirhamphus unifasciatus Ranz.

26. Mugil brasiliensis Agassiz.

27. Mugil incilis Hancock.

A single adult example with the scales noticeably smaller than in *M. brasiliensis*, and the vertical fins lower. Lateral line 43; 14 scales in a cross series. Longest dorsal ray less than half length of head.

28. Querimana harengus (Günther) J. & G.

29. Sphyræna ensis Jor. & Gilb.

30. Scomberomorus maculatus (Mitch.) J. & G.
31. Caranx caballus Gthr.
32. Caranx latus Ag. (=*fallax* C. & V.)
33. Caranx hippos (Linn.) J. & G.
34. Caranx setipinnis (Mitch.) J. & G.

This species has a well-developed series of spinous plates along the lateral line, as has been already pointed out by Bleeker and Steindachner. There seems to be no reason why it should not be referred to Caranx.

35. Selene vomer (L.) Lütken.
36. Oligoplites saurus (Bl. & Schn.) J. & G.
37. Trachynotus rhomboides (Bloch) Cuv. & Val.

Two small specimens, each 1½ inches long, differ from an example of the same size from Beaufort, N. C., in the much deeper body (depth 1⅖ in length), and in the greater development of all the spines. The triple spine at angle of preopercle is conspicuous, and the highest dorsal and anal spines are longer than the soft rays. Body thickly dusted with brown points; dorsal and anal blackish. D. VII–18; A. III, 17.

38. Centropomus armatus Gill.
39. Centropomus robalito Jor. & Gilb.
40. Centropomus unionensis Bocourt.
41. Centropomus undecimalis (Bloch) Lac.
42. Epinephelus sellicauda Gill.
43. Epinephelus analogus Gill.
44. Epinephelus multiguttatus (Günther) J. & G.
45. Serranus calopteryx Jor. & Gilb.

Two immature specimens from Panama and Pearl Islands respectively. Hitherto recorded only from Mazatlan and the Galapagos Islands (as *Prionodes fasciatus* Jenyns).

46. Lutjanus argentivittatus (Peters) J. & G.
47. Lutjanus guttatus (Steind.) J. & G.
48. Lutjanus novemfasciatus Gill.
49. Lutjanus aratus (Günther) J. & G.
50. Pomadasys pacifici (Günther) J. & G.
51. Pomadasys macracanthus (Günther) J. & G.
52. Pomadasys elongatus (Steind.) J. & G.
53. Pomadasys chalceus (Günther) J. & G.
54. Pomadasys brevipinnis (Steind.) J. & G.

This specimen extends the range of this species from Mazatlan to Panama. It may be noticed that the figure given by Dr. Steindachner (Ichthyol. Notiz., VIII, Taf. 5) is faulty in several respects. Thus the scales with their accompanying dark streaks are represented as oblique below the lateral line, whereas in reality they are horizontal. The inter-

Proc. Nat. Mus. 82——40   **May 22, 1883.**

maxillary processes are shown in the figure to project beyond the line of profile, while in the fish nothing breaks the evenly convex outline. The accompanying description does not countenance these errors of the artist. Dr. Steindachner has more lately (Ichthyol. Beit., II, 8) incorrectly identified *brevipinnis* with *Microlepidotus inornatus* Gill. The latter is a widely different species, with scaleless dorsal and anal and 14 dorsal spines.

55. Diabasis sexfasciatus (Gill) J. & G.
56. Diabasis scudderi (Gill) J. & G.
57. Diabasis flaviguttatus (Gill) J. & G.
58. Diabasis maculicauda (Gill) J. &. G.
59. Cyphosus analogus (Gill) J. & G.
60. Sciæna vermicularis Günther.
61. Sciæna chrysoleuca Günther.
62. Sciæna ophioscion Günther.
63. Larimus argenteus (Gill) J. & G.
64. Larimus breviceps C. & V.
65. Paralonchurus dumerili (Bocourt) J. & G.
    (*Genyanemus fasciatus* Steind.)
66. Isopisthus remifer J. & G.
67. Micropogon altipinnis Günther.

Very numerous immature specimens, showing: D. X-I, 20 or X-I, 21; and scales 6-45-13.

68. Cynoscion reticulatum (Gthr.) J. & G.
69. Cynoscion album (Gthr.) J. & G.
70. Upeneus grandisquamis Gill.
71. Polynemus opercularis (Gill) Gthr.
72. Gerres peruvianus Cuv. & Val.
73. Gerres dowi (Gill) Günther.

Three specimens, each about 6 inches long. Head $3\frac{2}{3}$ to $3\frac{1}{2}$ in length; depth $2\frac{4}{5}$ to 3. Eye 3 to $3\frac{1}{6}$ in head. Cheeks and sides without black specking.

74. Pomacentrus rectifrænum Gill.
75. Acanthurus matoides C. & V.
76. Gobius soporator Cuv. & Val.
77. Gobius paradoxus Günther.
78. Batrachoides pacifici Günther.
79. Thalassophryne reticulata Günther.

A single specimen about 12 inches long. Head $3\frac{1}{2}$ in length. D. II-25; A. 24.

80. Porichthys margaritatus Rich.

A specimen, $1\frac{1}{2}$ inches long, from Central America.

81. Scorpæna plumieri Bloch.

82. Scorpæna sp.

Four immature specimens, representing apparently two species, both distinct from *plumieri*, are in the collection from Panama and Acajutla. They are too small to permit identification.

83. Gobiesox adustus Jor. & Gilb.

Two specimens, in fine condition, are in the collection. They were obtained by Captain Dow on the coast of Central America. The species was hitherto known from Mazatlan only.

The following points were incorrectly stated in the original description: Width of head 2⅔ to 3 in length; pectoral one-third to two-fifths length of head; distance from base of caudal to front of dorsal, 2⅔ in length of body, 3½ in total, including caudal. D. 9 or 10; A. 7 or 8.

Emblemaria gen. nov. (*Blenniidæ.*)

Body slender, not eel-shaped, compressed, scaleless. Ventrals present, jugular, each of one spine and two soft rays. A single dorsal fin beginning on the nape and extending to the caudal, with which it is not confluent; no notch between spinous and soft rays. Head cuboid, compressed, narrowed anteriorly, with much the aspect of *Opisthognathus*. Symphysis of lower jaw forming a very acute angle. A single series of strong, blunt, conical teeth on each jaw, and on vomer and palatines. Vomer and palatine teeth larger, their series continuous parallel to the series in upper jaw. No cirri anywhere. Gill-openings very wide, the membranes broadly united below, free from the isthmus. Lateral line obsolete.

This genus bears some resemblance to *Blennius*, but the dentition is entirely different, approaching that of *Chœnopsis*.

84. Emblemaria nivipes sp. nov. (29,676).

Color in spirits: Sides dark brown, with 8 to 10 lighter vertical bars of variable width; body lighter below; obscure cross-bands on lower side of head. Dorsal blackish anteriorly, whitish behind, with membrane at intervals of every second, third, or fourth ray dusky; caudal light at base, its tip blackish; anal dusky-translucent; ventrals bright white, the basal portion dusky.

Head 3¾ in length; depth 7. D. XXIII, 14; A. 25. Body everywhere equally compressed, posteriorly tapering; head wider than body, of about equal depth, with very short, subvertical, sharply-compressed snout; eyes very large, approximated above, with some vertical range; orbital ridges sharply raised above, the interorbital region very narrow, channeled, about equaling diameter of pupil; eye 3⅔ in head. Gape very wide, horizontal, low, reaching much beyond eye, the maxillary about four-sevenths head, not produced beyond angle of mouth; intermaxillaries separated by a groove from the snout, this groove continuous for the entire length of the upper jaw, maxillary not evident, apparently adnate to the skin of the preorbital.

First dorsal spine inserted over margin of preopercle; spines all very slender and flexible, the posterior but weakly differentiated from the soft rays, the anterior portion of fin very high, the spines filiform, not exserted beyond the membrane; the longest dorsal spine about one-third length of body, the last spine about one-half head; membranes of last rays of both dorsal and anal slightly joined to base of caudal. Front of anal nearer snout than base of caudal by a distance equaling one-third length of head. Caudal three-fifths length of head; ventrals and pectorals slightly less.

A specimen 2 inches long, collected by Professor Bradley at the Pearl Islands, serves as the type of the species, and is numbered 29,676 on the register of the U. S. National Museum. Numerous smaller specimens are in the collection from the same locality.

85. Cremnobates monophthalmus Günther.

86. Salarias rubropunctatus C. & V.

Six specimens of this species, the longest 3 inches in length, were collected by Professor Bradley at the Pearl Islands. The fin rays, coloration, and proportions are those assigned this species by Cuvier and Valenciennes. In addition, there is a distinct jet-black spot behind the eye, with a narrow light edge anteriorly.

Head = depth, 4 in length (5 in total); eye $4\frac{1}{3}$ in head. D. XI, 16; A. 20. The teeth are somewhat less flexible than in *S. atlanticus*, and the canines in lower jaw are wholly wanting.

Specimens of the same species collected by Professor Bradley are in the collection from Callao. The species called by Kner (Novara Fische, 198) *S. rubropunctatus* seems to be different from this.

87. Dactyloscopus sp. nov.

?? Dactylagnus mundus Gill, Proc. Acad. Nat. Sci. Phila., 1862, 505, 506.

A specimen in the present collection, taken by Captain Dow on the coast of Central America, agrees well with the type of *Dactylagnus mundus*. It has, however, the dorsal beginning at the nape, and the pseudobranchiæ wanting. It is, therefore, a typical *Dactyloscopus*, and probably represents a species hitherto undescribed, but without further information we are not prepared to describe it as new, as it may be really identical with *Dactylagnus mundus*.

Color in spirits, light olivaceous, the edgings of the scales, some vermiculations on top of head, and the labial fringes, clear brown. Fins translucent; the caudal with a brown bar at base. Eyes dark.

Head $4\frac{1}{2}$ in length; depth $6\frac{2}{3}$. D. VI–38; A. II, 37. V. 3. Scales 6–51–5. B. 6. L. $3\frac{1}{4}$ inches.

Head and body slender, compressed, the greatest width at occiput four-ninths length of head; the greatest depth is immediately behind insertion of anal fin, thence tapering to a very narrow tail. Head narrow, cuboid, compressed, the upper surface nearly plane, the cheeks vertical. Eyes very small, superior, with little lateral range; diameter of orbit about $\frac{1}{15}$ length of head; snout very short, about equaling orbit. Anterior nostril in a short tube. Gape subvertical, the lower jaw very heavy, projecting, as in *Uranoscopus;* premaxillaries protractile, the processes reaching far behind orbits; lips fringed; both jaws with bands of villiform teeth; no teeth on tongue, vomer, or palatines.

Subopercle and interopercle very wide, flexible, striate, the latter overlapping throat and base of ventral fins, the former wholly covering base of pectoral fins; the striations of opercle terminate posteriorly in a wide, coarse, membranaceous fringe. Branchiostegal membranes not united, free from the isthmus. Pubic bones forming a sharp projection at throat. No pseudobranchiæ. Gills small, a round pore behind the fourth.

Dorsal beginning on the nape, its distance from snout about equaling depth of body. The first six rays are shorter than those following and not connected by membrane; as no traces of articulation can be found, they are probably flexible spines, but are not clearly differentiated from those immediately following. Origin of anal under fourth dorsal spine. Caudal distinct, narrow, short. Ventrals inserted under anterior margin of preopercle. Ventrals 2 in head; pectorals 1¼.

Scales large, with entire edges, wanting on head, breast, and region behind pectoral fins. Lateral line beginning at upper posterior angle of opercle, running parallel with the back on about 12 scales, then obliquely downwards to middle of body.

88. Fierasfer dubius Putnam.

(*Fierasfer arenicola* Jor. & Gilb., Proc. U. S. Nat. Mus. 1881, 363.)

Numerous specimens 3 to 4 inches long from Pearl Islands.

Head 6¾ to 7⅓; eye 4½ to 5 in head. Teeth in upper jaw small, acute, in a rather narrow band ; sometimes a few in the front of the jaw inconspicuously enlarged; those in lower jaw and on palatines conic, blunt, in somewhat wider bands, the outer series of lower jaw enlarged, canine-like ; vomer with a narrowly oblong patch of small, blunt teeth, surrounding a median series of 3 to 6 conspicuously enlarged, retrorsely curved canines, which are usually much the largest teeth in the mouth.

The original types of this species belonged to the present collection, having been sent by Professor Verrill to the Museum of Comparative Zoölogy. They are said to have been taken alive from the shells of pearl oysters. Our *Fierasfer arenicola*, from Mazatlan, is apparently not specifically different.

The generic name *Carapus* Rafinesque, has been lately substituted for *Fierasfer* by Professor Poey, following a suggestion of Dr. Gill (Proc. Ac. Nat. Sci. Phila., 1864, 152). This change does not seem to us justifiable, as it certainly is most undesirable.

The following is the original diagnosis of *Carapus* (Raf., Indice d'Ittiol. Siciliana, 1810, 57):

"XII. Gen. *Carapus*. Nessun' ala dorsale, ne caudale, un' ale anale, e due ale pettorali, mascella superiore più lunga dell' inferiore, coda nuda al disotto. *Osserv.* Differisce dal vero genere *Gymnotus*, che hà l' ala anale lunghissima, ricuoprendo il disotto della coda, e la mascella inferiore più lunga della superiore."

No species is here mentioned, but in the list of Sicilian fishes, on page 37, we find:

"272. Carapus acus. Raf., App. gen. 12 (Gymnotus acus Linn.) Carapo aguglia. *Anciduzza*."

We find that these two genera correspond to the first and second subgenera recognized under *Gymnotus* by La Cépède, the first (" *Gymnotus*") including *electricus*, *putaol* (*fasciatus*), and *albus;* the second ("*Carapus*") including *carapo, fierasfer* (=*acus* L.), and *longirostratus*. The name *Carapus* is evidently suggested by " *Carapo*," and the generic diagnosis of Rafinesque above quoted seems to be entirely extracted from Gmelin's description of *Gymnotus carapo* ("*Gymnotus* * * * dorso apterygio, pinna ani longitudine, * * * maxilla superiore longiore * * * ani pinnæ in caudæ apicem non excurrens, sed ante caudæ pinnam desinens"). The diagnosis does not apply to the species of Fierasfer, which have a distinct dorsal fin. It seems, therefore, proper to consider *Gymnotus carapo* L. the type of *Carapus* Raf., while *G. electricus* L. is evidently the type of Rafinesque's *Gymnotus*.

In the tenth edition of the Systema Naturæ, but two species are referred to *Gymnotus*, *G. carapo* and *G. asiaticus*, the latter not being a member of this group. If we date our nomenclature from this tenth edition, *G. carapo* L. must be taken as the type of *Gymnotus*, *Carapus* Raf. being a synonym of *Gymnotus*, while the name *Electrophorus* Gill should be used instead of *Gymnotus* for *Gymnotus electricus* L. (ed. xii).

89. Citharichthys spilopterus Günther.

90. Antennarius sanguineus Gill.

91. Antennarius strigatus Gill.

An adult example, 10 inches long, agrees in but few respects with the descriptions drawn, by Gill and Günther (*Antennarius tenuifilis*), from immature examples.

First dorsal spine elongate, filiform, twice the length of the second, with very slender dermal tip. Third spine more robust than second, wholly concealed in the skin, its length equal to that of first spine. Lips, maxillary, and a large transverse area behind second dorsal spine naked, each side of this area with a few spinous tubercles. Skin elsewhere covered with fine shagreen-like armature.

D. III–12; A. 7.

Color in spirits olivaceous everywhere on body, and on inside of mouth finely mottled with light olive brown; many irregular blackish areas on head and body, those on lower side of head showing a tendency to form concentric bars; some on sides forming irregular bars downwards from back; posterior portion of body not darker than the anterior; terminal parts of all the fins largely blackish. but with distinct black bars; some scattered round blackish blotches on sides, each consisting of a number of smaller black spots on an olive ground. Head and body with numerous pinkish and rose-red spots and bars, the latter sinuous, irregular, with wavy margins; a pinkish bar behind maxillary; a broad, saddle-like pinkish blotch across interval between second and third dorsal spines; a third bar from in front of origin of second dorsal downwards towards base of pectorals; a fourth across top of caudal peduncle. First dorsal spine narrowly barred with brown.

92. Balistes capistratus Shaw.

(Shaw Genl. Zool. V, pt. 2, 417, 1804 (based on *Balistebridé* La Cépède = *Balistes mitis* Bennett = *Balistes frenatus* Richardson.)

93. Balistes polylepis Steind.

94. Tetrodon angusticeps Jenyns.

(*Canthogaster lobatus* Steind., Ichthyol. Not. X, 18.)

This species is represented in the collection by two fine specimens from Panama, each about one foot long. They agree perfectly with Dr. Steindachner's Altata specimen (type of *C. lobatus*), but the nostrils are formed as in typical species of *Tetrodon, i. e.*, tubular, with two lateral openings near the summit.

Jenyns' description of *T. angusticeps*, from the Galapagos Islands, was evidently drawn from a specimen in poor condition. This would account for the alleged absence of prickles on the skin. In all other respects the description agrees with the specimens before us—the narrow, channel-like interorbital space, the minute papilliform protuberances on the skin, and the pair of fleshy flaps behind the nape being conspicuous features of the species.

95. Tetrodon politus Ayres.

96. Arothron erethizon, sp. nov. (29679).

Body all, except snout and caudal peduncle, thickly beset with long, robust, quill-like spines, which are longest and most numerous on the belly; these spines are concealed by the outer skin until the animal is inflated, in which case they protrude; under a microscope the skin is seen to be provided with innumerable minute protuberances, much as in *Tetrodon angusticeps*.

Snout short, cuboid, its length $1\frac{1}{2}$ times orbit; the upper profile slightly concave, interorbital space wide, slightly less than twice diameter of eye, strongly concave because of the elevated orbital ridges. Nostril tentacle bifid to the base, the divisions compressed, flap-like, without conspicuous openings; the inner surface of each division is thickly covered with minute, cup-shaped depressions, into which open the perforations of the tube. Distance from tentacle to eye but twice length of tentacle, which equals one-fourth diameter of orbit.

Caudal fin equal to length of caudal peduncle. Dorsal large, the base equaling three-sevenths height of fin.

Body without fleshy slips or folds.

Head 3¼ in length; eye about one-fourth head. D. 9 or 10; A. 10.

Color in spirits: Dark brown above, white below; entire upper parts including caudal fin, covered with round, white spots, most numerous on caudal peduncle, the largest much less than half pupil; a round black area surrounding base of pectorals, bounded by a white line; several parallel longitudinal black streaks below the pectorals; orbit with two concentric white rings.

Known from six specimens collected by Professor Bradley at Panama. The type is numbered 29679 on the register of the National Museum.

The following species are here for the first time recorded from Panama:

1. Ginglymostoma cirratum.
2. Urolophus halleri.
3. Syrrhina exasperata.
4. Stolephorus miarchus.
5. Ophisurus xysturus.
6. Ophichthys zophochir.
7. Sidera verrilli sp. nov.
8. Serranus calopteryx.
9. Pomadasys brevipinnis.
10. Gobiesox adustus.
11. Emblemaria nivipes sp. nov.
12. Salarias rubropunctatus.
13. Dactyloscopus sp. nov. (?)
11. Tetrodon angusticeps.
15. Arothron erethizon sp. nov.

INDIANA UNIVERSITY, *December* 1, 1882.

JUMPING SEEDS AND GALLS.[*]

By CHARLES V. RILEY.

Having recently received some fresh specimens of so-called "Mexican Jumping Seeds," or "Devil's Beans," as they are popularly called, I take occasion while yet they are active to exhibit them to the society. It will be noticed that these seeds are somewhat triangular, or of the shape of convolvulus seeds, there being two flat sides meeting at an obtuse angle, and a convex one, which has a median carina. They not only

---

[*]Read before the Biological Society of Washington November 24, 1882.

roll from one side to another, but actually move by jerks and jumps, and will, when very active, jump at least a line from any object they

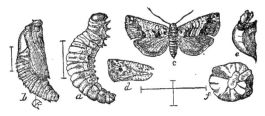

CARPOCAPSA SALTITANS: *a*, larva; *b*, pupa; *c*, imago—enlarged, hair-lines showing nat. size; *d*, front wing of a pale var.; *e*, seed, nat. size, with empty pupa skin; *f*, do. showing hole of exit.—(After Riley.)

may be resting on. The actual jumping power has been doubted by some writers, but I have often witnessed it. To the uninitiated these movements of a hard seed seem little less than miraculous. They are induced by a plump, whitish, lepidopterous larva which occupies about one-fifth of the interior, the occupied seed being, in fact, but a hollow shell, with an inner lining of silk which the larva has spun. The larva looks very much like the common apple-worm (*Carpocapsa pomonella*), and belongs, in fact, to the same genus. It resembles it further in remaining for a long time in the full-grown larva state before transforming, so that the seeds will keep up their motion throughout most of the winter months. When about to transform, which is usually in the months of January and February, it cuts a neat, circular door in the convex side of its house, strengthens the same with silk, spins a loose tube of silk within the seed, and therein transforms to the pupa state. The moth soon afterward pushes its way out from the little door prepared for it.

The moth was first described in 1857 as *Carpocapsa saltitans* by Prof. J. O. Westwood,* and afterward as *Carpocapsa dehaisiana* by Mons. H. Lucas.†

In regard to the plant on which these seeds occur there is much yet to learn, and I quote what Mr. G. W. Barnes, president of the San Diego Society of Natural History, wrote me in 1874 concerning it, in the hope that some of the botanists present may recognize it:

"ARROW-WEED (*Yerba de flecha*).—This is the name the shrub bears that produces the triangular seeds that during six or eight months have a continual jumping movement. The shrub is small, from 4 to 6 feet in height, branchy, and in the months of June and July yields the seeds, a pod containing three to five seeds. These seeds have each a little worm inside. The leaf of the plant is very similar to that of the ga-

---

* Proc. Ashmolean Soc. of Oxford, 1857, t. 3, pp. 137–8; see also Trans. Lond. Ent. Soc., ser. 2, 1858, t. IV, p. 27, 'and Gard. Chron. 1859, Nov. 12, p. 909.

† "Note sur les grains d'une Euphorbiacée de Mexique sautant au dessus du sol par les vibrations d'une larve de l'ordre des lepidoptères vivant en dedans."—(Ann. Soc. Ent de France, ser. 3, t. 6, Bull. p. 10, p. 33, p. 41, p. 44, 1859; t. 7, p. 561–566.)

rambullo, the only difference being in the size, this being a little larger. It is half an inch in length and a quarter of an inch in width, a little more or less. The bark of the shrub is ash-colored, and the leaf is perfectly green during all the seasons. By merely stirring coffee, or any drink, with a small branch of it, it acts as an active cathartic. Taken in large doses it is an active poison, speedily causing death unless counteracted by an antidote.".

In a recent letter he states that he is informed that the region of Mamos, in Sonora, is the only place where the plant grows; that the tree is about four feet high and is a species of laurel, with the leaves of a dark varnished green. "It bears the seeds only once in two years. The tree is called *Brincador* (jumper), and the seeds are called *Brinca-deros*. The seeds are more quiet in fair weather, and lively on the approach of a storm."

Professor Westwood mentions the fact that the plant is known by the Mexicans as "*Colliguaja;*" and Prof. E. P. Cox, formerly State geologist of Indiana, now living on the Pacific coast, informs me that the shrub has a wood something like hazel or whahoo; that the leaf is like a broad and short willow leaf. He confirms the statement as to its poisonous character; that a stick of the shrub, when used by the natives to stir their "penola" (ground corn-meal, parched), purges, and that the shrub is used to poison arrowheads. The plant is undoubtedly Euphorbiaceous.

The peculiarity about this insect is that it is the only one of its order, so far as we know, which possesses this habit, and it is not easy to conceive of what benefit this habit can be other than the possible protection afforded by working the seed, after it falls to the ground, into sheltered situations.

The true explanation of the movements of the larva by which the seed is made to jump was first given by me in the Transactions of the Saint Louis Academy of Science for December 6, 1875 (Vol. III, p. cxci).

The jumping power exhibited in this "seed" is, however, trifling compared with that possessed in a little gall which I also exhibit. This gall, about the size of a mustard seed, and looking very much like a miniature acorn, is found in large numbers on the under side of the leaves of various oaks of the White Oak group, and has been reported from Ohio, Indiana, Missouri, and California. It falls from a cavity on the under side of the leaves, very much as an acorn falls from its cup, and is sometimes so abundant that the ground beneath an infested tree is literally covered. It is produced by a little black Cynips, which was described as *Cynips saltatorius* by Mr. Henry Edwards. The bounding motion is doubtless caused by the larva which lies curved within the gall, and very much on the same principle that the common cheese-skipper (*Piophila casei*) is known to spring or skip. Dr. W. H. Mussey, of Cincinnati, in a communication to the Natural History Society of that city, December, 1875, states, in fact, that such is the case; though members of the California Academy who have written on the subject assert that the motion is made by the pupa, which I think very improbable. At all events the bounding motion is great, as the little gall may be thrown 2 or 3 inches from the earth; and there are few things more curious than to witness, as I have done, a large number of these tiny galls in constant motion under a tree. They cause a noise upon the fallen leaves that may be likened to the pattering of rain.

NOTE ON CLUSTER FLIES.

By W. H. DALL.

Having heard several years ago of a fly which was a great nuisance in the country houses near Geneva, N. Y., among members of my wife's family living there, I requested information and specimens when it should be convenient. Some time since a relative visited Geneva, and on his return brought me some of these flies alive in a bottle covered with gauze, which were exhibited at the last meeting of the Biological Society and turned over to Prof. C. V. Riley for identification. Since then a letter has been received, from which I make the following extracts:

"It is probably thirty years since the flies appeared in our neighborhood. I remember little about it except that they were at once a terror to all neat housekeepers, and from their peculiar habits a constant surprise. People soon learned to look for them everywhere; in beds, in pillow slips, under table covers, behind pictures, in wardrobes nestled in bonnets and hats, under the edge of carpets, and in all possible and impossible places. A window casing solidly nailed on will, when removed, show a solid line of them from top to bottom; they are uncanny. They like new houses, but are often found swarming in old unused buildings and go regularly to church, or perhaps only a few good ones abide in sanctuaries; any way they are there. Best of all they like a clean dark chamber seldom used, and if not disturbed form in large clusters about the ceilings. With them are usually found a number of purplish black hornets and some ladybugs (Coccinella). They are very cold and feel in the hand like small bits of ice. They are very oily; if crushed, leave on the floor a great grease-spot. I hardly think they breed in the houses, but do not know. About the 1st of April or as soon as the sun shines warm in the early spring they come out in the grass and fly up to the sunny side of the houses. Some possibly creep in open windows, or if the house is closed and sealed they have a faculty of going through any crack. They remain until some time in May, then disappear, and no more are seen until about September, when they come and remain as long as they are allowed to. They are very strong. A powder that suffocates common house flies has very little effect on them, and we attack them with ammonia and drown them with boiling water; even then are not sure they are 'kilt entirely.' Very few are found in the towns or villages; they live in country places altogether.

Words fail to describe their general depravity; it is beyond expression. If you wish to be happy, be sure you don't introduce cluster flies into your family."

The flies are also stated to be very sluggish—crawl rather than fly away when disturbed; hang from the cornice of a room in large clusters, like swarming bees, which can be brushed bodily into a vessel of boiling water; under buildings between earth and floor they are often found in incredible numbers; crawl in quiet, dark rooms between the sheets and under the pillows and vallances of made-up beds, and under the nailed-down edges of carpets, leaving nasty spots and a disagreeable smell wherever they go. If windows and blinds are opened and the room is occupied, they quietly vacate the premises in a little while unless they can crawl into some closet or wardrobe. There are in general appearance very like the common house-fly, but heavier-bodied, somewhat larger and more hairy—in short, coarser-looking.

Professor Riley writes as follows:

"So far as I have been able to investigate the matter your fly is the *Musca familiaris* of Harris (Ent. Corresp., p. 336), synonymous, without much question, with the *Musca rudis* Fabricius. It is not uncommon in New England in houses, nearly disappearing when *M. domestica* most prevails and found most in spring and fall. But I find no account of its abundance and annoyance in the manner you describe. It belongs to the genus *Pollenia* Robineau-Desvoidy."

"CLUSTER FLIES."

The fly presented by Mr. Dall at a previous meeting is the *Musca rudis* of Fabricius, a species known to be common to Europe and America * and redescribed, as *Musca familiaris*, in this country by Harris,† who says of it: "This species, not uncommon in houses in summer, nearly disappears the more abundant *M. harpyia* prevails. It resembles *M. rudis* Fabr., but is larger than the only specimen I have seen, and has the thorax much more densely clothed with fulvous hairs. From *M. harpyia* [*M. domestica* C. V. R.] it differs in the superior size, in having the eyes contiguous in the male, in the prominence of the front, in the hairiness of the thorax, etc. *M. obscura* of Fabricius is also synonymous, according to Meizers, who says of it (vol. V, p. 66) "Ein altes verwischtes Exemplar von *M. rudis*." It belongs to the genus *Pollenia* of Robineau Desvoidy, who made it, in fact, the type of that genus. This author in his "Histoire des Diptères des Environs de Paris," (vol. II, p. 600), mentions about 40 species of Pollenia, and says of *rudis*: "It becomes very common in autumn, and the first frosts compel it to take possession of our apartments. It here accumulates in numbers in the embrasures of windows and in the recesses of walls;

---

* Cf. Loew's note on this subject in his Ueber die Dipterenfauna des Bernstein's (translation in Sill. Journ. Sc. & Arts., vol. xxxvii, 2d ser., p. 318).

† Entomol. Corresp. of T. W. Harris, p. 336.

it then seems almost deprived of motion." He acknowledges that his *P. autumnalis* is also a synonym of *rudis.*

It will be seen from these facts that the species is not easily identified. This is accounted for on several grounds : 1st, the flies when they have frequented pollen-bearing flowers present a much brighter, yellowish appearance; 2d, the tufts of hairs which characterize it are very easily rubbed off; 3d, most of the insects of the family, as well as other Diptera of allied families, have a great tendency to grease, *i. e.*, they soon acquire in the cabinet a greasy, dark-colored aspect in which the characteristic markings are obliterated.

The genus, which is numerous in species and individuals, is chiefly distinguished by the bulging middle face; by the base of the antennæ being generally fulvous in color, and by the tufts of hairs at the sides of the thorax, to which last character the generic name alludes. The old genus *Musca* has been subdivided into numerous genera founded, as in this instance, on rather trifling characters, so that it becomes very difficult to separate some of them or to properly refer the species to them.

There are two authentic specimens of *Pollenia rudis*, determined by Baron Osten Sacken in my cabinet now in the National Museum, so that there can be no question as to the species.

In reference to the habits of the species it will be seen that what I have quoted from other authors corresponds very well with the facts as communicated by Mr. Dall, though I find no mention of any such unusual swarming in houses and working under bed clothing as communicated by his correspondent. The species is not infrequent in the fall of the year in houses in Washington and is readily distinguished from the common house-fly, even by an ordinary observer, by its larger size and more sluggish movements. The specimens submitted by Dr. Baker and received from Maine are specifically identical. Dr. S. W. Williston, of New Haven, Conn., writes me that he thinks he observed it in numbers clumsily crawling on the snow during mild weather in February and March.

Nothing definite is recorded of the larval habits and development of the species, though, speaking of the genus, Robineau-Desvoidy remarks that the eggs are laid in manure and in decomposing animal and vegetable matter. The larva doubtless lives in such decomposing substances.

It is not improbable that in parts of New York the species may have acquired more troublesome habits than it has elsewhere, for among the Diptera we have such instances of peculiar and injurious habits being locally developed, as in *Trypeta pomonella* Walsh, which in the West confines its work to the wild crabs and haws, while in the Eastern States it proves injurious to cultivated apples. *Lucilia macellaria* is a grievous pest in the Southwest, producing the well-known screw-worm so injurious to stock, whereas in the more northern States we never hear of such injury.

A REVIEW OF THE GENUS NOTURUS, WITH A DESCRIPTION OF ONE NEW SPECIES.

BY JOSEPH SWAIN AND GEORGE B. KALB.

The present paper contains the synonymy and diagnostic characters of the species of the genus *Noturus* Rafinesque, with short descriptions of those species in the accounts of which we have found confusion.

The material on which this paper is based is partly the same which served for Professor Jordan's account of the genus (Bull. U. S. Nat. Mus. X, pp. 96 to 103, 1877), with the addition of numerous specimens since received by him from different parts of the United States. Among these we find a single species which appears to be new to science.

We desire to express our indebtedness to Professor Jordan for the use of his collections and library, and for valuable suggestions.

ANALYSIS OF SPECIES OF NOTURUS.

* Premaxillary band of teeth with lateral backward processes. (*Noturus.*)

  a. Pectoral spine retrorsely serrate in front, almost or quite entire behind, its length 2 in head; body elongate; head about 4 in length; width of head about equal to depth of body, 5¼ in length; distance from snout to dorsal 3 in length; maxillary barbel reaching nearly to gill-openings; adipose fin deeply notched; anal rays 16. Color nearly plain yellowish brown, "in northern specimens blackish above, slightly mottled"; size large, reaching a length of more than a foot............................FLAVUS, 1.

** Premaxillary band of teeth without lateral backward processes. (*Schilbeodes* Bleeker.)

  b. Pectoral spine more or less dentate behind, serrate or somewhat roughened in front.

    c. Pectoral spine with posterior serræ weak, their length less than half the diameter of the spine; coloration almost uniform, margin of fins more or less edged with black.

      d. Pectoral spine very short and weak, about 3⅓ in head; adipose fin moderate, slightly notched; body robust; maxillary barbel short, 2 in head; jaws subequal; head not greatly depressed; anal rays 16.

                    ELASSOCHIR, 2.

      dd. Pectoral spine about 2 (1¾ to 2¼) in head; body elongate, especially in adult; head flat and thin; upper jaw more or less projecting; anal rays 14 to 16...INSIGNIS, 3.

    cc. Pectoral spine with its posterior serræ strong, spine-like, recurved, each little if any shorter than diameter of spine.

      e. Adipose fin large, deeply notched, but connected with caudal; pectoral and dorsal spines very strong; coloration much variegated with black and grayish; top of head, tip of dorsal, middle of adipose fin, and caudal black, with four broad cross blotches, one before dorsal, one behind it, one across adipose fin, and a small one behind it: anal rays 12 or 13.

                    MIURUS, 4.

      ee. Adipose fin entirely distinct from caudal fin, separated from it by a distance equal to the diameter of the eye; spines as in *miurus*, but somewhat weaker; head broad and flat; [anal rays, 11]; coloration little mottled. ....ELEUTHERUS, 5.

bb. Pectoral spine entire or grooved behind, never retrorse-serrate ; adipose fin
not at all or scarcely notched.

f. Head small and narrow, longer than broad ; its length about 4 in body,
its width 5¼ ; upper jaw projecting ; spines very short and weak, that
of the dorsal one-third height of fin ; pectoral spine slightly retrorse-
serrate without, rather obscurely grooved within, its length 3½ in head ;
anal rays, 14 ; color yellowish, rather dusky on head ; somewhat mottled ....LEPTACANTHUS, 6.

ff. Head short, broad, and deep ; pectoral spine without serration in front,
grooved behind, its length 2 in head ; jaws subequal ; barbels short,
maxillary barbel more than 2 in head ; color yellowish-brown, more or
less dusky, never blotched ; a narrow black lateral streak, and often
two dorsal ones ; anal rays, 15 or 16...................GYRINUS, 7.

1. Noturus flavus Rafinesque.

Noturus flavus Rafinesque, Am. Month. Mag. and Critical Rev., 41, 1818 ; Rafi-
nesque, Ich. Ohiensis, 68, 1820 (Falls of Ohio) ; Kirtland, Bost. Jour. Nat.
Hist., V, 336, 1846 (Mahoning River and Lake Erie) ; Storer, Syn., 406,
1846 (copied) ; Gill, Proc. Bost. Soc. Nat. Hist., 45, 1862 ; Cope, Journ.
Acad. Nat. Sci. Phila., 236, 1869 ; Günther, Cat. Fishes, V, 184 (Ohio) ;?
Uhler and Lugger, Fishes of Maryland, 151, 1876 (Potomac and Patapsco
Rivers) ; Jordan, Ann. N. Y. Acad. Sci., vol. I, No. 4, 1877, 118 ; Jordan, Ann.
Lyc. Nat. Hist. N. Y., 372, 1877 (Platte River to Saint Lawrence River,
Ohio Valley and N.E.) ; Jordan, Bull. Nat. Mus., X, 97, 99, 1877 (Vermont
and Canada to Va., Ohio Valley, and Missouri region) ; Jordan, Rept. on
Fishes of Ohio, 799, 1882. (Canada to Va., Mo., and Mou.) ; Jordan, Man.
Vert., 335, 1878 (St. Lawrence to Ky. and Upper Mo.) ; Jordan, Bull. Ills.
Lab. Nat. Hist., No. 2, 67, 1878 ; Jordan and Gilbert, Syn. Fishes N. A.,
100, 1883 (Ver. to. Va., Neb., and Tex.).

Noturus luteus Rafinesque, Jour. de Physique, 421, 1819.

Noturus occidentalis Gill, Proc. Bost. Acad. Nat. Hist., 45, 1862 ; Gill, Ichth.
Capt. Simpson's Rept., 423, 1876 (Platte River) ; Jordan and Copeland,
Check List, 160, 1876 (Platte River ; no description).

Noturus platycephalus Günther, Cat. Fishes, V, 104, 1864 (N. A.) ; Jordan and
Copeland, Check List, 160, 1876 (no description).

Habitat.—Vermont to Virginia and westward to Nebraska ; Lake
Erie ; Saint Lawrence. Ohio, Mahoning, Potomac, Patapsco, White,
Platte, and Missouri Rivers ; Swartz Creek, Michigan.

2. Noturus elassochir, *sp. nov.*

Habitat.—Illinois River.

Head, 4 ; depth, 5⅗ ; width of head, 5. D. I, 6 ; A. 16.

Body robust, somewhat elevated in the dorsal region. Head moderate. Mouth large ; its width 1⅔ in head. Interorbital space about equal
to the length of snout, which is 3 in head. Jaws subequal. Maxillary
barbel rather short, 2⅓ in head. Dorsal spine rather weak, but nearly
as long as the very short pectoral spine, which is 3⅓ in head. The serration is very similar to that of *insignis*, but much less distinct in front.
Humeral process obscure. Adipose fin long, low, with emargination
well defined ; caudal and anal fins large.

Color, in spirits, dark brown; vertical fins edged with darker; lateral line dark.

This species seems to be distinguished from *insignis* chiefly by the much shorter spines, by a slight difference in coloration, and by the more robust body. This species is described from a single specimen, 4¾ inches long, taken at Napierville, Ills., by Dr. Ernest R. Copeland. The type (No. 29677, U. S. Nat. Mus.) has been presented by Professor Jordan to the United States National Museum.

3. Noturus insignis (Richardson) Gill and Jordan.

> *Pimelodon livrée* Le Sueur, Mém. du Mus., V, 155, 1819.
> *Pimelodus insigne* Richardson, Fauna Boreali Americana, III, 132, 1836. (Name only; based on Le Sueur's description.)"
> *Noturus insignis,* Jordan, Bull. U. S. Nat. Mus. X, 97 and 100, 1877 (Penn. to S. C.); Jordan and Brayton, Bull. U. S. Nat. Mus. XII, 87, 1878 (Ohio and the Rivers James, Great Pedee, and Santee); Jordan, Bull. U. S. Geol. Sur., 414, 1878 (no description); Jordan, Man. Vert., 335, 1878 (Penn. to S. C.); Bean, Proc. U. S. Nat. Mus. 112, 1880 (James River, Potomac River, and Bainbridge, Pa.); Jordan and Gilbert, Syn. Fishes N. A., 100, 1883 (Penn. to Ga.)
> *Pimelodus lemniscatus* Cuv. and Val., Hist. Nat. Poiss., XV, 144, 1840; Storer Syn. Fishes, 405, 1846 (copied).
> *Noturus lemniscatus* Girard, Proc. Acad. Nat. Sci., 159, 1859 (no description); Gill, Proc. Bost. Soc. Nat. Hist., 45, 1862; Günther. Cat. Fishes Brit. Mus., V, 104, 1864 (N. A.); Jordan, Man. Vert., 303, 1876; Jordan and Copeland, Check List, 160, 1876 (no description).
> *Noturus occidentalis* Günther, Cat. Fishes, V, 105, 1864 (Platte River). (Not of Gill.)
> *Noturus marginatus* "Baird, MSS."; Cope, Jour. Acad. Nat. Sci. Phila., 237, 1869: Cope, Proc. Am. Philos. Soc., 484, 1870 (Catawba and Yadkin Rivers; no description); Jordan and Copeland, Check List, 160, 1876, (Ohio Valley to N. C.); Jordan, Ann. Lyc. Nat. Hist., XI, 372, 1877 (Ohio to Penn. and N. C.)
> *Noturus exilis* Nelson, Bull, Ills. Mus. Nat. Hist., 51, 1876 (Ills. River); Jordan and Copeland, Check List, 160, 1876 (Ills. and Wis.); Jordan, Ann. Lyc. Nat. Hist., Vol. XI, 372 (Ills. and Wis.); Jordan and Brayton, Bull. Nat. Mus. XII, 87, 1878 (Ohio and Ills.; no description); Jordan, Man. Vert., 335, 1878 (Ills. to Kan.); Jordan, Cat. Fishes, Ills. 67, 1878 (Ills., Wis., and Kan.); Bean, Proc. U. S. Nat. Mus., 112, 1880 (South Grand River, Mo.); Jordan and Gilbert, Syn. Fishes N. A., 100, 1883 (Wis. to Mo. and Kan.).

Habitat.—Pennsylvania to Georgia and westward to Nebraska; Delaware, Susquehanna, James, Great Pedee, Santee, Catawba, Yadkin, Saluda, Ohio, Illinois, South Grand, and Platte Rivers.

Head 3¾ to 4¼ in length; depth 5 to 6½ in length; width of head 4½ to 6 in length. D. I, 7; A. 14 to 17.

Body elongate in adult; head flat and depressed. Pectoral spine with well-developed serrations in front, posterior serræ weak, their length less than half the diameter of the spine, which varies from 1¾ to 2¼ in head. Upper jaw usually projecting. Dorsal spine about half the height of fin, 3 to 3½ in head. Adipose fin with slight notch. Maxillary barbel extending about to gill-openings. Coloration in spirits almost uniformly yellowish-brown; vertical fins yellowish usually, with a darker margin.

This description includes *N. exilis* Nelson, which is, in our opinion, not a distinct species.

4. Noturus miurus Jordan.

> *Noturus miurus* Jordan, Ann. Lyc. Nat. Hist. N. Y., 371, 1877 (Ohio Valley and S.W.); Jordan, Bull. U. S. Nat. Mus., X, 98 and 100, 1878 (Great Lakes, Ohio Valley, to Wis. and La.); Jordan, Rept. on Fishes of Ohio, 800, 1882. (Ohio Valley, Great Lakes, and southward to La.); Jordan, Ann. N.Y. Acad. Sci., 119, 1877 (White R., Wabash R., Ohio R., Tangipahoa R.); Jordan, Man. Vert., 336, 1878 (Ohio to Iowa and La.); Jordan, Bull. Ills. Lab. Nat. Hist., 68, 1878; Jordan and Gilbert, Syn. Fishes N. A., 99, 1883 (Great Lakes to Minn. and La.).
>
> *Noturus eleutherus* Jordan, Bull. U. S. Nat. Mus., X, 101, 1878 (foot-note: not text; Tar River, N. C.) (not type); Jordan and Gilbert, Proc. U. S. Nat. Mus., 363, 1878 (Neuse River; no description); Jordan and Gilbert, Syn. Fishes N. A., 99, 1883 (rivers of N. C. and E. Tenn.)

Habitat.—Great Lakes to Minnesota and Louisiana; Great Lakes; Ohio, Wabash, White, Blue, Tar, Neuse, Tangipahoa Rivers.

5. Noturus eleutherus Jordan.

> *Noturus miurus* Jordan and Copeland, Check List, 160, 1876 (French Broad; no description).
>
> *Noturus eleutherus* Jordan, Ann. Lyc. Nat. Hist. N. Y., 372, 1877 (French Broad River, Tenn.); Jordan, Bull. U. S. Nat. Mus., X, 101, 1877 (French Broad River); Jordan, Man. Vert.. 336, 1878 (French Broad River).

Habitat.—French Broad River, Tennessee.

Head, 3¾; depth, 5½; width of head, 4⅓. D. I, 7; A. 11.

Body rather robust, not elevated in the dorsal region. Mouth moderate, upper jaw much projecting. Interorbital space slightly convex, 3 in head. Maxillary barbel reaches about to gill-openings. Dorsal spine 2¾ in head; pectoral spine 1⅘ in head, with six large recurved teeth on inner edge, whose length is about equal to the diameter of the spine; outer edge obscurely serrated. Humeral process indistinct. Adipose fin low, distinctly separated from the caudal. Anal short and elevated.

Color, in spirits, brownish, dark above, becoming lighter behind and below; everywhere punctulate, except on belly.

This species is here described from the original type (No. 29678, U. S. Nat. Mus.), 3⅓ inches long. It was taken alive in the mouth of a water snake, by Professors Jordan and Gilbert, in the Big Pigeon River, a tributary of the French Broad, at Clifton, Tenn. It is the only representative of the species known, and may be an abnormal specimen of *N. miurus*. The specimens from Tar River, North Carolina, afterwards referred to this species by Professor Jordan, are large examples of *N. miurus*.

6. Noturus leptacanthus Jordan.

Noturus leptacanthus "Jordan, MSS, 1876"; Jordan and Copeland, Check List, 160, 1876 (Alabama River, no description); Jordan, Ann. Lyc. Nat. Hist. N. Y., vol. XI, 372, 1877 (Alabama River); Jordan, Ann. N. Y. Acad. Sci., vol. I, No. 4, 1877, 119 (Etowah River); Jordan, Man. Vert., 336, 1878 (Alabama and other Southern rivers); Jordan and Brayton, Bull. U. S. Nat. Mus., XII, 55 and 87,1878 (Alabama and Chattahoochee Rivers); ?*Hay, Proc. U. S. Nat. Mus., 514 and 515, 1880 (Enterprise, Miss.); Jordan and Gilbert, Syn. Fishes N. A., 98, 1883 (Ga. to Miss.)

Habitat :—Alabama, Etowah and Chattahoochee Rivers; ? Enterprise, Miss.

The specimens obtained by Professor Hay may perhaps represent a distinct species.

7. Noturus gyrinus (Mitchill) Rafinesque.

Silurus gyrinus Mitchill, Am. Mon. Mag., March, 322, 1818; DeKay, Fishes of N. Y., 186, 1842 (Walkill R.; copied.)

Noturus gyrinus Rafinesque, Jour. de Physique, 421, 1819; Ich. Oh., 68, 1820; Gill, Proc. Bost. Soc. Nat. Hist., 45, 1862; Cope, Jour. Acad. Nat. Sci. Phila., 237, 1869; Jordan and Copeland, Check List, 160, 1876 (E. Penn. and S. E. N. Y.; no description); Jordan, Ann. Lyc. Nat. Hist., vol. XI, 371, 1877 (Southern N. Y. to Penn); Jordan, Bull. Nat. Mus. No. 10, 102, 1877 (S. N. Y. to Penn.); Jordan, Bull. U. S. Geol. Sur., 414, 1878 (S. N. Y. and Penn.; no description); Jordan, Man,Vert., 337, 1878 (S. E. N. Y. and E. Penn. and N. J.); Bean, Proc. U. S Nat. Mus., 112, 1880 (near Piermont, N. Y.); Hay, Proc. U. S. Nat. Mus., 514, 1880 (Macon, Miss.); Jordan and Gilbert, Syn. Fishes N. A., 98, 1883 (N. Y., entire Miss. Valley, and Upper Lake Region.)

Schilbeodes gyrinus Bleeker, Ich. Arch. Ind. Prod., vol. I, 258, 1858.

Noturus flavus Jordan, Man. Vert., 303, 1876 (in part); Nelson, Bull. Ills. Mus, Nat. Hist., 50, 1876; Jordan, Proc. Acad. Nat. Sci. Phila., 46, 1877.

Noturus sialis Jordan, Bull. Nat. Mus. No. 10, 102, 1877 (Miss. Valley, Great Lake Region, and Red River of the North); Jordan, Bull. U. S. Geol. Sur., 414, 1878 (no description); Jordan, Man. Vert., 337, 1876; Jordan· Rept. on Fishes of Ohio, 801, 1882 (Miss. Valley to Red River of North); Jordan, Bull. Ills. Lab. Nat. Hist., No. 2, 68, 1878 (no description.)

Habitat.—New York and Pennsylvania and westward; Red River of the North; Walkill, Hudson, Chemung Ohio, and White Rivers. Pearl R. Miss.

---

* "The head is small and narrow, widening gradually from the narrow snout to the shoulders; the lateral outline of the head therefore straight; its length 4⅓ times in the body. Upper jaw projecting, *spines rather long and slender*, instead of being short as in the type, the pectoral spine being one-half the length of the head. *The color is quite dark.*" (*Hay.*)

Table of measurements.

Name of species	N. flavus	N. elassochir. (29677)	N. insignis	N. miurus	N. eleutherus. (29678)	N. leptacanthus	N. gyrinus
Locality	Ohio River.	Illinois River.	Susquehanna River.	Tar River.	French Broad.	Etowah River.	Illinois River.
	Inches and 100ths.	Inches and 100ths.	Inches and 100ths.	Inches and 100ths.	Inches and 100ths.	Inches and 100ths.	Inches and 100ths.
Extreme length	4.90	4.60	4.75	4.25	2.95	2.66	3.15
Length to base of caudal	4.20	4.05	4.00	3.55	2.60	2.30	2.60
(100ths of length)							
Body: Greatest height	19½	17	15	15	17	23	24
Head: Greatest length	25	25	27	30	28	26	28
Greatest width of orbital area	21	19	20	24	23	18	26
Width of snt	6	7	6	7	7	7	13
Length of snt	7	9	9	10	9	7	10
Length of maxillary barbel	15	10	14	26	18	12	14
Diameter of eye	4	4	4	4	5	3	3¼
Dorsal: Distance from snout	33	34	37	38	35	37	35
Length of base	11	13	10	13	12¼	11	14
Greatest height	11	11¼	14	16	16	10	15
Height of spine	7	6	8	12	12	7	12
Anal: Distance from snout	65	61	63	73	68	63	71
Length of base	18	23	19¼	16	15	18	18
Height at longest ray	11¼	12	13	12	13	11	16
Caudal: Length of middle rays	18	16	19	18	20	21	21
Pectoral: Length of spine	12	8	15½	19	17	6	15
Ventral: Distance from snout	49	46	46	51	46	47	48
Length	11	10	11	12	14	10	15
Dorsal, number of rays	I-7	I-6	I-7	I-6	I-6	I-6	I-7
Anal, number of rays	16	16	16	13	11	14	15

LIST OF NOMINAL SPECIES WITH IDENTIFICATIONS.

Noturus flavus Rafinesque	1818	Noturus flavus.
Silurus gyrinus Mitchill	1818	Noturus gyrinus.
Noturus luteus Rafinesque	1819	Noturus flavus.
Pimelodus insigne Richardson	1836	Noturus insignis.
Pimelodus lemniscatus C. & V	1840	Do.
Noturus occidentalis Gill	1862	Noturus flavus.
Noturus platycephalus Günther	1864	Do.
Noturus marginatus Baird	1869	Noturus insignis.
Noturus exilis Nelson	1876	Do.
Noturus eleutherus Jordan	1876	Noturus eleutherus.
Noturus leptacanthus Jordan	1876	Noturus leptacanthus.
Noturus miurus Jordan	1877	Noturus miurus.
Noturus sialis Jordan	1877	Noturus gyrinus.
Noturus elassochir Swain & Kalb	1882	Noturus elassochir.

INDIANA UNIVERSITY, *November 29, 1882.*

CATALOGUE OF A COLLECTION OF SAMPLES OF RAW COTTON PRESENTED TO THE UNITED STATES NATIONAL MUSEUM BY THE INTERNATIONAL COTTON EXPOSITION, ATLANTA, GEORGIA, 1881.

By S. M. INMAN.

LETTER OF TRANSMITTAL.

INTERNATIONAL COTTON EXPOSITION,
OFFICE OF THE TREASURER,
Atlanta, Ga., April 26, 1882.

DEAR SIR: Acting under instructions from the Executive Committee of the International Cotton Exposition, I take pleasure in forwarding to the Smithsonian Institution a collection of samples of the foreign cotton recently on exhibition.

*　　　*　　　*

Very truly yours,

S. M. INMAN,
Treasurer.

Prof. SPENCER F. BAIRD,
Washington, D. C.

CATALOGUE.

U. S. N. M. No.	Locality.	Grade on American standard.	U. S. N. M. No.	Locality.	Grade on American standard.
56, 093	The West Indies	Good, fair.	56, 111	Egypt	White, fair.
56, 094	Maceios, Brazil	Fair.	56, 112	......do	White, good, fair.
56, 095	......do	Good, fair.	56, 113	......do	White, good.
56, 096	Maranham, Brazil	Fair.	56, 114	......do	White, fine.
56, 097	......do	Good, fair.	56, 115	......do	Brown, fair.
56, 098	Pernambuco, Brazil	Fair.	56, 116	......do	Brown, good.
56, 099	......do	Good, fair.	56, 117	......do	Brown, fine.
56, 100	Paraiba, Brazil	Fair.	56, 118	Gallini, Egypt	Fair.
56, 101	Rio Grande, Bolivia	Fair.	56, 119	......do	Good, fair.
56, 102	Peru	Red.	56, 120	......do	Good.
56, 103	......do	Fair, rough. ?	56, 121	......do	Fine.
56, 104	......do	Good, fair, rough.	56, 122	Lagos, Africa	
56, 105	......do	Good, rough.	56, 123	Masandaran, Persia.	
56, 106	......do	Fine, rough.	56, 124	Kaukasus, Persia...	
56, 107	......do	Extra quality, rough.	56, 125	Taschkend, Persia..	
			56, 126	Bucharia, Persia....	
56, 108	......do	Fair, smooth. ?	56, 127	Bengal, India	Fair.
56, 109	......do	Good, fair, smooth.	56, 128	......do	Good, fair.
56, 110	......do	Good, good, smooth.	56, 129	......do	Good.

CATALOGUE—Continued.

U. S. N. M. No.	Locality.	Grade on American standard.	U. S. N. M. No.	Locality.	Grade on American standard.
56,130	Bengal, India........	Fine.	56,145	Rangoon, India.....	Fair.
56,131	Western India	Fair.	56,146	......do	Good, fair.
56,132	..... do	Good, fair.	56,147	Coimbatoor or Salem,	Good, fair.
56,133	......do	Good.		India.	
56,134	Tinnevelly, India...	Fair.	56,148	Hinghenghaut,India	Good.
56,135	......do	Good, fair.	56,149	Dacca, Bengal.India	Fine.
56,136	......do	Good.	56,150	Coconado, India	Fair, red.
56,137	Dhollerah, India....	Good.	56,151	Dharwar, India......	Good, saw-ginned.
56,138	......do	Fine.	50,152	Siam	Unginned.
56,139	Comrawuttee, India.	Good, fair.	56,153	China	Unginned.
56,140	......do	Good.	56,154	......do	Good, fair.
56,141	Scinde, India	Good, fair.	56,155	......do	Good.
56,142	......do	Good.	56,156	Nanking, China.....	Good.
56,143	Brooch, India........	Good, machine-ginned.	56,157	Fiji Islands	Rough stapled.
			56,158	..... do	Long stapled.
56,144	......do	Fine, machine-ginned.	56,159	Tahiti, Society Islands.	Fair.

DESCRIPTION OF TWO NEW SPECIES OF FISHES (MYROPHIS VAFER AND CHLOROSCOMBRUS ORQUETA) FROM PANAMA.

By DAVID S. JORDAN AND CHARLES H. GILBERT.

1. Myrophis vafer sp. nov. (29681.)

(*Myrophis punctatus* Günther VIII, 1870, 50. Jor. & Gilb., Bull. U. S. Fish Comm., 1882, 109: name only; not of Lütken.)

Body subterete anteriorly, compressed posteriorly, more robust and less vermiform than in *Myrophis lumbricus;* tail strongly compressed. Head comparatively large, its width posteriorly being greater than that of the body. Upper jaw considerably projecting; both jaws rather blunt. Eye moderate, considerably nearer angle of mouth than tip of snout, its diameter rather less than half snout. Gape rather long, about $3\frac{1}{3}$ in head. Head constricted behind the cheeks.

Teeth small, sharp, slender, hooked backward, apparently in one or two irregular series in each jaw, and a single long series, somewhat broken anteriorly, on the vomer.

Anterior nostrils with small tubes, posterior nostrils without tubes. Gill-openings moderate, oblique, placed in front of and below the bases of the pectorals, which are rather broader than the gill-openings.

Pectoral fins small, acute at tip, their length a little more than that of snout, and one-fifth to one-sixth that of head. Dorsal fin of moderate height posteriorly, its origin somewhat nearer gill-opening than vent, its distance from gill-opening a little more than length of head.

Head $8\frac{1}{4}$ in total length. Greatest depth of body about 28. Length of head and trunk $2\frac{1}{2}$ in total. Head $2\frac{1}{3}$ in trunk.

Color in life, light olivaceous; silvery on breast and belly; back and sides thickly dotted with fine, olive-brown specks. Snout somewhat dusky.

This species is very common in the rock-pools at Panama, where numerous examples (the types numbered 29681 U. S. Nat. Mus. register) were obtained by Mr. Gilbert, the largest $7\frac{1}{2}$ inches in length.

This species has been already noticed under the name of *Myrophis punctatus.* There is, however, no positive evidence that it is identical with the African species, so named by Professor Lütken. The Texan species of *Myrophis* (*lumbricus* J. & G.) and the Cuban species (*microstigmius* Poey) seem to be distinct from it.

Chloroscombrus orqueta sp. nov. (29165, 29278, 29285, 29343.)

 Micropteryx chrysurus Steind., Ichth. Beit. III, 61.
 Chloroscombrus chrysurus Jor. & Gilb., Bull. U. S. Fish. Comm. 1882, 110.

Proportions, fin rays, and coloration essentially as in *C. chrysurus*, which species it represents in the Pacific. From the Atlantic form, *C. orqueta* differs constantly in the much longer curve of the lateral line, and in the distinct armature along caudal peduncle.

Body ovate, strongly compressed, the edges trenchant. Dorsal and ventral outlines very regularly curved, the curve of the belly considerably stronger than that of the back, the axis of body much nearer the latter. Caudal peduncle very slender. Young less elongate than the adult, otherwise very similar in form.

Head small, rather pointed, the anterior profile nearly straight. Mouth small, very oblique, the lower jaw projecting. Maxillary broad, extending a little beyond front of eye, its length $2\frac{3}{4}$ in head, its tipe marginate; supplemental bone well developed. Teeth very small, those of the jaws not villiform, forming very narrow bands or single series in both jaws; villiform patches on vomer, palatines, and tongue. Adipose eyelid well developed. Preorbital very narrow, not half width of orbit. Gill-rakers numerous, very long, slender, close set.

Head nearly naked. Body covered with well-developed imbricated scales; the ventral ridge, and a narrowly triangular area forwards from front of dorsal naked.

Lateral line with a rather strong arch anteriorly, the chord of the curve being considerably longer than head, and $1\frac{1}{2}$ to $1\frac{3}{4}$ in the straight part. (In *C. chrysurus* the chord of the curve is about as long as head from tip of lower jaw, and $1\frac{2}{3}$ to $1\frac{3}{4}$ in the straight portion.) A distinct keel along caudal peduncle, the scales of lateral line enlarged and bony with bluntish tips. (In *C. chrysurus* the scales of lateral line are little if at all different from the other scales.) Dorsal and anal naked, the sheaths at their bases largely developed along the anterior half of each fin.

Antrorse dorsal spine concealed. Spinous dorsal persistent, the spines slender, the longest slightly shorter than anterior rays of soft dorsal, which are about half head. Soft dorsal and anal with anterior rays highest, the fins not falcate. Caudal widely forked, the upper lobe slightly longer than the lower, which about equals head. Anal spines strong. Pectorals very long, falcate, a little more than one-third length. Ventrals short, about two-fifths head.

Color somewhat darker than in *C. chrysurus*. In life, back green with blue reflections; sides and below silvery-white with bluish and purplish reflections; a distinct black blotch on upper angle of opercle, extending on shoulder girdle; inside of opercle, and skin lining shoulder girdle below, largely dusky. A quadrate black blotch on back of tail, extending backwards along bases of upper caudal rays. Fins light yellowish, the dorsal and anal edged with black; tip of upper caudal lobe black. Ventrals whitish. Tongue, base and roof of mouth, and skin of upper branchiostegals black.

Abundant at Panama; recorded by Dr. Steindachner from Magdalena Bay, Lower California. The types, numbered 29165, 29278, 29285, and 29343, were collected by Mr. Gilbert at Panama. It is known to the fishermen at Panama as Orqueta.

INDIANA UNIVERSITY, *November* 27, 1882.

DESCRIPTION OF A NEW EEL (SIDERA CASTANEA) FROM MAZATLAN, MEXICO.

BY DAVID S. JORDAN AND CHARLES H. GILBERT.

Sidera castanea sp. nov.
> (*Sidera panamensis* Jor. & Gilb., Bull. U. S. Fish Comm. 1882, 106 ; name only ; not *Murœna panamensis* Steind.)

Tail about as long as rest of body, or slightly longer. Head 2½ in length of trunk; cleft of mouth wide, 2⅓ to 2½ in head; teeth everywhere uniserial or nearly so, those on sides of mandible small, compressed, close-set, subtriangular, directed backwards, about 18 in number on each side; mandible with about four large canines anteriorly; upper jaw with the teeth partly in two series, some of the teeth being movable, the others mostly stronger, canine-like, especially anteriorly. Front of vomer with two very long slender canines, behind them a single series of small teeth; teeth all entire.

Eye large, slightly nearer tip of snout than angle of mouth, its diameter 2 to 2½ in snout; gill-opening one-third wider than the orbit; tube of anterior nostril short, less than half diameter of orbit; posterior nostril without tube; occiput not especially elevated, the anterior profile scarcely concave (perfectly straight in young 2 feet long).

Dorsal fin commencing much in advance of gill-opening, becoming unusually high posteriorly, where its vertical height is more than half greatest depth of body; the length of the longest ray more than greatest depth of body.

Color light brownish-chestnut, slightly paler on abdomen; no spots or bands anywhere; fins without dark margins; no dark spot on gill-opening or at angle of mouth; no black about eye; head without conspicuous pores.

The specimen here described is 44 inches in length; others about 2 feet in length agree very closely.

Sidera castanea is very common among the rocks about Mazatlan, where it reaches the length of about 4 feet, and is known to fishermen as *Anguila prieta*. It has not yet been observed elsewhere. The types numbered 28246, 29535, 29591, were collected by Mr. Gilbert.

INDIANA UNIVERSITY, *November 27, 1882.*

ON THE NOMENCLATURE OF THE GENUS OPHICHTHYS.

BY DAVID S. JORDAN AND CHARLES H. GILBERT.

The generic name "*Ophichthys* Ahl" has been adopted by Dr. Günther for a large group of eels, including numerous nominal "genera" of earlier writers. Whether this vast group will admit of further generic subdivision, we do not here propose to discuss. It is, in any event, divisible into subgenera, and for these subordinate groups we should adopt names in accordance with accepted rules of nomenclature. It becomes, therefore, important to ascertain what species should be taken as the type of *Ophichthys*.

As the original memoir of Ahl is not, as far as we know, in any American library, we have written to Dr. Lütken, of the University of Copenhagen, in regard to it. His answer to this letter is the source of the information given below in regard to the memoir in question. The following is the title:

I.X.Θ.Υ.Σ.

Specimen ichthyologicum *de Muræna et Ophichtho* quod seria exp. fac. med. Ups. præsid. *Carol. Vet. Thunberg. æquite,* etc.

Modeste offert.

Jonas Nicol. Ahl. 27 Jan. 1789. Upsalia.

The genus *Muræna* is in this paper divided into two, as follows:

Muræna.	*Ophichthus.*
"Animal apodum, pinnis ventralibus pectoralibus nullis. Membrana branchiostiga 10-radiata connata. Apertura branchiarum remota lateralis solitario."	"Animal apodum, pinnis ventralibus nullis; membrana branchiostiga 10-radiata connata. Apertura branchiarum remota lateralis ante pinnis pectoralis."

The species enumerated are:

MURÆNA.*	OPHICHTHUS.
1. *M. helena* L.	1. *O. ophis* (L.)
2. *M. nebulosa.*	2. *O. serpens* (L.)
3. *M. picta.*	3. *O. cinereus.*
4. *M. annulata.*	4. *O. myrus* (L.)
5. *M. fasciata.*	5. *O. conger* (L.)
	6. *O. anguilla* (L.)

The first species mentioned under *Ophichthus* is noticed as follows

"*O. ophis :* Cauda apterygia, corpore tereti, maculato. Hab. in Europeo mari et Indico.

"Synonym: Muræna ophis Linn. S. N. p. 425.
    " Serpens marinus maculosus Will. app. p. 19.
    "Houttyn Natural Hist. 1. D. p. 87.
    "Bloch, p. II, p. 35, t. 154."

As to the authorship of this paper, Dr. Lütken observes : " It is ques. tionable whether the dissertation should be ascribed to the 'Præsid.', Thunberg or to Ahl; you will see that the contemporaneous Vahl speaks of it as being of Thunberg, without phrase, and it is ascribed to Thun- berg also by Engelmann. Until a late time, in the Swedish universities, the dissertations were written by the professor and only ' defended' by the students whose name they bear. Thus often you will find that a page by Retzius or Linnæus was distributed to ten or twenty students, a sheet to each, for being defended, bearing these different names on their titles, but being afterwards collected and put together under the name of the *real* author. In other instances they were written by the student, when he was able to do it, and it is only to be seen from the paper itself whether it belongs to the student or to the master. In this special case it appears from the proëmium that Ahl really pretends to be the author, although probably he was not."

This appears to be the earliest attempt at subdivision of the genus *Muræna*, the name *Ophichthus* being intended for all eels with pectoral fins. For some group of these eels it must, of course, be retained.

In the diagnosis of the first three species the phrase " Cauda aptery- gia " occurs, and it is to eels thus characterized, that the name (more correctly spelled *Ophichthys*) is restricted by Dr. Günther. One of the three species, *ophis, serpens* and *cinereus*, must then be taken as its type. Two writers, Bleeker and Poey, have attempted further to restrict the genus *Ophichthys*. By a misapprehension, unfortunate, but easily ex- plained, Poey has considered *Muræna annulata* as the type of *Ophich- thys*. As we have seen, this species is explicitly excluded by Ahl, and

---

* This restriction of the Linnæan genus *Muræna* to *M. helena* and its supposed con- geners must, of course, take precedence over the restriction made by Bloch & Schnei- der in 1801, wherein *M. anguilla* was retained as the type of *Muræna*, and *M. helena* referred to a new genus, *Gymnothorax*.

the name *Ophichthys* cannot be used for the group (*Pisodontophis* Kaup) to which it belongs. It seems to us proper, with Bleeker, to consider *Murœna ophis* as the type of *Ophichthys*. It is the first species mentioned by Ahl, and for that reason it has already been taken by Bleeker as the type. It is also the species which suggested the generic name.

It is, however, not evident what this Linnæan *ophis* may be. It is based on a species of Artedi, which in turn rests on descriptions of Willoughby and Ray. To Ahl it was apparently known chiefly from the figure and description of Bloch. In any event, all the evidence points to a species allied to *Ophichthys triserialis, ocellatus,* and the like, and for this group we would retain the name *Ophichthys*. It would then be nearly equivalent to the genus *Oxyodontichthys* of Poey, and would probably, even if viewed as a subgenus only, include the following nominal genera: *Murœnopsis* Le Sueur; *Centrurophis, Pœcilocephalus, Microdonophis, Cœcilophis, Herpetoichthys, Elapsopsis, Scytalophis,* and *Leptorhinophis* of Kaup, and most likely several others of the same author.

The genus *Ophisurus* of La Cépède was originally based on two species, *O. ophis* La C. (not of L.) and *O. serpens* L. The first restriction of the name seems to be that of Swainson, in 1839, who removes *O. serpens* as the type of *Leptognathus* Sw., leaving the name *Ophisurus* for *O. ophis* La C. and its allies. Later (1856) Dr. Kaup gave to the latter group the name of *Pisoodonophis,* and made *O. serpens* the type of *Ophisurus*. The earlier restriction must take precedence and the name *Ophisurus* (or *Ophiurus*) must be retained for the species with granular teeth, if they be separated from *Ophichthys* proper. Whether these species again admit of subdivision, as suggested by Poey, does not now concern us, as the American species are typical *Ophisuri*.

The genus *Cœcula* Vahl has been adopted (Syn. fish N. A. 358) by the present writers, instead of *Sphagebranchus* Bloch. This name *Cœcula* occurs in a memoir (for a copy of which we are indebted to the kindness of Dr. Lütken) in the "Skrivter af Naturhistorie-Selskabet" 3d Bind. 2d Hefte. 1794, pp. 149–156, entitled "Beskrivelse af en nye Fiske-Slægt, *Cœcula,* af M. Vahl."

In this memoir, reference is made to Thunberg's separation of *Murœna* L., into *Murœna* and *Ophichthys,* and the generic name *Cœcula* is proposed for two species. The one is described in full and figured under the name of *Cœcula pterygera*. The other is the Linnæan *Murœna cœca,* a species unknown to Vahl, which he renames *Cœcula apterygia.* The genus *Cœcula* is thus characterized:

"Corpus teretiusculum, alepidotum, Branchiarum apertura collaris, linearis, Pinnæ ventrales & caudales nullæ. Oculi minutissimi."

The species are especially distinguished as *pterygera,* "pinna dorsali analique," and *apterygia,* "pinnis nullis."

This *Cœcula pterygera* is the only species of the genus known to Vahl from autopsy, and also the one placed first by him in his genus. It has been already (Syn. Fish. N. A. 358) considered by us the type of the genus *Cœcula*, and this restriction should stand. It is not quite certain what species Vahl had, and his type is not now to be found in the museum of Copenhagen. It is thought by Günther that *C. pterygera* is identical with *Sphagebranchus polyophthalmus* (Bleeker) Kaup. *Sphage-branchus rostratus* Bloch, the type of *Sphagebranchus* Bloch (1795) is also uncertain. It is, however, evident that the two species are closely related, and that both belong to Günther's subgenus *Sphagebranchus* "group **A**." *Cœcula* must therefore take the place of *Sphagebranchus*, of the still later *Dalophis* Rafinesque, and of *Lamnostoma* and *Anguisurus* Kaup. The two species of *Cœcula* described from the United States coast are not genuine members of that group, as they have the dorsal large, beginning in front of middle of head, and the gill-openings vertical and lateral, not oblique and ventral as in *Cœcula*. The genus *Callechelys* Kaup apparently coincides with this type, and may be accepted as a generic or subgeneric name for them.

The species of " *Ophichthys*" Günther known from American waters, north of the Tropic of Cancer, may be grouped as follows:

   a. No trace of fins anywhere.
1. *Apterichthys selachops* Jor. & Gilb.
   b. A high dorsal and no other fins.
2. *Letharchus velifer* Goode & Bean.
   c. Dorsal beginning on front of head; pectorals obsolete or nearly so; teeth small; pointed.
3. *Callechelys scuticaris* (Goode & Bean).
4. *Callechelys teres* (Goode & Bean).
   d. Pectorals developed, usually small; teeth granular.
5. *Ophisurus acuminatus* Gronow. (=*longus* Poey.)
6. *Ophisurus xysturus* Jor. & Gilb.
   e. Pectorals large; teeth all pointed.
   f. Teeth of each series subequal.
7. *Ophichthys miurus* Jor. & Gilb.
8. *Ophichthys triserialis* (Kaup).
9. *Ophichthys ocellatus* (Le Sueur).
10. *Ophichthys macrurus* Poey.
11. *Ophichthys chrysops* Poey.
12. *Ophichthys zophochir* Jor. & Gilb.
   ff. Some of the teeth strong canines.
13. *Ophichthys (Mystriophis?) schneideri* Steindachner.

INDIANA UNIVERSITY, *November* 15, 1882.

ON THE LIFE COLORATION OF THE YOUNG OF POMACENTRUS RUBICUNDUS.

By ROSA SMITH.

Hitherto only the adult form of this species has been known, and its uniform deep scarlet coloration has been considered to form a marked contrast to the coloration of the other species of *Pomacentrus*. I have lately secured numerous young specimens, and find their coloration quite different from that of the adult, and in general similar to that of the other members of this genus.

The ground color is dusky scarlet, with numerous markings of an intensely bright blue, which occasionally changes to bluish green. Two series of elongate spots form a blue stripe on either side of the median line, between tip of snout and beginning of dorsal fin; a line of blue on superior margin of iris is followed posteriorly by an irregular series of blue spots above the lateral line (the individual spots not quite equaling diameter of iris); the last of these spots is larger than those which precede it, being two-thirds of the orbital diameter, and extends up on the base of the dorsal fin at the posterior third of the spinous portion; thence very small blue dots continue to the end of the dorsal fin, describing a curve which exactly outlines the extent to which scales cover the base of the articulate dorsal rays; a conspicuous blue spot or bar crosses top of caudal peduncle close to posterior insertion of dorsal fin. One or more small blue spots at base of caudal. The spine and first ray of ventral blue; spines and tips of anterior rays of anal blue; a nearly round blue spot on posterior part of anal near its base. Sides of body more or less dotted with blue, as are also the cheeks and opercles. Pectorals and caudal semi-transparent, plain reddish. Ground color of anal bright red. Dorsal fin dusky, with minute blue dots anterior to the markings mentioned. Abdomen and under surface of head lighter, immaculate.

This description is made from specimens 1⅝ to 2 inches in length. Specimens less than 1 inch long have the spinous dorsal almost wholly blue and all the markings larger, while an individual 3½ inches long shows the markings similarly placed but relatively smaller, and the ground color is more olivaceous.

The fin-rays are proportionately higher and the eye relatively larger than in the adult. The suborbital and preopercle are without serrations. The small opercular spine, unlike that of the mature form, is smooth and wholly without denticulations. The greatest depth of the body is about half the length.

These brilliant little fishes inhabit only large, deep rock pools, hiding under the sea-weed of ledges, and frequently swimming out into the open water of the pool. They are accompanied by the adult, the usual uniform scarlet color of which appears a distinct lusterless yellow in the water.

The specimens described were taken at La Jolla, near San Diego. They have been sent to the United States National Museum.

SAN DIEGO, CAL., *November* 6, 1882.

ON A CINNAMON BEAR FROM PENNSYLVANIA.

By FREDERICK W. TRUE.

(Read before the Biological Society of Washington, October 27, 1882.)

1. In April of the present year Professor Baird received notice through the kindness of Mr. George Thurber, of New York, that a bear of peculiar color, which was said to have been killed in Pennsylvania, was exposed for sale at the commission house of Messrs. E. & O. Ward.

Professor Baird immediately effected the purchase of the animal, and in due time it arrived at the National Museum. It proved to be a very beautiful specimen of the Cinnamon Bear (*Ursus americanus* Pallas, *cinnamoneus* Aud. & Bachm.), a male about two-thirds grown.

The particulars of the capture being desired, the Messrs. Ward addressed a letter to the hunter, Mr. Seely Bovier, and received a reply, of which the following is an extract:

"ALBA, PA., *April* 20, 1882.

"E. & O. WARD

"GENTLEMEN: Yours of 18th just received. I would say that the bear was killed by myself on April 12, in Lycoming County, Monet Township, in this State, on what is known as the South Mountain ranges. I have hunted and trapped all my life and have never seen anything like this animal. All who have seen him are in doubt as to what species of bear he is. During all last summer in the back settlement near which I killed him, several of the men, women, and children were followed after night by what they called a panther. He would come very close and make an awful noise; sometimes he would be seen about dark in the buck field. I told the men there were no panthers in the county; that it must be something else. Undoubtedly it was this bear which followed them. I never saw him until the day I killed him. He was the most ferocious of all the bears I have ever killed. You will find that one ball went through his liver; that seemed only to increase his rage, however, and I was forced to put one into his brains. The spots on his head where the hair is off evidence the violence with which he 'tore around' after he was wounded.

"Such are the facts about the bear. To any one wishing further particulars about him I will cheerfully give them.

"Please have some one examine him very closely and tell me what species he is.

"Yours respectfully,

"SEELY BOVIER."

2. The following table shows the dimensions of the mounted specimen:

Table of actual superficial measurements.

Cat. No. 13455. Locality: Monet Township, Pennsylvania.

Measurements.	Centimeters.	Hundredths.
Tip of nose to base of tail	135. 25	100. 0
Tip of nose to occiput	36. 00	27. 0
Tip of nose to anterior margin of ear	27. 50	20. 0
Tip of nose to anterior margin of eye	12. 50	9. 2
Tip of lower jaw to corner of mouth	10. 00	7. 4
Breadth of head between eyes	10. 70	7. 9
Length of eye opening	2. 00	1. 5
Height of ear	13. 50	9. 9
Length of fore legs below body	46 00	34. 0
Length of hind legs below body	43. 50	32. 2
Length of fore feet (including claws)	19. 50	14. 4
Length of hind feet (including claws)	20. 20	14. 9
Length of tail with hairs	16. 50	12. 2
Girth of body ½ between fore and hind legs	105. 00	77. 6
Girth of fore leg at carpus	25. 00	18. 4
Girth of hind leg at tarsus	25. 30	18. 7

3. The hair, which is fine, is of two kinds, the longer straight, lustrous at the tips; the shorter crenulate, dull. The crenulate hair is absent on the feet and tip of the tail. It is also scarcely discernible on the anterior part of the head. The straight hair everywhere except on the head and back of the ears is of a medium sable color, the outer third having a pure golden lustre. On the anterior part of the head and on the backs of the ears the color changes to buff. The darkest color is on the cheeks and feet.

The crenulate hair is of a uniform, black-walnut color.

The straight hair measures 7.5 centimeters on the back and shoulders, where it is longest. It overtops the crenulate hair by about 2.5 centimeters. Its length at other points is as follows:

Centimeters.

Middle of the back 7. 5
Tip of the ear 1. 0
Middle of the forehead 2. 0
Tip of tail 9. 0

The eyelashes are 1 centimeter in length. The claws are pale horn-color at the base, but darker on the exposed portion. The skin is white. The lips and nostrils are of a dull-reddish or purplish brown. The soles of the feet are sooty. The eyes are brown.

4. The skull of the bear, which was in a badly broken condition when received, yields the following measurements:

Table of measurements of the skull.

No. 13455. Locality: Monet Township, Pennsylvania.

Measurements.	Centimeters	Hundredths.
Greatest length...	28. 20	100. 0
Proximal end of intermaxillary to surface of occipital condyle.............	26. 70	91. 1
Greatest width.....	17. 10	60. 7
Greatest height, not including lower jaw...................................	11. 45	41. 0
Least distance between orbits	7. 00	24. 9
Distance between orbital processes	9. 20	32. 6
Nasal bones—		
Length..	7. 45	26. 4
Width of both distally...	2. 60	9. 2
Narrowest part of muzzle behind canine teeth	5. 20	18. 5
Front margin of super. alveolus to first molar	7. 00	24. 9
Front margin of super. alveolus to posterior margin of palate.............	14. 30	51. 0
Distance between outer edges of the outer incisors	2. 80	10. 0
Length of super. alveolus occupied by molars ....—........	5. 40	19. 1
Least distance between inner edges of molars opposite sides	4. 10	14. 6
Distance from front of super. alveolus to proximal end of nasals..........	12. 60	45. 0
Distance from front of super alveolus to front edge of orbit..............	10. 70	38. 0

5. A number of bears from different localities in North America have been described under the name Cinnamon Bear, or *Ursus cinnamoneus.* Various views have been held regarding their taxonomic relations to the Black Bear, *Ursus americanus,* and to each other, the value of which can be determined only when a considerable number of skulls and skins shall be brought together.

Among the varieties of the Black Bear mentioned by authors is one called the "Yellow Bear of Carolina." No description of this animal occurs anywhere, so far as I am aware, except in Griffith's Cuvier's Animal Kingdom. For sake of comparison it may not be amiss to quote what is said regarding it. It is as follows: "The Baron [Cuvier] also thinks that the Yellow Bear of Carolina is a variety of the same species. This is scientifically termed *Ursus lutreolus.* We shall not venture to assert in contradiction to the Baron that this bear forms a distinct species, but assuredly it is a very strongly marked variety. Major Smith took a sketch of one at New York; the specimen was semi-adult. He does not consider that there is sufficient proof of its being a distinct species. In the specimen drawn by the major there was a greater convexity of forehead and a sharper nose than in the Black Bear. This comparison was easily made, as the two animals were chained very near each other. The ears of the Yellow Bear stood more back, were not quite so large, and the physiognomy was very different.* Both were remarkably tame. Although the Yellow Bear cannot be affirmed to be specifically different, yet it is certain that there is a distinct race of the animals. They were formerly common in Virginia, and they are still abundant in Northwestern Louisiana, where they are called White Bears, and are said to feed chiefly on honey, on acorns of a large size, wild berries, &c.

* It must be remembered that this specimen of the *Ursus luteolus* was but *semi-adult.* P[idgeon].

"The *Cinnamon Bear* in the Tower appears to be of the same race as this Yellow Bear."*

Richardson, writing in 1829, alludes to a Cinnamon Bear as follows:

" The *Cinnamon Bear* of the fur traders is considered by the Indians to be an accidental variety of this species [*U. americanus*], and they are borne out in this opinion by the quality of the fur, which is equally fine with that of the Black Bear."†

Audubon and Bachman, in their " Quadrupeds of North America," make the following allusion to a Cinnamon Bear:

"The Cinnamon Bear, so far as we have been able to ascertain, is never found near the sea coast, nor even west of the Ohio Valley until you approach the Rocky Mountain chain, and it is apparently quite a different animal,"‡ and again "sparingly found in the fur countries west and north of the Missouri, extending to the barren grounds of the Northwest."§

Other Cinnamon Bears were described by Professor Baird in 1859, from the copper mines of the Gila River, New Mexico. Regarding the specimens which he had under observation, he says: "Although about the size of the common black bear, *Ursus americanus*, or a little smaller, yet four skulls of all ages before me, when compared with a corresponding series of seven of *Ursus americanus*, exhibit such characteristic differences as to authorize the conclusion that the species is distinct."‖ In spite of these remarks, however, he places an interrogation mark after the name "*Ursus cinnamoneus.*"¶

DESCRIPTION OF A NEW PETREL FROM ALASKA.

By ROBERT RIDGWAY,

Curator, Department of Birds, United States National Museum.

An interesting collection of birds lately received at the National Museum from Mr. William J. Fisher, U. S. Tidal Observer at Saint Paul, Kodiak Island, Alaska, contains a specimen of a very handsome Petrel, which appears to be undescribed, and which, in honor of its discoverer, I propose to name and describe as

ŒSTRELATA FISHERI, sp. nov.   Fisher's Petrel.

SP. CH. *Adult ♂* (No. 89431, U. S. Nat. Mus.; collector's number, 54; Saint Paul, Kodiak Island, Alaska, June 11, 1882; William J. Fisher,

---

*Griffith. Cuvier's Animal Kingdom, II, 1827, pp. 228, 229.

†Richardson. Fauna Borealis-Americana, 1829, p. 15.

‡Audubon and Bachman. Quadrupeds of North America, III, 1854, pp. 126, 127.

§ l. c., p. 127.

‖BAIRD: Report U. S. and Mexican Boundary Survey, II, pl. ii, 1859, p. 29.

¶ See also COUES AND YARROW: U. S. Geog. Surveys W. of 100°, V, Zoology, 1875, pp. 66, 67.

HOFFMAN: Mammals of Grand River, Dakota. <Proc. Boston Society Natural History, XIX, 1876-'77, p. 99.

collector): Ground color of the head, neck, and lower parts pure white, but this unvaried only on the sides of the forehead, lores, malar region, chin, throat, jugulum, and crissum; feathers of middle portion of fore-head (longitudinally) and fore part of crown marked with a central spot of slate-color, these spots mostly approaching a lozenge-shaped form, but becoming gradually more transverse posteriorly, and at the same time paler in color; the terminal margin of the feathers grayish white; a distinct blackish spot immediately before and beneath the eye; sides of the breast washed with grayish; belly and flanks overlaid by a nearly uniform wash of smoky plumbeous, but the white showing through in places; many feathers of the sides barred with plumbeous-gray; anterior under wing-coverts dark sooty-gray or slate-color, those along the outer margin mainly of the same color; rest of under surface of the wing, including inner webs of primaries, uniform pure white, the latter having merely a narrow, but very abruptly defined, dusky stripe next the shaft, the white being margined for a short distance along the terminal portion with grayish; axillars mainly plumbeous, or barred with the same. Nape, back, scapulars, rump, upper tail-coverts, and middle tail-feathers, bluish plumbeous, darkest on the lower part of the rump, the feathers with distinct dusky shaft-streaks, except on the nape. Tail (except middle feathers) white, with very irregular transverse bars or vermiculations of plumbeous-gray. Lesser wing-coverts dark slate-color (many shades darker than the back); greater coverts, secondaries, and tertials plumbeous-gray (more silvery toward edge of wing), very distinctly edged with pure white; three outer primaries and primary coverts slate-black, the inner quills gradually more grayish, and narrowly bordered with white; bill uniform deep black; tarsi, most of basal phalanx of inner toe, and basal portion of webs, light brownish (apparently flesh-colored or lilaceous in life); rest of the feet dusky.

Wing, 10.15; tail, 4, slightly graduated; culmen, 1; depth of bill at base, .40; tarsus, 1.35; middle toe, 1.40.

This elegant Petrel, probably the handsomest of the genus, belongs to the delicately-formed, slender-billed group which includes Œ. cooki (Gray), Œ. gavia (Forst.), Œ. desolata (Gm.), and Œ. defillipiana (Gigi. & Salvad.). It is apparently most nearly allied to the last named, from which, however, it may be distinguished by the following characters:

Œ. fisheri. Lower parts chiefly smoky plumbeous on the surface, this color nearly uniform on the belly and flanks; greater wing-coverts, secondaries, and tertials silvery plumbeous, broadly edged with pure white, and in very conspicious contrast with the blackish slate lesser-covert. area; rectrices (except middle pair) white, transversely vermiculated with grayish. Wing, 10.15; tail, 4; culmen, 1; tarsus, 1.35; middle toe, with claw, 1.70. Hab.—Eastern North Pacific (off coast of Alaska); accidental in Western New York?

Œ. defillipiana. Lower parts pure white, tinged laterally with cinereous; greater wing-coverts, secondaries, and tertials dusky, edged terminally

June 26, 1883.

with grayish, and not contrasting noticeably with the lesser coverts; six middle rectrices uniform cinereous, the outer pair with exterior webs uniform white. Wing, 9; tail, 3.80; culmen, 1.04; tarsus, 1.07; middle toe, with claw, 1.40. *Hab.*—Eastern South Pacific (off coast of Peru).

The comparison with *Œ. defillipiana* resting only on the description and a colored plate, it may be, therefore, that some of the differential characters adduced in the above comparative diagnosis would not be found to hold good on actual examination of specimens. This is particularly liable to be the case regarding the coloration of the rectrices in *Œ. defillipiana*, which are not described with sufficient detail, while the figure may not be perfectly accurate so far as this feature is concerned.

The most nearly related species with which I have been able to compare *Œ. fisheri* is *Œ. gularis* Peale. The latter, however, is very distinct, the coloration being in almost every respect dissimilar, while the bill is much stouter through the base, and the tarsi and toes decidedly shorter.

A Petrel captured in Livingston County, New York, in April, 1880, described by Mr. Brewster in the Bulletin of the Nuttall Ornithological Club for April, 1881, and there referred to *Œ. gularis*, seems, judging from the description, to belong rather to *Œ. fisheri*. Should such prove to be the case, Mr. Brewster was evidently wrong in his determination. The specimen in question was compared with the type of *Œ. gularis*, and the differences of plumage ascribed to difference of age of the two specimens; but no fact in ornithology can be more thoroughly established than that, with the possible exception of the Albatrosses, *the Petrels have no distinct progressive stages of plumage*, the young assuming with their first feathers the fully adult livery.

DESCRIPTION OF A SPECIES OF WHITEFISH, COREGONUS HOYI (GILL) JORDAN, CALLED "SMELT" IN SOME PARTS OF NEW YORK.

By TARLETON H. BEAN.

Our attention has recently been called by the Rev. W. M. Beauchamp to a species of "smelt" in some lakes in New York, and finally Mr. J. C. Willetts has forwarded numerous specimens of this fish from Skaneateles. This is not an *Osmerus*, as the common name would imply, but a little-known *Coregonus*, and worthy of description.

The largest New York·specimen of this fish now in the collection is numbered 32162 in the National Museum Register; it was obtained in Seneca Lake, in June, 1878, by Prof. H. L. Smith, who sent it to the Museum. Seven additional examples were received October 2, 1882, from Skaneateles, N. Y., whence they were forwarded by Mr. J. C. Willetts. The catalogue number of these specimens is 32165. The individuals received from Mr. Willetts vary in length from 5½ to 6½ inches. Three of these examples have the air bladder much distended and filling the greater portion of the abdominal cavity.

The specimen received from Professor Smith, which we take as the basis of our description, is 10 inches long.

The species is most closely related to *C. artedi*, but differs from it and from all other species known to me in many important characters which have been only vaguely indicated in most of the published descriptions. It is much more widely separated from *C. artedi* than is the var. *sisco* of Jordan.

DESCRIPTION.—Body elongate, moderately compressed, slender. Head less compressed than body, its greatest width equaling one-half the distance from tip of lower jaw to nape; the lower jaw projecting considerably even when the mouth is closed. Mouth large, the maxillary reaching to the vertical through the anterior margin of the pupil. Preorbital bone long and slender, more than one-third as long as the head. Supraorbital as long as the eye, four times as long as broad.

The greatest height of the body is considerably less than the length of the head, and is contained five times in the total length without caudal. The greatest width of the body is less than one-half its greatest height. The least height of caudal peduncle equals the length of the orbit and about one-third of the greatest height of the body. Scales small, nine in an oblique series from the dorsal origin to the lateral line, eighty-two tube-bearing scales, and eight in an oblique series from the ventral origin to the lateral line.

The length of the head is one-fourth of the total length to the end of the lateral line. The distance of the nape from the tip of the snout is nearly one-third of the distance from the tip of the snout to the origin of the first dorsal. The length of the maxilla is one-third of the length of the head. The mandible is one-half as long as the head. Lingual teeth present. The eye is as long as the snout and one-fourth as long as the head. Gill rakers long and slender, the longest five-sixths as long as the eye; there are fifty-five on the first arch, thirty-five of which are below the angle. The insertion of the dorsal is nearer the tip of the snout than the end of the middle caudal rays. The longest ray of the dorsal equals the greatest length of the ventral and is contained seven times in the total length to the end of the middle caudal rays (six and two-third times in length to end of lateral line).

The length of the pectoral is one-sixth of the standard body length.

The insertion of the ventral is midway between the tip of the snout and the end of the middle caudal rays. When the ventral is extended the distance of its tip from the vent is only one-fourth of the length of the fin. In this respect this species differs widely from *C. artedi*.

COLORS.—Back grayish silvery; sides silvery; dorsal and caudal with darker tips.

Radial formula.—D. iii, 9; A. ii, 13; V. i, 12; P. i, 16; scales 9—82—8.

Measurements.

Current number of specimen	 32162	
	Milli-meters.	Hundredths of length.
Extreme length	253	....
Length to end of scales	217	100
Body:		
Greatest height	41	19
Greatest width	18	8
Height at ventrals	40	18½
Least height of tail	15	7
Head:		
Greatest length	52	24½
Distance from snout to nape	36	16½
Greatest width	20	9
Width of interorbital area	12	5½
Length of snout	14	6½
Length of operculum	13	6
Length of maxillary	18	8
Length of mandible	26	12
Diameter of eye	13	6
Dorsal (first)		
Distance from snout	112	51½
Length of base	20	9
Length of longest ray	33	15
Length of last ray	11	5
Anal		
Distance from snout	162	75
Length of base	24	11
Length of longest ray	20	9
Length of last ray	8	4
Caudal:		
Length of middle rays from end of scales	12	5½
Length of external rays	44	20
Pectoral:		
Distance from snout	52	24½
Length	36	16½
Ventral:		
Distance from snout	118	55
Length	32	15
Origin from anal origin	48	22
End of extended ventral to anal origin	15	7
Dorsal	iii, 9	....
Anal	ii, 13	....
Pectoral	i, 16	....
Ventral	i, 12	....
Number of scales in lateral line	82	....
Number of transverse rows above lateral line	9	....
Number of transverse rows below lateral line	8	....

NOTE ON A POTSDAM SANDSTONE, OR CONGLOMERATE, FROM BERKS COUNTY, PENNSYLVANIA.

By GEORGE P. MERRILL.

This sandstone is a coarse compact rock of a greenish gray color, though many of the included pebbles are of a rose-red tint. The cementing material, which is of a greenish color, shows under the microscope a fibrous structure and remains always light between crossed Nicols. It bears very many inclosures of rounded and angular grains of hematite, which by reflected light are of a bluish luster somewhat resembling menaccanite, but giving no distinct reaction for titanic acid when subjected to the proper tests. They are of all sizes up to a millimeter in diameter. A section through one of the rose-colored pebbles shows it to be traversed in all directions by numerous fractures in which are included, as if deposited by infiltration, innumerable minute blood-red particles or scale-like forms characteristic of red hematite When the light is shut off from below the stage of the microscope the quartz appears as a black opaque mass traversed by an irregular network of anastomosing red lines. The included scales are, apparently, sufficiently abundant and evenly disseminated to fully account for the red color of the pebbles. Besides the hematite the quartz grains contain numerous minute cavities, some of which are empty, while others contain a liquid and bubble. Numerous very small colorless needle-like crystals are also present, penetrating the quartz in every direction.

DESCRIPTION OF A NEW SPECIES OF ALEPIDOSAURUS (A. ÆSCU-LAPIUS) FROM ALASKA.

BY TARLETON H. BEAN,

Curator, Department of Fishes, U. S. National Museum.

The fish here to be described as the type of a new species was at first referred by me to *A. ferox.** It is number 27705 of the National Museum Register. Another example of the same species was previously taken at Unalashka by Mr. W. H. Dall. The type of the species was obtained at Iliuliuk, Unalashka, October 7, 1880, by Mr. Robert King, at his wharf. Mr. King first saw the dorsal fin of the fish emerging from the water, and this attracted his attention. The animal came up into shoal water, and acted as if it meant to go on the beach. Mr. King thrust a spear into it and thus secured it. In the stomach I found twenty-one individuals of *Eumicrotremus spinosus*, most of them adult, and one small squid. A cod-like fish was said to have been in the stomach also, but I did not see this. It is probable that the fish was driven ashore from the adjacent deep water by the torture of a parasite found in its flesh; this parasite has been identified with the genus *Tetrarhynchus* by Mr. F. W. True. It is said to be not an uncommon thing for the "wolf-fish," as this *Alepidosaurus* is styled, to throw itself on the beach at Iliuliuk.

It should be stated that the first notice of my species is published in Bulletin 16, U. S. National Museum, pages 888 and 889; this volume appeared early in April, 1883, but the original description was prepared much earlier than that date and the printing of it was delayed longer than was anticipated.

Alepidosaurus Æsculapius differs from *A. ferox* chiefly in the much shorter pectorals and ventrals and in the smaller number of ventral rays. Owing to the somewhat mutilated condition of the specimen, only the skin was preserved in alcohol after full measurements had been recorded.

DESCRIPTION.—The length to the origin of the middle caudal rays was 1,298 millimeters. The greatest height of the body (123 millimeters) is contained $10\frac{1}{2}$ times in the standard length. The depth at the ven-

---

* Proc. U. S. Nat. Mus. IV, p. 259, Dec. 24, 1881 (name only).

trals (105 millimeters) is contained 12⅛ times in the length. The least height of the tail is about equal to the length of the middle caudal rays. The species is much stouter in the second half of the body than our numerous examples of *A. ferox*. There is a well-marked fleshy keel along the median line, beginning a little in front of the ventrals and extending to the caudal.

The greatest length of the head (208 millimeters) is contained 6¼ times in the standard length. The width of the interorbital area (40 millimeters) is nearly equal to the diameter of the eye. The length of the snout is about twice that of the eye, which is contained nearly 5 times in the length of the head. The length of the intermaxillary (150 millimeters) is nearly ¾ that of the head; the bone extends behind the eye a distance equal to about ⅛ of the diameter of the eye. The length of the mandible is about 3 times the greatest width of the head. The nostril is nearly equally distant from tip of snout and the anterior margin of the eye.

The first dorsal was more or less broken, so that the lengths of its rays are not fully made out. The longest ray measured 235 millimeters. The distance of the first dorsal from the snout is about the same as the length of the head. The beginning of the dorsal, the posterior margin of the operculum, and the origin of the pectoral are in nearly the same vertical. The anterior edge of the first dorsal ray is very finely serrated.

The distance of the adipose dorsal from the snout is 5¼ times the length of the head. The length of the base of this fin is ⅝ of its width at the top.

The distance of the anal from the snout is somewhat more than 7 times the length of its base. The fifth, and longest, anal ray is ⅓ as long as the intermaxillary, and 3 times as long as the last anal ray.

The upper caudal lobe is imperfect, so that it cannot be known whether or not it was prolonged into a filament. The middle rays are equal to the least height of the tail.

The outer edge of the first pectoral ray is finely serrated. The distance of the pectoral from the snout (223 millimeters) is 3 times the length of the longest anal ray and 4 times the greatest width of the head. The length of the pectoral is less than that of the head by a distance equal to half the interorbital width.

The distance of the ventral from the snout (565 millimeters) equals 4 times the distance from snout to nape. The length of the ventral is a little more than twice that of the middle caudal rays. The first ventral ray is perfectly smooth (serrate in *A. ferox*).

Radial formula: B. 7; D. 39; A. 16; P. I, 12; V. I, 7.

Vertebræ, 50 (as in *A. ferox*).

Color.—General color dark gray, on the lower parts mingled with silvery; everywhere iridescent. Dorsal membrane black with steel-blue reflections. Adipose dorsal, pectorals and caudal black. Ventrals and anal silvery and gray. A row of small translucent spots on each side of the lateral line and keel.

MEASUREMENTS.

	Millimeters.
Length to origin of middle caudal rays	1,298
Body:	
Greatest height	123
Greatest width (near anal)	70
Distance of vent behind origin of ventrals	56
Height at ventrals	105
Least height of tail	36
Length of caudal peduncle*	130
Head:	
Greatest length	208
Distance from snout to nape	140
Greatest width	56
Width of interorbital area	40
Length of snout	85
Length of operculum	56
Length of intermaxillary	150
Length of longest palatine tooth	30
Length of mandible	165
Length of longest mandibulary tooth	16
Distance from snout to orbit	90
Diameter of eye	43
Diameter of iris	32
Dorsal (first):	
Distance from snout	210
Length of base	825
Length of longest ray †	235
Adipose dorsal:	
Distance from snout	1,098
Length of base	25
Length along anterior edge	45
Length along posterior edge	35
Width at top	35
Anal:	
Distance from snout	1,014
Length of base	142
Length of longest ray (5th)	75
Length of last ray	26
Caudal:	
Length of middle rays	37
Length of lower lobe	190
Pectoral:	
Distance from snout	223
Length	188
Ventral:	
Distance from snout	565
Length	77
Branchiostegals	7
Dorsal	39
Anal	16
Pectoral	I, 12
Ventral	I, 7
Number of vertebræ	50

* From end of adipose dorsal to origin of upper caudal lobe.
† Nearly all the rays are more or less broken.

ALPHABETICAL INDEX.

672   ALPHABETICAL INDEX.

O